T0398001

Handbook of Space Security

Kai-Uwe Schrogl
Editor-in-Chief

Maarten Adriaensen • Christina
Giannopapa • Peter L. Hays • Jana Robinson
Section Editors

Ntorina Antoni
Managing Editor

Handbook of Space Security

Policies, Applications and Programs

Second Edition

Volume 2

With 248 Figures and 47 Tables

 Springer

Editor-in-Chief
Kai-Uwe Schrogl
European Space Agency (ESA)
Paris, France

Section Editors
Maarten Adriaensen
European Space Agency (ESA)
Paris, France

Christina Giannopapa
European Space Agency
(ESA)
Paris, France

Peter L. Hays
Space Policy Institute
George Washington University
Washington, DC, USA

Jana Robinson
Space Security Program
Prague Security Studies
Institute (PSSI)
Prague, Czech Republic

Managing Editor
Ntorina Antoni
Eindhoven University of Technology
Eindhoven, The Netherlands

ISBN 978-3-030-23209-2   ISBN 978-3-030-23210-8 (eBook)
ISBN 978-3-030-23211-5 (print and electronic bundle)
https://doi.org/10.1007/978-3-030-23210-8

Introduction

Never before has security in space been more challenged.

Never before has space been more elaborately used for military and security purposes on Earth.

And never before was it more necessary to understand and to receive orientation in the policy area of space security.

This indeed is the purpose of this second edition of the *Handbook of Space Security*, which is addressed to all persons and institutions dealing with space security on a governmental, academic, societal, international, and diplomatic level. From now, the global future will depend on the secure use of outer space for all policy areas with particular stress on climate change and environmental monitoring, telecommunications, navigation, and cyber infrastructures as well as resource management. If space utilization as critical infrastructure is disrupted, our modern societies will break down. Space security is a key factor for survival. This is why space is also contested and space assets are vulnerable.

Consequently, we see a growing "securitization" of outer space. While some decades ago, there were rather clear distinctions between military and civilian uses of space with a smaller area in between called "dual-use." This zone is ever expanding under the label of "security" diminishing exclusively military and civilian use to small fringes. Just think of environmental security, cyber security, food security, water security, etc., and it becomes clear that we have to speak of a new paradigm also for the use of space. This can also be seen as the most prominent and accelerating development since the publication of the first edition of the Handbook in 2015.

Based on this, the definitional approach we use in this Handbook in space security can be drafted as:

"Space security" is the aggregate of all technical, regulatory and political means that aims to achieve unhindered use of outer space from any interference as well as aims to use space for achieving security on Earth.

This adds to an already existing abundance of definitions for space security. Since it is more than unlikely that a consensus would be reached, our definition stays in the practice of attempting to narrow general applicability and tailor-made approach. For us, it is a frame for the focus and the structure of the Handbook.

The structure of the Handbook comprises four parts. These four parts are edited by accomplished experts in the field. They are Peter L. Hays for Part I International Space Security Setting, Jana Robinson for Part II Space Security Policies and Strategies of States, Maarten Adriaensen for Part III Space Applications and Supporting Services for Security and Defense, and Christina Giannopapa for Part IV Space Security Programs Worldwide and Space Economy Worldwide. They have been assembling exclusive groups of contributors, which do not only reflect a broad geographic distribution (120 authors from 30 countries), but also offer an exciting diversity of practitioners and academics' approaches to the topics. An Editorial Advisory Board of 22 distinguished experts from governments and academia from some 20 countries and every continent (i.e., Africa, Asia, Europe, North America, and South America) assisted the editors in identifying and evaluating contributions to ensure a high-quality, coherent final product.

The Handbook applies political, legal, economic, and technology-oriented analysis to these topics and aims at providing a holistic understanding in each of the sections and for the theme of space security as a whole. It is a work of tertiary literature containing digested knowledge in an easily accessible format. It provides a sophisticated, cutting-edge resource on the space security–related policy portfolio and associated assets to assist fellow members of the global space community, academic audiences, and other interested parties in keeping abreast of the current and future directions of this vital dimension of international space policy. By analyzing the underlying developments in space environment and the linkages to space security from a wide range of disciplines and theoretical perspectives, this Handbook establishes itself as a leading work for reference purposes, as well as a basis for further discussion.

Furthermore, it is enriched by the account of numerous space operational applications that routinely deliver indispensable fast and reliable services for security and defense needs from and for space. The transformation of the space domain through new technological advances and types of innovation such as mega-constellations, machine learning, and artificial intelligence has resulted in major challenges. Finally, it examines how addressing these needs has led to space programs and the development of specific security space assets. In short, it examines the reciprocal relations among space policy objectives on one hand and operational capabilities on the other, allowing readers to understand the theoretical and practical interactions and limitations between them. It also features numerous recommendations concerning how to best improve the space security environment, given the often-competing objectives of the world's major space-faring nations.

The Handbook is available in both printed and electronic forms. Its success is documented by more than 100,000 chapter downloads for its first edition. The Handbook is intended to assist and promote both academic research and professional activities in this rapidly evolving field encompassing security from and for space. It aspires to remain a go-to reference manual for space policy practitioners and decision makers, scholars, students, researchers, and experts as well as the media. The Springer Publishing House has been an exemplary partner in this major undertaking. We would like to gratefully acknowledge the cooperation of Maury Solomon and Hannah Kaufmann (New York), Lydia Muller (Heidelberg), Juby George, and Sonal Nagpal (New Delhi). Ntorina Antoni, the Managing Editor of the Handbook, acted as the invaluable main liaison between our editorial team, contributors, and publisher throughout the project. It is the hope of all of us who joined together to prepare this second edition of the *Handbook of Space Security* that it will inspire those seeking to be active in shaping space security, on which we all depend.

September 2020

Kai-Uwe Schrogl
Editor-in-Chief

Advisory Board

Setsuko Aoki	Professor of Law, Keio University Law School, Tokyo, Japan Chair, United Nations Committee on the Peaceful Uses of Outer Space (UNCOPUOS) Legal Subcommittee, Vienna, Austria Vice President, International Institute of Space Law (IISL), Paris, France
Natália Archinard	Deputy Head of the Education, Science, Transport and Space Section, Directorate of Political Affairs, Federal Department of Foreign Affairs, Berne, Switzerland Chair, UNCOPUOS Scientific and Technical Subcommittee, Vienna, Austria
Frank Asbeck	Senior Fellow at the Prague Security Studies Institute (PSSI), Prague, Czech Republic Former Principle Adviser for Space and Security Policy, European External Action Service (EEAS), Brussels, Belgium Former Director of the European Union Satellite Centre (EU SatCen), Torrejon, Spain
Tare Brisibe	Senior Legal and Regulatory Counsel, Asia-Pacific, SES, Singapore Former Chair, UNCOPUOS Legal Subcommittee, Vienna, Austria Former Deputy Director (Legal) National Space Research and Development Agency of Nigeria, Lagos, Nigeria
Gerard Brachet	Space Policy Consultant, Paris, France Former Chair, UNCOPUOS, Vienna, Austria Former Director General, French Space Agency – CNES, Paris, France

Contents

Volume 2

About the Editor-in-Chief

Prof. Dr. Kai-Uwe Schrogl is currently seconded from the European Space Agency (ESA) to the German Federal Ministry for Economic Affairs and Energy in Berlin to support the preparation of the German Presidency of the Council of the European Union in the second half of 2020. Until 2019, Dr. Schrogl was the Chief Strategy Officer of ESA (Headquartered in Paris, France). From 2007 to 2011 he was the Director of the European Space Policy Institute (ESPI) in Vienna, Austria, the leading European think tank for space policy. Prior to this, he was the Head of the Corporate Development and External Relations Department in the German Aerospace Center (DLR) in Cologne, Germany. Previously, he also worked with the German Ministry for Post and Telecommunications and the German Space Agency (DARA) in Bonn, Germany. Dr. Schrogl has been a delegate to numerous international forums and has served from 2014 to 2016 as Chairman of the Legal Subcommittee of the United Nations Committee on the Peaceful Uses of Outer Space, the highest body for space law making, comprising 73 Member States. He also was chairman of various European and global committees (ESA International Relations Committee and two plenary working groups of the UNCOPUOS Legal Subcommittee, the one on the launching State and the other on the registration practice, both leading to UN General Assembly Resolutions).

He presented, respectively testified, at hearings of the European Parliament and the U.S. House of Representatives. Dr. Schrogl is President of the International Institute of Space Law, the professional association of space law experts from 48 countries, Member of the International Academy of Astronautics (recently chairing its Commission on policy, economics, and regulations) and the Russian Academy of Cosmonautics, as well as Corresponding Member of the French Air and Space Academy. He holds a doctorate degree in political science and lectures international relations as an Honorary Professor at Tübingen University, Germany. Dr. Schrogl has written or co-edited 17 books and more than 140 articles, reports, and papers in the fields of space policy and law as well as telecommunications policy. He launched and edited until 2011 the *Yearbook on Space Policy* and the book series Studies in Space Policy, both published by ESPI at Springer-WienNewYork. He sits on editorial boards of various international journals in the field of space policy and law (*Space Policy, Zeitschrift für Luft- und Weltraumrecht*, Studies in Space Law/Nijhoff; previously also *Acta Astronautica*).

About the Section Editors

Maarten Adriaensen currently works for the Procurement Department of the European Space Agency (ESA) in Paris, France. He was previously seconded by the ESA to the European Defence Agency (EDA) in Brussels, Belgium, assigned as Policy Officer Space. In that function, he provided strategic analysis and synthesis in support of EDA Directorates, internal coordination of EDA space activities, and support to cooperation on space activities externally. Prior to that, Maarten Adriaensen worked in the Procurement Department and Policy Department of ESA in Paris; the University of Leuven, Belgium; and the European Space Policy Institute (ESPI) in Vienna, Austria. He holds an M.Sc. in Space Studies (cum laude), M.A. in European Studies: Transnational and Global Perspectives (magna cum laude), M.A. in Modern History (cum laude), LLB, and an Academic Teacher Degree from the University of Leuven. He attended the 2012 International Space University (ISU) Space Studies Program (SSP) in Florida, hosted by the Florida Institute of Technology and NASA Kennedy Space Center. The content of Maarten's contributions to the Handbook reflect his personal opinions and do not necessarily reflect the opinion of the European Space Agency.

Dr. Christina Giannopapa works at the European Space Agency (ESA) since 2007. Currently, she is seconded from ESA to the Greek Ministry of Digital Policy, Telecommunications and Media as Special Advisor on high-tech and space applications related issues. Until February 2019, Dr. Giannopapa has been ESA's Head of Political Affairs Office in the Strategy Department of the Director General's Services in Paris, being responsible for providing advice and support on political matters to the Director General. From 2012 to 2015, she has been working on Member States Relations Department of ESA. From 2010 to 2012, Dr. Giannopapa has been seconded from the Agency as Resident Fellow at the European Space Policy Institute (ESPI) in Vienna, where she has been supporting the European Interparliamentary Space Conference (EISC) and lead studies on innovation, Galileo, Copernicus, and Africa. In the policy areas she also worked briefly in DG Research, European Commission. From 2007 to 2010, Dr. Giannopapa has been working in the Mechanical Engineering Department of the Technical and Quality Management Directorate of ESA in the Netherlands. Prior to joining ESA, she has worked as a consultant to high-tech industries in research and technology development. Dr. Giannopapa held positions in academia in Eindhoven University of Technology, the Netherlands, and in the University of London, UK. She has received 14 academic scholarships and awards and has more than 60 publications in peer-reviewed journals and conferences. Dr. Giannopapa holds a Ph.D. in Engineering and Applied Mathematics, an M.Eng. in Manufacturing Systems Engineering and Mechatronics, and an M.B.A. in International Management from the University of London, UK. Additionally, she is Assistant Professor in the Department of Industrial Engineering and Innovation Sciences at Eindhoven University of Technology. Dr. Giannopapa is the Chairperson of the Committee for Liaison with International Organisations and Developing Nations (CLIODN) and the Secretary of the Space Security Committee of the International Astronautical Federation (IAF). She is also the Director for Professional Development of Women in Aerospace-Europe (WIA-E).

Peter L. Hays is an Adjunct Professor at George Washington University's (GWU) Space Policy Institute and a Senior Policy Advisor with Falcon Research. He has been directly involved in helping to develop and implement major national security space policy initiatives since 2004 and serves as a senior advisor on governance, cadre, and strategic messaging issues. Peter served as a Staff Augmentee at the Office of Science and Technology Policy in 1988 and at the National Space Council in 1990. He served as an Air Force Officer from 1979 to 2004, flew C-141 cargo planes, and previously taught at the Air Force Academy, Air Force School of Advanced Airpower Studies, and National Defense University; he now teaches Air- and Spacepower Seminars at the Marine Corps School of Advanced Warfighting and the Space and National Security and the Science, Technology, and National Security Policy graduate seminars at GWU. Peter holds a Ph.D. from the Fletcher School and was an Honor Graduate of the Air Force Academy. Major publications include: *Handbook of Space Security*, *Space and Security*, and *Toward a Theory of Spacepower*.

Dr. Jana Robinson is Space Security Program Director at the Prague Security Studies Institute (PSSI). She previously served as a Space Policy Officer at the European External Action Service (EEAS) in Brussels as well as a Space Security Advisor to the Czech Foreign Ministry. From 2009 to 2013, Dr. Robinson worked at the European Space Policy Institute (ESPI), seconded from the European Space Agency (ESA), leading the Institute's Space Security Research Programme. Dr. Robinson is an elected member of the International Institute of Space Law (IISL) and the International Academy of Astronautics (IAA). She is also a member of the Advisory Board of the George C. Marshall Missile Defense Project of the Center for Strategic and International Studies (CSIS) in Washington, D.C. Dr. Robinson holds a Ph.D. in Space Security from Charles University in Prague and received two M.A. degrees from George Washington University's Elliott School of International Affairs and Palacky University in Olomouc, respectively.

About the Managing Editor

Ntorina Antoni is currently a Ph.D. candidate of strategic management in space security at Eindhoven University of Technology. Her research focuses on strategic decision-making under uncertainty and legitimacy in the context of high reliability organizations. She is also an Attorney-at-Law and member of the Athens Bar Association in Greece. Ntorina previously worked at the Strategy Department of the European Space Agency (ESA). In her role as Strategy and Policy Analyst, she got involved in the research and analysis of regulations, policies, and strategies of the ESA member states. Prior to that, Ntorina served as the Legal Counsel for Swiss Space Systems, a company which planned to provide orbital launches of small satellites and manned sub-orbital spaceflights. She was in charge of the legal and regulatory aspects of aviation and aerospace projects including aircraft purchase, certification of parabolic flights and suborbital flights, as well as regulation of small satellites. Ntorina is a Board Member of the Brainport TechLaw in Eindhoven and an elected member of the International Institute of Space Law (IISL). She holds a Law degree from the University of Athens in Greece, a Master's degree (LL.M) in International and European Law from Tilburg University in the Netherlands, and an Advanced Master's degree (LL.M) in Air and Space Law from Leiden University in the Netherlands.

Contributors

Olufunke Adebola Sam Nunn School of International Affairs, Georgia Institute of Technology, Atlanta, GA, USA

Simon Adebola Capacity Direct LLC, Atlanta, USA

Maarten Adriaensen European Space Agency (ESA), Paris, France

William Ailor Center for Orbital and Reentry Debris Studies, The Aerospace Corporation, El Segundo, CA, USA

Hamda Al Hosani Space Policies and Legislations Department, UAE Space Agency, Masdar City, Abu Dhabi, United Arab Emirates

Naser Al Rashedi Space Policies and Legislations Department, UAE Space Agency, Masdar City, Abu Dhabi, United Arab Emirates

Fatima Al Shamsi Space Policies and Legislations Department, UAE Space Agency, Masdar City, Abu Dhabi, United Arab Emirates

Jesse Andries Solar Terrestrial Center of Excellence (STCE) – Solar Influences Data Analysis Center (SIDC), Royal Observatory of Belgium, Brussels, Belgium

Ntorina Antoni Eindhoven University of Technology, Eindhoven, The Netherlands

Tal Azoulay Yuval Neeman Workshop for Science, Technology and Security, Tel Aviv University, Tel Aviv, Israel

Hanamantray Baluragi Indian Space Research Organization (ISRO), Bangalore, India

David Berghmans Solar Terrestrial Center of Excellence (STCE) – Solar Influences Data Analysis Center (SIDC), Royal Observatory of Belgium, Brussels, Belgium

Shamil Biktimirov Skolkovo Institute of Science and Technology, Moscow, Russia

Olavo de O. Bittencourt Neto Catholic University of Santos, Santos, Brazil

Nickolas J. Boensch Bryce Space and Technology, Alexandria, VA, USA

Ulrike M. Bohlmann European Space Agency (ESA), Paris, France

Bleddyn Bowen University of Leicester, Leicester, UK

Denis Bruckert EU Satellite Centre (EUSC), Madrid, Spain

Jean François Bureau Institutional and International Affairs, Eutelsat Group VP, Paris, France

Dean Cheng The Heritage Foundation, Washington, DC, USA

Jean-Pierre Darnis Université Côte d'Azur, Istituto Affari Internazionali, (IAI), Nice, France

Michael Davis Space Industry Association of Australia, Adelaide, Australia

Helene de Boissezon French Space Agency (CNES), Toulouse, France
Committee on Earth Observation Satellites (CEOS), Haiti Recovery Observatory, Port au Prince, Haiti

Philip De Man University of Sharjah, Sharjah, UAE
University of Leuven, Leuven, Belgium

Marta De Oliviera International Space University (ISU), Illkirch-Graffenstaden, France

Marco Detratti European Defence Agency (EDA), Brussels, Belgium

Davide Di Domizio European Defence Agency (EDA), Brussels, Belgium

Ferdinando Dolce European Defence Agency (EDA), Brussels, Belgium

Everett C. Dolman Air Command and Staff College, Air University, Montgomery, AL, USA

Andrew Eddy Athena Global, Simiane-la-Rotonde, France

Pascal Faucher Defence and Security Office, French Space Agency (CNES), Paris, France

Steven Freeland University of Western Sydney, Sydney, NSW, Australia

Daniel Freire e Almeida Catholic University of Santos, Santos, Brazil

Alexandru Georgescu National Institute for Research and Development in Informatics (ICI), Bucharest, Romania

Christina Giannopapa European Space Agency (ESA), Paris, France

Estelle Godard Institut d'études politiques de Paris, Paris, France

Elise Gruttner United Nations Assistance to the Khmer Rouge Trials, New York, NY, USA

Gaia Guiso Women in Aerospace-Europe (WIA-E), Milan, Italy

Xiaoxi Guo China Academy of Space Technology (CAST), Beijing, People's Republic of China

Peter L. Hays Space Policy Institute, George Washington University, Washington, DC, USA

Stacey Henderson Adelaide Law School, The University of Adelaide, Adelaide, SA, Australia

Yvon Henri OneWeb, London, UK

Per Høyland Department of Political Science, University of Oslo, Oslo, Norway

Irteza Imam Department of Defence and Strategic Studies, Quaid-i-Azam University, Islamabad, Pakistan

Isaac Ben Israel Yuval Neeman Workshop for Science, Technology and Security, Tel Aviv University, Tel Aviv, Israel

Anton Ivanov Skolkovo Institute of Science and Technology, Moscow, Russia

Nicole J. Jackson School for International Studies, Simon Fraser University, Vancouver, BC, Canada

Moriba Jah Department of Aerospace Engineering and Engineering Mechanics, The University of Texas at Austin, Austin, TX, USA

Jan Janssens Solar Terrestrial Center of Excellence (STCE) – Solar Influences Data Analysis Center (SIDC), Royal Observatory of Belgium, Brussels, Belgium

Hamid Kazemi Aerospace Research Institute, Department of Air and Space Law, Tehran, Iran

Ioannis Kechaoglou RHEA Group, Wavre, Belgium

Tanzeela Khalil South Asia Center, Atlantic Council, Washington, DC, USA

Ahmad Khan Department of Strategic Studies, National Defence University, Islamabad, Pakistan

John J. Klein George Washington University's Space Policy Institute, Washington, DC, USA

Alexandros Kolovos Automatic Control, AirSpace Technology, Defence Systems and Operations Section, Hellenic Air Force Academy, Athens, Greece

Tereza B. Kupková Prague Security Studies Institute (PSSI), Prague, Czech Republic

George D. Kyriakopoulos School of Law, National and Kapodistrian University of Athens, Athens, Greece

Ajey Lele Institute for Defence Studies and Analyses, New Delhi, India

Stijn Lemmens Space Debris Office, European Space Agency (ESA) – European Space Operations Centre (ESOC), Darmstadt, Germany

Francesca Letizia Space Debris Office, IMS Space Consultancy at Space Debris Office, European Space Agency (ESA) – European Space Operations Centre (ESOC), Darmstadt, Germany

Hongbo Li China Academy of Launch Vehicle Technology (CALT), Beijing, China

Dan Li China Academy of Launch Vehicle Technology (CALT), Beijing, China

Salvador Llopis Sanchez European Defence Agency (EDA), Brussels, Belgium

Lehao Long Chinese Academy of Engineering, Beijing, China

Zhuoyan Lu International Space University (ISU), Strasbourg, France

Holger Lueschow European Defence Agency (EDA), Brussels, Belgium

Tarlan Mammadzada "Azercosmos" OJSCo, Baku, Azerbaijan

Patrik Martínek The Prague Security Studies Institute (PSSI), Prague, Czech Republic

Jean-Christophe Martin CEO Marency SAS, Paris, France
CEO Marency SAS, Brussels, Belgium

Peter Martinez Secure World Foundation (SWF), Broomfield, CO, USA

Attila Matas Orbit/Spectrum Consulting, Grand-Saconnex, Switzerland

Robert Mazzolin RHEA Group, Wavre, Belgium

Wim Mees Royal Military Academy, Brussels, Belgium

Maria Messina Education Unit, Italian Space Agency (ASI), Rome, Italy

Edmondo Minisci University of Strathclyde, Glasgow, UK

Sebastien Moranta European Space Policy Institute (ESPI), Vienna, Austria

Elina Morozova International Legal Service, Intersputnik International Organization of Space Communications, Moscow, Russian Federation
International Institute of Space Law, Paris, France

Jean Muylaert RHEA Group, Wavre, Belgium

Annamaria Nassisi Strategic Marketing, Thales Alenia Space, Rome, Italy

Pat Norris VAPN Ltd, West Byfleet, UK

Francesco Pagnotta Presidency of the Council of Ministers – Office of the Military Advisor, Rome, Italy

Deganit Paikowsky Yuval Neeman Workshop for Science, Technology and Security, Tel Aviv University, Tel Aviv, Israel

Xavier Pasco Fondation pour la Recherche Stratégique (FRS), Paris, France

Isabella Patatti Strategic Marketing, Thales Alenia Space, Rome, Italy

Andrea Patrono EU Satellite Centre (EUSC), Madrid, Spain

Roberto Pelaez European Defence Agency (EDA), Brussels, Belgium

Regina Peldszus Department of Space Situational Awareness, German Aerospace Center (DLR) Space Administration, Bonn, Germany

Massimo Pellegrino Vienna, Austria

Joe Pelton International Association for the Advancement of Space Safety (IAASS), Arlington, VA, USA

Gina Petrovici German Aerospace Center (DLR), Bonn, Germany

Małgorzata Polkowska University of War Studies, Warsaw, Poland

Rajeswari Pillai Rajagopalan Nuclear and Space Policy Initiative, Observer Research Foundation, New Delhi, India

Marco Reali Presidency of the Council of Ministers – Office of the Military Advisor, Rome, Italy

Jana Robinson Space Security Program, Prague Security Studies Institute (PSSI), Prague, Czech Republic

Alvaro Rodríguez EU Satellite Centre (EUSC), Madrid, Spain

Luis Sanchez University of Strathclyde, Glasgow, UK

Chris Schacht Adelaide, Australia

Tommaso Sgobba International Association for the Advancement of Space Safety (IAASS), Noordwijk, The Netherlands

Lin Shen China Academy of Launch Vehicle Technology (CALT), Beijing, China

Byrana Nagappa Suresh Indian Space Research Organization (ISRO), Bangalore, India

Kazuto Suzuki Hokkaido University, Hokkaido, Japan

Mahshid TalebianKiakalayeh Graduate of Islamic Azad University, Tehran North Branch, Tehran, Iran

Leslie I. Tennen Law Offices of Sterns and Tennen, Glendale, AZ, USA

Jean-Daniel Testé OTA (l'Observation de la Terre Appliquée), Peujard, France

Jean-Jacques Tortora European Space Policy Institute (ESPI), Vienna, Austria

Maite Trujillo European Space Agency (ESA) – European Space Research and Technology Centre (ESTEC), Noordwijk, The Netherlands

Sufian Ullah Department of Defence and Strategic Studies, Quaid-i-Azam University, Islamabad, Pakistan

Cristina Valente Marketing and Sales, Telespazio, Rome, Italy

Petra Vanlommel Solar Terrestrial Center of Excellence (STCE) – Solar Influences Data Analysis Center (SIDC), Royal Observatory of Belgium, Brussels, Belgium

Massimiliano Vasile University of Strathclyde, Glasgow, UK

Nikita Veliev Skolkovo Institute of Science and Technology, Moscow, Russia

Xin Wang China Academy of Launch Vehicle Technology (CALT), Beijing, China

Jessica L. West Project Ploughshares, Waterloo, ON, Canada

Douglas Wiemer RHEA Group, Wavre, Belgium

Stefano Zatti European Space Agency (ESA) – Security Office, Frascati, Italy

Dong Zeng China Academy of Launch Vehicle Technology (CALT), Beijing, China

Vasilis Zervos University of Strasbourg and International Space University (ISU), Strasbourg, France

Shengjun Zhang China Academy of Launch Vehicle Technology (CALT), Beijing, China

Stefano Zatti has retired.

Part III

Space Applications and Supporting Services for Security and Defense

Introduction to Space Applications and Supporting Services for Security and Defense

37

Maarten Adriaensen

Contents

Abstract

This introductory chapter to the Space Applications and Supporting Services for Security and Defense provides the overview of subjects and areas covered in the Part 3. Apart from the introductory chapter, 20 contributions cover a wide spectrum of space-based and space-enabled services and applications. The section aims to cover all main applications areas in generic chapters as well as to provide information about niche areas in specific chapters. Together, the identified and covered themes and issues provide a comprehensive setting of space-based and space-enabled services and applications for security and defense.

Focus Areas

The security and defense interest in space came initially from the motivation to demonstrate ballistic missile capacities in the frame of nuclear deterrence. Subsequently, the operational use of space emerged considering the limitations of the ground and airborne assets used so far and the increasing added value from space-based services for Command, Control, Communications, Computer, Intelligence, Surveillance and Reconnaissance (C4ISR). In that frame, space enables services in

M. Adriaensen (✉)
European Space Agency (ESA), Paris, France
e-mail: maarten.adriaensen@esa.int; maarten.cm.adriaensen@gmail.com

© Springer Nature Switzerland AG 2020
K.-U. Schrogl (ed.), *Handbook of Space Security*,
https://doi.org/10.1007/978-3-030-23210-8_125

the fields of Satellite Remote Sensing (Earth Observation), Positioning, Navigation and Timing (PNT), Satellite Communications and Space Situational Awareness (SSA).

Numerous space operational applications routinely deliver indispensable fast and reliable services for security and defense. In terms of applications and services, space is indeed essentially the main global way of collecting, transmitting, and distributing information. Space-based services make a decisive contribution to the ability to conduct an effective foreign and security policy and to achieve whole-of-government security preparedness including defense aspects in a holistic approach. The relative dependency on space-based services in combination with the vulnerability of space assets drives countries to develop ways of protecting their critical space infrastructure, both civil and defense.

Satellite remote sensing information, based on the exploitation of imagery data derived from several category of sensors, electro-optical, radar, infrared (IR), multispectral, or laser, has important use in the security and defense domains. Earth Observation is an essential capability contributing to the intelligence picture at strategic, operational, and tactical levels and constitutes therefore a pivotal source of information for effective decision-making. A dedicated chapter on the role of satellite imagery in the frame of disaster management is included (H. de Boissezon and A. Eddy), as well as a chapter on Earth observation applications for security and defense purposes (F. Dolce, D. Di Domizio, D. Bruckert, A. Rodriguez, and A. Patrono). Complementing the chapters on specific remote sensing applications, a contribution on the regulatory aspects of remote sensing data completes the contributions on Earth Observation in this section of the Handbook (P. De Man).

Communication satellites are indispensable in support of command and control functions and are heavily used in military operations, requiring unique requirements for protected communications. The security operations also require such reliable communication services but without the same level of protection and thus rely on both commercial services and governmental satellites. The interpenetration of the military and civilian space communication services, however, increases with, on the one hand, the needs to connect mobile assets such as unmanned aircraft and network ground forces requiring large bandwidth and, on the other hand, the growth of the commercial market, driven by the demand for broadband Internet services and video distribution. A dedicated chapter on Satcom for security and defense is included in this section (H. Lueschow and R. Pelaez).

Positioning, Navigation and Timings (PNT) services and applications are crucial for security and defense purposes. GPS revolutionized military applications by providing accurate position, velocity, and time/synchronization information for all military assets worldwide, in all weather conditions, with a common time and geodetic reference frame supporting the essential interoperability requirement. Two chapters on GNSS for Security and PNT for defense are included in this section of the handbook (J. Martin; and M. Detratti and F. Dolce).

Space-based assets and services are facing multiple threats. A chapter on various threats to Space Systems provides a first overview of the threats faced by space-based assets and services, both encompassing the space, ground, and link segments (X. Pasco). Space services must be resilient to interference and must offer technical and procedural means to quickly remedy any interference occurring in services. Space system operators need to be able to identify the location and type of the interference or jamming source in order to take immediate and appropriate action. Besides electronic attacks, which interfere with the transmission of radiofrequency signals, cyberattacks target the data itself and the systems using data and information. A cyberattack on space systems can result in data loss, widespread disruptions, and even permanent loss of a satellite. One dedicated chapter addresses the setup of a cybersecurity space operations center (S. Llopis Sanchez, R. Mazzolin, I. Kechaglou, D. Wiemer, and J. Muylaert).

In the domain of SST and SSA, multiple chapters have been included in this section, considering the increasing global importance of SST and SSA, reflected in multiple programs developed in the USA, Europe, and other parts of the world. One chapter is dedicated to space object and event behavior quantification and assessment (M. Jah). Another chapter elaborates on STM through environment capacity from a holistic perspective (S. Lemmens). A specific chapter addresses the role of Artificial Intelligence in the frame of future SSA and STM (M. Vassile, L. Sanchez Fernandez-Mellado, and E. Minisci). The chapter on SST developments in Europe sheds a light on current development in the field of space surveillance capabilities in Europe (R. Peldszus and P. Faucher). Furthermore, this section includes a chapter on the status and evolution of space debris mitigation from a legal perspective (A. Nassisi, G. Guiso, M. Messina, and C. Valente). The risks to space-based services posed by space weather as well as the issue of Near-Earth Objects are addressed in this section with a specific chapter on space security in the context of cosmic hazards and planetary defense (J. Pelton) and a dedicated chapter on space weather services and applications (J. Janssens, J. Berghmans, P. Vanlommel, and J. Andries).

A dedicated chapter on space security and frequency management addresses the existing and upcoming challenges in that area (Y. Henri and A. Matas). The chapter on the use of space enabled services for food security focuses on the role of space assets and applications in the frame of agricultural post-harvest loss reduction in Africa (O. Adebola and S. Adebola). The chapter on a mega-constellation for space security provides a business case for commercial active debris removal (N. Veliev, A. Ivanov, and S. Biktimirov). Additionally, a specific chapter on security and defense applications from China features in this section (L. Shen, L. Hongbo, S. Zhang, and X. Wang). From the private sector perspective, a chapter is included on space security from the space operator perspective (J. Bureau).

Conclusions

Space is increasingly competitive, congested, and contested. The chapters in this section III provide a very broad scope of the space security and defense interests in space-based and space-enabled services and applications. Throughout the generic and specific chapters in this section III, the objective is threefold:

- Providing a coherent and representative picture of the generic security and defense interests for space-based and space-enabled services and applications
- Introducing niche areas through specific contributions that complement the generic overviews for various space subdomains (including Earth Observation, PNT, Satcom, SSA) and cybersecurity, with attention for future developments as well as legal and business perspectives
- Reflecting the increasingly competitive, congested, and contested nature of Earth's orbital space from the services and applications viewpoint

Disclaimer The contents of this introduction and any contributions to the Handbook reflect my personal opinions and do not necessarily reflect the opinion of the European Space Agency (ESA).

Earth Observation for Security and Defense 38

Ferdinando Dolce, Davide Di Domizio, Denis Bruckert,
Alvaro Rodríguez, and Andrea Patrono

Contents

Abstract

The contents reported in this chapter reflect the opinions of the authors and do not
necessarily reflect the opinions of the respective Agency/Institutions

Space-based Earth Observation is a consolidated capability providing added
value to reach information superiority, a crucial enabler for operations in both
security and defense domains. The availability and responsiveness of satellite
payloads, together with exploitation capacity, allow to plan, monitor, and inform
security and defense forces with performance not available with other means.

F. Dolce (✉) · D. Di Domizio
European Defence Agency (EDA), Brussels, Belgium
e-mail: ferdinando.dolce@aeronautica.difesa.it; davide.didomizio@eda.europa.eu

D. Bruckert · A. Rodríguez · A. Patrono
EU Satellite Centre (EUSC), Madrid, Spain
e-mail: Denis.Bruckert@satcen.europa.eu; Alvaro.Rodriguez@satcen.europa.eu; Andrea.
Patrono@satcen.europa.eu

© Springer Nature Switzerland AG 2020
K.-U. Schrogl (ed.), *Handbook of Space Security*,
https://doi.org/10.1007/978-3-030-23210-8_106

This chapter describes how the gap between security and defense domains is increasingly blurred and the capacity to exploit the "big data" made available by the satellite systems and other contributing missions is becoming a common technological and operational challenge.

Introduction

Space-based Earth Observation (SBEO) capabilities are one of the main data providers to imagery intelligence (IMINT) and geospatial intelligence (GEOINT) communities, since the technical and geographic information that can be derived from satellite systems through the interpretation or analysis of imagery is nowadays essential. However, SBEO products, including exploitation of imagery data derived from several categories of sensors, electro-optical, radar, infrared (IR), multi-spectral, or laser, can go well beyond IMINT/GEOINT domains and are used for both security and defense users for several purposes. Future SBEO satellites are providing big data from space and are building situation awareness, enabling the possibility to analyze the collected information, delivering products that will require strong optimization and improving in terms of delays in processing, interpreting, and disseminating to final customers. SBEO data/products, however, support also the monitoring phase, which relies on intelligence and is composed of two complementary functions: the early warning and the strategic surveillance. Furthermore, military planning, as well as geospatial support, also represents additional needs that can be accomplished through Space-based Earth Observation (SBEO) satellites' data and products at both political, strategic, and operational level.

In recent years, there has been an increase in the development of tools and techniques to improve the exploitation of collected imagery data also to face the proliferation of SBEO assets. However, it is judged that the security and defense communities have not fully benefitted from this development, and they will need tools and procedures to fully take advantage of these technologies and to increase the trust in such kind of future supporting capabilities. One of the main difficulties will be the need to better balance and leverage the skills of analysts and operators within effective and efficient operational workflows and trusted data exploitation algorithms.

This chapter is mainly focused on the analysis of current and future applications to support security and defense missions using Space-based Earth Observation sources.

Earth Observation Security and Defense Application Landscape

Earth Observation (EO) sensors mounted on space-borne platforms have now almost 50 years' life – successful – story. Three systems which represent the founding pillars of EO commercial satellites era: (1) Landsat-1 (1972), the first EO satellite to be launched to study and monitor the whole Earth's surface; (2) SPOT-1 (1986) that used a revolutionary commercial model for image distribution; (3) Ikonos (1999), the first commercial EO system capable to collect images with a ground sampling distance below 1 m (0.82 m) at Nadir (Denis et al. 2017). Meanwhile, US policy shift favored rapid market adoption for high-resolution satellite imagery anticipating a significant short- and long-term growth. Shortly after, DigitalGlobe launched QuickBird (2001).

In the last few years, there has been a proliferation of SBEO systems (archival, current, and planned – over 100 s of sensors (Committee on Earth Observation Satellites www.ceos.org)) and others are now planned up to 2030 and beyond. Performance of sensors and mission technology has progressed over the last two decades. Overall, missions experimented longer endurance than expected and both optical and SAR sensors meliorated their design increasing, e.g., sensing performance, positional accuracy, and platforms' agility. Moreover, satellite systems progressively moved from the single-sensor model to the constellation approach. Performances have been boosted as well by the progressive implementation of the "dual-use" systems concept that allow different user communities to manage and exploit them taking advantage of a synergetic approach (despite configurations and rules may vary from mission to mission). The most recent development is the launch of nano- and micro-satellites (with constellations that can reach 100+). Lowering the cost of access to SBEO, they are becoming increasingly more attractive than conventional satellites. As an overall consequence, availability and access to data obtained by space-borne missions are increasing – and will continue to – in an exponential way, offering better and truly affordable observation capabilities at a greater range of spatial, spectral, and temporal resolutions (Belward and Skøien 2015; Denis et al. 2016; Toth and Jóźków 2016).

Image analysis production based and organized as a sequential series of human interventions in a pipe way may soon get overwhelmed in the new scenario shaped by huge observation data handiness and increasing computing capability. Providers sitting on massive amounts of exploitable data and user communities progressively expanding their analytical appetite for new products and services need faster and further interactive production modalities. The increasing development of web-based solutions and cloud-based services has allowed better quality of online functionality and performance without having necessarily to host and manage the data. Fast access to extensive archives of data, integration of diverse workflows user-specific, qualification of providers and users to work in diverse but interconnected environments to consume data, provide services, generate information and distribute products, are step by step leading the way of EO exploitation and derived value-added production. Any implementation can/shall be adapted for ad hoc security environments, without implying different design but with enforcement of specific security protocols and restrictions – no misuse or free outflow (Holmes et al. 2018). As an example, NGA (former) Director Robert Cardillo, during his keynote at the 2018 GEOINT Symposium in Tampa, announced a new online platform for open collaboration and development of geospatial solutions.

An important component of EO in supporting the primary aims of the space and security and defense domain is the provision of image and geospatial intelligence products and services resulting from the exploitation of remotely sensed data acquired by sensors mounted on space-borne assets. The reflections or emissions measured by the different types of sensors are depicted in images that need to be converted into meaningful information. Observation data are currently showing a new unique scenario in terms of variety, volume, velocity, veracity, and value. Geospatial information and EO, together with modern data processing and big data analytics, offer unprecedented opportunities (Lee and Kang 2015; Nativi et al. 2015; Câmara et al. 2016). There are different application approaches to face this challenge and they mainly depend on the type of sensor used and the sort of information that needs to be extracted.

Historically, in the security-defense environment, information is derived through a subjective analytical approach principally based on the experience and the skills of the analyst who visually interprets the image(s). The spatial and contextual way to proceed varies and depends on the objective of the study. Spatial, pattern, texture, and, in general, spectral information is most of the time improved by standard image processing technics (i.e., image enhancement) for increasing the visual distinction between features. Different collateral/ancillary data, spatially and temporally corre-lated with the imagery, made available through different sources, may complement the analytical process providing worthwhile information, essential in helping, confirming, etc. the interpretation course and its inferences (Campbell and Wynne 2011).

When the analysis needs to cover large areas, perform quantitative investigation, implement complex monitoring, rapidly highlight features not detectable at first view, (semi)automation of the analytical process may facilitate the interpretation process, e.g., decreasing the analysis time span and the risk of poor detection rates when compared to only human, lengthy, scrutiny approaches.

The application of robust algorithms/models to transform spectral into "mean-ingful" information offers an invaluable support. Nevertheless, deterministic models have to be accurately parameterized according to the sensor performance, the nature of the analyzed variables, and the information to infer for a specific task (Adams and Gillespie 2006). Since this approach needs an exhaustive knowledge, testing, and repeatable conditions to establish firm physical relations (that not necessarily exist and that ideally should be supported by an extensive fieldwork activity that most of the time – in the security-defense domain – is unfeasible for the nature of the requests and/or its location), alternative ways to proceed are used to facilitate the analytical process.

The statistical analysis of the spectral information and its supposed relationship with the phenomena to be assessed is used to reduce or transform the dimensionality of the data and to increase either the computational efficiency of, e.g., an image classification or the understating and manual extraction of the analyzed features (Lillesand et al. 2014).

Spectral rationing with adequately chosen spectral areas and appropriate wavebands or combinations of wavebands may as well facilitate the depiction of specific information. They can be used to better reflect the image content and, as well, to further improve the performance of any of the hereby mentioned methods, including image fusion such as pan-sharpening techniques (Ghassemian 2016).

As temporal resolution of EO systems and constellations increased, multi-temporal data merging and change detection computing capacities augmented as well in terms of applicability and efficiency in supporting (semi)automatic monitoring of surface changes over varying time span intervals – including detection, estimation, and/or comparison of trends and dynamics (Fulcher et al. 2013; Hussain et al. 2013; Bovolo and Bruzzone 2015). This also simplifies the handling of the increasing load of imagery data, the controlling of alarms, and a better management of direct human involvement.

Where subjective, deterministic or statistical classic analysis become insufficient to identify relationships between the different pieces of available information – or simply are unknown or too lengthy, approximate, etc. to be established. Artificial Intelligence (AI) methods are progressively demonstrating the potential to get information out faster with more thorough and complete analysis. In recent years, neural network applications increasingly demonstrated better capability to automatically discover relevant contextual features in remotely sensed images (Arel et al. 2010; Long et al. 2017; Maggiori et al. 2017). Data volume and computational capacity increased exponentially, boosting precisely the application of neural network computing to satellite image (when compared to studies performed in the 1990s such as (Hepner et al. 1990) or (Atkinson and Tatnall 1997)). However, one of the major problems associated with precise recognition and extraction of objects from remotely sensed data is still the time and cost of wide-ranging training of algorithms, requiring experienced analysts (Ball et al. 2017), particularly when tasks to be undertaken are context specific and imply constant tailoring and precise knowledge, background, etc. as in the security-defense domain.

Collateral information gathered from social media are both worthy in supporting imagery analysis, and progressively more complex to use (i.e., floods of data, abundant, rapid, and accessible implying fast and qualified reactivity to provide the required situational awareness of relevant information) (Li et al. 2017). AI is as well improving the speed and accuracy of identifying enlightening evidences, allowing analysts to expand capacity, create new analytic products, etc. Reliable information gleaning has definitely progressed thanks to AI; nevertheless, it is still *in fieri* and constant adaptation and tailoring is often necessary to build up and maintain a knowledge data base, requiring expert interpretation processes to cope with uncertainty and/or incomplete information extraction.

The choice of analytic approach depends on the available data, the degree of understanding of the processes under examination, and the possible relationship between the EO data inputs and the goal of the analysis. In the security-defense domain, when the rather heterogeneous portfolio of possible EO-based services is considered, there is no rigid predefined approach to tackle any specific task. Experience, to be read as knowledge, understanding, mastering, etc., is at the core of any study and will guide the analyst to choose and combine, in an optimized way, any of the above-mentioned approaches, according to the context and the data availability. While operational use of EO keeps growing, gaps and opportunities for further development to tackle increasingly complex operational applications still exist and will always need adequate experienced human supervision throughout the entire analytical process.

Earth Observation Missions and Applications for Security and Defense

Space-based Earth Observation is now able to satisfy the growing needs of both security and defense entities and private customers coming worldwide since space systems are now becoming more and more numerous. Nowadays, it is expected that almost an infinite amount of information will be available, creating a high level of common awareness, while just a few years ago the prediction of a future information age was providing a different outlook. As matter of fact, the US commercial approach, known as "new space," is moving the market in the clear direction of an easier and cheaper access to space, reducing the life and dimensions of space missions and increasing the number of systems in orbit (Space Strategy for Europe – European Commission COM 2016). Governmental institutions and small countries can see micro-satellites' capabilities as the only opportunity to reach an independent, confidential, and trusted space-based capability due to the lower cost in development and launch phases they are promising.

On the other hand, in addition to real information, there is a lot of misleading information that can become a threat in modern warfare scenarios. In the past, such kind of information was not considered a relevant threat since they were limited to few numbers of potential events, while today it represents one of the most challenging threats to face. Criminal organizations can express their soft power generating misleading information, e.g., in the cyber domain. From this prospective, space-based information and communication services can represent a reservoir in terms of reliability and trustiness of the information more than other alternate sources.

In this congested and competitive space environment, EO products can certainly be derived by different platforms and the integration of the information coming from several sensor classes will represent the new bottleneck. With the availability of big data coming from space, such a huge offer of space imageries could move the equilibrium from the space to the ground segment. If yesterday access to space was the real challenge, and possibility to get access to space capabilities was the key enabling factor, now this is not anymore the case: the challenge will be the capability to acquire, store, manage, process, and deliver reliable and timely information, to be extracted by all essential data. Military will continue to define SBEO requirements in terms of accuracy and spectrum band; however, data fusion and integrated products merging different EO data, Positioning Navigation and Timing (PNT), and communication capabilities will be the key to deliver effective recognized pictures for defense operations.

Even considering that, the Ministries of Defences (MoDs) cannot certainly rely on commercial application to accomplish their task, especially if the data are provided by foreign companies, that are able to exercise a shutter control in certain specific time and area of interest (AoI). It is then easy to understand that space institutional flagship programs turn to be strategic as they provide not only a full set of information but also the control of the data acquisition, flow, policy, and security.

Space-based data moreover solve a key issue in terms of autonomy to the MoDs. In fact, one of the biggest strengths of the SBEO systems is that they are not affected by sovereign rights of States "overflown" by spacecrafts (United Nations Treaties and Principles on Outer Space 2002). This makes possible to obtain information about the area of interest through means regulated by agreed international laws, without any engagement of the States overflown by the spacecraft.

Even if the difference between defense and security domains is not easy to identify and both concepts could lead to misinterpretation, it could be summarized as the following: security's main task has to face with Member State's internal risks without a prerecognized enemy or attack to face, e.g., terrorism; on the other hand, defense's main task has to face with Member State's threats against an external identified enemy (Britz and Eriksson 2005; French white paper on Defence and National Security 2013). From this simple, but of course not exhaustive definition, it is clear that the capabilities required to deal with these two different scenarios are not necessarily equal. Nevertheless, the evolution of the global international scenario is generating boundaries that are quite often not clearly defined. The power's global model, in fact, is evolving quite rapidly moving from a clear unipolar international system after the end of Cold War, when some distinguished authors declared "The End of History" (Fukuyama 1989) to a more global and fragmented multipolar model, where the symmetry of previous scenarios is not anymore applicable. This asymmetry is certainly reflected into military operations, coping with a hybrid warfare scenario and threats that cannot be easily identified. In such conditions, the evolution of guerrilla environments led to an unclear definition and delineation of geospatial limits. The time when the Greek arena's competition model was applicable looks today as an ancient memory, while strategic models based on oriental philosophies, referring mainly to Sun Tzu's doctrine (Tzu 2007) where the art of camouflage is a key capability, are becoming more applicable to modern terroristic threats.

As a direct consequence of these new scenarios, the boundaries between internal and external activities are clearly not well identified, calling for an increasing application of defense capabilities for homeland security. Defense techniques, procedures, and expertise are now finding a great demand in the civilian and the security world (European External Acton Service 2016).

Nevertheless, there are still specific tasks related to defense domain that mainly stick with military operations and this is true also in the case of SBEO applications. In EU dimension, the taxonomy developed in the framework of the European Defence Agency, the "Generic Military Task List" (GMTL) clearly define some tasks that are not applicable to security dimensions. The GMTL, for example, refers to the conduct and synchronization of joint precision strike aimed to conduct efficient application of joint precision firepower. For such kind of tasks, SBEO data and products can play a key role. High-accurate weapons, in fact, are based on such kind of information that, if properly elaborated and ingested in the weapon system, produce a high added value. With the increase of revisit time and with the decreasing of processing time, also battle damage assessment (BDA), a typical military task could be supported by SBEO capabilities on top of more tactical vehicles, and a potential link between automatic change detection algorithms and tactical operational commanders could produce effective information (https://www.eda.europa.eu/what-we-do/activities/activities-search/persistent-surveillance-long-term-analysis-(sultan) 2019).

Furthermore, military planning is underpinned by a continuous process of information collection, military assessment, and analysis. The strategic planning, in particular, relies on information to be collected in conditions where forces are not yet deployed and the "expeditionary" characteristic of satellite systems, able to reach faraway points on the planet in a few hours and in the next future will be able to provide near-real time information with global coverage, are fundamental. On the other hand, geospatial support is a key enabler also for the planning and execution of military and civilian missions and operations, training, and exercises, and it is based on imageries also coming from space domain, supporting, in this case, tactical functions. Nowadays, geospatial support is essential in everyday life and hence it is even more necessary in security and defense operations (EU Capability Development Plan 2018). SBEO data are the pillars and the first layer to build on further information and to derive multiple products for multidomain assessments and to provide effective tools for decision making and military or mission commanders.

In addition to these specific military missions, in the domain of SBEO, there are three fundamental general requirements driving and steering the development of military space systems: availability, confidentiality, and integrity.

Starting from the integrity requirement and keeping in mind the disinformation threats are world-scale threats; it can be stated that only with an independent, well-defined, and verified information source, it is possible to implement armament control, confidence-building, and treaty monitoring, in particular in a framework of a common defense and security policy. To achieve this goal, MoDs shall have a reliable information source to reach a common situational awareness; otherwise, it will be difficult to set up a room to agree on a common foreign policy and to deal with common threats as well as to verify information accuracy. The point is, how such kind of requirement can influence the developments of future space-based reconnaissance systems.

In addition, SBEO applications present governance, data security, service continuity, and business model criticalities. For instance, the use of open-source applications not only involves criticality about the services themselves, but also allows to the service provider to gather and store key information about uses and users. The confidentiality is a general key issue for the future of information technology and this is particularly true for defense users, as revealed by recent application cases such as the application able to collect military positions around the globe through the use of connected fitness trackers (Fitness tracking app Strava gives away location of secret US army bases 2019). The same problem can be applied to commercial SBEO providers, where even only the information about the area and time of interest could represent an intelligence information, pointing out the importance of the confidentiality requirement. These issues have a direct impact in terms of SBEO needs for military missions. It raises the problem not only of the production and the availability of the information, but also the question of the control and the security of the data provided for the MoDs use. When imagery is obtained through commercial companies directly contracted by local MoD, the integrity of the information could not be guaranteed. Technically speaking, imagery data can be manipulated, even if such kind of theoretical operations could require some delay in providing the requested service. By building up its own fleet of satellites or strong restricted commercial licenses, including ground segments and processing, these potential concerns are not in place anymore.

Finally, also based on recent military operations' experience, where a coalition of States is involved, the same data might be needed by all of them at the same time, implying the requirements of the availability of the data. For this kind of issue, data exchange agreements must be addressed accordingly, leading in some cases to considerable additional costs and delays, while a broader and structured pooling and sharing approach would probably lead to more effective benefits for the coalition.

Security and Defense EO Application

Earth Observation from space in the defense sector was largely used historically for intelligence purposes, being considered as an extension of the capacity of spy aircrafts. In particular, the branch of intelligence dealing with imagery is known as IMINT.

IMINT is the technical, geographic, and intelligence information derived through the interpretation or analysis of imagery and collateral materials. It includes exploitation of imagery data derived from several categories of sensors: electro-optical, radar, infrared (IR), multispectral, or laser (US Joint Publication 2013).

The use of SBEO systems was initially devoted to specific strategic tasks (e.g., nuclear sites discovery). The current improvement of sensors' performance, the agility of the satellite platforms, and the possibility to integrate different datasets are important enablers allowing the use of SBEO also for more specific and repetitive tasks, even in direct support to missions and operations.

In this regard, system design parameters may however impose constraints on the ability to use SBEO satellites in military operations. The architecture of the mission and the choice of the orbit is one example of these constraints.

Traditionally, SBEO missions have been conceived with the use of low Earth sun-synchronous orbits. In this case, the complexity of system design was manageable thanks to the advantages of orbit stability, global coverage, constant sunlight on the platform, and of advantageous geometries for imagery collection. This type of orbit however limits the capacity of continuous observation (e.g., areas at equatorial latitudes are visited only twice a day), and moreover the satellite passes on target locations always at the same local time, reducing the possibility of discretional imaging. Constellations including several satellites, although improving the performance of continuous observation, would hardly be considered as sole source of information in the case of military operations.

Indeed, IMINT can be collected via satellites, but also with other assets: unmanned aerial vehicles, reconnaissance aircraft, and ground systems. These assets are not interchangeable and should be used in combination. A recent study conducted by the European Defence Agency evaluated the potential options to enhance collection capabilities in the area of IMINT through innovative and technologically feasible solutions, to meet the need of persistent surveillance of wide areas in defense and security operations (https://www.eda.europa.eu/what-we-do/activities/activities-search/persistent-surveillance-long-term-analysis-(sultan) 2019). To this extent, the analysis based on operational scenarios provided the respective merits of assets/systems based on geostationary earth observation satellite systems, constellations of optical and radar small/mini satellites in low earth orbit, High Altitude Pseudo-Satellite Systems (HAPS), and Remotely Piloted Aircraft Systems (RPAS). The quantitative analysis performed, while showing that the performance in resolution of geostationary EO satellites seems yet to meet the requirements of military operations, demonstrated a real complementarity between the LEO constellations and other technologies which are likely to be used concurrently or successively in order to achieve the objectives pertaining to a given phase of operations.

The intelligence communities are used to develop their activities on the basis of the so-called intelligence cycle. The IMINT cycle mirrors the intelligence cycle. The steps in this cycle define a sequential, interdependent process for developing IMINT. The management of operations of SBEO systems used to produce IMINT is typically harmonized with the steps of the IMINT cycle: tasking, collection, processing, exploitation, and dissemination processes (MCRP 2-10B.5 Imagery Intelligence – US Marine Corps).

Concerning the exploitation of imagery information, imagery analysts have a central role in this domain, especially taking into account the traditional approach mostly built on visual interpretation of satellite imagery.

In the above described framework of big data environment, the traditional analysts' task of building situation awareness and producing actionable intelligence is changing and needs to be supported by modern tools to obtain the promising enormous added value coming from such numerous amounts of data. In several cases, current tools are not able to adequately support analysis, producing delays in the processing and in the interpretation or not allowing to take advantage of the real potential of big data.

In the defense domain, the use of modern technologies might be hampered by the need to comply with security rules, to work on "closed" classified systems to protect the data and the information, not relying on the support of distributed resources normally available in large private networks or on the Internet.

In the last years indeed, we witnessed a large development of tools and techniques reaching a good level of maturity in providing useful information by exploiting collected imagery data. However, the military operational communities have not benefitted in full of this technology growth. For instance, although new techniques recently presented in the domain of big data analytics can provide added value for the security domain (Popescu et al.), a direct implementation in the defense applications needs to be properly addressed duly taking into account the still existing difficulties to put together the architectural elements of a cloud-based processing and the security constraints of classified systems. This does not mean however that defense imagery analysts are condemned to work with archaic tools.

As described previously, an important area of development is represented for instance by the future development of application of deep learning and artificial neural networks for imagery analysis. These capabilities will help to identify and refine the behavioral models by parsing and correlating the voluminous data streams available from space assets. Anomaly detection tools based on this concept are already available in Europe for the maritime domain with dual-use applications, valid both in defense (maritime situational awareness) and in security scenarios. Combining satellite radar imagery with Automatic Identification System (AIS) (IAC-14-B1.5.4 Cosmo-Skymed data utilization and applications), Vessel Monitoring System (VMS), coastal radars, and any available intelligence data provide useful information to build a database of normal behaviors concerning the vessel tracks in specific area. Any deviation from recognized track patterns might be considered as an anomaly to be further investigated.

This is one practical example of the use of Synthetic Aperture Radar (SAR) satellite imagery. This technology has become a consolidated asset of military SBEO in Europe, thanks to important satellite programs (ref. COSMO-SkyMed, SAR-Lupe, COSMO-SkyMed Second Generation, SARah). The evolution from the first generation of the years 2000–2010 to the one under development in these years is making available considerably larger amounts of data, thanks to the improved resolutions, larger swaths, and more imagery per orbit.

In this case, the challenges deriving from the increased amount of data are complicated by the inherent complexity of SAR data and by the preponderance of historically well-established procedures that make use of electro-optical images to support military operations and the decision-making process, relegating in several cases SAR imagery to a secondary source of information.

On the contrary, a thorough exploitation of SAR imagery strengths would enlarge the use of SAR imagery alone and/or in combined use with electro-optical images, thus taking full advantage of its unique 24/7 and all-weather characteristics, therefore raising the effectiveness of investments made by several European Ministries of Defence on SAR satellites.

Ongoing studies are investigating new techniques aimed at developing solid procedures in support of SAR imagery analysts, overcoming the inherent difficulties of interpretation of "salt and pepper" images and with the objective to reach high automation levels (https://www.eda.europa.eu/docs/default-source/eda-factsheets/2017-04-03-factsheet_react 2019).

The tasks can be performed by skilled analysts or by operators that might use tools developed for that purpose. In this regard, software exploitation tools for SAR images are available; however, the drawback is that those are not always able to extract and present the information that makes SAR images a product "easy and ready to use." In addition to this, the intrinsic peculiarities linked to the programming cycle of an SAR product and the lack of proper tools to assist the preparation of a task constitute an additional hurdle that limits the use of SAR images at operational level.

The procedures to analyses data are based on operational workflows. Those are defined as a series of activities that typically encompass several tasks: e.g., data preparation, data processing, visual interpretation. Operational workflows can be tailored on the basis of operational scenarios (ports, airfields, urban, lines of communications, industrial compound, etc.).

Data preparation are normally executed, thanks to the most common software functionalities already available in the market, e.g., co-registration, phase coherence extraction, geocoding, ortho-rectification.

Data processing would benefit from algorithms and tools available in the market or developed on purpose, according to the need of users, e.g., layover analysis, change detection (amplitude, coherent, or incoherent), edge detection and feature extraction.

The definition of workflows has a twofold advantage. First of all, the workflows become a guided process for imagery analysts through the complex applied physics of the SAR imagery interpretation. Secondly, in the near future, with application of deep learning techniques, it would be possible to train semiautomated systems to execute the workflows, requesting the intervention of the imagery analysts only in case of abnormal behaviors.

Military applications already investigated falls in the domain of damage assessment (Fig. 1), target analysis, monitoring, and military planning.

Significant elements characterizing defense-related SBEO applications have been described, also providing information on more recent developments in this domain.

Fig. 1 Multicoherence product from the execution of a workflow for damage assessment COSMO-SkyMed image © ASI 2017

The use of adequate satellite constellations with suitable architectural characteristics and possibly in combination with other collection sources is an important enabler. Furthermore, in order to be effective in current operational scenarios, military SBEO applications need to find the proper balance to use modern exploitation and analysis capacities and flexible dissemination chains with the constraints of secure environments typically set up to protect classified information.

Examples of EO Operational Tasks and Services for Security and Defense

Examples of EO Operational Tasks

The public domain has the perception of how SBEO works based on what they have seen in the movies rather than in the actual orbital dynamics that govern the movement of the satellites. The inescapable truth is that, once a satellite has been inserted into its orbit, there is not much we can do to control the moment at which it overflies our target of interest other than wait. This introduces a number of caveats that need to be carefully considered when using such systems for security and defense applications. Hence, the expression commonly used by image analysts who say that "when you need an image of a certain location the satellite is usually on the other side of the Earth; and when it finally reaches the desired coordinates, they are always cloud covered."

Fortunately, while this was usually the case two decades ago, the proliferation of satellite platforms that we have seen in recent years has somehow alleviated this limitation, increasing dramatically the number of passes/day over any given location. However, despite the efforts of some companies that claim to be able to provide imagery every 3 h, we cannot ignore the fact that a satellite does not and will not (for the time being) provide the same live feed as other systems such as RPAS or potentially HAPS, already mentioned above. Thus, although these are increasingly frequent, the views that they provide are still limited to particular instants in time. Thus, the image analysts have developed a series of skills over time that allow make assessments based on hypotheses developed using these views. It would be equivalent to try to understand a movie while only being able to see certain frames.

Image analysts call certain features that they use to elaborate these hypotheses "indicators." For instance, the sudden appearance of inflatable rubber boats at a makeshift illegal migrant camp located on a specific coastline is an indicator that, even if there are no departures visible on the image yet, there is a very high probability that launches will soon be taking place. Of course, the presence of indicators is very strongly associated to the identification of "patterns of life" or "patterns of behavior." And these, in turn, are associated with the continuous observation of a location of interest, or what is called "monitoring." Monitoring allows the analyst to establish a baseline, a visual understanding of the type and level of activity that is common at a certain location. When the analyst sees an event that departs from this usual activity, something that may be called an "anomaly," an alert can immediately be triggered and the level of surveillance be increased to identify the causes and possible consequences of such change. Of course, the reliability of the assessment is directly correlated with the duration of the monitoring period, meaning that longer baselines provide better results.

Some examples of this application are the monitoring of military installations, such as ports and airfields, that serve as a baseline for the detection of the deployment of certain types of weapons systems, troops, aircrafts, and vessels that may have strategic implications for the region: arrival/departure of aircrafts and vessels, deployment of SAM or SSM systems, improvement of facilities, development of new infrastructures, identification of the level readiness of the different units occupying the military installations, assessment of their operational status, estimation of their capability, etc.

Another example very commonly related to SBEO monitoring for defense is the field of treaty verification. This was in fact the origin of Open Skies, an initiative signed between the USA and the former USSR at the peak of the Cold War to guarantee support to the mutual assured destruction (MAD) doctrine by providing means to each of the parts to ascertain what the other was doing. Today, satellite imagery is used to monitor the development of nuclear weapons by measuring the level of activity taking place at well-known uranium mines, or monitoring the status of certain processing and enrichment plants or gauging the performance of certain nuclear reactors where plutonium is known to be produced, or assessing the results of nuclear detonations carried out at carefully concealed underground test sites. Monitoring is also the basis for the assessment of a country's strategic outreach in terms of its capacity to project power, either through the deployment of forces or the use of weapons of mass destruction (WMD) and their means of delivery. Other additional requested information, for example, are the capacity and status of their naval units: how many cruiser vessels do they have available; if they are building aircraft carriers: how many, when they will be operational; if they have ballistic missiles: how far they can go, from where are they launched; if it is likely to be another launch test soon: how accurate they are; where are their strategic bombers deployed; and so on.

Monitoring tasks generally account for a significant portion of SBEO applications for security and defense. There are other uses, however, for which intelligence derived from satellite imagery is also critical. One of this use is obviously military planning, an activity which occurs generally before actual events take place. The term coined for this in military parlance is "intelligence preparation of the battlefield (IPB)." There are numerous instances where products derived from images may support the IPB process: terrain reconnaissance, multicriteria cross-country mobility analysis (CCM), identification of Go/No Go areas, visibility analysis, analysis of critical infrastructures, route analysis, contingency planning, training, etc.

Other uses involve the assessment of a situation on the ground after a certain event has taken place, like an airstrike (BDA). Another very frequent post-event application of SBEO is the validation of intelligence obtained through other sources. In this regard, there is an increase of demands that deal with the investigation of illegal activities, including cross-border crime (CBC). A significant amount of these have to do with the trafficking of drugs or weapons, which pose an important security threat to

EU Member States. Most of them are related to the existence of vessels, aircraft, trucks, and other means of transportation and the need to confirm their presence at certain locations such as ports, airfields, or border crossing points (Figs. 2, 3, and 4).

The list of examples is obviously nonexhaustive and it leaves out some other plausible uses of SBEO for security and defense. However, we cannot close this section without mentioning one important security application which is the management of the crisis following natural disasters such as earthquakes, wildfires, or floods. In these cases, it is critical to have immediately after the event updated maps and spatial datasets of the theaters of operation which will most likely have changed significantly due to the unfolding of the disaster itself. These datasets will provide the rescue teams with the necessary information to establish priorities and make informed decisions on the ground as soon as possible even before arriving at the disaster area.

Security and defense operations and information managers will face a wide range of situations involving different requirements and end users. Industry and techno-logical innovation are developing at such a pace that the offer of SBEO services available is increasing exponentially. Now, more than ever, the GEOINT profes-sional needs to amplify his/her domain of knowledge in order to incorporate an understating of the different options available in order to choose that which better satisfies the needs of his/her customers. In most cases, the solution will consist of a mix of different tools, platforms, and sensors that, properly combined, will cover all the aspects of any given situation and provide the most efficient answer.

Copernicus SEA

Cooperation between the EU Satellite Centre and the European Commission (EC) is a key enabler for SatCen EO applications development. Such cooperation started more than 10 years ago with a strong involvement of SatCen in the EC research

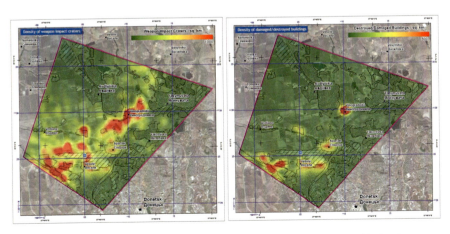

Fig. 2 Density maps comparing the weapon impacts visible on the image with the damage to buildings and infrastructures

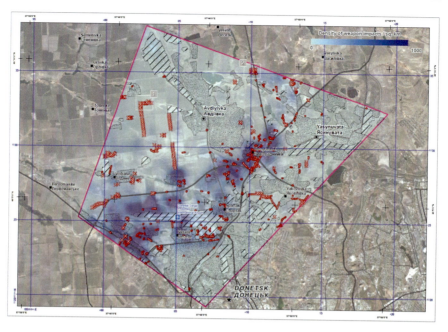

Fig. 3 Density map representing weapon impacts overlaid with the different military positions and equipment observed on the image

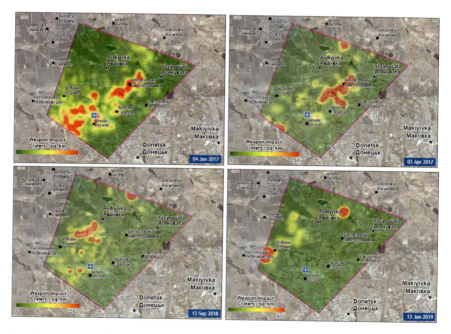

Fig. 4 A temporal series representing the evolution of the weapon impacts over the duration of the conflict

projects mainly in the areas of space and security and, in particular, through the FP6, FP7, and finally the H2020 Framework Programme.

The main element of this cooperation has been, and remains, Copernicus and several projects such as LIMES, GMOSAIC, G-NEXT, and BRIDGES that prepared the future operational role of SatCen in Copernicus, setting up the preoperational framework for the services that started in 2017.

Thus, Copernicus Support to EU External Action (SEA) is the result of many years of research and development by SatCen in partnership with the Industry under the European Union's Framework Programme for Research and Technological Development materialized by the transition of SEA from research and development and preoperational service provision to a fully operational mode.

Copernicus SEA is embedded in the Copernicus programme security component, therefore part of *"the world's largest single programme for observing and monitoring the Earth, for the ultimate benefit of all European citizens"* (Copernicus Support to Eu External Action Website) (Fig. 5).

Copernicus is composed of three components:

– *The space component*. This includes two types of satellite missions: Copernicus dedicated Sentinels and commercial or other space agencies' missions, called Contributing Missions (including very high-resolution satellite missions critical for security applications)
– In situ measurements (mainly ground-based providing information on oceans, continental surface, and atmosphere)
– Six services offered to authorized users and public

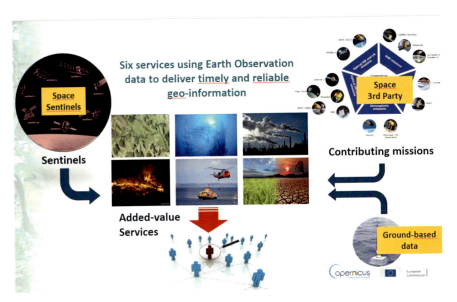

Fig. 5 Copernicus Programme structure – Source Commission DG-GROW (Presentation at SEA User Workshop – Paris)

The six services are land, marine, atmosphere, emergency, climate change, and security. Each service is delegated to different "entrusted entities."

Regarding the Governance, the EC has the overall responsibility of the program, and it is assisted by the Copernicus Committee including Member States, a Security Board (specific configuration of Committee), and a User Forum, as a working group to advise the Copernicus Committee on user requirements aspects (Regulation (EU) No 377/2014 of the European Parliament 2010).

The Security Board is involved in the management of information security for Copernicus and addresses issues such as the cyber security of the space and service infrastructures (Fig. 6).

Copernicus Security Services

The security service is to provide information in support of the civil security challenges of Europe improving crisis prevention, preparedness, and response capacities, in particular for border and maritime surveillance, but also support for the Union's external action, without prejudice to cooperation arrangements which may be concluded between the Commission and various Common Foreign and Security Policy bodies, in particular the European Union Satellite Centre (Regulation (EU) No 377/2014 of the European Parliament 2010).

In three key areas, i.e., Support to EU External Action, Border Surveillance and Maritime Surveillance, the security service is being implemented by the following entrusted entities: SatCen, FRONTEX, and EMSA. The operations started in 2016 for the Border Surveillance and Maritime Surveillance components of the security service and in May 2017 for the Support to External Action component.

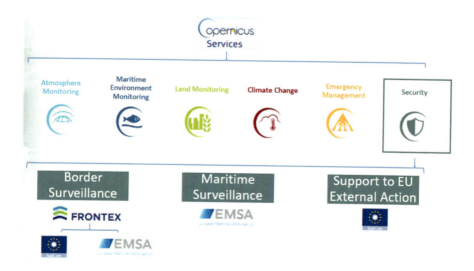

Fig. 6 Copernicus services (and components in security). (Source Commission DG-GROW)o

SatCen's main contribution is materialized by the role of entrusted entity for the operations of the Copernicus service in Support to EU External Action (SEA); SatCen also supports Border Surveillance through a Service Level Agreement with FRONTEX.

SatCen is thereby entrusted with the operational management of the Copernicus SEA service. Today, SEA addresses service production mainly through issuing and management of industrial service contracts such as a Framework Contract for "Geospatial production" but also the production of sensitive layers of information by image analysts and quality checks at SatCen. In addition, SEA implements user uptake activities mainly for the enlargement of the user base as well as service evolution activities taking benefit of state of the art in research and technological developments. For user uptake activities at least two workshops are organized per year. SatCen also implements a focal point for service's "Authorised Users" in the "SatCen Brussels Office." Security consideration regarding the requests is fully taken into account as each request is evaluated by the SatCen Tasking Authority (EEAS) from the sensitivity point of view. As Copernicus SEA does not currently manage EU Classified Information (EUCI) (2013/488/EU 2013), if a request is considered too sensitive and needs to be classified, it could be managed, if relevant, outside the perimeter of Copernicus as a SatCen classified task.

SEA's objective is to assist the EU and its Member States in civilian missions, military operations, and interests outside EU territory. It is designed to support the EU by improving the situational awareness of European Commission, European External Action Service, and Common Security and Defence Policy stakeholders including the Member States. The service can be activated to respond within very short timescales, as is necessary in cases of responses to crises such as political or armed conflicts. On the other hand, it is possible for the service to carry out monitoring campaigns over longer periods of time in order to develop a picture of how phenomena on the ground are changing. The primary target users are European entities, the EU, and Member State Ministries of Defence and Foreign Affairs as well as key international stakeholders, as appropriate under EU international cooperation agreements such as United Nations.

SEA Service Portfolio

After a ramping up of the service, SEA reached its full operational state in 2018 with, as mid-2019, more than 140 activations received from authorized users from EU Institutions, in particular EEAS and Member States. SEA products were built using mainly Copernicus Contributing missions as well as Sentinels satellites data as complementary sources (Fig. 7).

Mid-2019, the SEA service is mainly activated by the EU External Action Service: from the nine services of the portfolio, seven have been used so far (Fig. 8).

Analysis of EO data based on different techniques is used to identify patterns of illegal activity in an area of interest. Optical very high resolution (VHR) imageries are used to identify vehicles and infrastructure potentially suspicious. Radar Sentinel imagery interferometry techniques are used to identify the use of paths and roads during a time lapse.

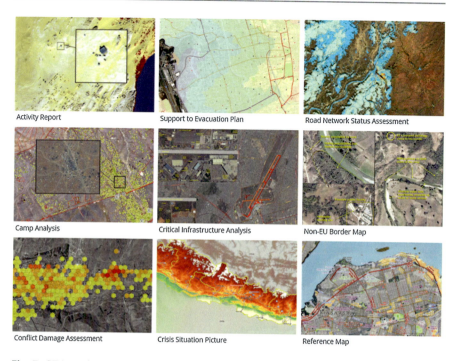

Fig. 7 SEA service portfolio. (Source SatCen (European Defence Action Plan – COM 2016))

Evolution of EO Services and Application at EU SatCen and Copernicus SEA

SatCen Service Evolution: Artificial Intelligence/Machine Learning

The concept of artificial neural networks and the theory of how these could be applied to a number of different applications, particularly in the field of EO and remote sensing, have been deeply described in the previous paragraphs. The development of the computing power necessary to drive this major breakthrough has reached critical mass, thanks to the continuous increase of chip capacity. Moreover, big data must be carefully stored over years of increasing generation and ingestion of information. But big data in itself is not useful. It only acquires a meaning if we are able to exploit it in such a way that it allows us to identify patterns, understand behaviors, bring to the surface the hidden structure of a certain phenomenon, and even predict what is going to happen next. It is particularly important in the field of Artificial Intelligence and Machine Learning (AI/ML) because for the first time SBEO service providers such as EU SatCen have accumulated enough data to train the algorithms to such an extent that they will provide meaningful, reliable, and actionable results. And once they are trained, the expectations are that these

Fig. 8 Example of an SEA activity analysis product for detecting smuggling and other illegal activity. (Source SatCen – SEA product portfolio (Copernicus website www.copernicus.eu))

algorithms will be able to breeze through the data and draw conclusions that would otherwise take an unfeasible amount of time for a human to reach.

There are numerous situations in the field of SBEO where AI/ML is already being used. Experience has shown that the algorithms are particularly efficient at performing repetitive tasks that may seem pretty straightforward in terms of complexity but often excruciatingly tedious for an analyst, such as scanning an image in search of changes or looking for certain objects like armored vehicles, aircraft, air defense sites, or other sorts of military equipment. At the EU SatCen, for example, it is not considered a future scenario in which the machine will eventually end up substituting the human analyst. There is a strong belief that certain traits which are common in successful image analysts, such as the capacity to unveil causal associations between elements on the image, or the ability to understand spatial relationships, or the facility to elaborate probable hypotheses to explain what is being observed, will very hardly, if ever, be outperformed by a machine. Thus, what is envisaged as a more likely scenario is one where the image analyst takes full advantage of the power of AI/ML to automate tasks such as automatic change detection and automatic feature detection and only intervene when the algorithms flag an alert, to alert that some relevant event has been found. This idea, which is sometimes known as a "tip and cue" approach, may fit surprisingly well with a hybrid SBEO collection plan which could include a mix of different sensors with complementary capabilities. As an example, to illustrate this, consider a situation where access is guaranteed to a constellation of microsatellites that provides fresh imagery at a medium spatial resolution but very high cadence, e.g., 3 m pixels every 2–3 h. The precision given by a 3 m pixel may not be enough to identify the type of equipment present on the ground, but if you know already what you are dealing with because of higher resolution imagery acquired at an earlier date, the medium resolution-high cadence imagery may be more than enough to highlight a change in the level of activity and trigger an alert. The analyst can then use the awareness of this event to tip off another constellation with higher precision sensors and program an acquisition with a better spatial resolution, and then may confirm the assessment. If the identification of the changes that triggers this mechanism can be done automatically by an AI/ML neural network, the analyst can significantly increase the area of surveillance and wait for these alerts to pop up, thus covering a larger surface and using his skills more wisely.

Copernicus SEA Service Evolution

Within the Copernicus Security Service component, the service evolution aims at promoting changes to the Service, aligned with the overall Copernicus strategy. The goal is to improve the existing portfolio of services by adding or modifying existing products or by implementing changes within the production or activation and delivery systems that improve the overall service experience to the users.

First, Copernicus SEA service is constantly adapting its response to the upcoming applicable policies, in particular those policies governing the EU External Action such as the EU Global Strategy for the European Union's Foreign and Security Policy and the Space Strategy for Europe, both issued in 2016. Any other relevant EU Policy will be considered as well.

Space Strategy for Europe (Space Strategy for Europe – European Commission COM 2016) states that *"Additional services will be considered to meet emerging needs in specific priority areas, including … (ii) Security and Defence to improve the EU's capacity to respond to evolving challenges related to border control and maritime surveillance with Copernicus and Galileo/EGNOS. This expansion will take account of new technological developments in the sector, the need to ensure adequate level of Security of the infrastructure and services, the availability of different data sources, and the long-term capacity of the private sector to deliver appropriate solutions."*

European Defence Action Plan (European Defence Action Plan – COM 2016): *"The Commission shall explore how Copernicus could cover further Security needs, including Defence. It shall strengthen Security requirements and will reinforce synergies with non-space observation capabilities in 2018."*

In the Space Strategy for Europe, additional services are considered in the area of security and defense. To some extent, the Copernicus SEA service could be considered as already implementing new services for defense and security, and therefore in line with the EDAP orientations (Member States defense users and CSDP military operations being part of the SEA users whenever they request to access the service within the context of the Common Foreign and Security Policy). The EDAP provides guidance on possible future evolution, in particular regarding the strengthening of security requirements and re-enforcing synergies with nonspace observation capabilities; this guidance shall be taken into account for the evolution of the Copernicus SEA service.

Nevertheless, SEA shall also be made available to new users having a bearing on the EU External Action. Copernicus SEA workshops, in particular the workshop organized in Paris at CNES (Centre National d'Études Spatiales) in December 2018, clearly highlighted that there are many potential new users in areas such as Ministry of Interior, Ministry of Foreign Affairs, maritime security actors, and agencies such as EFCA and EUROPOL that could get benefit from the service. Those users would need an easy access to the service, and this will have to be taken into account for its evolution. Regarding maritime security, it is worth mentioning the "European Union Maritime Security Strategy" (On 24 June 2014 the General Affairs Council of the European Union approved the "European Union Maritime Security Strategy" (EUMSS) 2018) endorsed by the EU Council. Its action plan revised in June 2018 specifically target *Support the conduct of CSDP missions and operations in the global maritime domain with EU maritime surveillance assets. ("In line with CISE* (Common Information Sharing Environment (A common information-sharing environment (CISE))), *ensure consistency and strengthen coordination between the existing and planned maritime surveillance initiatives on the basis of existing programs and initiatives by EDA, EFCA, EMSA, EUSC, FRONTEX, and other European agencies* (e.g. *ESA) as well as the Earth Observation programme (Copernicus), GALILEO/EGNOS (European Geostationary Navigation Overlay Service), and other relevant projects and initiatives. [MS/ COM/EEAS/EDA]")*.

Second, Copernicus SEA is strongly user driven and their requirements are fully taken into account both regarding the access to the service as well as the extension of the service portfolio.

Considering the rationale behind the Copernicus SEA, a set of predefined products has been defined and compiled in the Copernicus SEA portfolio, offering EU and international actors an initial pool of services that aim to tackle their needs in crisis situations or emerging crisis.

Service evolution is to bring new products to the users by extending SEA portfolio of services. Emerging requirements have been expressed, for example, in the areas of cultural heritage, illegal crop monitoring, security of EU/international events.

New products are achieved by finding new methods to exploit existing sensors by retrieving new types of information as well as exploiting new sensors and data. SEA service evolution demonstrated, for example, that the use of Copernicus Sentinels satellites was useful as complementary data based, for example, on the following capabilities: the revisit time of Sentinel-2, interferometry with Sentinel 1 to detect small changes in specific areas such as deserts, sea, etc.

Interagency cooperation is also a driver of innovation in this context, and it is worth mentioning the SatCen/EDA GeoHub project that is building a geo-spatial portal as well as the REACT project (briefly described previously (IAC-14-B1.5.4 Cosmo-Skymed data utilization and applications)) on the exploitation of SAR data. Both projects could be beneficial for SEA service evolution, as synergies are already well established.

Regarding the access to the service, SEA is currently benefiting from the infrastructure already in place at EU SatCen. The new developments planned for the infrastructure are aimed to provide the necessary hardware/software infrastructure to enable and optimize the management of the Copernicus SEA service, including activation workflow; seamless production and publishing; easier request and access to the products by the users. In the future, this infrastructure will need to be adapted to a considerable increase in data sources and volume, both for Earth observation and additional data, such as in situ, open source, etc. Additionally, the mentioned infrastructure must adapt to the need to "strengthen security requirements" and to "cover further security needs, including defense" (c.f. EDAP (European Defence Action Plan – COM 2016)) which might have an impact on the infrastructure in terms of the reinforcement of the capacity to process sensitive data.

Service evolution of this first phase of Copernicus SEA for the period 2014–2020 is currently extending the user community, the service portfolio, and is facilitating the access to the service.

SatCen is currently preparing with its partners the next phase of SEA within addressing "Copernicus 2.0" for the period 2021–2027, taking benefit of the results and lessons learned of service evolution during the first phase. A particular attention will be given to common requirements, interagency cooperation, interactive access through geo-portal, innovative tools such as artificial intelligence, and the availability of new space and nonspace sensors.

Conclusion

This chapter identified the current and future trend in the domain of Space-based Earth Observation (SBEO) from a security and defense perspective. Starting from a high-level state of the art, the current security and defense general needs have been described, pointing out how the future SBEO capabilities will be changed by the current new military scenario as well as the new space economy. In particular, the center of gravity will be more and more moved to the ground segment, always keeping in mind the specific military requirement of confidentiality, integrity, and availability of IMINT information.

Any SBEO capability shall be adapted for ad hoc security and defense environments, without necessarily implying different design but with enforcement of specific security standard protocols and restrictions, aiming to the interoperability and integration of different sources. The use of commercial and unsecured outflows can in any case represent a valid contribution that indeed needs to be properly balanced.

Considering the duality and increasing synergies between homeland security and external actions, the challenge will be in the implementation of a coordinated and holistic approach avoiding unnecessary duplication.

Some example of SBEO tasks and applications have been described, showing how the management of Artificial Intelligence and Machine Learning services will need to be properly customized to improve the inalienable analysts' skills, expensive, and precious resources that can be increased exponentially with tailored tools and related services.

Security domain, based on the experience of Copernicus Programme and EU SatCen services, nowadays is working with a cooperative model, delivering effective results in many applications.

This cooperative model has not yet reached the same level of maturity in the defense domain. However, significant efforts are conducted by national MoDs to cooperate on specific needs and activities. A further step forward might be a "pooling and sharing" model's application.

Furthermore, more support and contribution from EU institutions, eventually taking advantage of the security domain experience, tools, and facilities, might provide added value and cost benefit in the challenge of implementing a more structured and coordinated approach even in the defense domain.

The development of new common SBEO platforms/services could represent a first example (or the second one if we consider Galileo Public Regulated Service) of a European system to support defense needs of EU Member States.

References

2013/488/EU: Council Decision of 23 September 2013 on the Security rules for protecting EU classified information – Article 2: 'EU classified information' (EUCI) means any information or material designated by a EU Security classification, the unauthorised disclosure of which could cause varying degrees of prejudice to the interests of the European Union or of one or more of the Member States'

A common information-sharing environment (CISE) is currently being developed jointly by the European Commission and EU/EEA members with the support of relevant agencies such as the EFCA. It will integrate existing surveillance systems and networks and give all those authorities concerned access to the information they need for their missions at sea. (Source: https://www. efca.europa.eu/en/content/common-information-sharing-environment-cise)

Adams J, Gillespie A (2006) Extracting information from spectral images. In: Remote sensing of landscapes with spectral images: a physical modeling approach. Cambridge University Press, Cambridge, pp 1–38

Arel I, Rose DC, Karnowski TP (2010) Deep machine learning – a new frontier in artificial intelligence research. Comput Intell Mag IEEE 5:13–18

Atkinson PM, Tatnall ARL (1997) Introduction neural networks in remote sensing. Int J Remote Sens 18(4):699–709

Ball JE, Anderson DT, Chan CS (2017) Comprehensive survey of deep learning in remote sensing: theories, tools, and challenges for the community. J Appl Remote Sens 11(4):042609-1–042609-54

Belward AS, Skøien JO (2015) Who launched what, when and why; trends in global land-cover observation capacity from civilian earth observation satellites. ISPRS J Photogramm Remote Sens 103:115–128

Bovolo F, Bruzzone L (2015) The time variable in data fusion: a change detection perspective. IEEE Geosci Remote Sens Mag 3(3):8–26

Britz M, Eriksson A (2005) The European security and defence policy: a fourth system of European foreign policy? Polit europeenne 3(17):35

Câmara G, Ferreira Gomes de Assis LF, Queiroz G, Reis Ferreira K, Llapa E, Vinhas L, Maus V, Sanchez A, Cartaxo Modesto de Souza R (2016) Big Earth Observation data analytics: matching requirements to system architectures. BigSpatial16, Oct 31–Nov 03 2016, Burlingname. https://doi.org/10.1145/3006386.3006393

Campbell JB, Wynne RH (2011) Introduction to remote sensing. Guilford Press; 5th edition (June 21, 2011). ISBN-10: 160918176X ISBN-13: 978-1609181765. p 667

Committee on Earth Observation Satellites www.ceos.org

Copernicus Support to Eu External Action Website. https://sea.security.copernicus.eu/

Copernicus website. www.copernicus.eu

Denis G, de Boissezon H, Hosford S, Pasco X, Montfort B, Ranera F (2016) The evolution of Earth Observation satellites in Europe and its impact on the performance of emergency response services. Acta Astronaut 127:619–633

Denis G, Claverie A, Pasco X, Darnis J-P, de Maupeou B, Lafaye M, Morel E (2017) Towards disruptions in Earth observation? New Earth Observation systems and markets evolution: possible scenarios and impacts. Acta Astronaut 137:415–433

EU Capability Development Plan 2018

European Defence Action Plan – COM (2016) 950 final – Nov 2016

European External Acton Service – A Global Strategy for the European Union's Foreign and Security Policy, Jun 2016

Fitness tracking app Strava gives away location of secret US army bases. https://www.theguardian.com/world/2018/jan/28/fitness-tracking-app-gives-away-location-of-secret-us-army-bases. Last access Apr 2019

French white paper on Defence and National Security – 2013

Fukuyama F (1989) The end of history? Natl Interest (16):3–18, Summer

Fulcher BD, Little MA, Jones NS (2013) Highly comparative time-series analysis: the empirical structure of time series and their methods. J R Soc Interface 10:20130048

Ghassemian H (2016) A review of remote sensing image fusion methods. Inf Fusion 32:75–89

Hepner GF, Logan T, Pitter N, Bryant N (1990) Artificial neural network classification using a minimal training set: comparison to conventional supervised classification. Photogramm Eng Remote Sens 56(4):469–473

Holmes C, Tucker C, Tuttle B (2018) GEOINT at platform scale. In: The state and future of GEOINT 2018. Published by The United States Geospatial Intelligence Foundation USGIF, Herndon, pp 1–38

https://www.eda.europa.eu/docs/default-source/eda-factsheets/2017-04-03-factsheet_react. Last visit Mar 2019

https://www.eda.europa.eu/what-we-do/activities/activities-search/persistent-surveillance-long-term-analysis-(sultan) Last visit Mar 2019

Hussain M, Chen D, Cheng A, Wei H, Stanley D (2013) Change detection from remotely sensed images: from pixel-based to object-based approaches. ISPRS J Photogramm Remote Sens 80:91–106

IAC-14-B1.5.4 Cosmo-Skymed data utilization and applications

Lee J-G, Kang M (2015) Geospatial Big Data: challenges and opportunities. Big Data Res 2(2):74–81

Li J, He Z, Plaza J, Li S, Chen J, Wu H, Wang Y, Liu Y (2017) Social media: new perspectives to improve remote sensing for emergency response. Proc IEEE 105(10):1900–1912

Lillesand T, Kiefer RW, Chipman J (2014) Remote sensing and image interpretation. Wiley, Hoboken, p 704

Long Y, Gong Y, Xiao Z, Liu Q (2017) Accurate object localization in remote sensing images based on convolutional neural networks. IEEE Trans Geosci Remote Sens 55(5):2486–2498

Maggiori E, Tarabalka Y, Charpiat G, Alliez P (2017) Convolutional Neural Networks for Large-Scale Remote-Sensing Image Classification IEEE Transactions on Geoscience and Remote Sensing 55(2):645–657

MCRP 2-10B.5 Imagery Intelligence – US Marine Corps

Nativi S, Mazzetti P, Santoro M, Papeschi F, Ochiai O (2015) Big Data challenges in building the Global Earth Observation System of systems. Environ Model Softw 68:1–26

On 24 June 2014 the General Affairs Council of the European Union approved the "European Union Maritime Security Strategy" (EUMSS), Following the mandate by EU heads of state or government. The EUMSS action plan was recently revised (26 June 2018). This document provides guidelines regarding COM, EEAS and agencies (including SatCen) cooperation in the EUMSS framework. https://ec.europa.eu/maritimeaffairs/policy/maritime-security_en

Popescu et al. Cloud computing case studies and applications for the space and security domain. European Union Satellite Centre

Regulation (EU) No 377/2014 of the European Parliament and of the Council of 3 April 2014 establishing the Copernicus Programme and repealing Regulation (EU) No 911/2010

Space Strategy for Europe – European Commission COM (2016) 705 final – October 2016

Toth C, Jóźków G (2016) Remote sensing platforms and sensors: a survey. ISPRS J Photogramm Remote Sens 115:22–36

Tzu S (2007) The art of war. Filiquarian, First Thus edition. ISBN-10: 1599869772

United Nations Treaties and Principles on Outer Space (2002) United Nations, New York. United Nations publication, Sales no. E.02.I.20. ISBN 92-1-100900-6

US Joint Publication – Joint Intelligence – 22 October 2013

Satellite EO for Disasters, Risk, and Security: An Evolving Landscape

39

Helene de Boissezon and Andrew Eddy

Contents

Abstract

The number of EO satellites is growing rapidly, doubling in the last 10 years. Their data are increasingly integrated into comprehensive services such as the Copernicus Emergency Management Service, which offers full-cycle support for disasters in Europe and internationally. Nevertheless, widespread operational application of satellite EO to disasters concerns response and rapid mapping, focused on hazard information rather than risk reduction. New satellite missions offer increased capabilities, and organizations such as CEOS are initiating pilots with users for risk reduction and resilience building. Financing the value-added component remains a challenge.

H. de Boissezon (✉)
French Space Agency (CNES), Toulouse, France

Committee on Earth Observation Satellites (CEOS), Haiti Recovery Observatory,
Port au Prince, Haiti
e-mail: helene.deboissezon@cnes.fr

A. Eddy
Athena Global, Simiane-la-Rotonde, France
e-mail: andrew.eddy@athenaglobal.com

© Springer Nature Switzerland AG 2020
K.-U. Schrogl (ed.), *Handbook of Space Security*,
https://doi.org/10.1007/978-3-030-23210-8_93

733

Introduction

Over the past decade, the world of satellite-based Earth Observation (EO) has changed dramatically. From roughly 150 EO satellites in 2008, to over 350 in 2018, the number of observing sensors has more than doubled. This number excludes microsats and CubeSats, representing a further 330 (Pixalytics, 2018). Furthermore, that number is expected to double again in the next decade. Larger satellites continue to be replaced by more flexible constellations of smaller satellites, offering greater reach and scope, and greatly improved revisit time. Planet Labs alone now images the entire Earth every day. Even more striking is the range of observations that are being acquired. Whereas a decade ago, few satellites offered very advanced capabilities with sensors other than high-resolution optical or medium-resolution multispectral sensing, today, there are scores of satellites offering complex data requiring advanced processing and interpretation skills. Progress is also being made in making these tools and skills available to more people – through open data and licensing policies – and in bringing training on these easier-to-access tools to more people. In summary, the offering put forward by satellites has never been richer and more diverse, or more readily available.

Ironically, this enlarged offering brings an added complexity: choice. Not all sensors are equal; many applications require multiple sensors and different data sets. A whole research field has emerged in how to properly and most expertly fuse data for various applications. Furthermore, even finding the right data set with so much on offer is a challenge.

Gradually, users of satellite EO have enhanced their understanding of how satellites can be used, and new communities of users are emerging. When the French Space Agency CNES and the European Space Agency (ESA) created the International Charter Space and Major Disasters in 1999, providing near-real-time data during and immediately after major catastrophes was an innovation. Today, after some 20 years, 17 agencies regularly contribute to some 40–50 activations in any given year. Sister services have spawned in Asia (Sentinel-Asia) and Europe (Copernicus Emergency), providing strong regional hubs to address more events of regional and national impact.

The Global Facility for Disaster Reduction and Recovery (GFDRR), created in 2006, now boasts 37 member countries and 11 international organizations. Working in close coordination with the World Bank, GFDRR has greatly increased our understanding of how technologies and satellites in particular can be applied to risk reduction, in order to prevent natural hazards from becoming human catastrophes. Still more recently, Copernicus created the Risk and Recovery Program within its Emergency Management Services (EMS), which in addition recognizes that the specific needs of recovery can now also be met at an operational level by satellites. This requires effort to be applied to understanding user needs and offering pertinent EO-based services to those needs, and applying resources from the satellite community address them.

UNOSAT, UNITAR's operational satellite program, created in 2001, provides crisis and impact assessment and situational awareness during disasters but also offers expanded satellite-related capacity building and training throughout the developing world. UNOSAT aims to make satellite solutions and geographic information easily accessible to the UN family and to experts worldwide and more recently has provided operational support during Post-Disaster Needs Assessments (PDNAs), in conjunction with the EU's Joint Research Centre (JRC) and the World Bank, contributing to Disaster Recovery Frameworks.

Beginning in 2014, the Committee on Earth Observation Satellites (CEOS), an organization representing all the world's major space and remote-sensing agencies, created a permanent Working Group on Disasters with a view specifically to generating a comprehensive strategy for applying EO to Disaster Risk Management. The WG quickly established a series of thematic pilots aimed at exploring how satellite data could be applied to risk reduction more effectively, addressing one of the key priorities of the Sendai Framework. As the principal body bringing together the world's leading satellite agencies interested in EO, CEOS has a key role to play in encouraging the use of satellite EO for risk management and security.

The plethora of new resources applied to the satellite sector has concentrated on data provision according to increasingly demanding specifications, but very little dedicated funding has been applied downstream to making these data part of integral solutions for dedicated disaster and security systems. Until now, the burden of understanding what is available and integrating it into new and existing systems has been borne by the end user, with varying degrees of success. Recently, as the major satellite operators and distributors consider the next generation of services, leaders such as the European Commission have recognized the need for turnkey applications of satellite data tied to disasters and security. Trailblazing services such as the Copernicus EMS open new opportunities to grow awareness of satellite data use within user communities and, in turn, foster new service growth based on existing satellites.

This paper will review how satellite EO is applied to natural hazards in particular in the Copernicus EMS program, arguably the most mature service applying these technologies, and how these services are called to evolve to address the broader risk equation of hazards, exposure, and vulnerability. It will also review some emerging topics of interest to the disasters and security community, including new missions, and fora where leading organizations meet, as well as recent service innovations.

Satellite-Based Earth Observation: An Increasingly Critical Asset for Risk Management and Security

The Copernicus Emergency Management Service: The Hurricane Irma Example

As stated on the Copernicus website, Copernicus EMS provides information for emergency response in relation to different types of disasters, including meteorological hazards, geophysical hazards, deliberate and accidental man-made disasters, and other humanitarian disasters as well as prevention, preparedness, response, and recovery activities. While the service is focused on response, it is in fact a full-cycle service, offering components to address each phase of the disaster cycle. The emphasis is placed on information, rather than data. The service integrates data from a range of sources and provides information products relevant to end-user communities. The innovation in Copernicus EMS is that it is composed of an on-demand mapping component applied to all the aforementioned phases, as well as an early warning and monitoring components. This latter includes systems for floods, droughts, and forest fires. These early warning services are treated in section "Copernicus EMS Early Warning: EFAS-GLOFAS and EFFIS-GWIS."

In considering how satellite EO is applied to disasters and security, at first the rapid mapping component will be considered in detail and then the risk and recovery mapping component through the experience of the Irma Hurricane in France.

Copernicus EMS Rapid Mapping Service

The Copernicus EMS provides a complete portfolio of mapping products, including maps for reference, delineation, grading, and early monitoring. The response service is a 24/7 service operational 265 days a year, offering products to end users within a few hours of satellite overpass. It was this service that was activated in response to Hurricane Irma.

Hurricane Irma struck the French Caribbean Islands of St. Martin and St. Barthelemy on September 5 and 6, 2017. On St. Martin, the storm caused a crisis involving all areas of governance, including law and order, networks and essential infrastructure, information, and communications. St. Martin was hit by torrential rains and sustained winds of up to 297 km/h. Within the French Civil Security ministry (Ministry of the Interior), the Inter-ministerial Operational Center for Crisis Management or COGIC put itself on alert in "anticipation (of a) potentially catastrophic event" on September 4, as a result of the 5-day National Hurricane Center and Météo France forecasts. After Irma's landfall, the COGIC raised its alert level 1 to level 2 and then 3 (maximum mobilization), in order to manage an unprecedented situation in the French Caribbean (Fig. 1).

The Irma crisis affected all sectors of public activity. The electricity network was severely damaged: the exposed parts were destroyed and buried parts near the sea significantly damaged; the water network was out of order; the telecommunications network no longer functioned, whether via antennas or submarine cables. Roads were littered with trash, metal sheets, fallen trees, and other debris. To add anxiety to the desperation of the inhabitants, much looting and physical abuse followed this drama, which left a total of 11 victims. This lasted 4 days until the massive arrival of police. The relief operations were colossal, with nearly 2500 police personnel deployed to the scene. The COGIC, meanwhile, remained mobilized at full capacity on 24-h duty, nonstop during the September 6–20 period.

Support mechanisms for the crisis response by satellite imagery worked perfectly: Copernicus Emergency Rapid Mapping Service and the International Charter Space and Major Disasters were activated early on the morning of September 5 (1 day before the hurricane) by COGIC, French Authorized User, on St. Martin and St. Barthelemy. Thanks to their coordinated action, satellite imagery was acquired from the very first hours of the crisis, and cartographic products to support relief efforts

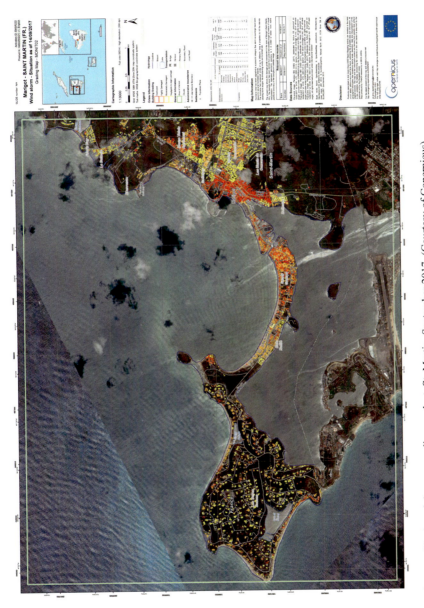

Fig. 1 Copernicus satellite-based damage grading product, St. Martin, September 2017. (Courtesy of Copernicus)

were delivered within 12 h of imagery reception for Copernicus products and 6 h for Charter products.

The French Space Agency (CNES) also decided to program Pleiades very high-resolution (70 cm) acquisitions beyond the standard Charter/Copernicus requests, guaranteeing daily acquisitions from September 7 to 16. The images were available 3 h after acquisition and immediately transmitted to the COGIC and the rapid mapping operator. Within hours, these images were analyzed to provide zoning for damage and impacts and to help organize relief (Fig. 2).

The Irma impact assessment information within COGIC was more than 95% based on satellite imagery and derived mapping. COGIC also produced an atlas at 1/5000 scale presenting "before/after" scenes based on Pleiades images and an exposure database with information on buildings, networks, etc. These same Pleiades images were also used by the Caisse Centrale de Réassurance (national public reinsurer) to provide a first quantification of the direct economic damage. Through the expertise of ICube-SERTIT, a French specialist in satellite-based analysis for crisis management support, the Caisse cross-referenced damaged buildings and insured exposure. The Copernicus and Charter products were also used by the

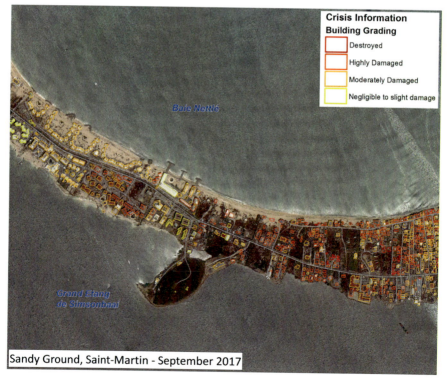

Fig. 2 Copernicus satellite-based damage grading product, St. Martin, September 2017, Sandy Ground Close-up. (Courtesy of Copernicus)

European Commission in the evaluation of the French request addressed to the Solidarity Fund of the European Union.

Copernicus EMS Rapid Mapping is a standard service provided by the European Commission. It has been activated more than 350 times since 2012.

Copernicus EMS Risk and Recovery Mapping Service

The Risk and Recovery Mapping Service addresses aspects outside the immediate response phase. It aims to provide generic risk-related information and specific products tied to recovery from disasters. It consists of the on demand provision of geospatial information. This information addresses prevention, preparedness, disaster risk reduction, or recovery phases (product delivery in weeks/months). The user may request products choosing from a pre-defined set of detailed topographic features and disaster risk information (hazard, exposure, risk) and/or describing in free text the information needs specific to the given situation and type of product wanted.

In the case of Hurricane Irma, beyond the crisis phase, satellite-based EO data has largely contributed to piloting and monitoring reconstruction post-Irma, through the comprehensive monitoring of the affected areas over many months. The main issues related to monitoring recovery are complex in a very densely populated territory, taking into account the risks, adaptation of infrastructures, and support to rapid economic recovery. The intensity of this Copernicus EMS Risk and Recovery postcrisis monitoring was unprecedented. On October 2, 2017, CNES decided to secure Pleiades acquisitions over St. Martin and St. Barthelemy for the following months and offered its support to the Inter-ministerial Delegation for Reconstruction and to the DEAL Guadeloupe (Regional directorate for equipment, planning and housing) for EO data to contribute to reconstruction efforts. The delegation, created by decree on September 12, 2017, had the mission to design and coordinate the actions related to the reconstruction of the two islands and their resilience to natural risks and climate change. The delegation and DEAL expressed an interest in continued access to Pleiades until 2020 on St. Martin, in order to precisely measure the necessary reconstruction effort, in support of the technical teams in charge of risk. All Pleiades images, orthorectified according to the national cartographic reference system, have been made available via the GeoPortail IGN (French National Geographic Institute) for the decentralized services of the state in Guadeloupe.

In February 2018, the Inter-ministerial Delegation for Reconstruction activated the Copernicus EMS Risk and Recovery Mapping (RRM) service over the two islands in order to obtain regular mapping, from March 2018 to September 2019, of building and storage area status, as well as rubble, shipwrecks, and improvised dumps. The products use images from Copernicus Third Party Missions, commercial data from partner states made available free of charge for Copernicus contractors, as well as for authorized users. In parallel, CNES decided to continue Pleiades programming on both islands (Figs. 3 and 4).

Copernicus RRM products were first produced at the end of March 2018. In order to contribute to the preparatory work for the fifth Inter-Ministerial Committee

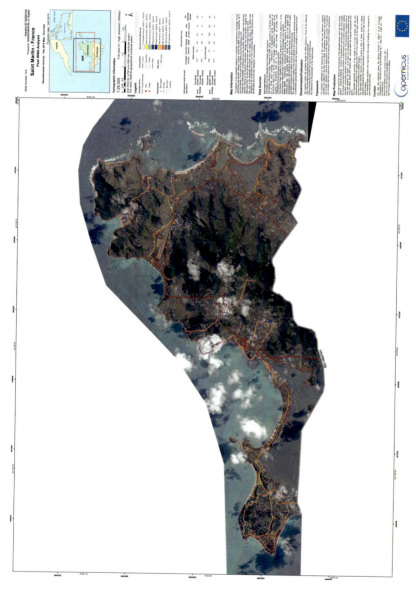

Fig. 3 Copernicus satellite-based monitoring product, St. Martin, December 2018. (Courtesy of Copernicus)

Fig. 4 Copernicus satellite-based monitoring product, St. Martin, December 2018, Sandy Ground Close-up. (Courtesy of Copernicus)

(March 12, 2018), CNES proposed a preliminary version of the Copernicus RRM cartographic products, produced under contract to CNES in time for the meeting. A set of cartographic products analyzing the evolution of affected buildings, rubble storage areas, landfills, and shipwrecks was generated from Pleiades images, providing a dynamic dashboard, building by building, and discharge by discharge, of the evolution of the situation as seen from space. Under the Risk and Recovery Program, pre-qualified consortia bid to provide a service, and once awarded a contract, the consortium has 20 business days to provide the full suite of products.

This exhaustive dashboard of the reconstruction of the islands was presented through a set of printed large-scale maps during the fifth Inter-Ministerial Committee meeting at Matignon, the French Prime Minister's offices.

The Copernicus project has taken over since the end of March 2018 for regular cartographic production, delivered in the form of directories of vector/raster files and reports. These results made it possible to write monthly briefing notes summarizing the Copernicus results between September 2017 and July, August, and September 2018 for the Elysée (French Presidency). This work served to prepare the Council of Ministers at the end of August 2018, dealing with the post-Irma situation in the Northern Islands and the first anniversary of the hurricane, in September 2018.

Once the work of the Inter-ministerial Committee was completed, the DEAL Guadeloupe became the reference partner for Copernicus at the end of 2018, with the strong involvement of their GIS unit. The DEAL felt the collaboration with Copernicus was the only mechanism allowing unfettered access to interpreted EO data, covering the whole territory, with an almost monthly update. These Copernicus-based results were the basis for official statistics on reconstruction progress, as they were the only source of comprehensive, repeated, comparable, objective, and available statistics.

The satellite-based databases provided:

- Estimated damages to buildings evaluated within the framework of refunds from insurers, immediately after the event
- Condition of buildings during imaging periods: post-Irma (September 2017 baseline) to February 2018, from February to April 2018, from April to June 2018, and from June to September 2018

While initially this quarterly frequency seemed particularly interesting to establish indices on dates of reconstruction, after 1 year, the frequency could have been reduced given the slowness of the reconstruction with St. Martin.

The data produced by processing Copernicus RRM products have been integrated into the GEOBASE as a QGIS project so that they can be viewed by all the services of DEAL Guadeloupe via a GIS cartographic tool. The raw data is also available on the data server, so it can be reused for further studies.

One concrete example of the use of Copernicus data by the GIS DEAL service is the location of possibly illegally reconstructed buildings in St. Martin. The problem of illegal, often fragile, constructions on the Guadeloupe archipelago as well as on the island of St. Martin is significant. The damage caused by Hurricane Irma highlighted the need for preventive measures. However, it is clear that the Natural Hazard and Risk Prevention Plan (PPRN), which defines zones where building is not allowed, is not always respected.

The availability of Copernicus data has allowed DEAL Guadeloupe to more precisely detect potentially problematic areas which orient controls in the field:

- Reconstruction in the red zone of the PPRN (construction in progress or completed);
- New construction in red zones where building is forbidden.

In this respect, the results of cross-referenced data made it possible to identify some 50 constructions that were presumed to be illegal because they were reconstructed in the red zone of the PPRN. This state of affairs represents a real challenge for various services in the DEAL (urban planning, sustainable housing policy, relocation of the victims, risk management). These data have allowed DEAL to have a relatively accurate and up-to-date knowledge of a territory without the need to mobilize a large number of field agents. From experience, they know that it is easier to stop a construction site than to destroy a completed construction. Finally, in terms of dissemination of post-Irma data: acquisition of images on both islands for the needs of DEAL 971 and all French public bodies such as the French geological institute BRGM, the biodiversity agency AFB, the forestry office ONF, the geographic institute IGN but also for the needs of the scientific community. Finally, the Ministry of Environment and CNES are working together to capitalize on feedback.

This activation demonstrates the crucial importance of satellite imagery during all phases of the risk cycle. The main benefits Copernicus Risk and Recovery brings to medium-term reconstruction monitoring include:

- Quick and frequent overview of the overall evolution of reconstruction (exhaustive, detailed);
- Figures used for Inter-ministerial Delegation for Reconstruction internal work, for information to French government, and for governmental communication (press releases, preparation of Presidential state visit, etc.).

The DEAL made the following recommendations for improvement:

- Closer links between the user and the RRM project officer, with integration of the user in the process;
- Shorter activation procedure;
- Direct link between the user and the RRM contractor, with regular user feedback;
- Operator support for technical issues encountered by the end users, depending on their technical equipment and skill level (data access, formats, GIS integration, raw images, etc.);
- Improvement of product accuracy over time: through user feedback and ground information (territorial data) potentially provided by end users;
- The Inter-ministerial Delegation would have appreciated to work in French for better expression of needs and easier interactions with the Copernicus project officer. They would recommend that national languages be used both for relations with Copernicus and its contractors and for product nomenclature, for optimal use at local level.

Thanks to satellite-based EO, St. Martin and St. Barthelemy have benefited from short- and medium-term recovery monitoring, providing an exhaustive overview of the territory that was frequently updated. Through Copernicus and Pleiades, updated geo-information of spatial origin have been regularly integrated into the geomatics repository of the authorities in charge of managing the reconstruction. Products derived from satellite imagery have increased and complemented the technical means, representations, documents, and standard information sources used by DEAL (official cartography, airborne means, drones and field photographs, administrative and technical documents of all kinds, etc.). They have been dynamically integrated into the reconstruction manager's geographic database, alongside other sources of information (Fig. 5).

The Copernicus Risk and Recovery Program was initially used mostly to analyze hazard components of risk outside the disaster period. It is only more recently that it

Fig. 5 Satellite-based overview of recovery status for St. Martin, December 2018. Product generated by SERTIT under CNES contract, based on Pleiades data acquisitions. (Courtesy of CNES)

has been actively used to monitor recovery. The project was launched in 2012 and has had some 61 activations. A new edition of this service will start in late 2019 or early 2020 and run for 4 years. It will significantly evolve, most likely proposing a standardized, rapidly triggered, component that could be well adapted to early recovery services, which would ensure continuity in image acquisition and geo-information products, after the response phase. A portfolio of standard products dedicated to these early stages of post-crisis will be available. This would allow recovery actors to activate the risk and recovery service (providing on-demand products) with the guarantee that the immediate postcrisis situation is being monitored.

Copernicus EMS Early Warning: EFAS-GloFAS and EFFIS-GWIS

Under the Copernicus Emergency Management Service (EMS)-Early Warning Component, flood and forest fire forecast and monitoring information is provided to the relevant end users through:

- The European and Global Flood Awareness Systems (EFAS and GloFAS)
- The European Forest Fire Information System (EFFIS) and the Global Wildfire Information System (GWIS)

These systems merge satellite data with relevant in situ data to generate information. While EFAS, GloFAS, EFFIS, and GWIS cover two different disaster types, they share many common in situ data requirements, including real-time and historical meteorological data and a range of geographical data sets. Flood forecasting requires access to real-time and historical river location and river flow data (hydrography and hydrology), and fire monitoring requires fuel (plant health, moisture) information.

The main flood warning system used at the European level is EFAS. It aims to deliver value-added information to the national hydrological services while at the same time providing a unique overview of actual and forecasted flood situations to the European Commission's Emergency Response Coordination Centre (ERCC). EFAS provides overview maps of flood likelihood up to 10 days in advance, as well as detailed forecasts at stations where the national services are providing real-time data. This system is in fact a comprehensive, integrated satellite and in situ network, made up of more than 30 hydrological and civil protection services. The network serves as a Copernicus EMS Rapid Mapping Service trigger for pre-tasking EO data coverage over areas that will potentially flood. Authorized users are made aware of the situation and can then trigger an activation.

On a global scale, the Global Flood Awareness System (GloFAS), jointly developed by the European Commission and the European Centre for Medium-Range Weather Forecasts (ECMWF), is a global hydrological forecast and monitoring system independent of administrative and political boundaries. This system offers a more comprehensive view of the state of the world's water cycle. It couples state-of-the-art weather forecasts with a hydrological model and, with its continental scale setup, provides downstream countries with information on upstream river conditions as well as continental and global overviews. In this sense, GloFAS is a complementary service to the more detailed EFAS. GloFAS has produced daily flood forecasts since 2011 and monthly seasonal streamflow outlooks since November 2017. Since 2018, GloFAS is a fully operational Copernicus Emergency Management Service component.

The European Forest Fire Information System (EFFIS) provides near-real-time and historical forest and forest fire information in the European, Middle Eastern, and North African regions. Fire monitoring in EFFIS comprises the full fire cycle, providing information on pre-fire conditions (fuel) and hot spots and assessing postfire damages (burned area mapping). EFFIS was established by the European Commission in collaboration with national fire administrations and constitutes a key data source to support protection of forests against fires in the EU and neighboring countries.

The Global Wildfire Information System (GWIS) is a joint initiative of GEO and Copernicus. GWIS builds on the ongoing activities of EFFIS by combining EFFIS outputs with the Global Terrestrial Observing System (GTOS), Global Observation of Forest Cover- Global Observation of Land Dynamics (GOFC-GOLD) Fire Implementation Team (GOFC Fire IT), and the associated regional networks. GWIS aims to bring together existing information sources at regional and national levels in order to provide a comprehensive view and evaluation of fire regimes and fire effects at a global level.

The existence of these operational services providing constant monitoring in the flood and fire fields provides a solid link to integrate satellite-based EO in full-cycle services geared toward relevant user communities. The existence of these services has made satellite observations much more critical, as they can be viewed and analyzed in a framework familiar to end-user communities. The early warning elements have also helped gain time in acquiring crisis time data over disasters.

Risk and Risk Mitigation

The discussion of the Copernicus activation for Hurricane Irma highlights the usefulness of a robust satellite-based emergency service, now available to global users, activated through the European Commission, and available free of charge. While this service now addresses the full cycle of disaster management, from early warning to recovery, it remains largely a hazard-focused service, aimed at disasters and their impact. As we will see below, satellites are now equipped to address the full range of natural hazards causing disasters. However, user-driven services are not yet exploiting on an operational basis the range of possible services to be exploited tied to risk or resilience. Risk covers not only the hazard but the exposure and the vulnerability of that exposure, where exposure is the assets that will be damaged by the hazard (e.g., population, buildings, roads, bridges, but also fields, crops, industries) and vulnerability the likelihood of damage as a factor of the characteristics of the exposed elements (e.g., population type, building material, building height, building age, types of crops). The following section examines how satellites can contribute to each of these three components of risk.

Hazards and Unrest

The Irma example demonstrates that satellite-based services can provide valuable warning information for hurricanes, as well as rapid damage assessment or situational awareness after an event. While not all data are integrated into operational services, there are many data streams relevant to hazards – not just hurricanes but the full range of possible hazards – and other causes of disaster such as civil unrest. The table below shows at a high level how satellite data can contribute to different hazards (Fig. 6).

The main challenge to accessing hazard data is related to understanding the source of the data, merging the data with other related data sets, and acquiring some data commercially. In some cases, there are regulatory constraints on who can access the data or where data can be collected. Most western commercial sensors cannot, for example, provide data to nationals of countries on specific lists or acquire very high-resolution near-real-time data over identified sensitive areas.

Exposure

Until now, the entire discussion has focused on hazard. Risk, however, is not just information about hazards. Satellites can provide detailed and up-to-date information about exposure, which is a critical factor for understanding risk. Exposure is

Hazard	Main Contribution	Main Data Sources VHR : very high resolution ; HR : high resolution; SAR : synthetic aperture radar
Floods	Flood extent in near-real time	VHR/HR SAR and optical data
Windstorms	Wind and rain damage	VHR optical
Wildfires	Burned area maps, « hotspots »	Medium resolution optical, thermal
Earthquakes	Rapid damage assessment to the built environment and landslide mapping, co and post seismic displacement	VHR optical, SAR interferogram
Volcanoes	Ground displacement, ash cloud monitoring, pyroclastic flow damage, main ash deposits	SAR data stacks, low and medium resolution meteorological satellites (geo and polar orbiting optical)
Landslides	2D mud and earth movement Damage to the built environment 3D ground displacement	SAR data stacks, VHR optical
Drought	Vegetation health; long-term plant health trends	Vegetation monitoring (NDVI or other index) from low and medium resolution optical.

Fig. 6 Satellite data applications for hazard analysis

defined as "the situation of people, infrastructure, housing, production capacities, and other tangible human assets located in hazard-prone areas." As stated in the UNISDR glossary, "measures of exposure can include the number of people or types of assets in an area. These can be combined with the specific vulnerability and capacity of the exposed elements to any particular hazard to estimate the quantitative risks associated with that hazard in the area of interest." In considering exposure, the usefulness of satellite data and the type of sensor to be used may vary significantly whether one is considering urban exposure (populations, buildings, networks) or rural exposure (population, crop types and values, land cover).

Detailed risk assessment requires an understanding of the impact of the hazard on built areas and other assets. In many cases, these data sets are years, if not decades, old. Very high-resolution satellite data can map at a building level or provide land-cover level assessments that place neighborhoods into a typology of affected areas. For some hazards, this work is fundamental to developing structural risk reduction measures. This is the case for slow onset flooding, for example.

Many practitioners use OpenStreetMap (OSM) as a base layer for exposure analysis, but as the example below over Jakarta demonstrates, even in areas where OSM is relatively densely populated with data, key information is often missing or misleading, and augmenting OSM with satellite data can be valuable in high-risk areas. In the example below, the first image is an OSM layer for Jakarta's port district. The information is up-to-date and provides a high level of detail.

Closer examination however reveals that the information is incomplete. The image below was produced in the RASOR project (www.rasor.eu) using a combination of OSM and detailed analysis of Pleiades satellite imagery. The imagery reveals further industrial infrastructure in the vulnerable port area. In areas where OSM is not as up-to-date, discrepancies might be much stronger (Figs. 7 and 8).

Vulnerability

The final component of the risk analysis is vulnerability. Vulnerability is defined by the UNISDR (International Strategy for Disaster Risk reduction, now UNDRR) as "conditions determined by physical, social, economic, and environmental factors or processes, which increase the susceptibility of a community to the impact of hazards."

Vulnerability can therefore be high or low depending on conditions, for example, people living in low-lying areas experience higher vulnerability to flooding than people who live higher up. There are different types of vulnerability, for example:

- Physical vulnerability: wooden homes which are less likely to collapse in an earthquake are more vulnerable to fire.
- Economic vulnerability: poorer families may live in squatter settlements because they cannot afford to live in safer (more expensive) areas.
- Social vulnerability: when flooding occurs some citizens, such as children, elderly, and differently able, may be unable to protect themselves or evacuate if necessary.

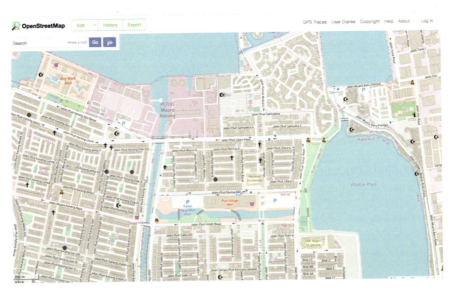

Fig. 7 OSM screen capture of Jakarta port area showing infrastructure and buildings. (Courtesy of OSM)

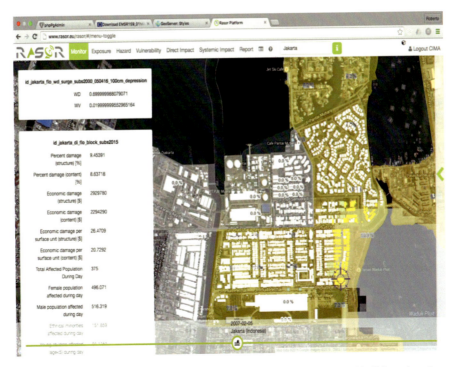

Fig. 8 RASOR screen capture of Jakarta port area showing infrastructure and buildings, based on merged OSM and Pleiades data. (Courtesy of RASOR)

- Environmental vulnerability: forested and deforested lands do not respond the same way to extreme weather events. Healthy ecosystems reduce environmental impacts from disasters.

There is significant overlap among these types of vulnerabilities, which is why when conducting vulnerability assessments, all these factors must be considered. However, when using satellite data to consider risk, practitioners are usually more concerned with physical vulnerability, as well as environmental vulnerability. Other types of vulnerability cannot be well measured by satellite. Indeed, traditionally, little data on vulnerability could be derived from satellites. Practitioners work with vulnerability curves developed initially in the USA such as HAZUS (GIS-based natural hazard analysis tool developed and distributed by the US Federal Emergency Management Agency). These curves are however adapted to US building types and do not export easily to Europe and even less so to developing countries. These curves must be adapted, and to adapt them, satellite information can occasionally provide useful information. This work is still not used operationally.

Understanding vulnerability is important to adopting effective risk reduction strategies, and satellite data could theoretically reduce the cost of collecting vulnerability information. In the case of both seismic and flood risk in urban areas, the structure, material and dimension of buildings, and their distance from other structures can be relevant. Understanding the plan, elevation, and structure of buildings allows the development of a building inventory which will allow a proper adaptation of the vulnerability curve to be applied to a given geographic area.

One of the most successful recent projects adapting remote-sensing data for vulnerability was the FP7 SENSUM project. SENSUM helped understand the changes in society's vulnerability and to integrate this into robust estimates of risk and of losses that follow an extreme natural event. This is especially important in countries where area-wide knowledge of the existing building stock is lacking and the urban environment is rapidly changing. SENSUM combined data from Earth Observation satellites and ground-based methods such as omnidirectional camera surveys. The methodologies developed through SENSUM demonstrate the usefulness of EO to supplement existing ground-based methods for vulnerability calculation but have not allowed a stand-alone satellite-based approach to emerge. The table below shows elements that can be derived through EO data, as proxies for in situ information, and associated vulnerability curves as modified (Fig. 9).

Evolving Supply and Hot Topics for Security

New Satellite Missions

The satellite world is quickly evolving. This discussion was opened by indicating that the number of EO satellites available to provide imagery for disasters, risk, and security has roughly doubled in the last decade and is poised to do so again. These missions bring more data of the types that have existed before (medium-, high-, and very high-resolution optical data, medium-, high-, and very high-resolution X, C,

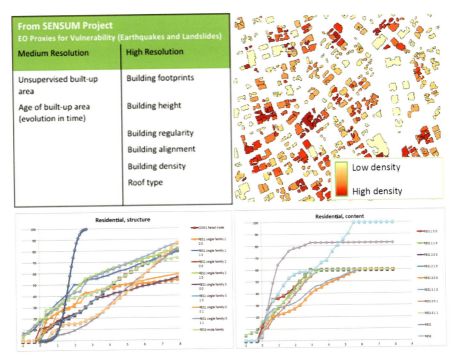

Fig. 9 EO proxies for vulnerability and resulting vulnerability curves. (Courtesy of SENSUM)

and L-band SAR data) but also new data types (S-band SAR, Ka-band SAR interferometry) and new monitoring configurations (e.g., high elliptical orbits or missions focused solely on water cover). It can be expected that these new data types will offer new applications, but the broader availability of data, and free and open data in particular, is enabling a range of new services not specifically tied to Copernicus but flowing from the data sets generated within the program. In this regard, it is worth citing the emergence of global volcano monitoring and rapid earthquake science products as two mature examples of how these data are being used.

Data Philosophies, New Sources, and Standards/Guidelines/Best Practices

The availability of free Copernicus data has been a major driver for the emergence of new services, such as risk monitoring. These services are user-defined and user-driven and often include new sources of complementary data that greatly increase the relevance of satellite data. Some noteworthy examples are the fusion of satellite images, drone data, and crowd-sourced information.

While drones alone can provide detailed surveys of affected areas, the use of drones in conjunction with satellites offers even more possibilities. Satellites detect affected areas, and the information can be relayed automatically to deployed drones in the zone. Drones are in fact a built-in component of the Copernicus EMS, but with limited means, the program cannot offer a wide enough array of drones to provide comprehensive use. One project trying to better exploit these synergies through use of the Internet of Things (IoT) is the Myriad project. Myriad triggers autonomous UAV surveys based on changes detected over Copernicus imagery. Monitoring large areas with drones is usually costly. Given the scale of the area to be surveyed, important changes in the landscape can go unnoticed, which may lead to environmental and security issues. Myriad aims to solve this problem using artificial intelligence to detect changes over Copernicus data sets from Sentinel-1, Sentinel-2, and Sentinel-3, sending those changes to a UAV that will then automatically survey the change. The acquired images will be further processed to confirm or discard the detected change. The whole service will be served through an online data viewer, where the user will be able to explore those changes in real time. The Myriad project suggests that the market for drones as a service may grow to as much as $US 18 billion by 2022.

The widespread adoption of mobile devices and social media platforms, coupled with the development of low-cost sensors, has made it easier for the public to contribute to and engage in scientific research and monitoring. This collaborative exchange with the scientific community is commonly referred to as crowdsourcing or citizen science. Numerous initiatives have emerged that actively involve citizens in environmental monitoring and stewardship, using citizen involvement to validate or complement EO data. The Group on Earth Observations (GEO) has dedicated a community activity to citizen science and lists a range of areas where citizens' observations, data, and information can complement official, traditional in situ, and remote-sensing EO data sources, including climate change, sustainable development, air quality monitoring, vector-borne disease monitoring, food security, flood, drought and natural perils' monitoring, and land cover or land-use change, among other topics. There is an enormous potential to use citizen-driven observations in combination with EO data from the Sentinel family of satellites, NASA Earth Observing Systems, and commercial imagery. Citizens provide in situ data for calibration and validation activities and for the integration of satellite and citizen observations to fill existing gaps. One of the leading applications of crowd-sourced data for disasters is Humanitarian OpenStreetMap (OSM) Team or HOT. HOT is regularly deployed during crisis situations to generate rapid situational awareness and provide a ready corollary for satellite-EO (www.hotosm.org). Given the strong focus of HOT on exposure and vulnerability, they are an ideal complement to satellite-based hazard mapping.

The European Commission (EC) has established Citizens' Observatories that are community-centric initiatives where citizens become more active in collecting and sharing environmental information, typically harnessing the latest technological advances (e.g., ubiquitous Internet connectivity, IoT, machine learning, social media, portable and inexpensive sensors). Citizens' Observatories empower citizens to get informed and actively participate in environmental decision-making, raise awareness about environmental issues, and help build more resilient societies. The seventh Framework Programme funded five Citizens' Observatories relating to various environmental issues such as air quality, flood and water management, coastal ecosystems, biodiversity, and odor annoyance. Horizon 2020 added four more: LandSense, Ground Truth 2.0, SCENT, and the GROW Observatory. Leveraging citizen input will enhance and augment the influence of existing Earth Observation monitoring systems, including GEOSS (Global Earth Observation System of Systems) and Copernicus.

In parallel to this, satellite data providers themselves are improving the specifications used to acquire and compile data for disaster users. Groups such as the International Working Group on Satellite-based Emergency Mapping (IWG-SEM) have drafted guidelines to set standards for the production of satellite-based maps, in an effort to ensure that satellite-based products meet clear quality thresholds to enable users to trust them.

According to the IWG-SEM Guidelines, the aim is to help support an effective exchange and harmonization of emergency mapping efforts leading to improved possibilities for cooperation among involved Emergency Mapping Organizations. This should facilitate the convergence of mapping procedures and thematic content across production teams in multiple response organizations, especially in the early response phases of disaster events. Easier information exchange allows for merging and quality checking of individual data/information layers generated by more than one Emergency Mapping Organization. The guidelines provide a framework, enabling the emergency mapping community to better cooperate during crisis times. The guidelines:

(a) Define fundamental principles;
(b) Establish a procedure for interactions and sharing of data, analysis, and mapping results;
(c) Organize mapping products, templates, and dissemination policies;
(d) Anticipate problems of uncertainty in communication;
(e) Commit to assurance of capacity and qualification;
(f) Prepare a glossary for emergency mapping vocabulary.

Additional separate, short, and thematically focused sub-chapters have been and are being published relating to specific disaster types.

In summary, the combination of satellite imagery, drones, and crowd-sourced data will offer a significantly more robust capability than has been available previously, supporting full-cycle risk management applications.

Initiatives and Services (Copernicus Security SEA)

In terms of relatively new services and service extensions, it is worth highlighting the Copernicus Service in Support to EU External Action (SEA), which has a security mandate, as well as emerging recovery-based pilots such as the CEOS-led Haiti Recovery Observatory, and new plans for the development of generic capabilities together with the EU, the World Bank, and UNDP.

The Copernicus Service SEA aims to support European Union policies by providing information in response to Europe's security challenges. It improves crisis prevention, preparedness, and response in three key areas: border surveillance, maritime surveillance, and support to EU External Action (SEA). Full-scale services are being developed to operationalize the use of satellites for border and maritime security. In parallel, the EU has established a service which brings satellite assets to bear for external crises when the EU seeks information to support policy decisions. As a global actor, Europe has a responsibility to promote stable conditions for human and economic development, human rights, democracy, and fundamental freedoms. In this context, the EU can provide assistance to third countries in a situation of crisis or emerging crisis and thereby help prevent global and trans-regional threats from having a destabilizing effect.

Pursuant to an agreement signed on October 6, 2016, the European Commission entrusted the European Union Satellite Centre (EU SatCen) with the SEA component of the Copernicus Security Service. In particular, the SEA component assists the EU in its operations, providing decision-makers with geo-information on remote, difficult to access areas, where security issues are at stake. It targets mainly European users, but it can also be activated by key international stakeholders, as appropriate, under EU international cooperation agreements. While still a relatively new service, the SEA security service has received positive feedback that demonstrates that the service responds to real needs, with ongoing user community enlargement.

Another example of innovative use of satellites is the Recovery Observatory effort led by CEOS with the World Bank, UNDP, and European Union. On a global scale, in both developed and developing countries, disasters strike regularly. Disasters sometimes take on monumental proportions, whether because of particularly vulnerable populations, a dramatic natural event, or exceptionally unfortunate circumstances. Hurricane Katrina, the 2010 Earthquake in Haiti, Typhoon Haiyan, or the Great Tsunami of Eastern Japan are examples of disasters that occupy a special place in our collective memories as mega-disasters after which people and governments take years to rebuild. Hurricane Matthew in Haiti was such a disaster. Since 2014, CEOS has been working on means to increase the contribution of satellite data to recovery from such major events. In December 2016, after the impact of Hurricane Matthew in Haiti became apparent, CEOS triggered the Recovery Observatory (RO). A project team made up of CEOS agencies, national partners, and international DRM stakeholders was established to oversee this project for a period of 4 years.

The aim of the RO is to:

- Demonstrate in a high-profile context the value of using satellite Earth Observations to support recovery from a major disaster:
  - Near term (e.g., support to PDNA process);
  - Medium term (e.g., recovery planning and monitoring, for up to 4 years).
- Work with the recovery community to define a sustainable vision for increased use of satellite EO;
- Establish institutional relationships between CEOS satellite data providers and stakeholders from the international recovery community;
- Foster innovation around high-technology applications to support recovery.

The main benefits of the RO include:

- Providing key information (analytical, geospatial) about the recovery to support end users in their decision-making processes and progress monitoring;
- Obtaining access to regular imaging of affected area over a long period, especially for higher-resolution data not typically available;
- Compiling in a single framework the key data sets (both satellite images and large number of other data) and use them seamlessly;
- Establishing a "real-life" demonstrator to identify where EO can bring useful information in the recovery phase and define "best practice" for the DRM community;
- Demonstrating usefulness of sat EO, together with other data sets, on a large scale for long-term recovery monitoring;
- Demonstrating applications tied to very high-resolution imagery and to high-frequency high-resolution images, to open the way to broader use of satellite EO after smaller events, and more regularly.

In May 2017, May 2018, and again in May 2019, the RO team made up of national champions; the French, and Italian space agencies; and international DRM stakeholders such as UNDP, World Bank/GFDRR, and EU delegation convened user workshops to bring together RO users around a list of main themes and to discuss RO products in each of these thematic areas (Fig. 10).

The table below summarizes the main thematic areas covered by the Haiti RO and the data served to develop the RO products (Fig. 11).

The RO platform is now up and running, and more data are being added regularly. It can be accessed at: https://www.recovery-observatory.org/

The CEOS RO partners, together with the World Bank/GFDRR, the United Nations Development Programme, and the European Union, are considering how

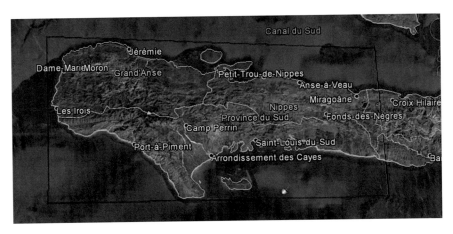

Fig. 10 Haiti Recovery Observatory area of interest. (Courtesy of Google Earth and CNES)

RO : Products Thematic Table (EN) April, 2019

	Produit	Utilisateur-clef	Elaboration	Données satellites
	Buildings Mapping	CIAT / Planning Ministry	CNES/SERTIT, Copernicus EMS R&R	Pléiades, WorldView-3
	Terrain Motion Change Detection	BME / URGeo	ASI, CNES/EOST	COSMO-SkyMed, Pléiades, Spot 6/7, TerraSAR-X
	Watershed / Flood	ONEV / Agriculture Ministry	ASI/CIMA Foundation	Pléiades, COSMO-SkyMed
	Agriculture	Agriculture Ministry	Copernicus EMS R&R	Sentinel-2, Spot 6/7, GeoEye-1, WorldView-2
	Macaya Park Monitoring	ANAP / ONEV / Environment Ministry	Copernicus EMS R&R, CNES/SERTIT	Spot 6/7
	Environmental Impact	ONEV / Environment Ministry	Copernicus EMS R&R	Sentinel-2, Spot 6/7, Pléiades, WorldView-2
	Land Use	All	CNIGS, CNES	Sentinel-2

Fig. 11 Haiti RO thematic products. (Courtesy of CEOS, CNES, and CNIGS)

to take the lessons learned from the Haiti RO and apply them to a generic RO concept that could be activated on a regular basis after catastrophic events.

In summer 2018, the Generic RO ad hoc Development Team was established to develop possible scenarios for such a generic, scalable, and replicable observatory. They are currently developing an advocacy paper to examine lessons learned through the application of satellite data to recovery challenges.

Conclusions

While in recent years, the application of satellite data to risk management and security issues has grown considerably and new services have emerged, there is every expectation that in the coming years, this trend is likely to accelerate. Not only are satellites more numerous than before and is satellite data more easily accessible, but the number of services integrating satellite data into operational security-related products is also growing. In Europe alone, the Copernicus program is likely to see its services augmented through a number of public good and commercial services spun off from the Horizon 2020 research program and related initiatives.

The real challenge for risk and security users is to understand the offering and identify services – or if not, data sets – that are adapted to their needs. For those integrating data into their own systems, there is a growing challenge of big data: very large data volumes required to drive new applications and the need for comprehensive processing capabilities to address these needs. In Europe, this data challenge is being addressed to some extent by the European Space Agency, which has foreseen the need for increased processing and integration capacity and has created a series of Thematic Exploitation Platforms or TEPs, many of which are directly related to security. Currently the TEPs address:

- Coastal themes;
- Forestry;

- Hydrology;
- Geohazards;
- Polar themes;
- Urban themes;
- Food security (under definition).

Similarly, new initiatives are coming forward to address clear needs. The experience in the Haiti RO is being expanded through collaboration from the World Bank, UNDP, and the EU to include a generic approach to accessing satellite-based recovery information. In the area of climate change, another new initiative, the Space Climate Observatory (SCO), aims to bring satellite data into play for situational awareness of climate impacts at the local level, offering a clear complement to global scale climate change models and observations. This new system may offer valuable data to climate-related security issues which will likely be a major theme of the next decade.

In many risk and security contexts, end users have access to classified data sources which are set up to provide a single stream of data and information. Being able to merge these classified data sources with a growing and increasingly rich set of public and commercial missions is not an insignificant challenge. In the increasingly complex network of systems and data streams, the ability to navigate, discriminate, analyze, and effectively add value will be a sought-after commodity among security users. Perhaps even more critical, the ability to do this in real or near-real time saves lives and protects property. As long as satellite data requires significant and sometimes expert human analysis, the cost of prevention and preparedness is prohibitive, and the ability to deliver in real or near-real time is illusory. The introduction of self-learning systems and other artificial intelligence in data interpretation will likely offer a major increase in mapping capacity and revolutionize the usefulness and hence the use of satellite data for risk and security in the decade to come.

References

http://effis.jrc.ec.europa.eu
http://gwis.jrc.ec.europa.eu/static/gwis_current_situation/public/index.html
http://sertit.u-strasbg.fr/RMS/
http://www.efas.eu
http://www.eohandbook.com/eohb2015/files/CEOS_EOHB_2015_WCDRR.pdf
http://www.globalfloods.eu/
http://www.hotosm.org
http://www.sensum-project.eu/
http://www.un-spider.org/risks-and-disasters/disaster-risk-management
https://emergency.copernicus.eu/
https://insitu.copernicus.eu/FactSheets/CEMS_Early_Warning
https://www.copernicus-masters.com/winner/myriad-satellite-triggered-drone/
https://www.pixalytics.com/eo-satellites-in-space-2018/
https://www.recovery-observatory.org/

Space-Enabled Systems for Food Security in Africa

40

Olufunke Adebola and Simon Adebola

Contents

Abstract

Africa would be able to fulfill only about 13% of its food needs by 2050. Systemic challenges such as postharvest loss cut across the agricultural value chain and represent an opportunity for integration and optimization, with space

O. Adebola (✉)
Sam Nunn School of International Affairs, Georgia Institute of Technology, Atlanta, GA, USA
e-mail: oadebola3@gatech.edu

S. Adebola
Capacity Direct LLC, Atlanta, USA
e-mail: space@capacity-direct.com

© Springer Nature Switzerland AG 2020
K.-U. Schrogl (ed.), *Handbook of Space Security*,
https://doi.org/10.1007/978-3-030-23210-8_142

technologies serving as a backbone for information generation, transmission, route optimization, and decision support. The chapter reviews space-based applications in support of decision-making processes of policy makers and presents a conceptual framework for their implementation in agriculture in general with postharvest loss reduction as an example.

Introduction

Space technologies play an important role in modern agriculture. They are deployed to strengthen agricultural practice by increasing the effectiveness of production and optimizing the efficiency of resource allocation at each stage of the agricultural value chain. There is a global consensus captured within the Millennium Development Goals (MDGs) and the succeeding Sustainable Development Goals (SDG) to leverage all necessary means to advance the state of humanity by accomplishing defined targets for bridging the development gap within the global community. This consensus has influenced the essential role that space technologies play in economic development, and their contribution towards achieving agriculture-related targets for nutrition, health, and environmental sustainability. This consensus to which space technologies contribute, has been reinforced by the principles, policy guidance, regional coordination, and domain-focused implementation efforts executed by the various United Nations instruments and organizations, including the UN-COPUOS, UNOOSA UNOSAT, WMO, and GEO, supporting the FAO and WFP in their mandate to end hunger globally (United Nations 2015). National and regional governments have recognized the role that space technologies play in agriculture and also justify investments in their space programs based on the enhanced capabilities they provide.

The technology triad of space-based imaging, geo-navigation, and satellite telecommunications enable the significant technological advances in agriculture that drive precision agriculture and are exponentially enhanced by the technologies driving the fourth industrial revolution. These disruptive technologies include big data analytics, the internet of things, artificial intelligence (AI), cloud computing, distributed ledger technologies, AI-powered robotics, and automation. Agriculture benefits from the multiple intersections of these technologies, and their convergence with the space triad on internet-enabled digital products, applications, processes, and platforms, is driving a new wave of advancements in precision agriculture, food processing, agriculture value-chain management, and commodity trading. Although primarily tied to private sector investments (Gagliordi 2018), the ensuing competitive edge gained by the early adopters of these technologies and approaches, exacerbates existing gaps in agricultural productivity and efficiency faced by developing countries. This chapter addresses that gap by highlighting the use of integrated systems that utilize space technologies in addressing one of the major problems facing agriculture (post-harvest loss), and that poses severe consequences for low- and middle-income countries.

Post-harvest Food Loss (PHL) is defined as measurable qualitative and quantitative food loss along the supply chain, starting at the time of harvest till its consumption or other end uses (De Lucia and Assennato 1994). It is also defined as the "losses due to spillage and degradation during handling, storage and transportation between farm and distribution (FAO 2011)." Losses can be quantitative – a reduction in the weight of edible grain or food available for human consumption. Losses could also be qualitative – reduction in weight due to factors such as spillage, consumption by pests and due to physical changes in temperature, moisture content, and chemical changes (Buzby and Hyman 2012).

Food loss has negative effects on the environment. Food in landfills decomposes anaerobically, yielding methane emissions, a gas more than 25 times as potent as carbon dioxide at trapping heat (FAO 2013). The Food and Agriculture Organization (FAO) ranks global methane emissions and places food wastage third after the United States and China. Communities affected by postharvest food losses lose out on the full benefits that agricultural products can offer in the form of nutrition, health, and wellbeing, and the use of plant products as an alternative to fossil fuels in energy generation. The use of "food" as biofuels continues to grow and may represent another demand factor adversely affected by food loss through wastage.

Agriculture is also a predominant sector of the economy and so reducing food waste could contribute to meeting the Sustainable Development Goals (SDGs). The agricultural sector accounts for 25% or more of gross domestic product (OECD and FAO 2016) and about 65% of jobs on the African continent (OECD/FAO 2016). It can be expected that a reduction in postharvest loss will increase household income and enable households to provide better education and healthcare for their families.

Global population growth could result in unmet demand for food supply which could result in political instability and unrest. This food insecurity has been blamed for protests such as those in North Africa in 2011 and the Arab Spring driven by the government's inability to address rising caloric demands and food prices (Johnstone and Mazo 2011). In addition to this, food insecurity would also increase migration, as people seek better opportunities for food. Similarly, ongoing conflicts and insurgency situations may be driven by the ability of terrorist organizations to "win the hearts of the people" through the provision of basic needs such as food and security (Mooney and Hunt 2009).

This chapter describes findings from a technology adoption study on postharvest loss in Africa (Adebola 2020). Field research was completed in Ghana during the summer of 2019 with questionnaires administered alongside focus groups and key informant interviews with stakeholders in the agricultural sector at multiple levels. This chapter provides lessons learned from the study and describes how space-enabled technologies can play a key role in improving the agricultural value chain and help address the problem of postharvest loss.

The Rationale for Improving System Efficiency, Integration, and Optimization

Solving the postharvest loss problem has long consumed the attention of actors in the engineering and development sectors. In the 1970s and 1980s, conventional wisdom offered modernization through technology adoption as the solution to the reduction of global postharvest loss. During this period, attempts to reduce postharvest loss have adopted a "one size fits all approach." Technologies such as improved seeds were deployed to "close the yield gap" and storage technologies were deployed to preserve the harvest. However, the effectiveness of these technology interventions has been inconclusive in Africa because of the heavy focus on the production and storage stages of food production.

Breaking away from a silo mentality in dealing with the problem of postharvest loss at single points within the agriculture value chain, this chapter proposes a systems approach, driven by space and other technologies in addressing the problem. According to Florkowski et al. (2009), there are three advantages to adopting a systems approach. Firstly, this approach offers the opportunity of providing improvements in the continuum and allows for predicting the impact of changes along the value chain without modifying the system. Furthermore, the systems approach can identify knowledge gaps and aid in prioritizing the efforts of researchers.

Food production is not a linear process but a chain that consists of a network of interdependent stakeholders involved in growing, processing, and selling the food to consumers. This network consists of:

- Food producers such as the farmers that trade, research, cultivate the food or raise animals for meat or milk
- Food processors that provide value-added services such as food processing and butchering, produce aggregators
- Food distributors, and those that market and sell food
- Food consumers
- Food regulators, including government and nongovernment organizations that monitor and regulate the entire food value chain from producer to consumer

According to the African Postharvest Loss Information System (APHLIS), about 18% of cereal harvested in Africa is lost before it is consumed (Africa Postharvest Loss Information System 2017).

Figure 1 shows the causal agents of postharvest losses across the cereal value chain. The magnitude and pattern of maize loss vary across the food supply chain.

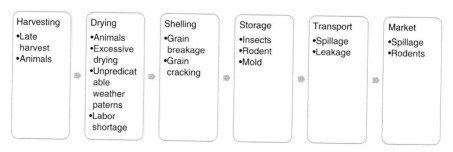

Fig. 1 Causal agents of Postharvest losses across the cereal value chain. (Source: Authors)

According to APHLIS estimates, about 6% of maize grown is lost at the harvesting and drying stages. At the harvest and drying stages, too early harvest will contribute to increased moisture content, making the commodity vulnerable to mold growth and invading insects. Too late harvest exposes the crop to birds and rodent attacks, and losses caused by natural disasters. Sun-drying also exposes the commodities to pests such as birds, rodents, insects, and other domesticated animals. Often, when mechanical dryers are used, the grains are not dried efficiently, leading to broken grains (Grolleaud 2002).

The traditional practice of harvesting crops by hand, using hand-cutting tools such as sickle, knife, scythe, and cutters as adopted by many smallholder farmers in Africa, has contributed to the loss of food crops at the harvesting stage (Kumar and Kalita 2017). Often, the manpower required for manual labor is inadequate because of increasing rural-urban migration, the prevalence of HIV/AIDS, and political conflicts. This has resulted in farmers delaying harvesting and subsequently, crops are exposed to pests such as birds (Paulsen et al. 2015). Furthermore, because of inadequate drying methods, most African farmers harvest their crops at physiological maturity, that is, when the moisture content is about 20–30%. The moisture content at physiological maturity makes the crops more vulnerable to pest attacks, mold growth, and other fungal contamination (Boxall 2002). Another 4% loss occurs during the drying phase. Harvested crops are usually sun-dried in developing countries (Hodges et al. 2010). Sun-drying is limiting because it is labor-intensive and weather-dependent. Farmers reported that they could predict the timing of the rains based on their experiences. This allowed them to prepare and plan to harvest in the driest period of the month of November. However, in recent times, the rains have been unpredictable with rain falling until the end of November, thereby hindering the grains from drying properly. The unpredictability of the rainfall patterns is further exacerbated by a lack of relevant weather information for agricultural practice.

Three percent of the harvest is lost during transportation to the farm because of the long distances between rural farms and the urban markets. Very often, these countries have poor road infrastructure and farmers do not have access to mechanical transportation, and so must rely on other modes of transportation such as bicycles, animal-drawn carts, and trailers. Food crops are also lost at this stage due to heat exposure, damage from pests, and theft (Ganpat and Isaac 2015).

Inadequate storage facilities contribute to a 4% loss of maize crops (World Bank 2010). Loss at this stage occurs because aggregators and traders take advantage of information asymmetry and imperfect market conditions and offer lower prices for agricultural produce. Farmers are encouraged to store their produce until they are sure they could get a higher price. In the absence of storage systems, farmers resort to storing their grains in their homes where the grains are susceptible to exposure to moisture, mold, rodent, and insect infestation. Two percent of harvested maize is lost during threshing and shelling processes. In developing countries, threshing is done through the traditional methods of manually trampling or beating the harvested crop to detach the grains from the panicles. Grain spillage, incomplete separation of the grain from the chaff, and grain breakage due to excessive striking are some of the major reasons for losses that occur during the threshing process (Kumar and Kalita 2017).

Losses could also occur when the grain is exposed to atmospheric and biotic factors during the process.

Space Technology Capabilities for Enhancing the Agriculture Value Chain

Information Generation

At different points within the agriculture value chain, information is created by different actors that: create and supply farm inputs, prepare for and assess readiness for planting, perform seedling and planting, monitor plant and crop health (and associated risks) throughout the different phases of the planting and harvesting season, carry out postharvest activities such as drying and shelling, and provide transportation services out of farms to market and storage locations. For each of these activities, space technologies provide useful information and services (United Nations 2015). Significant investments in space technology infrastructure such as the EGNOS, Galileo, GLONASS, and GPS geonavigation programs, and leagues of earth observation satellites, underscore the crucial importance placed on access to space-enabled capabilities by national and regional governments. The use of these technologies to advance agricultural practices that rely on the information that they provide, either individually or when integrated together with other data sources, is a key part of modern agriculture, with applications ranging from precision farming to animal husbandry.

Remote Sensing Imagery and Environmental Data

These are used in land allocation, land assessment, precision agriculture, monitoring weather and climatic events and patterns, and modeling to predict the effect of environmental factors on farming activities. Remote sensing also provides information on the effect of agriculture on the environment. Land use, land depletion, consumption of water resources, pollution from farm waste and effluents, and other ecological threats to pristine habitats are examples of factors that can be informed using remote sensing (United Nations 2015). Some of these factors exist at the meeting point of urbanization, agricultural land use, and environmental conservation, where a tenuous balance must be achieved between expediency and excess. This is all the more critical considering the responsibility of exploiting the environment for human uses but doing so in a responsible, equitable, and sustainable manner. Too much harm done to communities and long-term civilization has been overlooked due to the need to meet rising human demand, but it is all the more necessary that the pace and effects of agriculture on the environment be considered even as we improve our capabilities to plant and farm better using technology. Remote sensing technologies provide opportunities to track the suitability of the environment for agriculture and at the same time take a pulse of the sustainability of the environment in use (Weiss et al. 2020).

Geonavigation

In combination with mapping (Geographic Information System) and telecommunications technologies, geo-navigation provides valuable information for precision agriculture, logistics planning, and coordination in agriculture. Geonavigation data allows the overlaying of multiple types of sensor data collected at different times, about the same location (usually on the farm), so that the data can be analyzed using GIS to derive insights that are used in decision support. Also, because agricultural produce is perishable, maintaining its freshness or an unbroken cold chain between its point of production and its final destination is necessary for maintaining its nutritional and economic value. Geonavigation provides a foundation for location-based route-optimization to minimize the time spent in transit for food that needs to be delivered fresh. When combined with distributed ledger technology, geonavigation data can be used to create a tracking and quality assurance system for fresh produce that becomes crucial for maintaining an unbroken cold chain, and securing the integrity of automatic sensor readings and food inspector direct user input on the condition of agricultural products at each transit stop across the value chain. Collecting these data over time will provide reliable data for modeling the optimal conditions and transportation routes for food and other fresh agricultural products. By combining multiple data sources and technologies, a willingness to learn from failure, and applying advanced data analytics, agricultural systems can minimize losses and increase revenue while strengthening public confidence in the quality of food production (Tian 2017). This will also expand opportunities for the just-in-time production and delivery of goods for the "fresh or minimally-processed foods" industry. This is an important development for food safety and the food industry that is dependent on geonavigation data.

Another key aspect benefitting from the use of geonavigation technologies is the use of drones in agriculture (Anderson 2014). Drones are used in farm surveillance, irrigation, spraying of agrochemicals, and security. Their main benefit is the ability to replace a human presence in carrying out these activities, while serving as a low-cost, low-maintenance alternative in remote, dangerous, or expansive locations. However, geonavigation is needed to successfully deploy and manage drone technology. Precision is often required and so there may be a need to combine satellite navigation with other location sensors to provide the ability to determine farm boundaries and limit drone activity to defined areas within the farm.

Information Transmission

Space-enabled telecommunication is generally perceived to be useful for remote locations lacking in terrestrial network communications. This is still the case in many remote farming locations where access to telephone and broadband Internet access is still only possible via satellite telecommunications. The ability to integrate communications seamlessly across communications standards and protocols that combine space-enabled telecommunications with terrestrial network-based systems is important in agriculture. A driver for this is that agricultural produce is moved over long distances, using different transportation modalities and an array of tracking sensors deriving information on the agricultural product, its temperature, moisture content, packaging integrity, displacement, location, and other food safety, logistics, and business variables (Dlodlo and Kalezhi 2015). Tracking information received and processed in a timely manner is important in making health, safety, and business decisions relative to agricultural produce at multiple points within the agriculture value chain. A coordination point and information management system are needed to make sense of received information and tie that back to the right business or safety decisions that must be made.

Route Optimization

Finding an optimal route for the location and transportation of food and other agricultural produce is supported by space technologies such as geonavigation resources. Having accurate estimates of source and destination points, distance measures, potential routes, and navigation constraints are necessary in building a shortest path algorithm to drive logistics decision-making. Additional constraints include real time information on market prices at potential destinations and the precise timing of harvesting, processing, and transportation to the most profitable delivery point, to coincide with infrequent market days. To successfully accomplish this, logistics planning must be adept at integrating geonavigation technologies with route optimization resources, agriculture domain knowledge, and market experience. Route optimization is also used in planning the operation of farm-based machinery to achieve targeted data collection and application of farm inputs such as seeds and agrochemicals (Dlodlo and Kalezhi 2015).

Integrated Decision Support

The need for an integrated decision support system is to intelligently manage and use the varied data sources that can be combined to support multiple functions within agriculture. A system of this kind will coordinate end-to-end activities at each stage of the agricultural value chain (Dengel 2013). It will combine remote sensing and mapping technology, geonavigation, and satellite telecommunications technologies within a framework designed around the business rules and business processes that drive decision-making in agriculture. This is to ensure that farmers and other stakeholders have the right information to make decisions throughout the entire agricultural value chain, with the ultimate aim to reduce losses. As an integrated system of subsystems, it will provide a template for combining information flows, feedback loops, information staging, routing, and analytics (e.g., machine learning), originating from each process-based subsystem, into a composite, semantically interoperable, and synergistic whole.

Challenges and Solutions to System Integration in Agriculture

At the nexus of development problems and potential technology solutions, lies an opportunity to address a constellation of factors that act as forces that both shape and define the problem, as well as determine if the solution will succeed (Heeks 2005). We have seen that postharvest losses are encountered along the entire food production value chain (farm to table) and thus, Postharvest loss poses as a complex problem. Postharvest loss as a construct does not exist in isolation. It is variously influenced by the domains of agriculture, environmental sciences, transportation, commerce, and the attendant social and economic factors associated with these domains which include governance, political stability, and education. Describing these interlinkages and how distinct factors from each domain contribute to and drive the problem of postharvest loss is an aspect of its complexity that will be integrated into the conceptualization, design, development, and implementation of an integrated agricultural system of systems framework. The framework is aimed at providing decision support to aid the end-to-end functioning of the agriculture value chain. Understanding complexity enables comprehensive sense-making, problem solving, and systems-thinking which are key to defining the scope of interventions as well as guaranteeing long-term sustainability (Glouberman and Zimmerman 2002). These success factors are described below.

Knowledge and Skills

Decision support systems are an option for addressing and bridging the wide skills and knowledge gaps between successful programs and contexts, on one hand, and those struggling with reducing postharvest losses particularly in low-income countries, on the other. The goal in creating this integrated framework would be to abide by the five rights of decision support systems, which are to provide the right information, to the right person, in the right format, through the right channel, and at the right time in the workflow (Kaushal et al. 2003). However, to achieve this goal one must contend with knowledge and skills gaps, where they exist and could pose a barrier to engaging with and embracing decision support by farmers. These include gaps in improved scientific agricultural practices, technology applications in agriculture, precision farming, the use of information technologies, language barriers, collaboration, and negotiation skills. To the extent possible, an integrated system should account for and accommodate an end-user base with broad variation in skill sets and capabilities. In addition, training and end-user support can be provided to support the system's target user base.

Infrastructure Backbone

The right infrastructure is needed to support the operation of an integrated multi-subsystem decision support system for agriculture. The required hardware and software for data input, storage, management, analysis, exchange, and transmission should be included in designing an appropriate technical architecture for supporting all the data, process, and service components for each subsystem, as well as the flow of information between different subsystems (Uddin et al. 2017). Accurate and detailed systems documentation is needed in order to be able to perform this integration, especially if the decision support system is not being designed and developed in-house as an end-to-end system but built as a combination of existing software products from proprietary, customized and off-the-shelf sources. Similar considerations must be made in ensuring access to telecommunications services to support the implementation, monitoring, and tracking of agricultural processes, operating using varied forms of transportation on land, air, and sea. Access to geonavigation and real-time geolocation and mapping is needed to visualize and make intelligent assessments of agricultural goods in transit. In addition, the use of cloud-based services will provide a distributed infrastructure for data storage and management, global availability and access to information, reduce latency in information transfer, and increase system redundancy. The use of accompanying data management and advanced analytics capabilities on cloud platforms can be used to enhance the value of the data and develop a system with robust decision support capabilities. These will support the movement of agricultural goods across remote rural locations all through to their final disposition at processing sites, or to urban or other rural destinations for consumption.

System Interoperability

Drawing from space systems analysis and design, and Internet of Things (IoT) interoperability, the actual integration of subsystems would allow their component devices to communicate with each other using different interoperability standards and protocols (Noura et al. 2019). Specifically, data transport protocols such as TCP/IP will be used for moving data over the Internet, with options for securing data transfer implemented. Data formats for data exchange would include raw data and flat formats as well as structured formats such as JSON and XML for directly ingesting data into machines for analysis and automated decision-making. Application Programming Interfaces (APIs) will support data calls and responses between different subsystems. The use of an API-centric Service Oriented Architecture (SOA) for data exchange will permit the scalability and management of interoperability features across the system.

End-User Interfaces

An IoT-dense system that combines sensor data with other sources including space-enabled data, and exchanges these through APIs and an array of widgets and connectors designed to handle interoperability issues, will rely on a cross-platform web interface for deploying data visualization, information, and analytics insights

(Khattab et al. 2016). Similarly, enabling interaction with the system will support modalities such as key input, touch, voice, and motion detection. An interface, for example, should be intuitive, easy to interact with, support multiple languages, be accessible on different web-enabled devices, and have the resources for user training and support. Developing and deploying the end-user facing interfaces for the system should be done collaboratively with the end-users, system administrators, and product owners. This is to help gather their functional requirements that will then feed into the interface design, and subsequent product and user acceptance testing prior to deploying into production.

System Maintenance and Upgrades

Providing and maintaining good quality service for a system in production (operational use) requires budgeting for and providing resources to support regular system maintenance and upgrades. This should cover software version upgrades and patches, as well as repairs and replacement of failed hardware. A tracking system for reported failures and outages should be in place. The advantage of a sensor/IoT-dense system is the ability to collect data that can be used to automate issue reporting and tracking, as well as the predictive modeling of system failures, in order to enable intelligent system maintenance (Chuang et al. 2019). Regular feedback and iterative design sessions can be held with stakeholders, to ensure that the system as a whole continues to deliver useful information and updated services needed to remain effective and competitive.

An Example of an Integrated Agriculture Decision Support System

Examples have relied on combining inputs, models, feedback loops, and ensembles (Dengel 2013). This framework outlines the statistical, machine learning, and system automation approaches that will combine to integrate the varied inputs and separate models, into a functional system that captures the dependencies within the entire system. The level of description here simultaneously seeks to balance complexity with clarity in describing this integrated system (Table 1).

As described earlier, this is an IoT-dense system that combines sensor data with other sources including space-enabled data, and exchanges these through APIs and an array of widgets and connectors. The flow of information through the system is mediated through the different subsystems that manage the processes at each stage of the value chain, namely, planting, harvesting, processing (shelling, drying, packaging), storage, transportation, and market. The table above takes an integrated approach in giving examples of the different data elements, data types, and decision points relevant to each of these stages.

The flow of information seeks to enhance processes at each stage by making available information from a preceding or subsequent stage. These do not have to be

Table 1 Examples of variation in agriculture value chain data and integrated decision support algorithms

Capability	Data source	Agricultural value chain example data set (Postharvest loss example)
Sensing/imagery	Camera	Pest control, conflict hazards, type of crop, size of farm, estimating goods volume
	LiDAR/Radar	Precipitation, flooding, pollution, hydrologics
	Hyperspectral	Leaf color, NDVI, crop health, bush fires, soil temperature, air temperature, humidity
	Sensors	Grain moisture, aflatoxin and growth, infestation, contamination, packaging integrity
Geo-location	Localization	Storage and distribution infrastructure, market access
	Geonavigation	Planting and harvesting optimization, mixed crop harvesting, transportation
	Geo-optimization	Transportation, postharvest loss minimization, automated machinery
Telecommunication	Device	Bluetooth, infrared, NFC
	Network	TCP/IP, GSM, LTE, satellite communications
	Systems	RESTful APIs
System integration	Technical	Cost minimization, system logs
	Situational	Market regulations, supply conditions (i.e., competition), transaction costs
	Environment	Governance structure, regulatory compliance
Decision support	Product	Feedback from wholesalers and retailers, crop selection, future planting, crop rotation and substitution, production decision
	Logistics	Postharvest loss optimization
	Market/pricing	Profitability, expected return on investment, market forecasting, dynamic pricing

contiguous stages for shared information to be relevant or useful. The ability of an integrated system to store historical data for analysis endows the capability of providing both real-time data and historical insights as inputs to a decision-making process. Information about provisioning and procurement of resources for a downstream process would thus benefit from estimates and current data on the volume of produce to be anticipated from single or multiple farming locations. Similarly, the timing of harvesting (an upstream process) is informed by historical data on peak pricing, other market data, and the availability of transportation options for fresh produce. An exhaustive list of bidirectional information flows and the agricultural processes that they have the potential to inform will be determined over the course of interactions with stakeholders during the design phase of the system.

With the availability of historical data and real-time data, as well as a clear definition of agricultural processes that benefit from the availability of the data, statistical methods can be applied to develop diagnostic, predictive, and prescriptive models using these data. Understanding the details of the modeled agricultural processes is a prerequisite to successfully applying statistical machine learning in agriculture. Systems that collect, store, manage, and allow analysis, visualization, and integration of agricultural data inputs are essential for advancing the role of data as an agricultural resource, input, and asset. When machine learning models are stable in production and are successfully delivering gains in process improvements, they can then be allowed to drive agricultural process automation and inform process reengineering, where necessary. For example, the release of inventory stocks of raw produce into a food processing line based on forecasted demand to allow just-in-time delivery within the supply chain is a process that can be automated based on the success of machine learning models that draw on historical time-series data on demand patterns, as well as real time monitoring data on raw produce stocks in storage, and logistics data on transportation availability for the delivery of fresh products.

Similarly, a model for the pricing of goods at the market will need to input historical pricing data, demand forecasts, market competition data, real time inventory data, forecasts of anticipated commodity supplies, space-enabled environmental data predicting agricultural productivity, and industry financial targets, in determining where to set the price points. With additional learning, such a pricing model, that relies on the outputs of other statistical models with variable inbuilt uncertainty calculations and reliability, can be further enhanced through using artificial intelligence-based ensembles that can account for this wide variability and complexity. In addition, the pricing model can be further enhanced to maximize profit and minimize losses by varying the prices based on data on the location and timing of the sale, through using an optimization model for the setting of prices. This can then be automated as an AI-powered pricing system ensuring profitability even in the face of uncertainty which is a major factor in agricultural market risk modeling.

Agriculture is frequently described as a risky enterprise. With AI providing a way to intelligently deal with this uncertainty, a farmer can have at his disposal the tools he needs to minimize his risk in participating in farming. The same applies at every step of the value chain where actors in the agricultural value chain struggle with multiple points of failure in procuring, processing, and delivering to market various food products with different storage and processing requirements, within a highly regulated industry with stringent quality standards. AI will not eliminate all the risk in agriculture, but by linking and integrating varied data sources, and consistently doing this the right way over a long period of time, we will revolutionize the agricultural industry with significant impact on food production and availability, agriculture financing, commodity pricing, sustainable agriculture, nutrition, and hunger.

The planning and coordination of the integrated system will require skills in networked infrastructure to include the provisioning of hardware, networking, and software solutions for both back-end (resource management) and front-end (end-user facing) functions. User interface design and visual analytics skill sets will also be needed. Process automation, machine learning, optimization, and other advanced analytics skill sets will be required to turn the space-enabled and other data types into useful insights and intelligence for decision-making. Training, communications, and operational support resources are essential for successful deployment.

An Integration Model in Practice

Data governance should not be an afterthought when building complex systems, especially those derived from multiple sources, and those that may contain regulated data such as personally identifiable information (PII). Clear and detailed rules should be outlined to govern data storage, security, access, retention, sharing, and usage. Data has inherent value and so the ownership rights to the data should also be clear before, during, and after it passes through the system. It may be necessary to constitute a group that will be responsible for monitoring, managing, enforcing, and updating the data governance stipulations for the system.

In order to understand the complexity and depth of an integrated system for an agricultural problem, it is required to exhaustively explore and comprehensively describe the current state of data collection and knowledge management systems for tracking postharvest losses. Multiple data collection methods are needed to gather the needed data from the broad end-user base which will include individual farmers, their collectives (usually farmer cooperatives), food processing and transportation service providers, government, communities and other stakeholders including external donors, and agricultural technology manufacturers and vendors. The elicited findings will be used to illustrate cases that will inform functional requirements for how the conceptual framework will work in various process scenarios. In other words, their recommendations will be tailored to contextual circumstances. This will seek to analyze how efficiencies can be gained by implementing a context-sensitive and framework-based approach. This differs from current approaches that assume that one technological shoe fits all. A framework in addition to providing a template to compile lessons learned, also enables integrating data sources into meaningful information that can then be organized to provide decision support across the value chain. After exhaustively defining the data elements that feed into and that can be derived from the food distribution and postharvest loss across the food value chain, each data element is then further described in relation to the subsystems it supports, and how it is linked to other aspects of the system. This creates an understanding of how information flows within the existing system, and how this may be optimized to work better in an integrated decision support system.

When there is a high volume of information resources, information overload (with abundant false positives drowning out the actionable signal with unnecessary noise) could become a problem. To combat this, systems must be configured and trained to detect useful information and exclude unnecessary data. AI is a powerful tool for better defining the statistical boundary on which data classification is based, in order to improve the quality of actionable information. The goal is to work with only useful data and keep data products useful and relevant to the stakeholder and end-user base.

Another goal of this framework is to ensure that implementations of the integrated system result in intelligent systems with the ability to update and optimize decision support recommendations in response to changes in the data and other environmental factors. To the extent possible, it is also desirable that implementations carry a certain level of automation to further abstract their use from the inherent skills and expertise of the user. The ensuing postharvest loss optimized integrated decision support system (POIDS) will integrate different forms of data from a wide variety of sources, continuously streaming in, and optimize and learn from these data to automatically deliver value efficiently. This could serve as a framework for examining how a broad range of possible data elements representing widely varying upstream and downstream factors can be combined to achieve an advanced system of efficiency for agricultural and food distribution.

Legal, Political, and Economic Considerations

The success of the system will depend on the availability and sharing of satellite data. While Article 1 of the Treaty on Principles Governing the Activities of States in the Exploration and Use of Outer Space, Including the Moon and Other Celestial Bodies, states that "the exploration and use of outer space, including the Moon and other celestial bodies, shall be carried out for the benefit and in the interests of all countries, irrespective of their degree of economic or scientific development, and shall be the province of all mankind." The concept of benefit of all has continued to be ambiguous and a subject of legal contention (Aganaba-Jeanty 2015) particularly with relation to sharing satellite data. Also, being a soft law, the Treaty does not have the force of law, as such, States can decide the interpretation of common benefit it pleases and as such can decide not to share information for reasons such as national security. It appears that the wordings of The UNGA's 1996 Benefits Declaration that "States are free to determine all aspects of their participation in international cooperation in the exploration and use of outer space on an equitable and mutually acceptable basis" leave States to determine the extent and definition of international collaboration.

There is an obvious lack of agreement on the magnitude of the postharvest loss problem, even among highly resourced custodians of the problem in international development. Without a doubt, national governments have a role to play in the creation and validity of postharvest loss data. They, however, also face several significant challenges to meeting their governance obligations. Addressing the data availability and accuracy challenge would strengthen predictive models and better support decision-making to drive improvements along the food and agricultural value chain, to ultimately help reduce postharvest loss.

Investing in space technologies would have financial implications on the demand and supply sides. For example, on the demand side, there will be costs on the farmer for subscribing to the integrated support system service as they access information at the planting, harvesting, drying, and transportation stages. On the supply side, there are costs on the entity providing the integrated service for purchasing space data from private entities in developed countries and other sunk costs such as those of training the farmers on the most efficient utilization of the information provided and costs of business registration. It is expected that these costs will be shared among the stakeholders in the value chain. While the costs may be high in the early stages of the deployment, it is expected that over time, the service will reach economies of scale as more farmers subscribe to the service.

Conclusions

This chapter highlights lessons learned in addressing an agricultural problem that affects food security. It stresses the role of user communities in developing fitting solutions to user-defined problems while leveraging space-enabled technologies and other technology solutions.

An environmental scan of influencing factors in determining the success of integrated systems in agriculture would identify the different stakeholders, and their primary domains of interest in approaching the problem. The postharvest loss problem can be viewed as both a technology-focused or a development-focused problem. Each approach, to varying degrees, has been adopted as prevailing wisdom for resolving postharvest loss. It is, however, necessary to compare how the potential range of technology interventions have variously contributed to the success and in some cases the failure of programs aimed at improving postharvest loss. Research conducted by the authors in Ghana highlighted some issues of concern and core domains of interest as important to consider when integrating technology solutions into development programs.

African agriculture is often characterized by the low rates of adoption of Green Revolution technologies – such as high yielding crop-varieties; irrigation; and micronutrient fertilizers (Feder and Savastano 2017). Inappropriate land tenure, patriarchy, and ineffective government regulation of land use have been attributed as the cause of small land ownership in the continent (Melesse 2018). Small land ownership is broadly described as farmers farming on less than ten hectares of farmland (Samberg et al. 2016). Smallholder farmers are often described based on their resources using terms such as "small-scale," "resource-poor," and sometimes "peasant farmer" (Gininda et al. 2014). In effect, because of resource constraints, smallholder farming discourages farmers from investing in expensive technologies because the farms are not large enough to produce large outputs (Jayne et al. 2010).

It was observed that when acquisition costs for farmers and other stakeholders are too high then the adoption of new technological approaches will remain low (Kuehne et al. 2017). This is exacerbated by the gaps in purchasing power between the farmers on one hand, and those developing and producing the technology solutions that will be used for agriculture on the other. Similarly, access to maintenance services and the cost of maintenance, where available, could also pose a barrier to integrating technology solutions in an agricultural development program. An alternative solution to the acquisition and maintenance challenges would be the development of technologies and services locally and at a cost that will be affordable to a larger portion of the agricultural end-user base. Observations from the Ghana study show that locally produced technology is both feasible and available in practice but is prone to failure due to poor scalability where there is a lack of government or investor funding support.

Specifically, the use of space-enabled technologies has been on the increase, including among African countries, with Ethiopia recently joining South Africa, Egypt, Nigeria, Ghana, Morocco, Algeria, Kenya, Angola, and Rwanda on the list of African countries with space-based remote sensing or telecommunications assets. This should eventually lead to greater adoption of space-enabled services for agricultural purposes such as in reducing food loss, through the development and deployment of targeted, locally sensitive and context-specific solutions that embrace the challenges of their environment and provide options for disadvantaged farmers. Another option that has been tried is the provision of technologies at no cost to the farmers or their communities either through development grants or donor agency funding. It was observed that this often did not translate to immediate access for all those in need of the technology since such "gifts" still suffered from issues such as preferential allocation based on political affiliation, lack of skilled users, poor or absent maintenance, or inadequacy in meeting the need due to an insufficiently available number of the resource.

A potential approach would be to implement the concept of shared ownership of a technology resource. Ownership seeks to actively engage all stakeholders in the conceptualization, framing, and solution design of a shared problem. This allows a participatory approach, and a common sense of commitment to the success of the technology intervention. It also embraces a shared definition of success for the technology program which then allows each stakeholder to be represented in the allocation of resources relative to the intervention. The key then to increasing the adoption of an integrated system is the ability to demonstrate to farmers, at varying scales of farm size and technology advancement, how participation in the use of information technologies and taking advantage of decision support systems can enhance their farming knowledge and skills, increase their output, and profit them economically.

To further support the success of space-enabled technology interventions, the transfer of knowledge and skills to local product owners, system administrators, and end-users will serve to increase the long-term viability and sustainability of the programs. There needs to be a plan for not just immediate use of an application but also for significant technology leaps that will enable local ownership. This would solve the problem of technology interventions being abruptly disrupted due to cessation of development funding or the unexpected departure of a key resource that is difficult to replace. A previously deployed agriculture decision support tool, for precision farming and pricing, that was discontinued led to a distrust of such solutions by the farmers, and this may make it harder to convince them to adopt similar solutions in the future, especially if a cheaper alternative has been discovered. This could have been avoided if the technology intervention was deployed in a robust manner, with sufficient redundancy built into the system from a resource standpoint. It was observed that in some cases, the lack of knowledge was considered a problem at a communal level. To respond to this need, knowledge transfer can be done at a group level, over multiple iterations to ensure that knowledge stays fresh and practical. The integration of knowledge sharing platforms can further enhance access to information while providing peer support for troubleshooting and issue resolution.

Technology adoption does not happen instantly (Hall and Khan 2003). Rather, it is a process whereby some people are more willing and able to adopt innovation than others. The distribution of the actors within technology adoption models shows a bell curve where the spread of technology starts off slowly during the incubation period, accelerates during a period of rapid acceptance, and levels off as more people adopt the technologies (Caselli and Coleman 2001). Cultural factors remain a key driver of adoption and in determining if the impact will be realized from investments made in introducing technologies for development aims.

The chapter recommends pulling together varied data sources and integrating them into accessible decision support systems. These systems provide information for intelligent decision-making at the different intervention and leverage points across the agriculture value chain. Accuracy shall be refined in and through practice, and the systems should be set up to integrate new data sources, assumptions, constraints, and statistical models.

References

Adebola O (2020) Market-based approaches for postharvest loss reduction. PhD. Atlanta: Georgia Institute of Technology

Africa Postharvest Loss Information System (2017) APHLIS+. [online] Available at: https://www.aphlis.net/en/page/1/crop-tables#/datatables/crops-losses?metric=prc&year=2016. Accessed 10 Jan 2020

Aganaba-Jeanty T (2015) Common benefit from a perspective of "non-traditional partners": a proposed agenda to address the status quo in global space governance. Acta Astronaut 117:172–183

Anderson C (2014) How drones came to your local farm. [online] MIT Technology Review. Available at: https://www.technologyreview.com/s/526491/agricultural-drones/. Accessed 6 Mar 2020

Bharati P, Chaudhury A (2006) Studying the current status of technology adoption. Commun ACM 49(10):88–93

Boxall RA (2002) Storage losses. In: Golob P, Farrell G, Orchard JE (eds) Crop post-harvest: science and technology volume 1: principles and practice. Blackwell Sciences, Ltd., Oxford, pp 143–169

Buzby JC, Hyman J (2012) Total and per capita value of food loss in the United States. Food Policy 37(5):561–570

Caselli F, Coleman WJ (2001) Cross-country technology diffusion: the case of computers. Am Econ Rev 91(2):328–335

Chuang S, Sahoo N, Lin H, Chang Y (2019) Predictive maintenance with sensor data analytics on a raspberry pi-based experimental platform. Sensors 19(18):3884

De Lucia M, Assennato D (1994) Agricultural Engineering in Development: Post-harvest Operations and Management of Foodgrains. FAO Agricultural Services Bulletin No. 93. Rome: FAO

Dengel A (2013) Special issue on artificial intelligence in agriculture. Künstl Intell 27(4):309–311

Dlodlo N, Kalezhi J (2015) The internet of things in agriculture for sustainable rural development. In: 2015 international conference on emerging trends in networks and computer communications (ETNCC). Windhoek: IEEE, pp 13–18

FAO (2013) The food wastage footprint. FAO, Rome

Feder G, Savastano S (2017) Modern agricultural technology adoption in sub-Saharan Africa: A four-country analysis. In Agriculture and Rural Development in a Globalizing World (pp. 11–25). London: Routledge

Florkowski WJ, Prussia SE, Shewfelt RL, Brueckner B (eds) (2009) Postharvest handling: a systems approach. Amsterdam: Academic Press

Food and Agriculture Organization (2011) Global food losses and food waste: extent, causes, and prevention. Rome: FAO

Gagliordi N (2018) How self-driving tractors, AI, and precision agriculture will save us from the impending food crisis. [online] TechRepublic. Available at: https://www.techrepublic.com/article/how-self-driving-tractors-ai-and-precision-agriculture-will-save-us-from-the-impending-food-crisis/. Accessed 6 Mar 2020

Ganpat W, Isaac W (2015) Impacts of climate change on food security in small island developing states. Information Science Reference, an imprint of IGI Global, Hershey

Gininda PS, Antwi MA, Oladele OI (2014) Smallholder sugarcane farmers' perception of the effect of micro agricultural finance institution of South Africa on livelihood outcomes in Nkomazi local municipality, Mpumalanga Province. Mediterr J Soc Sci 5(27 P2):1032

Glouberman S, Zimmerman B (2002) Complicated and complex systems: what would successful reform of medicare look like? Commission on the Future of Health Care in Canada, Ottawa. Available from http://publications.gc.ca/collections/Collection/CP32-79-8-2002E.pdf

Grolleaud M (2002) Post-harvest losses: discovering the full story. Overview of the phenomenon of losses during the post-harvest system. FAO, Agro Industries and Post-Harvest Management Service, Rome

Hall BH, Khan B (2003) Adoption of new technology (no. w9730). Cambridge: National Bureau of Economic Research

Heeks R (2005) Implementing and managing egovernment: an international text. Los Angeles: Sage

Hodges R, Buzby J, Bennett B (2010) Postharvest losses and waste in developed and less developed countries: opportunities to improve resource use. J Agric Sci 149(S1):37–45. https://doi.org/10.1017/s0021859610000936

Jayne TS, Mather D, Mghenyi E (2010) Principal challenges confronting smallholder agriculture in sub-Saharan Africa. World Dev 38(10):1384–1398

Johnstone S, Mazo J (2011) Global warming and the Arab spring. Survival 53(2):11–17

Kaushal R, Shojania KG, Bates DW (2003) Effects of computerized physician order entry and clinical decision support systems on medication safety: a systematic review. Arch Intern Med 163(12):1409–1416

Khattab A, Abdelgawad A, Yelmarthi K (2016) Design and implementation of a cloud-based IoT scheme for precision agriculture. In: 2016 28th international conference on microelectronics (ICM). Giza: IEEE, pp 201–204

Kuehne G, Llewellyn R, Pannell DJ, Wilkinson R, Dolling P, Ouzman J, Ewing M (2017) Predicting farmer uptake of new agricultural practices: a tool for research, extension and policy. Agric Syst 156:115–125

Kumar D, Kalita P (2017) Reducing postharvest losses during storage of grain crops to strengthen food security in developing countries. Foods 6(1):8. https://doi.org/10.3390/foods6010008

Melesse B (2018) A review on factors affecting adoption of agricultural new technologies in Ethiopia. J Agric Sci Food Res 9(3):1–4

Mooney PH, Hunt SA (2009) Food security: the elaboration of contested claims to a consensus frame. Rural Sociol 74(4):469–497

Noura M, Atiquzzaman M, Gaedke M (2019) Interoperability in internet of things: taxonomies and open challenges. Mob Netw Appl 24(3):796–809

OECD/FAO (2016) OECD-FAO agricultural outlook 2016–2025. OECD Publishing, Paris. https://doi.org/10.1787/agr_outlook-2016-en

Paulsen MR, Kalita PK, Rausch KD (2015) Postharvest losses due to harvesting operations in developing countries: a review. In: American society of agricultural and biological engineers annual international meeting 2015, vol 1. New Orleans: American Society of Agricultural and Biological Engineers, pp 562–596

Samberg LH, Gerber JS, Ramankutty N, Herrero M, West PC (2016) Subnational distribution of average farm size and smallholder contributions to global food production. Environ Res Lett 11 (12):124010

Tian F (2017) A supply chain traceability system for food safety based on HACCP, blockchain & internet of things. In: 2017 international conference on service systems and service management. Dalian: IEEE, pp 1–6

Uddin MA, Mansour A, Le Jeune D, Aggoune EHM (2017) Agriculture internet of things: AG-IoT. In: 2017 27th international telecommunication networks and applications conference (ITNAC). Melbourne: IEEE, pp 1–6

United Nations (2015) Space for agriculture development and food security. [online]. United Nations Office at Vienna, Vienna. Available at: https://www.unoosa.org/res/oosadoc/data/docu ments/2016/stspace/stspace69_0_html/st_space_69E.pdf. Accessed 6 Mar 2020

Weiss M, Jacob F, Duveiller G (2020) Remote sensing for agricultural applications: a meta-review. Remote Sens Environ 236:111402

World Bank (2010) Missing food: the case of postharvest grain losses in sub-Saharan Africa. World Bank: Washington, DC

Satellite Communication for Security and Defense

41

Holger Lueschow and Roberto Pelaez

Contents

H. Lueschow (✉) · R. Pelaez
European Defence Agency (EDA), Brussels, Belgium
e-mail: Holger.LUESCHOW@eda.europa.eu; Roberto.PELAEZ@eda.europa.eu

© Springer Nature Switzerland AG 2020
K.-U. Schrogl (ed.), *Handbook of Space Security*,
https://doi.org/10.1007/978-3-030-23210-8_107

Abstract

The contents reported in this chapter reflect the opinions of the authors and do not necessarily reflect the opinions of the respective Agency/Institutions.

Today, coalition and national military operations are inconceivable without the support of space-based systems. Space-based assets and applications are essential to navigation; communication; meteorological, geospatial, and imagery services; early warning; and ballistic missile interception. They are a vital enabler for command and control (C2) and situational awareness through the provision of intelligence, surveillance, and reconnaissance (ISR) information.

Satellite communications play an indispensable role in security and defense-related governmental communication. They are used when other ground-based means of communication are not possible, reliable, or available. Satellite communications are critical assets in the ability of States to respond autonomously and in a timely manner to global defense, security, and humanitarian and emergency challenges.

Introduction - Historical Evolution

Communications have been critical to military organizations for centuries.

The year 1957 marks the beginning of the space age with the Russian SPUTNIK-1 being the first artificial satellite to go around the Earth. In 1962, the world's first active communications satellite Telstar 1 was launched. This satellite was built by AT&T and Bell Laboratories, USA. During its 7 months in operation, Telstar 1 provided to the world live images of sports, entertainment, and news. It was a simple single-transponder low-earth-orbit (LEO) satellite, but its technology of receiving radio signals from the ground, and then amplifying and retransmitting them over a large portion of the earth's surface, set the standard for all communications satellites that followed.

Since then the use of space has expanded constantly based on ever-growing demand for different types of space services, and SATCOM systems support a wide range of fixed and mobile telephone and data services including broadcast services. These are known under their official ITU definitions as the Fixed Satellite Services (FSS), the Broadcast Satellite Services (BSS), and the Mobile Satellite Services (MSS) for aeronautical, maritime, and land mobile applications. Over the years, a global change in telecommunications regulation and transition from monopolist communication providers to an open market has led to constant innovation and more competitive systems also in the satellite world.

During the 1960s and 1970s, satellite performances advanced quickly and a global commercial SATCOM industry began to develop. Focus in commercial SATCOM was for decades to support international and long-haul telephone traffic and to distribute broadcasting programs, and for many years, satellite broadcasting was the driver for the evolution of commercial SATCOM systems. Since those early days of SATCOM, commercial solutions have become more capable and could support higher data rates and more and more services. In parallel to this evolution, the user terminals have become smaller, cheaper, and easier to operate.

This development went hand in hand with the increased sophistication of military forces. Consequently, the opportunities offered by satellites for all types of military applications including communications have been understood, and it was natural that the military explored space to receive support to meet their communications requirements. This was the baseline for SATCOM as a vital enabler to support defense-related services.

Military organizations around the world began to look at ways how to use satellites to provide communications to their forces. Starting with the U.S. Department of Defense and its Initial Defense Satellite Communication System (IDSCS – a series of LEO satellites in a random orbit constellation that allowed for more or less global communications, although there were some periodic service interruptions due to gaps in the satellite coverage), a wide range of military communications satellites have been launched by a number of other defense forces around the world in the years and decades that followed. Over the last 60 years, a variety of satellites, deployed into different orbits, have been developed by many nations to meet their governments' and particular military communication needs.

As SATCOM systems come along with high costs and long planning and realization times, most of those military SATCOM projects and initiatives cannot meet all the communication requirements the military have, be it from a bandwidth or coverage perspective. Therefore, defense forces around the world have to rely on commercial SATCOM systems. Depending on the use case and application, the support by commercial satellite capabilities has always been necessary, especially when specific needs arise or military demand exceeds available military capacities.

Today, industry has a long experience in meeting the military user demands for telephone, data, networking, audio, and video services. SATCOM services and applications form the largest space market sector and are sold on a "global open market." Commercial SATCOM capacity is usually delivered by satellite operators who own a fleet of satellites which they procure, launch, and operate. Several different contract approaches have been used to obtain defense-related SATCOM capacity, and a whole industry has been developed around providing communications requirements for military purposes. In some cases, industry has provided the military user with full end-to-end (sometimes known as turn-key) services including the capacity leased from a commercial provider, terminals, other hardware, and operation of the full service to supplement the military capacity.

With the "new Space" and in particular the entry of new business actors from the "digital economy," the SATCOM market was faced with considerable changes and had to react to new requirements that currently lead to new or modified business models and SATCOM services.

High-throughput (HTS) and very high-throughput satellites (VHTS) as well as new mega-constellations of satellites represent the latest technology trends and promise for the future communications support for live video with low latency and at less cost compared to a decade ago.

SATCOM Frequency Bands and Orbits

From their initial stages, SATCOM services have quickly evolved to meet both civilian and military communication needs. The real drivers of this evolution were intercontinental and regional communication during the 1970s and the 1980s, broadcasts services, mainly TV, since the 1990s, and more recently multimedia applications.

An efficient management of the limited and scarce resource of frequency spectrum as well as of the different orbits (with a focus on the geostationary arc) is necessary to operate satellite services on a worldwide basis.

The International Telecommunications Union (ITU), a specialized organization of the UN, is executing this management and governing the use of the radio-frequency spectrum by the international "Radio Regulations" (RR) treaty. Within the RR, it has been agreed to allocate specific frequency bands to specific SATCOM services.

The RR distinguishes the following satellite services that are used for SATCOM:

- Fixed-satellite service (FSS)
- Inter-satellite service (ISS)
- Mobile-satellite service including (MSS)
  - Land mobile-satellite service
  - Maritime mobile-satellite service
  - Aeronautical mobile-satellite service
- Broadcasting-satellite service (BSS).

With the variety of satellite frequency bands that are allocated to above satellite services and can be used for SATCOM, designations have been developed so that they can be referred to easily. Traditionally, the C-band (3-7GHz), **Ku**-band (10–15 GHz) and **Ka**-band (17–43 GHz) are allocated to the fixed satellite services and the **L**-band (1–2 GHz) to mobile satellite communication. All those bands are heavily exploited by commercial industry.

While there is no explicit allocation in the RR reserved for governmental or military use, agreements between governments (specifically NATO nations) exist to use the **P**-band (200–400 MHz) and the **X**-band (7–10 GHz) as well as portions of the **Ka**-band (so-called military Ka-band) for governmental (military) purposes.

The P (or UHF)-band is dedicated to tactical voice communications and data links, although only providing a very limited throughput. Despite being an extremely limited service in terms of throughput, UHF has become an enduring technology for troops worldwide due to its utility for highly mobile, deployed forces. This utility is unlikely to change in the future despite the availability of handheld commercial SATCOM systems such as Iridium, Thuraya, and Globalstar.

The X-band is considered as ideal to establish secure and robust satellite communication links, for example to connect operational areas with the homeland HQs (reach-back), between theaters of tactical operations, maritime missions or over areas affected by a humanitarian crisis.

The military Ka-band (from 30 to 31 GHz and 43.5 to 45.5 GHz for the uplink and from 20.2 to 21.2 GHz for the downlink) provides a valuable spectrum resource in support of ISR (Intelligence Surveillance and Reconnaissance) missions. Providing high bandwidth for data-intensive applications over small terminals characterizes this band and makes it ideal for communications on the move.

The military users of governments traditionally focused on the geostationary (GEO) orbit for their SATCOM satellites. In this orbit, characterized by a flight height of almost 36,000 km above the Equator, a satellite needs exactly 1 day to fly around the earth and thus appears to be fixed above the same point on Earth. Because of this specific characteristic, a GEO orbit is very beneficial for supporting communications and earth stations are easy to install to keep the link between ground and satellite as they do not have to track the satellite's movement. This leads to lower terminal costs compared to more sophisticated terminals with the ability to track a moving satellite. Operation of terminals is also easier in GEO orbit SATCOM systems. A single satellite deployed into geostationary orbit provides visibility of almost a third of the earth's surface.

However, satellites in a geostationary orbit cannot cover the polar regions as they are invisible at latitudes higher than ~75°. Another disadvantage of GEO SATCOM solutions is the relatively high latency they impose on communication signals due to the long distance between a GEO satellite and the Earth.

Besides the geostationary orbit, there are several *non-geostationary orbits* (NGSO) used to provide SATCOM services.

Between approximately 500 and 2000 km above the Earth, *low earth orbits* (LEOs) are implemented. Due to their little height, they can better support applications that are latency sensitive.

However, their coverage is extremely limited and their visibility above a certain point on Earth is in the range of 15 min. As one LEO satellite can only provide this visibility every 90–120 min, usually constellations of LEO satellites are used to provide uninterrupted SATCOM services. This imposes some challenges to the ground system as a signal handover from one satellite to another is necessary.

Medium earth orbits (MEOs) describe the region of space between LEO and GEO. They are mainly used for navigational satellites but provide also communication services. Depending on the elliptical orbit MEO satellites are in, their visibility above Earth can vary from a few hours up to half a day. To provide a global coverage, a constellation of several satellites is therefore needed.

Besides those three classical orbits, more orbits of lower relevance are used by communication satellites as some of them are very suitable to support specific use cases.

Frequency regulation constraints and orbital reservation are very significant points to be tackled early in any SATCOM project.

SATCOM for Military

There is a general understanding that the provision of Satellite Communication (SATCOM) services for governmental use can be divided into three tiers which corresponds to different levels of information assurance although the exact definition of each tier may slightly vary.

Commercial SATCOM (COMSATCOM)

Mass market commercial SATCOM (COMSATCOM) is operated by private companies in a competitive market. Most of the COMSATCOM market is driven by television (broadcast), but there are also much more advanced systems that involve renting or buying dedicated communication stations (hereafter referred to as "terminals"). Governmental users are a small, though a high growth market for operators of COMSATCOM, the bulk of their turnover being linked to consumer multimedia services (TV and Internet).

In terms of communication security – in particular security mechanisms to counter threats such as anti-jamming, protecting against interception and demodulation, preventing unauthorized access, detection and neutralization of unauthorized activities – the level of protection currently implemented by COMSATCOM systems is generally considered as insufficient to meet the information security requirements for military and security use cases.

In addition, COMSATCOM systems usually don't offer specific guarantees in terms of access to the resource nor the system's vulnerability to external attacks.

Military SATCOM (MILSATCOM)

MILSATCOM satellites are primarily used for military missions at the national level, or in the framework of EU CSDP or NATO operations. These services are dedicated to the more critical applications requiring advanced protection (strong resistance to interference, military cryptography, resistance to nuclear events in orbit, undetectable communications, etc.).

MILSATCOM systems are technologically very similar to each other and to an extent interoperable. Security and technology are characterized as being highly specialized and largely sovereign in nature.

Governmental SATCOM (GOVSATCOM)

GOVSATCOM solutions are a new SATCOM service class that fits between mass market COMSATCOM services that do not offer specific features as regards security, robustness, and availability and MILSATCOM services that usually provide high security and guarantee levels, however, at much higher cost and thus difficult to implement for use in low- or medium-intensity crises.

GOVSATCOM means a highly available satellite communication, providing a level of security with some resilience, obtained using technological solutions available on the market with a minimum of changes.

Operational Needs

Flexibility is essential in modern conflicts.

Contemporary conflicts are usually small and asymmetrical with dynamic and often difficult-to-identify enemies typically characterized as insurgent forces. Military operations have become more international and are generally led by operational headquarters geographically distant from the active operation area.

This concept is called "reach-back" and refers to the situation where resources, capabilities, and expertise are at physical distance from the area of interest, supporting the people in the area to perform their task.

The most important advantages of this concept are the safety (by using reach-back facilities, less personnel has to be present in the area of operation), the mobility (ease staff members to move while keeping informed about the situation of the operation), the flexibility (in theater, military activities are hard to predict and this concept supports resilience in operations), the improvement in logistics (less logistical effort to deploy, maintain, and remove a command post), and the detection of the forces by the enemy (having smaller and agile centers of communication, the possibility to be detected by enemies decrease).

However, immediate disadvantage is the required information support. Constraints on bandwidth need to be overcome to ensure collaboration between units as if they were physically together.

SATCOM is the only viable means to overcome this constraint, and since the early 2000s alone, the EU led more than 30 "reach-back" missions and operations all over the world with many of them still ongoing. The geographical coverage of those missions and operations has greatly varied from Africa to the North-Eastern border of Europe and from the Mediterranean Sea and the Western Balkans to Eurasia into the Middle East and Asia.

The different scenarios that have to be supported by SATCOM range from "separation of parties by force" with large size formations and requiring a reaction time within 60 days over "conflict prevention" with medium personnel effort but requiring reaction time within 30 days to "evacuation operations" involving only small numbers of personnel and foreseen reaction time within 10 days. It is worth noting that most of those scenarios can be long-term missions (over 2 years).

All that gives an indication that the different natures of those operational needs derived from above scenarios requires different SATCOM service solutions. The military is dependent on having quick and easy access to fixed and mobile satellite services supporting tactical and strategical use cases and applications spanning all over the Earth.

Use Cases and Applications

As already addressed in the previous section, the SATCOM requirement for the "reach-back" infrastructure (strategic, operational, and tactical communication links) between tactical local networks, the deployed Force Headquarters (FHQ), and the (reach back) Operation Headquarters (OHQ) is dominant. Only with a secure and resilient SATCOM link, a command and control can be maintained functioning.

The following use cases and applications further define the military capabilities modern SATCOM systems must support:

- **ISR platforms** are increasingly utilizing high-definition optical cameras and/or high-resolution radars and other capabilities which create increasingly large amount of data that must be communicated from the remote location/platforms to locations in theater or OHQ securely, in a timely manner in order to inform operational decision making. These sensors are hosted on a range of mobile platforms, many of which are increasingly RPAS.
- **RPAS** are an increasingly deployed capability, used to extend the range of ISR and offensive capabilities by using systems which are autonomously operated or operated under human control, but which do not require human crewing. This allows them greater range and endurance. They are usually controlled and communicated by secure SATCOM links and they can carry ISR mission equipment.
- **Telemedicine** in remote and austere environments is more and more introduced in military operations. This relies on SATCOM to deliver health care services, such as access to specialists. With increasing medical capability that relies on ICT (information communication and technology), the volumes of data that need to be transmitted have risen markedly.
- **Logistics/Admin CIS** are keys to the success of modern military operations. The activities include planning for personnel deployment and cargo movements, via a complex mix of air, sea, and land routes; developing and managing cargo and asset tracking information; procurement and purchasing systems both in theater and at the rear echelons; providing in theater access to the military HR and admin systems.
- **Welfare** CIS in the digital world of military forces cannot be underestimated as their bandwidth and quality requirements will impose a challenge for SATCOM service provision. This use case embraces the provision of access to welfare services for deployed forces and supporting civilian/contract personnel. Extremely important to keep good morale, a wide range of services such as email, video conferencing, web browsing, and streaming services needs to be provided.
- **Command and Control links** commonly called C2, have to be supported in any military mission or operation. Without a safe and secure C2 link severe consequences for the conduct of any mission or operation can be expected.

Military SATCOM Solutions and Systems

The variety and different nature of above use cases imply that they request different levels of security and guarantees of availability.

Autonomy is an additional driver in defense systems. These all have led some countries to look for their own satellite communications solutions. In order to avoid dependencies and driven by the need to possess very capable and advanced own systems, several nations have generated the development of expensive multiyear programs for national (military) SATCOM capabilities to support military- and defense-related purposes. Most of those satellites meet specific security requirements such as anti-jamming and radiation hardening and operate mainly in the classical governmental frequency bands.

However, the vast majority of military users has no access to own SATCOM resources and is either dependent on SATCOM systems of partner countries through bilateral/multilateral agreements or has to rely on commercial SATCOM service providers.

The nature of support that is requested from SATCOM ranges from the provision of only power and bandwidth to end-to end (terminal to terminal) solutions and includes anchoring and backhauling.

United States

The United States currently owns the largest number of military SATCOM satellites, and it has developed into a nation with multiple constellations of satellites. These constellations tend to be frequency-band specific.

The United States has divided its SATCOM communications into four elements:

- Narrowband, unprotected communications using UHF – MUOS
- Wideband communications with limited protection features on X-band and Ka-band frequencies – WGS
- Protected communications with full hardening and survivability features using Ka-Band and SHF-frequencies (above 30 GHz) – AEHF
- Leased commercial satellite communications using L-band, C-band, Ku-band, and more recently X-band and UHF

MUOS (Mobile User Objective System) is today the most powerful UHF SATCOM constellation. It provides UHF secure voice, data, video, and network-centric communications in real time to US forces.

Most of the US military communications are supported by the **Wideband Global System** (WGS). Currently, nine satellites provide X-band and Ka-band communication services. Originally only build to support US forces and forces of allied partners Canada and Australia, the partnership over time has been extended and funding for the latest satellites has been provided also by Denmark, the Netherlands,

Luxembourg, and New Zealand in exchange for access to bandwidth from the entire global constellation.

The **AEHF** (Advanced Extremely High Frequency) SATCOM system is used to provide worldwide highly survivable and protected national communications capabilities, in support of strategic and tactical forces of the United States and its international partners.

The services provided by AEHF operate at Ka-band and SHF-band.

In summary, the WGS constellation represents the primary communications system for the U.S. Department of Defense operated alongside the specialized AEHF operated by the Air Force and the Navy's MUOS mobile communications system, while the use of UHF SATCOM solutions falls under the remit of tactical and highly mobile requirements.

United Kingdom

The United Kingdom, as the pioneer in European governmental SATCOM, has been a military SATCOM user since the late 1960s with the Skynet 2 satellites, actually the first communication satellites to be built outside either the United States or the USSR. The United Kingdom has always opted for multisatellite constellations. Skynet 4 was the last series of UK satellites owned and operated by the UK MOD, while the Skynet 5 fleet followed a private finance initiative (PFI) program with Airbus.

The system **Skynet 5** (four satellites with each UHF- and X-band capability) provides worldwide coverage and supports the UK military forces as well as NATO forces. It is expected that UK MoD will retake ownership of the satellites fleet by 2022.

To replace the current Skynet 5 military SATCOM capability, UK awarded a noncompetitive contract with AIRBUS DS to develop and implement Skynet 6 system, integrated by space crafts, service delivery elements to manage ground operations and an enduring capability program to provide future communication system capacity beyond the next decade. The first satellite Skynet 6A is expected to be fully operational by mid-2025.

France

France has been a member of the military SATCOM community since 1980 with Syracuse 1 and Syracuse 2 and currently operates the **Syracuse 3** constellation (2 satellites). The Syracuse 3 series is hardened and protected to meet NATO standards, similarly to the UK Skynet series, and concentrating on military X-band frequencies to support the French military. It is significant to note that France developed the highly resilient M21 modem, compliant with NATO STANAGS, as one of the key elements of the ground segment under Syracuse 3 program, which complements and supports tactical SATCOM networks.

With a look into future capabilities in SATCOM, in December 2015, France ordered the first two **Syracuse 4** satellites from Thales-Alenia Space and AIRBUS as the main elements of the future program COMSAT-NG. The new satellites, with specifications to be threat resistant to cyber-attacks, jamming, and HANE events, are expected to meet full operational capability from 2022.

The French position with regard to multinational SATCOM capabilities development is based on a vision of different layers that contribute to the overall SATCOM capability. The **SICRAL 2**, a joint Italian-French military SATCOM program, operates in X-band and augments the Syracuse system. This can be considered as core capacity for military use only. On the other side, as an extension of core capacity for use of military and government users, the **Athena-Fidus**, a French-Italian joint-venture, is a geosynchronous military and governmental Ka-band communications satellite capable of data transfer rates of up to 3 Gbps. Jointly procured by the French and Italian space agencies and defense procurement agencies, the system is intended to be used by the French and Italian armed forces as well as the civil protection services of France and Italy.

Besides those own resources, France has long-term contracts in place for leasing commercial SATCOM capacity that complements the national systems.

Germany

Germany has only recently entered the military SATCOM arena with its own dedicated assets after relying for many years on NATO capacity, intergovernmental agreements, and commercial leases.

The German Bundeswehr has launched **COMSATBw** 1 and 2 in October 2009 and June 2010, respectively, as part of its SatComBw-Stage 2 program and operates the two satellites with industrial support. Both satellites are identical carrying UHF- and X-band payloads. Besides those own resources, Germany has long-term contracts in place for leasing commercial SATCOM capacity.

Germany is currently defining requirements for the future systems replacing the existing SATCOMBw-Stage 2 capacities. COMSAT 1 and 2 satellites will reach their end of life by 2027. With this regard, the options include different models for acquisition of SATCOM capacity, from lease to cooperation with third party programs, also including future EU and NATO programs.

In addition, a recent contract has been signed between Germany's Deutsche Zentrum für Luft- und Raumfahrt (DLR) (on behalf of the German MoD) with OHB systems to develop a dual-use satellite using the SmallGEO platform, Heinrich Hertz, operating in Ku and military Ka bands. The satellite is expected to be fully operational by 2021.

Italy

Italy, with **SICRAL 1A**, launched the first military satellite communication spacecraft in 2001 into geostationary orbit, providing UHF and X-band capacity to Italian armed forces and also (jointly with France and the United Kingdom) to NATO forces. The second phase of the program began in 2009 with the launch of SICAL 1B.

The third phase of SICRAL program, the **SICRAL 2**, a joint Italian-French military SATCOM program, became operational in 2015. SICRAL 2 is a geostationary satellite that operates in X- and UHF bands, supports satellite communications requirements of Italian and French Armed Forces, and is able to enhance the SATCOM capabilities and provide backup to the Italian SICRAL 1 and French Syracuse 3 systems.

Italy is planning resources' allocation to maintain the capability of the SICRAL satellites, and further resources may be added by the National Space Programme for the acquisition of the SICRAL 3 system for strategic communications replacing or complementing the current systems.

Spain

Since 2005, Hisdesat (a company shared by the Spanish MoD and Spanish industrial players) is providing secure satellite communications services to the Spanish Ministry of Defence. With **SpainSat** and **Xtar-EUR,** Hisdesat has an innovative generation of satellites that provide satellite communication services in the X and military Ka bands.

Spain is currently working to replace the XTAR-EUR and Spainsat satellites that will reach the end of their final operational capability (after propellant savings) between 2022 and 2024. In this regard, the future satellite communications program, **SPAINSAT-NG**, has been approved in April 2019 and will encompass a space segment consisting of two highly protected threat-resistant satellites operating X-, military Ka-, and UHF bands, improved ground segment deploying new satellite terminals implementing EPM anti-jamming capabilities, and an additional enhanced anchor station. The SPAINSAT-NG will meet the operational capability by 2024.

Luxemburg

The youngest member of the European SATCOM-owning states launched its first satellite in early 2018. **GovSat-1** is a GEO multimission communication satellite operated by LuxGovSat, a public-private joint venture created in 2015 between the Government of Luxemburg and commercial satellite operator SES.

GovSat-1 was built by Orbital ATK and is offering 68 transponder-equivalent units of 36 MHz. Positioned at 21.5° East, it will provide satellite communications services within Europe, the Middle East and Africa, and enable operations over the Mediterranean, Atlantic, and Indian waters.

GovSat-1 was designed for dual-use to support both defense and civil security applications, including mobile and fixed communications. The satellite provides X-band and military Ka-band services on high-power and fully steerable mission beams. Equipped with anti-jamming features, encrypted telemetry, and control,

GovSat-1 will provide enhanced resilience capabilities to meet requirements of governmental and military users and targeting to support the future demands of NATO members.

Luxembourg has currently launched the initial phases to develop an additional GovSat-2 satellite to further enhance the available SATCOM capacities and services.

Other European Activities

Most European nations, other than the ones mentioned above, have neither the budget nor the depth and breadth of requirements to justify investment in own dedicated satellite capability. These nations have typically used intergovernmental agreements with their allies to gain access to protected communications (Germany did this with France for many years prior to launching the SatComBw programs).

When intergovernmental agreements are not possible, then long- or short-term lease contracts with commercial operators or service providers have often proved to be the vehicle of choice.

Nearly every nation has now leased one or more services from Inmarsat to include within its military portfolio for maritime or airborne communications, and this has been augmented over the last 5–10 years with leases of Intelsat, SES, or Eutelsat capacity and more recently with commercial X-band communications leased from either Paradigm or XTAR systems. These nations include, among others, Belgium, Czech Republic, Denmark, the Netherlands, Poland, Portugal, or Slovenia.

The launch of the Inmarsat GLOBAL Xpress but also other new satellite systems like high-throughput satellites will likely increase this usage.

In addition to above bilateral or commercial SATCOM service solutions, the 26 participating Member States of the European Defence Agency can meet their SATCOM requirements through two specific projects established in the Agency:

- **EDA GOVSATCOM Pooling and Sharing Demonstration Project (GSC Demo):** The main goal of this project is to prove and demonstrate the concept and benefits of a collaborative Pooling and Sharing model in GOVSATCOM and meet the **GOVSATCOM** demand of Member States and European CSDP actors through a pooled capability (bandwidth/power and/or services) provided by contributing Member States. By pooling the capability of the Member States, the project provides Governmental SATCOM services to the members, contributing to the overall operational efficiency. The project aims to demonstrate an efficient pay-per-use solution that does not impose any binding financial commitments beyond services requested. It is quick and flexible, establishing government-to-government agreements through EDA and so reducing the administrative burden for members who do not have to run their own GOVSATCOM capability. There are currently 15 contributing members to the GSC Demo project.
- **EDA – European Union Satellite Communication Market (EU SATCOM Market):** Mainly addressing COMSATCOM solutions, the EU SATCOM

Market (formerly known as European Satellite Communications Procurement Cell (ESCPC)) was launched in 2009. The project's aim to provide **commercially** available satellite communications (fixed and mobile) as well as related services (e.g., feasibility studies, maintenance, leasing of earth stations, SATCOM terminals, commercial crypto solutions) and options for other communication services for operations (e.g., theater local radio network, IT network backbone for the field headquarters) through the establishment of a number of Framework Agreements on behalf of the contributing members, to promote ease of access and improve efficiency. This project is now fully operational with a growing number of members and orders intake for long duration. The current contract will expire in January 2020 and will be renewed. There are currently 28 contributing members to the EU SATCOM Market including almost all CSDP military and civilian operations and missions.

NATO

For almost 20 years, NATO owned and operated the two NATO IVA and IVB satellites by itself. In May 2004, the NATO Consultation, Command and Control Agency (NC3A) decided to move away from owning and operating its own fleet of satellites and selected a multinational proposal to provide SHF and UHF communications. This program, entitled the NATO SATCOM Post-2000 (**NSP2K**) program, was based on the support of the French, Italian, and British governments to provide NATO with access to the military segment of their national satellite communications systems – Syracuse, SICRAL, and Skynet, respectively – under a Memorandum of Understanding (MoU). Compared to the previous NATO-owned capabilities, this policy change resulted in increased bandwidth, coverage, and expanded capacity for voice and data communications, including communications with ships at sea, air assets, and troops deployed across the globe.

Currently, the successor program is under discussion, under NATO Capability Package (CP9A0130) "Satellite Communications (SATCOM) Transmission Services." Through this CP130, the acquisition (common-funded) roadmap will address the renewal of allied SATCOM capacity for the period 2020–2034 across areas corresponding to the military SATCOM frequency bands (i.e., X-band, or SHF, EHF, UHF, and Military Ka-band).

Russia

Russia (former USSR) was the first country to orbit a satellite in 1957. It is reported that between 1960 and 1990, most Soviet satellites that were launched carried military payloads, even though until the last decade of the twentieth century there was no official acknowledgment of a military space program. During the first decade of the twenty-first century, Russia has continued its launch program and now identifies specific military satellites but with no specific information as to individual missions.

China

China launched its first satellite in 1970. Since then, its satellite activity has increased, particularly in the last decade of the twentieth century and the first decade of the twenty-first century. China has a large program of both reconnaissance and communications satellites utilized for military purposes.

China utilizes communications satellites for both regional and international telecommunications supporting both military and commercial users somewhat like several other countries.

SATCOM for Security

European Union (EU)

In December 2013, the Heads of States and Government of the EU met to discuss defense and the Common Security and Defence Policy. In the Conclusions, the European Council "*welcomed the Commission communication "Towards a more competitive and efficient defence and security sector.*" More specifically, the European Council welcomed plans regarding SATCOM: "*preparations for the next generation of Governmental Satellite Communication through close cooperation between the Member States, the Commission and the European Space Agency; a users' group should be set up in 2014; the European Council invites the Council, the Commission, the High Representative, the European Defence Agency and the Member States, within their respective spheres of competence, to take determined and verifiable steps to implement the orientations set out above.*"

That was a response to the growing needs identified in terms of secured and guaranteed access for institutional users and the starting point for several activities on European level to improve security and achieve guaranteed access to SATCOM for EU Member States and actors.

The Global Strategy for the EU Foreign and Security Policy (EUGS) issued in June 2016 stresses the importance of capabilities that should be developed with maximum interoperability and commonality and puts forward the objective of promoting the autonomy and security of European space-based services. "*European security hinges on better and shared assessments of internal and external threats and challenges. Europeans must improve the monitoring and control of flows which have security implications. This requires investing in Intelligence, Surveillance and Reconnaissance, including [. . .] satellite communications, and autonomous access to space and permanent earth observation.*" The EUGS claims that "*Defence policy also needs to be better linked to policies covering the internal market, industry and space.*"

The High Representative for Foreign Affairs and Security Policy (HRVP), through the EUGS, has thus emphasized that "*Member States need all major equipment to respond to external crises and keep Europe safe. This means having full-spectrum land, air, space, and maritime capabilities.*"

Building on the Global Strategy and in the frame of protection of the EU and its citizens, the Implementation Plan for the EUGS, issued in November 2016, states the EU can contribute to ensuring stable access to and use of the global commons, including the high seas and space. Taking forward the cross-cutting strategies in the domain of space (including in relation to the Copernicus and Galileo programs) and their links to CSDP, evolutions in the security and defense environment require the EU to reassess its space capabilities in areas relevant to Europe in the context of specific security needs. The Implementation Plan further stresses the needs to invest and develop collaborative approaches in satellite communications, autonomous access to space, positioning navigation and timing (PNT), and permanent earth observation.

The Space Strategy for Europe (SSE), issued by the European Commission in October 2016, describes space as a strategic asset for Europe. The SSE reinforces Europe's role as a stronger global player and is an asset for its security and defense. To increase security, satellite PNT, satellite communications, and Space-Based Earth Observation have been identified as crucial to contribute inter alia to detecting illegal immigration, preventing cross-border organized crime, and combating piracy at sea. The SSE describes emerging needs related to security and defense. "Space capacities are strategically important to civil-, commercial-, security-, and defense-related policy objectives. Europe needs to ensure its freedom of action and autonomy. It needs to have access to space and be able to use it safely. Due to growing threats emerging in space (from space debris to cyber threats), greater synergies between civil and defense aspects" will become "increasingly relevant" to reduce cost, increase resilience, and improve efficiency. Europe must draw on its assets and use space capacities to meet the security and safety needs of the Member States and the EU.

The draft EU Legislative Proposal for the EU Space Programme for the next multiannual financial framework (MFF 2021–2027), issued in June 2018, reflects the increased relevance of security and defense, synergies between civil and defense technologies, and applications of space-based assets and services and the potential applications resulting from the EU Space Programme components. EU GOVSATCOM, a new activity proposed by the European Commission, is one of the four Programme Components foreseen within the EU Space Programme. The European Commission intends to establish a Pooling and Sharing solution based on already available and foreseen commercial and governmental SATCOM resources.

European Space Agency (ESA)

The European Space Agency (ESA) has recognized that "the demand for secure satellite communications is increasing worldwide. Satellite communications have become part of the Digital Economy and are increasingly integrated into terrestrial solutions. In order to remain a reliable element in times of increasing cyber-threats, the satellite component needs to evolve and provide the resilience and cybersecurity expected in a commercial market. Furthermore, reliable and secured communications are more and more required in an institutional governmental setting. They support increasing societal needs such as for crisis management, maritime safety, and border control, which are also reflected in Europe's proposal for EU GOVSATCOM."

Therefore, and to underline ESA's contribution to the European goal to "ensure European autonomy in accessing and using space in a safe and secure environment," ESA is investing huge efforts in a variety of actions on Secure SATCOM for Safety & Security (4S), with the ESA GOVSATCOM Precursor activities as specific element responding to the growing need in Europe for secure communications for applications such as crisis management and maritime safety.

Conclusion, Outlook and Perspectives

SATCOM for defense and security will remain to represent of crucial support for a plethora of military and security applications.

The current activities, in particular at EU level, to achieve a GOVSATCOM capability for military and civilian security users describe already how the future of SATCOM for defense and security will look like.

Those users will more and more rely on solutions that are provided from both governmental and commercial resources. However, the commercial SATCOM service providers will have to proof that they will be able to meet the security requirements defined by the defense and security users and that they are also able to provide a resilient service whenever a need arises. Commercial SATCOM service providers have realized the challenge to improve the security within their satellite systems and activities are ongoing in this direction.

Satellite capacity should not be a scarce resource with the upcoming new commercial activities on (V)HTS. New applications like near-real-time video-on-demand and high-speed internet can create new use cases. In addition, the use of those commercial resources should be still affordable for defense and security users.

From a technical point of view, defense and security users will also benefit from ongoing innovations and implementations of advanced solutions for SATCOM systems.

GEO satellites will remain the main source of all SATCOM provisions. However, with the upcoming LEO and MEO constellations, new technical solutions will be offered and new use cases, which so far could not be supported by SATCOM due to specific latency requirements, are thinkable. The constellation approach is also promising in respect of cost reductions as standardization will lead to lower capital expenditures.

Further innovations that will contribute to military capabilities directly or indirectly are the use of electric propulsion that will allow to significantly reduce the satellite mass at launch allowing heavier payload and therefore more efficient satellites, having additional effects on cost and affordability and laser communications that will significantly increase security and data rates in SATCOM systems.

The commercial market and the defense and security users will therefore come even closer and classical high-security SATCOM solutions will become more and more the exemption, focused on very specific use cases and remain to be in the stock of only a few countries that again could make available those resources to partners.

References

A Global Strategy (GS) for the European Union's Foreign and Security Policy (2016) EEAS

EUROPEAN COUNCIL (2013) EUCO 217/13, EUROPEAN COUNCIL 19/20 DECEMBER 2013, CONCLUSIONS

EUROPEAN EXTERNAL ACTION SERVICE (2016) Shared Vision, Common Action: A Stronger Europe - A Global Strategy for the European Union's Foreign and Security Policy

EUROPEAN COMMISSION (2016) COM (2016) 705 final, Space Strategy for Europe

EUROPEAN DEFENCE AGENCY (2017) Governmental Satellite Communication (GOVSATCOM) - Feasibility Study, EUROCONSULT

EUROPEAN SPACE AGENCY, https://artes.esa.int/4s-govsatcom-precursor

Fritz DA et al. (2006) Military Satellite Communications: Space-Based Communications for the Global Information Grid, Johns Hopkins APL Technical Digest, 27(1)

Governmental Satellite Communication (GOVSATCOM) feasibility study (2017) EUROCONSULT

NATO (2014) Reference Catalogue of the Existing and Planned UHF, SHF and EHF Satellite Communications Equipment, NATO

Reference Catalogue of the Existing and Planned UHF, SHF and EHF Satellite Communications Equipment (2014) NATO

Space Strategy for Europe (SSE) (2016) European Commission

Position, Navigation, and Timing for Security

42

Jean-Christophe Martin

Contents

Abstract

Global Navigation Satellite Systems (GNSS) allow users to compute their position, velocity, and time anywhere in the world, anytime, and with a high accuracy. The best known and most popular GNSS is the US Global Positioning System (GPS), far in front of the Russian GLONASS system. However, due to the stnrategic importance of the GNSS, other powerful nations are developing their own global systems (GNSS): the European Union's (EU) Galileo and China's BeiDou, also known as KOMPASS.

Galileo may reach the FOC (full operational constellation) with 30 satellites (24 operational and 6 spare) in middle orbit in 2020/2021. BeiDou, the Chinese constellation, may also reach its FOC with five geosynchronous and

J.-C. Martin (✉)
CEO Marency SAS, Paris, France

CEO Marency SAS, Brussels, Belgium
e-mail: jcm@marency.eu; jcm.marencyconsulting@gmail.com

© Springer Nature Switzerland AG 2020
K.-U. Schrogl (ed.), *Handbook of Space Security*,
https://doi.org/10.1007/978-3-030-23210-8_21

797

27 satellites in middle orbit at the same moment. It means that each citizen on the Earth may be able to use the four constellations (GPS, Galileo, BeiDou, GLONASS), given that they have the proper receiver and chip. The combined capacity of the four constellations makes around 120 satellites around the Earth, and at least 15 satellites in view of each user begin in 2021. This makes a tremendous benefit and creates a multitude of opportunities for many applications and in particular for some secured applications, such as certain secured IoT (Internet of Things) or timing applications.

GNSS mainly offers two types of services: an open service, available to anyone, and an authorized service, providing better performance and available only to authorized users. The authorized services already support the defense military operations of the USA and Russia, while the open services have become instrumental for security in general and for civil security operations of any state supporting, for instance, police and civil protection. The fact that there is an opened service (OS) does not mean that secured applications cannot be developed relying on open services. The fact that a user can develop applications relying on four constellations (GPS, Galileo, BeiDou, and GLONASS) gives a huge number of applications for authentication, for example (applications on which the user is sure that he is using the right signals).

Galileo (and also the three other constellations) offers a specific service k called "Public Regulated Service" (PRS). Its "spectrum of applications" is broader than defense only but is "security" in a larger way. Each member state of the European Union, with the proper security organization, can use the PRS for its secured applications (police, special services, civil security, customs, etc.). This chapter addresses in particular these aspects, the use of the Global Navigation Satellite Systems (GNSS) for security applications.

Introduction

GNSS is the generic term for space-based systems that transmit signals that can be used to provide three services: position, navigation, and timing (PNT). The best known and most popular of the GNSS is the US Global Positioning System (GPS), although the Russian GLONASS system is regaining strength and other systems are being developed, such as EU's Galileo and China's BeiDou.

Some regional systems are also developed: IRNSS (Indian Regional Navigation Satellite System) by India and QZSS (Quasi-Zenith Satellite System) by Japan; the USA have their own SBAS system, WAAS (Wide Area Augmentation System), and Europe has developed EGNOS (European Geostationary Navigation Overlay Service). EGNOS is a geostationary system which, by improving the GPS signals over Europe, addresses very stringent users such as aeronautic or maritime users, for instance, or rail users, and surprisingly, agriculture, as they have very tough economic constraints. Some users have got precision constraints (the need to optimize the manure spreading of fertilizer in agriculture) or high constraints on integrity (rail), which is the parameter characterizing the trust the user can have in the signal (the signal reception is not always perfect, and the receiver often uses an algorithm called RAIM – receiver autonomous integrity monitoring – which assess the quality of the signal).

PNT applications are a domain in constant expansion with about 120 GNSS operational satellites (available in 2021), compared to the 30 satellites of GPS used by most applications in the year 2010. The development of IoT – Internet of Things– and connected device may rely for a part on GNSS concepts and PNT for logistics and scientific applications, for example. The consumption of the device, the receiver, is a major difficulty for the permanent use of GNSS, but IoT worldwide can use GNSS signals part time, for example, to improve the duration of the battery. As soon as you combine PNT device, a device which accesses to its PNT, and a telecommunication function, you open a huge number of applications, in particular secured ones.

The security applications for these technologies are extremely large. A wide security and civilian community can benefit from space services such as GNSS in support of strategic economic and commercial activities, strategic and critical transport, internal security (e.g., civil protection, firemen), law enforcement (police, professional mobile radio using TETRA or TETRAPOL standards), emergency services, customs, critical telecommunications, and critical energy.

Overview of Existing GNSS

Global Navigation Satellite Systems (GNSS) are experiencing a new era. The US Global Positioning System (GPS) now serves over a billion of users in a bewildering breadth of applications. The Russian system GLONASS is helpful but has never reached such figures (the order of magnitude of GLONASS users is of a few millions worldwide, mainly through chips in smartphones) but is increasing thanks to the mobile phones and their chips using all the available GNSS. China has invested and has developed its system known as BeiDou or Compass, with a constellation of 27 MEO satellites, 3 geosynchronous satellites, and 5 geostationary Earth orbit (GEO) with a global constellation which may be operational around 2021.

In addition, the European Union is developing the Galileo system, a GNSS that promises to place 24 satellites in medium Earth orbit (MEO) plus 6 spares in 2021. In 2019, 26 Galileo GEO satellites are in orbit and 24 GEO are operational (due to the fact that two satellites, which are active, were injected on a bad orbit in 2014 by a Proton/Soyuz launcher). "Initial services" became available on 15 December 2016. Then as the constellation is built-up beyond that, new services will be tested and made available, with system completion scheduled for 2020/2021.Once this is achieved, the Galileo navigation signals will provide good coverage even at latitudes up to 75 degrees north, which corresponds to Norway's North Cape – the most northerly tip of Europe – and beyond. The large number of satellites together with the carefully optimized constellation design, plus the availability of the three active spare satellites per orbital plane, will ensure that the loss of one satellite should have no discernible effect on the user.

Furthermore, in Europe, EGNOS (European Geostationary Navigation Overlay Service) is made of three geostationary satellites and a ground segment, which augments the existing GPS constellation by providing integrity and improved accuracy. EGNOS has been operational for aviation for more than 5 years. The USA has a similar operational augmentation system called WAAS (Wide Area Augmentation System). Furthermore, India and Japan are developing their own regional systems, respectively, IRNSS (Indian Regional Navigation Satellite System) and QZSS (Quasi-Zenith Satellite System), which will provide positioning and augmentation services (Fig. 1).

Description and Development of GNSS Systems

Detailed description of all GNSS has been published by the United Nations International Committee on GNSS (UN ICG). The UN ICG is an informal body of the UNOOSA (United Nations Office for Outer Space Affairs) with the purpose of promoting cooperation on civil satellite-based positioning, navigation, timing, and value-added services, as well as compatibility and interoperability among the GNSS systems, while increasing their use to support sustainable development, particularly in the developing countries. The participants are the GNSS providers (e.g., the USA for GPS and the European Union for Galileo) as well as various user communities. They have published different interesting documents to support space and GNSS (or Earth observation) in 2018, such as "European Global Navigation Satellite System and Copernicus: Supporting the Sustainable Development Goals. Building Blocks towards the 2030 Agenda" or "The Interoperable Global Navigation Satellite Systems Space Service Volume" in October 2018.

Fig. 1 Artist view of the ejection of the two first Galileo satellites with a Soyuz launcher

The US Global Positioning System

The GPS has been designed by the Department of Defense (DoD) in the 1970s to meet military requirements: at the beginning, it has been used to improve the navigation of the US ballistic missiles. The Navy, Air Force, and Army each came up with their own designs and ideas, and in 1973, a design was approved by the US government. It was also in parallel quickly adopted by the civilian world. The first satellite for the NAVSTAR GPS was launched in 1974, and from 1978 to 1985, another 11 were launched for testing purposes. The full nominal constellation of 24, which today allows navigation system to use worldwide GPS coverage, was completed in 1993. Currently, the GPS satellite constellation includes more than 30 operational satellites.

To understand the concept, the system comprises also the ground segment, with the ground control and its stations worldwide, which is key for the performance of the system, its security, and its availability (Fig. 2) (Kaplan and Hegarty 2017).

As said, initially, GPS was only intended for military use, even for very strategic applications linked to the performances of their ICBM (intercontinental ballistic missiles). **On 1 September 1983**, Korean Airlines flight KAL 007 from Anchorage to Seoul strayed off course into USSR airspace and was shot down by a soviet Su-15 fighter jet. All 269 passengers and crew were killed. Two weeks later, **US President Reagan suggested to use GPS for civilian purpose** to avoid further dangerous navigational mistakes.

The US government included a function called selective availability (SA) into NAVSTAR GPS that would degrade its accuracy (by a factor of 10) for civilian users to ensure no enemy or terrorist group could use GPS to make accurate weapons. It

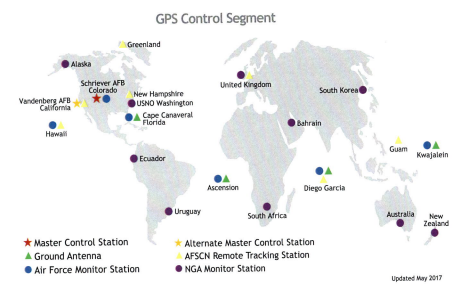

Fig. 2 Ground segment of the GPS system

has worked introducing deliberate errors into the data broadcast by each satellite. Military users could access the fully accurate system by using an encrypted signal that was broadcasted simultaneously but not available to unauthorized users. During the Gulf War, GPS became a strategic technology for the US military, which needed many more GPS receivers than it had. It is solved by using civilian GPS receivers; but to increase the accuracy of these devices, the SA function had to be temporarily disabled. Then, in 2000, US President Clinton announced that SA would be disabled completely, as US government "threat assessments" concluded that removing SA would have minimal impact on national security. He also said that the USA would still be able to "selectively deny'" GPS signals on a regional basis; it means the local jamming of a zone. This concept is known as "NAVWAR" in the military world.

In the civil world, this concept could be used – even if it has not been used yet – for security reasons, in case of a "hard riot," for example, in a town, to allow only security forces to use GNSS as authorized users. It is particularly interesting as this concept can give an operational superiority to police toward "hard rioting demonstrators" who are very mobile while using social networks.

GPS provides two types of services: a standard positioning service (SPS) and a precise positioning service (PPS). Authorized access to the PPS is restricted to the US Armed Forces (USAF), federal agencies, and selected allied armed forces and governments. The SPS is available to all users worldwide on a continuous basis and without any direct user charge. The specific capabilities provided by the GPS open service are totally opened. They are published in the GPS Standard Positioning Service Performance Standards (www.gps.gov/technical/ps/2008-SPS-performance-standard.pdf).

Likewise GPS, most space-based PNT systems will provide two types of services: an open service (OS) and an authorized service (AS). The UN ICG (see section "Description and Development of GNSS Systems") has proposed the following definitions:

- **Authorized Service**: a service which is specifically designed to meet the needs of authorized users in support of governmental functions (e.g., the PPS (P(Y) and the M-Code) for the military GPS and the PRS – Public Regulated Service – for Galileo authorized users by the governments); these services are encrypted with a high level of encryption, to be used by "authorized only" users; these types of signals are also deployed on the GLONASS and on the BeiDou system.
- **Open Service**: service (using one or more signals) provided to users free of direct user charges; the four GNSS have got such a service, available for all users; their detailed definition is free, opened, and published worldwide for the signal, in a specific document, called an ICD – (interface control document).

In addition to these two services, other types of GNSS services exist, such as the Galileo commercial service, which will provide added-value data to users who may have to pay for it. It will provide, for example, the capability of authentication of the signal which could be of interest for secured and legal applications. It means, for example, that users who need to be able to proof their position or timing on some applications, toward a tribunal, for instance, could use this encrypted commercial signal. Other applications even secured are foreseen for applications with needs which are commercially important but not as secured as for PRS applications.

GPS Modernization (Fig. 3)

The GPS modernization program upgrades the GPS space and control segments with new features to improve GPS performance, which include new civilian and military signals. GPS modernization is in particular introducing modern technologies throughout the space and control segments that enhance overall performance. The military are using the PPS (Precise Positioning Service), which is an encrypted signal that can only be used by military or users authorized by military such as coast guards. The USA has developed a new PPS signal, named M Code, which is also encrypted and dedicated to military users. The constellation (30 satellites) and the receivers are changed. But the user are "military only" (even if there are some exceptions such as the coast guards in the USA). Whereas for Galileo, the use of the PRS, which is foreseen and under the total control of the member states (who decide

Fig. 3 A GPS III-A, the last generation of GPS satellites

who their users are), is totally "security oriented" but broader than the PPS in terms of applications and users.

It should be pointed out that technical and political exchanges have permitted to define jointly the next generation of open signals, the GPS III SPS and the Galileo OS (Hein et al. 2001). Thanks to "the Signal Task Force," created in 2000 by the European Commission and the member states, ESA, and the Commission, following 4 years of technical exchanges with the USA, that have defined Galileo's signals (Status of Galileo Frequency and Signal Design 2002). GPS III's SPS signal and Galileo OS are therefore fully compatible, thanks to this excellent cooperation between the USA and the European teams.

The USA also developed a so-called satellite-based augmentation system (SBAS) named WAAS (Wide Area Augmentation System) over the US territory, relying on geostationary satellites, the system that improves the performance of the GPS constellation. The geo-satellites provide corrections to the user and to the signal propagation errors in the atmosphere and integrity monitoring. It is extremely useful for civil aviation over the US territory, principally for precision approach procedures down to 200 feet.

The Russian GLONASS System

Flight tests of high-altitude satellite navigation system, called GLONASS, started in October 1982 with the launch of the Kosmos-1413. The GLONASS system was brought into operational testing in 1993, and the whole orbit group of 24 satellites was formed in 1995. However, decrease in funding for space industry in 1990 led to the degradation of the GLONASS constellation. Russia has increased its contribution to the GLONASS constellation and could have spent around 10Md€ on its satellite navigation system on the period 2012–2020.

For the last 20 years, one of the major differences between GLONASS and both GPS and Galileo has been the choice of the management of the frequency bands and signals. The concept of sharing out frequencies is different for GLONASS than for GPS or Galileo. GLONASS uses one frequency for each pair of satellites (around the Earth at opposite sides). This leads to the use of 15 different frequencies for the constellation, the 30 satellites (so-called FDMA (Frequency division multiple access) techniques). There has been discussion between the European Commission and Russia in the year 2000: Russia wanted Galileo to use the FDMA concept. However, Galileo chose the CDMA (Code division multiple access: a multiplexing scheme is used to allow all the satellites of the constellation to use the same frequency. In fact, it is a bit more complicated as GPS and Galileo are using three different ranges of signals for different types of applications (see References 1 and 2 on E1/L1, E5, and E6 bands).) techniques, like GPS. Therefore, for the open service, GPS and Galileo are using one frequency for all their 60 satellites (Galileo uses three different frequencies and bands, like GPS, but two only are the same). This difference between the two modulations (CDMA versus FDMA) implies an interoperability complexity; combined GPS/GLONASS receivers were until recently more expensive and more complex. Thanks to improvements in semiconductor technologies over the past years, receiver manufacturers have managed to produce chips and receivers able to receive the four constellations – GPS, Galileo, GLONASS, and BeiDou – at a very reasonable cost and complexity level. It is in particular the case for smartphones such as iPhone or Samsung. Furthermore, GLONASS has conducted a modernization process, which has improved its interoperability with the other GNSS, through the choice of CDMA management of signals and frequencies. The interface control documents for GLONASS CDMA signals have been published in August 2016.

The first generation of GLONASS satellite was designed with a 3 years' lifetime, while the real operational life was of 4.5 years. In total, 81 satellites were launched. The second generation, known as GLONASS-M, was designed for 7 years' lifetime. The last recent generations of satellites, GLONASS-K1 and GLONASS-K2, are designed for a 10 years' lifetime.

Enhanced GLONASS-K1 and GLONASS-K2 satellites, to be launched in 2019 and after, may feature a full suite of modernized CDMA signals in the existing L1 and L2 bands, which include L1SC, L1OC, L2SC, and L2OC, as well as the L3OC signal. GLONASS-K series should gradually replace existing satellites. GLONASS-KM satellites may be launched by 2025. Additional open signals are being studied for these satellites, based on frequencies and formats used by existing GPS, Galileo, and BeiDou/Compass signals:

- Open signal L1OCM centered at 1575.42 MHz, close to GPS L1C and Galileo signal E1.
- Open signal L5OCM centered at 1176.45 MHz, close to the GPS L5 and Galileo signal E5A.
- Open signal L3OCM centered at 1207.14 MHz, close to Galileo signal E5B.

Users can access the GLONASS open signals. For classical applications, such as PDA devices, the use of the four constellations can be foreseen. It means that for some secured applications for worldwide users, the use of the four constellations (GPS, Galileo, BeiDou, and GLONASS) can be expected. Concerning trusted secured applications, the use of two constellations, GPS and Galileo, is easier to manage, in four bands of frequencies (E1/L1, E5/L5, E6, L2) which are complementary and have different constraints (e.g., L2 and E6 offer different solutions for compatibility with radar bands) (Fig. 4).

The Chinese BeiDou (Compass) System

In the year 2000, China has expressed its will to become a full partner of the Galileo program. They have offered 1 Md€ to become full member and in particular to access the Galileo PRS. The member states were not in favor of this option, leading to a strong willingness of China to develop its own program. The first studies have been launched around 2003. They have put a lot of pressure on their industry and have demonstrated their capability to be able to deploy a FOC (full operational capability) – around 2020/2021. On 14 April 2007, the first MEO satellite, named Compass-M1, was launched. On 15 April 2009, the first geostationary satellite, named Compass-G2, was launched. The Compass/BeiDou Navigation Satellite System was, as a first step, covering China and the nearby area in 2015. As of

Fig. 4 Artist view: last GLONASS-M satellite

December 2018, 15 BeiDou-3 satellites have been launched. This comprises five GEO satellites, 27 MEOs, and 3 in inclined geostationary orbit (IGSO). The launch occurs from spatial center XSLC of Xichang with a launcher CZ-3B, which puts in orbit two BeiDou-3 satellites each time for a total mass of around 2 tons. Designed by CALT (China Academy of Launch Vehicle Technology), a subsidiary of Aerospace Chinese group CASC, the CZ-3B belongs to the CZ-3A family launchers, which are usually used to put in orbit geostationary satellites. The current plan is to achieve a fully operational constellation (FOC) of 35 satellites by 2020/2021.

The Compass/BeiDou Navigation Satellite System in terms of functionality is very similar to the GPS or Galileo. BeiDou is able to provide two types of service worldwide: an open service and an authorized service. The open service provides, like GPS or Galileo, free positioning, velocity, and timing service. The authorized service should provide encrypted positioning, velocity, and timing services for authorized users (military users) (Fig. 5).

The European GNSS Galileo and EGNOS Systems

The Galileo program is Europe's initiative for a state-of-the-art global satellite navigation system, providing a highly accurate, guaranteed global positioning service worldwide. The program started in the European Commission, in 1998, with the so-called GNSS2 initiative. The system consists of 30 satellites, 24 operational and 6 spare, and the associated ground infrastructure. Galileo is interoperable with the GPS system (the signals have been commonly defined). Unlike GPS, Galileo is not a military system. It means that the system is not designed as a military system, with some nuclear resilience, for example. A Galileo satellite weight is about 600 kg versus 2 tons for a GPS satellite. However, Galileo can be used for security and

Fig. 5 Launch of BeiDou satellites

defense applications because each member state of the European Union is sovereign in choosing its users. When Galileo was designed, two main technical risks were identified: the "spatialization" of the atomic clocks and the management of the constellation (30 satellites). Even if there has been some technical difficulties on Galileo satellites, with the first generation of onboard European atomic clocks, this was probably the "price to pay" to develop an autonomous capability on such a critical technology in Europe. These technical risks have now been well identified and managed.

Galileo provides the European Union with an autonomous access to satellite navigation, a technology used in sectors that have become very important for its economy (about 7% of the EU GDD) and the well-being of its citizens. The market for satellite navigation services has been growing steadily and is expected to be worth €250 billion per year by 2022. Whereas EGNOS is operational since 2011, Galileo will become fully operational worldwide as GNSS in 2021 (FOC – full operational capability) (Fig. 6).

Independent studies show that Galileo may deliver around 90 billion euros to the EU economy over the first 20 years of operations, in the form of direct revenues for the space, receivers, and application industries and in the form of indirect revenues for society (more effective transport systems, more effective rescue operations, etc.).

A huge range of innovative applications is foreseen, such as improving air traffic management and securing applications for police, special forces, and government-controlled applications through the combined use of GPS and Galileo. As recalled in this paper, as GPS was already the "reference" for GNSS when Galileo was

Fig. 6 Launch of four Galileo satellites with an Ariane 5 launcher (2019) (artist view)

designed, the European GNSS is foreseen to complete the GPS and to be used in complement to improve its safety, integrity, and performances in general. As discussed also previously, for users who access the different open services of the four constellations (GPS, Galileo, BeiDou, and GLONASS), even some secured applications can be foreseen for specific needs (timing for critical infrastructure or secured communications, authentication, even navigation, and positioning, etc.). For each case, a specific threat/vulnerability and risks analysis should be conducted. In particular, the infrastructure plays an important role in the efficiency of the GNSS system as it includes the management of the security incident, the calculation of the ephemerids, or the ionospheric corrections. Such elements help enhance the efficiency for the user at the receiver level.

Here is a high-level scheme of Galileo's ground segment as deployed for initial operational capability (Fig. 7):

Prior to Galileo, Europe's first venture into satellite navigation was the development of EGNOS (European Geostationary Navigation Overlay Service), the European SBAS, and the equivalent of WAAS in the USA. As WAAS does over the US territory, EGNOS improves the performance of GPS over Europe, and, since 2011, EGNOS makes GPS suitable in Europe for safety critical applications such as flying aircraft or navigating ships through narrow channels. It should be noticed that EGNOS's design allows the augmentation in the future of GPS but also other GNSS constellations such as GLONASS or Galileo.

Fig. 7 High level description of Galileo ground segment for IOC

As a satellite navigation augmentation system, EGNOS includes a network of monitor stations across Europe and beyond that constantly monitor GNSS signals. Thanks to this monitoring functionality, EGNOS is able to correct GNSS orbit and clock estimation errors as well as signal propagation delays through the ionosphere providing a user positioning accuracy within 1–2 meters most of the time, a significant improvement compared to GPS alone, which may stay in the range of about 5 or more meters (depending mostly on ionospheric conditions).

EGNOS also provides verification of the system's integrity. Integrity is a feature which meets the demands of safety critical applications in sectors such as aviation and maritime, where lives might be endangered if the location signals are incorrect. Integrity mainly relates to the provision of timely warnings when the GNSSS system or its data should not be used for navigation. It also gives a measure of trust that can be placed in the correctness of the information supplied by GNSS.

Today, many GNSS receivers available on the market are also EGNOS enabled (Fig. 8).

Concept of GNSS Interoperability

The UN ICG define GNSS interoperability as the ability of global and regional navigation satellite systems and augmentations and the services they provide to be

Fig. 8 Galileo's infrastructure: Ground station in Fucino (Italy)

used together to provide better capabilities at the user level than would be achieved by relying solely on the open signals of one system. In practical terms, it means that:

- Interoperability allows navigation with signals from different systems with minimal additional receiver cost or complexity.
- Multiple constellations broadcasting interoperable open signals will result in improved observed geometry, increasing end user accuracy everywhere and improving service availability in environments where satellite visibility is often obscured such as forests and urban canyons.
- Geodetic reference frame realization and system time steerage standards should adhere to existing international standards to the maximum extent practical.

For a GNSS user, it means that in the future, his device will be able to process four GNSS constellations, not only GPS. These "combined receivers" will allow improved performance as compared to current GPS-only receivers.

The US-EU Agreement on GPS-Galileo Cooperation signed in 2004 laid down the principles for the cooperation activities between the USA and the European Union in the field of satellite navigation (Handbook of Space Security. Policies, Applications and programs 2015). In particular, this cooperation has led to the development of an interoperable and compatible OS signal design for the GPS and Galileo systems, as discussed previously, called the MBOC (multiplexed binary offset carrier) signal. The potential use of this signal by several GNSS is discussed between system providers, bilaterally and multilaterally within the UN ICG (Fig. 9).

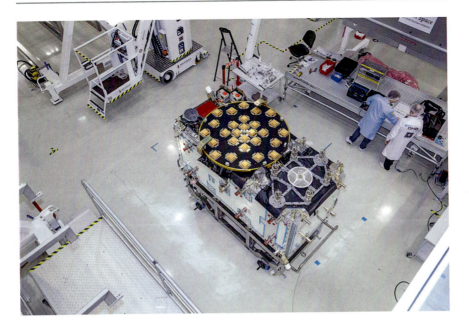

Fig. 9 Galileo's satellite assembly

A joint EU/US report, also derived from the US-EU 2004 agreement (EU-US Cooperation on Satellite Navigation 2010), has demonstrated the benefits of GNSS interoperability by showing the advantages of combining future GPS and Galileo open services. The report demonstrates and quantifies the improvements that can be expected when using GPS and Galileo open services in combination under different environmental conditions. Particularly, in partially obscured environments, where buildings, trees, or terrain block portions of the sky, the combined use of GPS and Galileo often allows a position fix that would have been impossible otherwise with only one system. The benefit of having at least 2 constellations of 30 satellites, meaning an operational use of 60 satellites, or possibly 4 constellations with a total of 120 satellites with secured access is widely recognized by the authorized users to be of a significant operational benefit.

Vulnerabilities of GNSS Services

Because of the increasing reliance of our societies to GNSS, a lot of work (In particular: Vulnerability assessment of the transportation infrastructure relying on the GPS, 29 August 2001, Volpe National Transportation Systems Center and Global Navigation Space Systems: reliance and vulnerabilities, The Royal Academy of Engineering, March 2011) has been done to assess the possible failure modes and vulnerabilities of these types of systems and the potential mitigation techniques.

The vulnerabilities of GNSS can broadly be classified into three main categories:

1. System vulnerabilities, including signals and receivers.
2. Propagation channel vulnerabilities (atmospheric and multipath) and accidental interference.
3. Deliberate interference.

1. GNSS have system-level vulnerabilities. For instance, GNSS satellites have on rare occasion broadcasted dangerously incorrect signals, or a reduced number of satellites visible because of a bad geometry could prevent the availability of a position fix, and GNSS receivers can incorrectly process valid signals to give incorrect results. In addition, GNSS signals are very weak, typically less than 50 watts transmitted from a distance of 23.000 km from the Earth. When received at the surface of the Earth, the signal strength may be as low as 10^{-16} watts, with a spectrum spread out effectively around ten times below the noise floor of the receivers.
2. Furthermore, signals are vulnerable to disruptions in the atmospheric medium they pass through, and receivers can also unintentionally lock onto reflections of the signals, known as multipath, giving unexpectedly large errors. They are also vulnerable to solar eruption and can be disrupted during this phenomenon. This causes can have quite different effects on users, such as partial or complete loss of the positioning and timing service; poorer accuracy; very large jumps in position, velocity, or time; and "hazardously misleading information" (HMI) that is to say, believable data that is dangerously wrong in safety critical applications like civil aviation. The aforementioned SBAS systems (EGNOS, WAAS) are able to detect HMIs related to the GNSS and the ionospheric propagation.
3. As mentioned supra, the GNSS signal is very weak at the surface of the Earth and can be easily jammed or disturbed internationally. This will be detailed here after.

Deliberate Threats to GNSS Services

Apart from the system-related vulnerabilities (over which the system providers have control), propagation channel errors (which are due to natural effects that can be modeled to some degree but that are by nature difficult to avoid), and accidental interference (which are difficult to anticipate), there are three distinct forms of deliberate man-made interference with GNSS signals: jamming, spoofing, and meaconing. These threats are increasingly worrying the GNSS stakeholders because of their possible safety and security consequences.

Jamming is the most likely threat and could impact the widespread use of the GPS. Jamming devices are radiofrequency transmitters that intentionally block, jam, or interfere with GNSS receivers. Criminal jamming is caused by people who are looking to defeat GNSS tracking systems. They may be car thieves, road toll evaders, and tracker evaders. It is illegal to use GNSS jammers in most countries.

However, in recent years, the number of websites offering "cell jammers" or similar devices designed to block communications and create a "quiet zone" in vehicles, schools, theatres, restaurants, and other places has increased substantially.

Meaconing (delaying and rebroadcasting of GNSS signals) and spoofing (transmission of a false GNSS signal) are more sophisticated and complex types of deliberate interference and therefore are for the time being less common. However, while a spoofer was a very bulky and expensive device some years ago, nowadays, spoofers can be portable and fit into a small box. While jamming intends only to disrupt GNSS service, meaconing and spoofing try to maintain GNSS service but with a false computed position.

The crudest form of jammer simply transmits a noise signal across one or more of the GNSS frequencies, to raise the noise level or overload the receiver circuitry and cause loss of lock. Circuits and assembly instructions for GPS jammers are widely available on the Internet, and commercial jammers can be bought for less than 20 euros. Commercial jammers are sophisticated and cheap: some are designed to fit into a pocket, some into car lighter sockets; most jammers are designed to block GPS, GLONASS, and Galileo. Powerful jammers are also commercially available, up to at least 100 W transmitted power.

There are concepts used in defense, known as NAVWAR (Navigation Warfare). The jammer or the spoofer can be used by military forces in certain areas of conflict, for example, in order to deny GNSS services to non-allies. In this case, only authorized users (allies) are allowed to have access to GNSS services, giving them operational superiority on the field.

One major threat to GNSS signals, even secured ones, is jamming. In this case, there are many ways to improve the performances of the receivers. One of them is to use specific antennas known as CRPA – controlled reception pattern antennas – which are able of digital beam forming and to create a "hole" in the pattern of the antenna in the direction of the jammer to cancel its action. It is de facto one of the best technologies to be used operationally in secured environments. To overcome the above-described potential risks of interference, there are several countermeasures at different stages of the receiver, some of which are mentioned here:

- Noise jamming can be overcome to some degree by CRPA and noise filtering in well-designed receivers.
- The use of two or more antennas can overcome the threat of spoofing by comparing the differential measurements obtained in the antennas with the real ones.
- The front end part of the receiver can incorporate between jamming and spoofing detection methods at the analog to digital conversion (ADC) of the signal, e.g., an indicator called jamming-to-noise ratio (J/N) is available in some receivers.
- Inertial measurement units (IMU) can be hybridized with GNSS receivers. Inertial sensors (i.e., accelerometers, gyroscopes, and gyrometers) are not impacted by external radiofrequency emissions and are therefore able to provide a valid position even in the presence of jamming and spoofing, at least for a

certain amount of time. This is probably the most robust countermeasure against the previous listed threats.

The proper combination of these different technologies can be, by design, the best GNSS security solution toward certain needs and in front of certain threats.

GNSS Applications Relevant for Security

Space has a strong strategic value. This allows countries to gain independence, scientific and technological prestige, and the capacity to act as a global actor. In fact, the development of space technologies has been often linked to a vision of worldwide strategic posture: that was the case, starting from the 1950s, in the USA, Russia, and France, where launchers and space assets were historically conceived as key elements of nuclear deterrence. But even if nuclear deterrence is put aside, space is synonymous of a whole chain of strategic technologies and activities: from launching to the establishment of satellite telecommunications, observations, meteorology, and navigation, space assets appear as a strategic set of infrastructures, meaning that they cannot be backed up by other types of ground networks and that their disruption would be critical to the whole society. Space assets should therefore be considered as "critical infrastructures," as their disruption would endanger both civilian and defense activities.

Space applications and technologies are best suited for dealing with an increasingly expanding concept of security. If, on the one hand, traditional customers are military users, on the other hand, a wider security and civilian community can benefit from space services which are being developed. This is particularly true for GNSS. Here are some users identified for some GNSS security applications (Fig. 10):

A critical component of any successful rescue operation is the knowledge of position and time. Knowing the precise location of landmarks, streets, buildings, emergency service resources, and disaster relief sites reduces that time and saves lives in case of natural or man-made disaster or any other type of crisis situation. This information is critical for the disaster relief teams and public safety personnel in order to protect life and reduce property loss. GNSS data (position and time) can contribute in every phase of the disaster management cycle (see the picture below) which is typically composed of three phases: (1) preparedness/prevention, (2) emergency response, and (3) recovery.

PMR: A Combined Use of GNSS and TETRA/TETRAPOL (Fig. 11)

In Europe, just a few companies can offer a European solution for the PMR device of which Airbus and Finmeccanica. In the USA, you can find other devices, such as Motorola. Airbus group has tried to sell its PMR activity in 2017/2018 but has finally kept this secured solution in Europe. The infrastructure of PMR is developed and maintained in Europe by many industrials like Airbus or Nokia, for instance. There

Fig. 10 PRS potential user communities and applications

are a lot of pressures for the development of the future of these technologies, with the arrival of 5G technology in particular. Huawei has probably invested more than 10 Md€ to develop it, whereas Nokia or Sony Ericsson may have invested around half of that amount. So, the future of the PRS for secured applications (no defense) is closely linked to the future of PMR. We could imagine that the marginal cost of the PRS, when it will be included in the PMR, could be in the order of magnitude of a few euros if included in hundreds of thousands of receivers, for police forces or customs, for example.

For the development of this market, the Commission has the full support of the GSA, the GNSS Security Agency, based in Prague (Czech Republic). This agency has an important role on the market analysis on GNSS and the support of the development of the user segment of GNSS and Galileo and its secured applications in particular. They are clearly key players for the future of secured GNSS and the PRS in particular.

Some specifications for applications require a high level of continuity and availability, in more of the security for the users. For example, rail users are the most stringent community for integrity, even more than aviation; so, some users belonging first to the European Union, may need to use the PRS, as long as their government and organizations have settled the proper organization. For example, the member states who want to use the PRS have to settle a CPA – competent PRS authority. This key authority, usually at a very high level in the states (Presidency in Italy or Prime Minister Services in France – the SGDSN), is responsible of many secured missions and tasks to secure the PRS (Decision 1104/2011 of the European

Fig. 11 Airbus TH9 Tetra
PMR (professional mobile
radio)

Parliament and of the Council of 23 October 2011 on the rules for access to the Public
regulated Service provided by the GNSS established under the Galileo program.).

Furthermore, the GSA have then conducted an important number of survey and
market analysis, published at end of 2018 (GSA 2018a, b, c), of whom I would like
to mention:

a) Report on maritime and inland waterways user needs and requirements (Reference: GSA-MKD-MAR-UREQ-229399, Issue/Revision: 1.0 Date: 18/10/2018)
   (GSA 2018).
b) Rail report on user needs and requirements (Reference: GSA-MKD-RL-UREQ-
   229496, Issue/Revision 1.0, Date: 18/10/2018) (GSA 2018).
c) Report on time and synchronization user needs and requirements (Reference:
   GSA-MKD-TS-UREQ-233690, Issue 1.0, Date: 18/10/2018) (for the next paragraph) (GSA 2018).

GNSS, Galileo, and Timing for Secured Applications

In Europe, some member states have conducted studies on their critical infrastructure, to analyze, in particular, the dependency of their infrastructure on GNSS. The

third report of the previous paragraph (report on time and synchronization user needs and requirements), from the GSA, stresses the fact that GNSS is not only strategic for Europe for navigation and positioning but that timing is probably as important as the others! It is key for the member states and for the banking sector to have a proper synchronization of secured telecommunications. For instance, many systems are using encryption and need a perfect synchronization mechanism to encrypt and decrypt signals.

The first discussions and negotiations between the European Commission and the US Department of States (DoS) in the year 2000 were focused on timing. The USA offered to deliver free GPS timing to Galileo. GPS uses a tremendous infrastructure in Washington, with more than 200 atomic clocks (a hundred in the year 2000), connected with Kalman filters and others, to deliver the GPS time. After discussions between the USA and Europe, the technical solution for the benefit of both parties has been that each party gives the other its timing reference. An exchange of the two times references (Galileo and GPS) through a facility called the GGTO – GPS Galileo Time Offset – provides a common reference of timing and significantly improves the resilience and also the performance of the systems. As an improvement, the critical infrastructure such as the PMR or banks, which rely for their timing in many cases on GPS, can easily rely also on Galileo timing and improve their reliability and resilience.

To conclude, the use of GPS and Galileo timing for critical infrastructures or sensitive and secured applications, such as telecommunications, banking, or commercial applications using encryption, is interesting to improve availability and resilience toward some incidents like signal disruptions. As timing is key for these secured applications, the GGTO and the use of at least these two constellations in the receivers improve the resilience of the systems using not only GPS but also GNSS. From an economic point of view, following threat and vulnerabilities analysis, when the use of GNSS is possible, the system designers could avoid using atomic clocks in many cases. Furthermore, there are huge opportunities for developing innovative applications for new developments like Internet of Things (IoT), as soon as the user needs a certain level of security for its assets.

The PRS and the Brexit

It is nearly impossible not to speak about the Brexit as the UK has been an essential contributor to the PRS on the Galileo program in the years 2010–2018. There are at least three major consequences in Europe, on Galileo and its member states, and in particular on the PRS and its secured applications:

- The security of the Galileo constellation is managed by a specific infrastructure called the GSMC – Galileo Security Monitoring Center. The main center is based in France, in Saint-Germain-en-Laye, whereas the backup center was based in NATS infrastructure in South UK, in Swanwick, between Portsmouth and Plymouth. Indeed, the first consequence of the Brexit is that the program has taken the decision to settle the new backup in Spain and to leave Swanwick.

- The second main consequence is that the UK, after the Brexit, would become a third state toward the European Union, as it is for the USA, Norway, or Switzerland, for example. They will have to sign a specific agreement with the European Union to access the PRS. And under the Decision N° 1104/2011, they will not have the same rights as the member states of the European Union.
- Furthermore, inside a PRS receiver, there is a specific module, called a security module (SM) (On the two last generations of military GPS, the PPS – Precise Positioning Service – and the M-Code, there is a similar SM which is called the SAASM – Selective Availability Anti-Spoofing Module. It has been designed in 1998 for some applications requiring protection from jamming or spoofing. SAASM allows satellite authentication, over-the-air rekeying, and contingency recovery.) in which the designer of the receiver hides secured information such as the cryptographic keys to access the PRS and decrypt its navigation messages. Should Brexit happen, the UK industry would not be allowed to build security modules of the PRS receivers (Legal consequence of the Decision N°1104/2011/ EU of the European Parliament and of the Council of 15 October 2011 on the rules of access to the public regulated service provided by the global navigation satellite system established under the Galileo program (Article 3.5 (b)).).

Conclusions

The revolution is there: about 120 GNSS satellites are in orbit around the Earth and nearly always 15 in view for any user worldwide. From civil to security applications, they give the users a wide array of potential new applications including timing, positioning, and navigation. Careful assessment of the threats and vulnerabilities and the needs of the different security users (police, customs, special forces, etc.) is essential to respond appropriately. For some applications, the users can develop applications relying on the commercial service of Galileo or on the open services of the four constellations, even hybridized with inertial components. In the field of security, the operational advantage of using encrypted GNSS for authorized users, in particular in times of crisis, when open signals can be jammed or spoofed, is huge. For the high level of security required by security users in Europe, the use of the Galileo PRS worldwide seems to be the best option.

Acknowledgments Frederic Bastide from the European Commission has contributed to the previous version of this article.

References

European GNSS Agency (GSA) (2018a) Report on location-based services user needs and requirements outcome of the European GNSS'. Gsa-Mkd-Lbs-Ureq-233604, (1.0)
European GNSS Agency (GSA) (2018b) Report on maritime and inland waterways user needs and requirements, 1–164. Retrieved from https://www.gsc-europa.eu/system/files/galileo_documents/ Maritime-Report-on-User-Needs-and-Requirements-v1.0.pdf

European GNSS Agency (GSA) (2018c) Report on rail user needs and requirements outcome of the European GNSS' User Consultation Platform, 80. Retrieved from https://www.gsc-europa.eu/system/files/galileo_documents/Rail-Report-on-User-Needs-and-Requirements-v1.0.pdf

EU-US Cooperation on Satellite Navigation, Working Group C, COMBINED PERFORMANCES FOR OPEN GPS/Galileo RECEIVERS, July 19, 2010. www.gps.gov/policy/cooperation/europe/2010/working-group-c/combined-open-GPS-Galileo.pdf

Handbook of Space Security. Policies, Applications and programs (2015) Springer. Kai-Uwe Schrogi and others. Volume 2. Part III Space applications and supporting services for security and defense, pp 609–630. Positioning, navigation and timing for security and defense. Jean-Christophe Martin and Frederic Bastide

Hein GW (Germany), Jérémie Godet (GISS), Jean-Luc Issler (France), Jean-Christophe Martin (European Commission), Rafael Lucas-Rodriguez (European Space Agency) and Tony Pratt (United Kingdom) (2001) The Galileo frequency structure and signal design. Proceeding of ION GPS 2001, Salt Lake City, Sept 2001, pp 1273–1282

Kaplan ED, Hegarty CJ (2017) Understanding GPS/GNSS: principles and applications, 3rd ed., Artech House Publishers, London, 31 May 2017, p 1064

Status of Galileo Frequency and Signal Design (2002) Günter W. Hein (Germany), Jérémie Godet (GISS), Jean-Luc Issler (France), Jean-Christophe Martin (European Commission, chairman), Philippe Erhard (European Space Agency), Rafael Lucas-Rodriguez (European Space Agency) and Tony Pratt (United Kingdom) Members of the Galileo Signal Task Force of the European Commission, Brussels

United States Government Accountability Office, Report to the Subcommittee on National Security and Foreign Affairs, Committee on Oversight and Government Reform, House of Representatives, GLOBAL POSITIONING SYSTEM, Challenges in Sustaining and Upgrading Capabilities Persist, Sept 2010

Internet Links

http://cospas-sarsat.org/
http://ec.europa.eu/growth/sectors/space/galileo_en
http://www.glonass-center.ru/en/
http://www.insidegnss.com
www.gps.gov/systems/gps
www.gps.gov/technical/ps/
www.gpsworld.com/gnss-system

PNT for Defense

43

Marco Detratti and Ferdinando Dolce

Contents

Abstract

Space-Based Position, Navigation, and Timing (PNT) services were conceived, designed, and developed in response to specific defense capability needs. PNT services are recognized as critical enablers for defense operations and as such must be available with the greatest possible robustness and dependability. Operational advantage of having access to such services is so evident that not only the military but also civilians started to exploit them up to the point that nowadays

---

The contents reported in this chapter reflect the opinions of the authors and do not necessarily reflect the opinions of the respective Agency/Institutions.

M. Detratti (✉) · F. Dolce
European Defence Agency (EDA), Brussels, Belgium
e-mail: marco.detratti@eda.europa.eu; ferdinando.dolce@eda.europa.eu

© Springer Nature Switzerland AG 2020
K.-U. Schrogl (ed.), *Handbook of Space Security*,
https://doi.org/10.1007/978-3-030-23210-8_105

space-based PNT have become a commodity significantly affecting not only defense operations but also the global economy. Threats and attacks to such services are for these reasons increasing, putting at risk the continuous availability required for military operations.

Introduction

Today, the secure service of the global positioning service (i.e., GPS) provided by the US government is an indispensable part of modern warfare for NATO allies, from strategic decision making or operational planning to the conduct of military operations. Since its conception, the main driver of space-based radio (or Satellite) navigation (nowadays addressed as a whole as Global Navigation Satellite Systems, GNSS) was to ensure that the service they provide could support the widest possible spectrum of military operations on a global scale (Fig. 1).

The general dependence of military tasks from such space services increases the number and typologies of attacks to space systems. From an operational point of view, such ubiquitous GNSS dependency, regardless the service considered (e.g., GPS, Galileo, or others), creates new weaknesses that adversaries could exploit easily and effectively. Indeed, despite their potential and strategic dimension, GNSS signals are vulnerable to several factors:

- Natural effects, such as multipath or ionospheric scintillations
- Signal deformations (GPS ringing phenomenon) and data corruptions, e.g., orbital and clock errors
- Nonintentional interference caused by radio transmitters, mobile communication networks, airborne navigation instruments, etc.
- Intentional interferences such as jamming and spoofing

Fig. 1 Prototype of 3D positioning system for soldier. (© European Defence Agency)

Within the current scenarios, intentional interference and threats (as well as technologies available to adversaries) are evolving in an extraordinary manner and EU Member States' Armed Forces need to face unprecedented challenges, whether stemming from modern high-tech warfare or more hybrid threats. As threats will continue to evolve quickly, so must PNT-dependent systems and platforms in response. Being able to rely on more secure and resilient PNT services for future operations in congested and contested scenarios is therefore a prerequisite which needs to be fully taken into account in the conception, design, development, and implementation of any PNT solutions for defense users. This has both political and operational consequences, rendering much more critical the identification and assessment of the vulnerabilities that adversaries could exploit.

As a matter of fact, space services and in particular GNSS could be considered, following Gen. von Clausewitz's theories (von Clausewitz 1832), as a center of gravity in the future fifth- or sixth-generation scenarios, where the global scale of future wars could be matched by the global coverage on Earth that only satellite-based services can provide. These considerations lead to multiple consequences and were probably also the basis of recent public political declarations from the EU Commissioner related to the need to start thinking on an EU Space Force (Teffer 2019) in parallel with the development of EU space capabilities such as Galileo, echoing the decision of the Trump Presidency to set up a plan for the creation of an independent US Space Force (Wall 2019).

Finally, it is important to underline how, after the transformation of satellite navigation in a commodity, a set of innovative technologies has emerged to improve and complement GNSS in any possible environment and overcome known weaknesses and vulnerabilities. These have to be fully considered and analyzed for the provision of highly robust and dependable PNT services for military forces. Even if GPS was in the driving position (and is going to remain there for a long time to come), PNT is today much more than GPS.

History of Space-Based PNT for Defense

Satellite navigation has its origins in the launch of the first artificial satellite by the Russians, the Sputnik, in 1957. After its launch, scientists in the US Johns Hopkins University discovered in 1958 that, due to the Doppler effect, the radio signals broadcasted by the satellite could be used to localize the satellite. This was used to reverse the problem and exploited to locate an object on the ground based on the knowledge of satellite position.

Based on this idea, in 1959 the U.S. Navy started the development of TRANSIT, the first navigation system to rely on satellites which became fully operational in 1964. Its main scope was to provide position information to the U.S. submarine ballistic missile force. It was not as accurate as today's satellite navigation systems (more than 20 m 2D accuracy, and performance greatly degrading as the speed of the platform increased), but introduced a set of innovations which are the basis of modern GNSS (Fig. 2).

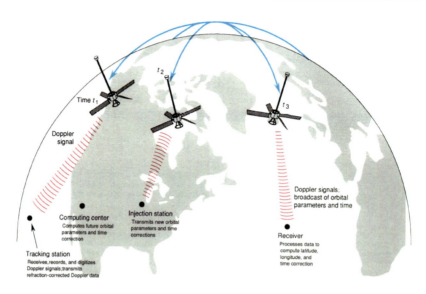

Fig. 2 The transit concept. (© The Johns Hopkins University Applied Physics Laboratory (Danchik and Lee 1990))

In parallel to TRANSIT development, a study was performed by the Aerospace Corporation for the U.S. Military to analyze tactical applications and utility of improved positioning accuracy. The study concluded with a proposal, in 1966, of satellites relying on highly stable (atomic) clocks to broadcast ranging signals continuously to receivers able to locate moving vehicles anywhere on Earth and in the air on a 24/7 basis. This was the beginning of the satellite navigation system as we know it today. After a series of other technical studies (among which some led to the identification of spread spectrum communication as the best way to transmit the ranging signals), the first four satellites constellation was developed leading to first demonstrations in 1974.

In 1978, the United States Department of Defense (DoD) started the launch of the first operational satellites (even if first Block I satellite with the first on-board atomic clock was launched in 1980) of the Navstar Global Position System, more commonly known as GPS, with the primary purpose to provide Position, Navigation, and Timing (PNT) information to defense users. DoD's primary purpose in developing GPS was to use it in precision weapon delivery, answering to the objectives of the US DoD second-offset strategy. As a second-order objective, such a space-based, all weather, and worldwide available and accurate PNT capability could address the needs of a broad spectrum of applications and would limit the proliferation of specialized PNT equipment supporting specific mission requirements reducing interoperability burdens, hence its almost immediate success.

Despite its early developments, it was only during the 1990s Gulf War (just after the first handheld GPS device for civilian applications, the Magellan NAV1000, was developed) that GPS demonstrated its full potential in operations. GPS navigation proved to be a crucial force multiplier for desert warfare. GPS satellites, even without a fully functional constellation (in 1991, there were only 19 GPS satellites in orbit, https://www.af.mil/News/Article-Display/Article/703894/evolution-of-gps-from-desert-storm-to-todays-users/), enabled forces to navigate, maneuver, and fire with unprecedented accuracy in the desert almost 24 h a day despite difficult conditions – sandstorms, no maps, no vegetative cover, few natural landmarks. GPS's fully operational capability was achieved in 1995 with the last of the first 27 operational satellites (including the spares) was launched. In the same year, the US DoD, fearing that adversaries could take advantage of the service, decided to decrease the accuracy of the openly available service through the activation of the selective availability.

The importance of this capability has been soon recognized worldwide also for civilian applications, as demonstrated by the development and deployment of other Global Navigation Satellite Systems (GNSS) especially after the end of the Selective Availability in 2000. Since then, GPS-based PNT has deeply changed the way many military operations are conducted by providing (an almost) continuous and ubiquitous precise positioning and timing for a vast variety of platforms at a reduced cost. The trend in the GPS/GNSS device market is not expected to decrease. Instead, it is expected to increase up to US$2.8 billion by 2027, at a CAGR of more than 2.7% (Market Research 2017). GNSS devices intended for munitions, soldiers, and ground platforms constitute the bulk of defense applications market. These are being acquired either as stand-alone devices or as part of soldier modernization programs. One of the most important factors driving the increase of the market is the fact that new constellations are becoming available and new technological advances are being integrated into PNT systems to increase their robustness by augmenting and complementing space-based capabilities.

GNSS PNT for Defense Users

Almost the totality of todays' PNT services rely, either directly or indirectly, on GNSS. Such services are key enabling capabilities in military operations contributing to all the military tasks, fundamental for the freedom of movement and acting as force multipliers. As such, PNT solutions must be secure and resilient. They have to be designed to withstand potential malfunctions and degradations and need to comprise adequate mitigation measures against complex attacks. This is achieved through the concept of PNT superiority against adversaries.

One of the key concepts of PNT superiority is commonly known as NAVWAR (Navigation Warfare). NAVWAR is defined as "the deliberate defensive and offensive action to assure and prevent positioning, navigation and timing information through coordinated employment of space, cyberspace, and electronic warfare. Desired effects are generated through the coordinated employment of components within information operations, space operations, and cyberspace operations, including electronic warfare, space control, space force enhancement, and computer network operations" (US FNP 2017). The underlying benefit of the NAVWAR doctrine for the military is to ensure military operations PNT superiority and advantage in the area of conflict without disrupting allied forces outside the theater of operations. This is substantially implemented by: "protecting authorized use of GPS; preventing the hostile use of GPS, its augmentations, or any other PNT service; and preserving peaceful civil GPS use outside an area of military operations."

Since its origin, GPS has become the "gold standard" by which other PNT solutions were (and still are) benchmarked. However, in order to improve the performance of navigation systems in cases of poor satellite coverage and low availability, new GNSS systems are being developed (notably EU is implementing the Galileo program, which is expected to reach full operational apability in 2020). Yet there are still a lot of concerns as space-based services can be denied or degraded in tactical environment; therefore, PNT superiority cannot be limited to NAVWAR. Given the reliance on PNT for operations, the challenge is to maintain a high degree of resilience and the highest possible confidence in any operational scenario even when no external aids to navigation and localization are available.

For the majority of defense forces' operations, the availability of a globally accurate, precise, and real-time location and timing information can provide a crucial advantage over adversaries. GPS first and other GNSS today are able to provide this capability in many operational conditions allowing, e.g., the effective engagement of opposing forces through accurate targeting, enhanced navigation, and maneuvering activities, and it helps preventing or minimizing collateral damage. This capability is used at several and different military levels such as a strategic analysis, which can take advantage of reliable and global positioning and timing information, or tactical operations, enabling the engagement of high-accuracy weapons's guidance. This needs to be clearly kept in mind, especially with regard to military rule of engagements that can effectively guide armed forces' actions and activities.

The revised EDA 2018 Capability Development Plan (CDP 2018) within the 11 identified EU capability priorities clearly reflects the indispensability of space-based communication and information services as an enabler for the defense systems, with a special emphasis on unmanned and autonomous systems (EDM 2018). Unmanned maritime high-end platforms, for instance, which have just been identified as a European priority to achieve maritime surface superiority through long endurance at sea, are only one example where support from space-based applications has become critical. If such systems do not have access to strong and resilient PNT support provided by satellites, they cannot be considered fully operational.

GNSS devices for defense applications are widespread across all operational domains and platforms: soldiers, vehicles, aircrafts, vessels, communications systems, and munitions are routinely equipped with GNSS systems to provide any combination on PNT information. Secure/encrypted GPS receivers for navigation and guidance solutions are available from a handful of manufacturers. Receivers are available either as stand-alone navigation devices or as embedded devices to be operated within a larger mechanical or electronic system.

Space-based PNT services used by European Union Member States and NATO Allies' armed forces are the US encrypted Precise Positioning System (P(Y) code). Today the de facto user equipment standard is the Defense Advanced GPS Receiver

Fig. 3 Left: Hand-held micro Defense Advanced GPS Receiver (DAGR). (© Collins Aerospace). Right: TOPSTAR M for avionics platforms. (Photo Thales © E. Raz)

(DAGR) used to provide precision guidance capabilities for vehicular, hand-held, sensor, and gun-laying applications. The latest generation being termed SAASM (Selective Availability Anti-spoof Module) with a quite small and low-weight hand-held form factor for war fighters, with an easy-to-use interface (Graphical User Interface and moving maps). Today there are models resembling conventional smartphone functionalities (including MP3 and camera), but there are several other form factor receivers matching different platform requirements (Fig. 3).

The availability of new GNSS could increase the robustness of the PNT services available to the defense users for the implementation of national or multinational operations. The defense sector should seek to maintain the right level of PNT capability in light of programmatic and technological opportunities and increasing threats to PNT information assurance. Key principles for the usage of space-based PNT solutions (mainly GNSS) as defense-enabling capability can be summarized as follows (ERNP 2018):

- The performance of PNT services in terms of accuracy, continuity, and integrity shall be commensurate to operational needs as defined by Member States in the Capability Development Plan.
- The delivery of PNT services must be subject to agreed governance arrangements and must be under full European control or dependably provided by an allied defense partner.
- Due to the worldwide extent of the EU's area of strategic interests, PNT services must ensure global coverage. Therefore, EU Member States shall have the right to unlimited and uninterrupted access to secured space-based PNT services worldwide.
- The PNT services for military use must ensure a high degree of resilience against all threats and risks; this should explicitly include all aspects of cyber warfare.
- The highest levels of PNT services availability should be sought. Space-based PNT should thus be highly resistant to disruption, denial, deception, and degradation.
- The use of PNT services must be accessible in contested and congested environments.
- It must be possible to deny the exploitation of the secure PNT services by adversary forces.

- Protection and robustness of the equipment must be adapted to the operational environments (e.g., physical security measures, antitampering).
- The interoperability of the PNT services must be ensured in areas of common interest negotiated with GNSS providers.
- The PNT services must be available with a high level of reliability in all operational environments (in particular urban).
- Augmentation systems, regional or local, may be considered in order to enhance available GNSS services.
- PNT services shall be workable from strategic to tactical level, and from the most complex weapon system down to, e.g., dismounted soldiers' equipment.

Space-Based PNT Systems Landscape for Defense Users

Until now, we have been focusing on GNSS as the source of PNT information for military forces. However, as a consequence of the increasing dependency and operational advantages associated to the mastering and control of GNSS systems, several techniques and devices have been developed that can severely degrade the performance of GNSS services. Such degradation, nevertheless, not always causes a complete denial of the service, but most often it causes misleading PNT information (spoofing). Such threats to GNSS might adversely affect various military tasks. In this context, PNT sources and systems can be identified to allow implementing diverse PNT architectures delivering different levels of performance according to the operational scenarios, threats, and missions.

GNSS Services and Systems

Broadly speaking, two main groups of satellite navigation services can be identified:

- GNSS open services: Provide positioning, velocity, and timing information that can be accessed free of direct user charge. The civilian services are accessible to any user equipped with a receiver, with no authorization required.
- GNSS-regulated and access-controlled services: These are robust services for the provision of PVT (positioning, velocity, and timing) information to authorized users. The regulated signals are typically designed with a focus on the robustness of its signal, which protects it against spoofing and makes it more resistant to jamming.

The only satellite navigation system currently used by military users in EU and allied forces is the US GPS system. However, in addition to GPS, there are three additional satellite navigation constellations in Medium Earth Orbit (MEO): one fully operational (the Russian Glonass), and the other two (the European Galileo and the Chinese BeiDou (BDS) navigation systems) under deployment. The four constellations transmit navigation signals in five different frequency bands featuring both open and regulated services according to the grouping below (Fig. 4).

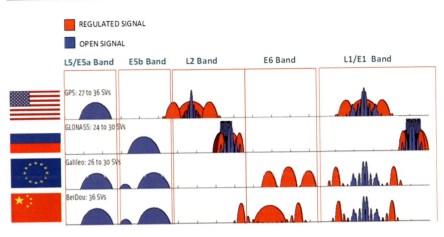

Fig. 4 MEO satellite navigation signals (red: regulated signals/services; blue: open signals/services). (Inside GNSS 2013)

In all the aforementioned satellite navigation systems with the exception of the European Galileo, access to regulated (i.e., encrypted) signals is reserved to users authorized by the Ministry of Defense, and therefore mainly targeting military uses and applications. However, albeit being considered and developed as a civil system under civilian control, Galileo will deliver the Public Regulated Service (PRS) which will offer strong and encrypted navigation signals. This secure service is restricted to governmental authorized users and is therefore suitable for services where robustness and complete reliability must be ensured, such as, but not limited to, military operations.

The access to PRS is regulated by Decision No 1104/2011/EU of 25 October 2011. According to it, the PRS is a service which is restricted exclusively to Member States, the Council, the Commission, and the European External Action Service ("EEAS") and, where appropriate, duly authorized European Union Agencies. It should also be possible for certain third countries and international organizations to become PRS participants through separate agreements. Decision No 1104/2011/EU mentions also that the PRS provide unlimited and uninterrupted service worldwide to PRS participants.

As soon as the Galileo PRS service will become operational, and after the availability of the associated defense user equipment, it could be expected that PRS will become an additional primary source of PNT for EU and allied forces. Indeed, EGNSS could increase resilience, availability, integrity of PNT information, and services for EU CSDP and MS operations.

Alternative Space-Based PNT

The military use of other satellite navigation services, like space-based augmentation (e.g., WAAS and EGNOS) and their benefit in terms of increased robustness in a military context is nowadays considered limited. The same applies to civil Differential Global Positioning Systems, RTK, and Post-processing Positioning Services (PPS), especially with respect to the robustness of a military system when relying on such civil augmentation systems. Currently, the main benefit of augmentation in a military context is for aviation applications, with dual-use of the civil system components, like SBAS receivers, and in parallel with military GNSS receivers.

GNSS (GPS, Galileo, Glonass, and Beidou) employ satellites in medium Earth orbit (MEO) broadcasting signals transmitted at an altitude of >20,000 km. This implies about 30 dB more path loss than signals transmitted from LEO satellites (at an altitude of less than 1000 km, like Iridium (https://en.wikipedia.org/wiki/Iridium_satellite_constellation)). Consequently, with similar transmit power available at the satellite, LEO signals are received at Earth's surface with a significantly higher power level than GNSS signals. For GPS and Galileo, the observed carrier to noise density ratio (C/N0) in good reception condition is around 50 dBHz, while, e.g., for Iridium, C/N0 values up to 80 dBHz are observed. The higher signal power levels lead to a much stronger jamming resistance of a LEO-based GNSS. Additionally, this allows the LEO signals to penetrate into difficult attenuation environments like deep indoors, where the reception of GPS or Galileo signals is not possible.

Another consequence of the lower altitude of the LEO satellites is a smaller antenna footprint on the Earth's surface, as illustrated in Fig. 5. In order to have one satellite in view at all times, only around ten MEO satellites are required, while at LEO rather one hundred satellites would be needed. On the other hand, while significantly more LEO satellites are required to provide the same coverage as MEO satellites, launching an LEO satellite is less expensive than launching an MEO satellite.

The lower altitude also leads to a shorter orbital period: for GPS and Galileo, the orbital periods are 12 and 14 h, respectively, while in LEO, the orbital period is around 100 min. Therefore, the LEO satellite geometry, as observed by a user on the Earth, changes much faster than for an MEO GNSS. This has several advantages: the user is observing larger Doppler shifts, which is beneficial for Doppler positioning

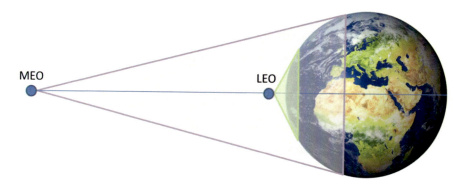

Fig. 5 LEO and MEO antenna footprints on Earth's surface. (© Reid et al. 2016)

(the same principle used in TRANSIT, see section "History of Space Based PNT for Defense"). Another positive impact of the fast changing geometry is that it whitens the multipath, which then averages out faster.

Currently, the only LEO constellation with global coverage that offers navigation capabilities is Iridium. Iridium/IridiumNEXT consists of 66 satellites at an altitude of 780 km and is used mainly for communication purposes and is available on a commercial basis. Since May 2016, Iridium offers a Satellite Time and Location (STL) service. Iridium uses overlapping spot beams and randomized broadcasts, which provides a mechanism for location-based authentication that is extremely difficult to spoof (Lawrence et al. 2017). Together with the high signal power levels providing jamming robustness and coverage in deep indoor environments, this antispoofing capability makes Iridium a very robust PNT source. However, Iridium does not yet achieve the accuracy offered by GPS or Galileo. After a convergence time of 10 min, the following performance has been observed in field tests: positioning accuracies between 20 and 35 m (1σ) and timing accuracy of 0.5 ms.

Even if the performance of Iridium is not comparable with those of the current GNSS, LEO-based PNT is a very dynamic environment, which has the potential for disruptive innovations in the near future that can complement classical GNSS considerably. In the near future, several other LEO-based constellations with possible navigation capabilities can be expected: OneWeb consists of a constellation of 648 satellites for broadband internet provision; the initial operational capability is planned for 2019 (Reid et al. 2016). SpaceX recently got approval from FCC to deploy its satellite internet constellation called Starlink, which adds up to more than 4000 spacecrafts (and up to 12,000) to be operational by the mid-2020s (Ralph 2019). Samsung proposed a LEO constellation of 4600 satellites, and Boeing published plans for a LEO constellation of 3000 satellites (Reid et al. 2018).

Additionally, the U.S. Defense Advanced Research Projects Agency (DARPA) already started to analyze whether military constellations (payloads: global surveillance, tactical communications, and PNT) in low Earth orbit are cheaper and nimbler alternatives to traditional military satellites. In 2017, DARPA launched the project known as Blackjack with the final goal to develop a low Earth orbit constellation to provide global persistent coverage for military operations (Erwin 2018). The project will aim to demonstrate an architecture showing the utility of a global LEO constellation for a wide variety of military payloads and missions.

In a further effort to enhance the PNT capabilities provided by MEO constellations (specifically GPS), the Air Force Research Laboratory (AFRL) in the US will launch in 2022 an experimental satellite in Geostationary orbit (AFRL 2017) called NTS-3 (Navigation Technology Satellite-3) as it follows NTS-1 and NTS-2 launched in 1974 and 1977, respectively, to test initial GPS functionalities). The experimental PNT satellite is intended to test new technologies and hardware to improve robustness and resilience of MEO-based constellations through a supporting layer of geosynchronous Earth orbit satellites. Technologies that will be tested include advanced antenna options, reprogrammable hardware, advanced clock technologies, and new signal structures. NTS 3 will also investigate secured-design technologies to enhance cyber-resilience, as well as modern cyber risks management approaches.

Modernization of Space-Based Secured PNT Services

In parallel to the completion of the deployment of new GNSS systems and the studies on possible alternative space-based PNT, possibly the most important aspect to be emphasized is the current modernization effort ongoing from US DoD to develop a new secure signal to improve the security features and jamming resistance properties of military navigation using GNSS, the M-code.

The motivations for such modernization can be found in the need to improve NAVWAR performance under the assumption that the threats against the military user (mainly driven by the hybridization of the warfare) are continuously evolving, resulting in an increasingly complex, more congested, and contested environment.

The M-code signal will provide better jamming resistance through much higher power transmission (up to +20 dB above current level, with the possibility through spot-beam transmission to direct the signal toward a specific area of interest) without degrading C/A-code or P(Y)-code reception and in openly available signals on L1 and L2 bands (Fig. 6).

The M-code design also features a more robust signal acquisition, more flexibility in its configuration and better security in terms of exclusivity, authentication, and confidentiality (hence better spoofing resistance), along with a simplified key

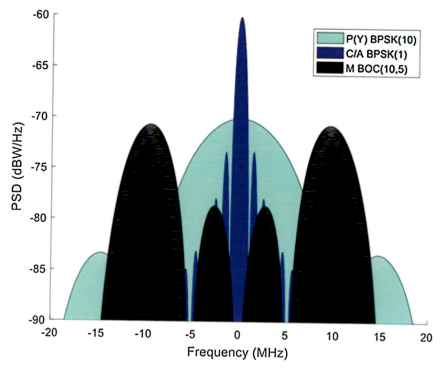

Fig. 6 M-code signal compared to legacy GPS signal around the L1 (1575.42 Mhz) frequency. (© GPSWorld)

distribution (Barker 2000). In particular, the signal is able to support NAVWAR activities: the energy signal is split into two lobes separated from the center frequency, enabling selective jamming of the open GPS (C/A code) without impacting military signal reception ("blue force jamming"). Currently, such blue force jamming is not possible with P(Y) code receivers, without also degrading the friendly force's receiver.

The M-code is designed to be autonomous, and so authorized users will be able to calculate their positions without requiring the use of other signals (e.g., the C/A-code in the case of the P(Y)), providing at least comparable performance to the P(Y)-code.

To the best of the authors' knowledge (GMV 2011), also within the Galileo program, in the frame of the design of its second generation (G2G), services are expected to deliver improved performance and features such as reliability, maintainability, availability, continuity, accuracy, and integrity. It is therefore auspicable that PRS performance will be improved in addition, likely to reflect the evolving threat landscape in which defense (a recognized user community of Galileo) will operate.

Technologies for Future Defense PNT Solutions

Multi-constellation Defense GNSS Receivers

Several techniques and technologies could be used to improve the robustness of GNSS receivers. Based on an analysis of publicly available information (F-DEPNAT), an overview of possible technologies and techniques for use in a military grade GNSS receiver is presented in Table 1.

In addition to the usage of multi-frequency (which is the de facto standard for military grade receivers), the use of multiple constellations allows for robust GNSS through improved availability of satellites with good geometry. Use of multiple constellations also mitigates against failure modes associated with a single constellation and supports diversity of signals across constellations and frequencies.

In a military context, the multi-constellation and multi-frequency receivers could make use of different secure, encrypted signals. In addition, a diversity of open signals may be used alongside advanced interference and spoofing mitigations to provide the highest levels of PNT security and robustness for users. For countries in NATO, the use of multiple secure signals will most likely mean GPS-PPS + Galileo PRS. The clear benefits of dual constellation GPS and Galileo receivers for employing encrypted signal services from different constellations are likely the key driver of the requests from United States (Gibbons et al. 2017) and Norway (De Selding 2015) governments to access the Galileo PRS.

Figure 7 illustrates a proposal for a dual mode receiver architecture employing GPS-PPS and Galileo-PRS. This architecture is based on separate security modules to process the encrypted PPS and PRS signals but maximizing the commonality of the remainder of the receiver (e.g., use of common frequency reference subsystem, radio front end). In the figure, a separate navigation processor is shown, but it is potentially feasible for this to be incorporated within one of the security modules.

Table 1 Technologies and Techniques for Military grade GNSS Receiver

Antenna & RF-Front-end technologies for military-type GNSS receiver	
Antenna type	Chip, helical, patch, choke ring, CRPA beamforming
GNSS bands signals	GPS L1, L2, P(Y), M-code (2020+), E6 PRS, E1 PRS (2020+)
Precorrelation BW	15–40 MHz
Sample quantization	2–8 bits/sample
Sample rate	Up to 200 MHz
Precorrelation RFI detection/ suppression	CW, swept CW, notch filter, dynamic chirp filter, pulse suppression
Reference oscillator	TCXO, OCXO,CSAC
Typical implementation (2018) w/o dc/dc converter	Single mini circuit board with 2–3 RF ICs and reference oscillator
Baseband processing techniques for military-type GNSS receiver	
Carrier tracking architecture	FLL-assisted PLL, PLL
Code tracking architecture	Carrier-aided DLL
Multipath mitigation technology	Narrow correlator, multicorrelator
Early-late correlator spacing	0.3–0.01
Other features	Massive parallel correlators for acquisition and tracking, multipath and RFI mitigation, sensor fusion capability
NavSEC/ComSEC	Separate security module for key generation/storage and message demodulation
Typical implementation (2018)	1–2 FPGAs and impeded or separate μ-processor, tamper protection, key loading module

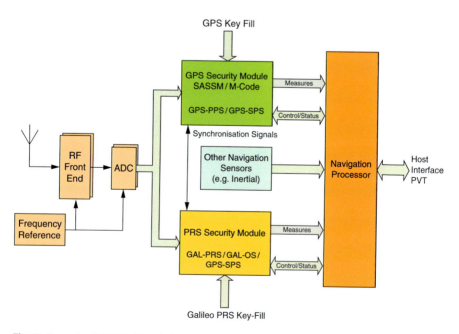

Fig. 7 Example of GPS/Galileo dual constellation receiver. (Courtesy of Nigel Davies © QinetiQ)

According to Davies et al. (2016), future challenges to be solved for dual constellation receivers include:

- Technical: Even if the development of operational PRS receivers is ongoing in Europe and subject to a number of initiatives, the development of the next generation of GPS-PPS, namely M-code, is more advanced.
- Security: There are separate US and European security rules and sensitivities. Therefore, security constraints arise for the key management and handling.
- Legal/political: The GPS-PPS security modules can be developed and approved by US contractors under the control of the US DOD and US National Security Agency (NSA). Similarly in Europe, the PRS access rules legislation requires that PRS security modules are developed and manufactured by authorized European manufacturers. Furthermore, use of GPS-PPS outside the US is limited to authorized nations for the purposes of defense under bilateral agreements with the United States.

These points, in addition to the agreement on which military uses could mostly benefit from a combined use of GPS- and Galileo-secured signals and the establishment of the relevant agreements, should be addressed adequately in order to ensure that the defense forces (in addition to the mass and automotive market which are already benefiting from open-service multi-constellation) would be able to fully exploit the benefit of multi-constellation performance.

Developments in the United States

The Military GPS User Equipment (MGUE) program started the development of M-code capable GPS receivers in 2013. According to the MGUE program schedule (Wilson 2015), GB-GRAM-M receiver kits are now available for integration in receiver housing and into military platforms, as shown in Fig. 8. In (Menschner 2018) the GPS enterprise, roadmap is shown for the full system, including also the MGUE Integration and test phases for the various user platforms which are expected to last until 2021.

Developments in Europe

The deployment of the Galileo Public Regulated Service user segment has been steered in recent years actively by the European GNSS Agency (GSA). One essential prerequisite for the future adoption of PRS by multiple user communities is the availability of receivers for different applications. Thereto, GSA has initiated the EU-funded projects P3RS-2 (awarded to an Italian/German consortium, (Leonardo 2014) and PRISMA ((GSA2015, which, based on publicly available information seems being awarded at least to French and German companies) for the development of the first generation of preoperational receivers and first prototypes. In addition to the prototype and proof of concepts developments driven by the GSA, several national development activities are running under the control of national defense agencies and competent PRS authorities. However, information on the expected availability date of the testing and integration of fieldable products are not publicly available.

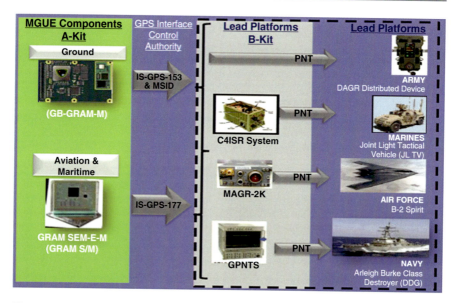

Fig. 8 MGUE Increment 1 M-code receiver development. (© GPS.GOV, Wilson 2015)

Other PNT Sources

Considering the strategic nature of space-based PNT infrastructures and in order to reap the most benefit from it for critical defense applications, its military potential needs to be recognized and fully understood. On this basis, resilience should be integrated to grant reinforced reliability and dependability to match the growing operational expectations following the evolution of modern warfare. For this reason, and despite not linked with space security, it is important to mention the fact that defense users are currently looking for alternative and autonomous (i.e., not dependent from GNSS or other PNT aids external to the concerned system) PNT sources to assure its continuity under all conditions.

Even if the widespread and worldwide adoption of GNSS in applications which were not even imaginable at its conception have implied the assumption that GNSS PNT is taken for granted, and the GNSS modernization in well underway with the progressive introduction of new constellations and signals, PNT and GNSS are not to be considered synonyms (a quite common mistake nowadays). With a host of technological evolutions on the horizon, PNT is much more than GNSS these days. Other technologies and techniques to augment and complement GNSS need to be studied to ensure PNT superiority to defense forces when GNSS is degraded (or simply not available for physical limitations). This is why several communities (laboratories, research organizations, civilian industry, and others) are being looked at and challenged by defense organizations to propose and explore novel and disruptive solutions. This is of utmost importance also considering the pace at which commercial innovation, without the constraints of traditional defense R&D, are progressing.

In particular, both hybrid and autonomous systems and components would need to be considered. As an example, it will be crucial to develop further Inertial Navigation Systems (INS), traditionally the optimal GNSS complement to guarantee and improve robustness of PNT information in GNSS denied or degraded environments. Work is ongoing to exploit advanced fiber optics (i.e., hollow core), quantum, or micro-PNT technologies (under development under multiple DARPA programs, McCaney 2015) to develop high-performing autonomous (i.e., not depending on external inputs) navigation sensors. In the area of quantum devices, Europe is also progressing with the development of the first commercially available inertial sensors (a quantum gravimeter, Fig. 9). However, current technological status is limited to quite big and static apparatuses with limited dynamic range and high sensitivity to environmental effects. Significant work is needed to develop solutions suitable for compact and dynamic environments and devices able to withstand defense operational and environmental conditions. Integration into defense platforms cannot be expected before 10–20 years.

GNSS evolutions will naturally benefit from related technological advances, some of which will make their signals less susceptible to interference and spoofing (e.g., through the usage of quantum cryptography). It is in fact important not to forget that, thanks to its worldwide availability and attainable cost/performance ratio, it is unlikely that GNSS will be replaced by another technology anytime soon.

Today's trend is in the provision of assured PNT relying on the integration of traditional PNT technology with nontraditional and emerging technology to improve the robustness and dependability of mission-critical applications in all the military

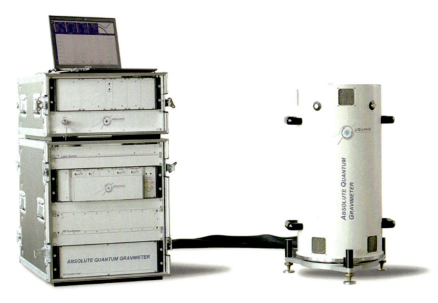

Fig. 9 First commercially available quantum inertial sensor. (Absolute Quantum Gravimeter, © Muquans, 2019)

operational domains. Fusing different sensor modalities to create a combined navigation solution is anything but a new idea. The benefits of combining GPS with an inertial sensor were recognized a long time ago, and this classic pairing continues to be the subject of research today. In particular, the deep integration produces an increasing of robustness to GPS jamming, compared with tightly coupled systems. However, technologies available to make PNT systems resilient are evolving and nowadays (and in the foreseeable future) a wide range of solutions to match different operational conditions and environments could be imagined.

By combining GNSS-based PNT equipment with detection and mitigation systems, we will continue to rely on trusted GNSS as the main source for PNT services. The combination of several PNT sources can generate different ways to deliver and to use the service. For example, based on the military mission or task, different PNT sources could be selected as main PNT provider, and standard operational procedures could be developed to support the use of different PNT source in case of NAVWAR environment. Such procedures could also be implemented in technical solutions, but in any case need to be flexible and adaptable to the operational tasks and the external threats.

PNT Superiority Impact on Military Tasks

The Generic Military Task List (GMTL) is a taxonomy agreed between EU Member States in the framework of the European Defence Agency to categorize the full spectrum of military activities divided into six domains: command, inform, deploy, engage, protect and sustain. Usually military systems are designed and developed to generate capabilities able to provide services in a subset of the above-mentioned domains.

Several systems can be operated to support one or more of these generic tasks; nevertheless, there are cross-cutting capabilities enabling a very large or, sometimes, the entire spectrum of GMTL tasks. Among those, PNT is one of the most critical and it gives a justification to why PNT superiority needs to be ensured. In Table 2, some examples of how PNT technologies can support the full spectrum of GMTL domains are reported.

Table 2 Examples of PNT support to military tasks

GMTL domain	PNT support
Command	Time and synchronization for C4I equipment, Blue Force tracking
Inform	Localization of jammers and spoofers, geographical information services
Engage	Mobility and personnel localization, weapons guidance, and precision engagement
Protect	Counter-IED support
Deploy	Platform mobility support
Sustain	Combat search and rescue

Timing sources are crucial for the synchronization and functioning of all networks and digital platforms at strategic, operational, and tactical levels. At tactical level, synchronization enables fundamental capabilities such as effective interoperability, and joint and combined operations that only a globally available service, such as GNSS, can automatically enable for an unlimited amount of users in different locations.

The dependence from PNT services is particularly critical for high precision engagement and missile defense, while a high level of dependence can be assessed for air defense and C4ISTAR services (ItAF 2014). As far as air high precision engagement capabilities, navigation, and guidance of weapons are highly impacted by a PNT-denied environments. As demonstrated by a NATO study (Schmidt 2013) through dedicated simulations, performance of precision-guided munitions can be degraded by a wideband GPS jammer close to the target up to a positioning error 10 times higher than in a noncontested environment. This can lead to catastrophic collateral effects due to the increased circular error probability (CEP), affected, at the same time, also by potential weapons' guidance systems' additional errors due to possible GNSS interferences. For these reason, in some circumstances, due to the rules of engagement, missions might be inexecutable in PNT-denied environments. In order to minimize such risks, air-launched precision-guided weapons, which rely on GPS during the navigation phase, can be complemented by laser inertial gyros to provide a precise guidance close to the target. Furthermore, once the weapon is very close to the target, an infrared receiver on the weapon compare the acquired imagery with the one memorized, providing high-level accuracy guidance in the last phase of the guidance (ItAF 2014).

There are air-launched unguided bombs that can be converted into precision-guided munition through an integrated GPS-aided INS guidance kit to improve the laser seeker and infrared technologies. The bomb could be used also with the laser seeker only; nevertheless, the accuracy degradation when GNSS is not available produces an increase of the target location error which in turn increases the probability of collateral damage (ItAF 2014). Also, as far as missile defense capabilities, there is a strong vulnerability of RADAR systems from GPS positioning information in the C2 management system (ItAF 2014) (Fig. 10).

Examples of these effects have been observed during the Iraq War in 2003, when the Iraqi Army used GNSS jammer to disrupt US GPS-guided missiles (Miles 2004). Several additional reports of GNSS jamming have been published about activities in Iraq and Afghanistan that have undermined military operations since then. Although in many cases such incidents can be quickly controlled in a military environment (they usually rely on a high power jammer which can be easily localized with modern equipment), it underlines the criticality of GNSS and the need not only to exploit space services but also to protect them.

Based on this information, it is clear that the improvement of GNSS services play a fundamental role. For example, the improvement of antijam capability of GNSS services is a clear requirement for military users. With the future generation of GPS Block III satellites, the accuracy of GPS service will increase from 3 m to 30–15 cm and in addition the new M-code in replacement of the current Y-mode will produce

Fig. 10 Pit drop trials during Joint Direct Attack Munitions (JDAM) integration on Tornado

Mon May 12 2003 11:23:49.098774 S

higher spoofing resilience. Antijamming capability will be increased with the increasing of the signal power, +20 dB above current level, with the possibility to direct the signal toward the area of higher interest.

Several techniques and devices have been developed and new ones will be certainly developed in the future that might severely degrade the performance of GNSS services, ultimately resulting in a complete denial of service, or even worse causing misleading PNT information (deception and spoofing). In particular, GNSS spoofing events have been regularly observed (C4ADS 2019) in the last few years in the Russian Federation, Crimea, and Syria, demonstrating how the military use and development of GNSS threats is growing to pursue tactical and strategic advantages both at home and in foreign territories.

All such threats to GNSS adversely impact various military tasks, such as those described above as some examples. For this reason, PNT superiority integration into military operations and systems is more and more relevant in view of ensuring force superiority in the battlefield of the future. This should be related to the understanding of the relevance of space domain and in particular the PNT sources in military operations. The threat scenario "A day without space," in fact, reveals a clear dependency (of current and future military operations) on space. At the moment, however, such a threat can be considered as an expression of the soft power (Nye 2005), even if it is clear that in a future strategic prospective the situation will evolve. As defined in "On War" (Von Clausewitz 1832) a center of gravity is "*the source of power that provides moral or physical strength, freedom of action, or will to act*," and in this sense, the space domain is already moving toward becoming a center of gravity for fifth-generation warfare.

A direct consequence of this qualitative consideration is based on the space threats proliferation that will generate the need to defend the space as a critical infrastructure, strategic nodes, or a military capability.

The real issue is that we are facing threats that were not planned to be dealt with, for technological but also mainly political reasons. Indeed, not only defense equipment is being targeted, but also the entire space segment as witnessed by several developments in this area. A recent report from the office of the Secretary of Defense (USDEF 2019) underlines how China is improving its counterspace capabilities. In addition to directed energy weapons, its antisatellite missile systems (following the tests in 2014) are being further developed. Even if the Chinese government has not acknowledged any specific antisatellite program, there are several publications from defense-funded academies that stress the necessity of "destroying, damaging, and interfering with enemy's reconnaissance…and communication satellites" suggesting that such systems (and the navigation ones) could be among the targets for attacks. The very recent Indian tests of antisatellite weapons (Foust 2019) as well as the decision to constitute a new military service dedicated to Space Force by U.S. President D. Trump (Wall 2019) are a clear signal toward this direction.

The peaceful use of space (UN 2008) would possibly become a right to defend and not an acquired status relying on international agreements. Space will follow the same path of the sea and air domains, started as a research and development environment when men were not able to navigate or to fly, and then transformed into capability domains to defend from external threats for the benefits of all the civil population. Rather sooner than later, the space domain will evolve to a sphere to be reached before the others, to a source full of enabling services. The above-mentioned recent anti-satellite's strike capabilities performed as test demonstrators are clear evidence that we need to start preparing to defend space from such kinds of threats.

In this view, the recent declaration of European Commission during the last European Space Policy conference emphasizing the need of start thinking to an EU Space Force (Teffer 2019) in parallel with the development of EU space capabilities, such as Galileo, looks as the product of a defense-oriented Strengths, Weaknesses, Opportunities, and Threats (SWOT) analysis performed in a process of strategic thinking. Indeed, as remarked in the EU Global Strategy, performant and robust PNT services are a key factor to enhance the responsiveness, the credibility, and the responsibility of the EU. In the end, denial or disruption of satellite navigation signals may heavily impact the effectiveness and the capabilities of EU military forces.

Conclusion

The assumption that space-based PNT services will be always and in any conditions guaranteed to EU Member States is debatable and the last EU Space Global Strategy clearly refers to this as a need of strategic autonomy. Such awareness in EU Member States is not homogenously spread. It could be erroneously considered that GNSS services are a given-capability, even without the standard requirements' definition process.

Recent events are confirming that space domain is already a congested and competitive domain, where several and new forms of threats will appear, operate, and evolve. PNT services will be affected, and considering the high level of military capability dependency from GNSS, a broader and comprehensive approach needs to be put in place supporting the development of dependable PNT services and sources capable of meeting European defense operational requirements in any of the envisaged scenarios.

All this, coupled with the almost ubiquitous dependency on space-based PNT services, require that defense planners and leaders understand that building more resilient PNT capabilities needs careful thinking and the implementation of architectures that transcend individual PNT-enabled systems, taking into account the need to protect not only the delivery of PNT services to users but also to render the space assets more robust and resilient to new types of threats.

A common definition and agreement of PNT requirements and related concepts of operations aimed to identify primary, alternate, and back PNT sources for defined military scenarios and platforms is indeed a crucial step to guarantee a holistic and robust PNT service to armed forces during military operations.

References

Air Force Research Laboratory Space Vehicles Directorate (2017) Navigation Technology Satellite-3 (NTS-3). https://www.kirtland.af.mil/Portals/52/documents/NTS-3-Factsheet.pdf?ver=2017-04-05-140456-050. Accessed online 12 Apr 2019

Barker B et al (2000) Overview of the GPS M-Code signal. In: Proceedings of the 2000 national technical meeting of the Institute of Navigation January 26–28, 2000

C4ADS (2019) Above us only stars. Exposing GPS spoofing in Russia and Syria

Danchik R, Lee P (1990) The navy navigation satellite system (transit). J Hopkins APL Tech Dig vol 11, pp 97–101 (ISSN 0270-5214)

Davies N et al (2016) QinetiQ, UK, D. Hagan, H. Mayoh, D. Mathews, Rockwell Collins, UK Towards Dual Mode Secured Navigation Using the Galileo Public Regulated Service (PRS) and GPS Precise Positioning Service (PPS), ION 2016

De Selding P (2015) U.S., Norwegian paths to encrypted Galileo Service Open in 2016. https://spacenews.com/u-s-norwegian-paths-to-encrypted-galileo-service-open-in-2016/. Accessed online 12 Apr 2019

EDA Capability development plan (CDP) (2018). https://www.eda.europa.eu/docs/default-source/eda-factsheets/2018-06-28-factsheet_cdpb020b03fa4d264cfa776ff000087ef0f. Accessed online 12 Apr 2019

EDA European Defence Matters (EDM) (2018) The future needs space, European defence matter issue #16, November 2018

Erwin S (2018) DARPA to begin new effort to build military constellations in low Earth orbit. spacenews.com/darpa-to-begin-new-effort-to-build-military-constellations-in-low-earth-orbit. Accessed Apr 2019

European GNSS Agency (2015) PRISMA – Development of low end operational PRS receivers including security modules architectures. https://www.gsa.europa.eu/prisma-development-low-end-operational-prs-receivers-including-security-modules-architectures

European radio Navigation Plan (ERNP) (2018). http://ec.europa.eu/DocsRoom/documents/33024. Accessed online 12 Apr 2019

Foust H (2019) India tests anti-satellite weapon. https://spacenews.com/india-tests-anti-satellite-weapon/. Last access 29 Apr 2019

Future Defence PNT (F-DEPNAT), EDA Procurement procedure 18.CAT.NP3.024. https://eda.europa.eu/procurement-biz/procurement/eda-procurement/18.cat.np3.024%2D%2Dfuture-defence-pnt-(f-depnat)

Gibbons G, Inside GNSS et al (2017) Delay continues for effort to add Galileo signal to U.S. Military receivers. https://insidegnss.com/delay-continues-for-effort-to-add-galileo-signal-to-u-s-military-receivers/. Accessed online 12 Apr 2019

GMV (2011) Galileo future and evolutions, Navipedia. https://gssc.esa.int/navipedia/index.php/Galileo_Future_and_Evolutions

Inside GNSS (2013) Something old something new. https://insidegnss.com/something-old-something-new/. Accessed online 12 Apr 2019

Italian Air Force (IAF) (2014) Spazio a support delle capacita' operative dell'Aeronautica Militare: Dipendenze, Vulnerabilita e Mitigazioni. Italian Air Force Publication, February 2014

Lawrence et al Innovation from LEO, GPS World, 30th June 2017

Leonardo (2014) Selex ES-led consortium wins €10.3M contract to hook EU governments up to Galileo satellite's secure navigation signals. https://www.leonardocompany.com/press-release-detail/-/detail/pioneer-ii

Market Research (2017) Global Military GPS-GNSS Devices Market 2017–2027. https://www.marketresearch.com/Strategic-Defence-Intelligence-v3944/Global-Military-GPS-GNSS-Devices-11068926/. Accessed online 12 Apr 2019

Mccaney K (2015) DOD puts emphasis on navigation warfare, accurate GPS signals, Defensesystems.com/articles/2015/02/09/dod-directive-navigation-warfare-pnt-tools

Menschner A (2018) GPS/PNT modernization progress: state of GPS III, MGUE, accelerating M-code, and resilient PNT, national space-based positioning, navigation, and timing advisory board meeting, May 2018

Miles D (2004) Iraq jamming incident underscores lessons about space. https://archive.defense.gov/news/newsarticle.aspx?id=25298. Accessed Apr 2019

Nye J (2005) Soft power: the means to success in world politics. PublicAffairs, New York

Ralph E. SpaceX Starlink becomes first US mega-constellation to gain FCC approval. https://www.teslarati.com/spacex-starlink-gains-fcc-approval/. Accessed online 12 Apr 2019

Reid T et al (2016) Satellite navigation from low earth orbit. SCNPT symposium 2016. http://web.stanford.edu/group/scpnt/pnt/PNT16/2016_Presentation_Files/S11-Reid.pdf. Accessed online 12 Apr 2019

Reid T et al (2018) Broadband LEO constellations for navigation. J Navig 65(2):205–220

Schmidt G (2013) Navigation sensors and systems in GNSS degraded and denied environments. NATO STO-EN-SET-197 2013

Teffer P (2019) EU commissioner floats idea for European space force, available through EU Observer. https://euobserver.com/foreign/143981. Accessed online 12 Apr 2019

U.S. Department of Defense, Department of Homeland Security, and Department of Transportation (2017) Federal radio Navigation plan (FRP) 2017. https://www.navcen.uscg.gov/pdf/FederalRadioNavigationPlan2017.pdf. Accessed online 12 Apr 2019

United Nations (UN) (2008) Treaties and Principle on Outer Spaces. New York. United Nations Publications. (ISBN 978-92-1-101164-7)

US Office of the Secretary of Defense (2019) Military and security developments involving the People's Republic of China, annual report to congress 2019

Von Clausewitz C (1832) On War, Revised edition with Introduction and Notes by Col. F.N. Maude. London. Kegan Paul, Trench, Trubner & C., 1918

Wall M (2019) Trump signs directive to create a military space force (Space Policy Directive-4). https://www.space.com/president-trump-space-force-directive.html. Accessed online 3 May 2019

Wilson J (2015) Military GPS User Equipment (MGUE), Lt Col J. Wilson, 29 Apr 2015. https://www.gps.gov/multimedia/presentations/2015/04/partnership/wilson.pdf. Accessed online 12 Apr 2019

Space Traffic Management Through Environment Capacity

44

Stijn Lemmens and Francesca Letizia

Contents

Abstract

Access to space is in principle regulated by international and national law. Once in orbit, however, the notion of national boundaries is not sufficient for space traffic management concepts as the physical reality dictates that actions of a single object in orbit has consequences on all its orbital neighbors. The space environment is therefore a limited shared resource, i.e., the ability to safely conduct operations. Based on reviewing the limitations of the current space debris mitigation guidelines, a natural extension is proposed by considering the capacity of the space environment to withstand risks associated with resource usage. To achieve this link, a metric capturing the resource consumption of an object in terms of collision risk induced on orbital neighbors is constructed. The integral risk over all actors in orbit is then used to quantify the notion of harmful interference with the environment. Managing this integral risk, i.e., the

S. Lemmens (✉)
Space Debris Office, European Space Agency (ESA) – European Space Operations Centre (ESOC), Darmstadt, Germany
e-mail: stijn.lemmens@esa.int

F. Letizia
Space Debris Office, IMS Space Consultancy at Space Debris Office, European Space Agency (ESA) – European Space Operations Centre (ESOC), Darmstadt, Germany
e-mail: francesca.letizia@esa.int

© Springer Nature Switzerland AG 2020
K.-U. Schrogl (ed.), *Handbook of Space Security*,
https://doi.org/10.1007/978-3-030-23210-8_109

environmental capacity, in the same way as is done for other resources linked with orbits, e.g., frequencies, leads to a dynamical framework for achieving space safety and security through space traffic management.

Introduction

The Introduction section will be used to briefly lay out the international legal and policy regime to deal with environmental issues such as space debris. The following section will describe the current global status in adhering to space debris mitigation guidelines put forward during the last decade, and highlight the various risks associated with both ignoring and adhering to them. The third section will go beyond the current guideline and introduce how a dynamics risk estimate, accounting for the design and operations of a space missions, naturally extends space debris mitigation and leads to the notion of the capacity of the space environment which can be managed as a natural resource in analogy with radio frequencies. The last two chapters mention some further applications and wrap-up with conclusions.

Space traffic management as a concept is not new. Various international fora, e.g., the International Institute for Space Law or the United Nations Committee on the Peaceful Uses of Outer Space (UNCOPUOS), have been discussing questions such as "how access to space should be regulated," "what are the rules of the road in orbit," or "how to integrate launch atmospheric re-entry traffic with air traffic" under various guises and with increased relevance since the 1990s. What has been an increased driver in the debate since the last decade is the focus on safety and security aspects, and the need to get the terminology agreed at international level. For the purpose of this work, the following definition of *space traffic management* is used: "The set of technical and regulatory provisions for promoting safe access into outer space, operations in outer space and returns from outer space to Earth free from physical or radio-frequency interference" (International Academy of Astronautics 2006), focusing on the physical interference.

Under the internationally widely adopted 1967 Treaty on Principles Governing the Activities of States in the Exploration and Use of Outer Space, including the Moon and Other Celestial Bodies (the "Outer Space Treaty") (United Nations 1966), and further derived treaties such as the 1972 Convention on International Liability for Damage Caused by Space Objects (the "Liability Convention") and 1975 Convention on Registration of Objects Launched into Outer Space (the "Registration Convention"), it is clarified that access to space is regulated at the State level, with only the common understanding of how to behave in orbit being laid out in those treaties themselves. In absence of fully internationally endorsed follow-on work, how the on-orbit behavior of actors can be made compatible with global safety and security objectives, it is important to recall the two main "hooks" that are available from a space traffic management perspective within the Outer Space Treaty, with a focus on environmental concerns. In particular, when it comes to what nowadays could be perceived as environmental concerns, the treaty does not offer a clear set of rights or obligations beyond Articles III and IX. The former concerns the need for State Parties, and hence the national actors for which they are responsible, to be acting in accordance to international law. The latter concerns the need of activities in space to avoid harmful contamination, which is one of the main purposes of space traffic management as defined for this work as well.

The currently most pressing environmental issue in outer space is the issue of how to deal with space debris, i.e., the nonfunctional objects of human origin left behind in orbit after six decades of spaceflight which carry the risk of, through collisions among themselves, creating orbital regions where space debris is so dense that spaceflight is impossible (Kessler and Cour-Palais 1978). The risk of triggerring this so-called *Kessler syndrome* is in essence a long-term risk. Studies that drive technical guidelines to avoid this fate generally focus on long-term simulations covering spaceflight activities 50–200 years into the future (Inter-agency Space Debris Coordination Committee 2007). However, there are already consequences of our limited stewardship of the space environment that present themselves to present day operators as ever-increasing short-term risks. In certain regions of Low Earth Orbit (LEO) and in the Geostationary ring, having a collision avoidance procedure in place and regularly exercising it has become a common practice. These short-term risks have been mostly caused by fragments from explosions of nonpassivated objects, left in orbit since the start of spaceflight activities, but collisions between space debris objects, including an active satellite with an intact abandoned satellite, have also occurred (European Space Agency 2019). Most recently, also near-term risks in the space environment have become apparent as actors in orbit start adapting to a congested space environment, due to regulations or out of self-interest. For example, a well-known space debris mitigation practice is limiting the orbital lifetime of a satellite to 25-years after the mission has ended. In the last 5 years, the increased adoption of this practice, combined with increased launch traffic into LEO of satellites without maneuver capability, has shown a build-up of object density around 600 km of geodetic altitude, which needs to be accounted for by objects passing through this region.

In order to address, among others, the issues posed by space debris on spaceflight activities, UNCOPUOS has taken the initiative to create a set of internationally agreed guidelines for the long-term sustainability (LTS) of outer space activities (United Nations 2019a). These guidelines contain recommendations on the policy and regulatory frameworks for space activities, the safety of space operations, rules of engagement for international cooperation, capacity-building and awareness, and scientific and technical research and development, but omitted security aspects. Under the assumption of an orbit, or orbital region, as a limited and shared resource, a contribution to space traffic management can be made by introducing concepts of top-down resource management compatible with the sustainability guidelines. At the hearth of this will be the quantification of the physical interference that spacecraft and launch vehicles, or collection of those as part of a single mission in the case of larger constellations, can have on the space environment itself and other operators.

Current Global Actions Undertaken to Preserve the Space Environment

Ever since the start of the space age, there has been more space debris in orbit than operational satellites. As space debris poses a problem for the near-Earth environment on a global scale, only a globally supported solution can be the answer. This creates the need for a set of internationally accepted space debris mitigation measures, in addition to national standards and license processes. A major step in this direction was taken in 2002, when the Inter-Agency Debris Committee (IADC) published its Space Debris Mitigation Guidelines (Inter-agency Space Debris Coordination Committee 2007). This document has since served as a baseline for nonbinding policy documents, national legislation, and as a starting point for the derivation of technical standards. The standardization of mitigation measures is important in order to achieve a common understanding of the required tasks leading to transparent and comparable processes. Even if a consistent set of measures is paramount to tackle the global problem of space debris, it is then up to the individual nations, operators, and manufacturers to implement them. The status of these national activities have been well documented by UNCOPUOS showing the increased uptake within its Member States (United Nations 2019b). In response to the LTS guidelines and its standardization role, the European Space Agency (ESA) publishes a supporting report to provide a transparent overview of global space activities, estimate the impact of these activities on the space environment, and quantify the effect of internationally endorsed mitigation measures aimed at improving the sustainability of space flight (European Space Agency 2019). In the spirit of transparency, this section is based on a recap of the cited report, in the public domain, with permission of the authors.

In more than 60 years of space activities, more than 5800 launches have resulted in more 44,000 tracked objects in orbit, of which more than 20,000 remain in space and are regularly tracked by the surveillance networks around the globe. About 26% of these so-called *catalogued objects* are satellites, or *Payloads*, and only a small fraction, about 2000, are still operational satellites today. About 17% of the catalogued objects are spent upper stages and mission-related objects such as launch adapters and lens covers. More than half of the population is made by fragments generated by more 500 breakups occurred in space, with the two major fragmentation events clearly visible as jumps in the population as shown in Fig. 1. Import to note is that the number of objects reflects the improvement in the capability of the space surveillance systems. When new objects are detected due to increased sensor performance, they can generally not be traced back any longer to an event or source and a growing category of "Unidentified" objects appear in the figure. This has legal repercussions, as there can be damage caused by well tracked and established space debris objects, for which an owner and hence a launching state cannot be identified.

To understand the active interest in space traffic management around the globe, one only has to note the remarkable change in the launch in traffic in Low Earth Orbit, both in terms of volume as in terms of actors, as shown in Fig. 2. The number of payloads has now reached four times the level of 10 years ago, with a steep

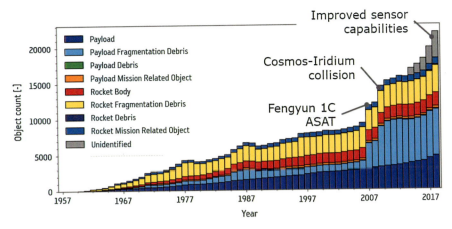

Fig. 1 Evolution of number of objects in geocentric orbit by object class. (With permission of the European Space Agency ©)

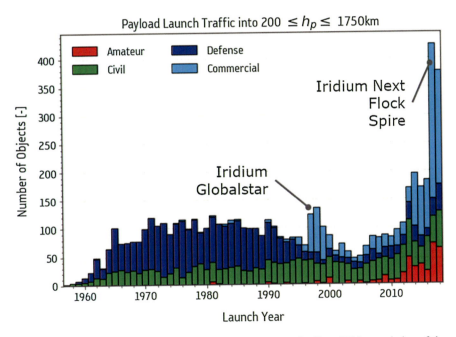

Fig. 2 Evolution of the launch traffic near LEO per mission funding. (With permission of the European Space Agency ©)

increase in particular since 2017. This growth in numbers is driven by the launch of ever smaller satellites, with around half of the satellites launched in the last 3 years having a mass smaller than 10 kg. This change in traffic is also related to a shift towards a more commercial exploitation of LEO and a diversification in the actors on stage. In 2019 alone, 180 satellites where injected in LEO as part of either the

Starlink, OneWeb, PlanetLabs, or Spire constellation. This number is expected to double in 2020. A similar trend was observed in GEO many decades ago and there it leads to collaboration among the different actors to ensure an effective exploitation of the available orbital slots. Similarly, in LEO, new collaborations among operators may emerge, as more and more operators take an active role in promoting best practices to limit the proliferation of space debris. Collaborations between some operators themselves is likely not sufficient, as the density of operations in LEO would imply that any operator can be affected. A global space traffic management system to coordinate these exchanged would be most efficient.

Counting and identifying launches and objects, and detecting and cataloguing space debris are needed to understand the global use of the environment. In a congested environment, equally important are the development of metrics that serve as proxies for the global adherence to space debris mitigation guidelines which have been put in place to protect the space environment from adverse effects such as the Kessler syndrome. These metrics, such as disposal strategies per object type or breakup events by underlying cause, consider both the historical evolution and the different performance achieved by different class of spacecraft. However, a caveat is that the analysis of the real progress made in the last years in terms of the compliance to space debris mitigation guidelines only appears after the operational life of a satellite. Hence the observed change in the traffic so far implies that in the future the signal could be dominated by the performance of large constellations.

As pointed out in the Introduction, the current short-term risk levels are dominated on-orbit breakups. Guidelines are clear: the potential for breakup should be minimized both during operational phases (for example, by a careful analysis of the failure trees) and after the end-of-mission, by releasing stored energy on-board, as the one in tanks and batteries. Intentional destruction and other harmful activities should also be avoided. Currently, on average 8 nondeliberate breakup events occur per year and this number has not improved in the recent years. One third of the events are related to failures in the propulsion system of the spacecraft. Even if the more systematic application of passivation strategies has contributed to reduce slightly this type of breakups, failures of the propulsion and of the electrical systems still represent a significant contribution to the population of fragments observed in Earth orbits.

It is well known that the distribution of these breakup events is not uniform across the population of objects in orbit, but rather some specific designs have exhibited over the time a higher tendency to fragment. This is particularly evident from the distribution of events by cause and by the time between launch and breakup. For "Anomalous" events, i.e., the separation at low speed of fragments from a parent object, the time appears to be rather uniformly distributed, whereas for certain classes of propulsion and electrical failures, the breakups are clustered around specific times, as show in Fig. 3. Design flaws can appear at different epochs from launch, but, for breakups in space, a higher incidence is observed during the first phase of operations, with half of the events occurring within 16 months from launch. This suggests that for the proposed large constellations operating at high altitude, where a repeating design is fundamental to the cost-model, the risk of breakups can

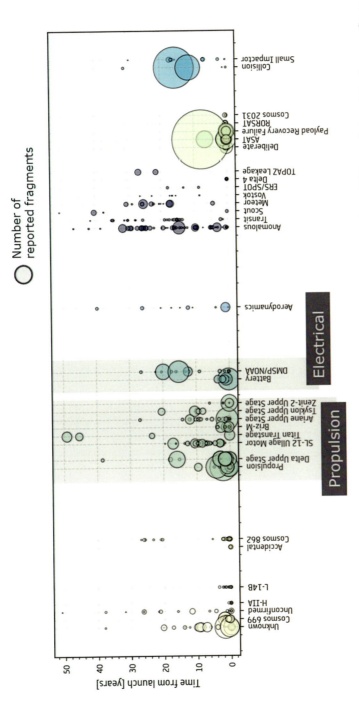

Fig. 3 Distribution of fragmentation events by estimated cause and elapsed time between launch and break-up. (With permission of the European Space Agency ©)

be mitigated by testing the system at lower orbits before moving to their operational orbit (Inter-agency Space Debris Coordination Committee 2017).

Whereas breakups in orbit are the current dominant source of space debris, the Kessler Syndrome implies that in the future this role would be taken by on-orbit collisions. The probability that these occur is essentially a function of the objects left behind on-orbit. As a mitigation measure, post-mission disposal guidelines have been formulated for two so-called *protected regions* defined by IADC, i.e., the Low Earth Orbit (LEO) and the Geostationary Orbit (GEO). Historically, these regions represent orbits where most operational spacecraft resides and where the collision probability is higher. These regions are protected because of their unique nature, which means that it is important to ensure access and operability in these regions for future missions and this requires defining their sustainable use with respect to debris generation. As mentioned in the Introduction, for objects in LEO, the recommended action is to accelerate their orbital decay such that their permanence in the protected region is limited to at most 25 years after the end of mission; for objects in GEO, it is recommended to move any spacecraft to a disposal orbit sufficiently above the GEO region and rocket bodies into orbits which don't intersect with it in the long term.

Disposal plans and their expected success rate are currently not systematically shared by operators but thanks to space surveillance data, the activity of a spacecraft can be derived and the orbital evolution predicted (▶ Chap. 50, "Space Object Behavior Quantification and Assessment for Space Security"). This create a transparent estimate on the behavioral state of any object, free from most biases, which could occur when considering intelligence sources. For objects in LEO, the residual orbital lifetime is estimated and compared to the 25 years mentioned in the guidelines. In this way, one can classify a spacecraft as compliant or not. For GEO objects, the orbital evolution over 100 years is checked to detect any return to the GEO protected region from the orbit where the spacecraft was disposed. In the LEO region, roughly 40% of the total number of payloads operate in orbits that naturally adhere to the space debris mitigation measures, i.e., they will re-enter in the Earth's atmosphere within 25 years from the end of their mission. In particular, around 78% of small payloads, i. e., below 10 kg in mass, operate in such regions. This means that still 22% of these spacecrafts are left in in potentially crowded orbital regions, without any maneuver capability.

For the objects that operate in non-naturally compliant regions in LEO, a low level of compliance is observed, with only around 15–25% of the payloads that have reached end-of-life during the 2010s attempting to comply with the space debris mitigation measures. This class of objects represents roughly 50% of the objects in LEO and around 60% of the total mass in LEO. They can become large contributors to a Kessler syndrome because of their mass (and so the number of fragments that they can generate) and because of their altitude (which results in a long residual lifetime). More optimistically, Fig. 4 shows that shift towards higher compliance rates is possible and already happening for rocket bodies in LEO. While around the year 2000, the attempts of disposal accounted to less than 20%, currently a value close to 80% is reached. Even better performances are reached in GEO, where the

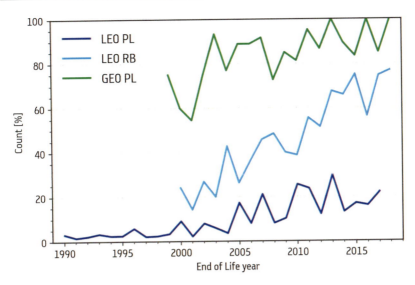

Fig. 4 Disposal attempt rate for non-naturally compliant objects such as Payloads/Satellites (PL) and launch vehicles/rocket bodies (RB) from the protected regions. (With permission of the European Space Agency ©)

disposal attempts have been consistently above the 80% level in the recent years. The case of GEO, where there is a clear commercial interest in keeping the operational orbits free from defunct of interference causing satellites, has a parallel for large constellations that will also have a similar interest in keeping their orbits clean. In case a collision, failure, or breakup occurs in a densely populated orbital region, the first operators to notice are the orbital neighbors that will have to adapt their procedures. In case of large constellations or GEO operators, these operators tend to form part of the same entity.

Whereas breakups and post-mission disposal address the long-term risk to the space environment, short- and near-term risk associated with space debris mitigation and environment congestion can be captured as well. The changing scenarios in terms of traffic and constellations has an impact on the operation of satellites that can be measured in terms of close approaches. As an experiment, one can assess the close approaches for ESA's fleet of satellites at lower orbits, i.e., below the congested debris regions, and note for these missions an increasing contribution coming from intact satellites in Fig. 5. Such intact, and in some cases operational, satellites include those belonging to constellations and general small satellites (mass lower than 15 kg). This trend, nearly certain to further increase, pushes operators to reconsider their current setup for the collision avoidance activities. Where a piece of debris is involved, collision avoidance processes only require the support of a space surveillance network to track, estimate, and predict the location of the chaser for a risk assessment. On the other hand, in case of operational chaser satellites, in addition also coordination is required to check their maneuverability status and whether any

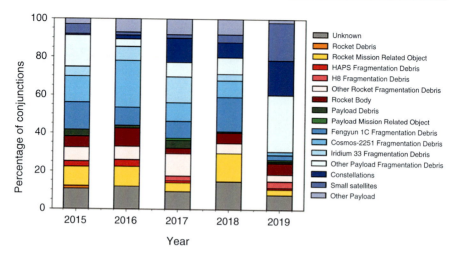

Fig. 5 Close approaches for ESA missions at low altitude LEO with various classes of chaser objects (both actively controlled as well as inactive). (With permission of the European Space Agency ©)

collision avoidance plan is in place on the other side. Manual coordination is no longer an option when considering a scenario with thousands additional operational satellites crossing operational orbits of others. There can be very valid reasons for doing such crossings, e.g., when performing a launch and early operations phase in orbits close to the Earth which are not debris dense before raising the orbit to the operational regime at 1200 km of geodetic altitude such as planned for the One-Web constellation in response to space debris mitigation actions (Inter-agency Space Debris Coordination Committee 2017).

A first step to ease the coordination among operators is to promote data sharing, for example, for what concerns the maneuverability of an object and its predicted ephemerides. A second step is to develop more automated systems for collision avoidance (Bastida Virgili et al. 2019). It is estimated that nowadays global satellite operators spend 14 million euro annually on debris impact avoidance maneuvers, but more than 99% of the conjunction notifications are false alerts. The changing scenarios, in terms of launch traffic, associated to small satellites and large constellations and in terms of improvements in the sensor capabilities, will generate a much larger number of collision warnings to deal with. The introduction of automated systems is not the final solution to the issue of collision avoidance, especially in the cases mentioned before where other operational satellites are involved, but it has the potential to ease the effort required for operations. More effective protocols for timely communication are needed in the future, also to ensure a smooth interaction between automated systems and systems with human in the loop, together with a set of well and unambiguously defined, and standardized, space traffic rules, including transparency in terms of the risk accepted by operators and on how reaction thresholds are defined.

The Limits of Space Debris Mitigation as We Know It and Beyond

In the previous Section, the state of the space environment and how to mitigate the adverse effects of space debris in the long- and short-term have been described. Many of the mitigation guidelines and standards in place world-wide have been based on the IADC guidelines (Inter-agency Space Debris Coordination Committee 2007), first released in 2002, which in turn were based on best practices available in its member agencies. The effectiveness of those guidelines was derived from space environment projection models. Such models define space traffic scenarios and on-orbit behavior to make a stochastic prediction on the evolution of the space debris environment. As a result of this process, the members of the IADC found that the 25-year post-mission disposal (PMD) rule was suited to reasonably bound the growth of object numbers provided that is has a high adoption and implementation rate.

Numerous studies of the long-term evolution by means of independent space environment projection models demonstrated that addressing breakups and post-mission disposals dramatically affects the growth rate of the debris population. For example, Fig. 6 shows the population of objects larger than 10 cm in LEO over 200 years. In the case where the current trends in terms of breakups and disposal continue are extrapolated to the future, an increase of more than three times in the number objects is predicted. The successful passivation of any spacecraft would already halve the final number of objects. If also a rate of 90% in the successful implementation of post-mission disposal is achieved, the increase of the number of objects over 200 years is only 30%.

This is an example of how, according to several consistent studies on the long-term evolution of the environment, the current level of compliance is not sustainable, in the sense that, if the same level is maintained also in the future, an exponential

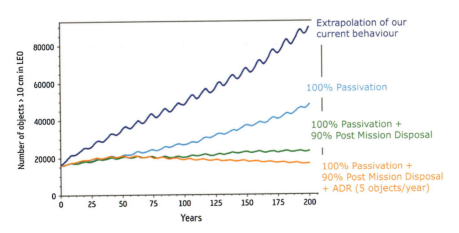

Fig. 6 Example predicted long-term evolution of the environment in different mitigation scenarios, under launch traffic conditions as observed during the 2000'ies as function of space debris mitigation scenarios including Active Debris Removal (ADR). (With permission of the European Space Agency ©)

growth of the population of objects is predicted, regular collisions will occur, and the associated consequences will be present in everyday operations. International standards and policy call for at least 90% success in post-mission disposal rates, and at most a 1 in 1000 change of accidental breakup (International Standards Organisation 2019). These target values are very far from, respectively, the observed 15% and 8-per-year events previously mentioned, so an important shift in how to deal with this sustainability target is still needed.

For reasons of practicality in view of implementation on national levels, all widely used guidelines and standards address space objects individually. This allows a convenient direct relationship between regulator and an individual space object, and leads to verifiable design targets such as the 25-year rule and 90% post-mission disposal success rate. However, the limitation of performing space debris mitigation in this way is the implicit dependency on a certain underlying space traffic (management) scenario assumed to guide effectiveness of the guidelines. For example, in the case when also the presence of large constellations in LEO are considered in analyses such as Fig. 6, the studies carried out within the IADC have shown how a reasonably stable evolution (but growing) of the environment is achieved only in the cases where not only the disposal success rate is at least 90%, but specifically for constellation objects is at least 95%. The 25-year post-mission disposal guideline can cope with drastic variations of space-traffic as long as it is almost perfectly adhered to (Inter-agency Space Debris Coordination Committee 2017), but technical systems will suffer from failures and a fulfilment of this guideline with a success rate of 100% is not achievable unless the object are inserted in orbits that is compliant to orbital lifetimes from the start. As noted earlier, this however creates a near term risk in form of an orbital shell around 600 km of geodetic altitude, which needs to be crosses by other operational satellites on the way up or down.

At latest with the start of the design of large constellations in LEO, it became clear that the space traffic has deviated dramatically from the assumptions made 20 years ago and space debris mitigation based on the one-requirement-fits-all-objects approach has reached its limits. On the other hand, it also serves to demonstrate the point that the space environment can de facto be seen as a limited shared resource, where the resource is the ability to conduct space operations, with interference being caused by the creation of debris that has to be mitigated. Indeed, considering the classical example for this type of interference one thinks of an object exploding or colliding in an Earth orbit: fragments will reach orbits that are 10s or 100 s of kilometers away in semi-major axis and quickly imply the need of regular collision avoidance maneuvers for the entire region affected once the fragments have been catalogued. A far larger amount of the debris created in this event will not be catalogued and can hence not be avoided, but will still have enough energy in a collision to affect the survivability rate of missions operating in this orbit (Krag et al. 2017). Less classically, releasing a constellation of 100 s of satellites without maneuvers capability in the space environment has environmentally the same effect as a cloud of tracked space debris and implies the same collision avoidance burden on nearby actors, but without the risk associated to undetected objects. The latter example is, however, more commonly interpreted as a space traffic management issue, but the interference nonetheless caused is the same as what is targeted by space debris mitigation.

To bridge the divide, it is of importance to establish a quantitative metric to describe the risk of interference a space object posed to the space environment, based on its physical characteristics only and in such a way that it can be integrated. A metric constructed in this way can capture the notion of resource consumption and hence quantifies the notions of harmful interference and damage as known from other global environmental issues. The stakeholder situation is diverse, as in absence of a comprehensive international legal regime to deal with space debris, various technical agencies, operators, and other stakeholder fora have been established which deal with recommending and enabling technical solutions to mitigate the proliferation of space debris (International Standards Organisation 2019) and to implement methods of exchange among actors in the space environment (The Consultative Committee for Space Data Systems 2013). The existence of standards and technical mechanisms to deal space debris have given rise to soft law practices by requiring space debris mitigation aspects to be considered as part of a national launch request (Tapio and Soucek 2019). To overcome this, the focus on using physical, or observable, parameters only for the quantification of the interference is importance to avoid misinterpretations. Indeed, as demonstrated by (United Nations 2019b), national regulators can also have different methodologies in tackling space debris mitigation. However, once in orbit the notion of national boundaries should no longer be the sole inspiration for space traffic management concepts as the physical reality dictates that actions of a single object in orbit has consequences on all its orbital neighbors.

When considering the "avoiding interference" part of space traffic management, an extension of the practice of space debris mitigation are on needs to address resource consumption questions such as "Which loss-of-mission risk is placed on other operators due to the behavior of a given mission?" and "What is the contribution of a given mission to the Kessler syndrome?".

Towards an Environment Capacity

As laid out in the previous sections, a quantification of the interference caused by an object in orbit to the space environment at large is required to bring the methods employed in defining space debris mitigation guidelines to the level of space traffic management. If successful, space traffic management in itself can contribute to a safe environment by strengthening existing space debris mitigation guidelines and requirements based on the actual interference levels in certain orbital regions.

Various attempts have been undertaken by many authors to formulate numerically what the impact is of an object on the space environment. Such approaches range from classical risk, i.e., probability times severity, analyses (Letizia et al. 2016) to purely environmental formulations based on modelling (Rossi et al. 2015), i.e., the likelihood of increasing the amount of objects in orbit. In general, the most simple approach, and common denominator among many said quantifications, describes in one way or another the impact of an object by the number of fragments released when involved in a collision or breakup (Utzmann et al. 2012). Considering these fragments, combined with the time they will remain on orbits gives a first quantification of how relevant an event is. Further extensions include accounting for feedback effects, i.e., accounting for the interaction of the released fragments with the existing environment (Letizia et al. 2019a), as having a fragment on orbit for a long lifetime is not necessarily a large issue if it has no probability of colliding with other object. In any case, the probability of an object to collide or breakup can be quantified as well and should be accounted for when quantifying interference, to avoid overly conservative estimates.

One metric has been developed for this use case: an index called Environmental Consequences of Orbital Breakups (ECOB) (Letizia et al. 2017). ECOB takes the form or a risk term, i.e., the severity of an event occurring multiplied by the likelihood of the event occurring. The form is shown in Fig. 7 and the main characteristics, relevant in this work, summarized below.

- The severity is assessed by simulating the fragmentation of the object studied and the effect thereof is measured in terms of the resulting collision probability for a set of targets representative of the operational satellites.
- This fragmentation can be due to collision or accidental breakup, with different probabilities and distinct fragments simulated in each case.
- The set or operational satellites in orbit change over time, and hence, the risk is a dynamical estimate, which can be reassessed when the space traffic conditions change.
- The state of the space debris environment at the moment of analysis is accounted for as part of the probability of collisional breakup for the object analyzed and the severity term is a proxy for the increased risk of collision for entire environment at

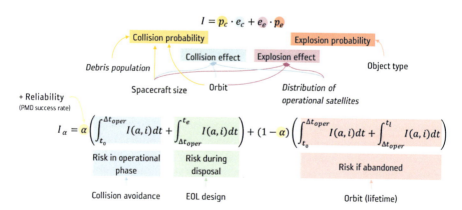

Fig. 7 Definition of the ingredients for metric underlying the risk quantification (EOL: End-Of-Life; PMD: Post-Mission Disposal). (With permission of the European Space Agency ©) (Letizia et al. 2017)

large. As such, the combined term is a proxy for the risk of triggering a local Kessler syndrome.

- The formulation accounts for mass, on-orbit exposed area, orbital ephemeris, and regime for a mission, the post-mission disposal and collision avoidance strategies, as well as the operations of the launch vehicle that brought the mission in orbit. Both for the object being evaluated as for other active missions in orbit.

There is also no need to limit the notion of ECOB to single object, as the results can be easily integrated over a group of object such as satellite and its launch vehicle, or a large constellation, or a satellite and objects intentionally released thereof, etc. The word *mission* is used to indicate an evaluation in such generality. The ECOB metric enables us to capture the three different risk perspectives of an object when interacting with the space environment at different time scales:

1. Short-term effect. The severity term is based on the effect of the potential breakup that a spacecraft or launch vehicles upper stage has on its orbital neighbors. On one hand, this includes the risk increase due to fragmentation debris on other active missions, but the ECOB of these other active missions can also be recalculated when a new mission is added to the environment and hence get a feedback on how the operational and disposal procedures would need to change to remain at the same level as without the additional mission.

2. Mid-term effect. Space debris policy measures can trigger behavioral changes of actors in the environment. E.g., the introduction of the 25-year PMD rule implies that nowadays favorable launch insertion is below this boundary, i.e., between 550 km and 650 km of geodetic altitude. By regularly updating the space debris environment model, as well as the representative list of active satellites, the ECOB formulation of environmental impact is dynamic and captures the changes in space traffic that take place.

3. Long-term effect. A risk formulation in terms of the feedback effect a collision or breakup in orbit has on other operational spacecraft is a proxy for the effect on the environment at large, which is currently still dominated by debris objects. As such, the integral risk over the population yields information on the status of the environment at large, and the triggering of the Kessler syndrome will show up as an environmental risk increase.

Environment capacity is defined as the term for resource usage of the space environment as quantified by the integral over all ECOB values of objects in the space environment that implies a sustainable future. There are various ways of defining a sustainable future scenario, e.g., ranging for those in which a net reduction in amount of space debris above a given size threshold is envisaged to those in which a maximum amount of yearly collision maneuvers is a given. Such environment trends, as computed by space environment projection models, can be evaluated in terms of their environment capacity, just like any individual mission would, and this value is set as maximum which can be in reality added to the environment across a period of time, e.g., a year. One natural candidate for a predicted future environment trend to follow comes easily to mind: one that consumes the same environment capacity as those predicted under the strict interpretation and application of the IADC space debris mitigation guidelines. The reason to base ourselves on an existing reference already applied to space debris mitigation is simply that the IADC guidelines have been demonstrated to be robust when adhered to stringently. The level of stringency required, e.g., in terms of postmission disposal success rate or accidental breakup probability tolerated, to follow the trend is automatically determined by the environment capacity to be achieved. A simplified example of how this environment capacity would practically work was played out for the first time in (Krag et al. 2018), with risk simply defined as the sum over all objects released into the environment as part of an event times their orbital lifetimes rather than ECOB.

The notion of thinking of a risk metric which determines the technical and operational constrains of a space mission, including the orbits it can access, is even less novel than the notion of space traffic management itself. In fact, the International Telecommunication Union's (ITU) constitution, which takes the form of an international treaty, via the radio regulation notes in Article 44 that "*In using frequency bands for radio services, Members shall bear in mind that radio frequencies and any associated orbits, including the geostationary-satellite orbit, are limited natural resources and that they must be used rationally, efficiently and economically, in conformity with the provisions of the Radio Regulations, so that countries or groups of countries may have equitable access to those orbits and frequencies, taking into account the special needs of the developing countries and the geographical situation of particular countries*" (International Telecommunication Union 2016). In case of the ITU and the frequencies they manage, the notion of harmful interference is the physical interference caused by electromagnetic waves within similar bandwidths which could render unambiguous reception and decoding of multiple signals impossible. Such interference can be predicted once the design and operational concept is known for a satellite, the interference determines how physically close the beams of satellite can be (e.g., in GEO orbits for telecommunication satellites, accounting for the fact they can go rogue), and the interference is easy to measure once it occurs in practice. Analogously and respectively, ECOB gives the risk of interference with the space environment based on the design and operational concept of a mission, the quantified risk can be used as an actionable value to decide if certain strategies are sustainable (e.g., how close can two large constellations operate together without causing collisions), and space surveillance networks track the on-orbit behavior of objects (including collisions and explosions).

A relevant overview of the full ITU process in this context can be found in ▶ Chap. 51, "Space Security and Frequency Management." In Table 1, a comparison is made between the legal regimes, broadly interpreted, that manage the frequency bands available as a resource and those in place for orbits as a resource. For the latter it is important to note that nowhere outside the ITU Constitution orbits as identified as such, and the Outer Space Treaty and common practices presented in this work indicate this. Nonetheless, it is clear that the commonalities with the far more regulated ITU

Table 1 Top-level differences between the legal regimes responsible for the management of the frequency and orbit resources

ITU constitution and regulations	Outer space treaty
ITU Constitution (Article 44): Radio frequencies and satellite orbits are limited natural resources	Outer space free for exploitation and use by all States Parties in conformity with international regulations;
The ITU Constitution creates, by means of an international treaty, an entire international body with regulatory capabilities. Implementation is a competence of the State Parties	States retain jurisdiction and control over objects they have launched into outer space
Regulations require a rational, efficient, economical use and equitable access	States Parties shall conduct exploration so as to avoid their harmful contamination and … and, where necessary, shall adopt appropriate measures for this purpose
Radio waves do not stop at national borders (Laws of physics)	Fragments of a an in-orbit breakup will not respect already occupied orbits (Laws of physics)
Interference-free is operations in the purpose of the Radio Regulations treaty	Interference free operations is the purpose of space debris mitigation policies
Allocation, power limits, and equitable use are achieved by clear rights and obligations: • Planning (block allocation) • Coordination (first come first serve)	Current level of international space traffic management is limited to: • Best effort coordination between operators, • "Soft Law" alignment across State on space debris mitigation

processes fall naturally into place with space traffic management when the notions of physical and radio-frequency interference are used as equivalences.

An example can be played out based on a top-down space traffic management framework modelled on the ITU with the physical interference defined as in this section. Most controversial could be the notion that orbits are then allocated based on the design and orbital parameters of a mission, and what can be accounted for in any given year. It is important to note that such an access to space policy based on environment capacity does not prohibit access to certain orbital regions per se. It might however imply stricter design requirements, e.g., a higher reliability on the mechanisms associated with post-mission disposal or a reduced orbital lifetime after the mission well below 25 years (Letizia et al. 2019a). The equivalent under ITU regulations, i.e., block allocation, can, in contrast, effectively block entire bands of the frequency spectrum for certain users. In practice, space traffic management based on environment capacity would imply the need for a simple scheduling and monitoring process:

1. Mission proposals need to be collected before launch with mission descriptions and planning in line with what is currently expected of Space Debris Mitigation Plans (SDMP, International Standards Organisation 2019). This ensures that the planning of available and consumed environment capacity can be monitored at short- and mid-term scales. This is analogous to the national and international coordination of frequency allocation for planned missions.

2. As the available capacity would be linked to a sustainable reference evolution of the environment, e.g., under hypothetical strict adherence to the IADC mitigation guidelines, there needs to be a limited time window in which proposals can be evaluated, in order to ensure that events happening in the environment can be accounted for, and orbits not blocked indefinably. The book-keeping of capacity allocation implies that ECOB's severity and probability terms can be regularly recomputed to track (planned) changes in the space environment. This is a common practice in planning-based resource management.
3. On a first-come-first-serve proposal-based system, if the capacity of a given launch year is nearly reached, more stringent design criteria would be required to achieve early launch or else an allocation would be postponed. Under-utilized environment capacity would become available in later years.

Technically, the implementation of such a space traffic management system could be as simple as a secure but open access web-based interface, which is synchronized with the ITU registry and linked with a credible and technical international body to ensure accurate modelling of the space environment. The same issues the ITU faces in terms of having reliable input in future behavior would, however, need to be tackled as well. A transparent space surveillance capability to validate that plans are executed as allocated is indispensable as well, e.g., to extend the analysis of on-going mitigation efforts such as shown in the previous section. Regulatory-wise, the implementation would not be complex per se. Indeed, as regulation currently targets enshrining mitigation requirements either by law, standards, or guidelines on the level of individual missions, and since current requirements need to remain active, this environment capacity approach could just be accomplished by an additional requirement. This additional requirement should ask to demonstrate that the space object fits within the overall capacity apportionment for the envisaged launch year, accounting for both its operational and disposal phase.

Applications Related to Environment Capacity

The notion of environment capacity is based on a quantified risk assessment that needs to be minimized. As this risk, i.e., the ECOB metric, is based on the design and operational concepts of a mission, it can effectively be used as design optimization tool for the sustainability of a mission as well (Letizia et al. 2019b). Among others, this allows the optimal space debris mitigation technology to be defined for various mission classes and orbital regimes. The concept can even be taken one step further, as working with quantified figures also enables new ways of benefiting from active debris removal activities or general trading schemes. E.g., an operator can opt to remove one of its previously abandoned satellites to increase its environment capacity allocation, or operators can trade unused allocation in case their mission performs more sustainable than planned.

At a practical level, the metrics to assess the impact of an object on the space environment have been included into larger environmentally oriented activities.

Transparent and procedure-based evaluations of space missions such as those described in the previous section can be included in life cycle assessments (Maury et al. 2019). Moreover, they have been adopted as an enabling component for a Space Sustainability Rating (Rathnasabapathya et al. 2019), aiming to positively identify those actors in orbit which take extra steps towards space sustainability beyond what is required by national regulations and space debris mitigation standards.

Conclusions

The access to space is currently regulated at the state level under the rules laid down in the outer space treaty, which over the years implied that states and international organizations adopted space debris mitigation policies, which have been based on internationally well-establish technical recommendations such as the IADC guidelines. The latter one is, however, a nonbinding legal instrument which can become applicable in law once a state adopts a derived technical standard, e.g., as part of a license to launch process. While such an update into legal regimes is an ongoing process, it is already clear that the global adherence to these guidelines, which were based on space traffic scenarios observed in the 1990s and 2000s, is currently not sufficient to achieve a sustainable space environment. With the recent advent of new actors in the space flight domain, and rapidly changing space traffic scenarios, these guidelines have also reached their limitations in applicability.

Space traffic management, when being based on the concept of avoiding physical interference which would be detrimental to actors in orbit and the space environment at large, offers a natural regime to put in place a top-down approach where orbits are considered as a limited shared resource and allocation is based on a quantified risk metric. To this end, a qualified risk estimator such as ECOB that captures the notion of the risk incurred by actors in the space environment due to the behavior, in terms of design and operational concept, of evaluated missions are tested to effectively quantify short-, near-, and long-term risks. This methodology is open and hence transparently available and compatible with present day space debris mitigation frameworks. The integral risk over all actors in orbit is the used environment capacity, i.e., the risk that we trigger a Kessler syndrome in the space environment. This value, including the predicted use by planned missions, can effectively be compared with space debris environment simulation models to identify sustainable trends. These trends can be tracked and accounted for on yearly basis. Where collision risk represents short-term physical interference with other actors, not tracking a sustainable trend represents the long-term physical interference with the environment.

In analogy with the modus operandi of the ITU, a space traffic management framework based on the philosophy of avoiding interference, based on the risk as defined in this work, would extend, rather than redesign, current national space debris mitigation frameworks into an internationally harmonized processes for space sustainability by ensuring space safety as the absence of harmful interference. Technological methods and processed to do so have been identified and prototyped. Adoption is pending.

References

Bastida Virgili B, Flohrer T, Krag H, Merz K, Lemmens S (2019) CREAM – ESA's proposal for collision risk estimation and automated mitigation. In: 1st International orbital debris conference, Sugar Land, TX, USA, 9–12 December 2019

European Space Agency (2019) ESA's annual space environment report. GEN-DB-LOG-00271-OPS-GR Issue 3. https://www.sdo.esoc.esa.int/environment_report/Space_Environment_Report_latest.pdf

Inter-agency Space Debris Coordination Committee (2007) IADC space debris mitigation guidelines. IADC-02-01, Revision 1, September 2007. https://iadc-home.org

Inter-agency Space Debris Coordination Committee (2017) IADC statement on large constellations of satellites in low earth orbit. IADC-15-03, Revision 0

International Academy of Astronautics (2006) Cosmic study on space traffic management

International Standards Organisation (2019) Space systems – space debris mitigation requirements. ISO 24113:2019

International Telecommunication Union (2016) ITU radio regulations. http://www.itu.int/pub/R-REG-RR/en

Kessler DJ, Cour-Palais BG (1978) Collision frequency of artificial satellites: the creation of a debris belt. J Geophys Res Space Physics 83(A6):2637–2646

Krag H, Serrano M, Braun V, Kuchynka P, Catania M, Siminski J, Schimmerohn M (2017) A 1 cm space debris impact onto the sentinel-1a solar array. Acta Astronaut 137:434–443

Krag H, Letizia F, Lemmens S (2018) Space traffic management through the control of the space environment's capacity. In: Proceedings of the 5th European workshop on space debris modeling and remediation

Letizia F, Colombo C, Lewis HG, Krag H (2016) Assessment of breakup severity on operational satellites. Adv Space Res 58:1255–1274

Letizia F, Colombo C, Lewis HG, Krag H (2017) Extending the ECOB space debris index with fragmentation risk estimation. In: 7th European conference on space debris, Darmstadt

Letizia F, Lemmens S, Virgili BB, Krag H (2019a) Application of a debris index for global evaluation of mitigation strategies. Acta Astronaut 161:348–362

Letizia F, Lemmens S, Krag H (2019b) Environment capacity as an early mission design driver. In: Proceedings of the 70th international astronautical congress

Maury T, Loubet P, Trisolini M, Gallice A, Sonnemann G, Colombo C (2019) Assessing the impact of space debris on orbital resource in life cycle assessment: a proposed method and case study. Sci Total Environ 667:780–791

Rathnasabapathya M, Wood D, Jah M, Howard D, Christensen C, Schiller A, Letizia F, Krag H, Lemmens S, Khlystov N, Soshkin M (2019) Space sustainability rating: towards an assessment tool to assuring the long-term sustainability of the space environment. In: Proceedings of the 70rd international astronautical congress

Rossi A, Valsecchi GB, Alessi EM (2015) The criticality of spacecraft index. Adv Space Res 56 (3):449–460

Tapio J, Soucek A (2019) National implementation of non-legally binding instruments: managing uncertainty in space law? Air Space Law 44(6):565–582

The Consultative Committee for Space Data Systems (2013) Conjunction data message. CCSDS 508.0-B-1

United Nations (1966) Treaty on principles governing the activities of states in the exploration and use of outer space, including the moon and other celestial bodies. RES 2222 (XXI). https://www.unoosa.org/oosa/en/ourwork/spacelaw/treaties/introouterspacetreaty.html

United Nations (2019a) Guidelines for the long-term sustainability of outer space activities. A/AC.105/C.1/L.366

United Nations (2019b) COMPENDIUM: space debris mitigation standards adopted by states and international organisations. https://www.unoosa.org/oosa/en/ourwork/topics/space-debris/compendium.html

Utzmann J, Oswald M, Stabroth S, Voigt P, Retat I (2012) Ranking and characterization of heavy debris for active removal. In: Proceedings of the 63rd international astronautical congress

Various Threats of Space Systems

45

Xavier Pasco

Contents

Abstract

For roughly two decades, orbital systems, beyond their traditional strategic value, have gained a pivotal role in modern conventional security and defense activities. As a consequence, they have been considered as possible new targets in military confrontations, and the recent years have indeed demonstrated a renewed activity in the field of antisatellite researches and tests. This piece attempts to put these efforts in perspective and detail their different forms. It appears that besides the traditional kinetic destruction of satellites, leading to uncontrolled long-lived debris, other threats may have equally destructive consequences with more limited side effects. Directed energy weapons in orbit or even cyberattacks may

X. Pasco (✉)
Fondation pour la Recherche Stratégique (FRS), Paris, France
e-mail: x.pasco@frstrategie.org

© Springer Nature Switzerland AG 2020
K.-U. Schrogl (ed.), *Handbook of Space Security*,
https://doi.org/10.1007/978-3-030-23210-8_9

become weapons of choice in the new space landscape. These likely perspectives must lead the international community to rethink the reality of threats related to space systems.

Introduction

Space systems have gained an increasing importance in the everyday life of the modern societies: telecommunications by satellite, broadcasting of television programs, observation of the Earth's surface and oceans, observation of the atmosphere for weather forecasts, navigation, and worldwide broadcasting of universal time have so many applications that they contribute intimately to the day-to-day making of our contemporaneous world.

Besides, the needs for the defense of States and for the security and safety of their citizen feed widely on data resulting from the use of observation, electronic intelligence, or early warning satellites. These have contributed in an essential way to producing a strategic piece of information during these last 50 years, helping in the prevention of the bipolar crises. Chastely qualified as "national technical means," observation, electronic intelligence, or early warning satellites became one of the touchstones of the strategic dialogue of the 1970s and 1980s. In this context, keeping space safe and preventing any evolution leading to putting space systems in jeopardy became a key word. In particular, American presidencies of the Cold War had effectively resigned themselves to this established fact. For decades, according to recently published official US documents, it was clearly recognized that any preparation of an antisatellite interception would have been contrary to the spirit if not the letter of the SALT (Strategic Armements Limitation Talks treaty signed in 1972 by Richard Nixon and Leonid Brezhnev) protection of "national technical means" with the risky perspective "to stimulate satellite interception since we are more dependent on intelligence from space sources and would have more to lose." (Memorandum from the President's Assistant for National Security Affairs (Scowcroft) to President Ford, Washington, July 24, 1976. For a more complete vision of the position of the US authorities at that time, refer more largely to the archives recently published under the direction of McAllister (2009).) In spite of two Soviet campaigns of antisatellite attempts during the 1970s which led to the US executive authorities to reexamine this position and realize a first antisatellite test in 1985, this particular form of militarization of the space was hardly pushed, the possible earnings remaining considered very thin with regard to the incurred strategic risks. The "stabilizing" function of these national technical means during the Cold War had been already well established and has been well informed since.

Considering this central aspect, the club of the space countries quickly agreed on the interest of keeping space free of weapons, in an explicit way or more implicitly. The text of the main legal body, the "treaty on principles governing the activities of States in the exploration and use of the outer space, including the Moon and the other celestial bodies," came into effect in 1967, has established the idea according to which the exploration and the use of the space are the privilege of the whole humanity. It has dedicated the freedom of research and circulation in space and has clearly indicated that the notion of State sovereignty cannot be extended in outer space or in the celestial bodies. Establishing the founding principle of the "peaceful uses" of outer space, the text does outlaw the deployment of weapons of mass destruction in outer space as well as any military activity on the Moon and on the other celestial bodies. (By the end of 2011, 100 countries had already ratified the Treaty, among which any major space nation.)

Nevertheless since approximately two decades, the international debate on the theme of the security of the spatial activities and more exactly on the militarization of the space returned to the front scene by becoming more radical. In the course of the transformations occurred during the 1990s, the initial preventions against a too extensive militarization of the Low Earth Orbit (LEO) have unmistakably weakened. Two main explanations can be called:

- The relative "downgrading" of the nuclear order as an international regulating principle and the consecutive "unbolting" of the debate on an increasing supposed vulnerability of the national spatial means: The United States in particular has mentioned the perception of an increased vulnerability considering the more and more central role played by satellites in the political, military, and economic life of most of the developed countries, with the United States in the first place.
- The emergence of new space actors, who may "threaten" to radically change the way space has been regulated under the auspices of a "club" of a few spacefaring countries, driving precisely these countries to anticipate this situation and bend over the elaboration of new international rules for the use of the space.

Change of Strategic Landscape: A Succession of Disturbing Events

More than in the 50 last years, this decade has known several events that have underlined the fundamental fragility of satellites. A series of destructions in orbit, deliberate or not, put space in full light, worrying the largest part of the diplomatic and military community. It came in a way to punctuate harder and harder debates in Geneva on the prevention of the arms race in space.

1. First of all, the shooting by China of a ballistic missile towards an old weather satellite on January 11, 2007, leading to its destruction and to the generation of a 3000 long-lived fragments on a very busy orbit, surprised the whole world. This test was the first of its kind since the one undertaken in 1985 by the United States which proceeded to the interception of one of their satellite by using a missile embarked under an F-15 fighter plane.

   At the very moment of the 2007 interception, the Chinese representatives were supporting without reserve the international efforts in the United Nations intended to limit the creation of space debris and opposed against the United States within the conference on disarmament in Geneva on the theme of the militarization of the space with a very proactive posture about prohibiting anti-satellite weapons.

2. Although they denied having had such intentions, the United States did not delay "answering" their Chinese counterparts by proceeding themselves on February 21, 2008, to the destruction of one of their military satellites in perdition. According to the American authorities, the point was to destroy a satellite which reentry was considered dangerous. Nevertheless, the successful attempt demonstrated, at least incidentally, the efficiency of one of the components focusing for the antimissile defense, whereas that it also meant the American intention "to mark" clearly its strategic territory. To complete the "state communication" picture, the American authorities did not miss to let know that this interception occurred at a much lower orbit than the Chinese interception, showing that this had been managed on the side of the United States in a more "appropriate way" by generating very short-term fragments of life. (Official US information has stated the figure of 175 detected debris (at the difference of 3037 for the Chinese event) with the last one reentered in the atmosphere by the end of October 2009.)

3. Less than a year later, on February 10, 2009, two satellites, one Russian (Cosmos 2251) and another one registered in the United States (a satellite of the Iridium constellation), collided and destroyed each other, generating some 1800 fragments on equally very frequented orbits. This collision, the first one in the history of space activity, was going to finish putting the question of space safety and security in the broad sense as one of the priority themes of the future space cooperation.

4. Finally, India performed an antisatellite test on March 27, 2019, using a two-stage ground based missile equipped with a terminal kill-stage that impacted a 750 kg Microsat-R satellite launched only about 2 months earlier in January 2019. This event was hailed by Prime Minister Modi as bringing "utmost pride" and having "a historic impact on generations to come." Communication was visibly prepared to avoid the level of criticism brought about in its time by the Chinese test. In particular, a FAQ document published by the Indian MoD immediately after the test underscored that "the test was done in the lower atmosphere to ensure that there is no space debris. Whatever debris that is generated will decay and fall back onto the earth within weeks." (See https://www.mea.gov.in/press-releases.htm? dtl/31179/Frequently+Asked+Questions+on+Mission+Shakti+Indias+Anti Satellite+Missile+test+conducted+on+27+March+2019 – accessed 29 March 2019.) Indeed, Miscrosat-R, supposedly an imaging satellite, was orbiting at about 280 km making it a "cleaner" target than the Chinese satellite, with debris supposed to burn in the atmosphere after only several months. It remains that this event has triggered the criticism of several operators of small low altitude satellites, such as Planet or Astroscale. However, it must be noted that the general reaction of governments, including China, has been limited to date.

Besides these well-known events, other recent disruptions in space have dramatized the space scene further, whether due to presumed cyberattacks (suspected in 1998 in the case of the US-UK-German satellite ROSAT recently reentered in the atmosphere), to laser blinding or tagging (as suspected from Chinese origin towards an US NRO satellite in October 2006), to interferences, whether purposeful as in the recent case of an Eutelsat satellite jammed from a source in the Middle East or accidental with the so-called zombiesat belonging to Intelsat and uncontrolled between April 2010 and January 2011 while emitting at full power and interfering during this period of time with a number of telecommunication satellites. As it will be explained below, these latest cases must also be considered as potential major sources of disturbances.

Early Armed Threats in Space

For a few years, the news has been dominated by controversies nourished by the supposed plans in a few countries of a possible deployment of weapons in the outer space. Such a subject is not new and has in fact been considered since the launch of the space activities. While no genuine "space arm race" has indeed been triggered during the Cold War, it is useful to remind nascent achievements in the 1960s, mainly carried out by the then USSR.

It must be noted that the *"weaponization"* of space has been considered very early in the history of space bipolar relationships. As early as February 1957, eminent US military officers did not hesitate to present space as a new "theater of operations": *"In the long haul, our safety as a nation may depend upon our achieving 'space superiority.' Several decades from now, the important battles may not be sea battles or air battles, but space battles, and we should be spending a certain fraction of our national resources to ensure that we do not lag in obtaining space supremacy."* (See excerpts of the famous 1957 speech by B. Schriever at http://www.af.mil/news/story. asp?id=123040817 (accessed August 2012).) A few weeks after Sputnik, that same year, the Air Force Chief of Staff, General Thomas D. White, reiterated this general assessment, ensuring that *"whoever has the capability to control space will likewise possess the capability to exert control of the surface of earth."* (Quoted in Stares (1985, p. 48). Military strategies would be also made public, for example, in a 338-page book, *The United States Air Force Report on the Ballistic Missiles* written by Colonel Kenneth Gantz (and forwarded by the well-known Generals White and Schriever). It was published by Doubleday and Comp in 1958. Besides the most common proposals aiming at developing antisatellite weapons, the US Air Force was proposing as soon as 1956 two different strategies for the military investment of space. One of those consisted in using a manned ballistic rocket (*Manned Ballistic Rocket Research System project*), while the other one (*Manned Glide Rocket Research System*) proposed the use of a reusable glide body launched from a main carrying rocket. If this latest project may recall the early NASA studies made about the shuttle at the end of the 1960s, this last project was purely military by essence as it envisioned the possibility to bomb the Earth surface since the altitude of 64 km! On its side, the Army, via the Army Ballistic Missile Agency (where Wernher Von Braun would ultimately help the United States to launch their first working satellite in January 1958), had the project of a super powerful rocket that would allow "colonizing" the Moon as well as other planets for military purposes. For a detailed expose of the military position at that time, see also Baker (1985, pp. 12–30)).

If many projects aiming at militarizing space *stricto sensu* have emerged in the United States, none of them were given real credits by the successive presidencies in this country. The political authorities were more inclined to capitalize on the nascent nuclear ballistic force to ensure the strategic balance with the USSR. However, first initial developments made in relation with the ballistic threat can be cited that paved the way for ground-based space weapons. A first "missile defense" capability was proposed in 1958 with the two-stage Nike-Ajax nuclear armed antiballistic missile, later on followed by the more powerful Nike-Hercules and Nike-Zeus. First ABM interceptions occurred in 1962, opening the way to newer ABM missiles, namely, the Sprint and Spartan version leading to the "Sentinel" and "Safeguard" program in 1969 with the objective to defend a limited number of strategic missile silos. It must be noted that as early as May 1962, the then Secretary of Defense McNamara allowed the conversion of the Nike-Zeus model into an ASAT program (Program 505) which led to simulated interceptions and then to a successful hit in May 1963 against a cooperative target. Another existing ASAT capability based on the THOR missile was also operational and led to the shutdown of the Nike-Zeus capability. The THOR capability would also be terminated a few years later in the mid-1970s.

The USSR gave itself the first role in developing threats actually *coming* from space orbital systems. A first series of "co-orbital" tests were indeed carried out starting from 1968 with a first alleged success in November that year and ended in 1971, obviously at a time when the new *"Detente"* was to be consolidated after the US-Soviet signature of Salt-1. (Signed in 1972 in Moscow, this test was incidentally pleading for the use of National Technical Means for treaty verification.) Realizing an alleged total of five successful interceptions during the first series of seven tests, the technique used by the USSR was the "co-orbital" explosion carried out by a specifically designed orbital system within a kilometer-wide radius of the target. A second series of similar tests was undertaken between 1976 and 1982, based on the advocated need for the USSR to respond to future presumed ASAT capabilities expected from the US space shuttle then in construction.

The Soviet activity in the field was then perceived as highly intensive, and President Ford directed the start of an equivalent ASAT program that would ultimately take the form of an airborne missile launched from an F-15 *Eagle* airplane. After a few test launches performed in 1982 and 1985, a third launch ended up with the interception of a US satellite target directly hit by the so-called Miniature Homing Vehicle (MHV), the third stage of the ASM-135 Vought missile. Again, the program was officially phased out in 1988, in a context when the strategic and budgetary soundness of such projects was questioned.

In any case, this early history amply demonstrates that initial ASAT programs had been envisioned as being possibly part of the global arsenals from a military perspective if not from a political one. Only the key role played by spy satellites in the mutual nuclear deterrence prevented weapons in space from becoming operational during the Cold War. This did not prevent national R&D projects to develop, paving the way for possible future threats in space.

A Generic List of Possible (Intentional) Threats in Orbit: Assessing Offensive Realities of Today

In today's completely renewed strategic context, these early antisatellite efforts have regained some momentum. Early programs have clearly served as a basis for more sophisticated projects allowed by technical advances, while new research domains seem to have emerged. The analysis of over more than two decades of R&D efforts lead to the following list of existing R&D orientations, possibly leading to actual space weapons:

- *Kinetic energy weapons* (KEW) implying a physical effect on the target, either by direct impact (so-called "hit-to-kill" techniques) or nearby explosion creating killing debris (such as in the case of the co-orbital Soviet systems)
- *High-altitude nuclear weapons* (EMP) creating ionization and/or electromagnetic effects on objects in the affected zone
- *Directed energy weapons* (DEW) mainly using laser or microwave techniques depositing energy on the target

Obviously other kinds of threats on space systems exist such as electronic warfare weapons (EW) using jamming techniques rendering communications impossible, or cyberattacks. Exactly like in the case of ground-based interceptors, such threats do not necessitate the use of space platforms to be effective. For this reason, such threats have not been treated as a key issue in the context of this chapter, as they do not define per se a threatening "space system." However, they shall not be discarded as their reality is largely tangible today as it will be explained further below.

- *Kinetic energy weapons*, while simple in their principle (physical collision), do not use simple techniques. They imply the use of maneuvering satellites as well as the mastering of precise "*rendezvous*" techniques, the least to achieve in case of "*hit-to-kill*" weapons! This can be related to techniques implemented by a number of existing systems, going from experimental surveillance satellites (or so-called "inspector" satellites) used, for example, to picture other orbital systems (such as in the case of the US XSS 11 and 12 or the Chinese SJ-12 or SJ-06F systems), to the European *automated transfer vehicle* (ATV) used for service and precise docking with the International Space Station (ISS). All these systems have in common highly maneuvering capabilities as well as precise terminal guidance systems allowing effective orbital "*rendezvous*." Mastering such technologies would theoretically allow developing kinetic energy ASAT. Protecting any satellite against the kinetic effect at orbital speeds becomes virtually impossible with pellets more than a few centimeters in size. As a matter of fact, protecting any satellite against a kinetic threat is almost paradoxical in itself. Satellite architectures are indeed based on the use of as light materials as possible involving some level of fragility. This is the case for satellite buses or for on-board solar arrays. "Armored" space systems are then hardly feasible and in any case would increase cost at all levels, from development to launch. Only some level of physical protection against small-sized debris (in the millimeters scale) can reasonably be applied nowadays.

However, if they can represent deadly threats, KE techniques remain highly costly in terms of energy (most notably when changes of orbits would be needed for performing an intercept) and, in a more sensitive manner, would create more debris that would add to the already rather congested orbital traffic. For sure, creating more debris would not account among the most preferred offensive strategies for most of the spacefaring countries whose space systems rely on an undisturbed and clean orbital space. Nations that do not intensively use space might possibly be less deterred from such actions.

- *High-altitude nuclear explosions* would make use of a nuclear bomb sent at an altitude of a few hundreds of kilometers with the objective to create highly intensive electromagnetic disturbances for Low Earth Orbiting (LEO) and even geostationary (GEO) objects. Cold War years were soon followed by the fear of an increasing nuclear proliferation that would make such a possibility more probable. Such an attack could indeed have an enormous effect on the whole activity in space, with, in the first place, the possibility for the attacker of annihilating a number of military systems precisely destined to warn against nuclear attacks, such as early warning systems, Earth observing, signal interception, or strategic communication satellites. In such a situation, most of the non-protected space systems would also be destroyed.

  Major studies (such as the HALEOS study published in 2001 under the auspices of the U.S. DoD Defense Threat Reduction Agency - DTRA) have shown that, compared with a terrestrial explosion, electromagnetic effects of a nuclear charge in space might be increased leading to potentially devastating impact beyond the only targeted orbits with short-term effects on the propagation of radio and radar waves, and longer-term effects involving the permanent excitation of Van Allen belts, with even the possibility of creating new magnetic belts resulting from the sudden expulsion of charged particles.

  While ionizing effects would be specific to such explosions, other effects, such as electromagnetic effects, would be no different from those created by *directed energy weapons* (DEW) using high-power microwaves (see below). As a consequence, protecting any system against such threats would mean protecting it partially from a major consequence of a high-altitude nuclear detonation. In other terms, the main characteristic of such a nuclear threat would remain its "nondiscriminatory" effects on the whole orbital population. In any instance, such an attack would mean that a situation of war would preexist. This makes the use of nuclear attack clearly different from other intentional actions that might take place in more ambiguous scenarios or even in a covert manner.

- *Directed energy weapons* (DEW) are sometimes perceived as presenting a coming threat for space systems. Indeed, this threat is theoretically characterized by some level of intensity leading to likely modulated effects on the target. It can be considered that DEW may have basically three classes of effects ranging as follows:

- Level 1: *A jamming effect*, i.e., time-limited disturbance of the satellite functioning that ceases when exposure to the weapon is over
- Level 2: *A disruption effect*, i.e., permanent disturbance (without definitive destruction) requiring an external intervention or reset
- Level 3: *An annihilation/destruction effect*, i.e., definitive disruption requiring an external replacement or repair at best. (This subjective scale can be paralleled to what has been almost theorized, or at least symbolized, in some US Air Force doctrinal documents using the infamous "5 Ds" to materialize the scale of gravity of any space attack: *"**D** eception, **D** isruption, **D** enial, **D** egradation, **D** estruction."* See USAF (2004), Counterspace Operations, Air Force Doctrine Document, 2-2.1.)

DEWs may have different effects according to their domain of functioning: For example, an intense laser ray has a thermomechanical effect on any material and as such can neutralize or destroy sensors or even some structures. By contrast, a microwave weapon would not have any thermal effect but would produce instead a high-power electrical effect on electrical components, whether directly or indirectly. Low-level components such as receivers or some class of sensors would prove particularly vulnerable to such threats. As envisioned by largely publicized projects very early on (such as the US Space-Based Laser project), equipping space platforms with powerful lasers for ASAT kind of activities might be theoretically possible with the objective to overflow or even destroy targeted sensors. However, aiming at sensors might not be an easy task, with the additional possibility of the development of self-protection devices for the most sensitive satellites.

The literature has frequently referred to powerful lasers in orbit, mainly inherited from the early R&D experiments engaged during the Ronald Reagan years under the auspices of the United States, so-called Strategic Defense Initiative (SDI) often dubbed *"Starwars."* (In this respect, it must be reminded that, at its apex, one of the several versions of this project was envisioning the deployment of many space and ground-based laser systems, possibly relayed by orbiting mirrors in order to destroy reentry nuclear heads. This complex network of sensors and effectors was considered as an addition to some more conventional 4000 intercepting "hit-to-kill" missiles or even satellites.) Laser-based ASAT developed under such concepts would be much more powerful with the objective to bring about mechanical destructions on the structure itself of space systems, most notably on deployed solar panels. Obviously, the development of such an armament would require much more energy generation that would make their development very problematic given the usual constraints applied to any space systems (size, weight, reliability). It is highly probable that these many technical constraints have largely put into question the development of such systems, even if it is probable that more or less secretive R&D has not ceased in this area. Following this logic, powerful microwave systems may represent a more threatening technology from an operational point of view than space-based lasers.

What Vulnerability, in Which Context? Very Different "Defensive" Situations

Of course, any offensive weapon will focus at the main vulnerabilities of spacecrafts. These vulnerabilities are usually related to support functions such as:

- Attitude control
- Tracking and telemetry
- Thermal management
- Power management

The dysfunctioning of any of these technical functions would generally mean a shutdown of the entire system in short or longer term. As a result, the attack modes may be very diverse, whether they involve the destruction of the solar arrays, the thermal increase of the satellite structure, or a cyber intrusion in automated management processes.

In addition, the vulnerability of any spacecraft can vary quite largely considering their very nature, the applied management processes, and even the very mission it has to fulfill. As an example, telecommunication satellites are controlled by multiple operators, private or public, which sell their services to many customers. In this particular case, many motivations can exist for attacking the space system, from a hostile action against a specific customer to a more "wide-range" terrorist-like attack. This means that the ways and means used for attacking the "satcom" function can be very different from an action to another, implying the need to protect many dimensions of a complex system. (Obviously, the uplink remains the targets of choice for any action against the satellite itself.)

As for the navigation satellite, their systemic redundancy makes them less an easy target. In this case, jamming may be used but this time with local effects, as it has been sometimes the case during the recent conflicts using GPS-guided munitions. In the case of Earth observation satellites, in addition to their highly critical pointing and control systems, their sensing payload and their downlink communication systems appear as high-value potential breaches. This vulnerability is indeed increased by their relatively few numbers and by the accessibility of the Low Earth Orbit (LEO) they usually make use of. Last but not least of this non-exhaustive list, the weather satellites, while mainly on the geostationary orbit, may also be vulnerable due to the reliance on the good functioning of their sensors as well as on their communication downlink capacity. It is obviously reinforced for those satellites that orbit on LEOs.

The Notion of "Space Threats" and Its Relevance for the Security of Space Activities

As just shown above, the notion of "space threats" as strictly defined by space systems posing a threat in orbit might not reveal itself as the most urging issue to tackle. Indeed, most of the space systems that might be considered as potential offensive candidates seem to remain fairly confined to the prospective horizon. From a technical standpoint first, using space systems as offensive weapons is not a simple operation. It involves relying on very demanding systems (in terms of sensing, maneuverability, energy management, cost, etc.) that may not make them so easy to produce and use. From an operational point of view also, this complexity may not be what a military user is looking for, notwithstanding the fact that, in the case of using offensive KEWs, the consequences of any attack will make no discrimination in the end between the victim and the attacker. For this reason, and from the policy perspective, it seems reasonable to put into question the very relevance of "threats in space" as a central notion for building the core of the future of space security. For sure, such a view does not imply that the international community should not pay attention to these developments. On the contrary, the fact that such techniques might be used one day should trigger a widespread awareness that in this field, earliest actions against the development of such weapons will be the most efficient. But in parallel, the rather prospective nature of these kinds of threats must not lead space-leading countries to underestimate the importance of other sorts of threats that may be much more meaningful on the shorter term. A brief (non-exhaustive) list of such threats may be recalled.

Ground-Based ASAT Tests

The most recent ASAT tests performed in 2007, 2008, and 2019, respectively, by China, by the United States, and most recently by India provide a good example of the practicability of and efficiency of ground-based ASAT missiles. As mentioned at the beginning of this chapter, the first one performed by China in January 2007 destroyed a decommissioned Chinese weather satellite on an 800 km circular orbit. A little bit more than 1 year later, the United States did hit a lower orbiting military satellite (246 km) with the stated goal to prevent an uncontrolled and dangerous reentry. In the first case, the interceptor used was a modified SC-19 missile, while the US military used the SM-3 sea-based intercepting missile developed for ABM purposes. Obviously, the proximity of anti-ballistic missile research and ASAT interceptors has been clearly apparent in all cases. While not completely known to date, the interceptor used by India has been officially acknowledged as a "DRDO's Ballistic Missile Defence interceptor (. . .) which is part of the ongoing ballistic missile defence programme" (Idem). It must also be recalled that China did several allegedly ABM high-altitude related tests in the aftermath of the 2007 ASAT experiment.

Even if it must be recalled that the targets were mainly cooperative and their trajectory well is known from their "attackers," these three cases have however amply demonstrated how much mastering space interception from the ground has become accessible to the most prominent ballistic and space powers.

Alleged Risks of "Cyberattacks"

Another type of risks, the "cyberattacks," has been alleged as becoming a major cause of concern for space systems. Cyberattacks can indeed take many forms and affect many elements of the entire space and control system. Tracking, telemetry, and control networks can be subject to such cyber-threat with the impossibility to transmit reliable data for the control of the satellite platform. As a consequence, any satellite can virtually be taken over by a non-authorized user who can force a system shutdown or a wrong maneuver leading the system to put itself in a safe mode or in any other uncontrolled mode. In theory, such a takeover can be implemented via cyber intrusions in the command center or through key ground stations. Awareness about the possibility of such attacks has increased over the recent years. In 2001, a NASA audit report pointed out that "six computers servers associated with IT assets that control spacecraft and contain critical data had vulnerabilities that would allow a remote attacker to take control of or render them unavailable." (The report goes on blaming that "moreover, once inside the Agency-wide mission network, the attacker could use the compromised computers to exploit other weaknesses we identified, a situation that could severely degrade or cripple NASA's operations." Source: NASA (2011).) These conclusions have been largely commented and have motivated the adoption of unprecedented protection measures for the Agency space systems.

A few recent cases have sometimes been cited that seem to sustain this assessment:

- Some reports have claimed that the German X-ray satellite ROSAT (made famous recently due to its uncontrolled reentry during the night of October 22–23, 2011) had been targeted in September 1998 by a cyberattack leading it to wrongly orient towards the Sun, ultimately causing its shutdown. This "wrong maneuver" (the cause of a loss of the satellite sensors) is reportedly related to a cyberattack carried out against computers of the Goddard NASA Center as unveiled in 1999 by one of the specialists in charge of the center computer services. At this time, the attack perpetrated against the X-ray department of the center was attributed to a Russian origin. However, those facts have never been confirmed, just a "troubling" coincidence between the move of the satellite and an intrusion in the computer system having been officially mentioned by the inquiry.
  Another satellite, INSAT-4B-S, this time a telecommunication satellite belonging to India has been mentioned as having been affected by a cyberattack (*Stuxnet Worm*) that would have caused a severe loss of power, ultimately leading to reduction of the telecommunication capacity of the satellite by more than 50%. (Again, this case has not been fully acknowledged, yet some other hypothesis (supported by ISRO) points out the loss of one of the solar arrays of the spacecraft. No official position about the incident has been confirmed up to this day.)
- Other examples have been cited in draft U.S. Congress reports citing interferences having affected the Earth imaging Landsat-7 satellite at least twice in 2007 and 2008, while another NASA EO satellite, TerraAM-1, experienced the same disruptions in 2008, for more than a single day in one occurrence. More recently, the U.S. National Oceanographic and Atmospheric Administration (NOAA) is reported having suffered a disruption of its Satellite Data Information System due to a severe hacking incident in September 2014. This implied for the Agency being denied sending weather forecast data for 48 h. (These episodes are

mentioned in Lewis (Patricia), Livingstone (David), "Space, the Final Frontier for Security," Research Paper, Chatham House, September 2016, p. 10. See also 2011 Report to Congress of the U.S.-China Economic and Security Review Commission, pp. 215–217 (https://www.uscc.gov/sites/default/files/annual_reports/annual_report_full_11.pdf).) Some other attacks occurred on NOAA satellite in October the same year, with press reports about Chinese hacking attempts. (See "Chinese Hack U.S. weather system- satellite network," Washington Post, 12 November 2014.)

Of course, another more generic risk is represented by a cyber intrusion in the information chain itself (data collection, processing, and dissemination) without affecting the satellite itself. This type of attack, even if indirect, may have consequences as serious as if the space segment itself was the target, for example, ending up with wrong data, unreliable imagery, and false alarms. While targeting only information, this type of intrusion can barely be deterred by strictly "space segment-oriented" defensive strategies and doctrines.

Cyberattacks will probably account for the most preferred offensive strategies when the objective will be to disrupt an entire "space system," especially when they are old generation, i.e., not protected against the latest software offensive devices. Here again, the capability to detect the origin of the attack and to attribute its responsibility will be the key for an effective deterrence strategy. At this level, there is no magic for space systems, and this type of vulnerability is essentially linked to a domain that remains partly external to the space sector itself.

The General Vulnerability of the Ground Segment

More generally, the ground segment represents a key node for ensuring the functioning of any space system. Losing the ground segment necessarily means losing the space segment. In theory, the consequences on the long term might be less definitive than when a spacecraft is destroyed, as regaining control on the operation of the space system might be possible once the functions of the ground segment have been recovered. Hence, losing the control of the ground segment might be considered as a reversible situation and might not imply the same kind of strictly deterring positions as in the case of the space segment.

However, the border between both situations may sometimes be very thin, as taught by a case occurred to Russian satellites more than 10 years ago, in May 2001. As reported at that time, a fire destroyed almost completely a main control station leading to a total loss of communication with four military early warning satellites placed on a highly elliptical orbit Podvig (2002). Only one satellite has been recovered after a while, the three others having derived well beyond their nominal position. Those remained well out of reach by their dedicated ground segment. This "ground" damage has then become irreversible for the space segment itself. It may even include risks for other spacecraft, proving at this occasion that the safety management of satellites may be key in collective space security.

Here again, protecting the ground segment against attacks or hostile actions does not directly imply the protection of the space segment only. It may involve some level of "systemic" thinking, some redundancies (ground and space), as well as some parallel hardening techniques (e.g., against high-power microwave devices or even to prevent possible EMP effects). In addition, the adoption of degraded modes must also be considered for any key node on the ground. It is important to note that such measures must apply to both military and civilian satellite owners to be fully efficient.

The Case of Orbital Hazardous Events: The Example of "Zombiesats"

A contrario, from April 2010 to January 2011, Intelsat, the largest telecommunication satellite operator, has lost control over Galaxy 15, one of about sixty satellites composing its geostationary fleet. This spacecraft has derived over a large portion of the geostationary orbit without offering any possibility for being recovered during that 8 month long time. This event has had a double consequence:

- An increased collision risk affecting the whole community of the satcom users, civilian, and military.
- A powerful jamming of satellite telecommunications as Galaxy 15 has kept on emitting at full power during the whole period of time. One of the most documented consequences was the loss of WAAS (the US regional GPS *Wide Area Augmented System* for improved satellite navigation) in Alaska.

The control of this satellite (quickly nicknamed "zombiesat" in the large amount of literature devoted to this case) has finally been recovered in January 2011 by Intelsat. However, this case has amply shown what kind of disturbances such an event can create with the necessity for operators to avoid possible collisions and interferences. (For example, it has been reported that, at this occasion, SES, the second largest geostationary satellite operator, had to proceed with many very precise maneuvers around some of its strategic orbital positions.) It shows how much non-intentional actions can also present serious threats to space security that do not clearly relate to deliberate actions. There may be a specific vulnerability in face of such "zombiesats" on the geostationary orbit due to the vicinity of the satellites around some key orbital positions. This must be taken into account as a complexity factor of the collective space security, as this makes disturbances rather quick to produce, intentionally or not, both for civilian and for military systems. It must be noticed that operators have seized the importance of such potential developments and have chose to share their knowledge by setting up a common database allowing them fostering early and precise coordination when needed. (Via the creation in 2009 of the Space Data Association, based on the Isle of Man. Obviously, considering the wealth of information contained in those databases, such a private initiative cannot be without consequences on the general management of international relations in space.)

The Jamming of Space Telecommunication from the Ground

Of course, last but not least, the simple jamming of space telecommunications by using ground-based devices must also be evoked in this list of "indirect" space threats. One of the most recent Iranian episodes (spring 2009) can be quoted as the Iranian government has decided to jam two satellites (*Hotbird 6/8 W6*, *Eurobird 9A/2*) managed by Eutelsat, one of the two major European telecommunication operators. The goal was then to prevent the broadcast of information perceived as contrary to the Iranian regime interest. The cost to access such technologies is relatively low at the level of a government, and these interferences remains sometimes hard to detect when they occur and in any case highly difficult to prevent. A contrario, the example quoted here, has shown that, for a time, operators themselves had been dissuaded to broadcast the controversial information (BBC and VoA notably).

All these examples show clearly that direct threats on space systems, as evoked in the first part of this chapter, do not represent the sole source of possible security breaches. They may not even appear as the most probable cause of space insecurity, at least for the short to middle term. The difficulty remains both the attribution of responsibilities and, more difficult even, the establishing of the intentional nature of any catastrophic event. Any questioning about the setting up of international regulation, whatever their form, or of some sort of "space deterrence" must take this complexity into account.

Some Effects on Space Deterrence: Protecting Against What Threat and/or Vulnerability?

In light of these possible developments, thinking about future threats on space systems means thinking about the probable nature of those treats as well as the kind of possible enemy using them. At first glance, the most developed spacefaring nations have used their space assets in a strongly asymmetrical context in which only a few countries were able to use similar orbital systems, possibly in a hostile way. However, it is probably necessary to take into account other kind of threats that countries on the verge of becoming space powers might likely use in case of political or military showdown.

Generally, deterring any threat to develop against space assets will imply a large appreciation of this diverse nature of possible threats, whether intentional or non-intentional. This approach will probably go through a few preliminary protective postures and actions:

- **Establishing the capability to attribute an effect to a certain cause:** This capability, addressing either intentional or non-intentional threats, relies on very specific technical capacities whether they aim at monitoring LEO or GEO orbits. But, in parallel, according to the nature of the threat in orbit (KEW, DEW, Jamming, etc.) or from the ground (using the same kind of techniques in a different way), very different means will have to be implemented to protect the satellites. Some strategies may envision having on-board devices allowing detection (and characterization) of laser attacks, for example. This may bring about a certain deterrent effect against an adversary who would rather have acted stealthily. Some other will possibly envision satellites more directly dedicated to detection and inspection. Of course, ultimately, these "defensive" systems may appear in reverse as potentially challenging this quest for permanent capacity to attribute any event to a certain cause. Indeed, by definition, such protective devices would make use of technologies that may allow discreet and more offensive actions. This is not the least of the paradoxes that such efforts would imply.

- **Creating a "red line" against any attack:** Provided the cause of any event solidly established, the difficulty remains to establish a sort of "red line" beyond which military protective action would be legitimate. First, characterizing between the intentional or the non-intentional move will be key in determining the reaction of the "victim." There probably lays the most difficult issue to tackle when it comes to ensuring a comprehensive protective posture (including military) against any threat on space systems. It must be noted that even in the case of a recognized intentional action, the possible "graduate" nature of the hostile action (from deception to *destruction* to recall the "5 Ds" approach) may render difficult any decision about the nature of the counteraction itself. This aspect may be at the center of the current effort to establish "rules of the road." No doubt that it will also raise expectations about the resistance capacity of the next-generation space assets. This is the approach followed for the hardening of the electrical components, for example, with two (possibly contradictory) principles. Making well known that the considered system has been hardened while, at the same time, keeping any possible adversary in the impossibility to determine the methods and the techniques used, as well as the very level of this hardening.

Conclusions

In any event, the road towards limiting by principle threats on space systems in a significant manner will probably remain quite bumpy for a while.

At this stage, satellites have remained vested with a highly symbolic value that continues to put them at the center of the current strategic relationships. The latest events (comprising the March 2019 Indian ASAT test as well as the Chinese ASAT in 2007 and in a way the US-made satellite destruction in 2008) have shown that affirming this kind of capability was also a part of "deterrence" postures or "state communication policies." It is well documented that satellites will become smaller and smaller, more and more able while less and less costly. The generalization of smaller high-performance spacecraft (whether military or civilian), possibly "launched on demand," announces the beginning of a new era for which a new equilibrium will have to be found. These progresses, sometimes promoted through concerted national efforts, are also a part and parcel of the "equation" aiming at balancing the protective approach with bolder technology-led solutions that are supposed to give an edge to the more advanced space countries. (Such as in the case of the US *Operationally Responsive Space* program, for example, even if this effort seems to remain in question nowadays.) Answering this question and finding a workable balance will determine the fate of our collective security against the threats on space systems as well as it will orient the future nature of a possible "space deterrence."

References

Baker D (1985) The history of manned spaceflight. New Cavendish Books, London

Caldicott H, Eisendrath C (2007) War in heaven: the arms race in outer space. The New Press, New York/London

Department of State (2009) Foreign relations of the United States, 1969–1976, vol E-3, documents on global issues, 1973–1976. United States Government Printing Office, Washington, DC

Gantz K (1958) The United States air force report on the ballistic missiles. Doubleday & Comp, New York

HALEOS, High-Altitude Nuclear Detonation against Low Earth Orbit Satellites (2001) Defense threat reduction agency briefing. http://www.fas.org/spp/military/program/asat/haleos.pdf. Accessed Aug 2012

McAllister WB (2009) Foreign relations of the United States, 1969–1976, volume E-3, documents on global issues, 1973–1976. United States Government Printing Office, Washington, DC

NASA (2011) Inadequate security practices expose key NASA network to cyber attack, office of audits, Washington, DC. For the complete audit document, see http://oig.nasa.gov/audits/reports/FY11/IG-11-17.pdf. Accessed 20 July 2012

Podvig P (2002) History and current status of the Russian early warning system. Sci Glob Secur 10:10–60

Schriever B (1957). http://www.af.mil/news/story.asp?id=123040817. Accessed Aug 2012

Stares P (1985) The militarization of space, U.S. policy, 1945–1984. Cornell University Press, New York

Stares P (1987) Space and national security. The Brookings Institution, Washington, DC

USAF (2004) Counterspace Operations. Air Force Doctrine Document 2–2.1. http://www.dtic.mil/doctrine/jel/service_pubs/afdd2_2_1.pdf. Accessed Jan 2010

European Space Surveillance and Tracking Support Framework

46

Regina Peldszus and Pascal Faucher

Contents

R. Peldszus (✉)
Department of Space Situational Awareness, German Aerospace Center (DLR) Space
Administration, Bonn, Germany
e-mail: regina.peldszus@dlr.de

P. Faucher
Defence and Security Office, French Space Agency (CNES), Paris, France
e-mail: pascal.faucher@cnes.fr

© Springer Nature Switzerland AG 2020
K.-U. Schrogl (ed.), *Handbook of Space Security*,
https://doi.org/10.1007/978-3-030-23210-8_104

Abstract

In order to safeguard European space infrastructure and contribute to global burden-sharing in the domain of Space Situational Awareness, the European Union set up a dedicated capability for Space Surveillance and Tracking (EU SST). Operational since 2016, EU SST employs a novel governance model that joins existing sensors of European Member States and provides collision avoidance, reentry and fragmentation analysis services to the European user community. This chapter provides an overview on the formation, internal governance, and operations of the current European Space Surveillance and Tracking Support Framework and offers a perspective on its future as part of the proposed EU space program.

Introduction

The capability of Space Situational Awareness (SSA) – monitoring and understanding real-time and foreseeable developments of activity in the orbital environment – has become a fundamental prerequisite for space security, both as a means for transparency, information advantage, and verification and as operational foundation for the resilience of space-based systems. While the major space actors maintain SSA capabilities to varying degrees (Lal et al. 2018), the US Strategic Command currently operates the most sophisticated SSA capability with dedicated data exchange arrangements to external state and non-state actors (West 2018).

Yet, not only is the safeguarding of the near earth environment as a global commons an international concern in the advent of an unprecedented scale and variety of activities across orbital regimes. The necessity for significant investment into development and operation of SSA sensor systems also lends itself to collaborative rather than unilateral efforts (McCormick 2013). In Europe, despite a range of sensor systems, governmental and commercial actors have been relying predominantly on SSA data provided by the United States. However, it has become paramount for Europe, particularly institutional actors such as the European Union, to protect their own space-based infrastructure, such as the Galileo navigation system. Furthermore, reaching a greater level of autonomy in space surveillance through a dedicated capability allows Europe to contribute to global burden-sharing in the domain of SSA and to enhance its position in international discussions (Pellegrino and Stang 2016, pp. 47–48; Dickow 2015, p. 123).

Hence, in recognition of the need for non-dependence in detecting risks in orbit and the opportunity offered by existing SSA legacies and advanced sensor capabilities in Europe, the European Union has since 2014 undertaken concrete efforts to consolidate a SSA capability. Initial focus was placed on one of the key elements of SSA (cf. European nomenclature (Rovetto and Kelso 2016)), Space Surveillance and Tracking (SST): sensor systems, such as radars, telescopes, and lasers, deliver data on the location and behavior of active and non-active spacecraft or debris, based on which risks can be evaluated on objects that are at risk of collision, have fragmented, or will reenter the atmosphere. By incentivizing the networking of hitherto discrete sensor assets already operated by separate entities in Europe, the European Space

Surveillance and Tracking Support Framework (EU SST) today constitutes the EU's primary SSA capability.

This chapter provides an overview about context, implementation, and perspective of the European Union's Space Surveillance and Tracking Support Framework. Following an outline of the political developments leading up to its formation, an overview on the internal governance of EU SST and its organization as a cooperation of EU Member States and a council agency is provided. Subsequently, its three operational functions are described that network SST sensors process SST data and form the basis for SST services to a varied user community. The chapter concludes with a perspective on how EU SST is bound to develop as part of the upcoming space program of the European Union and how it slots into to the emerging technical discourse on future space traffic management and coordination regimes.

Governance

Background

The notion of devising and implementing a dedicated European SST capability emerged in the decade preceding the adoption of the legal basis in 2014. European actors – with impetus from the new space policy mandate of the Lisbon Treaty – became either more aware of pending space security challenges in general or had already begun specifically to tailor, develop, or analyze possible responses to the safety and security challenges in the space domain. Increased non-dependence from US capabilities and contribution to global burden-sharing was already a central consideration, including the necessary trade-off between sovereign requirements and a degree of collective autonomy.

In response to a changing security environment after the Cold War, the European Union had begun to assess emerging drivers for future crisis management and related space operational requirements. In the string of critical space-based services including Earth observation (EO) and position, navigation, and timing (PNT), a space surveillance system was highlighted as a fundamental capability gap and priority of a future European space program (Commission of the European Communities 2003; SPASEC 2005). This was understood both in the context of more independently meeting the needs of burgeoning space efforts in Europe but in turn also to contribute to global burden-sharing, particularly with regard to efforts of the United States (Conclusion of the Workshop on Security and Arms Control in Space and the Role of the EU, 2007, cited in ESA 2008, p. 6). A Space Council Resolution then underlined the need to fill the capability gap of SSA, in order to accommodate safeguarding of space-based assets and infrastructure (navigation and observation) on the one hand and enable non-compliance and treaty verification on the other (Council Resolution 2008).

In view of this, the need for cooperation across multiple actors – rather than unilateral efforts – including across the civil-military intersection had already been recognized as a necessary key characteristic of emerging SSA efforts of European state actors and supranational and intergovernmental bodies (Spasec 2005). Of the European Union Member States, only France, the United Kingdom, and Germany had operational space surveillance capabilities at the time (Pagkratis 2011, p. 98; Veclani et al. 2014, p. 39 and 40), which were integrated or coordinated to some extent with European or transatlantic partners. A UK asset (Fylingdales) was part of the US Space Surveillance Network (US SSN). A bilateral Franco-German use of space surveillance and tracking radars (GRAVES and TIRA) had been underway since 2006 (Pasco 2009, p. 34; McCormick 2015, p. 46) and was explored as a potential model for the pooling of capabilities in a broader European framework (Robinson 2011, p. 21) or for possible approaches to data exchange or processing (European Commission 2013b, p. 13).

Meanwhile, the optional program for SSA at the European Space Agency (ESA) commenced work in 2009 (ESA 2008) borne out of earlier efforts in determining needs and requirements of a SSA system with a variety of stakeholders (Marta and Gasparini 2008, pp. 139–40) (Many of the participating countries were both ESA and EU Member States.). In this context, a range of studies explored possible approaches to a European SSA system: an architecture analysis examined three options of ground- and space-based SSA infrastructure, highlighting the merits of synergies of a comprehensive approach to including the monitoring of space objects and the space environment (Donath et al. 2009) (N.B. the different nomenclature of SSA in Europe to include SST, SWE, and NEO (the latter elements summarized as space environment in non-ESA contexts), in comparison to US nomenclature that included intelligence, surveillance, reconnaissance, environmental monitoring, and command and control (cf. Rovetto and Kelso 2016, p. 3)). Another comparative study on governance explored existing and potential models relating to organization and data policy, examining different models for decision-making including program-oriented and institution- or policy-oriented in the European Union (Pasco 2008). In trading off the constraints of national security interests with the potential of collective effort, it sketched the potential setup of a European supervisory authority overseeing the work of a managing organization, already accounted for different security requirements of the two central missions of space object surveillance (SST) and space environment monitoring (space weather and NEO).

At the same time, dialogue among the Member States of both EU and ESA was ongoing as to which body would be best suited to account for the political and practical security requirements of individual space actors involved in collective SSA efforts in Europe (Chow 2011, p. 11; Nardon 2007, p. 6). With the increasing involvement and consolidation of the European Union in space governance (Mazurelle et al. 2009, p. 18; Council Resolution 2010), additional EU bodies entered the debate. The EU's European Defence Agency (EDA) defined military user requirements (EDA 2010), which were later fused with civilian mission requirements in a dedicated set of high-level SSA civil-military use requirements (Council of the European Union 2011). Consultations between the EU and EU Member States' space agencies and ministries from 2009 onward had affirmed the emerging role of the EU as actor in space security matters and SSA in particular and were accompanied by further external studies on overall governance (European Commission 2013b, pp. 6–7).

As part of the continuing momentum for a European endeavor, a consensus emerged that an operational SST system be situated within the remit of an effort established and managed by the EU (Veclani et al. 2014, p. 40 and 74; Space Security Index 2012, p. 49–50). The de facto shift of the constituent national actors towards a setting as part of the European Union was formalized as part of the Space Council Resolution of 2010. It subsequently entailed a divestment from operational SST elements of ESA's SSA program (McCormick 2015, p. 48; ESPI 2018, p. 54), which continued with renewed focus on research and development for SST and on space weather and near-Earth objects (Flohrer and Krag 2017).

Interactions were also extended beyond Europe to inform and engage with regard to emerging governance options. Existing bilateral channels on SSA between EU Member States and their US counterparts (Robinson 2011) were complemented by dedicated interactions on the EU side. In 2012, the US Department of State hosted an EU-US workshop on critical infrastructure protection and SSA with representatives of the European Union (lead by Directorate General for Internal Market who was coordinating internally and other organs including the EEAS, EU Satellite Centre, and EDA, among others), as well as ESA on the US side including also the Department of Defense (Vittet-Philipp and Savova 2012, p. 4).

Discussions on developments continued in the EU community. A proposal for a SST framework was prepared (European Commission 2013a), which posited a governance model of European Member States and highlighted the potential benefits and security-related challenges. A dedicated impact assessment laid out the results of a consultation process and the conclusion already foreshadowed earlier that "European SST services" ought to be led by the EU rather than ESA (European Commission 2013b, p. 22). In exploring intergovernmental governance models other than those of the existing EU flagship programs, the assessment highlighted potential Member States for participation in a governance model (Next to FR, UK, and DE also IT and ES), who could bring sensors from the national side or developed as part of the framework of the ESA SSA preparatory program, but it also emphasized the need for dedicated networking and upgrading of hitherto separate capabilities (ibid, 2013b, p. 12–13). An initial appraisal of this work by the European legislative then considered different levels of engagement between Member States and EU organs (European Parliament 2013). The EU Satellite Centre, which was discussed as an entity contributing to service provision (European Parliament 2014, p. 11), had already been working on the exploration of potential elements related to the user interface of European SST services (Chatard-Moulin 2013, p. 12).

In view of a wider understanding of space-related risks, a nascent EU SST Support Framework was understood as a countermeasure to the emerging risk of space debris. It was added to the inventory of safeguards against risks to critical infrastructure (European Commission 2014a, p. 56) and highlighted as a necessity for the promotion of autonomy and security of European space-based services and for shared threat assessment (EU Global Strategy, 2016, p 42 and 45).

Legal Basis

Finally, in 2014 the European Parliament and the Council adopted the legal basis establishing a "Framework for Space Surveillance and Tracking Support" (European

Parliament and European Council 2014). Articulating the need for a safe, secure, and sustainable orbital environment and the need for resilience of European space-based infrastructure, the decision aimed at "ensuring the long-term availability of European and national space infrastructure, facilities and services which are essential for the safety and security of the economies, societies and citizens in Europe" (Art. 3). To this end, a "SST capability at European level and with an appropriate level of autonomy" was to be put in place (Art. 4). This was to include the establishment and operation of three functions, including a "sensor function consisting of a network of Member State ground-based and/or space-based sensors, including national sensors developed through ESA, to survey and track space objects and to produce a database thereof"; a processing function to "process and analyze the SST data at national level to produce SST information and services for transmission to the SST service provision function"; and a service function that would provide collision avoidance, reentry, and fragmentation analysis services (Art. 5.1) to entities of the European Union, its organs, Member States, and industry (Art. 5.2) (see section "Users").

Funding of 70 million Euros for a 5-year phase between 2016 and 2021 was set aside in parts from the Galileo and Copernicus funding streams for operational aspects, with additional funding earmarked for sensor upgrades from the EU research and development program, Horizon 2020, while significant previous investments of the participating Member States networking their assets were considered a prerequisite. The framework was designed not as an individual capability building exercise but as providing support to the networking and enhancement of existing and already emerging capabilities (That is until the start of the next so-called multiannual financial framework 2021–2027.).

A Consortium of European Member States in Cooperation with Council Agency

In order to participate as sensor and service providers, Member States of the European Union with ownership of or access to requisite SST sensors were invited to express interest and apply to the European Commission via their so-called national designated entities (in most cases space agencies) and, upon demonstrating eligibility, to form a Consortium.

Subsequently, in response to Decision 541 of the European Council and the European Parliament (European Parliament and European Council 2014) and the related implementing decision (European Commission 2014b) that set out formal application procedures, five Member States applied (France, Germany, Italy, Spain, and the United Kingdom). Each of their respective national designated entities – space agencies or their equivalents (CNES, DLR, ASI, CDTI, and UKSA) – submitted individual applications to the European Commission for joint participation in the SST Support Framework, demonstrating compliance with the criteria set out in Decision 541 and security aspects (ibid, p. 3). After an assessment by the European Commission, the five applications were deemed compliant and proceeded, according to the SST decision and implementing decision, to form a Consortium (European Commission 2014b).

A Consortium Agreement was signed by the heads of the participating space agencies in June 2015 (Via Satellite 2015; European Commission 2018a, p. 3). Encouraged by the decision, the partners of the SST Consortium collaborate with the EU Satellite Centre (EU SatCen), as an agency of the Council. An implementing arrangement was thus signed between the Consortium Member States and the EU SatCen in September 2015, thus formally constituting the SST Cooperation (ibid).

A tight timeframe between legal basis and proposed commencing of activities obliged the five-plus-one partners to proceed swiftly in setting up practical arrangements, applying formally for the requisite funding instruments earmarked by the European Union for this purpose, laying out technical activities for the initial services phase, and allocating resources among them in parallel (de Selding 2015). The activities were formally launched by the European Commission in January 2016 (European Commission 2018a, p. 3). This coincided with the aims set out in the 2016 Space Strategy, including a swift progression to reinforce and enhance the nascent SST activities further in view of the resilience of European space-based infrastructure (European Commission 2016, pp. 9–10).

A preparatory phase of six months focused on the fundamental coordination and preliminary joining of a distributed infrastructure, which involved the development of a service portfolio, an initial data and information policy, a model for internal burden-sharing in operations, the setup of front desk, and the acquisition of a first user cohort. Initial operations of the three SST services then commenced in July 2016 (European Commission 2018a, p. 4, 6; see also section "Services").

As early as spring 2016, additional European Member States expressed an interest in joining the newly formed Consortium and started interactions. With an implementing act for an additional cohort adopted in late 2016, proceedings for the accession of further Member States commenced in 2017. Eleven Member States entered into a dialogue with the Consortium, of whom three decided to apply to the European Commission to join: Poland, Portugal, and Romania. They were found eligible in 2018 and formally completed the accession process with a fresh set of agreements in 2019. Eight other MS expressed an initial interest in participation in the SST Support Framework but refrained from an application to the Commission (European Commission 2018a, p. 8).

Internal Governance

The SST Consortium is neither a legal entity nor an EU agency but constitutes a formal cooperation of the space agencies of the participating EU Member States. Despite different capabilities and legacies in the realm of SST, the partners agreed to commence their collaboration on quasi-equal footing with regard to voting rights and budget allocation, in order to cultivate a culture of cooperation and consensus necessary for joint efforts at the intersection of policy and operations.

The EU SST framework employs a novel governance model for space cooperation in Europe that differs from current governance models of communautized
efforts or previously trialed public-private partnerships, of the main EU flagship
programs for navigation, Galileo, and Earth observation, or Copernicus, which
represented either new capabilities that formerly did not exist or were not available
at the required scale.

The internal organization of the SST Cooperation (the Consortium and SatCen)
includes three layers – decision-making, management, and working levels – that are
jointly implemented by all partners in cross-agency teams. On the bottom level, the
working level is structured into operational activities (see section "Operations"). A
management layer coordinates the execution of the activities, handles administrative
matters relating to finances, and grants and reports on administrative matters to
European Commission and Research Executive Agency. On the decision level, the
governance of the Consortium is executed through three committees: the Steering,
Technical, and Security Committee. Each committee is formally staffed by two
delegates from each Member State, in some cases a civilian lead from the participating space agency and a second representative from the armed forces or ministries
of defense. The Security Committee handles all matters regarding data policy and
security assessment and oversees matters relating to operational risk. The Technical
Committee addresses all operational, research and development matters. Both committees provide direct input to the working and operational levels of SST that are
organized as working groups for specific activities (Gravier and Faucher 2018).

The highest-level decision-making body of the Consortium is the Steering Committee, which is responsible for all aspects of policy and strategy, decides on budget
allocation, and guides operational and technical activities. Its Chair and Secretariat –
who do not represent their Member State in order to afford a degree of neutrality –
are supported by a Co-Chair and maintain all external dialogue, both with the
European Commission and international partners, and represent the SST Cooperation formally in the SST Committee. (The first chairmanship term was served by
Germany from 2016 to mid-2017, co-chaired by the United Kingdom, followed
by France from mid-2017 to 2020, co-chaired by Germany and the United Kingdom
until end of 2018 and co-chaired by Germany since early 2019). The Steering
Committee also forms the core of a dedicated forum for exchange with the SatCen
(Coordination Committee) that addresses matters specifically to service provision
and front desk activities. A representative of the European Commission usually
observes the meetings of the Steering Committee. The committee meets usually
monthly for several days, in some phases quasi fortnightly or weekly. It takes
decisions by consensus but can vote with qualified majority to avoid impasse.

Several changes in the internal governance were initiated by the Consortium and
the European Commission in the course of the first two years of operation. Beyond
the formal accession procedure of new partners managed by the Commission, these
included measures countering the complex funding arrangements that proved admin-
istratively cumbersome for both the European Commission and the partners (i.e., the
regular reapplication for and simultaneous management of several 18-month-long
grants across three budget lines). Initiated internally by the Consortium, voting rights

in the governance committees evolved from unanimity to qualified majority. The Consortium also facilitated a greater involvement of representatives of the European Commission in internal governance as observers in the Steering Committee since early 2017 (European Commission 2018a) and fundamentally changed the operational setup of the service provision model (see also section "Services").

Interaction with European Union Stakeholders

The European Commission is formally responsible for facilitation of the implementation of the framework, interacting through regular meetings with the SST Cooperation, drawing up relevant coordination plans, and monitoring the execution of the grants in collaboration with the EU's Research Executive Agency (European Commission 2018a, p. 8). Due to the supplementary nature of the support framework – rather than a fully fledged program for now – the interaction between the SST Cooperation and its EU Stakeholders is not characterized by a classic customer relationship; notwithstanding, the SST Cooperation reports formally to the Commission and REA through the mechanisms associated with the funding instruments, as well as informally through regular, generally monthly meetings.

A further reporting line from the Consortium to its EU stakeholders – which surfaced as a fixture in the formal committee context when the Consortium became operational – is set up via the SST Committee. The SST Committee is a body of 28 EU Member States that monitors the implementation of the SST framework. Chaired by DG Grow, the committee serves as the forum for presentation, discussion, and adoption of proposals for the implementation of the SST framework. It usually involves briefings by the Commission, invites regular reports by the Consortium, and allows MS to discuss progress and initiatives. The forum regularly includes observers, both from EU organs (i.e., European External Action Service, European Defence Agency) or external entities in view of specific subject matter. Two further fora for SST experts and users allow for periodic discussion on specific topics relating to requirements and needs.

Transatlantic Relationship

Since the first early exchanges between European and US stakeholders on emerging SSA capabilities in Europe, US actors have been observing the developments of European SSA closely. From a transatlantic perspective, engagement between Europe and the United States is constituted on three levels.

The Member States of the Consortium have each retained – some long-standing – bilateral relations with the United States, both through their space agencies and, crucially for SSA cooperation on policy and operational level, through their ministries of defense and armed forces (e.g., for data sharing arrangements and liaison officers, see also section "Data Processing and Data Policy"). These bilateral relationships, in turn, extend to the multilateral level and form the engine for the transatlantic relationship between the SST Consortium as a whole and its US counterparts. As a multilateral group, the Consortium maintains a regular exchange on working level with representatives from the United States (cf. also section on STM). On supranational level, the EU-US space dialogue led by the European Commission does not yet regularly include EU SST representatives in its delegation but has grown to take into account the perspectives of EU SST for discussions related to SSA and STM, either by including relevant reporting through Commission officials or, more recently, through invitation of the chair of the Consortium per se.

Operations

After a preparation period of 6 months from the activation of the financial instruments and budget lines, EU SST started providing operational services as a joint effort on 1 July 2016. While its governance is managed as a hybrid of virtual and co-located interaction, EU SST service provision is managed as a distributed European ground segment system incorporating the major functional elements set out in the legal basis of Decision 541.

Service Provision Model Based on Internal Specialization

Operations consist of three main functions as per Decision 541, which make up the operational elements of the SST capability (i.e., aside from the governance elements described earlier): sensor, data processing, and service functions. SST sensors from all partners contribute data; this data is analyzed in the processing function and feed a joint database and ultimately a catalogue; from this, products are derived for three services that are generated by the operation centers and passed on to the users via a front desk.

In the interest of a swift progression to providing services to the EU user community after the entry into force of the legal basis in 2014 and the formation of the SST Consortium in 2015, the service provision model at the outset of operations in 2016 mirrored the equal footing approach of the participating partners: the CA service was operated on a fleet allocation principle, i.e., spacecraft or fleet was distributed to different operation centers (OCs), and services provided according a common guaranteed baseline with added-value elements at the discretion and capability of each operation center. At the time, RE and FG services were provided on a monthly rotational basis by all centers.

This provisional philosophy of each partner performing all functions was superseded in spring 2018 by an approach of specialization that is used to date, where each Consortium partner performs a pre-defined subset of the overall functions. Thus, pairs of OCs work in hot redundancy for the CA service, while all partners contribute offline and collaborate on post-analysis of events. The evolved service portfolio harmonizes products for all services. This internal burden-sharing constitutes a model of functional specialization, whereby specific partners are responsible for European level service provision (the French and Spanish OCs for collision avoidance service, Italian OC for both reentry and fragmentation) and data processing side (German OC), while all individual partners contribute sensor data.

The current EU SST capability uses existing assets (sensors and operation centers) that are virtually coordinated and complemented through joint efforts in combining discrete elements into an overall value chain. Across the three functions, operations are led by the operation centers of the participating Member States (COO for France, GSSAC for Germany, UKSpOC for the United Kingdom, ISOC for Italy, S3TOC for Spain, SSAC-PL for Poland, COpE for Portugal, and COSST for Romania). These are civilian, military, or civilian-military and may integrate capabilities or expertise of additional actors of the SSA ecosystem in Europe, including industrial subcontractors or scientific institutes for development, staffing, or operation of selected infrastructure.

Sensor Network

The EU SST sensor network has grown since initial operations, from 33 sensors in December 2017 of the initial Consortium of 5 partners (European Commission 2018a, p. 4) to a current total of 51 sensors made available for operations by its 8 partners. The network currently comprises of 12 radars (5 surveillance, 7 tracking), 35 telescopes (19 surveillance, 16 tracking), and 4 laser ranging stations. They provide coverage of all orbit regimes (LEO, MEO, HEO, and GEO) (European Commission 2018a, p. 4). The majority of sensors, including all radars and lasers, is located on European landmass (with the highest latitude site being Fylingdales in the United Kingdom); over a dozen additional telescopes afford coverage through locations in other geographical regions, including overseas territories or sites accessible through partners in the southern hemisphere. No space-based sensors are currently part of the system. Some Member State partners integrate assets or data from commercial or private entities (European Commission 2018a, p. 8). With the sensor network enlarged by additional Consortium partners, operational reviews are conducted annually to assess performance and contribution of each sensor.

The sensors operate at varying degrees of availability that are pre-defined and traded off among the partners. Control and tasking are retained by the respective Member States through their operation centers. On a network level, interactions are coordinated through the participating operation centers, which may send and receive tasking requests from other OCs and convene remotely for regular briefings.

The data provided from the sensors are shared either routinely in quasi real time or on request, depending on sensor type. They complement the data received through the 18th Space Control Squadron of the US Strategic Command in Vandenberg Air Force Base that are accessed through bilateral sharing agreements (for classified and nonclassified information). European measurements are thus systematically provided for the operation of each service, with dedicated campaigns performed for specific events: for collision avoidance services, in case of high-interest events

(HIE) and upon request of the nominal OC in charge of the service, all tracking radars and surveillance and tracking telescopes are activated to refine the orbit of the secondary object. For reentry services, upon request of the nominal OC, all tracking radars are activated to follow the reentry and provide European measurements; for fragmentation services, still upon request of the nominal OC, all radars and telescopes are tapped into according to their orbit regime.

Data Processing & Data Policy

The data derived from the contributing sensors is shared according to pre-defined principles. For surveillance telescopes, data is shared on a daily basis; for tracking radars, tracking telescopes and lasers in quasi real time (daily or on request); and for surveillance radars on a daily basis.

During initial services, data exchange was performed manually between the operation centers. Today, the growing network of sensors, and the increase in European data necessitated a dedicated platform for the ingestion and exchange of data and for further processing in view of setting up a European catalogue. Measurements from the sensors are hence fed into, stored in, and shared via a common platform so-called European database. After development work, the database went operational in April 2019 and is hosted, in line with the service provision model of functional specialization, by one of the operation centers (German OC).

The database constitutes the basis for building and maintenance of the precursor European catalogue currently in development. The operation centers providing the SST services will use the European precursor catalogue for service provision, in complementation of CDM received from the United States.

One of the main challenges posited for the Consortium at the outset was the design of an effective data policy due to the sensitive domain of SSA (Marta 2015, p. 9 and 10). The data currently handled on a multilateral level inside the Consortium is not classified, while any required filtering is performed on the sensor side. For the sharing and exchange of data for operational purposes, the Consortium operates on the basis of a dedicated data and information policy drawn up prior to initial operations in 2016 and reviewed in 2018 by the requisite internal body, the Security Committee. Since 2019, a fresh review has been underway to revise the documents in view of the requirements posed by an enlarged Consortium with additional security constraints of the individual partners.

In absence of a multilateral sharing agreement, the architecture of existing bilateral arrangements relevant to the exchange of SSA data are being taken into account. The data and information policy specifically address interactions for the purposes of EU SST by the Consortium partners. Initiating and concluding general SSA data sharing arrangements are not within the remit of the Consortium but the prerogative of the individual partners' military and national security stakeholders.

All Consortium MS have general security agreements for the protection of classified material with each other, constituting the prerequisite for any exchange of classified documents or other material. In the past few years, all partners concluded bilateral data sharing agreements with the United States for bilateral exchange of unclassified SSA data or are finalizing their respective arrangements (US STRATCOM 2019). France, Germany, and the United Kingdom have bilateral SSA data sharing agreements with the United States that also covers the exchange of classified SSA data, in addition to liaison officers of each of their respective Air Forces situated at the US Joint Force *Space* Component Command (*JFSCC*), formerly Joint Functional Component Command for *Space*, Vandenberg. Finally, two Member States of the Consortium, France and Germany, are currently the only partners to have concluded a bilateral SSA data sharing agreement with each other, covering the exchange of both unclassified and classified SSA data.

As approaches to data sharing are currently being further consolidated inside the Consortium, they must reconcile operational needs and individual security constraints with the features afforded by the evolving data processing function, and the general developments on a global level (i.e., developments on the US side with open architecture for data sharing; cf. also below for STM). Specific technical exercises on data sharing are therefore currently underway with relevant partners in the United States.

Services

Since 2016, the SST Cooperation has been providing three operational SST services as outlined in the Decision 541: collision avoidance service, reentry analysis, and fragmentation analysis service.

The collision avoidance (CA) service is provided by the French and Spanish OCs (COO and S3TOC) operating in hot redundancy. Satellites or fleets of satellites are allocated when a user registers for the service via the front desk (see section "Users"). The CA service constitutes an added-value service, whereby several sources of data are used: conjunction data messages (CDM) from the United States and CDM generated autonomously from European sensor data, in addition to ephemeris provided by the spacecraft operator. The EU SST service complements US CDM through automatic acquisition, checks, and analysis of all incoming information; it provides alerts based on a threshold pre-defined by the registered user on three levels of risk (high interest, interest, information); and issues recommendations for avoidance maneuvers. Observation campaigns are conducted for high-interest events (HIE). There is direct interaction with the spacecraft operator to support decision-making as to whether to take an avoidance action. For LEO and GEO, timelines of identification of incoming risks, tasking requests to EU SST sensors and EU SST CDM generation, and notification of operators range from 7 to 14 days prior to the time of closest approach (TCA). Products provided to the operator include information on events, objects, miss distance, etc. They are complemented where needed, based on a dialogue with users, by mitigation recommendations, maneuver design, and maneuver support.

The reentry (RE) service is under the responsibility of the Italian Space Operations Centre (ISOC). It consists of two main products, a list of upcoming reentries and reentry warning reports. Updated every 2–3 days, the list of upcoming reentries covers objects for which a reentry epoch within a period of up to 30 days has been computed, including object name, type, maximum latitude, size, and reentry prediction date range. Reentry warning reports are provided at least 3 days in advance of risky reentries and are updated as needed, including a final report to confirm the reentry. The user is actively notified and provided with information including details of the reentering object, estimated ground track, and uncertainty window.

The fragmentation (FG) service, also handled by ISOC, addresses fragmentations in orbit and consists of two products, a short-term and medium-term report. The short-term report is provided as soon as possible after a fragmentation event and includes the number of detected fragments and type of fragmentation. A medium-term report provides all additional available information including additional objects, orbit data, Gabbard diagram, and fragment cloud distribution and evolution. A current review of the fragmentation alert services foresees the addition of further features and a long-term analysis product of the event.

Users

The products related to all three services are generated by the operation centers and passed on to the user through the service provision portal operated by the SST front desk, which is under the responsibility of the EU Satellite Centre. The services are free of charge, are available 24/7, and are currently accessible to European users, including organs of the European Union (European Commission and Council of the EU, European External Action Service), public and private spacecraft owners and operators, and European Member States and their research institutions, civil protection authorities, and space agencies. In view of some interest expressed by specific non-European users, discussions are ongoing on the oversight level of the program on EU level for potential future inclusion of external users.

Since the start of operations in July 2016, user uptake has grown to a current total of 104 registered users from 60 organizations in 18 EU Member States as of September 2019. The collision avoidance service has 43 registered users from 21 organizations (whereby, a small number are still completing the interface control documents related to the registration process). EU SST hence currently protects 129 spacecraft across all orbital regimes (40 in LEO, 30 in MEO, 59 in GEO) from the risk of collision. These users include European constellations such as the Galileo satellite navigation fleet, as well as fleets by commercial communications providers, military assets, and spacecraft operated by governmental entities. While some users use the products to corroborate existing processes, most integrated the SST collision avoidance service as an integral part into the value chain of routine operations (Monham 2018). For reentry analysis, there are 71 users from 47 organizations and for fragmentation analysis 60 users from 40 organizations.

As part of a so-called user interaction mechanism overseen by the European Commission and handled by the EU SatCen, regular user feedback is integrated in service portfolio development.

Perspectives

Alongside the paradigmatic change in orbital utilization in the current decade, the global landscape of SSA is in the process of evolving significantly. As more actors bring various capabilities and interests to the table, elements such as governance and infrastructure come to the fore as key foci of international discourse in policy and operations (Lal et al. 2018). Security concerns have to be reflected on in light of post-Cold War practices of collaboration and transparency that have already overturned multilateral sharing practices in domains such as satellite imagery but must also cater to new risks and threats identified by major space-faring actors. Geographical opportunities for sensor sites and an array of cutting-edge and legacy sensor systems must be traded off with advanced operational needs of unprecedented orbital utilization, in order to define effective ground-based architectures. Moreover, the fusion, synthesis, and exchange of data and products from multiple sources need to be understood for SSA use cases that are likely to bifurcate broadly to include monitoring and coordination of large-scale orbital traffic on the one hand and the observation and verification of sophisticated proximity operations on the other.

At this point in time, European SST efforts also find themselves part of a dynamic debate close to home. As integration of SST capabilities is seen to take up speed (West 2018, p 11), the future of SSA capabilities in Europe and their operational, R&D, and political elements are being discussed on various levels including governmental (civilian and military programs and their bilateral partners), intergovernmental (European Space Agency), and – most crucially – supranational (European Union) levels.

Evolution as a Sensor Network

As part of ongoing programmatic activities inside EU SST, the three main functions of sensors, data processing, and services are complemented by dedicated efforts for the upgrade and maintenance of sensors, in line with a vision for an enhanced architecture for the timeframe until 2028. With the recent enlargement of the Consortium to eight Member States, the network of operational radars, telescopes, and laser stations has already grown to include wider global coverage. As further participants may join the effort in the short- and midterm, the integration and trade-off of additional sensor assets and the targeted use of funds to support further developments are being assessed.

A fundamental part of this work consists of architecture studies currently being undertaken by the Consortium on request of the European Commission, with a focus on the sensor layer of the EU SST ground segment, in order to plan structured upgrades of specific sensors and optimize their use in the evolving sensor network between 2021 and 2028 (2021–2027 constitutes the upcoming budgetary timeframe of the multiannual financial framework of the European Union.). To this end, sensors – both existing and under development – are being examined as part of an added-value analysis, in order to determine an optimal architecture in view of possible degrees of autonomy in different orbital regimes and prioritize sensors for upgrade accordingly. Specifically, the architecture studies involve the examination of several dozens of possible architectures until 2028, by simulating both sensor coverage and cataloguing performance, in view of a population of objects in orbit that evolves across time.

In consultation with the European Commission, these studies take into account the underlying philosophy of EU SST as a *support* framework that all but complements significant past and future investment of its Member States through funds for operation, maintenance, and upgrade, with limited EU co-funding assumed to consist of less than 50% in the future and two conservative budgetary scenarios of the upcoming EU space regulation (cf. next section). For GEO, a complete European autonomous surveillance capability for objects larger than 35 cm is feasible by 2028 with classical and wide field-of-view telescopes. For MEO, Europe could feasibly catalogue 80% of objects larger than 35 cm in the same timeframe. For LEO, Europe could be able to catalogue 9,000–11,000 objects larger than 7–10 cm by 2028 within the tentative budgetary constraints of the proposed EU space program.

Finally, while new sensor systems come online and legacy systems are due for upgrading in the short term, the Consortium is also set to lose valuable assets with the potential exit of the United Kingdom from the European Union. While this will signify the loss of a valued partner, the UK's interactions with the transatlantic SSA community have been long-standing. Since some partners of the SST Consortium actively engage in a range of multilateral space operations activities such as the Combined Space Operations (CSpO) initiative (US Air Force 2019), bi- and multilateral relations are likely to be maintained predominantly beyond the European Union SSA context.

Evolution in the EU Space Program

The ongoing architecture studies are linked to the opportunities posed by the proposed space program of the European Union (European Commission 2018b). While its draft legal basis is yet to be completed, the overarching context is all but decided: the draft regulation foresees a maturation of the current EU SST Support Framework into a sub-component of a dedicated SSA program, a notion already foreshadowed in the call for enhancing the SST framework into a fully fledged SSA program in the European Space Strategy of 2016 (European Commission 2016, p. 10). The SSA program will be situated next to the flagship programs for Earth observation (EO) and position, navigation, and timing (PNT), as part of a dedicated package of new "security components," which include SSA and Governmental Satellite Communications.

The SSA component is proposed to predominantly include SST, next to small-scale elements of space weather (SWE) and near-Earth objects (NEO), and will be furnished with a comparably minor share of the proposed overall 16 billion Euro budget. The current governance model of a Consortium or partnership of EU Member States – rather than a communautized program of all Member States or a public-private partnership – is being affirmed and carried forward, with an opportunity for the accession of additional Member States of the European Union. The philosophy and mechanisms of internal governance remain largely unchanged. Indeed, certain elements proposed and implemented under the first two leadership terms of the current Consortium are being explicitly prescribed: these include functional specialization of key capability elements such as services or catalogue by a specific actor or tandem of actors, mechanisms that strengthen increased transparency, and involvement of representatives of the European Commission.

Given the dual aspect and increased security angle of the program, the context of EU SST in the evolving European ecosystem will be further complemented through instruments for cooperation put in place by the European Union in the defense domain such as the European Defence Fund (EDF), its precursor European Defence Industrial Development Program (EDIDP), and Permanent Structured Cooperation (PESCO). It remains a subject of discussion as to how these will be utilized by European actors to address SSA elements while avoiding duplication of key capabilities. Further developments will need to be seen in context also of recent national policies, such as the 2019 French Defence and Space Strategy, which posits SSA as a key domain, or the impending space policy of NATO that may manifest a need for a recognized space picture across a similar but different range of partners.

Evolution in the Context of Space Traffic Management

In light of the emergence of large constellations with fleets of dozens, hundreds, or even thousands of spacecraft operated by private entities, the community of practice is increasingly discussing the notion of space traffic management (STM) or coordination. In absence of an unambiguous, shared definition of the concept among space-faring actors, the debate on STM gathers different perspectives that encompass technical and operational approaches linked to the exchange of data and information and to reflections on legal and regulatory aspects.

An emerging consensus in the discussion posits that on the one hand, STM activities will have to be addressed and implemented in the context of security political concerns that also impinge upon sovereign aspects (Becker 2019). On the other, future efforts in this evolving domain will also inevitably foot on a kind of multilateral or coordinated interaction between various global stakeholders (Lewis et al. 2018, p. 22 and 35). As the current SST framework is set to mature as part of the consolidation of the European space program in the coming decade, it will hence have to be seen in – and continue to consolidate its position within – the context of burden-sharing with transatlantic partners.

Recent developments in the United States saw the issuing of Space Policy Directive 3 (SPD-3) (Whitehouse 2018). SPD-3 proposes an integrated interagency approach towards STM. The US Department of Commerce is directed to act as a new civil focal point for public interface and on orbit SSA data sharing, while the US Department of Defense maintains the authoritative catalogue in view of its core national security mission. SPD-3 posits the main aspects of a space traffic management regime, including SSA data, STM services, and STM science and technology, and also addresses national orbital debris mitigation policy, as well as global engagement. While the latter two, national regulation for space debris mitigation and global engagement for the promotion of norms of behavior, remain the remit of individual Member States, in Europe to date, it is the aspect of technical and operational STM that resonates most with EU SST. Essentially, the activities of EU SST as a civilian SSA actor today correspond to the technical and operational aspects of the STM initiative in the United States in view of data sharing through a repository (open architecture data repository in the United States and European database in the EU) and the provision of basic and added-valued services (e.g., collision avoidance).

In response to the US developments, the dialogue in Europe on STM has picked up momentum. It increasingly forges links between the current body of thinking on potential regulatory needs, existing operational and technological capabilities of government actors, and commercial entities that explored operational and governance aspects on SST (Tortora 2019). Senior representatives of the European Commission have described the Union's SST activities to constitute the basis for a future European STM regime (Bieńkowska 2019).

Finally, the shift of responsibilities of the US Department of Defense to the Department of Commerce will nevertheless leave bilateral agreements between military SSA actors that allow the current sharing of US CDM with Consortium Member States untouched. Beyond these fundamental bilateral channels between the United States and the individual EU SST Member States, the Consortium maintains a regular dialogue with US stakeholders from the Department of Defense, Department of State, and Department of Commerce in mutual pursuit of a collaborative approach.

Conclusion

Space Surveillance and Tracking constitutes a fundamental capability to support and enable the operation of space-based services and assets. In view of foreseen activity in orbit of unprecedented scale, coupled with the growing criticality of – and risk to – space-based services, the domain of SSA will be salient in the mid- and long term. It will likely be pursued both by those state actors that operate their own maneuverable space assets and have a primary interest safeguarding them, and by state and non-state actors that can bring pertinent technology and infrastructural elements to the table.

In its comparatively short history, the implementation of the EU Space Surveillance and Tracking Support Framework has seen the convergence of a diverse group of stakeholders in a novel model for space cooperation in Europe. For some Member States, the inception of EU SST presented an opportunity to foster nascent engagement and interest in the domain of SSA; for others it serves as a multilateral forum to draw on existing rich legacies and make those available to the wider community. By employing a communal but tailored bottom-up approach to an SST system, EU SST allowed for a comparably rapid progression to fundamental operational services while honoring long-standing security constraints articulated by its key actors. A streamlined governance enables leverage for both the European Union and those state actors with long-standing operational and political investment in the SSA domain, and at the same time encourages increased participation by those state actors who begin to recognize SSA as an important field. It allows the contribution of industry actors to those parts of the SSA value chain deemed meaningful and effective by individual governmental actors while binding the latter in a joint effort of cooperation.

Borne out of the need to advance a strategic capability for the European community, EU SST hence constitutes an effort to consolidate and enhance capabilities of traditional and emerging space actors, as well as an exercise in transparency building within and beyond Europe. In fusing external data from the United States with that of an expanding distributed sensor network operated by European actors, SST services are provided free of charge to an increasingly large user community in Europe that includes commercial, governmental, and intergovernmental users who operate fleets of spacecraft or handle civil protection. Through an approach of layered dependencies among the Member States contributing to EU SST in Europe and as a European collective vis-à-vis transatlantic partners, the framework advances global burden-sharing of safeguarding strategic space assets. It also ties into the efforts of promoting the sustainability of the orbital environment and constitutes a building block for future operational activities in the area of space traffic management and coordination.

As integral component of the planned European Union space program, European SST activities are set to mature into a program in their own right, with an enhanced catalogue, services, and a globally distributed sensor architecture that affords greater levels of autonomy across orbital regimes. As such, the evolution of the SST capability of the European Union must be observed in the context of capabilities of global activity: like other major efforts, it faces the challenge and opportunity to slot into the global ecosystem of different governmental and commercial actors developing and operating existing and next-generation SSA capabilities; it contributes to the contemporary dialogue of global community on how to ingest of multiple sources of data and process for meaningful, actionable products; and must convene in organizational frameworks that take into account the different contributions and requirements of a highly diverse community.

References

Becker M (2019) Kein Krieg der Sterne, Internationale Politik, January/February 2019, 62–67

Bieńkowska E (2019) Keynote speech at 35th space symposium, Colorado Springs, 10 April 2019. https://europa.eu/rapid/press-release_CLDR-19-2012_en.htm. Accessed 17 July 2019

Chatard-Moulin P (2013) Horizon 2020 work programme, horizon 2020 space information day, Brussels, 11 December 2013. European Commission, Brussels

Chow T (2011) Space situational awareness sharing program. Secure World Foundation, Washington DC

Commission of the European Communities (2003) Space: a new European frontier for an expanding union: an action plan for implementing the European space policy (white paper), COM(2003) 673. Commission of the European Communities, Brussels

Council of the European Union (2011) High-level SSA civil-military use requirements. European External Action Service, Brussels

Council Resolution (2008) Taking forward the European space policy. Council of the European Union, Brussels

Council Resolution (2010) Global challenges: taking full benefit of European space systems. Council of the European Union, Brussels

de Selding PB (2015) A European space surveillance network inches forward. Spacenews, 17 June 2015 (online). https://spacenews.com/a-european-space-surveillance-network-inches-forward/. Accessed 17 July 2019

Dickow M (2015) The pursuit of collective autonomy? Europe's autonomy in "space and security" lacks a joint vision. Al-Ekabi C (ed) European autonomy in space. Stud Space Policy 10:113–123

Donath T, Schildknecht T, Martinot V, Del Monte L (2009) Possible European systems for space situational awareness. Acta Astronaut 66(2010):1378–1387

EDA (2010) EDA common staff target, 25 march 2010 (EDA SBD 2010–07). European Defence Agency, Brussels

ESA (2008) Space situational awareness preparatory programme proposal ESA/C(2008)142. European Space Agency, Paris

ESPI (2018) Security in outer space: rising stakes in Europe. European Space Policy Institute, Vienna

European Commission (2013a) Proposal for a decision of the European parliament and of the council establishing a space surveillance and tracking support programme. European Commission, Brussels

European Commission (2013b) Commission staff working document: impact assessment accompanying the document proposal for a decision of the European parliament and of the council establishing a space surveillance and tracking support programme, SWD(2013)55 final. European Commission, Brussels

European Commission (2014a) Commission staff working document: overview of natural and man-made disaster risks in the EU SWD(2014) 134. European Commission, Brussels

European Commission (2014b) Commission implementing decision of 12.9.2014 on the procedure for participation of the member states in the space surveillance and tracking support framework, C(2014)6342. European Commission, Brussels

European Commission (2016) A space strategy for Europe, COM(2016) 705. European Commission, Brussels

European Commission (2018a) Report from the commission to the European parliament and the council on the implementation of the space surveillance and tracking (SST) support framework (2014–2017), COM(2018)256. European Commission, Brussels

European Commission (2018b) Proposal for a REGULATION OF THE EUROPEAN PARLIAMENT AND OF THE COUNCIL establishing the space programme of the union and the European Union Agency for the space programme. European Commission, Brussels

European Parliament (2013) Initial appraisal of a European commission impact assessment: European commission proposal for a decision of the European parliament and of the council establishing a space surveillance and tracking support programme. European Parliament, Brussels

European Parliament (2014) Position of the European parliament adopted at first reading on 2 April 2014 with a view to the adoption of decision no .../2014/EU of the European parliament and of the council establishing a framework for space surveillance and tracking support, EP-PE_TC1-COD(2013)0064, European Parliament, Brussels

European Parliament and European Council (2014) Decision No 541/2014/EU of the European Parliament and of the Council of 16 April 2014 establishing a Framework for Space Surveillance and Tracking Support, Official Journal of the European Union, 27.5.2014, Brussels

Flohrer T, Krag H (2017) Space surveillance and tracking in ESA's SSA programme. In Proceedings of the 7th European conference on space debris, Darmstadt, 18–21 April 2017

Gravier A, Faucher P (2018) EU Space surveillance & tracking support framework: a consortium of member states safeguarding European space infrastructure & orbital environment, IFRI conference: La sécurité de l'espace dans les années 2020: perspectives transatlantiques, 27 November 2018, Bruxelles

Lal B, Balakrishnan A, Caldwell BM, Buenconsejo RS, Carioscia SA (2018) Global trends in space situational awareness (SSA) and space traffic management (STM). Institute for Defense Analysis, Washington D.C.

Lewis P, Parakilas J, Schneider-Petsinger M, Smart C, Rathke J, Ruy D (2018) The future of the United States and Europe: an irreplaceable partnership. Chatham House and Center for Strategic & International Studies, London/Washington D.C.

Marta LC (2015) The European space surveillance and tracking service at the crossroad. Def Ind 5:9–11

Marta LC, Gasparini G (2008) Europe's approach space situational awareness. In: Schrogl K-U, Mathieu C, Peter N (eds) Yearbook on space policy 2007/2008: from policies to programmes. Springer, New York, pp 138–151

Mazurelle F, Wouters J, Thiebaut W (2009) The evolution of European space governance: policy, legal and institutional implications. Int Organ Law Rev 6:155–189

McCormick PK (2013) Space debris: conjunction opportunities and opportunities for international cooperation. Sci Public Policy 40(6):801–813

McCormick PK (2015) Space situational awareness in Europe: the fractures and the federative aspects of European space efforts. Astropolitics 13:43–64

Monham A (2018) Securing EUMETSAT's mission from an evolving space environment 12th ESPI autumn conference, 27–28 September 2018. European Space Policy Institute, Vienna

Nardon L (2007) Space situational awareness and international policy (document de travail 14). l'Institut français des relations internationals (IFRI), Paris

Pagkratis S (2011) European space activities in the global context. In: Schrogl KU, Pagkratis S, Baranes B (eds) Yearbook on space policy 2009/2010: space for society. ESPI/Springer, Vienna, pp 2–132

Pasco X (2008) Study on suitable governance and data policy models for a European Space Situational Awareness (SSA) system. Fondation pour la Recherche Stratégique, Paris

Pasco X (2009) A European approach to space security. American Academy of Arts and Sciences, Cambridge, MA

Pellegrino M, Stang G (2016) Space security for Europe. EUISS report no. 29, July 2016. EU Institute for Security Studies, Paris

Robinson J (2011) Space security for the transatlantic partnership: conference report and analysis, report 38, November 2011. European Space Policy Institute (ESPI), Vienna

Rovetto RJ, Kelso TS (2016) Preliminaries of a space situational awareness ontology, advances in the astronautical sciences, preprint AAS 16–510

SPASEC (2005) Report of the panel of experts on space and security (issue 1.12). European Commission, Brussels

Tortora JJ (2019) Space traffic management: a long way ahead. J Space Saf Eng 6(2):69

US Air Force (2019) Multinational statement for combined space operations, 10 April 2019 (online). https://www.afspc.af.mil/News/Article-Display/Article/1810793/multinational-state ment-for-combined-space-operations/. Accessed 17 July 2019

US Strategic Command (2019) USSTRATCOM, polish space agency sign agreement to share space services, data, US strategic command public affairs, 11 April 2019, https://www.stratcom.mil/ Media/News/News-Article-View/Article/1811729/usstratcom-polish-space-agency-sign-agree ment-to-share-space-services-data/. Accessed 17 July 2019

Veclani AC, Sartori N, Battisti E, Darnis JP, Cesca E (2014) Space, sovereignty and European security: building European capabilities in an advanced institutional framework. Directorate-General for External Policies, European Parliament, Brussels

Via Satellite (2015) Five European countries sign space surveillance and tracking agreement, satellite today, 17 June 2015 (online), https://www.satellitetoday.com/government-military/2015/06/17/five-european-countries-sign-space-surveillance-and-tracking-agreement/. Accessed 4 July 2019

Vittet-Philipp P, Savova T (2012) International research update, issue 23, Brussels. European Commission, Brussels

West J (2018) Space security index 2018. Project Ploughshares, Waterloo

Whitehouse (2018) Space policy directive-3, national space traffic management policy. Office of the President of the United States, Washington DC

China's Capabilities and Priorities in Space-Based Safety and Security Applications

47

Lin Shen, Hongbo Li, Shengjun Zhang, and Xin Wang

Contents

Abstract

China's space industry has developed considerably since it was first established in 1956. Following a strategy based on self-reliance, China relies primarily on its own capabilities to develop space industry to meet the needs of modernization, based upon its actual conditions and strength. Significant achievements in the fields of Earth observation (EO), communication satellite system, positioning, navigation and timing (PNT), and space situational awareness (SSA) have been realized.

L. Shen · H. Li (✉) · S. Zhang · X. Wang
China Academy of Launch Vehicle Technology (CALT), Beijing, China
e-mail: tolinsh@sina.com; lhbspace@sina.com; zhangsj98@sina.com; wangxin.1122@163.com

© Springer Nature Switzerland AG 2020
K.-U. Schrogl (ed.), *Handbook of Space Security*,
https://doi.org/10.1007/978-3-030-23210-8_117

905

Introduction

China space industry has developed fast since it was established in 1956. Over the past 60 years, achievements include the development of atomic and hydrogen bombs, missiles, man-made satellites, and a lunar probe. China has also achieved manned spaceflight. China is unflinching in pursuing the road of peaceful development, while maintaining that outer space is the common wealth of mankind. While supporting all activities that utilize outer space for peaceful purposes, China actively explores and uses outer space and continuously makes new contributions to the development of manned space programs (The State Council Information Office of the People's Republic of China 2006).

China has contributed largely to some of the world's most advanced technologies in many important fields, including satellite recovery, multi-satellite launch, Cryogenic fuel rocket technology, cluster carrier rocket technology and geostationary Earth orbit (GEO) satellite launch, telemetry, and Tracking and Command (TT&C). China has also made significant achievements in the development and application of remote sensing satellites, communication satellites, and BeiDou Navigation Satellite System (The State Council Information Office of the People's Republic of China 2000). Space science, technologies, and applications have achieved fruitful results.

China's Space Safety and Security Policy

Space is the commanding point of international strategic competition. The weaponization of space emerged as a result of the development of space forces and means by some countries (The State Council Information Office of the People's Republic of China 2015).

China's policy advocates for the peaceful use of space and against the weaponization of space and the arms race in space. As such China is actively participating in international space cooperation. Closely following the space situation, the country monitors the threats and challenges of space security, while it safeguards the security of space assets, serves national economic construction and social development, and maintains space security.

China takes a defensive posture in the frame of its national defense policy, and its development does not pose a threat to any country. No matter how far China develops, it will never seek hegemony. The purpose of China's space development is to explore outer space and enhance the understanding of the Earth and the cosmos; to utilize outer space for peaceful purposes, promote human civilization and social progress, and benefit the whole of mankind; to meet the demands of economic, scientific, and technological development, national security, and social progress; and to improve the scientific and cultural levels of the Chinese people, protect China's national rights and interests, and build up its overall strength. (The State Council Information Office of the People's Republic of China 2016).

According to its policy, China's space industry is subject to and serves the national overall development strategy and adheres to the principles of innovative, coordinated, peaceful, and open development.

1. Innovative development. China takes independent innovation as the core of the development of its space industry. It implements major space science and technology projects, strengthens scientific exploration and technological innovation, deepens institutional reforms, and stimulates innovation and creativity, working to promote rapid development of the space industry.
2. Coordinated development. China rationally allocates various resources and encourages and guides social forces to take an orderly part in space development. All space activities are coordinated under an overall plan of the state to promote the comprehensive development of space science, space technology, and space applications and to improve the quality and efficiency of overall space development.
3. Peaceful development. China always adheres to the principle of the use of outer space for peaceful purposes and opposes the weaponization of outer space or an arms race in outer space. The country develops and utilizes space resources in a prudent manner and takes effective measures to protect the space environment to ensure a peaceful and clean outer space and guarantee that its space activities benefit the whole of mankind.
4. Open development. China persists in combining independence and self-reliance with opening to the outside world and international cooperation. It actively engages in international exchanges and cooperation on the basis of equality and mutual benefit, peaceful utilization, and inclusive development, striving to promote progress of space industry for mankind as a whole and its long-term sustainable development.

Since 2000, the Information Office of the State Council has published a space white paper every 5 years. So far, four space white papers have been published. The latest one was issued in 2016. The white papers offer a brief introduction about China's space development objectives, vision and principles, recent space progress, future space plans, development policies and measures, and international exchanges and cooperation.

Analysis of China's Capabilities in Space Safety and Security

Earth Observation (EO)

On April 24, 1970, China successfully developed and launched the first man-made Earth-orbiting satellite, Dongfanghong-1, and became the fifth country in the world to independently develop and launch man-made satellites. China is the third country in the world to master satellite recovery technology. The success rate of satellite recovery has reached the international advanced level. The use of the Fengyun (wind and cloud), Haiyang (ocean), Ziyuan (resources), Gaofen (high-resolution), Yaogan (remote sensing), and Tianhui (space mapping) satellite series and constellation of small satellites for environment and disaster monitoring and forecasting has been improved. (The State Council Information Office of the People's Republic of China 2016).

Fengyun Satellite Series

As early as the 1970s, China began to develop its meteorological satellites. Since then, 17 meteorological satellites have been launched. China is the third country after the United States and Russia to have polar orbiting meteorological satellites and geostationary Earth-orbit meteorological satellites at the same time (National Satellite Meteorological Center 2019).

Under the coordination and management of the World Meteorological Organization (WMO), all countries in the world have achieved the goal of exchanging meteorological observation data and acquiring global meteorological observation data. The Global Operational Meteorological Satellite Detection System (GOMSS) is a space-based detection system built through the joint efforts of countries launching and operating meteorological satellites under the coordination of WMO. China's operational meteorological satellites Fengyun-3, Fengyun-2, and Fengyun-4A have become important members of the global operational meteorological satellite detection system. They supplement each other in coverage and resolution and help China acquire global information and monitor regional disastrous weather and environmental and meteorological services, as well as Earth sciences.

Ocean Satellite Series

The first "Ocean" dynamic and environmental environment sensing satellite was launched in August 2011 (The State Council Information Office of the People's Republic of China 2011). It has the capability of all-weather and full-time observation in microwave region. The Haiyang-2 satellite is capable of all-weather, full-time, and high-accuracy observation of marine dynamic parameters such as sea height, sea wave, and sea surface wind. (The State Council Information Office of the People's Republic of China 2016).

China is developing three ocean monitoring satellites: color environment satellite (HY-1), ocean power ocean dynamic environment satellite (HY-2), and ocean radar satellite (HY-3). The system will realize product diversification, data standardization, application quantification, and operational operation and will meet increasing global requirements for ocean surveillance and monitoring. It also provides strong technical support for the implementation of marine development strategies. The system effectively implements marine environment and resources monitoring and provides services for the protection of marine rights and interests, disaster prevention and mitigation, and national economic construction by improving the accuracy and timeliness of marine environment forecasting and marine disaster early warning, (National Satellite Oceanic Application Center 2012).

Ziyuan Satellite Series

China has successively launched Ziyuan-1, Ziyuan-2, and Ziyuan-3. Until now, Ziyuan-3 01 and 02 stereo mapping satellites have achieved double star networking and operating. ZY-3 01 is the first high-resolution optical transmission stereo mapping satellite in China. It was successfully launched on January 9, 2012. It integrates mapping and resource investigation functions. ZY-3 02 was successfully launched on May 30, 2016. It is a high-resolution stereo mapping operational satellite. It is optimized on the basis of ZY-3 01 and carries payloads such as a three-line array mapping camera and a multispectral camera. The resolution of front and rear viewing cameras is improved from 3.5 m to better than 2.5 m. It also carries a set of experimental laser altimetry payload. The performance achieved is better than the ZY-3 01 satellite in image fusion ability and elevation measurement accuracy (Land Satellite Remote Sensing Application Center 2019).

Gaofen Satellite Series

China is building a high-resolution Earth observation system-of-system based on satellite imagery and stratospheric airships and aircraft, improving the corresponding ground systems and establishing data and application centers. Combined with other observation means, the system will form an all-weather, all-time, and global coverage Earth observation capability. By 2020, an advanced terrestrial, atmospheric, and oceanic Earth observation system will be built to provide services and assist in decision making processes for important fields such as modern agriculture, disaster reduction, resources and environment, and public safety.

The first Gaofen-1 satellite was launched on April 26, 2013. It combines high resolution and wide swath width, with a designated life span of 5–8 years (China National Space Administration 2019). The satellite is widely used in the fields of land and resources, environmental protection, precision agriculture, disaster prevention, and mitigation. Gaofen-2 achieves submeter optical remote sensing. Gaofen-3 is a SAR satellite and achieves 1 meter resolution. Gaofen-4 is China's first GEO high-resolution Earth observation satellite.

Environment and Disaster Monitoring and Forecasting Small Satellite Constellation

The HJ-1A/B/C environment and disaster monitoring and forecasting small satellite constellation includes two optical satellites, HJ-1A/B and one radar satellite HJ-1C, which can carry out large-scale, all-weather, and 24 h dynamic monitoring for ecological environment and disaster. These satellites are equipped with four remote sensors such as wide-coverage CCD scanner, infrared multispectral scanner, hyperspectral imager, and synthetic aperture radar, comprising a more complete Earth observation remote sensing series characterized by high and medium space resolution, high time resolution, high spectrum resolution, and wide coverage (China Centre for Resources Satellite Data and Application 2019).

Commercial Earth Observation

In 2015, the first satellite of China's first commercial remote sensing constellation, Jilin-1, was launched. In early 2018, four satellites of Gaojing-1 formed the first commercial remote sensing constellation with 0.5-meter resolution in China.

China established a ground data processing system for Earth observation satellites, common application supporting platform, and multilevel network data distribution system. Such capabilities greatly enhance China's abilities in data processing, archiving, distribution, services provision, and quantitative applications. Industrial application system-building is in full swing, having completed 18 industrial and two regional application demonstration systems, and set up 26 provincial-level data and application centers. An integrated information service sharing platform for a high-resolution Earth observation system has been built. Earth observation satellite data is now widely used in industrial, regional, and public services for economic and social development.

In the future, in accordance with the policy guideline for developing multi-functional satellites, and creating networks of satellites and integrating them, China will focus on three series of satellites for observing the land, ocean, and atmosphere, respectively. China is to develop and launch satellites capable of high-resolution multi-mode optical observation, L-band differential interferometric synthetic aperture radar imaging, carbon monitoring of the territorial ecosystem, atmospheric Lidar detection, ocean salinity detection, and new-type ocean color observation. China will take steps to build highly efficient capabilities of its own, comprehensive global observation and data acquisition with a rational allocation of low-, medium- and high-spatial resolution technologies and an optimized combination of multiple observation methods.

Communication Satellites

China is the fifth country in the world to independently develop and launch GEO communication satellites. Currently, China's aerospace industry has mastered key technologies in public platform of large-capacity GEO satellites, space-based data relay, and TT&T. The performance of satellites has been significantly improved, and the level of voice, data and radio, television communication, and broadcasting has been further improved. The successful launch of communications satellites such as Yatai and Zhongxing represent the completion of a fixed communications satellite support system whose communications services cover all of China's territory as well as major areas of the world. The successful launch and stable operations of the Zhongxing 10 satellite have greatly increased the power and capacity of China Telecom and Broadcasting Satellite. The Tiantong-1, China's first mobile communications satellite, has been successfully launched. The first-generation data relay satellite system composed of three Tianlian-1 satellites has been completed, and high-speed communication test of satellite-ground laser link has been crowned with success. The Shijian-13 satellite launched in 2017 will realize the broadband application of autonomous communication satellites and promote the development of

China's satellite communication industry. The Ka-band multi-beam broadband communication system on this satellite is the first application in domestic communication satellite. It can support multiuser and large capacity bidirectional using and download data at high speed through this satellite while supporting a large number of users to upload data at high speed in vast areas. The satellite breaks through a series of technical problems such as Ka multi-beam broadband system design, antenna reflector profile accuracy control and measurement, antenna pointing accuracy calibration, and so on. The related technology has reached the international advanced technology level (XINHUANET 2019).

By the end of 2016, China has developed three generations of Dongfanghong series communication satellite platform. Dongfanghong-4 platform is the main platform in service. The fourth generation Dongfanghong-5 super-large capacity platform is being developed.

Since the mid-1980s, China has developed satellite communication technology by using domestic and foreign communication satellites to meet the growing needs of communication, broadcasting, and education. For fixed satellite communication services, dozens of large and medium satellite communication Earth stations have been built throughout the country, with more than 27,000 international satellite communication lines connecting more than 180 countries and regions in the world. China has built a domestic satellite public communication network, with more than 70,000 domestic satellite communication telephone routes, and initially solved the communication problems in remote areas. Very small aperture terminal (VSAT) communication business has developed rapidly in recent years. There are 30 domestic VSAT communication business units serving 15,000 small station users, including more than 6,300 bidirectional small station users. At the same time, more than 80 special communication networks of finance, meteorology, transportation, petroleum, water conservancy, civil aviation, power, health, and journalism departments have been established, including tens of thousands VSAT. For satellite television broadcasting business, China has built a satellite television broadcasting system covering the whole world and a satellite television education system covering the whole country. Since 1985, China has used satellite to transmit radio and television programs. At present, it has formed a satellite transmission coverage network, which occupies 33 communication satellite transponders. It is responsible for transmitting 47 sets of central and local TV programs and educational TV programs, as well as 32 channels of internal and external broadcasting programs and nearly 40 sets of local broadcasting programs. Over the past 10 years since the launch of satellite education television broadcasting, more than 30 million people have received education and training in colleges and secondary schools. In recent years, the nation has built a satellite live broadcasting test platform, which transmits satellite TV programs from central and local governments to vast rural areas that are not covered by radio and television through digital compression, thus greatly improving the coverage of radio and television in China. There are about 189,000 satellite TV broadcasting stations in China. On the satellite live broadcasting test platform, a broadband multimedia transmission network of China's educational satellite has been established to provide comprehensive services of distance education and information technology for the whole country.

In the future, China will be oriented toward industrial and market applications and mainly operates through business models while meeting public welfare needs. China will develop both fixed and mobile communications and broadcasting as well as data relay satellites and build a space-ground integrated information network consisting of space-based systems such as high-Earth-orbit broadband satellite systems and low-Earth-orbit mobile satellite systems and ground-based systems such as satellite-access stations. TT&C stations, gateway stations, uplink stations, calibration fields, and other satellite ground facilities are to be built synchronously. These efforts are expected to bring about a comprehensive system capable of providing broadband communications, fixed communications, direct-broadcast television, mobile communications, and mobile multimedia broadcast services. A global satellite communications and broadcasting system integrated with the ground communications network will be established step-by-step.

Satellite Positioning, Navigation, and Timing System

Satellite navigation system is an indispensable information infrastructure for national security and economic and social development. The BeiDou Navigation Satellite System (BDS) has been independently constructed and operated by China with an eye to the needs of the country's national security and economic and social development. As a space infrastructure of national significance, BDS provides all-time, all-weather, and high-accuracy positioning, navigation, and timing services to global users (China Satellite Navigation Office 2018).

China attaches great importance to the BDS construction and development. In 2013, "the Medium and Long-Term Development Plan for the National Satellite Navigation Industry" was released, to make overall arrangement for medium and long-term satellite navigation industrial development and to provide the guidance of macro polices. In 2016, "China's BeiDou Satellite Navigation System," a governmental white paper, was released, to introduce the BDS development methods and policies.

Navigation satellite systems are public resources shared by the whole globe, while the multi-system compatibility and interoperability have become a trend. China has been applying the principle that "BDS is developed by China, and dedicated to the world," serving the development of the Silk Road Economic Belt and actively pushing forward international cooperation related to BDS. As BDS joins hands with other navigation satellite systems, China works with all other countries, regions, and international organizations to promote global satellite navigation development and make BDS further serve the world and benefit mankind.

BDS possesses the following characteristics: first, its space segment is a hybrid constellation consisting of satellites in three kinds of orbits. In comparison with other navigation satellite systems, BDS operates more satellites in high orbits to offer better anti-shielding capabilities, which is particularly observable in terms of performance in the low-latitude areas. Second, BDS provides multi-frequency navigation signals with which can be combined to improve service accuracy. Third, BDS integrates navigation and communication capabilities and has multiple service functions including real-time navigation, rapid positioning, precise timing, location reporting, and short message communication services.

BDS is mainly comprised of three segments: a space segment, a ground segment, and a user segment. The BDS space segment is a hybrid navigation constellation consisting of geostationary Earth orbit (GEO), inclined geosynchronous satellite orbit (IGSO), and medium Earth orbit (MEO) satellites. The BDS ground segment consists of various ground stations, including master control stations, time synchronization/uplink stations, monitoring stations, as well as operation and management facilities of the inter-satellite links. The BDS user segment consists of various kinds of basic BDS products, including chips, modules, and antennae as well as the BDS terminals, application systems, and application services, which are compatible with other systems. Relevant products of BDS have been widely used in transportation, marine fisheries, hydrological monitoring, meteorological forecasting, surveying and mapping geographic information, forest fire prevention, communication time unification, electric power dispatch, disaster relief and mitigation, emergency rescue, and other fields.

In the late twentieth century, China started to explore a path to develop a navigation satellite system suitable for its national conditions, and gradually formulated a three-step development strategy. The first step is to construct BDS-1. The project started in 1994 and was ready to operate in 2000, with the launch of two GEO satellites. With an active-positioning scheme, the system provided users in China with positioning, timing, wide area differential, and short message communication services. The third GEO satellite was launched in 2003, which further enhanced the system performance. The second step is to construct BDS-2. The project started in 2004, and by the end of 2012, a total of 14 satellites, including 5 GEO satellites, 5 IGSO satellites, and 4 MEO satellites, had been launched to complete the space constellation deployment. Besides a technical scheme which was compatible with BDS-1, BDS-2 added the passive-positioning scheme and provided users in the Asia-Pacific region with positioning, velocity measurement, and timing as well as short message communication services. The third step is to construct BDS-3. The project started in 2009, and by the end of 2018, a total of 19 satellites were launched to complete a preliminary system for global services. It is planned to comprehensively complete the deployment of BDS-3 with the launching of 30 satellites by around 2020. BDS-3 has inherited the technical schemes of both active and passive services, and can provide basic navigation (including positioning, velocity measurement and timing), global short message communication, and international search and rescue services to global users. Users in China and surrounding areas can also enjoy regional short message communication, satellite-based augmentation, and precise point positioning services, etc.

Until now, the construction BeiDou-2 system has been completed, with the networking of 14 BeiDou navigation satellites, officially offering positioning, velocity measurement, timing, wide area difference, and short-message communication service to customers in the Asia-Pacific region. BeiDou's global satellite navigation system is undergoing smooth construction. By the end of 2018, a total of 19 satellites were launched to complete a preliminary system for global services. Around 2020, the BDS-3 will be deployed with the launching and networking of 30 satellites.

In the future, BDS will continue to improve the service performance, to expand the service functions, and to enhance continuous and stable operation capability. Before the end of 2020, BDS-2 will launch 1 backup GEO satellite; BDS-3 will launch another 6 MEO, 3 IGSO, and 2 GEO satellites, to further improve the global basic navigation and regional short message communication service capabilities, and to realize the global short message communication, satellite-based augmentation, international search and rescue, and precise point positioning service capabilities, etc.

Space Situation Awareness

China advocates strengthening international space communication and cooperation and promoting inclusive development based on equality and mutual benefit, peaceful use, and common development. In order to achieve the goal of peaceful development and utilization of outer space and effectively safeguard space security, China will, on the basis of adhering to the principle of peaceful development and utilization of outer space and in full cooperation with the international community, especially developing countries, continue its opposition to the weaponization of outer space and the arms race in outer space, strive for the equal exploitation and peaceful use of outer space resources, and promote the establishment of a new international space order. It will make greater contributions to promoting economic development and social progress, safeguarding space security and world peace.

Space debris is the waste abandoned in space by human beings in space activities, also known as "space garbage." It mainly includes abandoned spacecraft and launch vehicle rockets, solid rocket burners, debris generated by spacecraft on-orbit operation and collision disintegration, etc. At present, there are hundreds of millions of space debris above the millimeter level, with a total mass of several thousand tons. The average impact velocity of space debris is 10 km per second. Space debris above centimeter level can lead to complete damage of spacecraft, while the cumulative impact effect of millimeter or micron level space debris will lead to performance degradation or function failure of spacecraft. According to statistics, China's on-orbit spacecraft has reached more than 270, with an average of more than 30 close-range dangerous rendezvous with space debris within 100 m per year. In addition, many space debris return to the atmosphere every year. These frequent meteoric events have posed a serious threat to the safety of ground personnel and property.

Space debris poses the greatest threat to space activities; thus, China will continue to strengthen space debris monitoring, mitigation, and spacecraft protection. China will develop technologies for monitoring space debris and pre-warning of collision and begin monitoring space debris and small near-Earth celestial bodies and collision pre-warning work. It will set up a design and assess system of space debris mitigation and take measures to reduce space debris left by post-task spacecraft and launch vehicles. It will experiment with digital simulation of space debris collisions and build a system to protect spacecraft from space debris.

In December 2008, the Key Laboratory of Space Object and Debris Observation of the Chinese Academy of Sciences was officially unveiled at the Purple Mountain Observatory (PMO) in Nanjing (Key Laboratory of Space Object and Debris Observation 2019). The laboratory set up a safety and early warning system for China in the field of space, in order to eliminate the threat posed by various human debris scattered in the vast space and protect the safety of space. The Key Laboratory of Space Object and Debris Observation will further make significant achievements in space target motion theory, space target and debris detection, spacecraft collision warning, upper atmosphere model, and so on. The laboratory will closely monitor space debris to reduce potential safety threat in space flight and detection. For the space debris discovered, the laboratory will track it in real time, search for the space garbage that has not yet been found, study the early warning technology for spacecraft launching and orbital operation, and then carry out risk assessment to resolve potential dangers.

At present, PMO has six outdoor observation stations: Qinhai Delinha Radio Observation Station, Jiangsu Xuyi Celestial Mechanics Observation Station, Jiangsu Ganyu Solar Observing Station, Heilongjiang Honghe Observation Station, Shandong Qingdao Observation Station, and Yunnan Yaoan Observation Station. Among them, Delinha Observation Station is the largest radio astronomical observation base in China. It has 13.7 m millimeter wave radio astronomical telescope, mobile submillimeter wave telescope, space debris detection optoelectronic telescope, and other large astronomical observation equipment. Xuyi Observation Station is the only celestial mechanics measurement base in China. It has the largest 1 m/1.2 m NEO detection telescope and space debris detection optoelectronic telescope in China (PMO 2019).

In June 8, 2015, the CNSA Space Debris Observation and Data Application Center was built relying on NAOC and is the technical support and daily operating organization of space debris affairs of CNSA. The establishment of the center is of great significance for promoting the development of space debris technology in China, enhancing the capacity of space debris management and service, ensuring the safety of spacecraft in orbit, supporting international space exchange and cooperation, and safeguarding the rights and interests of China's space development.

China has improved the monitoring and mitigation of early warning and protection against space debris and also enhanced standards and regulations in this regard. The monitoring of early warning against space debris has been put into regular operation, ensuring the safe operation of spacecraft in orbit. China has also made breakthroughs in protection design technologies, applying them to the protection projects of spacecraft against space debris. In addition, all Long March carrier rockets have upper stage passivation, and discarded spacecraft are moved out of orbit to protect the space environment. In the future, the standards and norms for space debris, near-Earth small objects, and space weather will be improved. In addition to this, China will establish and improve the basic database and shared data model of space debris; comprehensively promote the construction of space debris monitoring facilities, early warning, and emergency platform and network service system, strengthen the comprehensive utilization of resources; and further strengthen the spacecraft protection capability. China will improve the space environment monitoring system, build an early warning and forecasting platform, and enhance the space environment monitoring and disaster early warning capabilities. This chapter demonstrates the construction of monitoring facilities for small near-Earth objects and the improvement of their monitoring and cataloging capabilities.

International Exchanges and Cooperation on China's Space Safety and Security

The Chinese Government believes that international space cooperation should follow the basic principles set in the *Declaration on International Cooperation in the Exploration and Use of Outer Space for the Benefit and in the Interests of All States, Taking into Particular Account the Needs of Developing Countries*, adopted by the 51st United Nations General Assembly in 1996. In carrying out international space cooperation, the Chinese Government has consistently adhered to the following guiding principles:

• International space cooperation should aim at the peaceful development and utilization of space resources for the benefit of all mankind.
• International space cooperation should be carried out on the basis of equality and mutual benefit, complementarity of advantages, complementarity of strengths and weaknesses, common development and recognized principles of international law.
• The priority objective of international space cooperation is to jointly improve the space capabilities of all countries, especially developing countries, and to enjoy the benefits of space technology.
• International space cooperation should take necessary measures to protect the space environment and resources.
• Support the strengthening of the role of the United Nations Committee on Outer Space and support the United Nations Programme on Outer Space Applications.

China's international cooperation in space began in the mid-1970s. Since 2011, the Government signed 43 space cooperation agreements or memoranda of understanding with 29 countries, space agencies, and international organizations. It has taken part in relevant activities sponsored by the United Nations and other relevant international organizations, and supported international commercial cooperation in space. These measures have yielded fruitful results.

China participates in activities organized by the International Committee on Global Navigation Satellite Systems, International Space Exploration Coordination Group, Inter-Agency Space Debris Coordination Committee, Group on Earth Observations, World Meteorological Organization, and other intergovernmental international organizations. China has also developed multilateral exchanges and cooperation in satellite navigation, Earth observation and Earth science and research, disaster prevention and mitigation, deep-space exploration, space debris, and other areas. China's BeiDou satellite navigation system has become one of the world's four core system suppliers accredited by the International Committee on Global Navigation Satellite Systems and will gradually provide regional and global navigation and positioning service as well as strengthened compatibility and interoperability with other satellite navigation systems.

The cooperation between China and Brazil on Earth Resources Satellite is smoothly progressing. On October 14, 1999, China successfully launched the first China-Brazil Earth resources satellites. In addition to the Whole-Satellite cooperation, China and Brazil have carried out a number of cooperation in satellite technology, satellite applications, and satellite components. Following the successful launch of the Sino-Brazilian Earth Resources Satellite 02 in October 2003, the Chinese and Brazilian governments signed supplementary protocols on the joint research and manufacturing of satellites 02B, 03, and 04 and on cooperation in a data application system, maintaining the continuity of data of Sino-Brazilian Earth resources satellites and expanding the application of such satellites' data regionwide and worldwide.

China and the ESA have conducted the Sino-ESA Double Star Satellite Exploration of the Earth's Space Plan. China's relevant departments and the ESA have implemented the "Dragon Program," involving cooperation in Earth observation satellites, having so far conducted 16 remote sensing application projects in the fields of agriculture, forestry, water conservancy, meteorology, oceanography, and disasters.

Furthermore, a memorandum of understanding on technological cooperation in the peaceful utilization and development of outer space was signed between China and Venezuela. The two nations have established a technology, industry, and space sub-committee under the China-Venezuela Senior Mixed Committee. Under this framework, bilateral cooperation in communications satellites, remote sensing satellites, satellite applications, and other areas is promoted. China has exported satellites and made in-orbit delivery of Venezuela's remote sensing satellite-1.

China has exported satellites and made in-orbit delivery of Nigeria's communications satellite, Bolivia's communications satellite, Laos' communications satellite-1 and Belarus' communications satellite-1.

As one of the four major GNSS providers, BDS persists in open cooperation and resource sharing, actively carries out international exchanges and cooperation, and promotes the global satellite navigation development. China has taken part in international activities organized by the United Nations and other relevant international organizations, within the framework of relevant multilateral mechanisms. Under the framework of ITU, international frequency coordination activities have been conducted. China supported the extension of the radio-determination satellite service (space-to-Earth) allocations in the S-band and successfully pushed forward the S-band as another band for navigation satellites, with joint efforts with delegates from other countries. As members of the International Committee on Global Navigation Satellite Systems (ICG) and the ICG Providers' Forum, China actively participated in the meetings held by the United Nations Committee on the Peaceful Uses of Outer Space and the seminars organized by the United Nations Office for

Outer Space Affairs. The China Satellite Navigation Conference has been held annually, with more than 3,000 attendees every year. China actively established interaction mechanisms with navigation meetings of the United States, Russia, and Europe, participated in, organized, and hosted international academic exchange activities of satellite navigation, so as to strengthen international exchanges and attract global intellectual resources to jointly promote the development of satellite navigation technologies. Under the framework of the Asia-Pacific Space Cooperation Organization (APSCO), a number of cooperative projects are being implemented in the fields of monitoring and assessment, research and applications of the BDS/GNSS compatible terminals in disaster reduction, development of BDS/GNSS software receiver, as well as education and training on satellite navigation, in order to upgrade the technologies and to strengthen fundamental capacity building of the APSCO member states.

China actively participated in activities organized by the Inter-Agency Space Debris Coordination Committee, started the Space Debris Action Plan, and strengthened international exchanges and cooperation in the field of space debris research. The nation's independently developed space debris protective design system has also been incorporated into the protection manual of the Inter-Agency Space Debris Coordination Committee.

Conclusion

Currently, an increasing number of countries attaches importance to and takes part in developing space activities. Moreover, space technology is widely applied in all aspects of our daily life, exerting a major and far-reaching influence on social production and lifestyle and increasingly in the field of safety and security.

It is mankind's unremitting pursuit to peacefully explore and utilize outer space. Standing at a new historical starting line, China is determined to accelerate the pace of developing its space industry and actively carry out international space exchanges and cooperation, so that achievements in space activities will serve and improve the well-being of mankind in a wider scope, at a deeper level and with higher standards. China will promote the lofty cause of peace and development together with other countries.

References

China Centre for Resources Satellite Data and Application (2019) Retrieved 16 July 2019, from http://www.cresda.com/EN/

China National Space Administration (2019) Retrieved 16 July 2019, from http://www.cnsa.gov.cn/

China Satellite Navigation Office (2018) Development of the BeiDou Navigation Satellite System. Retrieved from http://en.beidou.gov.cn/SYSTEMS/Officialdocument/201812/P020190523251292110537.pdf

Key Laboratory of Space Object and Debris Observation, PMO, C (2019) Retrieved 16 June 2019, from http://www.sodo.pmo.cas.cn/

Land Satellite Remote Sensing Application Center (2019) Retrieved 16 July 2019, from http://www.lasac.cn/

National Satellite Meteorological Center. (2019). Retrieved July 16, 2019, from http://www.nsmc.org.cn/newsite/nsmc/channels/100003.html

National Satellite Oceanic Application Center (2012) National Satellite Ocean Application Service. Retrieved 16 July 2019, from http://www.nsoas.org.cn/NSOAS_En/index.html

Purple Mountain Observatory (PMO) (2019) Retrieved 16 June 2019, from http://www.pmo.ac.cn/

The State Council Information Office of the People's Republic of China (2000) China's space activities, A white paper

The State Council Information Office of the People's Republic of China (2006) China's Space Activities in 2006. Beijing. Retrieved from http://www.gov.cn/english/2006-10/12/content_410983.htm

The State Council Information Office of the People's Republic of China (2011) China's Space activities in 2011. Beijing. Retrieved from http://www.gov.cn/english/official/2011-12/29/content_2033200.htm

The State Council Information Office of the People's Republic of China (2015) China's military strategy. Beijing

The State Council Information Office of the People's Republic of China (2016) China's Space Activities in 2016. Retrieved from http://www.scio.gov.cn/wz/Document/1537091/1537091.htm

XINHUANET (2019) Retrieved 16 June 2019, from http://www.xinhuanet.com/

Cybersecurity Space Operation Center: Countering Cyber Threats in the Space Domain

48

Salvador Llopis Sanchez, Robert Mazzolin, Ioannis Kechaoglou, Douglas Wiemer, Wim Mees, and Jean Muylaert

Contents

The contents reported in the paper reflect the opinions of the authors and do not necessarily reflect the opinions of the respective agency/institutions.

S. Llopis Sanchez (✉)
European Defence Agency (EDA), Brussels, Belgium
e-mail: info@eda.europa.eu

R. Mazzolin · I. Kechaoglou · D. Wiemer · J. Muylaert
RHEA Group, Wavre, Belgium
e-mail: r.mazzolin@rheagroup.com; i.kechaoglou@rheagroup.com; d.wiemer@rheagroup.com; J. Muylaert@rheagroup.com

W. Mees
Royal Military Academy, Brussels, Belgium
e-mail: wim.mees@rma.ac.be

© Springer Nature Switzerland AG 2020
K.-U. Schrogl (ed.), *Handbook of Space Security*,
https://doi.org/10.1007/978-3-030-23210-8_108

921

Abstract

The focus of this article is to describe the rapid changing nature of space systems where the electromagnetic spectrum and cybersecurity aspects coexist, leading to a "change of course" to ensure secure space communications. Space systems must be designed to achieve cyber resilience. Existing vulnerabilities must be mitigated as resulted from a risk management methodology. To avoid compromises of security, a security monitoring mechanism and incident response for space systems have to be implemented. The authors provide an analysis of the functional capabilities and services of a "cybersecurity space operations center." System monitoring is essential in view of obtaining a cyber situational awareness and decision-making for space missions. Security operation center-related services could be built up in a permanent or in an ad hoc mobile infrastructure. An integrated cyber range capability to test, train, and exercise over modeled and simulated space systems will improve space operators' skills when confronted with time-critical interventions.

Introduction

National security is not restricted to securing the land, air, and maritime boundaries and pursuing strategic interests but encompasses all aspects that have a bearing on the nation's well-being. Outer space and cyberspace have emerged as the new enablers for nations, enhancing the speed and efficiency of national security and socioeconomic efforts and also providing novel applications in these areas. In an information-dominated world, they are instrumental in providing the competitive edge among the global community, strategic and tactical superiority in conflict situations, and projection of national power and influence. In addition to capability enhancement towards national aspirations, investments are also necessary for securing these capabilities against deliberate or unintentional intrusions or attacks and in ensuring safe and sustainable operations.

Assured access to cyberspace is a key enabler of national security. Two of the defining characteristics of a strong, modern, industrial nation are economic prosperity and a credible defense. The ability to use cyberspace has become indispensable to achieving both of these objectives. Business and finance executives, as well as senior defense leaders, rely on cyberspace for exactly the same thing – to get information, move information, and use information to make better decisions faster than the competition. The data- and information-rich environment of the modern battlespace presents a key area of both strategic opportunities and, equally, vulnerability. Both cyber and space technologies, in particular, are the new battleground for competing great powers who seek to limit opponents' access to information, analytics, and complex surveillance and reconnaissance information with which to inform decision-making.

With growing participation, commercialization has become an integral part of space operations and is receiving active governmental support. The peaceful use of space and the military significance of outer space continues to increase with some 60 countries currently utilizing it for peaceful purposes, for communications, banking, monitoring environmental and climate change, disaster management, e-health, e-learning and communications, and surveillance and guidance systems for military purposes. Regarding vulnerabilities, while space operations have always been vulnerable to natural forms of interference, progressive developments in the domain have also resulted in the emergence of unique novel challenges to space security and the sustainability of the environment. Greater participation in the domain and commercial prospects has more players vying for prime orbital slots, the radio frequency spectrum, and a larger share of the market. Overcrowding and increasing space debris are adversely affecting the survivability of satellites. The environment is therefore becoming more contested, congested, and competitive with the consequent increase in potential for disruption of operations. Space is being used extensively by advanced spacefaring nations for supporting military operations, and most new entrants would also leverage their access for these purposes. A corollary to this is that all assets in space providing a strategic or military advantage can be designated as valid targets in case of hostilities. Consequently, advanced nations are making efforts to dominate and control the environment to protect their interests and assured access to the realm, and the less capable ones would do the same to gain an asymmetric advantage through degradation and destruction of systems. Both these strategies demand development of counter-space threats and counter-cyber threats capabilities that would include risk mitigation actions. Such capabilities in the hands of rogue nations or non-state actors, who have limited interests in the space domain, could be extremely dangerous, even up to the catastrophic level of cascading satellite collisions. In recent years, national security challenges have necessitated a tacit acceptance of the use of the space domain for meeting national security objectives. The discussion about the creation of space forces in some countries – as a new service branch within the armed forces – to have control over military space operations is one example of the growing importance of the use of outer space.

Cyberspace is a man-made domain consisting of the interconnected networks of computing and communication devices and the information contained on these networks. Satellites and other space assets, just like other parts of the digitized critical infrastructure, are vulnerable to cyberattacks. When considering our daily lives, there is not an operation or activity conducted anywhere at any level that is not somehow dependent on space and cyberspace. This interdependency could be used to attack space assets from cyberspace. From a cyberspace perspective, it's irrelevant how high above the ground a computer is positioned. Cyber vulnerabilities associated with space-related assets therefore pose serious risks for ground-based critical infrastructure, and insecurities in the space environment will hinder economic development and increase the risks to society. Advancements in the cyber domain have not required a protracted thrust by governments towards its development as the transformational nature of the technology, its commercial potential and cheaper access, has caused a self-sustaining expansion of capabilities and capacities. Societies' increasing dependence on networks, however, has resulted in a surge in the number and sophistication of cyberattacks that exploit the hardware or software vulnerabilities of the networks with diverse motivations and consequences. Cyber intrusions could target government agencies and departments and private corporations or individuals, with diverse impacts on national security. These could be undertaken for espionage, to commit cybercrime or for denying or disrupting critical national infrastructure systems like power grids, telecommunication networks, transportation systems, water services, or financial and banking operations. They could be used for social engineering – spreading disinformation and molding public opinion with the intention to destabilize the internal security environment of the country, as is currently being seen as part of the current western electoral dialogue, and supporting evolving hybrid warfare techniques that see aggressive state-based actors conduct provocative activities that fall slightly below the threshold to provoke political sanctions or military response. As network centricity becomes integral to military operations, military equipment and operations could be attacked to gain strategic or tactical advantage. These attacks could even be used to cause kinetic effects and acts of sabotage or to hamper national response mechanisms, all of which would endanger lives, thereby impacting the credibility of governments and the underlying societal security frameworks. More widespread and systematic attacks could escalate tensions among states.

The perpetrators of these attacks currently range from an individual to hacking syndicates that work independently or are covertly supported by governments and corporations. States as well as non-state actors could seek an asymmetric advantage by employing such attacks to undermine an adversary's security and stability. With greater technological capability, the nature and scale of cyberattacks have continued to evolve. They are now more targeted and decisive, with clear political, economic, or military motivations and intentions. The traditional lines that would have earlier helped distinguish between the types of attacks and the motivation of their perpetrators have blurred. While the state actor could resort to an attack to undermine security or stability, a similar attack could now be undertaken by non-state actors for extortion or for obtaining information that could be sold to third parties. In that sense, a cyber threat intelligence capability is paramount to prevention and preparedness.

Networks continue to expand and become more complex, and their interdependencies continue to grow, further enhancing the vulnerabilities and increasing the difficulty in providing comprehensive protection. The environment is highly dynamic, and preventive and defensive countermeasures and reactive strategies, even with continuous efforts, are finding it difficult to keep pace with the rapidly evolving threats. Cyber situational awareness and the ability to monitor the domain, identify the vulnerabilities, and detect intrusions are still not sufficiently developed. Attempts at enhancement are hampered by concerns for privacy, freedom of speech, and the free flow of information. Even as detection rates have gone up through concerted efforts, cyber forensics need to be developed for attribution, as the attacker can easily hide his tracks in this intricate, borderless cyberspace domain. International legal regimes have failed to keep pace with the rapid technological advancements in this discipline. Advanced nations, whose critical dependence on space and cyberspace exposes them to asymmetric risks of disruption, are responding to these limitations by developing effective deterrence against misadventures, including counterthreat capabilities. Growing economies are investing heavily in computer networks and communication facilities to meet their aspirations and have received a further boost with the smartphone revolution, increasing the density and diversity of appliances used for access to the Internet. The diversity of machines makes it difficult to put in place comprehensive protection measures. Computers and networks rely mostly on foreign software and hardware, exposing them to risks associated with the global information technology supply chain. Most of the data generated in any country is exported and stored in foreign data banks using cloud computing infrastructure. Digital and smart city initiatives and the increasing involvement of the private sector in nation-building endeavors are progressive steps that are also increasing the scope and complexities of cybersecurity efforts and are further complicated by the Internet of Things (IoT). All these makes securing the domain an arduous task.

Clearly, it can be seen that most earth-based activities are touched in some manner by space when one considers the wide array of applications associated with satellite communications, precision navigation and timing, and earth and space observation. The contemporary nature of international strategic competition has reached an inflection point where national strategic security interests embrace a wide range of activities and communities ranging from traditional military to daily commercial interests. The integrity of the underlying technology base will be a central consideration to ensuring societal security and stability.

Description of the Cyber Threat Landscape Affecting Space-Based Information Systems

Space activities have obtained an increased focus in recent years. Within the space domain, space-based information systems (for the purpose of this publication, space-based information systems – referred as space systems in the further text – can be understood as the collection of devices and networks which permit the reception, processing, exploitation, and transmission of the information having their origin or destination in any type of space mission, e.g., navigation, telecommunication, earth observation, weather monitoring, etc.) use a digital infrastructure which connects primarily ground control stations with satellites. The architecture of satellite communications is composed of a ground segment or control segment, a space segment, and a user segment. Each segment has its own information exchange requirements and protection measures. Satellites can retransmit information to other satellites using relays without the intervention of ground assets. Ground control stations are connected with other ground tracking stations to exchange information. Ground stations can deliver commands through the uplink channel to the spacecraft and can receive data via the spacecraft downlink channel.

In cyberspace – beyond the computer networks, their connections through physical cables or fiber optics – there are other communication means which are often forgotten when a threat assessment is performed: the signals. Electromagnetic spectrum forms the invisible propagation medium through which the signals and their information flow. Electromagnetic spectrum has a remarkable influence in the space domain given the long distance between space assets and end users. The convergence between cyber defense and electromagnetic activities for space systems is even more significant. Reasons for that convergence reside in the need to keep control over space assets at any time as well as to ensure confidentiality, integrity, and availability of space operations. In the past, satellites were launched with an estimated life span of decades. Recognizing that satellites cannot be fixed easily when they are in orbit, the design of such satellites – expected to be fully operational for many years – foresaw minor and major software configuration updates to accommodate future enhancements. This contingency design might extend their estimated life cycle. In case of performing technical corrections to a satellite, stringent procedures must be put in place to prevent any disruption. The architectural design and engineering discipline of space systems are of the utmost importance. System engineering precepts for verification and validation testing (Byrne et al. 2014) expose security flaws which may not be previously identified by information security standard practices. The adopted cyber defense posture of space assets is a trade-off between a technological simplicity to permit defendable systems – which means fewer entry points known as attack vectors – and robust system architectures. Simplicity and robustness must go hand in hand, but it is not always feasible.

In this section, some cybersecurity topics relevant for space systems will be described with an expression of cyber threats and associated risks. The results will shed some light on challenges and recommended actions.

Electronic Warfare and Cyber Defense Convergence

It would be difficult to understand the digital battlefield as we know it today without the information services provided by space assets, e.g., geolocation. Thanks to a global coverage, space systems enable armed forces to operate in different scenarios across the globe. Space domain is more and more a contested and congested environment where armed forces aim to preserve freedom of operation. What is at stake is the ability of a military commander to plan and conduct military operations supported by a multi-domain command and control system (multi-domain command and control system refers to an emerging doctrinal concept which proposes the use of capabilities in concert across domains, e.g.. air, space, and cyber) which benefits from an integration with space systems. The global defense C4ISR (C4ISR stands for command, control, communications, computers, intelligence, surveillance, and reconnaissance) architecture leverages on space, especially because military operations are expeditionary. The armed forces are experiencing a digital transformation leading to a drastic modernization of command and control systems in all the traditional operational domains at land, sea, and air. Space systems contribute to a better integration and jointness of military operations.

The underlying physics of space communications could be altered by an adversary causing undesired events into a space mission. Space assets are equipped with thousands of sensors transmitting signals of a different nature and informing about parameters to a monitoring center. Signals may interfere with each other if they use close radio frequencies of which harmful effects depend – among other features – on the transmission power. To achieve a high level of cybersecurity in space, efforts must be pursued in promoting cyber resilience by hardening electromagnetic features of onboard space assets. Effects in cyberspace can be achieved by exploiting vulnerabilities in the electromagnetic spectrum and vice versa. Sophisticated electromagnetic techniques could make satellites vulnerable including ground equipment. Space systems must be protected against jamming or spoofing (electromagnetic jamming (United States Department of the Army 2017) is the deliberate radiation, reradiation, or reflection of electromagnetic energy, and spoofing (United States Defense Intelligence Agency 2019) is a technique that deceives a receiver by introducing a fake signal with erroneous information) and must have a capability to operate under degraded circumstances – which means to cope with limited communication capabilities.

Electronic warfare (EW) (electronic warfare (United States, Joint Publication 2012) refers to a military action involving the use of electromagnetic energy and direct energy to control the electromagnetic spectrum or to attack the enemy. EW consists of three divisions: electronic attack, electronic protection, and electronic warfare support) and cyber defense technologies need to keep pace with a fast technological development in space. Today the challenge is not to establish long-range communications that are almost taken for granted. The challenge is to secure such communications that the protocols that were born to ensure the information distribution are also valid to ensure confidentiality and resilience. Experience in this field is far from understood since the consequences of adversarial use are not publicly known. Cyber defense and EW operations are closely intertwined in a dynamic continuum like "two sides of the same coin." Ground-based EW systems considered as counter-space capabilities must be well taken into account when assuring the survivability of space-based tactical communications. Engineering design of space systems could advocate for a redundancy of key equipment in order to maintain service continuity, notably when a cyber-electromagnetic attack occurs or could enforce the use of radio waveforms less susceptible to external interferences. An increase of cybersecurity capabilities in space will allow governments to ensure their freedom of use, despite asymmetric capabilities possessed by an adversary. The duality cyber and EW and their effects cannot be neglected when approaching security solutions for space systems.

Space is a pivotal element of defensive capabilities for the military. Intelligence, surveillance, and reconnaissance (ISR) depend largely on space as well as positioning, navigation and timing (PNT). Situational awareness in space (SSA) helps to track and collect data related to space debris which is an important matter of concern for space operations. Reported cyberattacks on space systems including their lessons learned could work backwards to structure a cyber defense posture.

Managed Cybersecurity of Space Systems Is Key

It is impossible to fix a security vulnerability if neither monitoring nor detection capabilities are implemented into corporate networks of a given organization. This is a principle that should be obvious but pushes cybersecurity in space to become mainstream throughout all phases of space operations from the initial design, to launching and entry into service up to the final disposal phases. Like any other critical infrastructures do, space systems need to be monitored to detect malfunctioning or to correct deficiencies as they appear. But how big is this endeavor in the face of growing cyber threats affecting space systems? Can we extract some common ground to articulate a better readiness? The answers to these questions are not simple. Some ideas will follow to illustrate the complexity of the space security issues.

A few space-based cyberattacks can be found in the literature. A summary of cases grouped by the categories of the space missions affected is as follows (Zatti 2017):

- Missions on observation and exploration experienced incidents such as targeted interference and control take over.
- Missions on navigation experienced incidents such as a denial of service and spoofing.
- Missions on telecommunications experienced incidents such as deliberate jamming and unauthorized access.

Although, there are differences in the cyber threats affecting each category of space missions, a general classification can establish two main groups: (a) those focused on the *information*, e.g., infiltration in command and control networks, and (b) those targeting the space *infrastructure*, e.g., onboard components, payload, etc. The availability of cheap jamming devices has demonstrated the threat to global positioning system (GPS) applications (Cuntz et al. 2012). These threats may fall under the information category. GPS provides geolocation and timing information to a plethora of ground satellite receivers. GPS devices assist naval ships and drones to transit through international waters or to operate planned routes in the air, respectively. *GPS spoofing* (Falco 2018) has been revealed as a serious threat that could manipulate the GPS signal by injecting false data in the GPS receiver causing a miscalculation of the position – once again a clear example of the interrelated dependencies between electromagnetic signals and cyber defense. To reduce the impact of such attacks and minimize the risks, e.g., collisions at sea and deception, a backup to "classical" navigation systems could be an option besides demanding regulators to establish urgent measures, e.g., anti-jamming or anti-spoofing components or add-ons.

The commercial space sector is experiencing an explosion. The rapid emergence of low-cost *microsatellites* is characterized by a wide spreading of commercial-off-the-shelf (COTS) products (Falco 2018). New "CubeSat constellations" could increase the satellite's attack surface to potential hackers. Attention must be paid to ensure the adoption of minimum cybersecurity requirements before receiving an authorization to deploy in space. Policy recommendations would be focused on holding the owner of the satellite network liable for any damage that occurred and imposing a culture of "duty of care" for the space assets that private companies own, manage, operate, or develop. Another aspect of common concern which may fall under the infrastructure category is the *complex supply chains* when developing a satellite. Different suppliers may participate at various stages of the satellite's lifecycle. This situation introduces difficulties to manage accountability of the cybersecurity measures needed to be put in place. Therefore, system integrators must be empowered to certify a cybersecurity compliance – based on security standards – of every piece of hardware and software ever installed.

Challenges and Recommended Actions

To meet and outpace cyber threats in the space domain, it is recommended to create a cybersecurity space operations center (CySOC) as the focal point to handle cybersecurity incidents for space missions forming part of a wide area network with other security operation centers (SOC), computer emergency response teams (CERT), and security monitoring centers (SMC). CySOC is a threat-driven response with responsibilities to detect and support the mitigation of cybersecurity incidents, train space operators, and provide expertise about the convergence of cyber defense and electromagnetic activities to counter-cyber threats. An integral part of the CySOC would be a full-fledged cyber range for space systems. A cyber range would provide training, education, exercise, and testing functionalities including the promotion of a system engineering discipline aiming to build cyber resilient architectural designs of space systems avoiding architectural points of failure. Security certification and accreditation of space systems shall be governed by and construed in accordance with an overall cybersecurity certification framework.

A cyber range could simulate space scenarios including the environmental conditions present in space. Combining intelligence with cyber operations as part of the operational activity at CySOC could prevent the materialization of future attacks. An information sharing network (ISN) would allow the exchange of information of cyber threats to other space security centers. The CySOC must provide a cyber situational awareness capability for space systems contributing to a SSA. A digital forensic capability as part of an incident handling would permit to conduct forensic investigations appropriate to support attribution. An interesting concept would be the development of a mobile deployable CySOC which could complement the work of a permanent center in situations where it is necessary to conduct on-site activities, e.g., spaceports and military operations overseas.

Functional Capabilities for Implementing a Space Operation Center

Baseline Functions of a Cybersecurity Space Operations Center (CySOC)

The need for a cyber defense situation awareness capability (CDSA) is well documented in cybersecurity-related literature and results from extensive studies in the domain. CDSA itself is rooted in the application of foundational situation awareness (SA) concepts as applied to the cyber domain. Although there may be variants, the work of Mica Endsley (Endsley 1995a) and her decision-making model is still widely accepted as the foundation, including three levels of situation awareness (SA): perception, comprehension, and projection. Fundamental to development of any level of cyber operator SA is implementation of an appropriate means to monitor the cyber environment through collection of cybersecurity-relevant data and transforming this into information useful for the operator understanding. A CySOC provides a focal point for this essential capability of data collection and information transformation.

The CySOC focuses on the early detection of an event that may reveal an incident. It provides the necessary tools to help the analyst not only to timely classify the event but also to identify effective mitigation measures. In an effort to enhance its proactive capabilities, it will utilize any available input including a direct user input, through a call center. In addition, various threat intelligence feeds are a valuable asset to identify threats and develop mitigation measures.

Since the security analysts are not normally in control of the operational systems, the CySOC team works closely with the monitored system's administrators who implement the resulting mitigation measures. This close collaboration is also essential for continuous security protection optimization and sensor tuning to enhance the cybersecurity resilience.

As a baseline, the functional role of the CySOC is a derivation of cyber situation awareness, providing:

(a) Knowledge and understanding of the current state of the monitored systems
(b) Timely and accurate assessment of potential threats

Moreover, the CySOC provides incident management support and post-incident analysis capabilities to support forensic investigation of incidents. The scope of the investigation will focus on the:

(a) Circumstances (how and why) that allowed this incident to occur
(b) Identification of affected assets
(c) Evaluation of the extend of the compromise
(d) Identification of indicators of compromise (IOC)

Finally, the CySOC can produce meaningful reports containing situational awareness data, system statistics, and regulatory compliance status as deemed necessary per network/mission.

As already introduced above, the application of resulting mitigation measures is out of the scope of the cybersecurity operations. The role of the CySOC is strictly advisory to the monitored system's administrators who are responsible for the final decision and application of a mitigation strategy.

Commonly, cybersecurity operations center services are divided into the following three service packages to cover different needs of the monitored entities based on their internal cybersecurity capabilities and risk assessment:

1. Managed Threat Detection: The service is limited to detection and report of probable incidents,
2. Managed Detection and Response Support: Building upon detection and report, the service expands to incident response support in collaboration with the monitored entity.
3. Specialized Services: It contains additional supporting services that may be utilized on demand based on specific needs.

Managed Threat Detection (MTD)

Managed threat detection service is limited to monitoring and triage of events informing the monitored entity for probable incidents that require attention. Normally it includes the following functions:

- Call center: In the form of a call center, utilizing telephone, email, and ticketing system, the CySOC is receiving input from the users of the monitored system as an extension to the deployed sensor. The user reports aim to enhance the capability of early incident detection covering advanced threats able to evade the monitoring infrastructure.
- Proactive real-time monitoring and prioritization: Aggregating and analyzing all security events from IT and operation technology (OT) systems, as well as any necessary preprocessing and visualization to provide an overview of the current situation. This encompasses immediate analysis of real-time events and data feeds such as alerts from various intrusion detection systems or other system logs. The objective is to spot probable immediate threats or incidents that require the attention of the monitored entity.
- Reporting: Production of meaningful reports containing system statistics and regulatory compliances status as deemed necessary per network/mission.

Managed Detection and Response Support (MDRS)

The service aims to support the monitored system's incident investigation and response capabilities. The involvement of the CySOC does not stop on the reporting of a probable incident (like in MTD service) but continues with the comprehensive investigation of the event, in collaboration with the monitored entity providing mitigation strategy advices monitoring the effectiveness of the applied measures.

Managed detection and response support will normally include the following functions:

- Incident and threat analysis: Involving the in-depth analysis of events that are categorized as an alert indicating a probable incident. This capability usually involves detailed analysis of various data artifacts from different sensors and systems.
- Situational awareness: Providing knowledge and understanding of the current state of the IT/OT systems, which can provide timely and accurate assessment of the effectiveness of the mitigation strategy.
- Incident response support and collaboration: A key capability providing the tools (both technical and administrative) to support an effective response by the monitored entity to a successful cyberattack. The CySOC will collaborate with the monitored entity in order to identify in a timely manner the most effective mitigation measures to contain and minimize the impact of a cyberattack. This involves working with affected system owners to gather further information about an incident, understand its significance, and assess the potential mission impact.
- Reporting: Expands on the MTD service reports to include development of targeted threat intelligence data for the space sector.

Specialized Services

The specialized services utilize the advanced capabilities of the CySOC to support specific needs of the monitored entity. Those services can be provided on demand and may be totally independent from the other services or build upon those:

- Forensic investigation/analysis: Providing tools, means, and forensic investigation capabilities to analyze cybersecurity incidents in detail and to provide intelligence on cyberattacks in support of MDRS service or as part of post-incident analysis investigation. This encompasses analysis of digital artifacts (including various storage media, network traffic, memory analysis, etc.) to determine affected assets, identify the attack vector, produce IOCs, and establish a timeline of events.
- Threat intelligence: Ingesting institutional, governmental, or commercial threat indicators to provide timely intelligence to use against cyberattacks relevant to the monitored entity in the form of security feeds.
- Security protection optimization recommendations and system tuning: Including analysis of the monitored system's capabilities to identify required improvements, updates, upgrades, or tuning in support of the monitoring operation. This applies only to security-related devices utilized by the CySOC including, but not limited to, intrusion detection and prevention systems (IDS/IPS), firewalls, log servers, and endpoint protection solutions. The security operation team passes recommendations to the systems' owners leaving to them the implementation responsibility. The service does not provide system vulnerability assessment.

Concept of Operations

The cybersecurity operations team is organized into three teams, called Tiers, Tier-1, Tier-2, and Tier-3, organized under a technical leader and managed by a service manager.

Tier-1 and Tier-2 analysts compose the proactive detection and analysis team. The team is responsible for identifying threats having the Tier-1 analyst to perform continuous monitoring of the alert queue (system generated or by the call center) and triaging security alerts escalating them to Tier-2 analyst for further investigation and deep-dive incident analysis. The team is also responsible for identifying mitigation measures under the supervision of the technical leader supported, when required, by the advisory team. They coordinate with the asset owners under threat for a timely and effective response providing recommendations only (no actions upon the monitored system) and situational awareness reports. In the case of MTD service, the team is limited to triage of alerts leaving the further investigation and response to the asset owners.

Real-time monitoring is dependent on the necessary presence of cybersecurity-related sensors, log collection and aggregation, event correlation, and operation of security information and event management (SIEM) tools.

Tier-3 analysts form the security advisory team. The team is composed by specialists that possess in-depth knowledge in one or more of the following fields, network security, endpoint security, threat intelligence, forensics, and malware reverse engineering, as well as the functioning of specific applications or underlying IT/OT infrastructure, acting as an incident "hunter," not waiting for escalated incidents and closely involved in developing, tuning, and implementing threat detection analytics. Their functions include:

- Forensic investigation/analysis: In support to the forensic investigation/analysis service.
- Space threat intelligence feed production: Leveraging commercial and open threat intelligence feeds, as well as internal findings, the team produces space-specific threat intelligence. The threat intelligence results are used to improve CySOC operations (e.g., feeding the system with IOCs) and are reported to the monitored entity as part of the reporting function of MDRS service.
- Reporting: In addition to the required reports for the monitored entity, the team produces reports related to the CySOC itself. The reports include metrics on the effectiveness and efficiency of the operations, mean response time, etc.
- Monitoring system optimization: In support of the security protection, optimization recommendations, and system tuning service including also the CySOC system itself. The identified improvements for the CySOC system are implemented directly in contrast to the monitored systems in which the implementation falls under the responsibility of the system owners.
- Tool improvement and engineering (to improve effectiveness and efficiency): Encompassing improvements of existing supporting tools and engineering of new to improve the effectiveness and efficiency of the CySOC operations.

• Proactive detection and analysis team support: As subject matter experts, they support in incident response operations.

The above functions of Tier-3 analysts are aided by cyber simulation technologies. A fully functioning cyber range provides the means to analyze current and potential threats evaluating security feeds and hunting potential threats. The cyber range having the capacity to represent the monitored systems at high fidelity offers the optimal means to conduct the Tier-3 functions.

The service manager has a complete view and responsibility for the state of the service leading the daily operations. The technical leader is the senior security analyst of the proactive detection and analysis team. He has the full view of the service and the service operations status and is responsible for the coordination of CySOC operations.

Additional Capabilities of a Cybersecurity Space Operation Center (CySOC)

The previous section described the baseline functions of a cybersecurity operation center. However, as previously described in this article, the realities of threats to space missions go beyond the traditional monitoring and analysis of IT/OT systems and comprise other threats including (a) those focused on the *information*, e.g., infiltration in command and control networks, and (b) those targeting the space *infrastructure*, e.g., onboard components, payload, etc. As has been described, threats against the command and control networks may include specialized threats in the electromagnetic spectrum, while threats against the onboard components may take the form of kinetic threats from other space objects.

Therefore, it is recommended that the concept of CySOC should encompass not only traditional cybersecurity operations capacity but be augmented as an all-hazards threat assessment and mitigation service as a supplemental capacity to existing space operations. While the existing roles for space operations, including signal interference detection and response or SSA, should remain as the primary operational capacity for ensuring space operations against natural or random occurrences, these operations are not designed nor equipped to assess the potential of these effects being part of a larger, intentional, threat campaign. Meanwhile, a properly equipped CySOC, having access also to signal interference analysis capabilities and SSA information, could develop a larger threat and risk-based view of SA also involving indicators of compromise (IOC) from terrestrial networks, social engineering attacks, and others.

As an added perspective, to augment the role of the CySOC towards a more complete space mission assurance capability, it is recommended that the CySOC should also analyze information from ground-based physical security surveillance systems, enabling a more complete all-hazards risk assessment capability. Certainly, in all cases, CySOC operations would remain limited to the role of analysis and response assessment, offering advisory services to space operations, network and system operations or physical security operations, with each having their respective roles as final decision authority for implementation of the recommended mitigation response.

Importance of Collaborative Information Sharing and Incident Management

It is already understood that effective cybersecurity operations have a high dependency on collaborative information sharing of cyber-relevant information. Advisory services providing reference information related to vulnerability of systems, availability of malware exploits and resulting IOCs, and threat awareness information are essential aspects of cybersecurity analysis and response assessment. However, creation of effective information sharing communities for the purpose of coordinating cyber incident response remains an elusive goal. While communities of cyber adversaries will actively collaborate to develop comprehensive attack campaigns through coordination of social engineering attack specialists, information reconnaissance and data collection, and the development of target-specific malware to penetrate systems, defenders are hampered by ineffective collaboration tools to allow secure information sharing of cyber incident information across organizational boundaries.

Fortunately, changing regulations regarding responsible disclosure, improved attitudes towards information sharing, and improved technologies to support secure and selective information sharing of cyber incident information are increasing the potential for developing a coordinated incident response capability. For example, within Europe, four pilot projects, sponsored by the European Commission, have been established to develop an EU network of cybersecurity competence centers. This initiative represents the largest ever investment into creation of a collaborative cybersecurity information sharing capacity. The current initiative includes various studies and system development activities to create secure means for partners to share information among a wide scope of trusted relationships. Instances of the proposed CySOC services should participate and collaborate among such a network of cybersecurity competence both to leverage on the derived knowledge and to contribute to improvements in cybersecurity competence. In this way, the CySOC can take advantage of knowledge coming from dependent sectors and thereby improve the effectiveness of mitigation response to threats posed to space systems.

The Role of Education and Training for Developing a Skilled Space Workforce

Introduction

Space agencies are facing the challenge to educate and train a workforce that will be required to operate in an increasingly contested and complex space environment that may be affected by deliberate adversary cyber operations (Martin and Inspector

General 2012). Space operations are moreover increasingly reliant on information system technology, and even though increasingly efforts are being dedicated to better securing networks and systems, no system of systems will ever be entirely cybersecure. Therefore, a paradigm shift is needed from information assurance to mission assurance, which has to be considered in the context of complex systems that comprise not only physical and information assets but also cognitive and social domains. Subsequently solutions need to be designed for cyber resilience with the human operator as an integral part of the solution to ensure mission control in a degraded information environment.

At this time few opportunities exist for space operators to develop a clear understanding of how cyberattacks impact the systems they manage or a space mission as a whole. When cyber events are considered in a training exercise, they are typically developed in an isolated storyline out of fear that the potential cascading effects might interfere with the other training objectives. Given the principle that you "fight like you train," it is however essential to incorporate realistic cyber scenarios into a significant part of the education and training exercises, in order to produce a workforce that is capable of mitigating the effect of cyber incidents in an operational space system environment.

The design of cyber injects for training non-cyber operators can be performed in two possible ways (McArdle 2019). The "systems engineering approach" consists in first identifying an attack vector (weaponized file, social engineering, hardware of software supply chain, etc.) and subsequently the target of the attack (organization, mission, network, system, etc.). Next the specific effects of the identified attack opportunity are estimated, such as induced system failures, denial of service, and data exfiltration, and this effect is then simulated to the trainees.

The second approach is an "information assurance approach," which is based on the confidentiality, integrity, and assurance (CIA) triad. By applying the loss of one or more of the CIA attributes to a specific system, the result on the broader encompassing systems of systems, on the mission or on the organization, is estimated and presented to the trainee during the training. Certain observable effects of cyberattacks may be difficult to be distinguished from more innocent technical failures; however, both situations require a different response. Therefore, it is important that trainees learn to recognize a cyberattack in order to trigger the appropriate response and recovery. This is therefore also a training objective.

Training Environments

When training a space workforce, the option of training cyber incidents with live systems is not an option. This can be circumvented through the use of a high-fidelity synthetic training environment, where the operators interact with a virtual environment that runs as much as possible the real software systems combined with models and simulations of the physical phenomena the systems normally interact with. Synthetic systems exist for specific simulations or trainings. However, they are often tailored to a specific task and siloed, which makes it hard to integrate them into a complete synthetic training environment. It is therefore necessary to encourage simulation system developers to open up their systems and to develop federation technologies to interconnect them. It is essential to be able to run multi-domain simulations, covering all ground segment systems and radio links, as well as all aspects of the space segment ranging from the power management subsystem, attitude determination and control, communications, computing and data handling, to all possible payloads (Cayirci et al. 2017). This will make it possible to design realistic scenarios that confront the trainees with their day-to-day user interfaces and with a realistic baseline system behavior.

The goal of the non-cyber workforce is not just to train them for identifying cyber incidents but also for training the interaction with the cyber workforce, so they know the procedures, the contact points, the information to provide, how not to destroy valuable forensic information, etc. Finally, the ultimate goal is to create a digital twin of the systems that the trainees are being trained for. A digital twin is a virtual representation of a system that has been loaded with the real-world inputs and environmental parameters that were acquired on the real-world system so that it can be used to accurately simulate the real-world system's behavior.

Designing Training Scenarios

Attackers have an interest in being unpredictable. Indeed, high-threat potential actors will deliberately misrepresent their capabilities, having a stockpile of zero-day exploits at hand. They are typically capable of deploying varying tactics, techniques, and procedures (TTPs), for instance, by relying on different hacker groups. Finally, any up-to-date detailed threat intelligence covering adversaries' capabilities will likely be classified and therefore difficult to include in training scenarios. Therefore, it is impossible to design training scenarios that cover the exact incidents the trainees will face during normal operation. Nevertheless, using the available doctrinal documents from potential adversaries and drawing inspiration from documented past attacks, it will still be possible to develop relevant scenarios.

As a baseline reference for designing adversarial cyber operations, the following aspects should be considered. Adversaries will be especially interested in high-value targets, such as ground stations, communication satellites, etc. They will furthermore try to undermine trust in the information that is produced, for instance, by manipulating sensor data that is acquired by a payload or telecontrol information from the satellite itself in order to mislead the operators. Finally, they can also resort to blended strategies, where they combine kinetic attacks, like jamming or sabotage, with non-kinetic cyberattacks that reinforce and prolong the effect of the physical attack.

Evaluating Training Results

The education and training processes will have to be periodically evaluated and improved in order to ensure that they achieve the objective of a cyber-skilled

workforce. Alongside the typical evaluation approaches for more scholarly education as well as for operational training activities, it will be necessary to ensure that
the cyber-related objectives for the training have been achieved. More in particular
the question that needs to be answered is whether the staff members, both in cyber
and in non-cyber positions, are able to take the appropriate decisions with respect to
cyber incidents and events.

A valuable approach for measuring in an objective way whether the trainees have
observed and appropriately processed the necessary information to support their
decisions is the situation awareness global assessment technique (SAGAT) (Endsley
1995b) that was developed by Mica Endsley and is based on her decision-making
model with three levels of situation awareness (SA): perception, comprehension, and
projection. SAGAT is an on-line approach for real-time, human-in-the-loop trainings
that freeze the simulation at specific moments during the training and query the
subjects on specific information elements situated at the three levels of SA. Unfortunately, the fact of freezing a complex training scenario does have an effect on
trainee performance (Matthews et al. 2000); therefore, it may be more appropriate to
use real-time probing by sending SA questions to the trainees as a part of the normal
exercise communication (Jones and Endsley 2000). Since the probing should not
disturb the normal operation of the trainee too much and therefore the sampling rate
and size shall be limited, it may be useful to complement it with nonintrusive expert
behavior observations, for instance, based on the situation awareness behaviorally
anchored rating scale (SABARS) approach (Strater et al. 2001).

It must finally also be noted that cyber range-based trainings could be used for
selecting operators or for verifying the required skills for candidate operators before
they start a long and expensive training, in a similar way as Liu et al. did to select
astronauts to be trained into qualified robotic operators (Liu et al. 2013).

Conclusions

Space is changing from a selective preserve of wealthy states or well-resourced
academia, into one in which market forces dominate. Current technologies bring
space capability into the reach of states, international organizations, corporations,
and individuals that a decade ago had no realistic ambition in this regard; and
capabilities possessed a few years ago only by government security agencies are
now in the commercial domain.

Space and cyber, both technologically intensive domains, need to be harnessed
optimally for national security. Formulation and articulation of an all-encompassing
national security policy would help define domain-specific strategies and roadmaps.
There is a global trend towards increased instability in these domains as nations
develop counter-space threats and counter-cyber threat capabilities. Consequently,
space has been labeled as the fourth, and cyber the fifth, dimension of warfare.

The threat manifests itself in a variety of forms on a daily basis. Lacking a
coherent way forward, the collective accumulation of the variety of impacts from
the myriad of cyberattacks across the breadth of critical systems could serve to drain
western innovation, economy, and commerce without reaching the threshold of

triggering meaningful government and military and commercial engagement and response. There has never been a more important time or imperative for us to act upon this issue given the increasing potential existential threats posed to our societies and make this a more central element of our public agenda. The potential exists for a revolution to drive space-based security.

Disclaimer This article is a product of the authors. It does not represent the opinions or policies of the European Defence Agency or the European Union and is designed to provide an independent position.

References

Byrne DJ, Morgan D, Tan K, Johnson B, Dorros C (2014) Cyber defense of space-based assets: verifying and validating defensive designs and implementations. Proc Comput Sci 28. http://www.sciencedirect.com/science/article/pii/S1877050914001276

Cayirci E, Karapinar H, Ozcakir L (2017) Joint military space operations simulation as a service. In: Proceedings of the 2017 winter simulation conference. IEEE Press

Cuntz M, Konovaltsev A, Dreher A, Meurer M (2012) Jamming and spoofing in GPS/GNSS based applications and services – threats and countermeasures. In: Aschenbruck N, Martini P, Meier M, Tölle J (eds) Future security 2012, Communications in computer and information science, vol 318. Springer, Berlin/Heidelberg

Endsley MR (1995a) Toward a theory of situation awareness in dynamic systems. Hum Factors: J Hum Factors Ergon Soc 37(1):32–64

Endsley MR (1995b) Measurement of situation awareness in dynamic systems. Hum Factors 37 (1):65–84

Falco G Job one for space force: space asset cyber security; cyber security project, Belfer Center for Science and International Affairs. Harvard Kennedy School, July 2018. https://www.belfercenter.org/publication/job-one-space-force-space-asset-cybersecurity

Jones DG, Endsley MR Can real-time probes provide a valid measure of situation awareness. In: Proceedings of the human performance, situation awareness and automation: user-centered design for the new millennium, Savannah, GA (2000)

Liu AM et al (2013) Predicting space telerobotic operator training performance from human spatial ability assessment. Acta Astronaut 92(1):38–47

Martin PK, Inspector General Nasa cybersecurity: an examination of the agency's information security. US House of Representatives, Feb (2012)

Matthews MD et al (2000) Measures of infantry situation awareness for a virtual MOUT environment. In: Proceedings of the human performance, situation awareness and automation: user centred design for the new millennium conference

McArdle J (2019) Rethinking cyber training for the non-cyber warrior, conference summary and conclusions, Pell center for international relations and public policy. https://pellcenter.org/wp-content/uploads/2019/01/McArdle-Rethinking-Cyber-Training-Final.pdf

Strater LD et al (2001) Measures of platoon leader situation awareness in virtual decision-making exercises. No. SATECH-00-17. TRW INC Fairfax VA Systems and Information Technology Group

United States Defence Intelligence Agency, Challenges to Security in Space, January 2019

United States Department of the Army, FM 3-12 Cyberspace and electronic warfare operations, April 2017

United States, Joint Publication 3–13.1 electronic warfare, 8 February 2012

Zatti S, The protection of space missions: threats and cyber threats; information and systems security. In: 3th international conference, ICISS 2017, Mumbai, India, 16–20 Dec 2017, Proceedings

AI and Space Safety: Collision Risk Assessment

49

Luis Sanchez, Massimiliano Vasile, and Edmondo Minisci

Contents

Abstract

The New Space environment will introduce notorious changes in the space environment during the next decades with the irruption of mega-constellation, the boost of small satellites, and the generalization of low-thrust engines. Current safety strategies and collision avoidance procedures will no longer be capable to deal with the increase on conjunction alerts. Artificial intelligence appears to be the best strategy to cope with this new situation, thanks to its ability to perform faster than physical models and make decisions based on a wider range of parameters than human operators. This results in better performances when more data are available, as the situations will present on the coming years.

L. Sanchez (✉) · M. Vasile · E. Minisci
University of Strathclyde, Glasgow, UK
e-mail: luis.sanchez-fdez-mellado@strath.ac.uk; massimiliano.vasile@strath.ac.uk; edmondo.minisci@strath.ac.uk

© Springer Nature Switzerland AG 2020
K.-U. Schrogl (ed.), *Handbook of Space Security*,
https://doi.org/10.1007/978-3-030-23210-8_136

941

Introduction

The space environment is under a radical transformation that affects technologies, use of the space, mission concepts, and operations. Electrical propulsion, once proved its reliability and capabilities, has started to be used during the last decade on commercial and scientific satellites, both in Low Earth Orbit (LEO) and Geostationary Earth Orbit (GEO), and its use is expected to grow. In the late 1990s, technology improvements have resulted in miniaturization of space components that eventually have allowed satellites to reduce its size. Since 2003, when the first CubeSat was launched, the use of such small satellites by universities or for commercial usage has continuously increased, and it is expected to keep growing during the next decade. Along with this increase on small satellites, a higher rate of launches per year and new countries and private actors entering the scene are also expected. Among these new actors, maybe the most relevant due to its impact on the orbit environment, will be swarms and constellation mission. Along with small satellites, mega-constellations will represent the most profound change in the LEO regime during the next decade. Several of these constellations, each of them compounded by thousands of satellites, are planned and some of them have already started the deployment stage. It is expected that in the next years, the number of satellites in orbit will multiply by several times. Bearing in mind that the current number is slightly below 2,000, it will push the figures to tens of thousands of operational satellites in orbit at the same time (Lewis et al. 2017).

Space debris also increase notoriously the problem, since it is the most common type of object orbiting the Earth. Space debris refers to all man-made objects in space apart of operational satellites as well as micrometeoroids captured by the Earth's gravity. It includes upper-stage rocket bodies, inoperative satellites remaining in orbit, objects left by missions, and fragment from old satellites due to fragmentation or collision. From the beginning of the Space Era in 1958, the number of space debris objects has kept growing to reach the current state where there are in orbit more than 34,000 objects bigger than 10 cm, more than 900,000 between 1 cm and 10 cm, and millions of them even smaller (ESA Report 2019). These numbers are also expected to increase in the following years, not only linked to the increase in space traffic but also due to improvements in the current tracking techniques. New infrastructures are expected to start their operation in the next decade allowing the detection of smaller objects, which have not been possible to track until now. While this increase in the cataloged objects does not mean an increase in the actual number of objects since they are already in orbit, it will boost the number of conjunction alerts experienced by satellite operators (Haimerl and Fonder 2015) (Fig. 1).

What does it mean for safe operation of satellites? First of all, the collision of operational satellites between them or with space debris will become a more tangible threat, and the very single event of this characteristic ever happened can repeat on a more congested future space environment. In the second place, these new satellites will operate in already congested areas, some of them with limited propulsion system or with an electric one, what means more complicated and longer operations will be

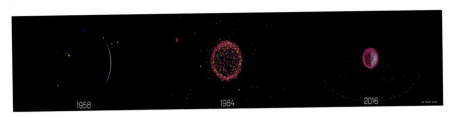

Fig. 1 Evolution of the number of objects in orbit: 1958–1984–2016. (Credits Dr. Stuart Grey)

associated to them, with an impact on the planning of collision avoidance maneuvers (CAM), more time crossing critical regions, like ISS and human space flight regions or already populated orbits. This highlights the impact of future New Space's satellites and the precariousness of the system at critical stages when agreeing a common avoidance strategy between operators, regarding that already a CAM has had to be implemented between an ESA satellite and another from Starlink mega-constellation, where this one is not even completely deployed. On the third place, the expected increase on cataloged objects can collapse the STM system, since linked to it, an increase on conjunction alerts can be expected. Despite most of these alerts does not mean a collision is going to happen or a CAM will be needed, time and effort has to be put on evaluate the risk link to each of them. If besides the probability of collision, the metric used for evaluating events as high-risk or low-risk conjunctions, presents important limitations (note that the aforementioned Iridium-33/Cosmos-2251 collision event presented a probability of collision not classified as high risk by several operators), the system leads to a catastrophic result unless major renovations are implemented (Peterson et al. 2018).

Among those renovations, automation is a major one. A shift from a system in which each satellite is operated by several agents to a system where only one operator can manage several satellites is desirable. However, such a situation is not possible with the current system structure, especially considering the expected traffic growth. It is at this point where artificial intelligence (AI) plays a crucial role. AI techniques can operate faster than current models and can take decision considering a wider set of parameters than human operators and have the capacity to perform better when the available data increases, which is the scenario expected for the next years in space. If a certain set of reliable data is provided, AI systems are able to learn directly from them and predict accurate results without the need for any physical model. In a scenario where more and more data will be available and when time is a critical resource, using the surrogate model these techniques provides can be the key for the automation of the Space Traffic Management system. While only a few examples of AI applied to Space Traffic Management can be found, they have been successfully used for predicting events, classification, and decision support in other engineering fields, including space and air traffic management. This allows thinking that AI systems are a promising trend for the next years.

The rest of the chapter deepens on the application of AI in space engineering in general, and in Space Traffic Management in particular. Beginning with, a summary of the current situation of the Space Traffic Management (STM) and Space Situational Awareness (SSA) systems is presented, highlighting the critical situation for the future regarding the expected increase in space traffic. An overview of studies about AI in the field of traffic management, collision avoidance, and space engineering is then presented, followed by a survey of the main works on the application of AI on the STM system. Finally, some challenges to be addressed for a good implementation of AI techniques are stated.

AI and Space Safety: Collision Risk Assessment

Space Safety System

A fundamental concept in space safety is Space Traffic Management (STM), which is defined as "the set of technical and regulatory provisions for promoting safe access into outer space, operations in outer space and return from outer space to Earth free from physical or radio-frequency interference" by the IAA (International Academy of Astronautics) in the Cosmic Study on Space Traffic Management (2006, 2018). This concept includes a wide field where different knowledge areas play a role on space safety. On one hand, there are the rules, standards, and recommendations related to the satellite operations, maneuvers, conflict resolution, and collision avoidance. This group also includes the protocols to be implemented if a conjunction between two operational satellites is reported as well as the good practices on sharing satellite's operations information. On the other hand are the technical aspects whose aim is the implementation of the previous protocols and good practices for the safe operation of the satellites, including tracking of space objects, conjunctions detection, and risk assessment as well as action for the mitigation of the risk of collision.

Another concept related to space safety is Space Situational Awareness (SSA) that involves the actions, techniques, and technologies for the tracking, orbit determination, and calculation of ephemerides of the space objects. Both SSA and STM are closed-related since the STM system needs the knowledge provided by SSA about the state of the satellites to provide conjunction alerts and perform the correct collision avoidance maneuvers (CAM) if needed. Combined, these two systems create a more complex one that involves information of thousands of space objects; requires the coordination of different operators, satellite owners, and teams; and provides alerts and recommend actions to be taken whose consequences have to be managed in a short interval of time.

Continuous monitoring of all the trackable space objects around the Earth, both operational and nonoperational satellites, is carried out by SSA service providers. The main actor is the USSTRATCOM, although commercial companies and other states are getting more relevant in the last years. When a potential encounter between one operational satellite and a piece of space debris or another operational satellite is detected when propagating the observable states, a Conjunction Data Message (CDM) is created and sent to the operators in charge on the involved satellites. Since information collected by the SSA system (USSTRACOM) presents low quality, especially for space debris objects, the observable state and the propagated one are affected by uncertainty. A more dedicated following can be carried out to reduce this uncertainty. With all the available CDMs associated with a single event, a conjunction risk assessment is executed by the operators' CARA (conjunction assessment risk analysis) team to determine if the event represents a true threat for the satellite and the space safety or not. In the case of a high probability of collision associated with the event, a complex process starts. The first step, if the event involves two operational satellites, is agreed a common strategy, relying on manual communication between operators that delay the process. The common procedure (the two satellites moves, the biggest one moves, the one with propulsion system moves) is then analyzed with the payload team, flight dynamics team, and ground stations to come up with a possible collision avoidance maneuver strategy. This step requires a lot of coordination, time, and workload as it is critical for the success of collision avoidance. Secondly, the proposed strategy is then evaluated to ensure that the risk of the current event is reduced and no future possible collision arises: with the same object (secondary collisions) or with other bodies (tertiary collisions). Eventually, once the maneuver is approved, the event is closely monitored, and 1 or 2 days before the Time of Closest Approach (TCA), as long as the risk remains high, the CAM is performed. After the CAM is executed, the state of the satellite should be monitored again to check the maneuver has been correctly performed.

It can be seen how many critical points the performance of a single CAM associated with only one event presents. In the first place, information of conjunction is provided just a few days in advance what gives a tight interval of time for the whole process to be performed. Since CDMs are available for the last 7 days before TCA, this is the time window operators have. However, the actual time interval is shorter since first CDMs present high uncertainty and better quality data are usually required. In the second place, some of the steps involved in the process are computationally expensive and time-consuming for operators, which is added to the tight time window where the process is carry out. If better quality data are demanded for the conjunction risk assessment, sensors require time for providing accurate orbit determination information. Not only that but also accurate orbit propagation is time-consuming, and it is an operation that has to be implemented in several stages: for using actual orbits for obtaining the risk of collision, for evaluating different CAM proposals, or for assessing future collisions once the CAM is implemented. It is not just a time issue. Besides, coordination effort is a key aspect of the process. Flight control, flight dynamics, ground station team restrictions, and mission requirements have to be considered when evaluating the possible collision avoiding strategies. The coordination tasks would be even more critical if the potential collision involves two operational satellites when teams of both missions have to agree a common strategy, a problem that worsens due to the lack of protocols and specific regulations (Peterson 2019).

STM is also responsible for managing all the conjunction alerts received before the collision risk assessment process stipulates the event that represents high probability of collision or not. All those alerts that do not need a conjunction risk assessment and those that after the assessment do not require a CAM are considered as false negatives. They do not give any information about real collisions but increase the operators' workload. There is a greater number of non-actionable alerts than actual high-risk events, which means that an important part of the resources is spent on events that are not relevant for space safety. Contrary to these false alarms (false positives), there is the possibility of false negatives to occur. False negatives are those high-risk events that are misclassified, which can lead to collision or risky events not noticed by operators in advance. As mentioned before, the collision between Iridium-33 and Cosmos-2251 was a situation like this (Peterson et al. 2018). The root of these events resides, partially, in the bad quality of initial position data, especially for the space debris objects, what makes the acquisition of better quality information essential, bringing more information to be managed by the system.

Note that the situation presented shows the current state of the system, where the traffic of space objects has not experienced the expected next year's growth. The implementation of a CAM explained above involves only a pair of space objects; however, it involves multidisciplinary teams to coordinate a lot of information in a very constrained interval of time. False alerts and false positives mentioned in the previous paragraphs currently happen. The increase on launch rate programmed for the next decade leads to the question of the scalability of the system; the final issue STM and the space safety system will face scalability. The increase on space traffic will make operators struggle in managing all the available information, and sub-optimal decisions are likely with effects on space safety. If currently, hundreds of alerts are triggered, the future space environment will push this number to limits that the system may not cope with. Since more resources should be put on filtering false alarms, the assessment of collision risk and mitigation strategies will suffer from this increase of alerts. Besides, future operators' systems would not be based on a team formed by several operators taking care of one or a few satellites, but smaller teams controlling a whole constellation with several satellites each. Such a situation is not possible unless a greater level of automatizing is implemented, and the use of a decision support system is used to replace most of the operators' tasks (Nag et al. 2018) (Fig. 2).

SSA, which is also a fundamental part of the system, also introduces critical points to the process. It is responsible for obtaining position information about all the objects orbiting the Earth, satellites, rocket bodies, or pieces of debris, as well as the ephemerides of those bodies. As can be expected due to a large number of objects orbiting the Earth, the amount of information the system has to deal with is already enormous, not counting for the expected growth of space traffic of the next decade. Furthermore, the next years will witness the start of the Space Fence system operations, the new SSA system developed by the USA for improving the monitoring of space objects. The expected increase in sensitivity will allow the track of smaller objects, invisible at the moment, including bodies in the range of 1–10 cm, making it possible to include up to 200,000 orbiting objects in the catalogs. It means that information related to space objects position and ephemerides will increase even more at a rate much bigger than the numbers of launches since most of the new objects that will enter in the catalogs are already in orbit. Another contributor to the

Fig. 2 Operative, on deployment, and planned constellation on LEO region

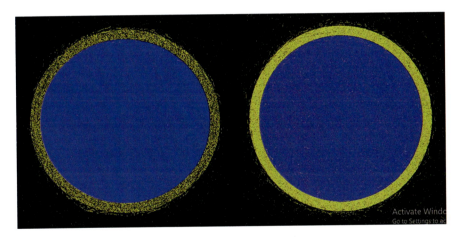

Fig. 3 Current LEO catalog versus expected catalog when Space Fence is operative

SSA system is the commercial providers that independently to the traditional sources carry out track campaigns, whose information has to be merged with those from the US and agencies' catalogs (Crosier 2016) (Fig. 3).

The New Space environment presents similarities with Air Traffic Management (ATM) and Unmanned Air Traffic Management (UTM) systems, where the increased on traffic population have forced them to adapt themselves to the new circumstances. ATM system is a well-established system which has coordinated an increasing air traffic population for several years. Key aspects of this system is the clear distribution of responsibilities, a proper set of protocols and common practices, and the effectiveness to control several objects under only one control center, which has facilitated the automation of activities previously carried out by human operators. However, as in STM, the population growth, especially in certain regions, has forced actors involved on the system to develop a more automatic system (Kochenderfer and Chryssanthacopoulos 2011). UTM is another example where automation has been implemented to handle the rapid traffic increase and the necessity of a quick decision-making process. Some studies show the efficiency of implementing automation on the UTM and the possibility to adapt the proposed system structure to the STM (Murakami et al. 2019). Furthermore, decision support systems (DSS) based on AI has started to be implemented on unmanned aerial vehicles (UAVs) control systems not for automatizing but for supporting operators on the decision-making stage. These approaches use AI, fuzzy logic, and other related techniques for rapidly taking into account a wide set of parameters and compute a ranking of the best options to implement under a conflict, automatizing tasks previously done by operators and speeding up the whole process. Operators can then select the appropriate actions based on the ranked list of alternatives and based on certain criteria, reevaluating alternatives under new criteria or recomputing the list if more information is available. While STM presents its particularities with respect to UTM or ATM, it is clear that when traffic management systems have experienced the congestion of the environment, they have tended to the automation of the system, usually relying on AI techniques.

The successful examples of applying AI on other engineering fields, including space engineering and traffic management have boosted the interest on using these techniques on the STM system for automatizing tasks, speeding up the process, or supporting operators on taking optimal decisions in an environment that is overcoming the capacity of human operators since more and more data and variables have to be considered. Among the actors interested on implementing AI for STM, ESA and NASA can be named, both with programs to study the availability and applicability of AI methods onto real missions and scenarios (Benjamin Bastida et al. 2019; Mashiku et al. 2018). ESA has identified three main issues to be addressed by AI for facing the population increment on the orbital environment: reducing operators' workload (automation), lowering the decision-taking time on risk conjunction assessment and collision avoidance planning, and scaling down the number of false alerts.

Artificial Intelligence in Engineering

Artificial intelligence is referred to as the ability of computers to learn from data and reasoning, acquire knowledge, react to the environment, and correct themselves to imitate human intelligence or behavior without being specifically programmed to do it. It is a wide knowledge area including machine learning, natural language representation, computer vision, and data mining, among many others (Russell and Norvig 2009). It has been studied for some decades, but only during the last years, with faster and more capable computers and the availability of big dataset, it has been possible its implementation into real applications in a broad range of disciplines, including engineering (Fig. 4).

Among some of the applications of AI in engineering, one interesting field related to space safety is traffic management and collision avoidance. An important trend in recent years is the application of AI on autonomous cars. Image recognition, intelligent decision systems, and autonomous collision avoidance are issues presented in this field and addressed by AI. However, the applicability of those techniques to a completely different environment as it is space is not a straightforward task, and it is currently under research. The development of robotics has also brought some improvements in autonomous collision avoidance algorithms. Regarding the increasing autonomous of satellites, bringing them closer to the general idea of what a robot is, some attempts of extrapolating those algorithms to the space environment have been analyzed, and it is an interesting research area where promising results are expected on the next years.

The See Traffic Management (SeeTM) system presents also some examples of the application of artificial intelligence on collision avoidance. While space and maritime environment presents notorious differences, there is also some similarities, like an initial sparse and wide operational space which has experimented an increase on traffic density, having led to the necessity of implementing autonomy on the traffic management systems, or regions where this density is reaching current system limits, like ports on the see environment and the LEO region in space. In this sense, it is interesting the work presented in Statheros et al. (2008). There, an intelligent ship collision avoidance system is implemented, combining both dynamic and AI

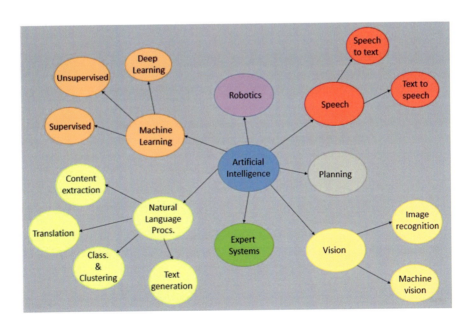

Fig. 4 Artificial intelligence areas

surrogate models. Such a system structure could be implemented, and it is on process to be implemented, on STM.

In the field of traffic management, there are also examples of using AI in the Air Traffic Management (ATM) system and Unmanned Air Traffic Management (UTM) systems. Autonomy, both on the vehicles and on the operators' activities, is spread on these systems, although not necessarily by using AI. Nevertheless, the increase in air traffic and the irruption of commercial UAVs interacting with the convectional aerial traffic have forced the system to implement AI techniques for supporting the operators on the management of the system (Kochenderfer and Chryssanthacopoulos 2011; Julian and Lopez 2016; Ramirez-Atencia et al. 2017).

The space sector is getting interested in AI too, having incorporated techniques and methods in different areas. Natural language processing (Berquand et al. 2018), knowledge representation, automated reasoning, computer vision (Jasiobedski et al. 2001), trajectory optimization and navigation (Izzo et al. 2019), satellite autonomy (Anderson et al. 2009), and robotics are some of the fields in space engineering where AI have made interesting contributions.

Artificial Intelligence in Space Safety

It seems clear that there is a well-established field of research and application of AI in different fields of engineering, including dealing with conflict, managing traffic, and supporting decision-making. Based on the studies presented in the previous section, it is reasonable that AI and machine learning (ML) methods can be applied also in STM. As mentioned previously, space agencies and other actors involved on space safety have started to implement lines of investigation on this direction, and it is worth to explain in more detail the three main issues AI is expected to solve according to ESA (Benjamin Bastida et al. 2019):

- Reducing the tasks operators currently carry out by implementing automation. Future increase in space traffic will translate in growth on the time and effort operators will spend just dealing with alerts, classifying events, performing detail conjunction risk assessment, planning and executing maneuvers, collecting better data, or managing end-of-life strategies. Currently, some of these activities present a certain degree of automation, while others require several dedicated hours. Investing in the automation of most of these activities will allow operators to focus on the decision-making stage, on the nominal operation of satellites or the handling of more satellites simultaneously. Another important area where automatizing can liberate much of the operator's time is on the coordination between teams and other operators in the event of a conflict, switching for the current manual procedures for a much more automatic one, with clear protocols and standardized steps.
- Lowering decision-taking time. Automation of operators' tasks will allow them to spend more time and effort on the critical steps of decision-taking in collision risk assessment, collision avoidance maneuvers, or disposal strategies evaluations.

However, the expected rise in space population will imply the number of satellites to be controlled, and the amount of information to be considered will exceed human operators' capacities. AI-based systems for supporting on the decision-making stages, like DSS agents, will be able to handle all this information and propose alternative strategies to operators in much lesser time than current approaches taking into account a wider range of variables. Besides, surrogate models provided by AI techniques for skipping computational expensive propagator or dynamical models or the uses of databases with predefined maneuver examples to automatically find the optimal one are other AI-based options for reducing time in the future STM system.

- Reduce false alarms. Currently, the vast majority of conjunction alerts reported to operators correspond to events that do not require any additional action (neither avoidance maneuvers nor a more detailed evaluation). While triggering alerts, this kind of events do not imply true collision scenarios but consumes time and resources unnecessarily. In the next decade, when smaller objects can be tracked and more satellites will be in orbit, the number of such events will boost and more resources would be needed only to filter the actual collision encounters or high-risk events from all the non-actionable cases. Correctly selecting events without missing the high-risk ones (false negatives) nor wasting resources on false alerts (false positives) will be as essential as challenging for future STM, regarding current databases are dominated by those less interesting low-risk events.

While AI has been used in other areas of space engineering, its application on Space Traffic Management and Space Situational Awareness is limited. However, it is possible to find some pioneer works on this subject. While scarce, they cover different aspects of the STM and SSA system, addressing some of the previous aspects highlighted by ESA as priorities.

Some of those works are focused on improving orbit determination by the implementation of ML. In Peng and Bai (2018a), support vector machine is used for reducing the positional error of satellites after orbit determination and orbit propagation processes. In Peng and Bai (2018b), they continued with this line of research, switching from SVM to artificial neural networks (ANN). What they proposed in those works is the use of ML for improving orbital determination parting from the idea that classical models keep unused certain embedded information from historical data. Using both SVM and ANN, they tested the models for predicting a satellite's position and velocity error caused by measurement and dynamic propagation model limitations. Using the historical information of a certain resident space object (RSO) during an interval of time, they expected to find the relation between them and the aforementioned error in three circumstances: for the same RSO in the same interval of time of the historical data, but at epochs not including in the training set, for the same RSO but for times after those included on the historical data, and for near RSOs, both in the same interval as the training data and posterior epochs. They demonstrated by different numerical experiments the possibility of using ML for reducing orbit determination error and thus, improving orbit position knowledge. The benefits of this method for the SSA system are clear. While SSA is responsible to keep track of all the thousands of RSOs orbiting the Earth, the accuracy of observations and models is restricted due to the great number of objects and limited knowledge of the environment when building the models. Being able to correct the errors associated with them, especially derived from imperfect modeling of the dynamics (drag, solar activity. . .) and limitations of the observation sensors will automatically provide a better position for detecting conjunctions and evaluating their risk. Nevertheless, the same authors are aware of some of the limitations affecting this approach as reported on another publication (Peng and Bai 2017). Lack of real data for propagating, time window limitations on the predictions, and restricted generalization to other objects different than the ones used for training are some of them (Fig. 5).

The previous approach corrects orbit determination and propagating errors, but it is still limited to the orbit propagation of each of the object of interest, which is time-consuming when several bodies are considered. Sanchez et al. (2019) propose an ML-based system for predicting collision encounters by using a set of ANNs for predicting the equinoctial parameters of a satellite during an interval of time, by providing exclusively the initial Keplerian parameters. By comparing the predicted orbits of a couple of objects, the equinoctial parameters obtained during the whole interval of time are good enough for estimating potential conjunctions by calculating the impact parameter (B-parameter) between the two bodies. In the end, the proposed method is not anything else but a surrogate orbit propagator that substitutes the dynamic models by a

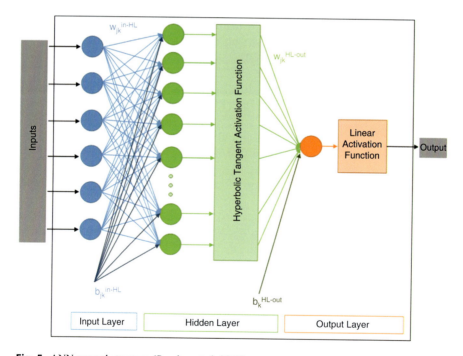

Fig. 5 ANN general structure (Sanchez et al. 2019)

surrogate model based on data. Some interesting aspects explain the importance of this approach and summarize some of the general advantages of ML. First of all, this method provides a surrogate model of the underlying problem (the two bodies perturbed movement) that does not rely on any dynamic model nor uses any integration method (nor analytical nor numerical). Since no integration is involved in the propagation, it performs faster. We are moving toward an environment where thousands of pieces can be a threat to the operational satellites and where operators will be responsible not only for one but for several of them, including constellations. Moreover, the tracking system will struggle on providing good positional data from every piece of space debris at any time. Possessing a fast and accurate model able to compute the propagate orbit of these thousands of satellites becomes crucial for the future of STM. The second advantage this approach presents is that the model relies on the data used for training. As in Peng and Bai (2018a, b), dynamic model errors are avoided since ML does not use any physical model but builds one based on the available data. In this way, by using the historical real position data, the uncertainties associated with drag, solar radiation pressure, and any other physical effects difficult to model simply do not influence the final result. As can be seen on the results proposed in Sanchez et al. (2019), the error is not dependent on the closeness to the initial epoch, as it usually happens on dynamic based orbit propagators, since an independent set of six ANN has been trained for each epoch based on the real orbital parameters of the training RSOs. This work is presented as a first step toward the use of ML in STM and, therefore, also presents some limitations: data used for training (assumed as real position) comes from a virtual database obtained by using a high fidelity propagator, and the conjunction events prediction is made assuming the Keplerian propagation of one of the satellites involved on the conjunction. However, despite these limitations and using a relatively simple ANN model, it can provide accurate results for equinoctial parameters and detection of conjunction events for RSOs different from those used during training. In addition, it performs quickly compared to orbital propagators when several object's orbits are propagated. Despite providing preliminaries results, it sets a promising path for using ML in orbit determination and orbit propagation (Fig. 6).

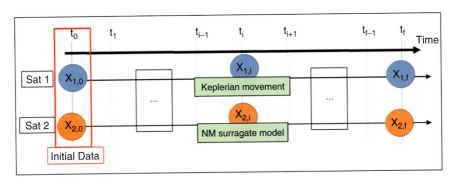

Fig. 6 ANN for orbit propagation and conjunction event prediction introduced in (Sanchez et al. 2019)

Other approaches have been followed for applying AI in STM. In Sanchez et al. (2020), ML algorithms have been tested for classifying conjunction events based on a new approach for evaluating the risk assessment. The new approach pretends to overcome some limitation of a common risk assessment metric, probability of collision, by using the belief and plausibility concepts coming from evidence theory, accounting thus for epistemic uncertainty on collision risk assessment. This approach takes the conjunction geometry between two objects involved in a conjunction and includes the uncertainty from the point of view of the evidence theory. Assuming one or more sets of statistical distributions, each parameter defining those distributions is provided by the different sources (i.e., sensors or experts) as an interval or intervals, without assuming any distribution of the parameters, but only the true value is included on one of them. Each of these intervals is associated a basic probability assignment accounting for the reliability of each source, which allows taking into account aleatory and epistemic uncertainty independently. The classification criteria proposed for conjunction events are then based on the time to the encounter and belief and plausibility thresholds. Some ML methods, like artificial neural networks, random forests, support vector machine, and k-nearest neighbors, have been tested for creating two different intelligent classification systems, one using as input values of belief and plausibility as well as time and the other, considering time and geometry, allowing skipping the time-consuming step of computing the belief and plausibility curves. Each of the classes is related with an action that would be suggested to operators in the decision-making process. The results proposed in this work show the potential of using ML for supporting decision-making. Another intelligent decision support system is presented in Vasile et al. (2017). The idea of the proposed method is supporting operators in the planning and implementation of collision avoidance maneuvers when needed. An interesting contribution of this work is the creation and exploitation of a database of possible predefined maneuvers to be implemented in a conjunction event scenario. A virtual satellite position dataset was created to obtain conjunction events and later, computing the optimal maneuvers, which were stored in a new database. The new orbits generating after the CAM were also stored in the initial database and analyzed for detecting future encounters and thus, obtaining a wider range of CAMs for feeding the ML algorithms. The availability of a database with these characteristics, with a broad variety of possible maneuvers, provides fundamental information to an intelligent decision system for providing alternative proposals based on certain criteria. The criteria selected on this work considered the risk of not executing the collision compared with the risk associated with future possible collisions. The ideas presented on those works lay the foundations for future intelligent DSS for supporting operators. Other criteria can be implemented on the AI-based DSS to elaborate more sophisticated ranked lists of proposals to the operator like confidence on the sources, the inherent risk of executing a maneuver, the cost of the maneuver versus the cost of the satellite itself, restriction due to mission requirement, or fuel usage limits. Sophisticated DSS system accounting for several variables and proposing alternatives in relative short interval if time has already proposed in other fields of traffic management (Ramirez-Atencia et al. 2017).

There is another aspect of the satellites' mission crucial for space safety: the disposal and reentry stages. The end of life of a satellite affects space safety in several aspects. First of all, it is essential for decreasing the rate of space objects in orbit, since it is the easiest way of removing bodies from space; second, during the decay stage, satellites have to cross highly populated regions, something that will become more critical when mega-constellations are completely disposed of. Finally, it is an extremely uncertain stage since atmosphere drag starts to be the dominant effect and density models are imprecise, solar activity is still not well modeled, and knowledge on the behavior of satellites during reentry is hard to predict. Minisci et al. (2017) presented a study for uncertainty propagation during the last stages of a GOCE mission. Besides the uncertainty quantification and characterization study, the use of high-dimensional model representation (HDMR) methods and the creating of large databases have set the path for the future use of AI on reentry time windows prediction. In the same work, meta-models based on AI were preliminary studied for mapping initial stated and model uncertainties to reentry time windows. Initial results suggest that it is possible to estimate the reentry window by this method and without any propagation. Further analysis is being carried out for a better implementation of this idea, and results suggest the potentiality of this approach.

Some other works and studies relating AI with space safety, STM, and SSA have been carried out recently. Furfaro et al. (2019) used recurrent neural network (RNN) and convolution neural networks (CNN) for classifying and characterizing RSOs based on their curve of light for STM. In Mashiku et al. (2019), supervised and unsupervised ML algorithms and fuzzy logic have been implemented for predicting close approaches by using not only the classical probability of collision but other parameters as well. Finally, Shabarekh et al. (2016) use ML approach for predicting where and when will maneuver be executed in the future to improve SSA capabilities.

Challenges for the Future

AI is a promising approach for being implemented in STM to face the challenges of the new space environment expected for the next years. Space agencies, operators, and commercial agents have shown interest in these techniques to ensure the future of space satellites, and there are already ongoing researches for addressing the issues. However, being a new approach means that it has to face several challenges before we can talk about a space safety system based on AI.

As can be seen from current and past studies, a common problem is the lack of appropriate datasets for training the models. AI techniques are based on the availability of enough information to fit the models, extract information or capture the patrons relating data. However, actual information from real satellites is not always available in the desired format or with the required quality. Indeed, orbiting satellites are periodically tracked, allowing accessing to a great amount of historical data; however, some of these objects are not tracked with good enough quality to allow AI techniques to extract reliable information or more accurate results than traditional methods. On the other hand, some information is not available at all, like maneuvers implemented by satellites, or the information is not enough to allow the AI models extracting patterns. Therefore, current AI techniques relay on simulated databases that have the advantages of creating a broad casuistic. However, an important challenge for the coming years regarding the implementation of AI is the creating of databases with information coming from real scenarios: real CDMs, information about implemented collision avoidance maneuvers, uncertainties associated with measurements, and state propagation.

Artificial intelligence involves a wide set of branches. So far, space safety has just scratched the surface on the application of those techniques in STM. Most of the methods implemented and studied are centered on the machine learning branch, more specifically on supervised learning. However, there is a wide range of possibilities in AI where STM can take techniques from. Intelligent Problem Solving, including Evolutionary Computing and Constraint Satisfaction Programming, can be an interesting branch for DSS development along with fuzzy logic, automating reasoning, or knowledge representation. Computer vision and image recognition are also open areas where STM can benefit from, besides data mining. The implementation of AI in STM is still a new research area, but the potential for solving some of the problems already identified is huge. The advantage is that AI is a more tested technology in other fields, including engineering. As has been seen, traffic management has already benefited from AI, and space engineering has already used AI techniques for some years. The current state of the art allows STM system to use these technologies and their advantages for solving its own challenges, automating tasks and speeding up processes.

There is still another challenge to face, as it is the implementation of these kinds of techniques onto real applications. The work carried out so far is focused on proving the capability of these techniques to improve the STM system and ensure space safety in the oncoming scenario. However, there is still a long way for being able to implement those techniques on real missions or in the actual system. More research has to be done for really understating the relation between training AI models and the physical laws ruling the data; more detailed studies for optimizing techniques should be performed as well as adapting the system for gradually incorporating the proven methods. It is now a perfect time for testing new approaches since the space environment is changing, and new techniques are not advisable but mandatory for the sustainability of the system, but at the same time, it is critical to implement reliable methods in order not to collapse the system. This leads to the last and main challenge the implementation of AI in STM has to face: the lack of standards on STM. Several AI-based approaches can be suggested, but as long as there are no protocols of actuation and standardized actions in conflict situations, the problem of a congested space will still be there. AI techniques as a way for supporting operators and moving to an automated scenario will work as long as a set of common rules and practices are shared by the different agents using the space.

Conclusions

New Space will bring great challenges to space safety in the next decades. The implementation of new technologies, new concepts of satellites, and new kinds of missions, like low-thrust engines, small satellites, or mega-constellations, will push the limits of the space system to its limits. On top of all of this, the problem of space debris, which is going to become worse with the increase in space traffic, will make it completely necessary to carry out drastic changes on the system in order not to collapse it.

Although these changes can come from different approaches, there is a consensus on the space community that automation of the Space Traffic Management and Space Situational Awareness systems is one of them. To achieve the required level of automation, AI techniques arise as the most promising tool due to a series of factors. Their ability to deal with huge amount of data, and not only that but also learning from them and improving performances when more information is available, the advances on computer systems that allow its implementation both in the ground segment and in-orbit computers, the wide range of fields of application and task they can be applied to or the possibility to speed up the process where they are used, and the capacity for automation and decision-making support are just some of their advantages.

While used in other engineering fields, like traffic management or computer vision among many others, the application in space engineering started near in the past, focused on image recognition, autonomous navigation, satellite autonomy, orbit trajectories, or robotics. However, it is only in recent years where space safety has started to implement AI techniques, where only a few promising studies have been carried out. However, the trend followed by agencies and space actors points in an increasing relevance of AI for STM, and since it may be the only tool able to handle all the information the congestion environment expected for the next decade will generate.

Three main issues are expected to be addressed with the implementation of AI on space safety, space traffic management, and collision avoidance: automation of certain task to reduce operator's man workload, minimize time between decisions (conjunction risk assessment or collision avoidance planning and implementation), and reduce the number of false alerts in relation of potential high-risk conjunction events.

However, as a starting technique on the field, there are still some challenges to overcome. A common limitation already faced is the lack of proper database based on real scenarios. AI techniques are based on the availability of representative data. The creation of appropriate databases with information coming from real satellites, events, and scenarios, or at least, a database of virtual scenarios closely similar to real situations is vital for obtaining the better performance of these techniques. AI is a wide area with several fields. At this moment, only some of them have been preliminary studied, mainly focused on the machine learning area. Studying different approaches and performing analyses to determine the best AI branch to solve each problem related to space safety are highly recommendable to obtain the maximum benefits from AI. Finally, lack of protocols and standardized practiced is a drag for obtaining the best performances of some of these methods. A promising area on AI is the development of intelligent agents or intelligent decision support systems. However, these methods required a series of clear rules to provide the appropriate advice to operators. Agreeing on common rules and practices for all space actors is essential for the proper implementation of AI in space safety.

References

Anderson JL, Kurfess FJ, Puig-Suari J (2009) A framework for developing artificial intelligence for autonomous satellite operations. In: Proceedings of the IJCAI-09 workshop on artificial intelligence in space, Pasadena, 17–18 July 2009

Benjamin Bastida V, Flohrer T, Krag, H, Merz K, Lemmens S (2019) CREAM – ESA's proposal for collision risk estimation and automated mitigation. In: 1st international orbital debris conference, Sugar Land, 9–12 December 2019

Berquand A, Murdaca F, Riccardi A, Soares T, Gerené S, Brauer N, Kumar K (2018) Towards an artificial intelligence based design engineering assistant for the early design of space missions. In: 68th international astronautical conference, Bremen, 2018

Crosier C (2016) United States Strategic Command space situational awareness sharing program update. United Nations Committee on the Peaceful Uses of Outer Space, Scientific and Technical Subcommittee, 54th session, Vienna, 3 February 2016

ESA Space Debris Office (2019) ESA's annual environment report. Technical report, ESA. https://www.sdo.esoc.esa.int/environment_report/

Furfaro R, Linares R, Reddy V (2019) Space objects classification and characterization via deep learning and light curves: applications to space traffic management. In: Space traffic management conference, Austin, 26 February 2019

Haimerl J, Fonder G (2015) Space fence system overview. In: Advanced Maui optical and space surveillance technologies conference proceedings, Maui, Hawai, September 2015

Izzo D, Märtens M, Pan B. (2019) A survey on artificial intelligence trends in spacecraft guidance dynamics and control. Astrodyn 3, 287–299 (2019). https://doi.org/10.1007/s42064-018-0053-6

Jasiobedski P, Greenspan M, Roth G (2001) Pose determination and tracking for autonomous satellite capture. In: 6th international symposium on artificial intelligence, robotics and automation in space, Montreal, June 2001

Julian K, Lopez J (2016) Policy compression for aircraft collision avoidance systems. In: IEEE/AIAA 35th digital avionics systems conference (DASC), Sacramento, September 2016

Kochenderfer MJ, Chryssanthacopoulos JP (2011) Robust air-borne collision avoidance through dynamic programming. Lincoln Laboratory, Massachusetts Institute of Technology, Lexington, Massachusetts

Lewis H, Radtke J, Rossi A, Beck J, Oswald M, Anderson P, Bastida Virgili B, Krag H (2017) Sensitivity of the space debris environment to large constellations and small satellites. J Br Interplanet Soc 70:105–117

Mashiku A, Frueh C, Memarsadeghi N (2018) Supervised-machine learning for intelligent collision avoidance decision-making and sensor tasking. 2018 NASA Goddard workshop on artificial intelligence, Greenbelt, Maryland, November 2018

Mashiku A, Frueh C, Memarsadeghi N, Gizzi E, Zielinki M, Burton A (2019) Predicting satellite close approaches in the context of artificial intelligence. In: AAS/AIAA astrodynamics specialist conference, Portland, 11–15 August 2019

Minisci E, Serra R, Vasile M, Riccardi A, Grey S, Lemmens S (2017) Uncertainty treatment in the GOCE re-entry. In: 1st IAA conference on space situational awareness, Orlando, November 2017

Murakami D, Nag S, Lifson M, Kopardekar P (2019) Space traffic management with a NASA UAS traffic management (UTM) inspired architecture. AIAA SciTech Forum, San Diego, 7–11 January 2019. https://doi.org/10.2514/6.2019-2004

Nag S, Murakami D, Lifson M, Kopardekar P (2018) System autonomy for space traffic management. In: IEEE/AIAA digital avionics and systems conference, London, September 2018

Peng H, Bai X (2017) Limits of machine learning approach on improving orbit prediction accuracy. In: Advanced Maui optical and space surveillance technologies (AMOS) conference. Maui Economic Development Board, Wailea Marriott, September 2017

Peng H, Bai X (2018a) Improving orbit prediction accuracy through supervised machine learning. Adv Space Res 61(10):2628–2646. https://doi.org/10.1016/j.asr.2018.03.001

Peng H, Bai X (2018b) Artificial neural network–based machine learning approach to improve orbit prediction accuracy. J Spacecr Rocket 55:1248–1260. https://doi.org/10.2514/1.A34171

Peterson G (2019) Establishing space traffic management standards, guidelines, and best practices. Center for Space Policy and Strategy, The Aerospace Corporation, El Segundo, California

Peterson G, Sorge M, Ailor W (2018) Space traffic management in the age of new space. Center for Space Policy and Strategy, The Aerospace Corporation, El Segundo, California

Ramirez-Atencia C, Bello-Orgaz G, R-Moreno MD, Camacho D (2017) Solving complex multi-UAV mission planning problems using multi-objective genetic algorithms. Soft Comput 21(17):4883–4900. https://doi.org/10.1007/s00500-016-2376-7

Russell S, Norvig P (eds) (2009) Artificial intelligence. A modern approach, 3rd edn. Prentice Hall Press, Upper Saddle River

Sanchez L, Vasile M, Minisci E (2019) AI to support decision making in collision risk assessment. Paper presented at the 70th international astronautical conference, Washington, DC, 21–25 October 2019

Sanchez L, Vasile M, Minisci E (2020) On the use of machine learning and evidence theory to improve collision risk management. Paper presented at the 2nd IAA international conference in space situational awareness, Washington, DC, 14–16 January 2020

Shabarekh C, Kent-Bryant J, Keselman G, Mitidis A (2016) A novel method for satellite maneuver prediction. In: Advanced Maui optical and space surveillance technologies conference, Maui, September 2016

Statheros T, Howells G, McDonald-Maier K (2008) Autonomous ship collision avoidance navigation concepts, technologies and techniques. J Navig 61(1):129–142

Vasile M, Rodriguez-Fernandez V, Serra R, Camacho D, Riccardi A (2017) Artificial intelligence in support to space traffic management. Paper presented at the 68th international astronautical conference, Adelaide, September 2017

Space Object Behavior Quantification and Assessment for Space Security

50

Moriba Jah

Contents

Abstract

As the spacefaring community is well aware, the increasingly rapid proliferation of human-made objects in space, whether active satellites or debris, threatens the safe and secure operation of spacecraft and requires that we change the way we conduct business in space. The introduction of appropriate protocols and procedures to regulate the use of space is predicated on the availability of

M. Jah (✉)
Department of Aerospace Engineering and Engineering Mechanics, The University of Texas at Austin, Austin, TX, USA
e-mail: moriba@utexas.edu

© Springer Nature Switzerland AG 2020
K.-U. Schrogl (ed.), *Handbook of Space Security*,
https://doi.org/10.1007/978-3-030-23210-8_103

quantifiable and timely information regarding the behavior of resident space objects (RSO): the basis of space domain awareness (SDA). Yet despite six decades of space operations, and a growing global dependence on the services provided by space-based platforms, the population of Earth orbiting space objects is still neither rigorously nor comprehensively quantified, and the behaviors of these objects, whether directed by human agency or governed by interaction with the space environment, are inadequately characterized.

Key goals of advanced SDA are to develop a capability to predict RSO behavior, extending SDA beyond its present paradigm of catalog maintenance and forensic analysis, and to arrive at a comprehensive physical understanding of all of the inputs that affect the motion of RSOs. Solutions to these problems require transdisciplinary engagement that combines space surveillance data with other information, including space object databases and space environmental data, to help decision-making processes predict, detect, and quantify threatening and hazardous space domain activity.

Introduction

This chapter presents an introductory overview of space object behavior quantification and assessment through the lens of more known functions such as surveillance, tracking, and information fusion for space domain awareness. The presumption is that humanity as a whole will be more efficient, protected, and successful in their future space domain activities and dependencies if a common operational perspective can be achieved in the space domain. Without a common perspective of the space domain, serious operational weaknesses may result when space services and capabilities are degraded or denied to any entity by either natural or human-made causes. With the rapid assimilation of information technology globally and associated applications to the modern space domain, it becomes imperative that the world holistically maximize its total space domain awareness.

Space capabilities and services have been essential supporting utilities underpinning much of the global economy and technology for many years. However, the threats to those capabilities and services, along with the consequences of their loss or degradation, have only recently become a global concern given the proliferation of plausible threats and the growing dependence upon space throughout the world. All space capabilities and services are of individual national origin. The protection of space systems (satellites and controlling ground infrastructure) that provide the capabilities and services to users and customers is a singularly sovereign responsibility. In fact, not only the operation of those systems but the collection/acquisition and dissemination of information on and about the space domain are also inherently sovereign in nature. Since space has become more important as a contested domain extending human activity, this increasingly poses a dilemma for achieving global space security, safety, and sustainability.

Historically, two of the more pervasive challenges of integrating multi-entity-sourced data and information that contributes to space domain awareness are (a) overcoming the national and commercial sensitivities of data sharing and (b) the accurate technical fusion of such data. Those challenges extend to the effective achievement of space domain awareness for the world. It is posited that the following three types of data and information may be more likely to be shared than other more sensitive information (e.g., space-related intelligence): (1) space surveillance and tracking, (2) space environment, and (3) radio-frequency interference experienced by space communication links. Thus, these three areas of space domain awareness are used within this chapter as the exemplars for characterizing a likely initial global space domain awareness picture.

The space domain can be defined as all conditions, areas, activities, and things terrestrially relating to space, adjacent to, within, or bordering outer space, including all space-related activities, infrastructure, people, cargo, and space-capable craft that can operate to, in, through, and from space.

Space domain awareness, in this context, can similarly be defined as the effective understanding of anything associated with the space domain that could impact the security, safety, economy, or environment of space systems or activities, globally. The definition acknowledges the supportive activities and threats related to land, maritime, air, and cyber regimes relevant to space operations. It requires the combination of space situational awareness foundations of detecting, tracking, and environmental monitoring, along with space intelligence foundations of characterizing normal behavior and sensitivity, to detecting change to know when an event or process has or is predicted to occur. A purpose of SDA is to provide decision-making processes with a timely and actionable body of evidence of behavior(s) [predicted, imminent, and/or forensic] attributable to specific space domain threats and hazards. Although there are no limits on what constitutes space domain awareness data, it is essential to initially address, at a minimum, space weather and environmental reporting, space object tracking and characterization/classification, as well as radio-frequency interference characterizations and attributions against satellite control links and communication services.

To date, SDA has lacked credible scientific and technical rigor to quantify, assess, and predict space domain threats and hazards. The current state of the art suffers from a number of inadequacies: there are no standard definitions of elements in the space domain; descriptions of space objects and events are limited; no standard method of calibrating sensors and information sources has been developed; tasking is addressed to individual sensors for specific data rather than to a more holistic system for information required to address needs and requirements; there exists no rigorous understanding of space environment effects and impacts on space objects; there is no framework that encourages and enables big data analysis and supports an investigative "from data to discovery" paradigm; we lack a consistent method to understand all of the causes and effects relating space objects and events.

The need to address these concerns has never been greater. On-orbit collisions, natural or intentional, are a global concern that threatens the long-term sustainability of our space activities and environment and worsens the impact of the space debris population growth in critical mission-dependent orbital regimes. It accounts for an increase in the useless space object population of about 1% annually (with isolated events contributing spikes upward of 20% population growth) and jeopardizes the livelihoods of tens of millions of people who depend on critical space capabilities and services.

Traditionally, efforts to develop and maintain awareness of all trackable space objects have relied upon the USSTRATCOM's Space Surveillance Network (SSN). But these sensors are often prohibitively expensive for even the richest of nations, and the space domain is too vast for traditional space surveillance, ground- or space-based, to be truly effective by itself. Protecting important space assets, especially those that provide critical services and capabilities such as communication, weather, bank routing, position, navigation, and timing, requires a new approach encompassing twenty-first-century technology and a fundamental understanding of the processes governing the behavior of objects in space.

It is in this context that the following sections in this document are given and seek to place the characterization and behavior of space objects on a rigorous scientific footing. Until now, the global approach to space operations has been largely reactive, following the latest commercial exigency or governmental demand signal of the day. By contrast, the fundamental work required should lead to new ways to understand, measure, and predict behavior in space. In turn, that work will underpin the development of best practices in space traffic management and inform efforts to improve mission assurance and mitigate the effects of space debris hazards.

Space Domain Awareness Goals: Knowing and Predicting Events and Processes

The set of all space domain events and processes, as a whole, is unknowable for many reasons. Within this whole set, we have a subset of events and processes that we believe have and are occurring. Not all of these beliefs are measured and for those that haven't been measured, we can refer to them as "hypothesized knowledge." There is also a subset of space domain events and processes that have been measured, and this is what we call evidence. However, not all evidence has been processed or used to extract knowledge. So, for evidence that has yet to be processed and either used to generate a belief or confirm or refute one, we call these "latent knowledge." Lastly, where beliefs and evidence are not mutually exclusive, we call this "inferred knowledge." The subset of space domain events and processes for which we have neither evidence nor beliefs, we acknowledge as the Arcana or "Ignorance" (Fig. 1).

Space domain awareness (SDA) is the actionable knowledge required to predict, avoid, deter, operate through, recover from, and/or attribute cause to the loss and/or degradation of space capabilities and services. The main purpose for SDA is to provide decision-making processes with a quantifiable and timely body of evidence of behavior(s) attributable to specific space threats and/or hazards. SDA encompasses all activities of information tasking, collection, fusion, exploitation, quantification, and extraction to end in credible threat and hazard identification and prediction. Understanding the synergy between the space environment, the interaction of this space environment with objects (astrodynamics), the effects of this space

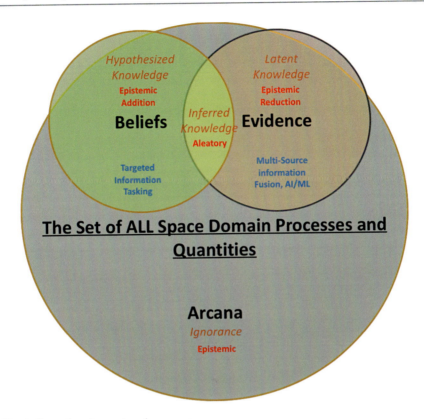

Fig. 1 Space domain events and processes

environment on objects (operational and not), and the available sensors and sources of information is critical to meaningful SDA. Included in the SDA purview is collecting raw observables, identifying physical states and parameters (e.g., orbit, attitude, size, shape), determining functional characteristics (e.g., active vs. passive, thrust capacity, payloads), inferring mission objectives (e.g., communications, weather), identifying behaviors, and predicting specific credible threats and hazards. Intuitively, SDA is a natural "big data" problem, drawing from existing and potential metadata and data sources. The problem at hand is (a) how these articulated needs can be rigorously addressed using first principles; (b) what methods, techniques, and technologies must be leveraged from other fields or targeted for development; and (c) what sensors, phenomenology, sensor tasking, or additional data are needed to support the SDA mission.

Existing research and technology focuses largely on collecting observables, identification of physical states and parameters, and determining functional characteristics. Advances include extracting observations and new information from non-traditional sensors, improving track association and initiation using admissible regions, using Finite Set Statistics methods to improve detection and tracking, and classifying space objects using ontology and taxonomy approaches.

The intent of any operational component is to predict resident space object (RSO) behavior with quantified uncertainty in order to provide decision-makers with timely warnings of specific hazards and threats. Behavior prediction must take into account the behavior of other RSOs, physics, and indirect information gleaned from non-standard sources.

To achieve this, let us focus on the foundational philosophy with which the SDA problem space should be engaged. The most important ingredient in SDA is evidence. Ideally, it all begins with that. To know something, one must measure it; to understand something, one must predict it. Both of these require observations, one to abduct knowledge and the other to demonstrate understanding. As long as one has a hypothesis that explains all past observations and can predict future ones, to within a quantifiable measure of precision, it cannot be proven that such a hypothesis is false. However, the truth about all observers is that they all lie. There is no such thing as an observer that reports the truth, because all observers (physics- and human-based) are corrupted by noise and/or bias and are of finite resolution, precision, and accuracy.

Most people attempt to force the data to answer a very specific question (i.e., estimate a specific state), yet no one has ubiquitous observations. In other words, the set of data available to the analyst is by and large almost certainly incomplete. Yet, decisions are pervasively made based upon incomplete data, and one tends to find ad hoc (and oftentimes simplifying) assumptions in place in order to arrive at such decisions. Even when the answer is obviously "insufficient information to decide," many analysts make a decision anyway without truly understanding the consequences of deciding outside of the available information content or inferable knowledge. Another way to state this given Fig. 1 is that most decision-makers make decisions based upon hypothesized knowledge (beliefs unsubstantiated by evidence).

Robust and meaningful space domain awareness requires the analyst to embrace the complexity of the problem at hand and seek to abduct knowledge from the available data and assume nothing external to what the data allow one to assume. In other words, one must mine all of the possible and available evidence, refraining from prejudice.

I call this the "Moriba Jah Refrigerator Approach" to analysis, because it is likened to how a refrigerator cools an object. Cooling is not achieved by directly cooling the object but rather by indirectly doing so via the elimination or removal of heat from the object. Therefore, for space domain awareness, the approach should be to remove ambiguity from the system instead of attempting to estimate a sole and specifically posed state. In other words, the analyst must seek to be prejudice-free in the inference process and hypothesize as much as possible and only use the available data to discard a hypothesis that cannot explain the evidence. In so doing, all surviving hypotheses have some non-zero likelihood of being true. The analyst should make decisions based upon all the possibilities that survive the scrutiny of evidence. Not doing so exposes the end user to unquantified or erroneous risk. Imagine that in the beginning, the analyst has a body of hypotheses much like a block of marble of arbitrary shape. As data are gathered and analyzed, the analyst must subject each hypothesis to predict what has been observed, and if a hypothesis fails to do so within acceptable and quantifiable precision, then it should be discarded and likened to chipping away pieces of marble, with the ultimate goal to reveal the statue residing within the block itself. Additionally, it would be of favor if the analyst could use data alone to form hypotheses and, in this way, not subject a priori knowledge (bias) to the analysis. This may be a role for methods such as machine learning and artificial intelligence, but these should be invoked judiciously and not in the absence of known laws of physics.

In this text, so-called hard inputs will refer to information sourced from physics-based sensors such as radars and telescopes, and so-called soft inputs will refer to information sourced from human observations or interpretations. In general, hard inputs will be numerical-valued functions, and soft inputs will be semantically valued. But this is a generality and not universally true. What follows are finer descriptions of these two sources of information.

Human-Based (Soft Inputs)

Of importance to space domain awareness, which is really "decision-making knowledge" for the domain, is context. This context cannot result solely from an interpretation of physics. Although most of the tracked resident space objects (RSOs) are defunct objects, debris, there is a subset of the population that are actively controlled by humans. Moreover, humans have valuable information to provide into a system that is attempting to quantify, assess, and predict the behavior of objects in space.

Much can be learned about an upcoming launch from human-based information, such as semantic corpora like "tweets" and other online media outlets. Detecting and tracking newly launched RSOs, for example, has proven to be challenging even for the most expert space surveyors. This could be made easier if one can fuse or couple physics-based inputs (i.e., measurements from physical sensors such as radars and telescopes) with human-based inputs (e.g., an opinion or human-made observation). One example resides in the use of so-called two-line elements or TLEs. These are provided by USSTRATCOM but have no measure of precision or uncertainty associated to them. Are TLEs useful? Indeed they can be, as long as they are treated as a human-based input and not like a physical sensor measurement. Prior work by Delande et al. (2018) demonstrates how these two sources can be indeed fused to achieve improved insight into RSO trajectories. It is beyond the scope of this chapter's section, but please refer elsewhere for more details. In essence, it is possible to model the uncertainty of a RSO driven by systematic effects or ignorance (epistemic) rather than assume all uncertainty is driven by randomness (aleatory).

One of the salient challenges in exploiting human-based information is that it must be properly curated and normalized. This can be readily understood in the following example:

Let there be an unmanned aerial vehicle (UAV) in an enclosed room with a ranging sensor in one corner but this ranging sensor is unable to detect objects in the whole room. Moreover, the room contains obstacles represented as the furniture.

Now, assume that there are also three windows allowing three humans to see into the whole room. Each human has their own microphone to provide input into the UAV guidance, navigation, and control, along with the physics-based (range) sensor. At a given moment, the UAV is on an almost assured path to colliding with a chair in the room. The range sensor can weakly observe the UAV, the first human says "watch out," the second says "ummmmm, that's going to hurt," and the third says "you have an obstacle 3 meters ahead of you." The range sensor's input is straightforward to implement but what about the three humans? Not only do they have information that could be useful to the benefit of the UAV, but they say three very different things which all attempt to convey the same predicted event, a collision. So, these human data need to have an associate measure of uncertainty, and these semantic inputs need to be mapped to a lingua franca or "normalized" in order to minimize redundancy and confusion.

While a radar and telescope are examples of physics-based sensors, natural language processing (NLP) is an example of a human-based sensor as its objects are semantic. Finding relevant information content in a physics-based sensor can be "trying to find a needle in a haystack," but finding relevant information content in human-based inputs can be "trying to find the needle in the needlestack."

Structured and Unstructured Information

Without spending too much time in this section, for the intents and purposes of this text, structured data are those that can be found or had in specific formats, oftentimes repeatable and defined by standards (e.g., ISO). These may be ideal in terms of how they are formatted because one can develop software and algorithms to consistently and appropriately interpret and exploit these data. However, as for unstructured data, one can think of these as serendipitous, not necessarily repeatable, or even with any metadata or ancillary information to provide more meaningful context. For example, a query "scraping" the Internet for semantic data regarding launches will result in some tweets, some online media articles, etc., and these do not have a common format or structure to them, making them more difficult to interpret. Moreover, as in the example of the UAV, many different things can be alluding to a common event or process, and this must also be inferred.

Standards, Calibration, and Metadata

To several points made earlier, true and meaningful SDA cannot occur in the absence of properly engineered, modeled, and curated data. Something that makes this more straightforward to achieve is the use and adherence to standards. For space we have quite a few, such as ISO, CCSDS, RINEX, EOSSA, and others. An interpreter can be developed for each, so that when data are provided in any of them, the data can be readily "ingested" and mapped/transformed to a common and appropriately labeled framework (e.g., a graph database) that can in turn be queried by a variety of users. If users have to interpret each type of data to achieve their space domain decision-making knowledge needs, it is impossible to have this be timely, actionable, consistent, etc. Instead, what is desired is to make this interaction with "raw" information sources transparent to the user and simply provide the user with an ability to query some form of database or knowledge graph that already contains the relevant and salient information in a way that quickly and readily provides the information required to perform analytics or other. Standards make this process simpler.

Information content can only exist in the presence of differences or rates of change. How does one know whether or not a given information source is "credible" or not? Biases are commonplace among information sources, and some are additionally noisy and corrupted by systematic or random errors. If all one has is a single clock, it is quite challenging to actually know the local time. However, if one has a time reference for comparison, one can quantify how accurate and precise one's clock is. This underscores the importance of calibration as a frequent process within SDA. Think of information as evidence and each piece of evidence needs to be assessed for its value, relevance, and accuracy/precision. We must know how to interpret and "weigh" each source. This is the goal of calibration, and it is a critical element of meaningful data curation.

Regarding the interpretation of information from any given source, it is insufficient to simply have measurements absent context. For example, simply having several thousand observations from an unknown telescope or radar makes it almost surely impossible to exploit these data. In order for the data to be actionable, they require details regarding where they were collected, the precision of the sensor, and perhaps other information such as how they were collected or assumptions made, etc. This context or ancillary information is oftentimes referred to as "metadata." The word "meta" means beyond. Datum is simply an assumption or premise from which inferences can be drawn. Data is the plural of datum. Hence, metadata would be "beyond the information" but really is context or information regarding the information. Metadata are critical in providing the SDA practitioner with guidance on how to use the data.

The Importance of "Independent Observations" and Big Data

Related to the previous section, how does one know if one has the most accurate clock in the world? One must have access to as many clocks as possible, of all types, so that one can weigh all the independent evidence and find its "barycenter." (Think of this term as the "center of mass" of multi-source information or evidence, when all are appropriately weighed and compared.) Then and only then can one know because then one can compare his or her clock to this information barycenter. How precise is your clock? Well, along with determining the information barycenter once can also determine how scattered or spread apart these clocks are from each other and this "variance" can provide insight on our clock's precision. Let's explore why independent observations are critical for SDA in an example. Let's assume that you have a satellite near the GEO region and have a genuine concern if any other RSO is within 1 km of your satellite. What if I told you that I predict that another RSO will be within 100 m of your satellite this time tomorrow. Would you maneuver your satellite to avoid the risk? What if I then told you that I arrived at that conclusion based upon a single sensor. Would your decision change? You might want another source (independent) of information that can either confirm or refute my hypothesis, right? In order for evidence to be the most actionable possible, it must be corroborated evidence, and this can only happen if there is at least one additional and independent source of information observing the same or common event or process. It's of little use to the global community if India makes one of their satellites explode on orbit stating that the resultant debris will all reenter the Earth's atmosphere in just over 1 month, and the United States states that they are tracking debris in higher orbits that will take decades to reenter. We have two entities with opposing hypotheses and no ability to corroborate either of these publicly. This is a global problem that must be addressed because it is only a matter of time before two entities have a dispute to resolve. What will be the body of evidence required to satisfactorily resolve the dispute?

Moreover, having lots of telescope data to observe the debris would be welcomed. Yet it is much more useful to have a lot of disparate data, so, for example, having several telescopes, radars, observers from the ground, observers from space, etc. all adding to the pool of evidence. A lot of data is just that, a lot of data. However, a lot of disparate data is so-called big data. Why are big data important? Because it provides additional perspectives and "observability" into a common event or process via mutual information that a single type of data cannot provide due to limitations in the sensor itself. No single type of sensor can observe in all wavelengths and frequencies and resolutions. Is it more powerful to see a hazard or does hearing it and perhaps even smelling it also help make a decision? Disparate sensing allows for more quickly abducting the underlying system or behavior from the data. It helps eliminate wrong answers more quickly. We should always evaluate evidence not only for its ability to support a hypothesis but perhaps even more so for its ability to discard a hypothesis.

Space Domain Information Fusion: A Model

Data on the space environment and objects in it, imported into the SDA process, come from a disparate variety of sources and sensors. To maximally exploit the information, we must in some sense fuse the data. In this context, the concept of "data fusion," which is so often only vaguely defined, means that we seek quantitative answers to specific questions with the lowest uncertainty permitted by all the available data. For example, "Where will this object be next Tuesday at 3 o'clock?" or "What is the likelihood that my on-orbit network capability will be disrupted by space debris within the next 2 years?" To address this challenge, we have defined a Space Domain Information Fusion (SDIF) model, illustrated in Fig. 2.

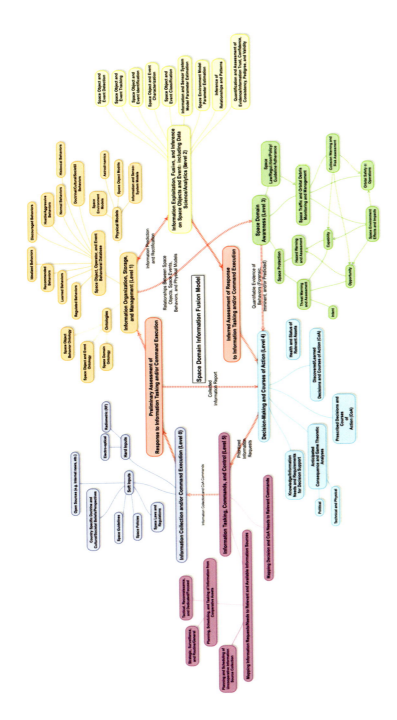

Fig. 2 Overview of the space domain information fusion model

We wish to maximize the mutual information between the evidence and our beliefs. Note that this is very different from a conventional data processing approach, which seeks to make the output (e.g., an image) "look like" the scene. A caveat is that arriving at an equivalent measure of information for soft inputs such as opinions is still a tremendous challenge.

The SDIF model is designed to demonstrate a system of systems that accomplishes a series of tasks:

- Facilitate the gathering of information from a system, driven by the specific needs of a given user.
- Autonomously determine how to weigh, trust, and process new information and evidence into the system.
- Provide a rigorous and physically and semantically consistent picture of the space domain via hard and soft input information fusion.
- Discover previously unknown elements of space objects and events via the leveraging of knowledge graphs and ontological frameworks (See http://astria. tacc.utexas.edu/AstriaGraph.).
- Provide space object behavior and event predictive capabilities that are credibilistically quantifiable.
- Demonstrate the art of the possible in terms of decision-making processes and enabling command and control products and services.

The SDIF model provides a closed-loop information framework that can satisfy a variety of user needs, with a broad range of operational concerns, where the knowledge of the space domain is common. The framework provides a common operating picture that is consistent for all users. The model consists of six main levels as shown in Fig. 2 and described in summary as follows:

Level 0

Here, raw (instance) data enter the system. These data sources include hard inputs from a variety of sensors and historical surveys, as well as soft inputs such as United Nations guidelines, European Union codes of conduct, country-specific doctrine and cultural beliefs, press announcements, and other open-source literature. Both are important in predicting, quantifying, and assessing space object behavior including space threats and hazards. The focus should not be constrained to information that may only seem to be relevant to the space domain, but all information, as numerous and as disparate as possible. The reason for this is that the Space Domain Information Fusion model, coupled to the exploitation of knowledge graphs and ontologies, will see to discover things regarding the behavior of space objects and events via correlations with any seemingly unrelated information. For instance, what if there is movement of certain space objects prior to each G7 summit? One would never know this unless these events could be easily linked and correlated. The following are examples of information sources:

- Information relevant to space object operational status
  - Owner/operator telemetry
- Information relevant to space object and event behavior and dynamic models
  - Thruster activity
  - Thermal profile (radiation and energy balance)
  - Solar/Earth radiation and Earth albedo
  - Outgassing
- Information relevant to non-dynamic models
  - Sensor performance (noise, biases, and latencies)
  - Sensor locations and Earth orientation parameters (EOPs)
  - Media corrections (atmospheric refraction, signal delays, etc.)
- Space object and event background, context, and sensor/metadata
  - JSpOC or country mission catalogs: TLEs, VCMs, state vectors, conjunction assessments
  - Observations from ground- or space-based sensors
  - Commercial data providers (e.g., AGI, GMV, ISON, Airbus, Zodiac Systems)
- Space object and event historical and country-specific behavior
  - Known break-up events
  - Known anti-satellite (ASAT) weapon tests
- World history, news, and geopolitical information
  - Councils, summits, wars/armed conflicts, etc.
  - Financial trends, events, etc.

Level 1

This forms the heart of the system: it is the foundational piece that must be correct. The space domain is described through knowledge graphs with relationships between objects described by a set of ontologies (schema). A behavioral database, models of the physics, and other information about the space domain "universe" are at this level. This is where all incoming information and evidence is stored, before and after processing, and where past, current, and predicted knowledge and beliefs about our "logosphere" reside. The fundamental function of this level is to go "from data to discovery": it is designed to leverage big data science and analytic methods. Whenever anyone wants to know something regarding space objects and events, this is where that information will be drawn from and, if absent, will generate an information request that will be sought to be achieved. This graph-based database also contains and maintains various representations of uncertainty and ambiguity associated with the data. At this level, no judgements are made regarding the behavior of space objects and events. Here are examples of categories and mechanisms of information to be stored and managed:

- Knowledge graphs and ontologies
  - Consistent and rich representation of space objects and events that facilitates linking of large and disparate sources of information

- Behavioral context and history
  - Cultural/societal perspectives
  - Known past behavior and events
  - Geopolitical positions
- Physics-based models
  - Space environment
  - Space objects
  - Astrodynamics
  - Sensor and information systems
- Information storage and management
  - Dynamic nature
  - Concurrent access and sharing

Level 2

This is where our beliefs and knowledge in Level 1 are subjected to critical scrutiny. Here also we assess the degree to which any new evidence can be trusted and, if the evidence indicates that our beliefs should change, to what extent we allow that change to be made or our confidence in our belief to be adjusted. What do we do when evidence seems to conflict each other? How do we quantify, incorporate, and fuse the information we might find in someone's opinion? So far, no specific questions have been asked of the information; the intent is simply to update knowledge of the "logosphere" as described to the extent possible given the evidence provided. No judgments are made.

Once our beliefs have been rectified (confirmed, changed, or neither because any new evidence was unrelated), any changes are mapped back to Level 1 to bring our knowledge up to date.

Level 3

Here is where we ask specific questions about things in our "logosphere" and where the tools of our analyses are brought to bear to make judgments about those things and their relationships. Users will supply their own questions and decision-making criteria. For example, to one user an object 1 km from a specific space asset may be threatening. Another may be comfortable with a separation as small as 100 m. Level 3 takes the knowledge from Level 1 and assesses it against user-defined criteria.

By keeping Levels 2 and 3 separate from Level 1, users can apply different evidence and judgments to the information without changing how space objects are defined and represented. In this paradigm, the picture of the space domain is consistent regardless of the specific user.

Level 4

At Level 4, decisions are made by addressing questions such as "Should I do something?", "If I do this, what is the expected effect?", and "What other information do I need to decide between these three courses of action?" Some courses of action might be predetermined by the user, and others not. The user may simply be looking for a body of evidence of something occurring in the space domain that concerns them.

A not-so-subtle issue at this level is the notion of confidence. If the user is concerned with having any other space object within 1 km of their own, how accurately must we know the answer? If our answer were "object X will be within 1 kilometer of object Y tomorrow at noon UCT," the customer will ask "what's the error on that?" If our reply is "+/− 10 kilometers," the customer may likely choose to do nothing because the level of uncertainty may not warrant the effort and risk. Thus, it is critical to have not only the customer's criteria for warning or notification but also the level of knowledge required to enable a decision. This is part of what is meant by providing "actionable" knowledge. There is work that aims to quantify this measure of confidence regardless of whether from hard or soft inputs (Bever et al. 2019).

Level 5

Any output from Level 4 that leads to a requirement for further information passes to Level 5 where sensors and information sources are tasked to collect new information. Other non-information gathering actions may also be tasked. A prioritized list of actions is established and executed. The user has a lot of flexibility into what happens at this level.

Space Surveillance and Tracking

As defined by the United States Strategic Command (USSTRATCOM), space surveillance involves (but is not limited to) detecting, tracking, cataloging, and identifying man-made objects orbiting Earth, which include active/inactive satellites, spent rocket bodies, debris, and fragments. Space surveillance accomplishes the following:

- Analyze new space launches and evaluate orbital insertion.
- Detect new man-made objects in space.
- Chart present position of space objects and plot their anticipated orbital paths.
- Produce and maintain current orbital data of man-made space objects in a space catalog.
- Inform NASA and other government entities if objects may interfere with the orbits of the Space Shuttle, the International Space Station, and operational satellite platforms.
- Predict when and where a decaying space object will reenter the Earth's atmosphere.

- Prevent a returning space object, which to radar looks like a missile, from triggering a false alarm in missile-attack warning sensors of the United States and other countries.
- Determine which country owns a reentering space object.
- Predict surface impacts of reentering objects and notify the Federal Emergency Management Agency and Public Safety Canada if an object may make landfall in North America or Hawaii.

It is important to note that USSTRATCOM developed and implemented this process and sensor network since the launch of Sputnik. As such it has been improved in an evolutionary process and does not represent the art of the possible in terms of space surveillance capability. If one were to develop a space surveillance and tracking network and system at present, it would probably not look (or operate) like the USSTRATCOM Space Surveillance Network (SSN).

SSN Sensors (Taken Directly from a USSTRATCOM Fact Sheet)

The SSN uses a series of sensors to achieve its mission. Below is a brief description of each type of sensor:

Phased-array radars can maintain tracks on multiple satellites simultaneously and scan large areas of space in a fraction of a second. These radars have no moving mechanical parts to limit the speed of the radar scan – the radar energy is steered electronically. A detection antenna transmits radar energy into space in the shape of a large fan. When a satellite intersects the fan, energy is reflected back to the detection antenna, where the location of the satellite is computed. Two examples of these radars include Cavalier AFS in North Dakota and Eglin AFB in Florida.

Conventional radars use moveable tracking antennas or fixed detection and tracking antennas. A tracking antenna steers a narrow beam of energy toward a satellite and uses the returned energy to compute the location of the satellite and to follow the satellite's motion to collect more data. These radars include the Altair complex at the Reagan Test Site in the Kwajalein Atoll and the Haystack Millstone facility at the Massachusetts Institute of Technology.

Electro-optical sensors consist of telescopes linked to video cameras and computers. The video cameras feed their space pictures into a nearby computer that drives a display scope. The image is transposed into electrical impulses and recorded on magnetic tape. This is the same process used by video cameras. Thus, the image can be recorded and analyzed in real time.

Midcourse Space Experiment (MSX) satellite is a low-Earth orbiting satellite system with a payload containing a variety of sensors, from UV to very-long-wave IR. Originally a platform for Ballistic Missile Defense Organization (now known as the Missile Defense Agency), the MSX was moved to the SSN in 1998.

Ground-based electro-optical deep space surveillance sites assigned to Air Force Space Command (AFSPC) play a vital role in tracking deep space objects.

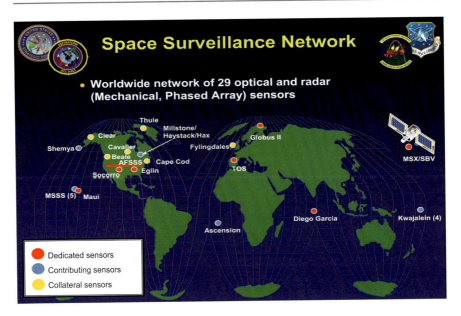

Fig. 3 Space surveillance network

Between 2000 and 2500 objects, including geostationary communications satellites, are in deep space orbits more than 22,500 miles from Earth.

The SSN sensors are categorized as dedicated (those with the primary mission of performing space surveillance) or contributing and collateral sensors (those with a primary mission other than space surveillance). Combined, these types of sensors take between 300,000 and 400,000 observations each day (Fig. 3).

Space Object Tracking

Without loss of generality, tracking an individual in a population implies an ability to "tag" (read uniquely identify) the individual and monitor this individual through time/space/frequency with quantifiable ambiguity or uncertainty, evaluating the interaction of the individual with others and its environment. However, if an individual cannot be physically tagged (or labeled) in a uniquely identifiable way, this poses serious limitations and challenges to space surveillance and tracking, which is indeed the case we face.

Tracking a space object means that one can identify this object with quantifiable and acceptable ambiguity and reconstruct and predict its behavior (usually referring to its location or motion). When this is constrained to the object's trajectory or flight path, this process is more commonly known as orbit determination and prediction as shown schematically in Fig. 4.

Orbit determination (OD) is the process of adjusting trajectory models to best match the observed tracking data and quantify the error associated with the trajectory estimated. The collected tracking data are the actual or *Observed* measurements. The

Fig. 4 Orbit determination
process

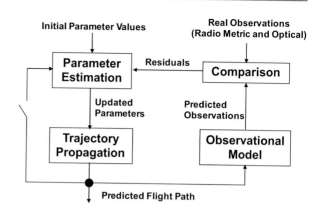

trajectory models produce predicted or *Computed* measurements. Then, what are termed Data *Residuals = Observed–Computed* measurements. The OD method typically aims to minimize the residuals by adjusting the trajectory models. These residuals are minimized in a weighted least-squares sense. The OD process accounts for measurement accuracies and accuracies with which parameters were known before taking measurements (a priori *uncertainty*). The OD produces (a) an updated trajectory estimate and (b) an estimate of error associated with current trajectory prediction. The various forces influencing the motion of the space object must be understood. An example list of these forces follows:

- Gravitational forces
  - Dominant body force (dominant body is treated as spherically symmetric; produces pure Keplerian motion)
  - Non-dominant body forces (third body forces)
  - Dominant body gravity field asymmetries
  - General relativistic effects
- Non-gravitational forces
  - Thruster activity
     Trajectory correction maneuvers or orbit trim maneuvers
     Attitude control system
     Angular momentum desaturations (AMDs)
  - Solar radiation pressure and Earth albedo/radiation
  - Thermal radiation
  - Aerodynamic effects (Drag)
  - Gas leaks (real or compensative)
     Propulsion system
     Outgassing
     Unknown/unmodeled accelerations

OD, especially for uncooperative space objects, requires scientific detective work. Applying the scientific method as an ongoing process is what successful

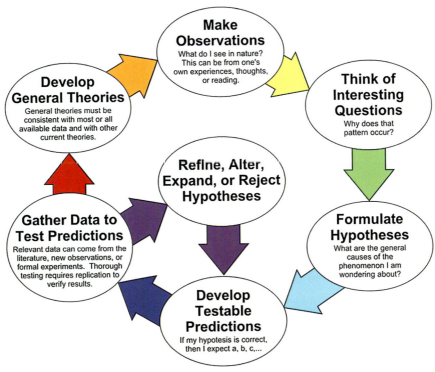

Fig. 5 Scientific method as an ongoing process

OD requires (refer to Fig. 5). The OD process is subjective in that the result is not unique given a variety of assumptions on the space environment models, the astrodynamics models, and the sensor or observation system models. Moreover, the result will also differ depending on what states and parameters are estimated and the assumed prior uncertainty on these. There is a finite amount of information contained within any given set of data, and the resulting state estimate greatly depends upon what is asked of the data.

Determining a space object's orbit is typically much easier than predicting it. In order to best predict a trajectory, the OD process must attempt to not only reconstruct the trajectory of the space objects but also infer or refine knowledge of key model parameters, dynamic and non-dynamic. The only way to demonstrate that one truly understands the space object's behavior and its interaction with the space environment is via the ability to accurately predict its behavior as corroborated by future observations.

Trajectory prediction involves accurately modeling and estimating all past forces and events, as well as predicting all forces and future events. This includes the current *Estimated* trajectory error, as well as all future, or non-estimated errors that can also contribute (*Considered* error sources). More specifically, there is a need to consider the error contribution due to any uncertainty in model parameters that cannot be estimated in the OD solution. For maneuvering space objects, future thrusting events are uncertain (even if predicted) and must be included as potential uncertainties that cannot be estimated (i.e., there is a random error in every thrusting event). Many times the orientation, size, and material properties of the space object are unknown, and their uncertainty should be considered upon the influence and uncertainty in the predicted trajectory as well.

OD cannot be absolutely validated because the collected data do not have *observability* into all components of state. There are several indicators of solution quality: Regarding quality of fit, are the data residuals mean zero with no systematic trends? Regarding estimated parameters, are estimates realistic, within a priori uncertainties? The solution quality can be trended by comparing various solution strategies that are (a) data span dependent, (b) data type dependent, (c) sensor dependent, and (d) model assumptions dependent.

One of the most challenging tasks in space surveillance is in associating detections to unique objects. Many studies and analyses regarding space surveillance make two fundamentally flawed assumptions: (1) if the space object is in the sensor field of view, it is detected with a probability of one; (2) if a space object is detected, it is known which space object generated the detection. The problem with the first assumption is that the probability of detection is never exactly one; several things contribute to the total probability of detection. The probability of detection is comprised of three components: (1) a sensor-dependent component, (2) a dependence on everything between the sensor and the object, and (3) an object-dependent component. Rarely, if ever, are all three components known with absolute certainty. The problem with the second assumption is that there is always some level of ambiguity in the detection-to-object assignment unless the object is transmitting/transponding a known frequency.

This was the challenge raised earlier in the context of being able to uniquely tag and track individuals in the population. To highlight the extent of this issue, please refer to Fig. 6 which shows a plot of detections made in a single night by the Space Surveillance Telescope (SST) located near Socorro, New Mexico. Every dot is a detection generated by a space object in various orbital regimes: Molniya (highly eccentric orbits or HEO), Mid Earth Orbit (MEO) with Global Navigation Satellite Systems (GPS, etc.), and the near geosynchronous regime (GEO). All of the dots that are black are detections that were associated/correlated to unique known objects. All of the other dots are detections generated by space objects that are unknown. At least 50% of all detections are generated by unknown (read untracked) objects. This is a major unresolved issue.

In essence, the problem is as follows: given a series of observations (tracking data), determine which detections belong to unique objects and compute their trajectories. One mechanism for performing this is via Joint Probabilistic Data Association (JPDA) (Stauch et al. 2017). A simpler approach is nearest neighbor (NN) which is somewhat captured in the schematic that follows, but this simpler

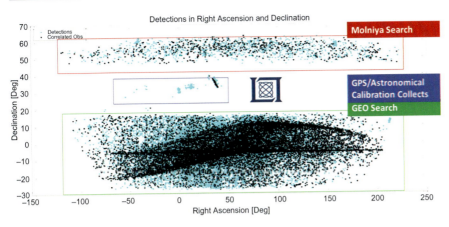

Fig. 6 Correlated versus uncorrelated detections from the Space Surveillance Telescope (SST)

approach tends to suffer from higher false positives (i.e., associating data from objects that are not the same, assuming a unique object).

Figure 7 shows a series of unassociated detections at an initial time, and then a set of hypotheses are computed from these. These hypotheses are compared to future data, and the hypothesis with the highest likelihood agreement (exploiting the Mahalanobis distance) is assigned to the detection in question. Hypotheses that are the most unlikely tend to be pruned.

The goal of multiple hypothesis testing is to converge on the correct hypothesis by pursuing an abductive/inductive reasoning approach whereby the data are used for their ability to identify and remove the wrong or unlikely hypotheses. This was referred to earlier as the "Moriba Jah Refrigerator Approach." Any surviving hypotheses at any given time have a statistical likelihood of explaining past observations. The following provide more insight into the difficulties of data association (Fig. 8).

Let us examine the concept of the JPDA. Assume we have two known space objects that we are tracking and we have just received two detections (measurements) as in Fig. 8b. The ellipses surrounding each object represent the object's uncertainty projected (or transformed) into the reference frame where the measurements are represented, assuming that the uncertainty can be represented as a Gaussian probability distribution. There are many occasions when this is not the case, and thus care must be taken in considering realistic measures of uncertainty. The next, below, shown a comparison of orbital probability mass distribution against uncertainty represented as a Gaussian assumption versus one that allows the representation to adapt to the underlying propagated errors called AEGIS (DeMars et al. 2013). It is seen how the Gaussian error ellipses do not conform to the shape or density distribution of error in the orbital plane. All decisions are based upon the assumed uncertainty, so if it is grossly inaccurate, then only flawed decisions can be the result.

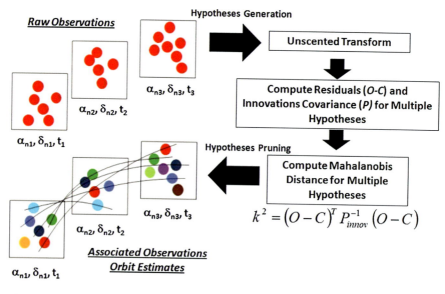

Fig. 7 Data association

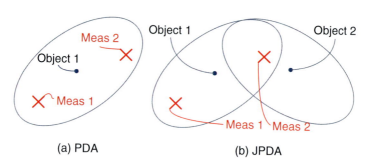

(a) PDA

(b) JPDA

Fig. 8 Probabilistic and joint probabilistic data association

What are the possible joint events or hypotheses that could explain the scenario in the Fig. 9 with multiple hypotheses? Table 1 enumerates these, and all of them could be possible and thus must be considered in the tracking framework. As the number of objects and detections increases, it can be seen that the combinatorics involved invokes the need of computational capacity and efficiency to say the least.

Not considering any of these joint events could lead to errors in the future because we may be missing important information and this could lead to a degraded tracking capability. To this point, there are actually more joint events to consider in the prior scenario that we did not include such as the joint event of measurement 1 having originated from object 2 which performed an unknown (therefore unmodeled) propulsive maneuver. Other domains such as air, ground, etc. may not require an accurate knowledge of the physics because they tend to be data-rich environments

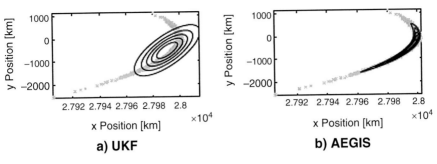

Fig. 9 Comparison of orbital probability mass distribution against uncertainty representation

Table 1 JPDA scenario example

Joint event	Marginal events	Object # associated with measurement:		Event Description
		1	2	
Θ_1	$\theta_{0,1}, \theta_{0,2}$	0	0	Meas. 1 & 2 from clutter
Θ_2	$\theta_{1,1}, \theta_{0,2}$	1	0	Meas. 1 from Obj. 1, Meas. 2 from clutter
Θ_3	$\theta_{0,1}, \theta_{1,2}$	0	1	Meas. 1 from clutter, Meas. 2 from Obj. 1
Θ_4	$\theta_{0,1}, \theta_{2,2}$	0	2	Meas. 1 from clutter, Meas. 2 from Obj. 2
Θ_5	$\theta_{1,1}, \theta_{2,2}$	1	2	Meas. 1 from Obj. 1, Meas. 2 from Obj. 2

(i.e., there are many measurements available). However, the space tracking problem, specifically space surveillance, tends to be data-starved, and thus the physics must be relied upon to properly predict the motion in between sparse observations.

Summary

In order to develop and maintain a database or catalog of space objects and events, and perform space surveillance and tracking successfully, all of the elements presented in this paper must be brought together very skillfully and effectively. Data must be properly curated, collected, transformed, stored, managed, rectified, fused, exploited, disseminated, etc. Great care must be taken in making assumptions and avoiding the temptation to assume more than what the data and information available allow or indicate. Being successful requires an ability to do the proper "detective" work and learn as much as possible from the data for the purpose of improving one's predictability of future behavior. Many people can reconstruct events and trajectories, but few can predict them because prediction requires one to know and understand the underlying system. The SDIF model is the framework that drives the entire process, and it should be driven by user needs and requirements. Without the SDIF model, the output of the tracking and surveillance will be less useful to user needs or may not satisfy them entirely. Of great importance is the ability to quantify and realistically represent the uncertainty of the system. Most people make the assumption that all of the errors are Gaussian, but there is substantive evidence that this assumption is oftentimes flawed, dependent upon the scenario. The focus should be on uncertainty realism versus blindly constraining oneself to Gaussianity. Last but not least, one must understand the data that one is collecting, receiving, and processing. Many errors in what is inferred from the space surveillance activity can be attributed to exploiting measurements under invalid assumptions.

References

Bever M, Delande E, Jah M (2019) Outer probability measures for first and second order uncertainty in the space domain. IAA-AAS SciTech-040, Moscow, Russia, June

Delande E, Houssineau J, Jah M (2018) Physics and human-based information fusion for improved resident space object tracking. Adv Space Res 62(7):1800–1812. https://doi.org/10.1016/j.asr.2018.06.033. Elsevier

DeMars KJ, Bishop RH, Jah MK (2013) Entropy-based approach for uncertainty propagation of nonlinear dynamical systems. J Guid Control Dyn 36(4):1047–1057. https://doi.org/10.2514/1.58987

Stauch J, Bessell T, Rutten M, Baldwin J, Jah M, Hill K (2017) Joint Probabilistic Data Association and smoothing applied to multiple space object tracking. J Guid Control Dyn (Special Issue on Space Domain Awareness):1–15. http://arc.aiaa.org/doi/abs/10.2514/1.G002230

Space Security and Frequency Management

51

Yvon Henri and Attila Matas

Contents

Abstract

The International Telecommunication Union (https://www.itu.int) (ITU) is unique among UN-specialized agencies in having a mix of public and private

Y. Henri
OneWeb, London, UK
e-mail: yhenri2017@gmail.com

A. Matas (✉)
Orbit/Spectrum Consulting, Grand-Saconnex, Switzerland
e-mail: am@orbitspectrum.ch

© Springer Nature Switzerland AG 2020
K.-U. Schrogl (ed.), *Handbook of Space Security*,
https://doi.org/10.1007/978-3-030-23210-8_91

985

sector members. In addition to the 193 Member States, the organization includes also over 800 members comprising the world's leading ICT operators, equipment manufacturers, software developers, service providers, R&D organizations, and local, regional, and international ICT bodies which are approved by the Member State concerned. ITU has three main areas of activity organized in "sectors" which work through conferences and meetings. The Radiocommunication Sector (ITU-R) coordinates this vast and growing range of radiocommunication services, as well as the international management of the radio-frequency spectrum and satellite orbits. The Telecommunication Standardization Sector (ITU-T) is in charge of developing ITU standards (called Recommendations), fundamental to the operation of today's ICT networks. Last but not least, the Telecommunication Development Sector (ITU-D) is in charge of initiatives for ITU's internationally accorded mandate to "bridge the digital divide." For over 150 years, ITU has worked alongside the industry, building global consensus, reconciling competing interests, and forging the new technical standards that have served as the platform for the development of what is now the world's most dynamic business sector.

Introduction

The International Telecommunication Union (https://www.itu.int) (ITU) is unique among UN-specialized agencies in having a mix of public and private sector members. In addition to the 193 Member States, the organization includes also over 800 members comprising the world's leading ICT operators, equipment manufacturers, software developers, service providers, R&D organizations, and local, regional, and international ICT bodies which are approved by the Member State concerned. ITU has three main areas of activity organized in "sectors" which work through conferences and meetings. The Radiocommunication Sector (ITU-R) coordinates this vast and growing range of radiocommunication services, as well as the international management of the radio-frequency spectrum and satellite orbits. The Telecommunication Standardization Sector (ITU-T) is in charge of developing ITU standards (called Recommendations), fundamental to the operation of today's ICT networks. Last but not least, the Telecommunication Development Sector (ITU-D) is in charge of initiatives for ITU's internationally accorded mandate to "bridge the digital divide." For over 150 years, ITU has worked alongside the industry, building global consensus, reconciling competing interests, and forging the new technical standards that have served as the platform for the development of what is now the world's most dynamic business sector.

Concerning space services, during the last 56 years, from the first ever World Administrative Radio Conference (WARC-63) (ITU 2019) (https://www.itu.int/en/history/Pages/RadioConferences.aspx?conf=4.89) in 1963, up to and including the forthcoming World Radiocommunication Conference, in 2019 (WRC-19)

(ITU 2019) (https://www.itu.int/en/ITU-R/conferences/wrc/2019/), many ITU conferences have addressed the regulation of spectrum/orbit usage by stations of the space radiocommunication services. The ITU Member States have established a legal regime which is codified through the ITU Constitution/Convention (ITU 1992) (Constitution of the ITU, 22 DEC 1992, (CS); Convention of the ITU, 22 DEC 1992, (CV)), including the Radio Regulations (ITU 2016) (ITU Radio Regulations, Edition of 2016) (RR). These instruments contain the main principles and lay down the specific regulations governing the following major elements:

– Frequency spectrum allocations to different categories of radiocommunication services
– Rights and obligations of Member administrations in obtaining access to the spectrum/orbit resources; International recognition of these rights by recording frequency assignments and, as appropriate, orbital information for a space station on-board a geostationary satellite or for space station(s) on-board non-geostationary satellite(s), used or intended to be used in the Master International Frequency Register (MIFR) or by their conformity, where appropriate, with a Space Plan

The above regulations are based on the main principles of efficient use of and equitable access to the spectrum/orbit resources laid down in No. **196** of the ITU Constitution (ITU 1992) (Article **44** of the ITU Constitution), which stipulates that *"In using frequency bands for radio services, Members shall bear in mind that radio frequencies and any associated orbits, including the geostationary-satellite orbit, are limited natural resources and that they must be used rationally, efficiently and economically, in conformity with the provisions of the Radio Regulations, so that countries or groups of countries may have equitable access to those orbits and frequencies, taking into account the special needs of the developing countries and the geographical situation of particular countries."* As indicated in the above provision, further detailed regulations and procedures governing orbit/spectrum use are contained in the Radio Regulations (RR), which is a binding international treaty (ITU 1992) (Article 31 of the ITU Constitution).

Specific procedures have been established to ensure international recognition of the frequencies used and to safeguard the rights of administrations when they comply with those procedures.

The fact that the ITU Constitution and Convention and the RR that complement them are *intergovernmental treaties ratified by governments* means that those governments undertake:

• To apply the provisions in their countries
• To adopt adequate national legislation that includes, as the basic minimum, the essential provisions of this international treaty

Organizational Structure of the ITU

The ITU Plenipotentiary Conference consisting of the representatives of the Member States of the Union is the highest policy-making body of the ITU. Held every 4 years, it is the key event at which ITU Member States decide on the future role of the organization, thereby determining the organization's ability to influence and affect the development of information and communication technologies (ICTs) worldwide. The conference sets the Union's general policies (adoption of the 4-year strategic and financial plans; election of the senior management team of the organization (Secretary General, Deputy Secretary General, and Directors of the 3 Bureaux (Radiocommunications, Standardization, and Development), the Member States of the Council (48 Members states representing the five geographical regions (Americas, Western Europe, Eastern Europe and Northern Asia, Africa, Asia, and Australasia)), and the 12 members of the Radio Regulations Board. The last ITU Plenipotentiary Conference met for the 20th time in Dubai, United Arab Emirates, in November 2018.

The Council, on the other hand, acts as the Union's governing body in the interval between Plenipotentiary Conferences. Its role is to consider broad telecommunication policy issues to ensure that the Union's activities, policies, and strategies fully respond to today's dynamic, rapidly changing telecommunications environment. ITU Council also prepares a report on the policy and strategic planning of the ITU and responsible for ensuring the smooth day-to-day running of the Union, coordinating work programs, approving budgets, and controlling finances and expenditure.

The Radio Regulations Board (RRB) is composed of elected members who are recognized by their qualifications in the field of radiocommunications and their practical experience in the assignment and utilization of radio frequencies. The duties of the RRB are defined in Article **14** of the Constitution and include, inter alia, the approval of the Rules of Procedure (RoP) (ITU 2017) (https://www.itu.int/en/publica tions/ITU-R/pages/publications.aspx?parent=R-REG-ROP-2017&media=electronic) in conformity with the RR and with decisions by radiocommunication conferences, the examination of any other issue that cannot be resolved through the application of the RoP, and any appeals against decisions made by the Radiocommunication Bureau (Bureau) regarding frequency assignments.

Major Principles

In the process of establishing the ITU's space-related regulations, emphasis was laid from the outset on *efficient*, *rational*, and *cost-effective utilization*. This concept was implemented through a "first come, first served" procedure. This procedure ("coordination before use") is based on the principle that the right to use orbital and spectrum resources for a satellite network or system is acquired through negotiations with the administrations concerned by actual usage of the same portion of the spectrum and orbital resource. If applied correctly (i.e., to cover genuine requirements), the procedure offers the means of achieving efficient spectrum/orbit management; it serves to fill the gaps in the orbit(s) as needs arise. On the basis of the RR, and in the frequency bands where this concept is applied, Member administrations designate the volume of orbit/spectrum resources that is required to satisfy their actual requirements. It then falls to the national administrations to assign frequencies and orbital requirements, to apply the appropriate procedures (international coordination and recording) for the space segment and earth stations of their (governmental, scientific, public, and private) networks, and to assume continuing responsibility for the networks.

The progressive exploitation of the orbit/frequency resources and the resulting likelihood of congestion of the geostationary satellite orbit (GSO) prompted ITU Member countries to consider more and more seriously the question of *equitable access* in respect to the orbit/spectrum resources. This resulted in the establishment (and introduction into the ITU regulatory regime) of frequency/orbital position Plans in which a certain amount of frequency spectrum is set aside for future use by all countries, particularly those which are not in a position, at present, to make use of these resources. These Plans, in which each country has a predetermined GSO orbital position associated with the free use, at any time, of a certain amount of frequency spectrum, together with the associated procedures, guarantee for each country equitable access to the spectrum/orbit resources, thereby safeguarding their basic rights. Such Plans govern a considerable part of the frequency bands available for the space communication services.

During the last 56 years, the regulatory framework has been constantly adapted to changing circumstances and has achieved the necessary flexibility in satisfying the two major, and not always compatible, requirements of efficiency and equity. With the dramatic development in telecommunication services, increasing demand for spectrum/orbit usage for practically all space communication services has been observed. This increase is attributable to many factors. These include not only technological progress, but also political, social, and structural changes around the world. Those factors also include their impact on (i) the liberalization of telecommunication services, (ii) the introduction of non-geostationary satellite orbit (non-GSO) satellite systems for commercial communications (large non-GSO constellations as well non-GSO systems with short duration mission), (iii) growing market orientation, (iv) the change in the way this widening market is shared between private, and (v) state-owned service providers and the general globalization and commercialization of communication systems.

Frequency Allocation Structure

The Table of Frequency Allocations (Table) (ITU 2016) (Article **5** of the RR) and associated principles represent a basis for the planning and implementation of radiocommunication services. The current approach is based on a block allocation methodology with footnotes. The regulated frequency band (8.3 kHz–3000 GHz) is segmented into smaller bands and allocated to over 40 defined radiocommunication services (ITU 2016) (Article **1** of the RR). The radio services are identified as *primary* or *secondary* (the latter shall cause no harmful interference to, or claim protection from, the former) and footnotes are used to further specify how the frequencies are to be assigned or used. The Table is organized into three Regions of the world and is supplemented by assignment and allotment Plans for some bands and services, and/or by mandatory coordination procedures.

Using the Table as a starting point, the frequency spectrum management authority of each country selects appropriate frequencies with a view to assigning them to stations of a given service. Before taking the final decision to assign a frequency to a station in a given radiocommunication service in a given frequency band and to issue an appropriate license, the authority concerned should be aware of all other conditions regulating the use of frequencies in the band concerned, e.g.:

- Are there other mandatory RR provisions governing the use of the frequencies?
- Is the band concerned subject to a pre-established international assignment or allotment Plan? Are the characteristics of the assignment in accordance with the appropriate entry in the Plan? Is there a need to apply the Plan modification procedure prior to issuing a license?
- Is there a need for effecting the coordination procedure prior to notification of the concerned assignment to the Bureau or prior to its bringing into use?
- Is the procedure mandatory or voluntary? Is the procedure specified in the RR or in a special agreement?
- Is there a need to notify the frequency assignment to the Bureau, when should such notification be effected, which characteristics are to be notified, what action should be foreseen after the recording or otherwise of the frequency assignment concerned?

Regulations Applying to the Use of Frequencies and Orbits by Satellite Networks

The specific procedures setting out the rights and obligations of the administrations in the domain of orbit/spectrum management and providing means to achieve interference-free radiocommunications have been laid down by successive WRCs on the basis of the two main principles referred to above: efficient use and equitable access. In order to put these principles into effect, two major mechanisms for the sharing of orbit and spectrum resources have been developed and implemented:

- A priori planning procedures (guaranteeing *equitable access* to orbit/spectrum resources for *future use*), which include:
  - The Allotment Plan for the fixed-satellite service using part of the 4/6 and 10–11/12–13 GHz frequency bands contained in Appendix **30B** (ITU 2016) (Appendix **30B** of the RR)
  - The Plan for the broadcasting-satellite service in the frequency band 11.7–12.7 GHz (Appendix **30**) and the associated Plan for feeder links in the 14 GHz and 17 GHz frequency bands (Appendix **30A**) (ITU 2016) (Appendix **30/30A** of the RR)

- Coordination procedures (with the aim of *efficiency* of orbit/spectrum use and interference-free operation satisfying *actual requirements*), which include:
  - Geostationary satellite networks (in all services and frequency bands) and non-geostationary satellite networks in certain frequency bands governed by the RR No. **9.11A** procedure, which are subject to advance publication and coordination procedures
  - Other non-GSO satellite networks (all pertinent services and certain frequency bands), for which only the advance publication procedure is required before notification

Procedures Applying to Non-planned Space Services

The procedures for non-planned space services are contained primarily in Article **9** of the RR "Procedure for effecting coordination with or obtaining agreement of other administrations."

The coordination procedure is based on a "first come, first served" principle. Successful coordination of space networks or earth stations gives an international recognition to the use of frequencies by these networks/stations. For such frequency assignments, this right means that other administrations shall take them into account when making their own assignments, in order to avoid harmful interference. In addition, frequency assignments in frequency bands subject to coordination or to a Plan shall have a status conditioned by the application of the procedures relating to the coordination or associated with the Plan. The relevant provisions involve three basic steps:

- Advance publication information (Section I, Article **9**) (ITU 2016) (Article **9**, Section I of the RR)
- Coordination request (Section II, Article **9**) (ITU 2016) (Article **9**, Section II of the RR)
- Notification (Article **11**) (ITU 2016) (Article **11**, Section I of the RR)

Advance Publication Information (API) Procedure

For a satellite network or a satellite system not subject to the coordination procedure, an administration shall send to the Bureau a general description of the network or system for advance publication in the International Frequency Information Circular (BR IFIC) (ITU 2019) (https://www.itu.int/ITU-R/go/space-brific/en) not earlier than 7 years and preferably not later than 2 years before the planned date of bringing into use of the network or system (see also No. **11.44** of the RR). The characteristics to be provided for this purpose are listed in Appendix **4** of the RR (ITU 2016) (Appendix **4** of the RR). Upon receipt of the advance publication, administrations should check whether the planned system is likely to affect their existing or planned systems or stations and both administrations shall endeavor to cooperate in joint efforts to solve any difficulties, with the assistance of the Bureau, if so requested.

The first two steps of the procedure (advance publication and coordination) have been streamlined for GSO networks and non-GSO networks and systems subject to coordination by WRC-15. In the case of GSO networks and non-GSO networks and systems subject to Section II, Article 9 coordination, no requirement to send to the Bureau the advance publication information in addition to the coordination request ones. Upon receipt of the complete coordination information, the Bureau shall publish, using the basic characteristics of the coordination request, a general description of the network or system for advance publication in a Special Section of the BR IFIC.

For some space services commonly used in non-GSO satellite networks not subject to coordination as Earth exploration-satellite service, meteorological satellite service, space research service, space operation service, etc., the administration has to submit advance publication information to the Bureau before the notification procedure, the regulatory final step for the recording of the frequency assignments in the MIFR.

The date of receipt of the advance publication or coordination request information as applicable starts of regulatory 7-year time limit for the frequency assignments of the system to be notified and recorded in the MIFR and brought into use.

Procedure for Effecting Coordination of Frequency Assignments

Coordination is a further step in the process leading up to notification of the frequency assignments for recording in the MIFR. The coordination procedure is a formal regulatory obligation both for an administration seeking to assign a frequency assignment to its network and for an administration whose existing or planned services may be affected by that assignment. An agreement arising from this coordination confers certain rights and imposes certain obligations on the administrations concerned; as such, coordination must be effected in accordance with the relevant regulatory procedures laid down in the RR and on the basis of technical criteria either contained in Appendix 5 of the RR (ITU 2016) (Appendix 5 of the RR) or otherwise agreed to by the administrations concerned.

For most coordination cases related to space services, the responsible administration shall send to the Bureau the request for coordination together with the appropriate information listed in Appendix 4 of the RR. On receipt of the request for coordination, the Bureau will promptly examine the information in terms of completeness and conformity with the Convention, the Table of Frequency Allocations and other provisions of the RR (see RoP under No. 11.31). The Bureau will then examine the information received in order to identify any administration with which coordination may be needed based on frequency assignments affecting or being affected, as appropriate, and relating to the threshold levels and conditions given in Tables 5-1 and 5-2 in Appendix 5 of the RR.

Finally, the Bureau will publish the complete information (Appendix **4** information and, as appropriate, the names of identified administrations and the specific satellite networks or earth station for coordination between GSO networks with which coordination may need to be effected), in a special section of its BR IFIC.

When a coordination request is received, an administration studies the matter to determine the level of interference likely to be caused to frequency assignments of its networks or stations or caused to assignments of the proposed network or station by its own assignments (No. **9.50** of the RR). Within a total period of 4 months from the date of the publication of the request for coordination in the relevant special section or the date of dispatch of the coordination data, as appropriate, it shall:

- Communicate its agreement to the proposed coordination (Nos. **9.51** and **9.51A** of the RR) or
- Provide to the notifying administration (with a copy to the Bureau) the technical data upon which its disagreement is based, along with its suggestions for resolving the problem (No. **9.52** of the RR).

The Bureau's assistance can be requested at the coordination stage of the procedure, by either notifying or objecting administration, with a view to resolving any difficulties which may arise.

As indicated above, there is an obligation for the notifying administration to coordinate with any administration which has initiated the coordination process at an earlier stage. However, there is also a provision (No. **9.53** of the RR) stipulating that both the notifying administration and the objecting administration shall make every possible mutual effort to overcome any difficulties which may arise in a manner acceptable to the parties concerned. The intent of this provision is to facilitate the entry of the newcomer and, even though an administration was first in line, encourage concessions to that end on the basis of mutual cooperation.

Non-GSO FSS Satellite System: Particular Features

There is no regulatory definition in the RR related to non-GSO system in the Fixed-Satellite Service (FSS), called time to time also "Mega or Large Constellations," and intended to provide high-speed, low-latency broadband services, including Internet connectivity, throughout the world, including locations which cannot be reached using GSO satellites. A non-GSO FSS system can be considered as "a constellation comprised of a group of satellites, operating in the frequency bands allocated to fixed-satellite service, with similar characteristics and functions, operating in similar or complementary orbital planes under a shared control for coordinated ground coverage."

Typical non-GSO FSS satellite system consists of one or several space stations located on low-Earth orbit (LEO), medium-Earth orbit (MEO), or highly elliptical orbit (HEO) and several gateway stations. The gateways stations purpose is to connect the non-GSO FSS satellite system with terrestrial networks, to provide each user an access to the private or public networks.

The sharing relationship between GSO networks and non-GSO systems in the FSS and the broadcasting-satellite service (BSS) is covered by the provisions No. **22.2** of the RR (ITU 2016) (Article **22** of the RR), when not otherwise specified in RR:

> No. **22.2** § 2 1) Non-geostationary-satellite systems shall not cause unacceptable interference to and, unless otherwise specified in these Regulations, shall not claim protection from geostationary-satellite networks in the fixed-satellite service and the broadcasting-satellite service operating in accordance with these Regulations. No. **5.43A** does not apply in this case. (WRC-07)

In view of the shortage of suitable frequencies, and in order to take advantage of existing space infrastructure and to allow a fair sharing between non-GSO systems and GSO networks in portions of frequency bands allocated to FSS and BSS, WRC-97 and WRC-2000 adopted equivalent power-flux density limits (EPFD) by which an administration operating a non-GSO system in the FSS in compliance with these limits (RR Nos. **22.5C**, **22.5E**, and **22.5F**) shall be considered as fulfilling its obligations under RR No. **22.2** with respect to any GSO networks, provided that operational and additional operational EPFD limits (Tables 22.4A, 22.4A1, 22.4B, and 22.4C) are not exceeded.

There are single-entry limits applicable to non-GSO FSS systems in certain parts of the frequency range 10.7–30 GHz to protect GSO FSS satellite networks operating in the same frequency bands. **RESOLUTION 85** (WRC-03) (ITU 2016) (**RESOLUTION 85** (WRC-03) of the RR) deals with the application of Article **22** to the protection of GSO FSS and BSS satellite networks from non-GSO FSS satellite systems. In accordance with this Resolution and availability of the EPFD validation software, the Bureau is currently able to verify compliance with the limits in Tables 22-1A, 22-1B, 22-1C, 22-1D, 22-1E, 22-2, and 22-3 and to determine the coordination requirements under Nos. **9.7A** and **9.7B** (specific large earth station). Additional protection GSO FSS and BSS from all non-GSO FSS system is described in **RESOLUTION 76** (REV.WRC-15) (ITU 2016) (**RESOLUTION 76** (REV.WRC-15) of the RR) which ensures protection of GSO FSS and BSS networks from the maximum aggregate EPFD produced by multiple non-GSO FSS systems in frequency bands where EPFD limits have been adopted.

This Resolution is instructing that all administrations operating or planning to operate non-GSO FSS systems, individually or in collaboration, shall take all possible steps, including, if necessary, by means of appropriate modifications to their systems, to ensure that the aggregate interference into GSO FSS and BSS networks caused by such systems operating co-frequency in these frequency bands does not cause the aggregate power levels given in Tables 1A to 1D (see RES-76) to be exceeded and in the event that the aggregate interference levels in Tables 1A to 1D (see RES-76) are exceeded, administrations operating non-GSO FSS systems in these frequency bands shall take all necessary measures expeditiously to reduce the aggregate EPFD levels to those given in Tables 1A to 1D (see RES-76).

Notification and Recording in the MIFR

The procedure for notification and recording of space network frequency assignments in the MIFR is described in Article **11** of the RR. The MIFR represents one of the pillars of the international radio regulatory setup as it contains *all frequency in use or plan to be used notified to ITU*. It should be consulted before selecting a frequency for any new user. For these reasons, *notification of frequency assignments to the Bureau, with a view to their recording in the MIFR, represents the ultimate goal of the frequency registration process and a major obligation for administrations, especially in respect to those frequency assignments that have international implications.*

According to Article **11** provisions, any frequency assignment liable to have an international implication has to be notified to the Bureau:

- If the use of that assignment is capable of causing to or suffering harmful interference from existing or future stations in another country; or if that assignment is to be used for international radiocommunication
- If that assignment is subject to the Article **9** coordination procedure or is involved in such a case
- If it is desired to obtain international recognition for that assignment
- If it is a non-conforming assignment and if the administration wishes to have it recorded for information

For that purpose, administrations submit the relevant characteristics of the frequency assignments to be notified, as specified in Appendix **4** of the RR, to the Bureau. The Bureau shall publish the information in PART I-S of the BR IFIC, thereby ensuring that all administrations are informed of the use of the assignments and that they are taken into account in any future planning conducted at the national, regional, or international level.

The subsequent processing of a notice varies according to the frequency band and service concerned. Each notice is first examined with respect to its conformity with the Table and the other provisions of the RR (regulatory examination); this examination consists in checking that the assignment (frequency, class of station, notified bandwidth) does indeed correspond to an allocation in the Table or the footnotes thereto and, where appropriate, that it complies with other technical or operating conditions laid down in other articles or appendices of the RR (power limits, authorized classes of emission, minimum elevation angle, etc.). If the result of this examination is *unfavorable* and the administration concerned has not explicitly undertaken that the assignment shall be operated subject to not causing interference to assignments operating in conformity with the RR, making reference to No. **4.4** of the RR, the examination stops there and the notice is returned to the notifying administration after publication of the finding in *PART III-S of the BR IFIC*.

When the result of the first examination (under No. **11.31** of the RR) is *favorable*, the assignment is *recorded in the MIFR*, or examined further, if appropriate, from the viewpoint of its conformity with the coordination procedures (No. **11.32** of the RR) or with a world or regional allotment or assignment Plan (No. **11.34** of the RR).

Following such examinations, the assignment is either recorded in the MIFR and published in *PART II-S of the BR IFIC* (if the finding is *favorable*) or is published in PART III-S of the BR IFIC and returned to the administration (if the finding is unfavorable). The administrations are normally advised to complete the coordination procedure with the identified administrations, or to apply the relevant Plan modification procedure. However, in some specific cases, an administration may resubmit the notice without completing the coordination or Plan modification procedure and the concerned assignment may be recorded in the MIFR under specific conditions.

Bringing into Use of a Satellite Network

One major element of the satellite registration procedure is the respect of the regulatory time limit for bringing a satellite network or system into use (BiU) and submitting notices for recording in the MIFR. No. **11.44** of the RR stipulates that the notified date of bringing into use of any assignment to a space station of a satellite network shall be no later *than 7 years* following the receipt of the advance publication information by the Bureau. WRC-12 defined further bringing into use of a GSO satellite network as contained in RR No.**11.44B** which requires that the "frequency assignment to a space station in the geostationary-satellite orbit shall be considered as having been brought into use when a space station in the geostationary-satellite orbit with the capability of transmitting or receiving that frequency assignment has been deployed and maintained at the notified orbital position for a continuous period of ninety days. The notifying administration shall so inform the Bureau within thirty days from the end of the ninety-day period." WRC-15 instructed then the Bureau that upon receipt of bringing into use information and whenever it appears from reliable information available that a notified assignment has not been brought into use in accordance with the regulations, the Bureau shall consult the administration for clarification as prescribed in RR No. **13.6**.

Regarding non-GSO systems, no provisions in the RR specifically address the bringing into use issue. It has been the practice of the Bureau to declare their BIU successfully completed when one satellite is deployed into a notified orbital plane and capable of transmitting and/or receiving those frequency assignments. This practice, reflected for FSS and MSS non-GSO systems in section 2 of the RoP for RR No. **11.44**, has been used for a number of years. Furthermore, it has been used irrespective of the number of satellites or of the number of orbital planes indicated in the notification information provided under RR No. **11.2**.

In order to clear the notion of bringing into use for non-GSO systems, the ITU-R has been tasked to study the issue and proposed draft elements in the Conference Preparatory Meeting 2019 (CPM-19) Report for administrations to take a decision at WRC-19, both on the bringing into use of frequency assignments to non-GSO systems and the possibility of adopting a milestone-based approach for the deployment of non-GSO systems composed of multiple, multi-satellite constellations, in particular frequency bands.

The first general conclusion is that the bringing into use of frequency assignments to non-GSO systems should continue to be achieved by the deployment of one satellite into one of the notified orbital planes within 7 years of the date of receipt of the advance publication of information (API) or request for coordination, as applicable. This conclusion applies for frequency assignments for all non-GSO systems in all frequency bands and services. However, three options are proposed with respect to the minimum period during which a satellite has to be maintained in a notified orbital plane: 90 days (as currently required for fixed-satellite service (FSS) and mobile-satellite service (MSS) non-GSO systems in the RoP for RR No. **11.44**), some period less than 90 days, or no fixed period.

The second general conclusion is that a new WRC Resolution should be adopted to implement a milestone-based approach for the deployment of non-GSO systems in specific frequency bands and services. This milestone-based approach would provide an additional period beyond the 7-year regulatory period for the deployment of the number of satellites, as notified and/or recorded, with the objective to help ensure that the MIFR reasonably reflects the actual deployment of such non-GSO systems. Several options are proposed with respect to the number of milestones, the milestone periods, the required percentage of satellites deployed to satisfy each milestone, the consequences of failing to meet a milestone, and appropriate transitional measures to fairly and equitably address the case of the recorded frequency assignments to non-GSO systems already brought into use, and that have reached the end of their 7-year regulatory period, but where the non-GSO system has not been fully deployed.

The final decision on the bringing into use of non-GSO systems will be taken at WRC-19.

Responsibilities of the Notifying Administration After Recording in the MIFR

Recording in the MIFR does not mean the end of activities for the notifying administration as regards the concerned frequency assignment. The notifying administration should remain in close cooperation with the licensing authority and satellite operator and any change in the characteristics of the concerned assignment would have to be notified to the Bureau so as to be reflected in the MIFR, if necessary following additional coordination with the administrations of other countries concerned.

The notifying administration has also to respond to coordination request of any administration which has initiated the coordination process at a later stage with the objective, on the basis of mutual cooperation, to overcome any difficulties which may arise in a manner acceptable to the parties concerned, as stipulated under RR No. **9.53**.

Furthermore, the notifying administration should remain in close contact with the monitoring authority so as to check whether the concerned frequency assignment is operated in compliance with the notified characteristics and whether other elements (e.g., frequency tolerance) are kept within the limits prescribed by the RR. The notifying administration should also initiate appropriate monitoring programs with a view to detecting any operational or technical irregularities in the operation of frequency assignments pertaining to other administrations, and to initiate appropriate actions in this regard, so as *to ensure interference-free operation* for stations under its jurisdiction.

Non-GSO Satellites with Short Duration Mission (SDM)

There is currently no regulatory definition for non-GSO SDM satellites. The RR is recognizing only GSO and non-GSO satellites. However, WRC-15 adopted **RESOLUTION 659** (WRC-15) (ITU 2016) **(RESOLUTION 659** (WRC-15) of the RR) Studies to accommodate requirements in the space operation service for non-geostationary satellite with short duration mission, and invited the ITU-R to study the spectrum requirements for telemetry, tracking, and command in the space operation service for the growing number of non-GSO satellites with short duration missions. A new term "short duration mission" (SDM) was used for the first time in the ITU-R, which refers to a non-GSO satellite system having a limited period of validity of not more than typically 3 years. The Resolution invited also WRC-19 to consider the results of ITU-R studies and take necessary action, as appropriate, provided that the results of the studies are complete and agreed by the ITU-R study groups.

Two Agenda Items of WRC-19 were agreed on the non-GSO SDM systems, Agenda Item 1.7 (to study the spectrum needs for telemetry, tracking, and command in the space operation service for non-GSO satellites with short duration missions, to assess the suitability of existing allocations to the space operation service and, if necessary, to consider new allocations, in accordance with **RESOLUTION 659** (WRC-15) and Agenda Item 7 – Issue M – Simplified regulatory regime for non-GSO satellite systems with short duration missions.

In response to the first issue, the CPM-19 Report for administrations to take a decision at WRC-19, developed four methods, and associated regulatory texts to satisfy this agenda item:

- Method A proposes no change to the RR.
- Method B1 proposes a new Space Operation Service (SOS) (Earth-to-space) allocation for non-GSO SD systems in the frequency range 403–404 MHz.
- Method B2 proposes a new SOS (Earth-to-space) allocation for non-GSO SD systems in the frequency range 404–405 MHz.
- Method C proposes to use the SOS allocation in the frequency band 137–138 MHz for downlink and the band 148–149.9 MHz for uplink and to provide appropriate associated regulatory provisions in the RR for telecommand links of non-GSO SD missions.

Regarding WRC-19 Agenda Item 7 Issue M, the studies recognized the specific nature of non-GSO SDM satellite systems being developed by academic institutions, amateur satellite organizations, or by developing countries that are using these satellites to build their expertise in space capability and proposed in the CPM-19 Report for administrations to take a decision at WRC-19 on a draft Resolution – Simplified regulatory regime for non-GSO SDM satellite systems for non-GSO SDM satellite systems, operating under any space radiocommunication service not subject to the application of Section II of Article **9**. This draft Resolution recognizes the specific nature of the spectrum allocated to the amateur-satellite service that shall operate in accordance with the definition of the amateur-satellite service as contained in Article **25** (ITU 2016) (Article **25** of the RR) of the RR, resolves the total number of satellites in a non-GSO SDM satellite system constellation shall not exceed [10], and that the maximum period of operation and validity of frequency assignments of a non-GSO SDM satellite system shall not exceed 3 years from the date of bringing into use of the frequency assignments, which is equal to the satellite launch date, without any possibility of extension, after which the recorded assignments shall be cancelled.

BSS Plans and Their Associated Procedures (Appendices 30/30A)

Appendices **30** and **30A** to the RR contain Plans for the broadcasting-satellite service (BSS) in the 12 GHz band and the associated feeder-link Plans in the fixed-satellite service (FSS) in the 14 and 17 GHz bands. These Plans are occasionally referred to as the "BSS and the associated feeder-link Plans" and were established with a view to facilitating equitable access to the GSO for all countries.

The BSS and associated feeder-link Plans are presented in a tabular form in Articles **10** and **11** of Appendix **30** (hereafter referred to as AP30) and Articles **9** and **9A** of Appendix **30A** (hereafter referred to as AP30A), respectively. The regulatory procedures associated with the Plans are contained in the Articles of those Appendices. They apply to Plan implementation and modification as well as sharing with respect to terrestrial and other space services in the frequency bands of AP30/30A. Several technical annexes exist containing sharing criteria, calculation methods, and technical data relating to the Plans.

Characteristics of the national assignments, such as nominal orbital position, the service area defined by the ellipse parameters and e.i.r.p. values, and the channel numbers (Frequency assignments) are contained in Articles **10** and **11** of Appendix **30** and Articles **9** and **9A** of Appendix **30A**.

FSS Plan and Its Associated Procedures (Appendix 30B)

Appendix **30B** of the RR contains the Plan for the fixed-satellite service (FSS) in the 6/4 GHz frequency bands and in the 13/10–11 GHz frequency bands. This Plan is also referred to as the "FSS Plan" and was established with a view to facilitating equitable access to the GSO for all countries.

The FSS Plan is contained in Appendix **30B** (hereafter referred to as AP30B) together with its associated regulatory procedures. Several annexes exist containing criteria, calculation methods, and technical data relating to the Plan. The FSS Plan is an allotment plan. Each allotment in the Plan comprises:

- A nominal orbital position
- A bandwidth of 800 MHz (uplink and downlink) as listed in paragraph 1 above
- A service area for a national coverage

Characteristics of the national allotments, such as nominal orbital position, ellipse parameters, and power-density values, are contained in Article **10** of AP30B.

Administrative Due Diligence

Following one of the recommendations in the report by the Director of the BR on **RESOLUTION 18** (Kyoto 1994) (ITU 2016) (**RESOLUTION 18** of the RR), WRC-97 adopted **Resolution 49**, which has been modified by subsequent WRCs, on the administrative due diligence applicable to some satellite communication services as a means of addressing the problem of reservation of orbit and spectrum capacity without actual use. This resolution will apply to any satellite network of the fixed-satellite service, mobile-satellite service, or broadcasting-satellite (except in 21.4–22 GHz band) service in frequency bands subject to coordination under Section II of Article **9**, as well as modifications of the Appendices **30** and **30A** Plans and additional uses in the Appendix **30B** planned services.

For the above cases, an administration shall send to the Bureau due diligence information relating to the identity of the satellite network (name of the satellite, notifying administration, reference to the special section publication, frequency range, name of the operator, orbital characteristics) and the spacecraft manufacturer (name of the manufacturer, date of execution of the contract, delivery window, number of satellites procured); this information is to be submitted as early as possible before bringing into use, but must in any case be received before the end of the 7-year period established as a time limit for bringing into use a satellite network. Before notifying its satellite network for recording in the MIFR, the administration shall also send to the Bureau information relating to the launch services provider (name of the launch provider, date of execution of the contract, anticipated launch or in-orbit delivery window, name of the launch vehicle, name and location of the launch facility).

After verifying its completeness, the Bureau will publish the information in a special section of the BR IFIC. Should an administration fail to supply the complete required due diligence information in time, the networks concerned shall be cancelled (cancellation of the coordination request or modification to the Plan or entry in the MIFR) and shall not be recorded in the MIFR.

RESOLUTION 552 (WRC-12) (ITU 2016) (**RESOLUTION 552** of the RR) contains due diligence procedure for BSS in the band 21.4–22 GHz. The Resolution is entitled "Long term access to and development in the band 21.4-22 GHz in Region 1 and 3." The content of this resolution is somewhat similar to **RESOLUTION 49** (ITU 2016) (**RESOLUTION 49** of the RR) and new data elements are required to be submitted by administration under this Resolution. Under this resolution, administrations have to submit due diligence information not only when the space station is brought into use for the first time but also information about any further change, like deorbiting of the satellite or moving of the satellite to another orbital location. Further, this Resolution requires ITU to provide an ITU-ID for each of physical satellite network brought into use in this band. This satellite ID remains the same for the whole life time of the satellite irrespective of the orbital location of the satellite or its responsible administration until it is deorbited.

Preventing Harmful Interference to Satellite Systems: Non-interference as a Norm

In recent years, an increasing number of cases of harmful interference have emerged, including deliberate ones with the intention of disturbing or preventing the reception of signals, which particularly affect telecommunication satellites. In some cases, instances of harmful interference have targeted radio navigation-satellite service (RNSS) signals used by civil aviation, threatening international air traffic with dire consequences including potential loss of life.

A primary objective of ITU is to ensure interference-free operations of radio-communication systems. This has been emphasized at ITU World Radio-communication Conferences, as citizens of every country around the world depend on terrestrial and space radiocommunication systems for the provision of reliable telecommunication and broadcast services. As the leading United Nations agency for management of the radio-frequency spectrum and satellite orbits – and hence responsible for resolving instances of intentional or unintentional harmful inference – ITU is extremely concerned about the growing number of satellite networks which are the targets of deliberate harmful interference. Although currently some mechanisms exist to resolve harmful interference between the parties concerned, a resolution of this nature is often an expensive and lengthy process.

ITU's response so far has been limited to appealing to all parties that may be involved to exercise the utmost goodwill and to provide mutual assistance in settling issues of harmful interference. In such cases, ITU applies the provisions enshrined in Article **45** (ITU 1992) (Article **45** of the ITU Constitution) of the ITU Constitution and Section VI of Article **15** (ITU 2016) (Article **15** of the RR) of the RR. In most instances, the information provided to ITU by an affected administration includes evidence on the location of the source of harmful interference. However, as the information often comes from a single source, and in the absence of proper means to investigate or corroborate the information, irrefutable evidence cannot be provided to the administration under investigation to assume responsibility for stopping the interference.

Within this context, ITU has signed memoranda of cooperation (MoC) with administrations and organizations that have the capacity to monitor the use of spectrum allocated to satellite services in order to assist us in performing measurements related to cases of harmful interference. A memorandum of cooperation has also been signed between ITU and the International Civil Aviation Organization (ICAO) regarding cases of interference involving the RNSS on board civil aircraft. Such agreements, along with continued work in improving the technical and regulatory environment, are expected to help in the timely settlement of harmful interference cases, including RNSS, which would have a profound impact on improving aviation safety.

To better respond to the satellite interference issue, the Bureau is providing administrations, satellite operators, space agencies, and other space stakeholder member of the ITU-R Sector with an ITU online application "Satellite Interference Reporting and Resolution System" (SIRRS) (ITU 2019) (**SIRS**: https://www.itu.int/en/ITU-R/space/SIRRS/Pages/default.aspx), to be used for reporting and exchange of information concerning cases of harmful interference affecting space services. This online application has been developed in response to **RESOLUTION 186** (Busan 2014) (**RESOLUTION 186** (ITU-PP-14)) and in line with Annex 2 to Decision 5 (Rev. Busan 2014) on modern electronic communication methods. The objective of this system is to facilitate the communication between the parties concerned in case of harmful interference and to assist them in the identification of sources of interference and their prompt elimination in accordance with the provisions of Articles **15** and No. **13.2** of the RR (Articles **13** of the RR) (ITU 2016). The system allows to capture information in accordance with Appendix **10** of the RR and to upload additional information in the format of Report **ITU-R SM.2181** (ITU 2010) (Report **ITU-R SM.2181** https://www.itu.int/pub/R-REP-SM.2181-2010), Recommendation **ITU-R RS.2106-0** (ITU 2017) (Recommandation **ITU-R RS.2106-0** https://www.itu.int/rec/R-REC-RS.2106-0-201707-I/en) or any other standard format.

Security Aspects and Protection of Frequency Assignments Recorded in the MIFR

According to Article **8** of the RR (Article **8** of the RR), the international rights and obligations of administrations in respect of their own and other administrations' frequency assignments shall be derived from the recording of those assignments in the MIFR. Any frequency assignment recorded in the MIFR with a favorable finding under No. **11.31** shall have the right to international recognition. For such an assignment, this right means that other administrations shall take it into account when making their own assignments, in order to avoid harmful interference. If harmful interference to the reception of any station whose assignment is in accordance with No. **11.31** is actually caused by the use of a frequency assignment which is not in conformity with No. **11.31**, the station using the latter frequency assignment must, upon receipt of advice thereof, immediately eliminate this harmful interference.

In accordance with No. **4.4** of the RR, administrations shall not assign to a station any frequency in derogation of either the Table of Frequency Allocations or other provisions of the RR, except on the express condition that such a station, when using such a frequency assignment, shall not cause harmful interference to, and shall not claim protection from harmful interference caused by, a station operating in accordance with the provisions of the Constitution, the Convention, and the RR. The Regulations recognize therefore the operation of non-conforming stations, however, under strict conditions of non-interference against stations in conformity with the RR, for information purposes only.

Following the RoP on No. **4.4** adopted by the RRB, the scope of No. **4.4** is therefore limited to derogations to the Table of Frequency Allocations. In particular, administrations intending to authorize the use of spectrum under No. **4.4** still have the obligation, under Sections I and II of Article **9**, Nos. **11.2** and **11.3**, to notify to the Bureau "any frequency assignment if its use is capable of causing harmful interference to any service of another administration." The Board also concluded that administrations, prior to bringing into use any frequency assignment to a transmitting station operating under No. **4.4**, shall determine:

(a) That the intended use of the frequency assignment to the station under No. **4.4** will not cause harmful interference into the stations of other administrations operating in conformity with the Radio Regulations
(b) What measures it would need to take in order to comply with the requirement to immediately eliminate harmful interference pursuant to No. **8.5**

When notifying the use of frequency assignments to be operated under No. **4.4**, the notifying Administration shall provide a confirmation that it has determined that these frequency assignments meet the conditions referred to above in item a) and that it has identified measures to avoid harmful interference and to immediately eliminate such in case of a complaint.

The provisions relating either to frequencies or bands to be used for safety and distress communications or allocated for passive usage prohibit any other use:

(a) Provisions relating to safety and distress communications – in Appendix **15** (Global Maritime Distress and Safety System – GMDSS) (Appendix **15** of the RR), Tables 15-1 and 15-2: frequencies marked with an asterisk (∗) to indicate that any emission causing harmful interference to distress and safety communications is prohibited
(b) Provisions relating to passive usage – No. **5.340** of the RR (Provision No. **5.340** of the RR)

The Board considers that, in view of this prohibition, a notification concerning any other use than those authorized in the band or on the frequencies concerned cannot be accepted even with a reference to No. **4.4**; furthermore, the administration submitting such a notice is urged to abstain from such usage.

Conclusion

"With a concerted effort, we can reduce, and to the extent possible remove, all obstacles impeding the development and bringing into operation of new satellite networks and systems; we have to think carefully about how we can continue to foster peaceful cooperation among nations through the equitable sharing of global resources and use and improve satellite access to help connect the unconnected, and make the world a better and a fairer place for all" (Henri 2015).

References

Busan (2014) ITU Plenipotentiary Conference Busan (2014). https://www.itu.int/en/history/Pages/PlenipotentiaryConferences.aspx?conf=4.294

Henri (2015) Orbit/Spectrum international regulatory framework. ITU. www.itu.in/en/ITUD/Regional-Presence/AsiaPacific/Documents/Events/2015/October-IISS-2015/Presentations/S1_Yvon_Henri.pdf

ITU Constitution, 22 DEC 1992, Collection of the Basic Texts of the International Telecommunication Union adopted by the Plenipotentiary Conference. https://www.itu.int/pub/S-CONF-PLEN-2011. ITU. Last accessed 09 May 2019

ITU (2010, 2017) 2017 Rules of Procedure, approved by the Radio Regulations Board. https://www.itu.int/pub/R-REG-ROP/en

ITU Radio Regulations (2016). http://www.itu.int/pub/R-REG-RR/en. ITU. Last accessed 09 May 2019

ITU-R Space Services Department (SSD) http://www.itu.int/ITU-R/go/space/en. ITU. Last accessed 09 May 2019

Kyoto (1994) ITU Plenipotentiary Conference (Kyoto, 1994). https://www.itu.int/en/history/Pages/PlenipotentiaryConferences.aspx?conf=4.15

Space Weather: The Impact on Security and Defense

52

Jan Janssens, David Berghmans, Petra Vanlommel, and Jesse Andries

Contents

Abstract

Space weather refers to variations induced by the Sun of the Earth's space environment and the impact that these variations can have on technological systems and human health. During space weather events, technology, such as radio communication and GNSS positioning, can be seriously affected. Space weather can cause the loss of satellites, increase radiation levels at aeronautical flight levels and on the ground, and has the potential to catastrophically damage

J. Janssens · D. Berghmans (✉) · P. Vanlommel · J. Andries
Solar Terrestrial Center of Excellence (STCE) – Solar Influences Data Analysis Center (SIDC),
Royal Observatory of Belgium, Brussels, Belgium
e-mail: David.Berghmans@sidc.be

© Springer Nature Switzerland AG 2020
K.-U. Schrogl (ed.), *Handbook of Space Security*,
https://doi.org/10.1007/978-3-030-23210-8_94

power grids. We review the space weather cause-effect chains from the source to the affected technologies with special attention to the impact on security and defense.

Introduction

Besides being an exciting scientific discipline, space weather is increasingly recognized as a source of risks to critical infrastructures. Governmental and commercial organizations become progressively aware of their vulnerabilities to the impact of space weather on technological systems and human health. Space weather thereby evolves along the same path as meteorological weather: end users require to be informed on the impact the environment has on their activities, both on average over long time scales as well as operationally in real time.

Most space weather is ultimately due to events in the solar atmosphere such as solar flares, solar energetic particle events, coronal mass ejections, and coronal holes. The impact of these events propagates through interplanetary space and can directly affect technologies and astronauts deployed in space, or indirectly through disturbances of the Earth magnetosphere and ionosphere. Space physics is however only part of the story; the final impact experienced by the end user is critically dependent on the details of the technology and the local environment. Space weather is therefore undeniably multidisciplinary, spanning different space and geophysics as well as engineering domains.

The difficulty with setting up a user-oriented service is that the required space weather and technology expertise is scattered over various disciplines, over many organizations (government organizations, research institutes and companies) in different countries. Several international organizations are therefore taking initiatives to create collaborative frameworks to bring together the required expertise. The World Meteorological Organization (WMO, a specialized agency of the United Nations) has created the Inter-programme Coordination Team on Space Weather (ICTSW) that is tasked to worldwide standardize and enhance the exchange of space weather data and services. In the past years, the Space Situational Awareness (SSA) program of the European Space Agency (ESA) brought together many European space weather assets in a coordinating framework. At the end of 2019, ESA regrouped its space weather activities in a "Space Safety & Security" pillar at the ESA Ministerial Council "Space 19+".

Section "Space Weather from the Sun to the User" of this chapter provides an overview of space physics phenomena and the potential impact they have on technologies. This is further illustrated in section "Historical Space Weather Events with Defense and Security Impact" with a sample of historical space weather events selected for their relevance to security and defense. Section "Space Weather as a Challenge" addresses the challenges space weather poses for (observational) research, for societal risk handling, and finally as a security and defense issue. The last section provides the conclusions on this chapter on space weather.

Space Weather from the Sun to the User

The Sun is a variable star at only 150 million km from the Earth. It is a plasma ball, i. e., a hot ionized gas where the interaction of its electrical and magnetic energy makes its dynamics and variability. Its variability is characterized by the 11-year cycle observed in the counts of the sunspot index (Fig. 1). Sunspot are dark features appearing in groups (Figs. 2 and 4) where the solar magnetic field pierces through the visible surface of the Sun. Higher up in the solar atmosphere, in the solar corona, the sunspots expand in so-called active regions composed of magnetic loops that connect with nearby sunspot groups (Fig. 7). In contrast, in "coronal holes," the magnetic field is not bundled in strong coronal loops, nor does it reconnect to other place on the solar atmosphere, but expands outwards and fills most of the interplanetary space with the (fast) solar wind. The Sun rotates around its axis in roughly 28 days, with the solar poles taking a few days longer for a complete rotation than the solar equator. Also, the (fast) solar wind is swept around as a fire-hose stream passing by the Earth with 28-day recurrence.

With increasing complexity, the magnetic active regions in the solar corona tend to become unstable to so-called magnetic reconnection thereby producing energy emissions, called solar flares, over the entire electromagnetic spectrum. Solar flares are categorized according to their emission in soft X-rays using a logarithmic scaling (Fig. 8). The most violent class of flares (X-class) amounting to an increase of a factor 10,000 in solar X-ray emission with respect to the background level (A-class)

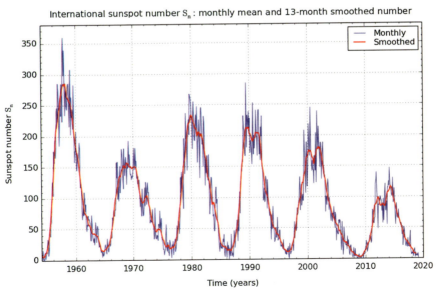

SILSO graphics (http://sidc.be/silso) Royal Observatory of Belgium 2019 March 1

Fig. 1 The international sunspot number characterizing the 11-year magnetic activity cycle of the Sun

Fig. 2 White light drawing from the Mount Wilson Observatory on 19 September 1941, showing the sunspot group responsible for the brilliant aurora the night before. Note the remark from the observer-on-duty in the lower right coroner ("Aurora last nite"). The Mt. Wilson 150-Foot Solar Tower is operated by UCLA, with funding from NASA, ONR and NSF, under agreement with the Mt. Wilson Institute (http://obs.astro.ucla.edu/intro.html)

observed on quiet days. A similar magnetic reconnection process can lead to coronal mass ejections (CMEs) whereby a large blob of mass is expelled from the solar atmosphere into the solar wind. If the propagation of the CME is faster than the ambient solar wind, then the CME fronts can steepen into shock waves. Solar plasma particles can be accelerated to near-relativistic speeds, either in solar flares or in shock waves. The particles escape away from the Sun into space. Moving through space, these electric particles are forced to follow the magnetic field lines present in the space. These fast particles are called "Solar Energetic Particles" or in short SEPs.

Given its proximity, the Sun drives space weather around the Earth on three timescales:

1. Within 8 min, the electromagnetic radiation emitted by flares and travelling at the speed of light
2. Within an hour, SEPs travelling at near-relativistic speeds from flares or CME shocks
3. Within a day, the bulk mass of the fastest CMEs (~2000 km/s)

Life on the Earth's surface developed experiencing little or no influence from the above-mentioned solar drivers of space weather because the Earth is surrounded by different protecting layers. The magnetic field of the Earth expands outwards in space and creates a magnetic bubble (the magnetosphere) from which the solar wind is deflected. Under the pressure of the solar wind, the magnetosphere is deformed on the sun-light "day" side but remains big enough in normal conditions to expand beyond the orbit of geostationary satellites. Under the impact of CMEs or fast solar wind stream however, the magnetosphere temporary further deforms and later resettles. This process is called a geomagnetic storm and is often associated with aurora near the magnetic poles. In the northern hemisphere, these polar lights are most often seen from Scandinavia, Canada, Alaska, and Siberia. By creating large-scale induction currents, geomagnetic storms can also lead to power grid collapse. On longer timescales, geomagnetic induction currents through oil pipelines lead to enhanced corrosion.

The top layer of the Earth atmosphere (roughly from 60 to 1000 km) protects us from extreme ultraviolet (EUV) and X-radiation from the Sun and is called the ionosphere. The ionosphere reflects HF (high frequency) radio waves and is therefore important for radio communication. The ionosphere is a reactive environment which can be strongly influenced by solar flares, by geomagnetic storms and by solar energetic particles. Disturbed ionospheric conditions can lead to HF radio blackouts, degraded satellite communication, and GNSS (Global Navigation Satellite Systems) positioning errors. Under the influence of solar radiation, satellites in low Earth orbit can experience enhanced drag at times the ionosphere is "thickened."

Solar energetic particles reach the Earth environment at near-relativistic speeds along the interplanetary magnetic field embedded in the solar wind. They are an immediate risk to astronauts and spacecraft electronics. They are deflected by the Earth magnetosphere towards the magnetic poles of the Earth where they collide with atmospheric particles creating a shower of secondary radiation. The most energetic SEP events can however strike anywhere on the Earth with their shower of secondary radiation reaching the ground (so-called ground-level events, GLEs). During such strong SEPs, airplane crew and passengers risk ionizing radiation, and notably during polar flights.

Not unlike earthquakes, space weather events are exponentially distributed with small events being much more numerous than large events. The largest events in recorded history are referred to as extreme space weather. Among geomagnetic storms, the classical reference is the "Carrington event" in 1859 which made telegraph systems, the most advanced technology at the time, fail all over Europe and North America. If such an event would happen again, modern technology would be catastrophically affected. The chance for such extreme events to occur again in the next 10 years is only 12% (Riley 2012). Cases of extreme space weather are thus high impact-low probability events.

Historical Space Weather Events with Defense and Security Impact

In this section, we first give an historical account of some noteworthy space weather events with military impact.

Polar Lights as the Ultimate Weapon

During the Second World War, both belligerent parties made use of the bright shine of the polar lights to attack critical assets of the adversary. A noteworthy example took place in September 1941 (Love and Coïsson 2016). Greenwich sunspot group 1,393,703 appeared from behind the eastern solar limb on 10 September. During its 2-weeks transit over the solar disk, it continued to grow in size and complexity, making it all the way into the Top 50 of largest sunspot groups of the last 170 years! The region was very flare active, and a particularly strong eruption over this sunspot group was observed with the Greenwich Observatory spectrohelioscope in the morning of 17 September. Barely 20 h later, magnetometers on Earth went haywire when what is now known as a coronal mass ejection (CME) struck the earth's magnetic field. Starting around 09:00 UT on 18 September, severe to extremely severe geomagnetic storming levels were continuously recorded for a period of 24 h, a level of geomagnetic activity that has not since been matched (Cliver and Svalgaard 2004). The resulting aurora was observed as far south as New Mexico and California (Fig. 2). In Chicago, motorists parked on the highways had caused a traffic jam as they sought a clear view of the celestial spectacle. The press used literary one-liners such as "celestial pyrotechnics," "neon lights," or "ethereal blitz" to describe the impressive event. Some citizens even wondered if a new type of anti-aircraft search battery was being tested.

The true effects on war took place in Europe, where the British Royal Air Force carried out a raid on a German supply base in the Baltic Sea, whereas the Germans bombarded Leningrad, each under the lights of the aurora borealis. In the North Atlantic, German U-boats were operating to sink eastbound ships supplying Great Britain. During the night of 18–19 September, the captain of the U-74 recorded in his war diary that the conditions were "as bright as day." An allied convoy consisting of cargo ships and accompanying anti-submarine warships ("corvettes"), which normally would have been hidden in the dark of night, was detected. Just a few hours later, a decisive torpedo was fired hitting the corvette HMCS Lévis, nearly cutting the vessel in two. Shortly afterwards, the HMCS Lévis sank leaving many casualties.

The Disappearance of the HMS Acheron

Since the start of the measurements in the early 1940s, only 72 GLEs have been recorded, qualifying them as rare. The strongest GLE occurred on 23 February 1956, and increased radiation levels on several locations by several thousands of percent (Bieber et al. 2005; Fig. 3).

Coincidentally, the HMS Acheron, a submarine of the Royal Navy, was performing arctic trials in the Denmark Strait between Iceland and Greenland. The 1,123-ton Acheron was a sister ship of the Affray, which sank in the English Channel a few years earlier, leaving many casualties. When the Acheron did not respond to a routine radio check on 24 February at 10:05 a.m., the Admiralty flashed a "sub-sunk" order, signaling an immediate search with all available ships and planes. Royal

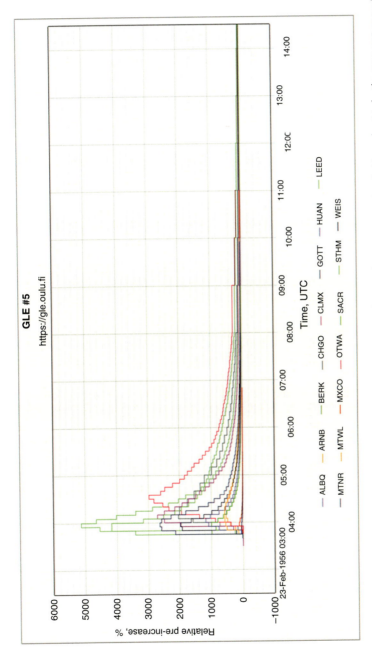

Fig. 3 On 23 February 1956, neutron monitors around the world recorded a strong increase in neutron counts, what turned out to be the strongest ground-level enhancement ever recorded. A GLE is a rare event with only 72 registrations since the start of the observations in the early 1940s. Most GLEs do not reach 100% in neutron counts increase, corresponding here to halfway between 0% and the first gridline (200%). This testifies of the formidable strength of the 1956 event. Plot from the Oulu Cosmic Ray Station of the University of Oulu, Finland (https://gle.oulu.fi/#/)

Air Force planes roared off for Reykjavik, Iceland, to set up a base for search operations. U.S. Air Force units on Iceland already were standing by. Ships steamed out from Scotland and Iceland. Three hours later, the British minesweeper Coquette radioed that she had made "visual contact" with the sub in gale-swept seas. The Acheron then proceeded to Iceland, and the Admiralty called off the search for the British submarine, which was feared lost for nearly 6 h. It was quickly pointed out that the unusual sunspot activity over the past 2 days might have been the cause of the radio blackout, as gigantic explosions on the Sun had bombarded the Earth with cosmic rays, interfering with communications. In Copenhagen, the Danish government's telegraph authority said no radio messages had been received from Greenland stations since early morning on 23 February. "Frankly," a spokesman for the authority said, "we cannot see how a vessel could get signals through while we cannot receive a word from powerful land stations." (from the "Amsterdam Evening Recorder," 24 February 1956).

Jamming Missile Warning Systems

In 1967, the Cold War between the Soviet Union with its satellite states (USSR) on the one hand and the United States and its allies on the other hand was in full swing. The USA had a stockpile of more than 30,000 nuclear warheads, while the USSR was making a recovery effort to exceed that number. Ballistic missiles that could carry such warheads were deployed at an increasing pace, and radar systems to detect such missiles were being operated. Tension was high and pushed even further by the ongoing race to the Moon, the continued launch of spy satellites, and other conflicts such as between Egypt and Israel. It was against this backdrop of geopolitical and military turmoil that a large and complex sunspot group appeared at the Sun's east limb on 17 May. Further increasing in size during the next days, the region started strong flaring activity from 21 May onwards, which would last for a full week. The strongest flare (X6) took place on 23 May around 18:46 UT, which was at sunset for European countries, but near local noon for the central United States (Fig. 4). It was followed by another strong flare (X2) at 19:53 UT. These extraordinary solar eruptions manifested themselves very strongly over all portions of the electromagnetic spectrum, and extreme, hours-long solar radio bursts (i.e., burst in the radio part of the solar spectrum) were recorded. At frequencies of 606 MHz, peak flux densities reached 373,000 sfu (solar flux units), making it the strongest solar radio burst observed up to that time, and of the entire twentieth century. For comparison, typical values for the "undisturbed" flux density at this frequency are somewhere between 30 and 45 sfu.

   The solar radio bursts significantly disturbed the United States' Ballistic Missile Early Warning System, BMEWS for short (Knipp et al. 2016). This radar system operated at 440 MHz from sites in Alaska, Greenland, and the United Kingdom. During the 23 May event, all the BMEWS radar systems had a good view on the Sun, with the Greenland radar particularly well aligned. In response to the solar radio bursts, the radar screens showed many "impacts" which were subsequently interpreted as

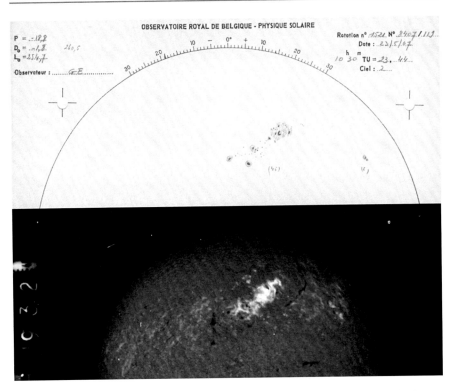

Fig. 4 The famous sunspot group responsible for the very high solar activity on May 1967. On top, a white light drawing by the USET solar telescope (Royal Observatory of Belgium, Brussels, http://www.sidc.be/uset/) on 23 May (10:30 UT). The bottom picture is an H-alpha image taken on 19:32 UT at Sacramento Peak when the X6 flare was still in progress. The size of the bright flare ribbons, covering most of the sunspot group, testifies of the strength of the event (Image from the Flare Patrol H-alpha instrument, NSO/AURA/NSF, ftp://nispdata.nso.edu/flare_patrol_h_alpha_sp/movie/56/)

jamming by the operators who had never witnessed such an intense radio event. Logically, Cold War military commanders viewed full-scale jamming of surveillance sensors as a potential act of war and positioned their bombers in a ready-to-take-off position. The decision to launch or not to launch is well worth noting, because in view of the tense political situation in May 1967, a full-scale aircraft launch by the allied forces could have been very provocative and, just as importantly, difficult (if not impossible) to abort in view of the impaired radio communications. Fortunately, and despite the limited data available at the time, solar forecasters from the USAF Air Weather Service were able to extract sufficient information from solar and radio observations to convince high-level decision makers at NORAD (North American Air Defense) that the Sun was the likely culprit in contaminating the BMEWS radar signals. This key element defused the critical situation, and the decision was taken to return aircraft and alert status to their normal levels. Further details on this event and its space weather impacts, as well as on the effects from the related extremely severe geomagnetic storm, can be found in Knipp et al. (2016).

Unexpected Detonation of Sea Mines

Over the years, the solar storm of 4 August 1972 has reached a legendary status among space weather scientists. This storm was very similar to the famous Carrington event in 1859, except for the aurora which was not seen from low geomagnetic latitudes. The main eruption took place during the morning of 4 August and was accompanied by a powerful proton event that reduced the life expectancy of solar panels aboard satellites with no less than 5 years. Yet, things could have been much worse. Indeed, the storm occurred right between the Apollo 16 and 17 missions. If astronauts had been walking on the Moon when this proton event took place, then they would most likely have suffered radiation sickness (Cucinotta et al. 2010; Fig. 5). The CME accompanying the eruption still holds the record for being the fastest CME travelling the Sun-Earth distance. No CME has ever done better than 14.6 h (Cliver and Svalgaard 2004). This achievement was most likely the consequence of eruptive activity during the hours and days prior to the main eruption, clearing the path for the powerful 4 August CME. When it arrived at Earth, the magnetopause, usually at about 10 Earth radii distance, was pushed back to the Earth to only about 5 Earth radii, suddenly exposing several satellites to the wrath of the disturbed solar wind. As the subsequent geomagnetic storm unleashed all its power, some of the magnetometers on Earth went off-scale, and aurora was bright enough to cast shadows. As just recently revealed by the scientists (Knipp et al. 2018), this severe geomagnetic storm had also another effect which had been buried in the military archives for more than 40 years. Indeed, back in 1972, the United States were at war with Vietnam. In an attempt to isolate North Vietnam from

Fig. 5 Astronaut Eugene A. Cernan checking out the Lunar Roving Vehicle (LRV) during the Apollo 17 mission in December 1972 (Credits: NASA). Just a few months earlier, energetic particles released during a strong solar eruption would have caused radiation sickness if any astronauts had been walking around on the lunar surface at that time

the rest of the world, magnetic-influence sea mines ("Destructors") had been dropped into the coastal waters of North Vietnam just 3 months prior. On 4 August, aircrews reported the sudden detonation of some two dozen of sea mines near Hon La in just 30 s. Aerial observations indicated evidence of some 4000 additional detonations along the North Vietnamese coast during the first weeks of August. The US Navy quickly concluded that the magnetic field variations were the cause of these detonations, in line with measurements from magnetometers in nearby locations such as Manila, the Philippines. This conclusion led to the radical decision to replace all the magnetic-influence-only sea mines with magneto/seismic mines, meaning there were now two triggers needed before the sea mines could detonate.

The Battle of Takur Ghar

So far, the cited examples took place during record setting space weather storms. However, the following case illustrates that even mild disturbances, enhanced by a series of unfavorable conditions, can also lead to insecure situations.

Operation Anaconda took place in Afghanistan from 1 to 18 March 2002. It was a large-scale international military campaign led by the United States aiming at the destruction of Al Qaeda and Taliban forces. An intense battle took place during the morning hours (before dawn) of 4 March on Takur Ghar, a 3191-m high mountain top, as US Special Operations Forces came under heavy fire from the Al Qaeda and Taliban forces. A Chinook helicopter was directed to rescue the team. However, in view of the rocket-propelled grenades and heavy machine guns that the insurgents were using, a SATCOM (Satellite Communications) message was sent to the Chinook to avoid the mountain top. Unfortunately, and despite repeated attempts, the Chinook helicopter never received that critical message. It landed on the top of Takur Ghar and came immediately under intense fire, resulting in several casualties.

A subsequent analysis of the incident blamed the radio outage on poor performance of the UHF (ultra high frequency) radios on the helicopters as well as on terrain radio interference. However, in 2014, Michael Kelly and his team of researchers from the John Hopkins University came to a different conclusion, offering a viable alternative for the outages (Kelly et al. 2014). Utilizing a model that uses UV (ultraviolet) data from the TIMED spacecraft to retrieve the 3D electron density, they came to the conclusion that a combination of ionospheric disturbances with multipath effects (multiple radio reflections from the mountainous terrain) could also have caused the decreased communication links (Fig. 6).

The main cause of the ionospheric unrest is the presence of equatorial plasma bubbles, i.e., depletions of electron density in the ionosphere. Their number correlates with the solar activity level, and they also are more numerous during the equinoxes (spring and autumn) than during the solstices (summer and winter). They usually form after sunset at the bottom of the F-region (main ionospheric layer), where small low-density irregularities can grow into turbulent bubbles. The bubbles have a typical size of about 100 km and their effects usually end around midnight. They can occur during relatively minor levels of geomagnetic activity,

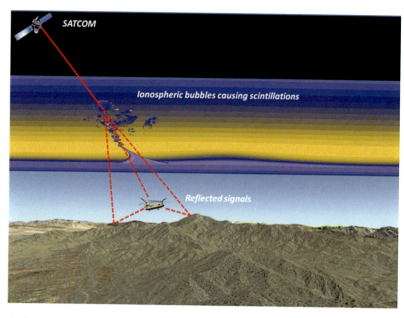

Fig. 6 Sketch prepared by ROB/GNSS (Dr. Nicolas Bergeot) with Dr. Yokoyama's model (NICT/AERI) as base. It shows how a combination of ionospheric disturbances ("scintillation") with reflected signals from the surroundings can weaken and effectively blacken out radio signals from space- or ground-based transmitters

especially during solar cycle maximum. Radio wave propagation can be severely affected in terms of power and intensity as these waves travel through small-scale structures in the ionosphere (i.e., scintillation of radio waves).

At first sight, one would expect relatively strong geomagnetic activity to explain the ionospheric disturbances, but this was not the case on 3–4 March. Quiet to unsettled geomagnetic conditions were observed, with a single active episode during the 21:00–24:00 UT interval on 3 March. According to Kelly and his team, this suppressed the generation of the evening-side depletions and delayed the onset of the ionospheric bubbles until after midnight.

Moreover, analysis of the data from the TIMED satellite led Kelly and his team to conclude that the plasma bubbles affecting the Takur Ghar war theater did not have very steep density gradients, and thus would have resulted only in mild ionospheric disturbances. Normally, this mild ionospheric "scintillation" is not a problem for SATCOM, but the intensity of the already weakened radio signals could have further been reduced by the multiple reflections from the surrounding mountains. Hence they concluded that "... the destructive multipath interference from complex terrain reflections coupled with scintillation could cause a signal blackout. ...". The take-home message here is that even mild disturbances, under a set of unfavorable conditions, can create dangerous situations and that users should always be very attentive.

Solar Flares Hampering Hurricane Relief Efforts

To conclude the summary of historical space weather impacts on the military operations and technology, we discuss this last recent case to illustrate that sometimes, it's just a matter of unfortunate timing.

The 2017 Atlantic hurricane season was one of the deadliest and most catastrophic hurricane seasons in recent history, with a damage total exceeding $200 billion (USD). One of the strongest hurricanes was hurricane Irma, ravaging the Caribbean island chain from the northern Leeward Islands, via Puerto Rico and Cuba to Florida from 6 till 11 September. Coincidentally, from 4 till 11 September, Active Region 2673 (Fig. 7) developed on the solar disk into a complex and very active sunspot group, producing no less than 27 medium and 4 extreme solar flares, including the two strongest flares of the entire solar cycle 24, resp. an X9 on 6 September and an X8 on 10 September (Fig. 8). The numbers of the active regions are determined by the National Oceanic and Atmospheric Administration, NOAA.

These strong flares induced a rapid ionization of the equatorial upper atmosphere, resulting in a disruption of HF communications while emergency workers were struggling to provide critical recovery services to the Caribbean communities (Redmon et al. 2018). The Hurricane Weather Net (HWN) reported that the 6 September solar flare caused a near-total communications blackout for most of the morning and early afternoon. The French Civil Aviation authorities also reported that HF radio contact was lost with an aircraft off the coasts of Brazil and French Guyana

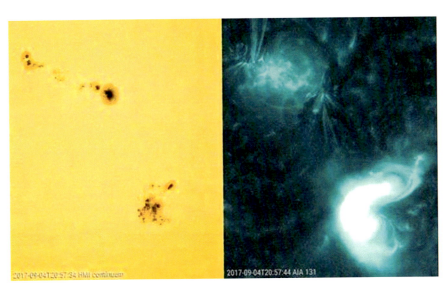

Fig. 7 A picture of Active Regions NOAA 2674 and NOAA 2673 (right) on 4 September 2017, as imaged by the Solar Dynamics Observatory (SDO) in white light (left) and in extreme ultraviolet (right). A medium-class flare was in progress in NOAA 2673 at that time. Image courtesy NASA/ SDO, AIA and HMI science teams. SDO is the first mission for NASA's Living With a Star (LWS) Program

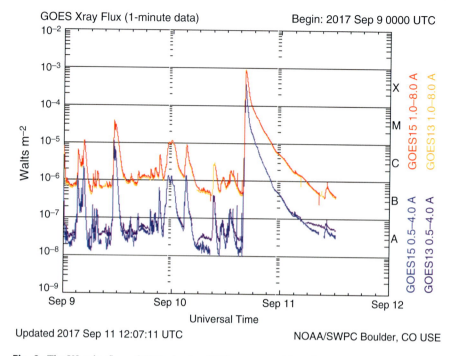

Fig. 8 The X8 solar flare of 10 September 2017 as measured in soft X-rays by the GOES13 and GOES15 satellites. Flares are categorized as A, B, C, M, or X-class flares. Each class ranges from 1 to 9. Image courtesy NOAA/GOES

for approximately 90 min, triggering an alert phase until a position report was received by New York radio. The 10 September flare also severely disrupted HF communication, with a widespread communication blackout lasting for nearly 3 h, which basically could not have happened at a worse time. The researchers conclude that "... These solar eruptions led to geoeffective space weather impacting radio communications tools used in the management of air traffic as well as emergency- and-disaster assessment and relief, temporarily complicating an already extreme terrestrial weather period." (Redmon et al. 2018). Further reading on Hurricane Irma is found in ▶ Chap. 39, "Satellite EO for Disasters, Risk, and Security: An Evolving Landscape".

Space Weather as a Challenge

The Research and Observation Challenge

The above two sections illustrate how the cause-effect chains are coupled over different physical domains from the Sun to the Earth. Many links in these cause- effect chains are poorly understood and space weather, or solar-terrestrial physics, remains thus an active field of research. As compared to meteorological weather, progress in space weather is hampered by the difficulty of obtaining constraining measurements. Measurements of space weather conditions start from observing the Sun, which are necessarily performed from a distance. These "remote sensing observations" (images or timelines) can only be obtained from ground-based telescopes during daytime in wavelengths not hindered by the Earth's atmosphere (radio and visible parts of the spectrum). Studies of the sources in the solar corona require observations in X-rays or EUV, which can only be observed by satellites outside the Earth's atmosphere. Even then, only one perspective on the solar globe is obtained and a full 360° view requires several more satellites in deep space observing the Sun from other viewpoints.

The propagation of solar wind plasmas and energetic particles throughout the interplanetary space on their way to the Earth can be confirmed by in situ measurements, but the only stable place between the Sun and the Earth is the Lagrange L1 point where the Sun's and Earth's gravitation balance. Unfortunately, following the relative difference in masses, this L1 point is at only 1% of the Earth-Sun distance from the Earth, providing very little warning time for upcoming solar wind disturbances. Further near-Earth in situ space measurements are required to track the state of the magnetosphere and to measure energetic particle fluxes at the top of the atmosphere. The Earth ionosphere is traditionally observed with ionosondes from the ground or through analysis of signals from GPS satellites. Finally, additional observations from the ground are required to measure deflections of the Earth magnetic field during geomagnetic observations. GLEs are confirmed with neutron monitor measurements.

The above list of measurements is only the tip of the iceberg and many additional measurements are possible and probably required to improve scientific understanding. A COSPAR working group has produced a road map for space weather. This report also highlights the importance of another set of measurement data, those of the impacts within the technologies. Measurements of radiation doses at flight altitudes during solar storms or induced currents in the power grid are much less readily available than the corresponding geospatial space weather measurements. Also, in the realm of observed impacts on satellite operations, data are generally not released publicly.

The Societal Challenge

Space weather poses risks to global society, with some regions (e.g., high latitude zones) and some sectors (e.g., those dependent on high precision GNSS) more affected than the others. A proper response to the involved risks will thus differ from region to region and from sector to sector but will require in any case the following elements:

- Understanding of the vulnerabilities, including impact and likelihood
- Preparedness through improved engineering of the affected systems

• Maintaining awareness of the current state of the space environment through observations and analysis in real time

Understanding of the Vulnerabilities

The first step towards a proper handling of the space weather risk within a sector is to have a proper understanding of the (potential) impacts and their likelihood. It is important to note that the exact impact of the space weather phenomena on technological systems goes well beyond space physics and is to a large extent an active separate area of (technological) research. While the physical principles are understood and often historical examples are known to exist (see above), it is much harder to accurately estimate the exact magnitude of the impacts and, more importantly, the associated cost and consequences.

Also, in the case of extreme space weather, one faces the problem of high impact-low probability events which are so rare that proper statistics on impact, cost, and consequences cannot be accumulated (Eastwood et al. 2017). Nevertheless, over the last decades, the understanding of the vulnerabilities has grown substantially in many sectors. Several countries around the world have also explicitly included space weather within their national risk assessments and have accounted for space weather phenomena within their national and sectoral risk management strategies.

Preparedness Through Improved Engineering

When potential impacts are known, there are broadly two venues in order to mitigate the impacts. The first is an engineering solution where the design of the technology is altered such as to become immune or at least less prone to impacts caused by space weather. For example, with sufficient "shielding," damage to spacecraft electronics by Solar Energetic Particles can be strongly reduced. GNSS positioning errors due to ionospheric variability can to some degree be addressed with dual frequencies or augmentation systems. Such engineering solutions are typically relatively successful for most mundane space weather events but become prohibitively expensive (or even physically impossible) for the most seldom but extreme space weather events.

Maintaining Awareness

When improved engineering or system hardening is not possible, the second strategy is to learn to live with the impacts but reduce their consequences through maintaining active awareness of the current and anticipated space weather conditions, and by actively feeding such information into the operation procedures within each affected sector. As just one example, the International Civil Aviation Organization (ICAO, a specialized agency of the United Nations) has recently identified world space weather information providers that will issue from autumn 2019 onwards so-called advisories to inform aviation actors on the ongoing space weather impacts on HF communication, GNSS positioning, and radiation levels at flight altitudes.

If this second mitigation strategy is followed, one must be aware that the observational challenge highlighted above becomes yet more stringent as the data must be acquired and processed in real time, which is not the case when the observation data are used for research and model development. Given the implied costs, there is a strong case to be made for international cooperation, both for ground-based observations and for observations from satellites. In the case of ground-based observations, there are additional logistical reasons to make observations from multiple sites across the global. Ground-based solar observations require observatories in different time zones and ionosphere and surface geomagnetic field data require spatial resolution.

International Coordination of Space Weather Services

But maintaining awareness of space weather conditions requires more than just having the observational data available. In addition, models to forecast (near) future behavior and human-based assessment of the meaning of the incoming are needed. It also includes deriving more tailored and focused services for specific users based on the observational data, the models, and the forecaster experience.

On a global scale, the International Space Environment Service (ISES, http://www.spaceweather.org) provides daily space weather monitoring and forecasting and brings together space weather monitoring centers from 19 countries around the world. This organization is the main international body devoted to the promotion of such services, facilitating the exchange of data and the exchange between those centers of best practices in providing space weather services. The NOAA Space Weather Prediction Service (SWPC) in the USA is the best-known member. In Europe, space weather services are offered among others through the ISES Regional Warning Centers in the UK (MetOffice), Belgium (Royal Observatory of Belgium), and Poland (Space Research Center). In Europe, the European Space Agency (ESA) has federated many European space weather assets in its Space Situational Awareness (SSA) network with specialized entities (the Expert Service Centers) explicitly focusing on a specific physical domain. The ESA/SSA network concept is still in development and is to be considered preoperational.

The need for globally coordinated observations has recently been channeled through the World Meteorological Organization where a dedicated work stream (Inter-Programme Team on Space Weather Information, Systems and Services) is in the process of including the corresponding observation requirements within the WMO databases regarding observation requirements.

Specific Defense Challenges

Space is sometimes considered the fifth warfare domain. It is at least a critical enabler for land, air, maritime, and cyber operations. Governments and armed forces strongly rely on space-based capabilities and services to fulfill their missions, especially in the frame of expeditionary and intelligence operations. Satellite communications, Global Navigation Satellite Systems (GNSSs), space-based Intelligence, Surveillance and Reconnaissance (ISR), and environmental monitoring are paramount for the success of operations. Therefore, timely reliable space weather information is essential.

The defense sector may be impacted by space weather effects on different levels. At the first level, one might expect military technology to be affected by space weather effects much like the civilian technology as discussed above. Obviously, details of the vulnerabilities of military technology are not readily available to space weather researchers. Here we discuss additional challenges faced by the defense sector beyond the direct impact on technology.

The first specific defense-related challenge is the relation between the needs for space weather services by military users and the capabilities and governance of currently developed space weather service systems. These space weather service systems are currently being developed out from the research community. Consequently, space weather warnings and alerts are typically formulated in scientific jargon describing the state of solar-terrestrial system as a physical system. Such information is not directly actionable by a military user in the field but needs to be translated in terms of potential impact on specific equipment. As this translation requires knowledge of the military equipment in use, it must be done by experts within the defense sector itself. These technical experts need to be trained to understand the scientific jargon and to be able to translate it into actionable information.

In this perspective, we notice huge disparities between the NATO Nations. In fact, only few NATO Nations are quite advanced while the majority has recently envisaged to bridge civil space weather services to the needs by military users. In America, the USAF (US Air Force) already works for years in close collaboration with the Space Weather Prediction Center (NOAA) and fully relies on its experience and knowledge in solar science to produce space weather forecasts. In Europe, the UK Met Office produces space weather forecasts for national civilian and military customers. Also, the Joint Meteorological Group (JMG) relies on the Solar and Terrestrial Center of Excellence (STCE) of Belgium for the provision of space weather data and has developed its own "Space Weather Impact Matrix" depicting the impact of space weather events on Dutch weapon systems.

In NATO, space weather has been recognized a specific matter and belongs to the METOC (Meteorological and Oceanographic) field of expertise. The United States is currently assuming the role of "Assisting Nation" (AN) for space weather. Last year, an ad hoc group has been created to work on standardization and harmonizing in the field of the space weather support to NATO operations.

In Belgium, the Meteo Wing of the Belgian Air Force (BAF) has started a collaboration with the STCE, especially in the field of education and training. What space weather products are concerned, the BAF currently relies on the US being the Space Weather AN. Nonetheless, the BAF strongly envisages extending its collaboration with STCE through the development of specific space weather products to support Belgian military assets and the support of the STCE in distributing and validating space weather advisories for the ICAO.

Another specific defense challenge is to distinguish a space weather impact from sabotage or jamming from the enemy. The same space weather source can affect a wide range of different technologies all over the planet. As we have discussed above, the impact includes reduction in communication possibilities, deterioration of positioning services, and unexpected malfunctioning. Confusion with a coordinated sabotage activity is therefore not excluded. In order to distinguish between a space weather impact or deliberate action, it is necessary to maintain an awareness of the state of space weather.

As mentioned, the – mostly – civilian space weather systems are currently being developed by the research community, although armed forces throughout the world are becoming increasingly active in the field of space weather, for instance, the USA and the UK. This implies that much of the existing space weather services are dependent on research observational infrastructure, both on the ground and in space. This observational infrastructure typically has poor long-term perspectives, very little redundancy and security measures, and even its own functioning might be strongly affected when strong space weather occurs. An independent, fully operational infrastructure for space weather observations is required to maintain serious "space situational awareness." This would involve many ground-based observatories (per time zone, redundancy for weather conditions) and/or many satellite observatories (remote sensing of the Sun from different perspectives, as well as in situ solar wind observations) and becomes therefore quickly prohibitively expensive for both civilian and military programs.

A final specific challenge for defense is at the extreme side of space weather events. Such extreme events have been observed in the past (e.g. 1859) but not in modern times when society has become fully dependent on power grids and satellite communication. The most extreme space weather events have the potential to damage the major transformators of the power grid worldwide (Schrijver et al. 2014). Such large-scale damage would require months to years for repair resulting in serious problems for basic services such as hospitals and water ans food distribution and might ultimately lead to collapse of the civilian society. The original space weather driver of such unrest will have faded away in at most a few weeks but the military might be called upon to maintain homeland security for much longer.

Conclusion

Space weather as driven by the variable activity of our nearby star, the Sun, has been around since before the dawn of humankind. Our increasing dependence on technology in space and on the ground makes us however increasingly vulnerable to the impact of space weather on technological systems. Whereas – in most sectors – everyday space weather can be handled through improved system hardening, the catastrophic impact that seldom but extreme space weather events can have, cannot be fully excluded. Given its dependence on global communications, accurate GNSS positioning, and satellite operations, also the defense and security sector must take the space weather risks into account. In this chapter, we have listed several historical space weather events that have had surprising impacts on military operations. We advocate that maintaining a "space situational awareness" requires structural communications between space physicists providing real-time analysis of the space phenomena, with technology expert operators that can estimate the impact of these phenomena on the technological systems that the end user cares about.

Acknowledgments The contribution by Capt d'Avi Damien Lebrun (Meteo Wing, Be) to section 4.3 is much appreciated.

References

Bieber JW, Clem J, Evenson P et al (2005) Largest GLE in half a century: neutron monitor observations of the January 20, 2005 event. In: Sripathi Acharya B, Gupta S, Jagadeesan P, Jain A, Karthikeyan S, Morris S, Tonwar S (eds) Proceedings of the 29th international cosmic ray conference, August 3–10, 2005, Pune, India, vol 1. Tata Institute of Fundamental Research, Mumbai, p 237

Cliver EW, Svalgaard L (2004) The 1859 solar-terrestrial disturbance and the current limits of extreme space weather activity. Sol Phys 224(1–2):407–422. https://doi.org/10.1007/s11207-005-4980-z

Cucinotta FA, Hu S, Schwadron NA et al (2010) Space radiation risk limits and Earth-Moon-Mars environmental models. Space Weather 8: S00E09. https://doi.org/10.1029/2010SW000572

Eastwood JP, Biffis E, Hapgood MA et al (2017) The economic impact of space weather: where do we stand? Risk Anal 37(2):206–218. https://doi.org/10.1111/risa.12765

Kelly MA, Comberiate JM, Miller ES et al (2014) Progress toward forecasting of space weather effects on UHF SATCOM after Operation Anaconda. Space Weather 12(10):601–611. https://doi.org/10.1002/2014SW001081

Knipp DJ, Ramsay AC, Beard ED et al (2016) The May 1967 great storm and radio disruption event: extreme space weather and extraordinary responses. Space Weather 14(9):614–633. https://doi.org/10.1002/2016SW001423

Knipp DJ, Fraser BJ, Shea MA et al (2018) On the little-known consequences of the 4 August 1972 ultra-fast coronal mass ejecta: facts, commentary, and call to action. Space Weather 16 (11):1635–1643. https://doi.org/10.1029/2018SW002024

Love JJ, Coïsson P (2016) The geomagnetic blitz of September 1941. Eos 97. https://doi.org/10.1029/2016EO059319

Redmon RJ, Seaton DB, Steenburgh R et al (2018) September 2017's geoeffective space weather and impacts to Caribbean radio communications during hurricane response. Space Weather 16 (9):1190–1201. https://doi.org/10.1029/2018SW001897

Riley P (2012) On the probability of occurrence of extreme space weather events. Space Weather 10 (2):S02012. https://doi.org/10.1029/2011SW000734

Schrijver CJ, Dobbins R, Murtagh W, Petrinec SM (2014) Assessing the impact of space weather on the electric power grid based on insurance claims for industrial electrical equipment. Space Weather 12(7):487–498. https://doi.org/10.1002/2014SW001066

Further Reading

Large Solar Event Detected During Irma. NCEI/NOAA, 14 September 2017. https://www.ncei.noaa.gov/news/large-solar-event-detected-during-irma

Missing British Sub Feared Lost, Safe; Search Called Off. Amsterdam Evening Recorder. LXXVII (158). Amsterdam, New York. 24 February 1956. p 1. https://en.wikipedia.org/wiki/HMS_Acheron_(P411)#cite_note-4

Odenwald S (2015) Solar storms: 2000 years of human calamity. CreateSpace Independent Publishing Platform. ISBN-10: 1505941466

Space Security in the Context of Cosmic Hazards and Planetary Defense

53

Joe Pelton

Contents

Abstract

When most people think of space security, and especially "space safety," they tend to think of space safety for astronauts and successful launch of satellites, or strategic systems related to missile defense. Some may even think of the threats that could come from orbital space debris. This chapter, however, examines space security from yet another perspective. It addresses cosmic hazards that humanity faces from asteroids, comets, solar storms, and even shifts in the Earth magnetosphere. These space hazards are very real threats to global security and modern civilization. Too often we tend to forget that our world travels through the hostile environment of outer space and is indeed quite vulnerable to cosmic hazards of various types.

This chapter focuses on the fact that humans live on a six sextillion ton planet that orbits the sun once a year and travels millions of kilometers through a quite dangerous cosmic environment as it speeds through space at close to 100,000 km/h.

J. Pelton (✉)
International Association for the Advancement of Space Safety (IAASS), Arlington, VA, USA
e-mail: joepelton@verizon.net

© Springer Nature Switzerland AG 2020
K.-U. Schrogl (ed.), *Handbook of Space Security*,
https://doi.org/10.1007/978-3-030-23210-8_89

Chapter 11 in Part 1 of the HOSS discusses "space safety" from the perspective of astronaut and spacecraft risk and reliability, and also considers the problems of sustainability of space, orbital debris, and space traffic management.

This chapter, however, addresses "space safety" from the perspective of asteroids and comets that are potentially hazardous and could impact Earth in various ways – in the air, on land, in the oceans, or just offshore. It also addresses the hazards that come with solar storms that include high-energy X-ray flares and coronal mass ejections (CMEs). Further the chapter even addresses emerging problems associated with the Earth's reversing magnetic poles. The now clearly detected shift in the geomagnetosphere is a concern. This is because during this time of reversal, the Earth's natural protective shielding against the ions from the sun that come with CMEs are greatly diminished. This means that humanity and our electrical power grids, pipelines, communications networks, and satellites will be much more vulnerable to violent solar storms during the time of this magnetic reversal.

The structure of this chapter is that it first addresses the threat to the world from potentially hazardous asteroids and comets. It then addresses the issue of solar radiation flares and coronal mass ejections (CMEs), and then finally it addresses the complication that comes from the reversal of the Earth's magnetic north and south poles.

Keywords

Anthropocene Age · Comets · Coronal mass ejection (CME) · European Space Agency · Geomagnetosphere · International Asteroid Warning Network (IAWN) · Mass extinctions · NASA · National Atmospheric and Oceanic Administration (NOAA) · Near-Earth objects · Potentially hazardous asteroids · Solar flares · Solar storms · Space Mission Planning Advisory Group (SMPAG) · Space weather · UN Office of Outer Space Affairs · UN Committee on the Peaceful Uses of Outer Space · Van Allen Belts · X-Ray flares

Introduction

The dinosaurs that were wiped out by the so-called K-T Mass Extinction some 65 million years ago really never knew what hit them. The Earth's various plant and animal life forms have to date experienced five mass extinctions that have each occurred hundreds of millions of years apart. Four of these event have come from thermal shift's in the Earth's atmosphere, but the most recent one, the K-T event, came when a 5-km-wide asteroid crashed into the coast of Mexico and the Caribbean Sea and created a massive cloud that blocked out the sun, killed off the dinosaurs and three-fourths of all species on Earth, and led to the rise of mammalian life forms over the life forms that became extinct (The K-T Mass 2018).

And killer asteroids are not safely things of the past. In May 2018, two potentially hazardous asteroids, one the size of a city block zoomed by Earth within the lunar orbit. Indeed the largest one passed only 114,000 km away (Close Encounter 2018).

In February 2018, a smaller asteroid came within 70,000 km of smashing into our world (Bob King 2018). In cosmic dimensions, these "near misses" are considered uncomfortably close. Although the millions of asteroids in our solar system are largely concentrated in the asteroid belt, there are perhaps tens of thousands of asteroids that are close enough and large enough and travelling fast enough that they are of a serious concern. The potentially hazardous asteroids that are tracked by ground observatories or infrared space telescopes are generally grouped into four categories. These are pictured in the graphic below. The most common are Apollo (representing some 62% of the inventory) and Amor (representing 32% of the inventory) The other types are characterized as Aten (representing some 6%) and finally Atiras which are the Inner Earth Objects (IEOs). Atiras, which are the least common, currently represents a total population of only six potentially hazardous asteroids. Recently, a Trojan asteroid that follows the same orbit as Earth around the Sun has been located to create yet another category (See Fig. 1) (Jet Propulsion Laboratories 2018).

There is a concerted worldwide effort to detect and track potentially hazardous asteroids. These near-Earth asteroids (NEAs) and/or near-Earth objects (NEOs), also known as potentially hazardous asteroids (PHAs) are of special concern. NEA or PHAs are defined as those objects whose orbits come within 0.05 astronomical units (or 7.5 million kilometers) of our planet and have a magnitude (H) of 22 or brighter. When an asteroid actually impacts Earth, its mass and velocity can do great damage (See Fig. 2).

And unfriendly asteroids are not the only cosmic hazard that threaten our planet. In 1859, the so-called Carrington event, a coronal mass ejection from the Sun, bombarded Earth with trillions of high-speed ions. This led to telegraph offices catching on fire and the "Northern Lights" descending all the way done to Cuba and

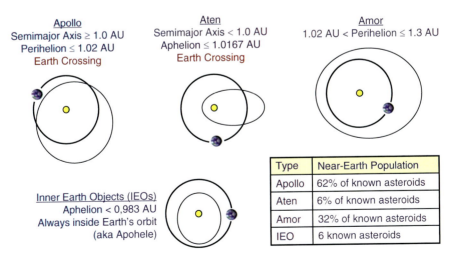

Type	Near-Earth Population
Apollo	62% of known asteroids
Aten	6% of known asteroids
Amor	32% of known asteroids
IEO	6 known asteroids

Fig. 1 The types and population frequency of potentially hazardous asteroids. (Graphic courtesy of the Jet Propulsion Labs)

Fig. 2 Potentially hazardous asteroids streak past Earth with regularity. (Graphic courtesy of NASA)

Hawaii. Records in China suggest that a similar solar storm event occurred in the early 1700s. More recently, there was the 1989 Montreal event that wiped out many transformers from Chicago to Montreal. In 2003, there was the Halloween event that impacted the electrical grids in Scandinavia. In areas exposed by the so-called Ozone holes in the polar regions, there are also particular concerns with solar flares represented by high-energy x-rays and ultraviolet radiation. These flashes of radiation, if not blocked by the Ozone layer of Earth's atmosphere, can expose animals and people to genetic mutation and elevated risk of skin cancer. Any serious examination of space security and "space safety," must therefore consider the risks that come from cosmic hazards and consider the extent to which today's and tomorrow's space programs could mitigate these risks through the use of space systems and technology either today or in the future. We are now beginning to develop new and sophisticated space systems and technology that could protect are small and fragile planet against such cosmic hazards.

It is ironic that as we develop new types of space systems that could protect Earth from these cosmic hazards, that there are real threats to humanity originating hear or Earth. The exponential growth of global human population, the spread of industrial pollution, and other types of human activities are now beginning to threaten the world's biosphere via climate change and over consumption of the Earth's limited resources. This growth of the human impact on our planet has led geologists to proclaim that we now live in the Anthropocene Age. This means that human activity is the major shaper of the world's geology and ecology. Some suggest that these trends might even lead to what might be called the "Sixth Mass Extinction" or "Anthropocene Mass Extinction." Apparently, today's world needs protection from both cosmic hazards as well as human expansionist trends and climate change here on Earth (Mooney and Dennis 2018).

Today the risk that comes from potentially hazardous asteroids or comets is now quite severe. This is because of potential loss of life, devastation to vital infrastructure, and environmental impacts that could occur. Yet, although the risk of devastation is large, the probability of a near-term collision is thought to be remote. Indeed, the probability is much higher of Earth being hit by a deadly solar storm that could disrupt the lives of a huge number of people in modern society. Indeed, it is far more likely that a massive coronal mass ejection (or solar storm) could provide major disruptions to our modern infrastructure of electrical power systems, pipelines, satellites, and industrial control systems on which we now greatly depend. If we picture Earth as an apple, then our atmosphere that protect us would be represented by the skin of the apple. Today there is a wide range of activities with regard to sustainability of space, defense of Earth against asteroids and comments via the International Asteroids Warning Network (IAWN) and the Space Mission Planning Advisory Group (SMPAG), and activities related to research and protective programs related to extreme space weather events and the Earth's magnetic fields. It is not possible to list all of these activities in a comprehensive manner in this one article. Even if an attempt were made, the listing would be soon out of date. It is recommended that the UN Office of Outer Space Affairs website be consulted to obtain useful updated information by visiting the following and related websites: http://www.unoosa.org/documents/pdf/smpag/st_space_073E.pdf

Cosmic Hazards from Potentially Hazards Asteroids and Comets

The threat, or risk level, from potentially hazardous asteroids is not easily assessed. When NASA was first asked by the U.S. Congress about the cosmic hazards there was difficulty finding a good answer as to the extent of the planetary threat level. NASA first suggested that a comprehensive inventory of asteroids that could threaten Earth might be set at 1 km in diameter. In subsequent discussions with Congress with NASA, it became clear that 1 km was not sufficient. When the George Brown Act part of the NASA Authorization Bill was passed in 1993, it was agreed to refine the research parameters. Thus they reduced the size of PHAs for the NASA search inventory for threatening asteroids to 140 m or larger. NASA, under this legislation, was to have completed this inventory with a comprehensive list of all such PHAs by now but still have not been able to finish this search process. Although over 90% of PHAs over 1 km in size have been identified, the inventory of asteroids down to 140 m is still far from complete. The NASA NEOWISE program has helped to identify numerous NEAs, but this infrared space telescope program has now finished its mission. NASA has plans for a new infrared telescope to assist with this discovery process, called NEOCAM, but it is still in a Phase 1 definition and would not be launched before 2022 or 2023 (Maizer et al. 2015). Based on a so-called "debiased" sampling of the results achieved with the NEOWISE asteroid search program, it was estimated that only 25% of asteroids greater than 100 m have now been identified. The estimate as to the likely this unidentified population of potentially hazardous asteroids above 100 m in size is estimated to be somewhere between 3200 ando 6200 (or 4700 plus or minus some 1500) (Ibid.). This is a rather frightening number when it is realized that 35 m asteroids could be "city killers" and that the estimated number of unidentified PHAs of this size could possibly be in the tens of thousands.

The problem is that the experience that comes with the study of past asteroid strikes such as the Tunguska strike in Siberia, Russia, in 1908 (estimated to be some 40 m in diameter) that asteroids that are only 30–40 m in diameter could destroy an entire city anywhere on Earth. Indeed the area of devastation from the Tunguska strike covered an area equivalent to the size of the San Francisco Bay area. There have been others that such as the B612 Foundation that has suggested that an infrared space telescope could be designed with the discriminatory capability to detect potentially hazardous asteroid down to 35 m in size. This project that was called Sentinel would be designed to detect "city killers" that could still create massive destruction should it actually make a direct hit on a major city. In short, it would be able to detect many more PHAs and of much smaller size than NASA's intended infrared space telescope that is designed to meet the 140 m radius standard officially set by the U.S. Congress.

Figure 3 shows in details the relationship between the diameter of asteroids, their typical kinetic energy, their numbers (or frequency of their occurrence), and the likelihood of their hitting Earth. Specific examples of major asteroid strikes that occurred in the past are noted in the case of Chelyabinsk, Russia, that impacted with an energy level of about 100,000 tons of TNT and the case of Tunguska, Russia, that impacted with an impact energy level of about 1 megatons of TNT.

This type of chart suggests that large asteroid or large meteor strikes impact with great energy, but that such impacts are fortunately also rare. Indeed this is not only because larger asteroids are fewer in number but also because we are also, in a sense, protected. The enormous gravitational fields of the Sun and the Jupiter are far more likely to scoop up such large space rocks before they crash into Earth. The unfortunately part of the equation is that there are so many asteroids and even comets out there. Many millions are safely in the asteroid belt, yet there are still tens of thousands of potentially "city killing" asteroids that could still hit Earth and release terawatts of kinetic power as they crash into earth with the force of a nuclear explosion (Pelton 2015).

It is important to note that the number of asteroids of a larger size are exponentially less in number. Alternatively, this means that as the size of asteroids decreases, the number of them increases exponentially. Larger and therefore more deadly

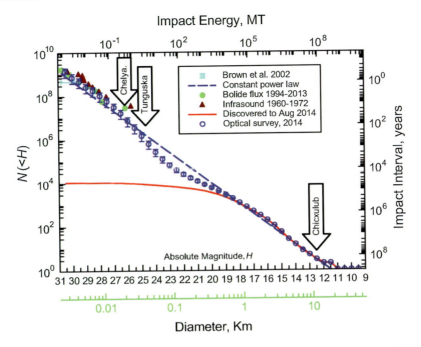

Fig. 3 Different sizes of near-Earth asteroids and impact energy in equivalent megatons of TNT. (Chart courtesy of NASA)

asteroids, as well as comets, which travel at even more deadly speeds, are fortunately likely to be captured by other bodies in the Solar System such as the Sun or Jupiter since this objects have much greater gravity wells than Earth. But what might be misleading is that the frequency of occurrence does not mean that the next deadly asteroid strike can be counted on to be thousands or even millions of years in the future. In truth, a major strike could threaten within the next year. Further the projected deaths shown in Table 1 are not truly foreseeable since it would depend on where and how it landed. If the 44 m asteroid indicated in the chart landed in a totally isolated area, such a Tunguska, it might result in no deaths, but if it landed directly on Moscow, New York, London, Beijing, or Mexico City, then the deaths could be in the many thousands or even millions. As the Earth becomes more populated and the number of megacities with a population of 10 million or more continues to rise, the risk of a catastrophic impact from an asteroid also rises. (See Table 1)

The other possibility, of course, is a comet impact. Comets that could threaten Earth are fortunately much rarer than potentially hazardous asteroids. The speed and size of comets could make their impact far more devastating. The Shoemaker-Levy comet cluster impacted Jupiter in giant pieces up to 2 km size between 16 and 22, July 1994. Each of these pieces of the comet, which were traveling at speeds of about 60 km/s, could have created incredible damage if they had smashed into Earth and threatened the existence of all living things on our planet including humanity.

Table 1 Key estimated impact figures for potentially hazardous asteroid

Asteroid damage parameters

Ground impact	Probability/ year	Diameter	Blast radius	Deaths	Ground damage	Interval within Lunar Orbit (i.e., within 400,000 km)
5 megaton event	1.67E^{-03}	44 m	About 20 km	About 200	About 1200 km^2	1/20 years
100 megatons	1.00E^{-04}	120 m	About 60 km	About 19,500	About 10,000 km^2	1/80 years
1000 gigatons	2.00E^{-06}	3 km	About 500 km	Unknown	About 800,000 km^2	1/600 years

Note: Figures are estimates that would vary significantly depending on actual point of contact

Currently, there are efforts to address the threat of Earth being hit by either an asteroid or comet in future years. The U.N. General Assembly has sanctioned the creation of the an International Asteroid Warning Network (IAWN) whose work is coordinated by the U.N.'s Office of Outer Space Affairs (International Asteroid Warning Network).

The Minor Planet Center in Cambridge Massachusetts plays a key role in cataloging all of the near-Earth asteroids as reported by infrared space telescopes and by at least 30 ground observatories and space-based infrared telescopes (The International Astronomical Union (IAU) Minor Planet Center 2018). A listing of the 30 ground-based observatories and their asteroid discoveries can be found at the following website (List of Asteroid-Discovering Observatories). The greatest number of asteroids in recent years has come from the NEOWISE (Near Earth Orbit Wide-range Infrared Surveyor Explorer). This IR Space Telescope during the period of its reactivation between 2013 and 2016 discovered 541 NEOs and 99 comets (Howell 2016). This capability will in theory ultimately be replaced by the NASA NEOCam mission. Currently, during the Phase A study authorized in January 2017 will consists of an infrared telescope and a wide-field camera operating at thermal infrared wavelengths. The NEOCam telescope has been funded for an extended Phase A study by NASA in the Planetary Defense Coordination Office will, however, be limited under current guidelines to searching for asteroids of a size of only 140 m and larger (NEOCam 2018).

This worldwide effort is also assisted by the Spaceguard Foundation in Italy (The Spaceguard Central Node 2018), The International Astronomical Union (IAU), the International Academy of Astronautics (IAA), the Spaceguard Foundation, the various national and regional space agencies, plus nongovernmental such as the Planetary Society, the B612 Foundation, and the Secure World Foundations (SWF) are all very much involved in considering how to better identify threats from near Earth objects (NEOS). Scientific bodies and NGOs have for instance suggested that search strategies should be seeking to identify asteroids that are smaller than 140 m in size and promoted the idea of launching new infrared space telescopes that are able to identify NEOs that are as small as 35 m in diameter. Indeed the B612 Foundation has defined such a space mission, named Sentinel with an estimated cost of $450 million that could find NEAs down to this size. When NASA ended matching funding in 2015, the B612 Foundation end this program and is now exploring a smaller constellation of small sats that might be able to complete an inventory of threatening asteroids of smaller size (B612 2017).

When the U.N. General Assembly in 2013 sanctioned the creation of the IAWN, they also sanctioned the creation of the Space Mission Planning Advisory Group (SMPAG) to address strategies and even actual missions to address identified cosmic threats. The work of the IAWN and the SMPAG are both overseen by the U.N. Office of Outer Space Affairs (OOSA). The responsibility of the IAWN is to create an international warning network to identify the threat. The responsibility of the SMPAG would be to consider actual strategies that might be employed to ward off an asteroid or comet that might threaten a direct impact with Earth in the future (UN Asteroid Defense Plan 2013). It is recommended that one visit the website of the U.N. Office of Outer Space Affairs to learn more about the most recent activities in these areas.

Protective Strategies for Planetary Defense Against Asteroids and Comets

There have been a wide range of possible strategies that have been discussed to address what to do if a definitive threat to Earth by an asteroid or comet is identified. These strategies have varied from longer term strategies that are consistent with years of advance warning. These activities might include such efforts as undertaking the deployment of so-called "Laser Bees." This would be a small sat mission where lasers activated from small sats would create small jket streams of ejected mass from asteroids so as to divert the asteroid's path (Laser Bee Project 2014). There are other concepts that have been identified that involve placing a satellite of some mass in close proximity to the threatening NEO so that its gravitational effects (i.e., to become a "gravity tractor") would over time move the NEO into a less threatening orbit (Sohn 2018). These strategies would likely be more suited to an asteroid that a very fast moving comet.

Another type strategy might involve either a kinetic impact between a spacecraft – perhaps equipped with a bomb – and a PHA in order to divert its path. This also might be done via a particle beam system that would create momentum transfer to divert the asteroid's path. One version of this approach would be for enterprises engaged in "space mining" to engage in a "two-fer." Thus they would divert a NEO into Lunar orbit so as to extract resources from the asteroid over time. They would, however, equip the asteroid with thrusters that would allow a collision between the threatening asteroid and the "mined asteroid" in lunar orbit so as to prevent either space object colliding with Earth.

There are other versions of such shorter-term strategies. This might including creating a directed high energy beam to blast the PHA long enough to divert its path.

Or, as has been depicted in films, there might be a mission to as asteroid in order to detonate a bomb or even nuclear device to blow the asteroid apart. There are currently various types of research programs around the world to better understand the composition and shape of asteroids and to examine which of these strategies would be most effective and which might be best used with longer, medium, or short-term warning of the threat (Atkinson 2011).

There are efforts such as the European Union-led NeoShield initiative. But such efforts are today only modestly funded by a few millions of dollars. Only a truly well-funded program involving billions of dollars could develop the needed capabilities – especially for short reaction time programs (http://www.neoshield.eu/neoshield1-summary/neoshield-1-team/).

There have been nearly 10,000 NEOs identified. NASA has currently identified some 1400 asteroids that are considered the most serious threats as potentially hazardous asteroids that are plus 1 or plus 2 on the Palermo Threat Scale. Some asteroids are projected to be in orbits that could come quite close to Earth in the next few years. Aphopis, for instance, is now being tracked in an orbit that will come dangerously close in 2029 and 2036, but it is not expected to actually hit Earth in those years. Some have suggested that there are large asteroids concealed in the Taurid Meteor Shower that is associated with Comet 2P/Encke, and others suggest that a better inventory is needed of all the periodic meteor showers that bombard Earth during yearly cycles to determine there are not deadly space rocks concealed in these meteor clouds

The thing that is most clearly true is that the inventory of potentially hazardous asteroids is still far from complete and that an accounting of asteroids that is limited to those that are larger than a 140 m threshold is inadequate. The second conclusion that is not as clear. We should seek to develop planetary defenses against comets and asteroids. But it is unclear as to what systems or technology can create the best space guards against cosmic hazards. We still do not know what are the most effective and cost-efficient strategies that might be used to defend Earth against impending threats from a near Earth asteroid or comet? Although the UN-sanctioned International Asteroid Warning Network (IAWN) and the Space Mission Planning Advisory Group (SMPAG) are a start, we are still a long way from knowing exactly what and how to defend Earth against such hazards and who should be in charge of such activities.

Solar Weather and Ionic Storms that Threaten Planet Earth

The two main cosmic threats differ greatly in terms of their frequency of occurrence. During solar max in the 11-year cycle of solar activity, there can be several coronal mass ejections a day and solar radiation flares are twice as common as the Ionic blasts. In short, solar storms of significance are much more common that major strikes by asteroids or comets. Yet largecosmic bodies slamming into Earth will typically do much more damage. The bottom line, however, is that asteroids, comets, solar radiation flares, or coronal mass ejections can all do significant harm. A major coronal mass ejection that creates major problems, such as the so-called Carrington Event of 1859 are projected to occur every 150 years while a major devastating asteroid hit seems to space millions of years apart. The difficulty is that "average frequency of occurrence" does not mean that a catastrophic event might not occur tomorrow. The concern level as to a major coronal mass ejection strike should be high for several reasons: (i) It has been some 160 years since the last truly major coronal mass ejection; (ii) the only truly significant result of the Carrington event of 1859 was telegraph offices catching on fire, but with today's world, the risks are enormous in terms of possible adverse affects on electrical power grids, pipelines, industrial control systems, telecommunications and data networks, and satellites; (iii) the world's population since 1859 has grown from 1.5 billion to some 7.5 billion and that population is 53% concentrated in urban centers hugely dependent on modern lines of supply and service-based jobs that are heavily dependent on electrical power grids, telecommunications and data networks, pipelines, satellite systems, and other infrastructure vulnerable to a CME containing trillions of ions blasting Earth with incredible force.

There is a constant stream of so-called solar wind that streams from the sun but typically this "normal" space weather is warded off by the Earth's geomagnetosphere with little or no affect on our daily lives.

During Solar max, there are perhaps 10–12 X-Class (or the very highest energy X-ray emissions) solar flares and about half the time these will be accompanied with tremendously powerful coronal mass ejections that send trillions of ions at millions of kilometers/hr outward from the sun. About ever dozen years there is an EMP that will reach Earth and do damage to electrical power grids and information networks. The Montreal event of 1989 and the Scandavian Halloween event of 2003 represent examples. Fried electrical transformers are the most common result of such an event, but many types of infrastructure on Earth, in the skies, and in orbit are potentially at risk (See Fig. 4).

The smashing into the Earth's electromagnetic shields by high energy ions traveling at millions of kilometers/hour is a tremendously disruptive force that stresses the magnetic fields so they stretched over 30 times the Earth's diameter before they resume their normal form that are seen in the usual shaping of the Van Allen belts. Without this protective shielding, Earth would not have its life-giving atmosphere.

Mars, which lacks an iron core such as that possessed by Earth, does not have a magnetic field to protect it. The solar wind thus continually strips away Mars' atmosphere. Then coronal mass ejections blasts away virtually all of the modest atmosphere that Mars' gravity manages to maintain. Thus Mars is only able to sustain about 60 milli-bars of atmospheric pressure.

When solar wind reaches Earth, the magnetic field pushes off the solar ions to the North and South Poles and it is this solar phenomena that create the aurorae that can be seen there. Sometimes Earth's location is called a "Goldilocks" planet since our world is just the right distance away from the Sun in terms of temperature and radiated light and its gravitational field is ideal for human physical characteristic. One should also note that its geomagnetic shielding is just right to sustain an

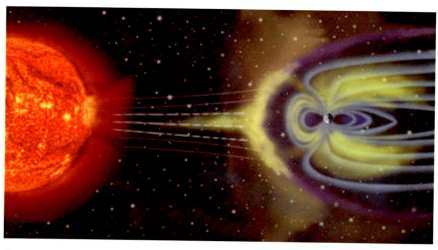

Fig. 4 Depiction of a coronal mass ejection showing impact on Earth's magnetic fields. (Graphic courtesy of NASA. Note: Graphic is not to scale)

atmosphere that is able to sustain life and protect against harmful radiation and solar ions. The Earth's Van Allen Belts and ozone layer are critical to sustaining life in our "Goldilocks" world. Without this protective shielding, the radiation from the sun would create enormous problems due to skin cancer and genetic mutation would lead to birth defects and other problems.

The natural protections of our planet have been sufficient to sustain life and allow human civilization to develop for millions of years. The twenty-first century that has seen humanity to grow to unparalleled numbers that might reach 10–12 billion by 2100 and lead to 80% urbanization with heavy dependence on modern infrastructure could lead to a new level of vulnerability in many ways. Modern jobs and urban living are now dependent on electrical power systems, pipeline networks, telecommunications, data and transportation networks, and satellite systems. If solar storms could knock all of these systems out of operation, then billions of people could be put at risk in terms of having food to eat, jobs to support economic viability, and protection against genetic mutation. The unprecedented growth of humanity from 800 million in 1800, to 1.8 billion in 1900, to nearly 7 billion in 2000 now exposes humanity to risk due to over development, global warming and climate change, and major disruptions if modern infrastructure is somehow disabled (Pelton and Marshall 2010).

It is for the above reasons that the threat from solar storms should be taken quite seriously. Currently vital governmental units, such as the U.S. National Oceanic and Atmospheric Administration (NOAA), which is a part of the U.S. Department of Commerce, are providing a real-time monitoring capability of space weather. There is a web-based space weather dashboard that reports on many different aspects of space weather. There are special networks to providing warnings – especially with regard to X-Class solar radiation flares and high-energy coronal mass ejections

(CMEs) that could threaten everything from electrical power grids to everything that depends on electrical power to sustain their operation. Defense ministries have their own monitoring and alert systems as well.

These efforts are currently aimed to provide systems alerts. There are some efforts to create more robust protective circuit breaker systems and more accurate and rapid warning systems, but the next step to consider what form of protective shielding might be created in space to provide some form of shield against major solar storms is largely in the realm of science fiction. It is now time to recognize the extent of the danger that solar storms present to the global economy and to explore what types of protective steps might be taken.

Solar Shields and Other Planetary Defense Strategies

The natural location to consider for a protective solar shield is clearly Lagranean Point 1. This is the location about 1.5 million kilometers out in space – well beyond the Lunar orbit – where the gravitational field of Earth and that of the Sun are more or less in balance or a form of homeostasis. This is actually more of an "area" than a point as demonstrated in Fig. 5.

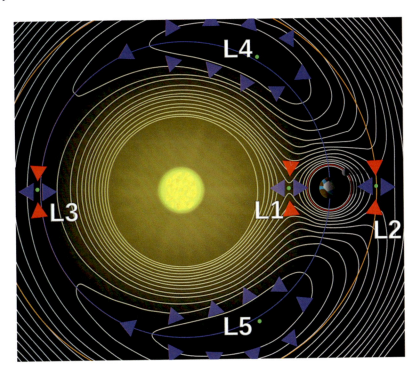

Fig. 5 The Lagrange Point L1 where Earth and Sun gravitational fields balance. (Graphic Courtesy of NASA)

The key would be to create a large enough magnetic field, both in magnitude and geographic size such as 2–4 Teslas covering a 1000 km^2 such that a blast of CME ions would be diverted to the sides so that the ion stream would miss Earth and not "tail" around the space shield and come back together so as to still impact Earth.

The idea of creating a solar shield, sometimes called a LAPSE (LAgrange Protection for Solar Ejections), has two significant elements to consider. One element is that a LAPSE could be created not only for Earth but for Mars as well. Mars has no natural magnetic protection and accordingly has its atmosphere stripped away. A LAPSE for Mars has the potential to create a significant atmosphere that could sustain life in the future. The projections have been made to suggest that Mars atmospheric density could increase from 60 milli-bars to ten times this density over time (Green et al. 2017) (See Fig. 6).

The other element is that a Lagrange Point 1 (LP1) facility could be expanded to undertake additional function such as Solar System wide transit point with virtually no take-off gravitational limitations to reach any point of destination. Further this location would also be idea for creating a solar radiation modulation system that could provide some temporary limits to global warming due to climate change while longer term solutions are devised.

There is a particular reason why the idea of a LAPSE Solar Shield is a concept of some urgency to be considered and implemented for Earth as soon as it could be designed and implemented. This reason is the current reversal of the world's magnetic poles. This reversal could weaken in a significant way the natural geo-magnetic shield that Earth currently enjoys.

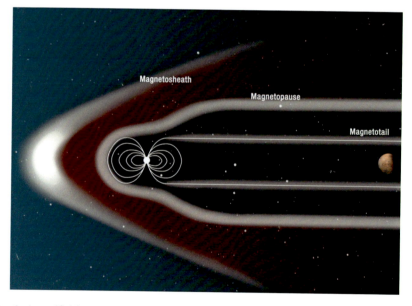

Fig. 6 An artificial magnetosphere of sufficient size generated at L1 allows Mars to be well protected by the magnetotail. (Graphic courtesy of James Green, NASA)

The Complication of the Earth's Shifting Magnetosphere

Currently, there are two satellite missions measuring the Earth's change magneto-sphere. One is operated by the European Space Agency and is a three-satellite constellation called Swarm. The other is the NASA mission that is called the Magnetospheric Multiscale (MMS) constellation of four satellites flying in tightly configured constellation. Both mission are confirming that the world magnetosphere is changing and that the Earth' magnetic poles are shifting. Figure 7 below shows this shifting magnetosphere and how magnetic North has moved down to Siberia and the magnetic South has traveled upward as well.

There is now evidence of an irregularity in the Earth's iron-based core under Africa may be responsible for this shift to occur over long spans of time (Pappas 2018). The timing of this shift is not known, nor is the impact of this shift on the natural protective shielding of Earth from coronal mass ejections. There has been some computer modeling that suggest that the future results could be quite daunting. Some modeling as shown in Fig. 8 suggest that the natural shielding could be reduced down to 15% of the current shielding as defined by the current conventional shaping of the Van Allen belts as shown in the right and the quite distorted shaping during the reversal process.

The future might be envisioned as a "perfect storm" of adverse factors. This future of expanding global population, higher levels of urbanization, greater

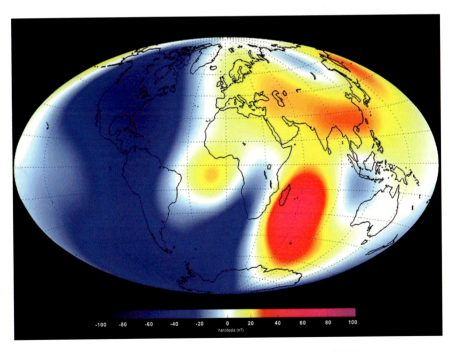

Fig. 7 Survey of Earth's changing magnetosphere. (Graphic Courtesy of ESA)

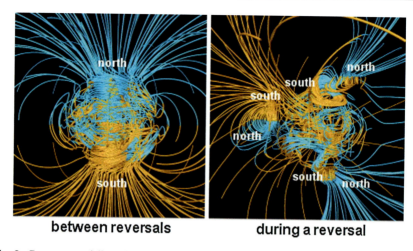

between reversals   **during a reversal**

Fig. 8 Computer modeling of changes to the Earth's magnetic field during magnetic reversal

dependence on electrical power grids, telecommunications and data networks, industrial control systems, and satellites that are more and more vulnerable to solar storms as the magnetic poles shift and natural shielding is eroded more and more.

Conclusions and Strategies for the Future

The twenty-first century world and human civilization is vulnerable to significant cosmic hazards. These include potentially hazardous asteroids, comets, solar flares, coronal mass ejections, and eroded natural shielding as the Earth's magnetic field reverses. There is a need for agile and innovative adjustment to these conditions by global policy makers. The space agencies of the world need to adjust their mission and priorities. Space exploration and missions to the Moon and Mars are worthwhile objectives, but if the viability of human survival race and vital infrastructure on which the world depends becomes clearly and unequivocally at risk, then innovative space design and construction projects such as solar shields, improved IR space telescope to identify asteroid and comet threats, and space systems to cope with NEO hazards.

Space security and safety is more than astronaut safety, missile defense, reliable satellites or coping with the hazards of orbital space debris. The world of space is on the edge of a new era where mega-construction projects in space are technically and intellectually possible. The key to such progress forward is perhaps first and foremost a clear vision and a setting of new priorities for space agencies for the next stage of the twenty-first century.

References

Atkinson N. Every way devised to deflect an asteroid. Universe Today, 9 Nov 2011. https://www.universetoday.com/90798/every-way-devised-to-deflect-an-asteroid/

B612 studying smallsat missions to search for near Earth objects. Space News. 20 June 2017. http://spacenews.com/b612-studying-smallsat-missions-to-search-for-near-earth-objects/

Close encounter: asteroid as big as a football field will whiz right by Earth this week. Tribute, 13 May 2018. https://www.sanluisobispo.com/news/nation.world/world/article211064469.html#storylink=cpy; https://www.sanluisobispo.com/news/nation-world/world/article211064469.html

Green J et al (2017) A future mars environment for science and exporation. Planetary Science Vision 2050 workshop, (LPI contribution no. 1989)

Howell E. WISE: NASA's infrared asteroid hunter. Space.com, 4 Aug 2016. https://www.space.com/33659-wise-space-telescope.html

International Asteroid Warning Network. http://iawn.net/about.shtml

Jet Propulsion Laboratories. http://neo.jpl.nasa.gov/neo/groups.html. Last accessed 6 Oct 2018

King B. Asteroid 2018 CB Zips by Earth. Sky and Telescope, 7 Feb 2018. https://www.skyandtelescope.com/astronomy-news/asteroid-2018-cb-graze-earth-friday-watch-online/

Laser Bee Project, 14 May 2014. http://www.planetary.org/explore/projects/laser-bees/

List of asteroid-discovering observatories. https://en.wikipedia.org/wiki/List_of_asteroid-discovering_observatories

Maizer A et al (2015) Space-based Infra-red discovery and characterization of minor planets with NeoWISE. In: Pelton JN, Allahdadi F (eds) Handbook of cosmic hazards and planetary defense. Springer, Cham, pp 583–590

Mooney C, Dennis B. Climate change warning is dire. Washington Post, P A1 and A22. 8 Oct 2018

NEOCam: finding asteroids before they find us, JPL NASA. https://neocam.ipac.caltech.edu/. Last accessed 8 Oct 2018

Pappas S (2018) Electro-Blob under Africa may be 'Ground Zero' for Earth's magnetic field reversal, 1. Space.com, 10 Mar 2018. https://www.space.com/39942-africa-blob-earth-magnetic-flip.html

Pelton JN (2015) Introduction to the handbook of cosmic hazards and planetary defense. In: Pelton JN, Allahdadi F (eds) Handbook of cosmic hazards and planetary defense. Springer, Cham, pp 583–590

Pelton JN, Marshall P (2010) MegaCrunch: ten survival strategies for 21st century challenges. Emerald Planet, Washington, DC

Sohn E. Gravity tractor as asteroid mover. Science News. https://www.sciencenewsforstudents.org/article/gravity-tractor-asteroid-mover. Last accessed 8 Oct 2018

The International Astronomical Union (IAU) Minor Planet Center. https://www.minorplanetcenter.net/iau/mpc.html. Last Accessed October 15, 2018

The K-T mass extinction event, exploring Earth. https://www.classzone.com/books/earth_science/terc/content/investigations/esu801/esu801page01.cfm. Last accessed on 10 Oct 2018

The Spaceguard Central Node. http://spaceguard.rm.iasf.cnr.it/. Last accessed 10 Oct 2018

UN Asteroid Defense Plan, LA Times, 28 Oct 2013. http://www.latimes.com/science/sciencenow/la-sci-sn-un-asteroid-defense-plan-20131028-story.html

Active Debris Removal for Mega-constellation Reliability

54

Nikita Veliev, Anton Ivanov, and Shamil Biktimirov

Contents

Abstract

Provided that the hazard of space debris in orbit can pose threats to space exploration missions and, thereby, influence the redundancy of Earth observation and telecommunication constellations, this chapter addresses the case for mega-constellation reliability. The space security challenge in this case does not only relate to the regulatory and legal framework thereof but also to the business development of technical solutions for space security. Although the current level of technologies enables active debris removal (ADR), its business applicability remains to be investigated. In this study, a multiparametric mega-constellation model has been developed to take into account orbital motion, coverage, ground communication, reliability, collision risks, and service consumption in the global

N. Veliev (✉) · A. Ivanov · S. Biktimirov
Skolkovo Institute of Science and Technology, Moscow, Russia
e-mail: n.veliev@skoltech.ru; a.ivanov2@skoltech.ru; shamil.biktimirov@skoltech.ru

© Springer Nature Switzerland AG 2020
K.-U. Schrogl (ed.), *Handbook of Space Security*,
https://doi.org/10.1007/978-3-030-23210-8_138

telecommunication market. The research and simulations performed on the model allowed for the analysis of possible financial metrics (revenue, cash flows, total replenishment cost) of the company who operates the ADR, as well as replacement scenarios and weak points of the mega-constellation. All combined, the chapter provides insights into the market that exists for ADR technologies, by demonstrating the ADR business applicability for mega-constellations.

Introduction

Since the inception of mega-constellation projects, the space debris problem got worse. Despite the fact that the history of spaceflight has witnessed cases of collisions between operating satellites and space junk, space debris objects have not been seriously considered in the business of aeronautics. Provided that the space telecommunication market utilizing low Earth orbit (LEO) and medium Earth orbit (MEO) becomes larger, the demand for new satellite systems and constellations, respectively, grows. Hence, the density of the satellites and other objects in orbit is growing as well. Eventually, by 2030 the number of the manmade objects in LEO and MEO is expected to grow ten times bigger than it is right now (European Space Operations Centre 2019). This definitely has repercussions for the market and as it constitutes a high financial risk.

Accordingly, this chapter examines the problem of space debris and how to potentially tackle it. The solution to the problem is what has been discussed for at least a decade: active debris removal (ADR), which is based on the mechanical process of returning the space object to the Earth atmosphere. Along with a passive debris removal, which is based on the phenomenon of atmospheric drag, active debris removal is one of the main space debris mitigation methods. The approach of this chapter is based on the premise that a satellite constellation operator and a company, aiming at the development of active space debris removal, could create a mutually beneficial situation. This would allow the operator to lower the risks of collisions, increase the stability and quality of the service, and improve its financial indicators while the ADR company would position itself in the market. At the same time, ADR can be of benefit to all mankind by ensuring the sustainably of the outer space environment.

Additionally, the analysis was based on the creation of a simulation environment that could facilitate in general the analysis of business reports for the whole telecommunication market. In particular, the simulation environment could become a quite effective instrument that enables risk assessment, market benchmarking, market specification, and selection of the appropriate business strategy. Hence, the simulation of in-orbit processes and the assessment of the data links were selected as the methodology of this chapter. For this purpose, the sustainable and stable simulation environment was created with valid and verified models. This environment has included different time scales, operating scenarios, constellation types, and satellite types to estimate financial metrics of the operator. As a final stage, the study applied the simulation environment developed to the specific business study. It was produced for the ADR company working along with the first echelon of the SpaceX Starlink constellation.

The results of the simulations and the results of the analysis showed that the loss of satellite could significantly influence the quality of the service reducing the coverage rate up to 20% and lead to extremely high financial losses for the operator (up to one billion dollars for a half of the lifetime of the satellite). That said and taking into account space technology readiness level and existing and developing business strategies, the ADR could successfully enter the space market and become profitable.

Simulation Model for Satellite Mega-constellation Reliability X

Concurrent Engineering Approach for SpaceX Starlink

The first baseline of this study was the research made by the Skoltech Space Center, which applied concurrent engineering methods in practice. The object of the research was the SpaceX company Starlink (Kharlan et al. 2018). The work included assessment of the statements of the company (displayed in Table 1) over its work, methods, service, and implementation process. Thus, the tasks of the team included validation and reverse engineering of the technical part.

The concurrent engineering approach assesses the applicability, the main technical parameters, and the evaluation of potential financial metrics of the company, as well as possible bottlenecks of the project (Shishko 1995). Accordingly, a team of ten Space Center researchers and students conducted the breakdown of the SpaceX Starlink satellites to subsystems (Kharlan et al. 2018). The process of concurrent engineering design consisted of seven sessions and five iterations based on the use of a special software. For the purposes of Skoltech Concurrent Engineering Design Laboratory, the CDP4-IME software created by RHEA Group was selected as the best solution because of its open-source code that supports the needed functionality. The research was finalized in a month with results described in the article prepared for the IAC conference in 2018 (Kharlan et al. 2018). Some particular research outcomes are displayed in Fig. 1 exposing the mechanical part 3D model and cost distribution.

Reliability Simulation Model for Satellite Constellation

Accordingly, the rationale for the simulation model was to understand whether the satellite is operational or not over time. Hence, the evaluation of a satellite's possible lifetime, which is generated based on the reliability distribution, becomes the most

Table 1 SpaceX statements regarding service and project characteristics (SpaceX 2017)

Parameter/unit	Value
Mass of the satellite, kg	400
Overall cost of the project, billion US$	<10
Connection speed, MB/s	>512
Service cost, $	<300 per subscription

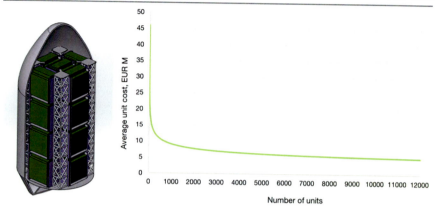

Fig. 1 Results of the reverse engineering of the SpaceX Starlink project: mechanical 3D modeling and composition (left), average cost distribution chart (Wertz et al. 2011) (right)

crucial part of the model. The main challenge presented in the process of sampling the lifetime of the satellites is the reliability distribution as the function of time, as that becomes the statistical task. The solution can be found by compiling the data of the launch date and the failure date (if any) of all satellites available in the database of the space objects currently located in orbit. The statistical learning based on the data for satellites and space debris is listed in the SpaceTrack closed database (Castet and Saleh 2009). The overall approximation numerical formula which can be used for the calculations is (Castet and Saleh 2009)

$$R(t) = 0.000120114 * \frac{\exp\left(0.000265681 * t^{0.4521}\right)}{t^{0.5479}} \tag{1}$$

where t is time in sec.

The next step of the reliability assessment was to generate the array of satellite states – the matrix describing the status of the satellite during whole lifetime was "1" for operating, "2" for interrupted, "3" for failed, "4" for being on replacement, and "0" for not working. This provided for a simple and obvious understanding of the status of the satellite during every step of the simulation.

Propagation

This study has considered part of the upper echelon of SpaceX non-geostationary orbit (NGSO) satellite system to demonstrate economic feasibility of various ADR strategies. The SpaceX satellite system was expected to be deployed during the first deployment phase (see Table 2). The echelon comprises of 4,425 spacecraft operating in the Ku and Ka bands. According to SpaceX (SpaceX 2017), upper echelon consists of five sub-constellations corresponding to different orbit altitudes and

Table 2 Orbital parameters of the SpaceX NGSO satellite constellation (SpaceX 2017)

Parameter	Initial deployment (1,600 satellites)	Final deployment (2,825 satellites)			
Orbital planes	32	32	8	5	6
Satellites per plane	50	50	50	75	75
Altitude	1,150 km	1,110 km	1,130 km	1,275 km	1,325 km
Inclination	53°	53.8°	74°	81°	70°

inclinations as shown in Table 2. During this research, the first sub-constellation – corresponding to altitude of 1,150 km above the Earth and 53 degrees – was considered because of the fact that it is planned to be deployed earlier than others. This helps demonstrate operation performance of satellite communication systems. Therefore, it is worth modeling possible advantages and risks of active debris removal technologies at this step.

The satellites in NGSO are located in circular orbits and evenly distributed according to the Walker constellation design pattern (Larson and Wiley 1992). This allows placing the satellite within the constellation in a way that ensures evenly distributed Earth coverage and avoids possible collision between satellites. The main parameters of the constellation according to the Walker design pattern are the total number of spacecraft T, number of orbital planes P, and phasing parameter F (Larson and Wiley 1992). Due to the lack of information regarding the phasing parameter, it was set equal to one, which is suitable for constellation of this size.

The propagation model enables the prediction of satellite's position and velocity at a required time (Larson and Wiley 1992). Each spacecraft in constellation is equipped with propulsion system for phasing maneuvers and orbit maintenance (SpaceX 2017). Therefore, the effects of atmospheric drag and solar radiation pressure can be omitted because they mostly influence the shape of the orbit, but not the precession (Larson and Wiley 1992). Thus, the perturbation caused by Earth's oblateness is taken into account. Right Ascension of the Ascending Node (RAAN) velocity is

$$n_\Omega \approx -\frac{3J_2\mu_G^{1/2}R_E^2}{2R_0^{7/2}}cosi \qquad (2)$$

where $R0$ and i are radius and inclination of the orbit, $RE = 6378.245$ km is the Earth's mean equatorial radius, $\mu_G = 3.986*105$ km^3/s^2 indicates the gravity parameter of the Earth, and $J2 = 1.082626*10^{-3}$ is the first zonal harmonic coefficient.

As it was mentioned earlier, satellites in the constellation have circular orbits. Therefore, in order to describe satellite location in orbit, it is more convenient to use the argument of latitude denoted. The latter is equal to the sum of the argument of latitude and true anomaly. To describe time derivative of argument of latitude, the following expression is used (Vallado 2001):

$$\omega_D = \omega_o \left[1 - \frac{3}{2} J_2 \left(\frac{R_E}{R_0} \right)^2 \left(1 - 4 \cos^2 i \right) \right]$$ (3)

where $\omega_D = 2\pi/T_D$; T_D is the period of satellite's revolution around the Earth, also called as draconic period; and ω_0 is mean motion of a satellite.

The example of 1-day propagation of the upper echelon satellite is exposed in Fig. 2.

To model the ground-track and Earth coverage of a satellite within the constellation, a projection of the satellite position was made. It was calculated in Earth-centered, Earth-fixed (ECEF), coordinate system to Earth sphere. All calculations of satellite motions made according to the previous paragraph are conducted in inertial reference frame such as Earth-centered inertial (ECI). Therefore, it should be converted to ECEF in order to calculate ground track and coverage. Transformation of coordinates is performed according to the International Earth Rotation and Reference Systems Service (IERS) 2010 conventions (Gérard and Luzum 2010) where such effects as precession and nutation of Earth rotation axis are considered.

Simulation Process

The process of simulation starts when the majority of data is prepared for the processing. First the lifetime statuses of the constellation satellites are estimated, then the ABGN is created, and, last, the propagation matrix is ready. The initial parameters of the simulation are:

1. Simulation related parameters:
   a. The length of the simulation period
   b. The size of the timestep of the simulation – which definitely is supposed to be selected corresponding to the parameters of the preprocessed files

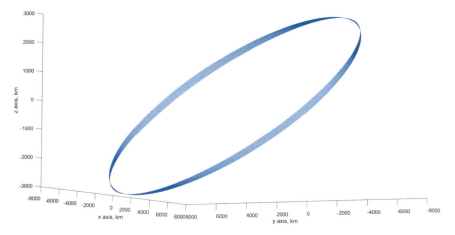

Fig. 2 One-day propagation of the satellite taken from the upper echelon (Larson and Wiley 1992)

2. Constellation-related parameters:
   a. Altitude
   b. Inclination
   c. Accuracy of the ABGN grid
3. Satellite-related parameters:
   a. Mass
   b. Volume
   c. Antenna field of view
   d. Coverage rate
4. Spare strategy type:
   a. Two options are available: "none" for no strategy and "lod" for launch-on-demand strategy.

The process of the simulation itself consisted of a method that reads the states of the satellites and is based on information that reveals the coverage of the constellation. Moving forward, the method estimates the possible revenue and the costs for the specific timestep and finally calculates the amount of space debris in the orbit. The simulation requires much resources, including random access memory (RAM), computational power, and hard drive (HD) memory. With all the optimization, the simulation parameters were set to be constant (displayed in Table 3).

Replenishment Scenarios

The approach of the simulation considered two types of interactions with spare issues: no strategy at all and launch-on-demand (LOD) strategy assuming that every time the satellite fails in the orbit, the other one is supposed to be launched on its place in the shortest time possible. In that case, the ADR effectiveness is supposed to be evaluated based on satellite costs and the risk of in-orbit collision assessment, which is performed using the growing collision probability formula:

$$PC = 1 - \exp\left(-SPD * VR * AC * t\right) \tag{4}$$

where SPD is spatial density (n/km^3), VR is relative velocity (km/s), AC is aerial collision cross section (km^2), and t is time (sec).

In Fig. 3, the collision probability along with the growing density of the debris during the 1st year of constellation operation is displayed. As it can be clearly seen,

Table 3 Optimal simulation parameters based on parameter-sensitivity analysis by the authors

Parameter, units	Value
CPU count, units	40
Step size, s	100
CPU frequency, GHz	3.2
Chunk size, steps	3942
RAM, Gb	200

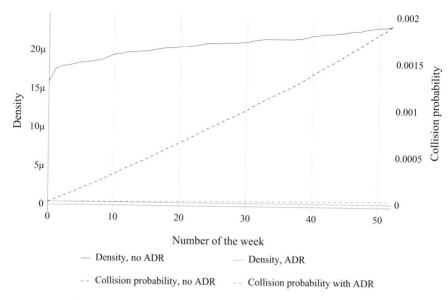

Fig. 3 Space debris density distribution and collision probability distribution based on simulation results of the authors

the ADR is supposed to line up the risk management and lower the probability of the chain reaction.

Marketing Model for Satellite Mega-constellation

As long as the satellite constellation model was created and it became clear how each satellite in the constellation moves and what was the coverage of each satellite, it became available to assess how each satellite influences the connectivity and, consequently, the company operator's financial metrics. For that purpose, it was necessary to assess the market and allocate marketing data to the coordinates on the Earth surface.

Obviously, as an understanding of the possible business applications of the ADR is based on the financial metrics of the operators, the marketing analysis and model are taking the most important part in the research. It facilitates the calculation of possible revenue for the constellation through the implementation of economics and marketing.

The underlying reasoning of detailed market study was to understand the number of subscribers in every point of the Earth globe and according to the pricing estimations to calculate the potential positive and negative cash flows as a benchmark. The created marketing model consists of two major blocks:

- Pricing model
- Population model and market penetration

Pricing Model

The study considered two different types of pricing currently being used by the telecommunication companies worldwide: all-flat and traffic-based tariffs (Deloitte 2019). For the flat plans, the price of the service was calculated as the average price of the flat tariff in the particular country. Figures 4 and 5 show the statistics used for the calculations of the traffic-based subscriber contract.

Both calculations were made and verified using the open-source data for telecommunication companies' statistics in Russia, Statista Inc. database. The average errors for both methods are shown in Table 4.

Population Map and Market Penetration

The population data is based on the information prepared by the NASA Socioeconomic Data and Applications Center (SEDAC) (Doxsey-Whitfield et al. 2015) based on the gridded population of the world (GPW). The fourth version of the data is the distribution of the human population across the Earth. The statistical data is generated based on the Earth observations and provides globally consistent data for any type of the researches and studies.

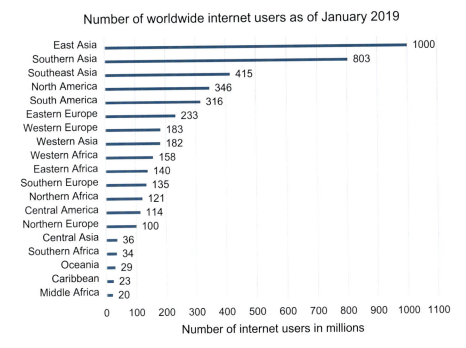

Fig. 4 Internet user number distribution by region (Statista Inc. database 2019b)

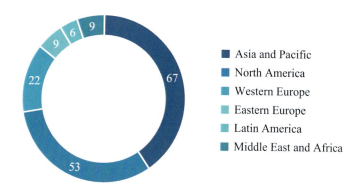

Fig. 5 Internet traffic consumption per month by the region (Statista Inc. database 2019a)

Table 4 Verification error for different pricing models, based on the simulation assessments of the authors

Method	Error rate, %
Flat	28
Consumption-based	13

The Earth observation data was transferred to the population data values; then the data passed the process of normalization using the official statistic of the countries. This work uses data for 2015 with a resolution of 1 degree. However, increasing the accuracy significantly decreases the calculation speed and affects the overall simulation time, according to Formula 5 (representing the needed number of calculation steps in order to facilitate the whole grid for a single timestep):

$$N_{steps} = \frac{180deg * 360deg}{A} \tag{5}$$

where N_{steps} is the number of steps, A is accuracy (step size, deg^2), and 180*360 is altitude-longitude degree grid.

The population is not the only thing that is necessary for the calculations though. The other important elements are market penetration and target audience. Both parameters are limited to the number of users interested in the service and able to pay for it. The generic coefficient determining the part of the population that supposed to use the service of the exact provider can be estimated by the formula:

$$K = MS * I * IP \tag{6}$$

where MS is the market share, I is income availability, and IP is Internet penetration.

Theoretical three-parameter model describes three sides of the approach of estimation of the amount of the target audience: economic, marketing, and technical.

The market share of the company is determined dynamically as a function of time, describing the entry of the company to the market with boundary conditions. This is described in the Federal Communications Commission (FCC) request of SpaceX technical information in 2017, as well as in official forecasts of the SpaceX Starlink stating 40 million subscribers and 30 billion US dollar revenue by 2025 (Harris 2019). As a market acquisition model, the Gompertz curve (Zlatić and Štefančić 2011) has been selected, because the growth speed of the curve is pretty similar to the market share speed. The approximation was used to describe the mobile phone penetration to the population (Islam et al. 2002). The exact formula is presented in Eq. 7:

$$MS = 0.2158 * e^{-3.0716 * e^{-0.00000354 * t}}$$ (7)

where t is time in sec.

Income availability is the parameter describing price availability of the product on the market. This parameter is being calculated based on the price of the product and the distribution function of the income per capita in each country. According to the fact that the Internet connection payments on average are holding 10% of all spends per month (Visser 2019), selecting the tariff price gives an understanding of the number of people that are able to pay. For most countries, the shape of the curve is described with different data. However, for those countries with no data available for a "rich-poor" curve, the average worldwide curve was used. This curve is displayed in Fig. 6 as an example (OurWorldInData Inc. database 2014).

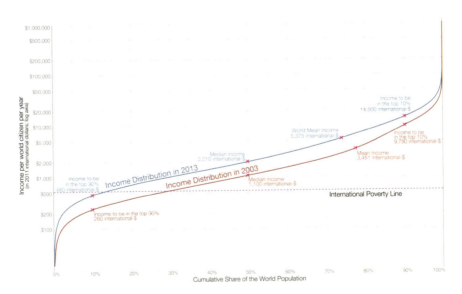

Fig. 6 The world population income distribution in 2003 and 2013 (OurWorldInData Inc. database 2014)

Agent-Based Ground Network

Considering the output of the economical modeling process, the agent-based ground network (ABGN) has been created. The term ABGN as well as the grid itself was created during the research specifically for the simulation to simplify the process of markets of the Earth-based consumers. The ABGN is the operational set of agents, or in this case subscribers of the Internet service, integrated to the peer-to-peer-linked network distributed by the Earth globe and having a set of parameters, enabling the assessment of the entire system. In the case, when the subscriber is a single user or a household, the agent network could be represented as the grid with number of the subscribers in surrounding area as the nod value.

Each agent can be represented as an object of the "subscriber" class with a set of attributes. This is based on a self-made Python class of objects containing a set of parameters with relative information regarding the selected type of the subscriber. The attributes are:

1. Position. The positional argument, describing the position of the subscriber (or a set of subscribers) on the Earth globe. The distribution could be random, functional – set up with a function of time, evenly weighted – normally distributed nods of the grid, setting up a single subscriber or number of subscribers in the surrounding area, or single-located, the array of coordinate of the subscribers.
2. Money capacity. The parameter of money capacity is being calculated based on the amount of the target audience in the nod and the price of the service. This parameter is the permanent base for the calculations of the money flows of the operational company and dynamic coverage methods.
3. Traffic demand. The parameter is calculated based on the amount of target audience in the nod and the traffic demand in the location per capita. This parameter hardly influences the link budget and dynamic coverage methods.

The agent-based ground network is able to represent not only the large number of users but the single user as well. In other words, this system works with both the business to consumer (B2C) and business to business (B2B) strategy of operation problems. That, in particular, allows to solve the static and dynamic tasks (or a combination of such). This advantage allows the model to be applicable to other scenarios for various companies. The system works on a plug-and-go basis meaning that setting up the type of the ABGN does not require changing code in the core of the simulation.

Model Validation

As soon as the environment is ready, the verification takes place in order to determine the applicability of the model to real scenarios. The validation has been delivered as a two-step process: the first one was the validation of the model itself, performed with running the simulation for the constellation that already exists; the second was the case validation, determining whether the model is applicable to the particular case study.

In the first part, the simulation process was run with the characteristics of and information about the Iridium Inc. constellation (from annual report to Stockholders in 2009) that has been firstly launched in 1997 and consists of 75 active satellites with a coverage of 100% of the Earth globe. The idea was to compare the results of the simulation with company open data (including revenue). The outputs are displayed in Fig. 7, with the error made up to 7%.

The case validation was based on the open data of the SpaceX company official statements claiming the amount of revenue by 2025 (claiming 30 billion US$ of revenue by the time) (Mosher 2019). The error appeared to be 122%, as the results display in Fig. 8. The big error is corresponding with lack of data.

Commercialization of ADR and Insurance Strategy

According to the baseline of the simulation, the propagation and coverage footprint were calculated. Based on the coverage and reliability model, the overall constellation coverage was calculated. Along with created ABGN model, the possible market coverage of the operator was calculated for the selected satellite constellation formation. The data for each timestep gave an opportunity to understand main

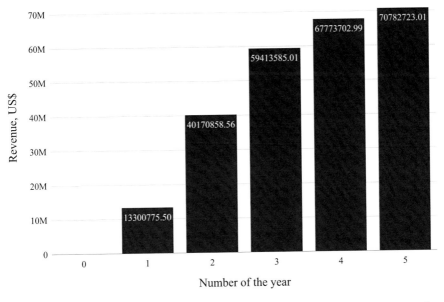

Fig. 7 Annual yearly revenues for the Iridium Inc. simulated, where year 1 is the 1st year after main launch in 2003 based on simulation results of the authors

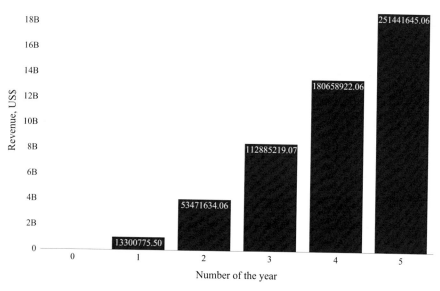

Fig. 8 Annual yearly revenues for the SpaceX simulated, where year 1 is the 1st year after main launch in 2020 based on simulation results of the authors

financial metrics of the operator (such as positive cash flow from the servicing, income, and operational expenses).

This, consequently, enabled the research group to run some basic assessment scenarios of the telecommunication segment. The environment and the models have been verified using the existing cases of constellations and open information of the company at hand.

Afterward, the environment was used several times to assess the satellite service to be provided and supported by the SpaceX Starlink company for the simulation periods equaled to 1 month, 1 year, and 3 years. The output data is represented as a time evolution of several important parameters: revenue flow, costs, space debris density in orbit, and coverage. The results are exposed in Figs. 9, 10, and 11.

Following the assessments made in the beginning of the work, losing the satellite led to several huge impacts on the constellation operational indicators. It can be seen in Fig. 9 that operating with no replenishment strategy could be followed by revenue decrease (around 1.5 mln USD for the 1st year and nearly 1 bln USD for the first 5 years). In Fig. 10, the same data showed in percentage which is up to 6% revenue loss for the 1st year. Figure 11 also shows that the malfunction satellite can also lead to a coverage loss leading to the quality issues.

The assessment showed that active debris removal is critically important for the company-operator as it allows to increase revenues and decrease the risks of collisions. The main result of the outcome analysis is the fact that ADR is not only practically necessary to tackle the challenges of the mega-constellations, but it can also be commercialized.

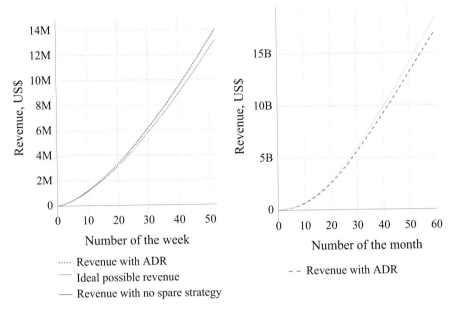

Fig. 9 Revenue time evolution for the case study: simulated for 1 year (left) and for 5 years (right) based on simulation results of the authors

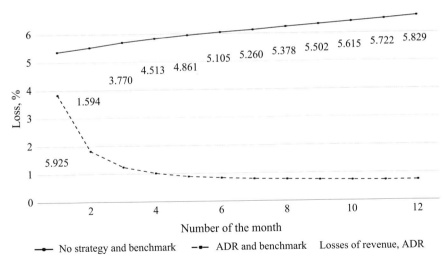

Fig. 10 Revenue losses for the consequent periods based on simulation results of the authors

Based on the research outcomes, a marketing analysis was followed to create a sustainable business strategy for the ADR company. For that purpose, the company was evaluated from a business perspective. The analysis showed the specifics of the business and possible strategy options. Three major business plans

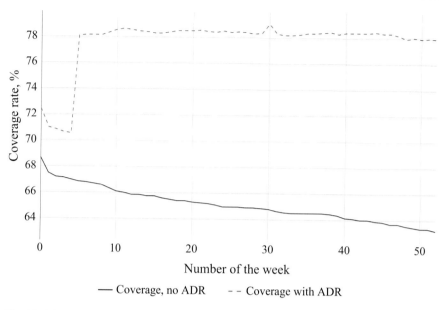

Fig. 11 Time evolution of the coverage based on simulation results of the authors

were assessed: the "flat insurance," the "dynamic flat insurance," and the "pay-as-you-go tariffs." For each of the strategy, a separate simulation turn was run in order to understand the applicability of the strategy, its profitability, and positives and negatives of its use.

The flat insurance strategy is based on the business plans of the insurance and reinsurance companies providing the full insurance. This includes the complete ADR services, for a fixed reward. After simulating, this plan, however, appeared to be noncompetitive since the reward calculated is quite high and appears to be not advantageous for the operator.

The pay-as-you-go tariff, on the contrary, means to implement the system of rewarding the ADR company every time the satellite replacement takes place. However, in that case, things can get worse for the company itself. Since the satellite failure is highly connected with the reliability of the satellite, it seems that the company's financial behavior can turn to be unpredictable. This can subsequently lead to investment overlaps and breaks – the incomes of the company are going to be strictly connected to the satellite loss, which is not equally distributed over time – the cash flow becomes "jumping."

The compromise lies in merging both of two options and adjusting the details of the strategy. The dynamic flat insurance implements the dynamic rewarding methods and selective ADR use. The rewarding technology is connected with the distribution of the satellite reliability over time, taking into account any kind of collision risks. In Fig. 12, the rewarding scheme is presented. The rewarding scheme based on the business plan enabling dynamic price change; in that case, the price depends on the overall reliability of the constellation. It means that the more time the constellation

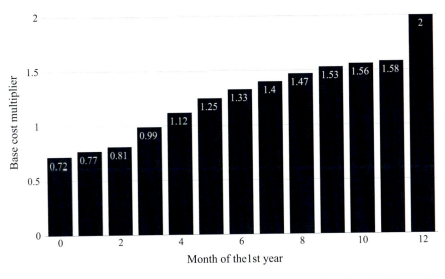

Fig. 12 Reward strategy: base cost multiplicator distribution over time based on simulation results of the authors

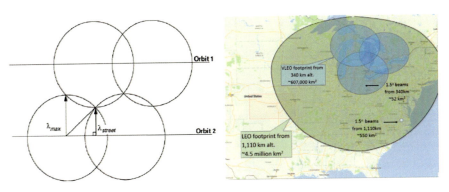

Fig. 13 Footprints of SpaceX lower echelon (SpaceX 20,187) (right) and "street size" description (left)

lives, the less its reliability, the more money the customer supposed to pay to cover the ADR service.

Additionally, the simulation showed that the loss of one individual satellite has approximately no influence on the service quality, while the failure of two coherent satellites leads to 1.5 times as big revenue losses. This actually is explained by the fact that the satellite coverage footprints overlap as it can be clearly seen in Fig. 13 (Kharlan et al. 2018) – the blue circles are displaying some of the first echelon satellite footprints. The picture shows that the loss of one satellite leaves relatively small piece of Earth surface, which can lead to predictable and short connectivity losses. At the same time, losing several consequent satellites significantly increases the uncovered surface bringing continuous and unbalanced connectivity failures.

Conclusions

The objective of this chapter was to analyze a business case connected with an active debris removal (ADR) process. It aimed to understand its applicability and prove the market existence for ADR. In order to pursue this investigation, the analysis was based on the creation of a simulation environment for mega-constellations, the Starlink SpaceX first echelon constellation. The results of the study give insights into the success of an ADR company, being interconnected with the role of operators of these large constellations.

[Can you summarize here a bit more about the results? I think they are very important to highlight in relation to the mega-constellation challenge].

Despite the importance of the results, the limitations of the method are evident due to the study performed for the specific large case of mega-constellations. In order to address these limitations, future research could look into the following proposed directions:

1. Perform a study for different constellation with different orbital parameters.
2. Perform a study for different altitudes (MEO and GSO).
3. Perform a study for other markets, such as Global Navigation Satellite System (GNSS), Earth observation (EO), or defense applications.
4. Apply different optimization methods.
5. Assess the replenishment times and the constellation sizing factors.

References

Castet JF, Saleh JH (2009) Satellite and satellite subsystems reliability: statistical data analysis and modeling. Reliab Eng Syst Safe 94(11):1718–1728

Deloitte Center for Technology, Media and Telecommunications (2019) Telecommunications industry outlook. https://www2.deloitte.com/content/dam/Deloitte/us/Documents/technology-media-telecommunications/us-telecom-outlook-2019.pdf

Doxsey-Whitfield E, MacManus K, Adamo SB et al (2015) Taking advantage of the improved availability of census data: a first look at the gridded population of the world, version 4. Pap Appl Geogr 1(3):226–234

European Space Operations Center (2019) ESA's annual space environment report, 5, pp 1–77

Gérard P, Luzum B (eds) (2010) IERS conventions. Technical note no. 36. Bureau International des Poids et Measures, Frankfurt am Main

Harris M (2019) The space-wide web. New Sci 242(3228):44–47

Iridium Communications Inc (2009) Annual report to Stockholders, Accessed 28 Feb 2019. https://investor.iridium.com/download/IRDM_10-KWrap2009.pdf

Islam T, Fiebig DG, Meade N (2002) Modelling multinational telecommunications demand with limited data. Int J Forecast 18(4):605–624

Kharlan A, Ivanov A, Veliev N et al (2018) University-based facility for evaluation and assessment of space projects Alexander Kharlan. In: 69th international astronautical congress, Bremen, pp 1–5

Larson WJ, Wiley J (eds) (1992) Space mission analysis and design, 3rd edn. Microcosm Press, Torrance

Mosher D (2019) SpaceX may be a $120 billion company if its Starlink global internet service takes off, Morgan Stanley Research predicts, Business Insider. https://www.businessinsider.com/spacex-future-multibillion-dollar-valuation-starlink-internet-morgan-stanley-2019-9. Accessed 28 Sept 2019

OurWorldInData Inc (2014) Global income distribution. https://ourworldindata.org/global-economic-inequality. Accessed 4 Mar 2019

Shishko R (1995) NASA systems engineering handbook. National Aeronautics and Space Administration, Washington, DC

Singh N, Browne LM, Butler R (2013) Parallel astronomical data processing with Python: recipes for multicore machines. Astron Comput 2(8):1–10

SpaceX (2017) Spacex V-band non-geostationary satellite system attachment a technical information to supplement schedule S. International Bureau Application Filing and Reporting System, FCC. Available at: https://licensing.fcc.gov/myibfs/download.do?attachment_key=1190019

Statista Inc (2019a) Global consumer internet traffic by region. https://www.statista.com/statistics/267199/global-consumer-internet-traffic-by-region/. Accessed 15 Mar 2019

Statista Inc (2019b) Number of internet users worldwide. https://www.statista.com/statistics/273018/number-of-internet-users-worldwide/. Accessed 15 Mar 2019

Vallado DA (2001) Fundamentals of astrodynamics and applications, 2nd edn. Microcosm Press, El Segundo

Visser R (2019) The effect of the internet on the margins of trade. Inf Econ Policy 46(3):41–54

Wertz JR, Everett DF, Puschell JJ (2011) Space mission engineering: the new SMAD. Microcosm Press, Hawthorne

Zlatić V, Štefančić H (2011) Model of Wikipedia growth based on information exchange via reciprocal arcs. Europhys Lett 93(5):58005

Space Security and Sustainable Space Operations: A Commercial Satellite Operator Perspective

55

Jean François Bureau

Contents

Abstract

The increasing need of commercial Satcom services (for governments, economy, social, and cultural purposes) augments the call for a coherent "space security" discussion which will lead to the development of rules and guidelines for sustainable space operations. Two main items of utmost importance to commercial space operators are elaborated in this chapter: the management of space operations in increasingly "crowded orbits" and the protection satellite services should receive, both in space and on Earth. During the last two decades at least, guidelines, best practices, agreed principles, and "soft law," have been the pragmatic answer to move forward. However, the exponential increase of space-based services will call for a "governance framework" which should strengthen the principles of the Outer Space Treaty and ensure that the Treaty will remain effective in the coming decades.

Views expressed in this article are the author's one only, and cannot be Eutelsat ones.

J. F. Bureau (✉)
Institutional and International Affairs, Eutelsat Group VP, Paris, France
e-mail: jbureau@eutelsat.com

Introduction

Since the United Nations Group of Governmental Experts (GGE), established in 2011 (GA Res A/RES/65/68, January 5th 2011) and concluded its report on "Transparency and Confidence Building Measures in Outer Space activities" (A/68/189 of July 29, 2013) (TCBM) in 2013, the discrepancy between space security requirements and space risks and challenges have significantly increased.

The TCBM proposal identified by the UN GGE has been largely ignored in terms of implementation, and the failure of the EU Code of Conduct proposal as well as the deadlock on the discussion of the long-term sustainability (LTS) guidelines created a sense of "no hope" on the diplomatic discussions. At the same time, the large and rapid changes that have taken place in the last 5 years in the field of space-based services and applications, as well as the disruptive innovations that supported a lot of new projects especially in low orbits (including the so-called mega-constellations), have increased the urgency for a more space security organized environment, considering the overall sustainability of space activities.

The 50th anniversary of the Outer Space Treaty in 2017 was not an occasion for many events. This low attention probably reflected a more substantial question about the extent to which the Treaty and its framework of international conventions (first of all the International Liability for damage caused by space objects Convention of 1972 and the Registration of objects launched in outer space Convention of 1975) would still shape the coming developments of human activities in and from space in the future. In addition, how the Treaty and related Conventions provisions would be resilient enough and inclusive enough to shape and regulate all coming developments which are under preparation.

The nonappropriation of outer space, along with the freedom of exploration, the liability for damage caused by space objects, the prevention of harmful interference with space activities, and the obligation to notify the international community and to register space activities are all legally acknowledged principles. (But, during the last period of time, different projects seem to assume that the nonappropriation of Outer Space could allow for the privatization of the mining of asteroids.) Yet, they are challenged, either because they are bypassed, or because they are discreetly ignored, or even openly violated.

Last but not least, and even if the prohibition of placement of nuclear weapons or other weapons of mass destruction in outer space seems to be still in force, the number of warnings that a new round of technologies could prepare for a militarized space is obvious. To quote the famous assessment from the US National Security Space Strategy of 2011, "space, a domain that no nation owns but on which all rely, is becoming increasingly congested, contested and competitive." (US Office of the Director of National Intelligence: "National Security Space Strategy," January 3, 2011.)

All these trends are feeding the sense that space could become an out-of-rule domain, a kind of jungle of the twenty-first century which will undermine many coming projects, or even destabilize the most established space-based services.

Because a part of the new revolution links together space-based systems and commercial services, in an increasing number of applications, be they civilian, military, or dual, commercial operators (be they "old" or "new" space) perceive the forthcoming environment as challenging. The concerns include not only an increasing lack of regulation, but also essential space dependency for many more activities on earth. According to the US Commerce Secretary Wilbur Ross, "today, the global space economy is roughly 400 billion dollars, about 80% of which is commercial activity" but "Morgan Stanley projects the global space industry could reach 1.1 trillion dollars by 2040." (Remarks by US Commerce Secretary Wilbur L. Ross at the US Chamber of Commerce Space Summit, December 6th 2018. Office of Public Affairs, US Department of Commerce.) As an observer remarked, "so the department (of commerce) is trying to deregulate the industry, making it easier for entrepreneurs to jump in, and to lure capital from venture capital firms, hedge funds, sovereign wealth funds and even mainstream pension funds." (Tett 2018; see as well, Donohue 2018.)

As space becomes a truly essential service base for a large spectrum of human activities on Earth, security of all the stakeholders (States, private, and international organizations), resilience of the regulations which allow and organize those services, sustainability of the infrastructure deployed in space, and finally, predictability of the upcoming rules based on a truly world-wide consensus constitute the ingredients of a safe and secure space for the future. No need to say that to overcome this challenge, which is multilateral by nature, requires a lot of effort and willingness from all stakeholders.

From a satellite operator perspective, three sets of key issues can be listed, which could, altogether, jeopardize the future of its activities, in terms of development, sustainability, and affordability. In this time of "crowded orbits," (refer to Moltz 2014) (1) it is tempting to identify these challenges at the geostationary orbit; (2) at the juncture between the geostationary orbit on one side, and the medium-earth orbit and low-earth orbit on the other; and (3) finally on earth as presented in the following sections.

Space Operations in GEO

As already suggested, the most crowded and congested orbit is the geostationary orbit where 548 of the 1886 existing satellites are positioned, 1186 being in LEO orbit and 112 in MEO. (Data provided by the Union of Concerned Scientists "Satellite database" in its update of August 10th 2018, Accessed 12 Dec 2018.) It is interesting to notice that in 2012, 1050 satellites were in operation, among which 432 were in geostationary orbit, 73 in medium-earth orbit, and 503 in low-earth orbit. (The figures are provided by J. C. Moltz, in "Crowded orbits," op.cit pp. 20–23.) These figures deliver quite a strong message: the GEO orbit is now growing very slowly in terms of new satellite populations, as opposed to LEO that is witnessing a dramatic increase of resident space objects.

GEO orbit is crowded and congested to such large extent that the management rules for the fleets raise more constraints. Yet the core of the International Telecommunications Union (ITU) regulation of the space-based services for telecommunications is still mainly based today upon the noninterference principle as it was at the beginning of the 1970s, when the management of the orbital positions was far less constrained. The change is such that, at that time, the management of the orbital positions was also complemented by another implicit rule – first arrived, first served – which cannot any longer be implemented. Of course, the principle of noninterference is of much importance as it allows for proper technical functioning of the satellites at close orbital positions, through coordination agreements among operators, but appears too limited to address all potential situations.

Among them, the dynamic management of fleets, the appearance of a graveyard orbit, and the emergence of "clinging satellites" are quite troubling and may illustrate the extent to which existing regulations are not able to face the coming challenges. Basically, the need for enduring transparency about what is happening at the GEO orbit and "who" does what may have to be on the top of the to-do list the international community could set up. It is assumed that one key obstacle of such transparency has been, in the past, the importance of military services which were provided by military satellites at this orbit, especially for nuclear testing detection, early warning of missiles launching, or detection of preparations for a military offensive. Such missions are for sure still needed even if the geopolitical and strategic environment is nowadays profoundly different from the Cold War one, and the satellites associated with them may legitimately receive a different treatment.

However, managing transparency at the activities which take place at the GEO orbit, from the satellites which are registered by the UN Register now ratified by 68 nations and some intergovernmental organizations (like Eumetsat, Eutelsat, ESA, or Intersputnik) at the end of 2018, would also be the beginning for more security in orbit, keeping in mind that, according to the UNOOSA, 91% of all satellites, probes, landers, crewed spacecraft, and space station flight elements launched into Earth orbit or beyond have been registered with the Secretary-General. (UN Office for Outer Space Affairs (UNOOSA) website, Register of Objects Launched into Outer Space, Accessed 12 Dec 2018.) From that standpoint, it is important to notice that the Registration Convention is still attracting new signatures, like the Luxembourg one, a nation which has recently decided to join the Registration Convention. However, registration should probably be based on a wider basis than a very administrative process. Like ships and planes have received an identification once the management of their movements was obviously needing a way to track them to prevent fatal collisions, providing an ID to the registered satellites may be the next needed step to ensure that the Registration Convention will still be a significant tool for space governance.

As we shall see, absence of even minimal rules related to transparency at the GEO orbit is providing an avenue to behaviors which are obviously dangerous, and which could become common because of the impunity the players which are behaving so do feel. The fact that impunity prevails over responsibility and accountability is obviously not a good situation.

Because of the saturation at the GEO orbit, operators need to develop a very dynamic "fleet management" which includes increasingly frequent satellites movements along the orbit. (As an example Eutelsat which owns close to 40 satellites has had to manage an average of 5–6 moves each year since 2011 (however of very different magnitude), and there is no doubt that the frequency of these moves has increased along these years.) For that reason – ensuring that close movements by different satellites will not create harmful interferences or even collision risk among them – the satellite telecommunications operators have established a shared and common database of technicalities associated with each satellite they operate, which can facilitate safe movements and close locations, and by itself could illustrate the kind of best practice and transparency which could be extended to all registered satellites. (The database is managed by the "Space data association" (SDA). SDA "is to seek and facilitate improvements in the safety and integrity of satellite operations through wider and improved coordination among satellite operators and to facilitate improved management of the shared resources of the space environment and the radiofrequency spectrum" (SDA website consulted December 13, 2018). SDA is now looking at enlarging its database to other orbits than GEO.) From a regulatory perspective, and in order to comply with ITU rules related to the orbit management, satellite operators have noted that a frequency assignment can be brought into use only if a satellite is maintained at the orbital position for a continuous 90-day period so fighting against "paper satellites," and that from another standpoint, a frequency assignment cannot be suspended for a period exceeding 3 years. All these rules which ITU has developed in order to better manage the orbit and allow access to it to all newcomers are also aimed at limiting the cases where "force majeure" is claimed. It is fair to say that these rules, because they are more and more constraining, are also creating business opportunities among satellite operators, which will need each other more often to ensure the continuity of service to their customers, when the scarcity of orbital positions and frequency assignations do not allow for mismanagement of these "rights."

It must be added that national legislations can also contribute to this transparency. In the case of France, satellites movements are notified 1 month in advance to the French space authorities, as an obligation deriving from the French Space Operations Act of 2008. However, there is no binding international rule creating the same obligation at the moment, even if recommendations to notify such movements have been adopted by UN member states. In a nutshell, guidelines based upon transparent behaviors and explicit rules for the orbit management to be followed by all parties to the Register Convention will be ever more needed, and in the interest of all stakeholders, will be.

National Space Laws (like the 2008 French one) also include provisions aimed at ensuring the protection of the space environment during the satellites' life cycle, which means proper control of the satellite during its in-orbit life (station-keeping, relocations) and deorbiting at the end of its operational cycle in ways which will minimize the risks for health and the environment. Those rules are directly derived from some guidelines and best-practices recommendations established under the UNOOSA via its COPUOS technical works, as part of the space debris management issue. However, for this last part, the guideline is based upon the principle that the satellite must be able to reach a position (the "graveyard orbit") located at 300 km above the GEO orbit, and that, once at that final position, the satellite has been totally passivated and will no longer be a threat of any kind (energy, mechanics, radio-frequency, health). For sure, this deorbiting rule is providing a rule aligned with the fact that those satellites, beyond any kind of reach from earth at 36,000 km, cannot be of any use. However, it should be recognized that we are only at the starting point of this "graveyard policy" and that many satellites launched since the beginning of the last decade has still to reach the graveyard. Hence, there is some doubt that this rule will be sustainable in the long term, as the number of out-of-cycle satellites will quickly increase in the next 5/10 years because satellites launched in the 2000s will have to be replaced, in order for operators to keep the orbital positions they have been authorized to use and from which they provide services.

Finally, the principle of "noninterference" does not prevent "passive hostile" attitudes like the situation by which an unidentified satellite is coming so close to your satellite that it enters the "box" where this last one is supposed to be maintained in order to serve properly and in due compliance with the noninterference principle, and this unidentified satellite clings to your satellite for a period of time which can be a matter of months, and not only of hours or days. Despite the obvious danger such a behavior entails, this passive hostile attitude does not allow for complain, in absence of interference. However, a satellite which is stationed for a long time close to another one which has the full rights to stay at its orbital position, in its "box," is potentially dangerous, even from an orbit management perspective, and finally, could be considered as a serious threat, and as such in breach of the basic rules of peaceful uses of outer space. Every behavior of such nature should be accountable, and transparency rules should request for compulsory statements by which the nation responsible would have to explain the reasons of such close presence at an orbital position (the "box") which is not supposed to be the harbor of a passive, but potentially hostile, clinging satellite. It is obvious that along the last years this story of clinging satellites has expanded to a point that, after US officials raised the issue during the Space 50 Conference, the French ministry of Armed Forces also made it publicly when she revealed that a French-Italian satellite (ATHENA-FIDUS) had been spied by a Russian satellite (Luch (Olymp-K)) without any kind of govern-mental comment or official justification. It is well known that other Russian satellites have behaved in the same way along the last years with different commercial satellites owned by several different satellite operators. That the existing rules, or more precisely their weakness, can allow such movements and "passive aggression" demonstrate the magnitude of the gap which is now created between what could be considered as a responsible management of the GEO orbit and what is taking place. The next step in such an escalation of dangerous behavior will obviously be that a commercial satellite is facing a situation where its integrity is at stake and takes initiatives for movements which could, at the end, turn into a collision between the two objects. Even if the article IX of the Outer Space Treaty calls for a conduct of all space activities "so as to avoid harmful contamination" and asks countries to notify other countries before engaging in any activity that might cause "harmful interference" with activities of others, there is no regulation, procedure, and even concrete sanctions that could limit such attitudes and even prevent them. If there is a will to restore a safe and secure use of the space domain, the most efficient way to prevent such an escalation is to make such attitudes public (as the US and French officials did recently), make them transparent, and strengthen the rules which must prevent them. The shame of the present situation is that such behavior is even not clearly in breach of the rules, despite the very significant danger posing for the stability and safety of space-based activities and services. An increase of global space situational awareness (SSA) capabilities and a "naming and shaming policy" should put an end to the impunity that some stakeholders make use of.

Space Operations in LEO, MEO, and Transit Orbits

It has already been noted that one of the most disruptive changes taking place at the moment is the quick and significant development of projects planning to use the low-earth orbit (LEO), and to a lesser extent the medium-earth orbit (MEO). The magnitude of change is such that some consultants foresee 6500 smallsats to be launched in LEO orbit before 2027. (NSR press release about its report "Small satellite markets," 5th edition, Accessed 28 Nov 2018.) Other figures have even been mentioned, like 11,943 satellites, authorized by the FCC, which are planned by M. E. Musk constellation "Starlink," (Wall Street Journal (online), Accessed 18 Dec 2018) or the 3500 satellites One Web is intending to deploy. Should all the projects already made public by developers and startups be implemented, a total of 18,000 satellites could reach the LEO orbit before the end of the next decade.

Obviously, such an order of magnitude is creating a lot of unknown challenges, but the significance of this change can be measured against the fact that the world aeronautics industry looks for more than 36,000 planes in service in the world in 2032 (against 17,740 in 2013) which will transport 6.3 billion travelers. In other words, and assuming that most of the current projects will be implemented, where we were seeing roughly 1 satellite against 10 planes in the mid of the 2010s, the ratio could become 1 satellite against 2 planes in the mid of the 2030s. In addition, it must be noted that, in many cases, these new low-orbit constellations will aim at providing connectivity in planes, like many key geostationary already decided (e.g., Eutelsat VHTS) will do. This new connection between telecommunications satellite deployments and air fleets development could be of structural consequence for both domains in the next future.

These disruptive figures may give a sense of one new and key issue which will have to be faced: the organization of the management of the different orbits, and the notion that the different orbits will have to be understood in their dynamic interaction.

At the moment, the main regulatory guideline, issued by the ITU in 2007 along with the "noninterference principle," states that the signal coming from the GEO satellites shall not be interfered by signal from satellites at lower orbits, the GEO signal being far less powerful when it reaches the earth than the other signal, especially those emitted from LEO satellites. These rules have been set up for the Ku and Ka bands which are providing the most significant services (Ku for television by satellites, Ka for internet and connectivity by satellites).

The wording of the ITU stands as follows (Radio Regulation No 22.2, as decided by WRC-07):

> 22.2§ 2*Non-geostationary-satellite systems shall not cause unacceptable interference to and,* unless otherwise specified in these Regulations, *shall not claim protection from geostationary-satellite networks* in the fixed-satellite service and the broadcasting-satellite service operating in accordance with these Regulations. No. 5.43A does not apply in this case.

These rules may have to be strengthened. However, new parts of the spectrum, like Q- and V-bands, will be extensively used by satellites systems, notably to connect the earth gateways and the satellites in the coming decade. Most of the "very high throughput satellites" (VHTS), like those designed by Thales Alenia Space and/or Boeing, will make use of these Q- and V-bands.

Those issues are listed on the WRC19 agenda (Agenda Item 1.6 of WRC19). In order to prepare for that discussion, the WRC15 (2015) asked to conduct and complete in time for WRC-19:

> 1 *studies of technical and operational issues and regulatory provisions for the operation of non-GSO FSS satellite systems* in the frequency bands 37.5–42.5 GHz (space-to-Earth) and 47.2 48.9 GHz (limited to feeder links only), 48.9–50.2 GHz and 50.4–51.4 GHz (all Earth-to-space), *while ensuring protection of GSO satellite networks* in the FSS, MSS and BSS, without limiting or unduly constraining the future development of GSO networks across those bands, and without modifying the provisions of Article 21; (...).

In addition to the large and (maybe) numerous constellations in LEO related to telecommunications services, the coming Internet of Things (IoT) constellations and the maturation of the observation market which will drive the deployment of dedicated constellations, like Planet6Labs which plans for 140 low orbit satellites, must be considered. These two kinds of applications will, for sure, include an increasing number of artificial intelligence (AI) assets, which will be on-board small (like shoe boxes) and unexpansive satellites. (The New York Times International was referring to satellites of 7000 $ a piece, Accessed 12 Dec 2018.)

With these new developments, space-based services are reaching a kind of industrial age which they ignored until now, despite the more than 42,000 TV channels currently broadcasted globally by satellite. AI and IoT will for sure drive a standardization process, in terms of production, and a development of satellite-based services unknown until now. In addition, new assets, like balloons, high altitude pseudo satellites (HAPS), and drones, will become part of the connectivity networks deployed in space. As well, suborbital flights will have to be considered in terms of legal and regulatory terms, and it will have to be decided whether they should be regulated by the air-space rules or by the space management principles, both systems needing to be (at least partially) aligned to be manageable.

The industrial age of the space-based applications will entail new services like in-orbit services. These services intend to extend the life cycle of the satellites by refueling them (when their propulsion is mainly chemical) and/or repairing some of their major components like solar antennas. A few companies are currently planning for such in-orbit services. However, this next step is already raising a lot of questions: if space services ability to extend the life cycle of the satellite can be affordable, do we need to plan for 15 years of life cycle for the telecommunications satellite, as it was done until recently or should we think about less duration for the satellites (8–10 years for example) with a better ability to adapt the services to the consumers expectation, when planning for the market in 2030 is so difficult?

Such in-orbit services would entail an increasing level of space movements and traffic which would need specific regulations but are still to come. It has been said that soon the first "mission extension vehicle" (or "space tug") will be launched to extend the life duration of Intelsat 901. Assuming that this interest in life-extension of the satellites could be shared by other satellite operators, what will be the legal status of these objects? Should they be registered like satellites? What is the consequence if such an object fails to provide the service it was assumed to? What is the status of such an object if left in space? Is it a debris, and if so, could the Liability Convention apply? All these questions need to be addressed quite urgently as there is no common rule already agreed. Furthermore, because of the growing number of LEO projects, which will entail hundreds of launches, the overall effect of this change will be an increasing number of debris, even in absence of military activities aimed at testing anti-satellite weapons.

Space surveillance, including space surveillance and tracking, will therefore receive a growing attention from all space users. SSA and Space Traffic Management (STM) are, for the time being, sovereign missions, developed by governments or groups of governments (the EU plans to develop its own capacities). The condition under which these data could be shared with commercial operators, will have to be decided. At the commercial level, the SDA has outsourced a capacity of that kind, adjusted against commercial needs (of course very different from the military needs, and much more limited than them) and allow for access to data to the contributing operators. In the future, the growing importance of such SSA/STM data is so great that the US government has decided that the US Department of Commerce would become the interface with the commercial satellite operators in terms of SSA data sharing. This recent change could also open the door to established commercial SSA and STM services which would complement the governmental ones. If that is the case, they need to decide how these SSA/STM data from different sources will be shared, and along which rules these will be of much importance.

Two questions will be of key importance in order to ensure a sustainable space operations environment:

- How will all stakeholders of the space-based infrastructures, among them the private satellite operators, have access to the space surveillance awareness data? It can easily be assumed that not all data will be accessible, but according to the "need-to-know" principle, there is room to decide which of them are of interest for the commercial operators.

- To which extent the dedicated tools, implemented and managed by governments or groups of governments, will be phased with the pace of the commercial services development? If it is recognized that the SSA data could be shared if and when a denial of service could be intended, an obvious threat to some assets could be experienced, or more broadly, in order to better understand the new developments which could put satellite operators at risks, then SSA policies would strengthen private policies aimed at ensuring continuity of service and agreed service levels.

For sure, satellite operators will consider SSA and STM as key strategic issues which should be inclusively designed. The way to proceed and the roadmaps to de defined will become major issues of the relationship these commercial operators will have with governments. However, if we take into account GEO management challenges, LEO/MEO developments, and multiplication of space objects (like space tugs), there is no doubt that the entire Space Situational Awareness and/or Space Surveillance and Tracking purposes and methodologies have to be redesigned in accordance with the disruptive changes which are going to take place in the way all space stakeholders will make use of this common good.

Protection of Satellites Services from and at Earth

Since the core business of telecommunications by satellites has moved from direct-to-home (DTH) TV broadcasting to internet and broadband services, the value of the different components has faced a major shift: when the DTH broadcasting was mostly an investment in the space segment in the 1990s, which represented more than 90% of the total (space + terrestrial) investment, the space segment is about 60% of the total investment when it comes to broadband services by high throughput satellites (HTS) which are based upon a multispots coverage. In other words, around 40% of the investment is nowadays related to the terrestrial segment of the space-based system. This terrestrial segment is mainly distributed between gateways which connect the final user equipment (antenna + modem) to the internet, through the satellite. The network of gateways, installed in different places (nations) throughout the coverage of the satellite, constitutes a distributed hub managing the traffic and the spectrum allocation to and from different satellite spots.

This architecture which is becoming the standard of connectivity provision in the GEO high and very high throughput satellites (VHTS) will become even more essential in the LEO constellations in order to ensure continuity of service when the constellation "flies" over a location.

Hence, protecting the terrestrial gateways and networks from interferences, be they technical (deliberate or not), legal, or even political, is crucial. This matter is mostly regulated by nations, according to their conceptions of "internet freedom," the resilience of the technical solutions selected by the internet service providers (ISP), the security regulations which apply to these networks, and more widely, to the broadband policies decided by governments.

The freedom of information has been defined in 1948 by the UN Charter on Human Rights as: "Everyone has the right to freedom of opinion and expression; this right includes freedom to hold opinions with-out interference and to seek, receive and impart information and ideas through any media and *regardless of frontiers.*" (article 19).

The UN Covenant on Civil and Political Rights (1966) developed its consequences as follows:

> Article 19.1. Everyone shall have the right to hold opinions without interference.
> 2. Everyone shall have the right to freedom of expression; this right shall include freedom to seek, receive and impart information and ideas of all kinds, regardless of frontiers, either orally, in writing or in print, in the form of art, or through any other media of his choice.
> 3. The exercise of the rights provided for in paragraph 2 of this article carries with it special duties and responsibilities. It may therefore be subject to certain restrictions, but these shall only be such as are provided by law and are necessary: (a) For respect of the rights or reputations of others; (b) For the protection of national security or of public order (ordre public), or of public health or morals.

The fact that both of these documents state that freedom of information must not be limited by frontiers receives a special meaning in the case of internet by satellite, which is by nature, the most efficient tool to disseminate information "regardless of frontiers." However, the worldwide trend seems to look for more constraining rules, in many cases for security reasons. How to ensure free provision of space-based services and freedom of information at the internet age, on one hand, and security requirements, on the other hand, at the same time needs to be considered not only at the level of the governments, in charge of setting national rules, but also at the international level, which is appropriate to set rules aimed at implementing the principles as stated by the UN Covenant, which is binding for ratifying states. Previous attempts to move forward in this direction have failed in the past, when TV dissemination was the main satellite service. However, the new development of space-based services will call for more determined actions to update the implementation of these lasting principles which have to be protected in this new situation, as well.

From another standpoint, the new era of space-based services will look at the convergence between telecommunications and navigation services (like GPS and Galileo). If connectivity has to support mobility, both kinds of services will be needed at the same time, and the satellite will be, again, an indispensable tool because of its territorial coverage, when terrestrial telecommunications network (i.e., fiber, 4G) basically address urban and concentrated populations. With the 5G coming, the end user will ask for a seamless connectivity, every time and everywhere. This expectation, which will ensure continuity of service, will request hybrid and complementary networks, terrestrial and space together providing a resilient service based on a large interoperability of the networks, able to match any kind of unexpected situation as the satellite will also provide the back up to the terrestrial network in case of need, or in case of urgency. Such architecture will allow for satellite to serve remote, distant areas, and mobility needs when terrestrial networks will first of all provide services in highly dense and urban areas. Furthermore, the resilience of the overall supply will extend the spectrum of applications, especially to the very demanding services related to defense and security, as we already noticed.

Along this change, the clear-cut separation between defense/security needs and commercial needs will continue to blur. There is already a lot of security/defense needs which are fulfilled by commercial objects and operators; military planners are now including the commercial assets in their assessment of the resources which could be mobilized in case of need by governments (because of the flexibility provided by large commercial fleets), and commercial operators consider the governmental needs as a key driver for their future development (i.e., EU Govsatcom in Europe which should include services provided by commercial operators; DISA reform in the United States by which the Air Force Space Command will "oversee management of nearly all military and commercial SATCOM for the DoD" (US Air Force Space Command press release, Accessed 12 Dec 2018). Furthermore, disruptive technologies like software-driven satellites will allow the best provision of the power and of the spectrum based on end user needs. (The "Quantum" satellite, to be launched in 2019, and built by Airbus in UK, will be the first 100% software-driven satellite.)

If governments and operators together develop collaborative policies and solutions to fulfill the security needs, more robust rules will have to ensure that hostile actions against commercial satellites will be treated in the same way as hostile actions against sovereign satellites. However, the requisite for such approach is the ability to designate the origin of the hostile action, in other words the ability to attribute. Here, the challenge is probably of the same nature as it is when it comes to cyber-attacks. In both cases, satellite jamming and cyber-attacks, the actors bet on the impunity they can expect from the difficulty to identify their behavior and to attribute the unlawful practice. Hence, and even if new satellite technologies allow for anti-jamming equipment and geo-localization mechanisms on-board the satellite, which become quite conventional on commercial satellites, it seems that satellite manufacturers need to invest in research and development to ensure that those which are tempted to make use of cyber-attacks or jamming of the signal will be deterred from such behavior because of the increasing risk of being identified. Again, being able to "name and shame" the origin of the infringement and the identity of the rule breaker, like a whistleblower, is a must. To the extent that such infringement is facing a sanction!

Government's responsibility in case of jamming of the satellite signal falls under the ITU rule: however, despite recent progress from the ITU Pleny Potentiary of 2014 (resolution to set up a database of the geo-localized jamming), the fight against deliberate jamming (for political reasons, to prevent the reception of a signal in a territory) must be strengthened. Western nations are currently the only ones which may be ready to recall the principles, and more important, to ensure that they are still implemented by the community of nations which are UN members.

Space security must be understood in a comprehensive way: it is not only about protection of the space segment. A space-based infrastructure is a system combining a space asset and a network of terrestrial gateways/infrastructure which are as important as the space segment in order to provide the service. This terrestrial component of the space-based system has an increasing strategic value as it is the service provider component to reach the end user: we can observe an increasing pressure from governments to receive a right to have access to or even to be able to control the flows of data coming from (or going to) the satellite. There are good reasons for that (e.g., fight against terrorism), but there are also very serious threats resulting from that trend (reduction or even suppression of freedom of information). The international community has established strong grounds which have supported the development of the space applications (like TV and internet services provision), which is the freedom to access to information "without borders consideration"; (art 19 UDHR; art 19 of ICCPR; EU HR Chart): these principles are obviously challenged in an increasing number of situations.

Space operations and space services are at the core of the discussion about cybersecurity in space. They all need very dynamic and robust cryptology methods in order to ensure that the very quickly increasing number of services based upon space infrastructure will be resilient to adverse behaviors. New technological developments like laser transmission of data (EDRS in EU; NASA next generation relay satellite/post TDRSS) is giving the space-based solution some kind of advantage to ensure the security of the data transmissions: the space community (governments, industry and operators) should identify the contribution the space-based solutions can bring to a more secure cyber environment, and set up the rules which will strengthen this key advantage at the moment when satellite services are even more needed.

Conclusion

Along the last decade, it has been expected that this need for a more regulated environment for space-based activities could be fulfilled by guidelines, best practices, and agreed principles. This set of rules will be even more needed in the coming future than it already is. As the governmental, economic, and social value of the space-based applications multiply, such rules will only protect the proper management of the humanity "common good" which is the space domain.

If there is no deployment of offensive weapons systems (like ASW) in space, this "soft law" recommendation should concentrate on two main issues, which unite most of the described situations: service (and access to) denial; attribution of action (transparency/responsibility/liability).

However, it should be recognized that, with its code of conduct proposal, the EU tried to make steps forward towards a more sustainable space environment. The failure of such proposal to reach the consensus shows how important it is to strengthen the efforts. It seems that far from being the "common good" subject to peaceful activities, as described by the Outer Space Treaty of 1967, it moves towards a kind of jungle where survival and security of the services, whatever their nature, will not be granted.

Looking at the plans to develop new activities related to space exploration and the prospect of "celestial commercialization," a commentator was stressing that "the time has come to clarify international space law and allow commercial ventures to go ahead subject to sensible safeguards." Noting that "a full-scale revision of the Outer Space Treaty might be desirable but is not necessary," this comment was suggesting that "a governance framework agreed by all spacefaring countries would do the job." (See Financial Times, "The world should update its laws on outer space," Accessed 28 Dec 2018.) This cautious approach was taking stock of the very difficult challenges a revision of the Treaty would entail at a moment of hardened competition among the space nations, and despite the benefit all humanity could find from an updated treaty which could decide for the regulations of the coming and extended space activities. However, far worse, deciding for a revision of the Treaty could open a Pandora box which could, finally, undermine the principles which the Treaty has recognized and which are still very meaningful for the future. Should a Treaty revision be decided, it should aim at strengthening those principles, and set regulations which would help to manage the coming challenges, not destroy the "space order" which is still based upon the Treaty.

Obviously, and in absence of a "consolidated" Treaty, the governance framework which is required is more than urgent, and not only because of the asteroids commercialization. Space is at the juncture of sovereign and commercial activities, national and international projects, closely related to the emerging data and digital needs and technologies, close to extend significantly the number of industries and services which will rely on it on a constant basis. More than an infrastructure, space-based activities are becoming the most ubiquitous domain which most not to say all future human development plans will require. Absence of governance, allowing all these plans and services to find their path will become a challenge from which, in absence of significant progress, the mere future of this attractive and even exciting new area could collapse without delivering its promise.

International organizations, governments, and private actors must now bend their efforts to establish this governance framework which is so much needed.

In that respect, commercial operators should build upon their own experience of space-based services to further contribute to shaping this framework. Among the different items they could raise, some seem more urgent.

A first priority could be to develop the *transparency* of movements in space. As an example, adding a satellite identification to the registration obligation could help monitoring movements at the different orbits, facilitate their notifications during the 15 years of the life cycle of many satellites, and discriminate satellites from space tugs and high altitudes balloons or drones, for example. History shows that such transparency (based on an ID) was the condition for a safe and secure use of the airspace by planes, and seas by ships. It can be doubtful that the huge increase of space objects in the near future will allow for an enduring regime of quasi-secrecy about the movements in space, even more if collisions between space objects were to remain nonliable.

The second aspect could be to develop a *responsibility* scheme where international norms, best practices, and/or national space laws, financial incentives finally could combine to promote a sustainable use of space. The launching industry has demonstrated a significant capability to reduce the risks of failures in the launching business because of the attention the operators paid to the insurance cost of the launchings they ordered. The need for an insurance could be extended to more space activities, and could act as an incentive to develop designs which could reduce the number of debris, extend the life time of space assets, or reduce the collision risks (especially at low orbits).

The third aspect should tackle the absolute need to ensure the *continuity* of service which end users are expecting from the space rules and actors. The guarantee of service continuity is the condition for an area of large recourse to space-based services, especially if these services deal with connectivity and mobility. It should encompass the prevention of deliberate interferences, the technical developments which will ensure the robustness of the signals against cyber-attacks, the ability to identify and designate the parties which are threatening the continuity of service. A lot of investment (technical and financial) is paid to achieve this service continuity objective, but it needs to be backed by rules which will penalize the actors undermining it.

A final dimension could set up the appropriate forum where all issues related to space sustainability and governance could be discussed between international organizations (CD, COPUOS, ITU...), governments, and private actors (the list of which is extending quite quickly). At the moment, all concerned entities develop their "own" framework for discussion, but these different discussions are not coordinated along with a *common agenda* which could have been agreed by all stakeholders. The need for an agreed "space agenda" which would answer the questions – what are the issues at stake? what are the key priorities to address? to which extent proposals and identified solutions can be implemented by the actors, need "soft law" solutions (best practices, standards...), or need an urgent "hard law" (conventions, revised OST) – all these issues need to be addressed in a forum which could report to the UN, as the ultimate responsible for a safe and secure space.

Security and sustainability of space-related activities is facing a very serious challenge. Commercial space operators have an insight on technological developments (which they very much drive when ordering the space-based and ground-related infrastructure), affordability of coming services (especially in terms of connectivity and data management), added value of space-based services (by comparison to terrestrial solutions), and regulations which could support the space based economy. All these issues cannot be solved by one government, or even one international organization. They are multifaceted, evolving, and pressing at the same time.

References

Donohue TJ (2018) CEO of the US chamber of commerce. Space, the new economic frontier. Space News, December 17, 2018

Moltz JC (2014) Crowded orbits. Conflict and cooperation in space. Columbia University Press, New York

Tett G (2018) Space: the final frontier for finance. Financial Times Magazine, December 19, 2018

Space Debris Mitigation Systems: Policy Perspectives

56

Annamaria Nassisi, Gaia Guiso, Maria Messina, and Cristina Valente

Contents

A. Nassisi (✉)
Strategic Marketing, Thales Alenia Space, Rome, Italy
e-mail: annamaria.nassisi@thalesaleniaspace.com

G. Guiso (✉)
Women in Aerospace-Europe (WIA-E), Milan, Italy
e-mail: gaiaguiso@gmail.com

M. Messina (✉)
Education Unit, Italian Space Agency (ASI) , Rome, Italy
e-mail: maria.messina@est.asi.it

C. Valente (✉)
Marketing and Sales, Telespazio, Rome, Italy
e-mail: cristina.valente@telespazio.com

© Springer Nature Switzerland AG 2020
K.-U. Schrogl (ed.), *Handbook of Space Security*,
https://doi.org/10.1007/978-3-030-23210-8_143

Abstract

The preservation of the space environment against space debris threats requires the development of a legal and regulatory framework. The sustainability and ongoing security of the space environment requires an agreement at international level. The study argues that in order to develop a set of specific rules for the governance and regulation of space debris it might be helpful to draw on similar contexts. First, maritime law on salvage on the high seas, developed by the International Maritime Organization (IMO), can facilitate the approach to abandoned spacecraft. Second, the role of international organizations such as the International Telecommunication Union (ITU) can serve as a point of reference in the development of binding space debris mitigation rules. The development and implementation of such rules will contribute to the successful sustainability of outer space activities.

Introduction

The chapter provides an analysis of the legal and policy issues concerning future operational debris mitigation systems. This is an essential step in preparation of specific rules, agreed at national and international level, for a legal and regulatory framework.

The analysis performed focuses the attention on these main three topics:

1. Space debris problem and need for regulation
2. Maritime law as a reference to address the rules for abandoned spacecraft/space debris
3. ITU as a reference organization that can adopt binding rules

What we need is a binding international legal instrument. Yet, as it might take too long before we have the next treaty, addressing soft law and rules of the road could be relevant. Maritime law can offer some frame of reference to solve this issue. Specifically, the international law of salvage in high seas, as outlined in the International Convention on Salvage, IMO 1989, may help to establish a similar regime, in space, for abandoned spacecraft.

In this regard, the pivotal role of the International Telecommunication Union (ITU) in setting a regulation for the Active Debris Removal missions (ADR), in the years to come, should not be underplayed. Since this agency is in charge of promoting and coordinating the definition of technical standards, as well as of assigning satellite orbits, it might be able to establish itself as the referral organization for the issuing of finally binding space debris mitigation rules. As in the case of frequency regulation, also in the case of debris mitigation and removal, it is crucial to have an institution capable of supporting the multiple stakeholders and security constraints. So much of the future of space activities will depend on how these rules will succeed in limiting the growing amount of floating debris in space. Are the "Space debris" a threat for Space Exploration? This causes legal uncertainty.

The tendency to overcome limits confirms an unavoidable desire of knowledge, inherent in all human beings. Despite the new technology innovations, the ambitious plans relating to exploration and commercial exploitation of extraterritorial resources implies the need of a legal certainty at national and international level. Protecting outer-space from damages caused by negligence or breach of duty should foster new legislative efforts and finally a new instrument of hard law. The sovereignty aspect continues to guide the international community toward flexibility and escape clauses. If we contextualize Durkheim's thought, the French sociologist who lived in the mid-nineteenth and twentieth centuries, we would say that managing debris as a social phenomenon involves measurement of its detrimental effects, and subjugating individual interests to collective's one.

In accordance with the principles contained in the "Treaty on Principles Governing the Activities of States in the Exploration and Use of Outer Space, including the Moon and Other Celestial Bodies" (Outer Space Treaty 1967), governmental organizations aim to pursue "the benefit and the interests of all countries, irrespective of their degree of economic or scientific development." This provision is essential for a comparative analysis between Space Law and the principles governing the Law of the Sea, especially on the sustainability and protection of natural resources.

The lack of ratification of the "Agreement Governing the Activities of States on the Moon and Other Celestial Bodies" (Moon Agreement 1979) demonstrated the difficulty to further develop the legal framework regulating activities on the Moon or other celestial bodies. Following the new geopolitical order of the 1970s, the negotiation of the Moon Agreement occurred at the same period with United Nations Convention on the Law of the Sea (1982), both echoing the intention of States to define responsibility in a newly international cooperation framework. Although United Nations Convention on the Law of the Sea (UNCLOS) and Moon Agreement are similar in terms of legal nature, the same cannot be said for the ratification process. Although the Moon Agreement has only been ratified by 13 states, it took the UNCLOS 10 years of negotiations to be ratified by 164. The Law of the Sea therefore represents an example for the progressive development of international space law.

Current Situation of Space Debris and Risk Posed to Long-Term Sustainability

The debris population could be generated from accidental case or voluntary case. The well-known antisatellite test (ASAT) of China is an example of voluntary case where a defunct weather satellite, called Fengyun-1C, was destroyed by a missile that imparted an estimated 350 joules per gram of its mass. Another well-known collision that we can classify as accidental case, is the 2009 collision between the Iridium33 and Kosmos2251 satellites. Both cases have produced an increment of debris population, a risk for operational satellites and a threat of sustainability as highlighted in the report of the International Interdisciplinary Congress on Space Debris: "Guidelines for the Long-term Sustainability of Outer Space Activities of the Committee on the Peaceful Uses of Outer Space (2019)(*UN A/74/20*)."

A lot of commercial initiatives consider solutions for space debris removal. As such, ensuring that sustainability of outer space can be realized through the promotion of progressive development of international space law. Promoting awareness of this issue should be sought through a plurality of legal and scientific efforts with a motivated incentive to open data.

Drawing Analogies Between Outer Space and Maritime for the Regulation of Space Debris

The roadmap for the common spaces – the Earth and the seas – have always been a good arena for all policy makers, scientists, and economic stakeholders to discuss. Both outer space and the seas have posed significant challenges at international level. Although in both cases, the idea to set up a supranational legal framework was suspicious of any effort to control and limit freedom. As it will be indicated below, particular attention was paid to environmental management from the point of view of safety policies and protection of the marine environment: hence, the importance of drawing analogies between outer space and maritime for the regulation of space debris.

The adoption of the 1989 International Convention on Salvage and the "no cure no pay" rule (art. 12. 2 "no payment is due under this Convention if the salvage operations have had no useful result") marked a step forward in the management of maritime environment (IMO 1989).

The International Maritime Organization (IMO) Convention calls upon States Parties:

- To carry out the salvage operation with due care (article 8)
- To take measure to protect coastline or related interest from pollution or threat of pollution following upon a maritime casualty (article 9)
- To give a reward to encouraging salvage operation (article 12)

Any analysis seeking to guarantee such protection in outer-space led to non-legally binding results. That is why the Space Debris Mitigation Guidelines of the Committee on the Peaceful Uses of Outer Space (COPUOS) an example of many voluntary political engagements that have become well-established practice. The guidelines describe space debris as "*all man-made objects, including fragments and element thereof, in Earth orbit or re-entering the atmosphere, that are non-functional*," and emphasize that rapid growth increases the probability of collision. It is stated that "*the prompt implementation of appropriate debris mitigation measures [is meant] towards preserving the space environment for future generations*." The use of words such as "*minimize,*" "*avoid,*" and "*limit*" contained in Space Debris Mitigation Guidelines underpin that the guarantor of their application is the recipient itself.

Even up to today, the only binding policy provisions related to the return of objects launched into outer space are contained in the Agreement on the Rescue of Astronauts, the Return of Astronauts, and the Return (The Rescue Agreement) of objects launched into outer space of 1968, but with the sole aim of recognizing duties of each contracting party and launching authority.

Nonetheless, article 5, paragraph 4 of the Rescue Agreement takes into consideration the issues of environmental protection in outer space. It is based on the provision that it is extremely important to react and eliminate any space object or its component that may be considered as potentially harmful. This interpretation could be a bit of a stretch: the agreement is mainly guided to promote international cooperation and provide all possible assistance to astronauts. Lastly, in accordance with article 6, the launching authority should pay the costs incurred to recover and return space object but, unlike the Convention on the International Maritime Organization – adopted 1948 and entered in force in 1958 – there is no reference to a possible reward. Thus, although it suggests to act and protect outer space environment, it's not clear to what extent though. The International Convention on Salvage, IMO 1989, is crucial therefore in reference to the liability and compensation mechanism, helping states to face the challenges that remain unresolved.

Need for New Rules Ensuring Sustainable Development for the Evolution of Satellites

In a world that is becoming increasingly sustainable, a new set of rules should acknowledge such a paradigm shift in outer space as well. While there have been some attempts to create a widespread endorsement of evaluation methods for the footprint of substances used in the space industry, in this regard, the Intergovernmental Panel on Climate Change (IPCC) and the European Commission's Joint Research Centre's Life Cycle Assessment (LCA) can be cited. Currently the legislation producing the most impact on the production chain of substances susceptible of harming the planet and the space environment, include the Montreal Protocol on Substances that Deplete the Ozone Layer (1987), the European Commission's Restriction of Hazardous Substances (RoHS) directive (2003), and the Registration, Evaluation, Authorization, and Restriction of Chemicals (REACH) regulation (2006) – all placing restrictions on particular substances used in space programs (ESA Safety). Also in maritime law, similar agreements have been reached to protect the maritime environment: for instance, the Protocol on Barcelona Convention concerning Cooperation in Preventing Pollution from Ships and, in Cases of Emergency, Combating Pollution of the Mediterranean Sea (January 2002), replacing the Protocol on Barcelona Convention concerning Cooperation in Combating Pollution of the Mediterranean Sea by Oil and Other Harmful Substances in Cases of Emergency (Barcelona 16 February 1976)

It is very interesting to note that a major impulse to incorporate sustainable materials in the development of new satellites is coming from what has been labelled as "ecodesign" (Design for Innovative Value). In the context of space policies, *ecodesign* would mean that, the approach to the building of satellites and orbiting space facilities should focus on "sustainability by design," to use a terminology borrowed from the contemporary concept of "privacy by design" (Regulation (EU) 2016/679). The goal of manufacturing sustainable spacecraft is linked to the shaping of policies for deorbiting management. Since sustainable is something that allows a process to endure indefinitely, minimizing or eliminating its negative impact on the environment, then sustainable production implies taking advantage of compostable materials. In this respect, national and international space legislation should place limits to the type and nature of materials to be used in space, by setting a percentage of compostable composition depending on the space item. This would enable de-orbiting only for those (smaller) portions of spacecraft that cannot biodegrade in space.

Ongoing Search for Space Authority

The lack of binding international agreement on space debris is, in particular, related to political and regulatory aspects. Just the absence of an authority for the regulation of space debris creates that gap which results in inability to reach a common agreement at the international level to address such a crucial issue.

The matter needs to be carefully unraveled step by step, starting from fundamental questions that are to be posed: should the above said authority be a new one, explicitly created for this purpose, or should it be an already existing institution appointed? And, if it be an already existing institution, which one? Would it be suitable to promote, drive, and guarantee the legal certainty and standardization, if necessary, coercing states into it, to safeguard the long-term sustainability of outer space activities, and the balanced exploitation of its resources?

During UNISPACE+50, the Committee on the Peaceful Uses of Outer Space (COPUOS) identified seven thematic priorities and evaluated potential initiatives introducing legal mechanisms necessary to achieve the safety and sustainability of space operations. The Chair of the Working Group on the Status and Application of the Five United Nations Treaties on Outer Space noted the lack of exhaustiveness relating to semantic richness of some notions. Actually, there is not one responsible authority for tracking space debris internationally but several ones. The United States does track space debris to protect European satellites, while within Europe some of that information is shared with the rest of the world or within the SST Framework Program. In addition, the Inter-Agency Space Debris Committee (IADC) is an international association involving various space agencies from around the world with the aim of addressing the problem of space debris.

In this regard, the "Space2030" agenda, encouraging a collaboration between the Office for Outer Space Affairs and industry entities, called for efforts aimed at promoting transparency and sustainability of space activities through the pillars of space economy, space diplomacy, and space accessibility. In line with a view to straighten a space global governance, the General Assembly looked at "Space 2030" as a strategic overview within UNISPACE+50 conference which engaged all key stakeholders in the space arena (A/AC.105/1166).

ITU Role in Leading Regulation on Space Debris

In the current global panorama, the International Telecommunication Union (ITU) could play a role in fostering a common understanding on security challenges in space. Founded in 1865, the ITU is the United Nations agency tasked with allocating "global radio spectrum and satellite orbits, [developing] the technical standards that ensure networks and technologies seamlessly interconnect, and [striving] to improve access to ICTs to underserved communities worldwide."

When addressing the matter of space debris and de-orbiting, in the recent past and hitherto, the ITU has mostly issued recommendations. One of the most notable recommendations has been, for instance, the December 2010 Recommendation ITU-R S.1003-2, concerning the "Environmental protection of the geostationary-satellite orbit." In here, the role of the agency comes into play because of the increasing threat posed by spacecraft fragments, resulting in multiplying debris crowding the orbit at stake.

ITU-R S.1003.2 provides guidance about disposal orbits for satellites in the geostationary satellite orbit (GSO). In this orbit, there is an increase in debris due to fragments resulting from increased numbers of satellites and their associated launches. Given the current limitations (primarily specific impulse) of space propulsion systems, it is impractical to retrieve objects from GSO altitudes or to return them to Earth at the end of their operational life. A protected region must therefore be established above, below and around the GSO which defines the nominal orbital regime within which operational satellites will reside and maneuver. To avoid an accumulation of non-functional objects in this region, and the associated increase in population density and potential collision risk that this would lead to, satellites should be maneuvered out of this region at the end of their operational life. In order to ensure that these objects do not present a collision hazard to satellites being injected into GSO, they should be maneuvered to altitudes higher than the GSO region, rather than lower.

The recommendations embodied in ITU-R S.1003.2 are:

- Recommendation 1: As little debris as possible should be released into the GSO region during the placement of a satellite in orbit.
- Recommendation 2: Every reasonable effort should be made to shorten the lifetime of debris in elliptical transfer orbits with the apogees at or near GSO altitude.
- Recommendation 3: Before complete exhaustion of its propellant, a geostationary satellite at the end of its life should be removed from the GSO region such that under the influence of perturbing forces on its trajectory, it would subsequently remain in an orbit with a perigee no less than 200 km above the geostationary altitude.

- Recommendation 4: The transfer to the graveyard orbit removal should be carried out with particular caution in order to avoid radio frequency interference with active satellites. The document, only one of a series in this decade, aims to ensure "the rational, equitable, efficient and economical use of the radio-frequency spectrum by all radiocommunication services, including satellite services, and carry out studies without limit of frequency range on the basis of which Recommendations are adopted,"[3] providing then direction and guidelines to national and international space operators.

By definition though, an act of recommendation is not legally binding, and this explains the "softer" capacity of its provisions. For instance, the one expressed therein, that cites: "before complete exhaustion of its propellant, a geostationary satellite at the end of its life should be removed from the GSO region such that under the influence of perturbing forces on its trajectory, it would subsequently remain in an orbit with a perigee no less than 200 km above the geostationary altitude," or even "the transfer to the graveyard orbit removal should be carried out with particular caution in order to avoid RF interference with active satellites."

ITU Role in Guaranteeing Space Security Against the Debris Threat

As mentioned before, one of the duties of the ITU is to oversee the definition of technical standards for the use of radio waves, and so, directly and indirectly, the functioning of, and the proper access to, the interconnection between the ICT infrastructures all around the world. This means that, through one of its three internal sectors, the ITU-R, the assignation of satellite orbits is already part of its statutory activities. Furthermore, since the 1973 ITU Convention, article 33, defined the Geostationary orbit (GEO) as a "limited natural resource" (Hacket 1994), many decades passed, and ITU continued its rationalization attempts of the orbit use for telecommunication purposes, thus shaping, in the form of resolutions and recommendations, an incomparable technical and political legacy that, as extremely valuable is today, even more will be in the future, with additional private operators joining orbits exploitation. As things stand, it would be a natural consequence to just broaden ITU mandatory sphere of action to the debris issue, rather than deferring it in the expectation that, eventually, an agreement over the institution of a new ad hoc regulatory body will be reached soon.

The second reason for believing that ITU would be the most logical option for a leadership role lies in its very structure: in fact, it has the peculiar characteristic of being the sole UN agency to enlist both public and private stakeholders. With its 193 Member States, about 700 tech companies, ICT regulators and leading academic institutions, ITU is perfectly capable of establishing itself as the global funnel for the different stakeholders, and for all space players' interests. Furthermore, as noted by Antoni et al, "the ITU working groups which are a great example of effective mechanism to overcome conflicting interests among national authorities, and allow for coordination at international level and the bottom-up development of binding instruments" (Antoni et al. 2020). The efficiency and immediacy of the dialogue between governments and private entities, in particular, would benefit from this choice, and also from the technical asset that the ITU is the bringer and guarantor of.

Discussion: Fairway Charges in Space, a Starting Point for Space Debris Regulation?

A last question still remains. How may international law support settling on the ITU as the organization of reference for space debris mitigation rules? Since the dawn of the space era, jurists have examined maritime law to find elements of similarity that facilitate the development of rules for outer space activities. A number of provisions are even similar, both in phrasing and in content (adapted, of course, to the different contexts), with the first one serving as a model for inspiration for the latter. Especially in relation to the high seas discipline, this becomes more evident. Therefore, a study on the applicability of the fairway charges rulebook in space, albeit with some variations, could be the starting point for consistently assuring the efficient use of the spectrum/orbit resource.

In maritime law, the en route rights are governed by the 1972 IMO Convention concerning International regulations for preventing collisions at sea (COLREGs 72), which was drafted and adopted as an updated version of the 1960 Collision regulations. The Convention, and its subsequent amendments, lays down rules on a crucial issue in the maritime environment, that of the naval traffic in the sea, aiming at reducing the risk of collision between vessels, through the introduction of traffic separation schemes, and in particular through the establishment of priorities between routes and actions to be taken, for instance in case of overlapping trajectories.

A parallel drawn between vessels and spacecraft as well as routes and orbits would be an interesting exercise. The latter could indeed bear some interesting outputs, benefitting the management of spacecraft especially in the most crowded orbits afflicted by the increasing presence of debris. Provisions such as the one that cites: "a vessel which was required not to impede the passage of another vessel should take early action to allow sufficient sea room for the safe passage of the other vessel. Such vessel was obliged to fulfil this obligation also when taking avoiding action in accordance with the steering and sailing rules when risk of collision exists" (Rule 9) could shape the basis for an obligation, upon the launching states, to take all possible and preventive action not to impede the passage of other spacecraft in orbit. Extensively, this would result in de-orbiting measures to be mandatorily planned in advance, and set in motion before the spacecraft's operational life comes to an end, in order to clear the path as required by the rule.

Moreover, Rule 17 of COLREGs 72 states that a stand-on vessel may "take action to avoid collision by her manoeuvre alone as soon as it becomes apparent to her that the vessel required to keep out of the way is not taking appropriate action in compliance with these Rules." Transposed for our purposes, that would pose states which command operative spacecraft, risking collision with a "dead" one, on an upper decisional level than the state that owns the latter, broadening the borders of their maneuver power. The same would go with responsibility issues, where Rule 18 "includes requirements for vessels which shall keep out of the way of others." Indeed, this would open new scenarios as for the framework of international liability and responsibility for damage between two or more launching states, with the rules currently in force being called into question.

Conclusions

By way of conclusion, maritime law can really be a platform where space law can grow. It remains to be seen to what degree, and by what means, the international institutions and operators will want, and be able, to draw on analogies from the same successful maritime role model in space.

In addition, the ITU can be the right reference to contribute to the sustainability of outer space that can be guaranteed from regulation to the support of commercialization. Finally, it is important to create an international consciousness and awareness on the importance of solving debris legal aspects.

References

Antoni, N., Giannopapa, C, Schrogl, KU (2020) Legal and policy perspectives on civil-military cooperation for the establishment of space traffic management. Space Policy J, In press, JSPA_101373

Design for Innovative Value. Towards a Sustainable Society: Proceedings of EcoDesign 2011: 7th International Symposium on Environmentally Conscious Design and Inverse Manufacturing BY Mitsutaka Matsumoto, Yasushi Umeda, Keijiro Masui, Shinichi Fukushige, Springer Science & Business Media, 3 Apr 2012

ESA Safety and Security. https://www.esa.int/Safety_Security/Clean_Space/ecodesign

Guidelines for the long-term. sustainability of outer space activities, Committee on the Peaceful Uses of Outer Space, Fifty-ninth session, Vienna, 8-17 June 2016

Guidelines for the long-term sustainability. of outer space activities, Committee on the Peaceful Uses of Outer Space, Sixtieth session, Vienna, 7-16 June 2017

Hacket GT (1994) Space Debris and the Corpus Iuris Spatialis, Forum for Air and Space Law, vol 2. Editions Frontieres

http://spacenews.com/op-ed-chinas-well-crafted-counterspace-strategy/

http://www.esa.int/About_Us/ECSL_European_Centre_for_Space_Law/About_space_law

http://www.imo.org/en/About/Conventions/Pages/Home.aspx

http://www.un.org/depts/los/convention_agreements/convention_overview_convention.htm

http://www.unoosa.org/oosa/en/ourwork/spacelaw/treaties/introliability-convention.html

http://www.unoosa.org/oosa/en/ourwork/spacelaw/treaties/intromoon-agreement.html

http://www.unoosa.org/oosa/en/ourwork/spacelaw/treaties/introrescueagreement.html
https://treaties.un.org/pages/ViewDetails.aspx?src=TREATY&mtdsg_no=XII-1&chapter=12&
    lang=en
Regulation (EU) 2016/679 of the European Parliament and of the Council of 27 April 2016 on the
    protection of natural persons with regard to the processing of personal data and on the free
    movement of such data, and repealing Directive 95/46/EC (General Data Protection Regulation)
Thematic priority. 2.Legal regime of outer space and global governance: current and future
    perspectives, Committee on the Peaceful Uses of Outer Space, Note by the Secretariat, A/
    AC.105/C.2/2019/TRE/L.1.

Security Exceptions to the Free Dissemination of Remote Sensing Data: Interactions Between the International, National, and Regional Levels

57

Philip De Man

Contents

Abstract

International rules on remote sensing are generally silent on restrictions to remote sensing data dissemination in favor of broadly phrased principles of free collection and exchange of information. At the same time, those States and regional

P. De Man (✉)
University of Sharjah, Sharjah, UAE

University of Leuven, Leuven, Belgium
e-mail: philip.deman@kuleuven.be

© Springer Nature Switzerland AG 2020
K.-U. Schrogl (ed.), *Handbook of Space Security*,
https://doi.org/10.1007/978-3-030-23210-8_119

groups that have adopted specific remote sensing legislation typically focus on the security implications of such free data dissemination when applied to sensitive information. This chapter aims to give an overview of the meshwork – at the same time overlapping and partially lacunal – of international, regional, and national rules specifically adopted for the regulation of remote sensing activities, with a view of distilling the general benchmarks of a regime for the dissemination of remote sensing data, and the acceptable limitations for national security concerns. In so doing, the chapter aims to keep a clear focus on specific regulations rather than abstract tendencies. The discussion of every level therefore centers on a legal document deemed particularly emblematic of the regulation of remote sensing at that level. Reflecting this, the chapter is divided into the following sections: (I) the international level, analyzed through the lens of the 1986 United Nations (UN) Remote Sensing Principles; (II) the national level, illustrated by the comprehensive remote sensing regulation in the German Satellite Data Security Act; and (III) the regional level, where the interactions between the international and the national remote sensing rules are exemplified by the regulations of the European Union (EU) on data and information disseminated through the Copernicus program.

Introduction

As a legal matter, checks on the dissemination of remote sensing data for reasons of national security combine issues governed by, inter alia, the law on outer space and telecommunications, human rights law, and trade and export control regulations. They also stand at the crossroads of intersecting regulations at the international, national, and regional levels. The laws on outer space and space communications deal with issues that are intrinsically international in nature and their general principles are laid down in a limited set of multilateral conventions and resolutions. These typically focus on the freedom of States to use outer space for activities such as remote sensing, including dissemination of their data. Matters of national security are, by definition, eschewed in multilateral negotiations so as to allow States to retain a high level of discretion for their regulation at the municipal level. It should not be surprising, therefore, that, while international rules on remote sensing are generally silent on restrictions to the free dissemination of remote sensing data, the focus of national regulation lies rather on the security implications of such freedom when applied to sensitive data.

This shift in focus is also due in part to the distinct legal environment in which national legislators operate, where general principles on remote sensing agreed to at the international level must be integrated in an overlapping patchwork of rules on such diverse matters as national licensing of space activities, trade, and export control. In this respect, national export control rules still often focus heavily on trade in satellite technologies and ground control hardware, leaving the dissemination of the collected data to other areas. (See Schmidt-Tedd and Kroymann 2008, 103 and 107; noting that the German Satellite Data Security Act was necessary to close a potential legal loophole in export control regulations, which typically only apply to satellite technologies but not to the distribution or transfer of data generated by these satellites.) And while States are increasingly active in adopting national space legislation, cases where the dissemination of remote sensing data are explicitly regulated at the domestic level remain rare. Further, the relative novelty of highly detailed remote sensing imagery collected, stored, and disseminated on a large scale by private actors means that, even if specific rules of national space law have been adopted, detailed regulation of the security implications of remote sensing data dissemination may still be missing.

In light of this situation, the present chapter wishes to give an overview of the meshwork – at the same time overlapping and partially lacunal – of international, regional, and national rules specifically adopted for the regulation of remote sensing activities, with a view of distilling the general benchmarks of a regime for the dissemination of remote sensing data, and the acceptable limitations for national security concerns. In so doing, the chapter aims to keep a clear focus on specific regulations rather than abstract tendencies. The discussion of every level therefore centers on a legal document deemed particularly emblematic of the regulation of remote sensing at that level. Reflecting this, the chapter is divided into the following sections: (I) the international level, analyzed through the lens of the 1986 United Nations (UN) Remote Sensing Principles (Principles Relating to Remote Sensing of the Earth from Outer Space, UN Doc. A/RES/41/65 of 3 December 1986 (hereinafter: "Remote Sensing Principles" or "RS Principles")); (II) the national level, illustrated by the comprehensive remote sensing regulation in the German Satellite Data Security Act (Act to give Protection against the Security Risk to the Federal Republic of Germany by the Dissemination of High-Grade Earth Remote Sensing Data of 23 November 2007, 2590 *Federal Gazette (BGBl.)*, Year 2007, Part I, No. 58, 28 November 2007 (hereinafter: "Satellite Data Security Act" or "SDSA")); and (III) the regional level, where the interactions between the international and the national remote sensing rules are exemplified by the regulations of the European Union (EU) on data and information disseminated through the Copernicus programme (see, in particular, Commission Delegated Regulation (EU) No. 1159/2013 of 12 July 2013 supplementing Regulation (EU) No. 911/2010 of the European Parliament and of the Council on the European Earth monitoring programme (GMES) by establishing registration and licensing conditions for GMES users and defining criteria for restricting access to GMES dedicated data and GMES service information, *OJ L* 309/1 of 19 November 2013).

International Level

Introduction

The activity of remote sensing in its broadest form can be defined as any act of gathering information about an object without making physical contact. A more narrow approach, however, defines the notion as the act of collecting and distributing information about phenomena on Earth with the aid of satellites. As such, it is subject, in the first place, to the rules of international law regulating activities in outer space, in particular the Outer Space Treaty (Treaty on Principles Governing the Activities of States in the Exploration and Use of Outer Space, including the Moon and Other Celestial Bodies of 27 January 1967, 610 *UNTS* 8843 (hereinafter: "Outer Space Treaty" or "OST")). The provisions of the Outer Space Treaty avoid any references to the collection of data from outer space, or indeed any other specific type of space activities, in favor of a general set of principles built on the fundamental notion of free exploration and use outer space. However, existing practice at the time of negotiations of the Outer Space Treaty indicates that States generally agreed on the legality of remote sensing activities as non-intrusive means of verification, transparency, and confidence-building. (The US Corona program, aimed at photographing the territory of the former USSR, was launched as early as June 1959 and lasted until 1972. For more on the early days of remote sensing, see Bosc 2015, 557–563.) Despite lacking, therefore, a legal basis explicitly confirming the legality of remote sensing, be it for scientific, commercial, or military reasons, the absence of any specific prohibition of an on-going activity should be taken as an indication that remote sensing is covered, in all its varieties, by the general freedom of States to explore and use outer space in Article I, para. 2 OST. At the same time, the codification in binding language of the principle that outer space is not subject to national appropriation by any means whatsoever in the same convention effectively deprived sensed States from invoking sovereignty as a legal basis for objecting to data collection of their territories from space. (Art. II OST. This distinguishes remote sensing in space law from air law, where the recognition of complete and exclusive sovereignty of States over the airspace above their territory effectively prohibits States from remotely sensing the territories of other States without their permission. See Art. 1 of the Paris Convention relating to the Regulation of Aerial Navigation of 13 October 1919, *Br. Treaty Series* 1923, No. 14; Art. 1 of the Chicago Convention on International Civil Aviation of 7 December 1944, 15 *UNTS* 295. Art. 36 of the Chicago Convention further provides that each contracting State "*may prohibit or regulate the use of photographic apparatus in aircraft over its territory.*")

Though the activity of placing satellites in an orbit around Earth with the purpose of collecting, for military reasons or other, data about the territories of States and areas outside national jurisdiction, was hence accepted, the related issue of the processing and dissemination of the data collected in that process continued to raise sensitive questions rooted in sovereignty and national interests. After a lengthy process of negotiation, marked by disagreement over both content and form, the Legal Subcommittee (LSC) and the plenary Committee on the Peaceful Uses of Outer Space (COPUOS) finally reached consensus on 15 Principles Relating to Remote Sensing of the Earth from Outer Space, which were subsequently adopted, also by consensus, by the UN General Assembly (UNGA) in 1986. (The item on remote sensing was first introduced in the agenda of the LSC in 1972: see Report of the Legal Sub-Committee on the Work of its Eleventh Session (10 April–5 May 1972), UN Doc. A/AC.105.101 of 11 May 1972, 3, agenda item 4 (c). For the history of the negotiations of the RS Principles, see Smith and Reynders 2015, 86–87) The discussions mainly centered on the legal form of the document and the extent of the application of the accepted principle of States' permanent sovereignty over natural resources to data about these resources, and pitched developed countries against developing countries on issues of prior consent and exclusive access. (See the UNGA Resolutions concerning Concerted Action for Economic Development of Economically Less Developed Countries, UN Doc. A/RES/1515 (XV) of 15 December 1960, and Permanent Sovereignty over Natural Resources, UN Doc. A/RES/1803 (XVII) of 14 December 1962.) Though ambiguous at points in their formulation and nonbinding in nature, the 1986 Principles were successful in ending a laborious process of negotiations and consultations by consensus, and as such should be the first stop in any analysis aiming to clarify the limits, if any, to remote sensing data collection and dissemination on security grounds.

Scope of Application and Security

One of the main points of contention that arose during the negotiations for the RS Principles concerned the scope of application of the 15 principles. As noted, remote sensing is not necessarily restricted to data collection of Earth from space, and different States had different ambitions for the instrument they were negotiating. Ultimately, the RS Principles Resolution defines its own scope of application both narrowly and broadly, in that it only applies to remote sensing of the Earth from outer space when performed for certain purposes, while including all phases of the data collection and dissemination process in the definition of these activities. As such, the RS Principles provides that the term "remote sensing activities," when used in the document, refers to *"the operation of remote sensing space systems, primary data collection and storage stations, and activities in processing, interpreting and disseminating the processed data."* (Principle I (e). According to Principle I (b), (c), and (d), "primary data" *"means those raw data that are acquired by remote sensors borne by a space object and that are transmitted or delivered to the ground from space by telemetry in the form of electromagnetic signals, by photographic film, magnetic tape or any other means; ['processed data'] means the products resulting from the processing of the primary data, needed to make such data usable; [and 'analysed information'] means the information resulting from the interpretation of processed data, inputs of data and knowledge from other sources."*) However, the term "remote sensing" in this definition exclusively refers to *"the sensing of the Earth's surface from space by making use of the properties of electromagnetic waves emitted, reflected or diffracted by the sensed objects, for the purpose of improving natural resources management, land use and the protection of the environment [...]"* (Principle I (a) RS Principles).

If the peculiar approach to delineate the scope of the RS Principles appears to undercut its relevance for the purpose of space security, a proper understanding of the document's genesis and the nature of remote sensing will suffice to reject this notion. The fact that the instrument defines the term "remote sensing" only with respect to natural resource management, land use, and the protection of the environment does not mean that the activity of remote sensing may only be performed for these three purposes, and that remote data collection of other State's territories for other aims such as security would be disallowed under international law. As the phrase "for the purpose of these principles" makes clear, the only implication of the definition as provided is that the 1986 Resolution will simply not, as such, apply to remote sensing activities performed for reasons other than those listed in the definition. (Compare Lee 2001, considering this a "creative" interpretation, though ultimately appearing to argue in favor of it.) Indeed, the definition of the term "remote sensing" in the RS Principles merely reflects the primary concerns of the States gathered to negotiate a principles document in a forum – COPUOS – that eschews discussions on national security interests and military activities in favor of principles of a general application or, in this case, tailored to natural resource management and related purposes. The definition should also be read against the background of the Outer Space Treaty, which does not distinguish in the formulation of its principle of free exploration and use between activities performed for military or other purposes.

The above does not mean that the RS Principles have no legal implications for remote sensing activities performed for military reasons, for two reasons. First, it will often be difficult to determine, in practice, whether a remote sensing activity is solely carried out for one or more of the purposes mentioned in the 1986 Resolution due to the inherently dual-use nature of many remote sensing satellite activities. The GMES programme on Global Monitoring for Environment and Security, developed jointly by the European Union and the European Space Agency and later renamed Copernicus, is a prominent example of a remote sensing project set up for purposes that, at the same time, fall within and outside of the scope of the RS Principles (see section "Copernicus Regulations: Scope" on Copernicus). Further, even when a program as such was not originally set up "for the purpose of" dual-use, data collected by satellites through purely civilian programs may also be used for military purposes, while conversely remote sensing operations of the military may result in civilian applications for the management of natural resources.

At this point, the RS Principles' narrow definition of "remote sensing" must be read in conjunction with the instrument's broad approach to "remote sensing activities." As noted, the latter concept includes the operation of space systems and the collection and storage of primary data, as well as activities in the processing, interpretation, and dissemination of such data. The sweeping nature of data collected by modern remote sensing activities means that it is often only at the stage of analyzing and interpreting the collected data with the aim of using and disseminating them that the nature of the activity may be classified as either for security or other purposes. It follows from a good faith interpretation of the RS Principles that they will need to be taken into account for the regulation of remote sensing activities whose design or implications combine one or more of the three purposes covered by the definition of "remote sensing" in the Principles with other purposes, including security, at least as far as the phase of data collection, storage, and, possibly, processing is concerned. It is only when the data collected is then analyzed, interpreted, and disseminated for purely security purposes that the activity would arguably no longer be covered by the RS Principles.

Secondly, and more importantly for the present chapter, it is not because a program of remote sensing is performed solely for the purpose of one of the three goals enshrined in Principle I (a) of the 1986 Resolution, and that the processed data and analyzed information following from that program is only in fact used for civilian purposes, that their dissemination would necessarily be bereft of any security implications. Such dissemination is undoubtedly and even under the strictest inter-pretations of the definitions of "remote sensing" and "remote sensing activities" covered by the RS Resolution. Hence, it is clear that an answer to the question of the precarious balance between free access to and dissemination of data and information collected by remote sensing programs for civilian purposes, and restrictions thereto for reasons of national security, will at least in part depend on the content of the RS Principles. The answer will ultimately remain ambiguous, however, for the princi-ples, while acknowledging the freedom to use outer space for the purpose of remote sensing, are silent on considerations of national security that may limit this freedom.

Free Use and Dissemination

Though the preamble of the Remote Sensing Principles does not explicitly refer to the Outer Space Treaty, its contents are clearly inspired by this convention. As such, the Principles are founded, in the first place, on the freedom of States to operate remote sensing satellite systems, and freely store, process, interpret, and disseminate data collected in the process (Achilleas 2008, 2). This follows from a combined reading of the definition of "remote sensing activities" in the 1986 Resolution with the explicit reformulation of and reference to Article I OST in its Principles II and IV. The latter Principle confirms the legality of remote sensing activities by noting that they "shall be conducted in accordance with the principles contained in [Article I OST], which, in particular, provides that the exploration and use of outer space shall be carried out for the benefit and in the interests of all countries, irrespective of their degree of economic or scientific development, and stipulates the principle of freedom of exploration and use of outer space on the basis of equality." Principle II para-phrases the remaining paragraph of Article I OST by providing that "[r]emote sensing activities shall be carried out for the benefit and in the interests of all countries, irrespective of their degree of economic, social or scientific and techno-logical development." To placate the developing countries, the Principle adds that the needs of these States must be taken "into particular consideration," without, however, identifying specific ways, including limitations to the above freedom, in which this should be done.

Other principles echo different provisions of the Outer Space Treaty. Principle III of the RS Resolution restates and clarifies Article III OST by requiring that remote sensing activities shall be conducted "in accordance with international law, includ-ing the Charter of the United Nations, the [Outer Space Treaty], and the relevant instruments of the International Telecommunication Union." Principle XIV is notable as well, for it confirms the application of Article VI OST to *"States operating remote sensing satellites,"* which shall hence assure that their activities, even when performed by nongovernmental entities, *"are conducted in accordance with [the RS Resolution] and the norms of international law."* Commercial owners and operators of remote sensing satellite systems must hence also abide by all applicable rules of international law by virtue of their State's responsibility for their activities. The reference in this context to "norms of international law" appears to expand the text of Article VI OST, which only refers to the *"provisions set forth in the [Outer Space Treaty],"* but ultimately is the logical result of a combined reading of this provision with Article III OST. More notable is that the phrasing of Principle XIV, in apparent deviation from the general scope of the RS Principles, only confirms the international responsibility under Article VI OST for the activities of *"States operating remote sensing satellites,"* rather than States performing "remote sensing activities." When read in conjunction with the requirement of authorization and continuing supervision by States of the activities of their nationals, Principle XIV could thus be interpreted restrictively as merely confirming the international responsibility of States for the operation, by nongovernmental entities, of remote sensing space systems, but not for the other aspects of the remote sensing activity as defined in the 1986 Resolution, i.e., the collection and storage of primary data and, crucially, the dissemination of the data resulting therefrom (compare Mantl 2015, 177–178). Regardless of this interpretation, however, States in their national laws will often go further and also require their remote sensing data providers to obtain permits for disseminating data, at which point a sensitivity check for reasons national security will be performed. (See infra on the German Satellite Data Security Act of 2007. It could be argued that States would still be responsible for the dissemination of data collected from remote sensing operations of their nationals under Art. VI OST, for that provision more generally refers to "national activities in outer space" and "the activities of non-governmental entities in outer space." However, such argument would rest on the extension of the broad definition of "remote sensing activities" in the RS Principles to the term "(national) space activity" in the Outer Space Treaty, which should not necessarily be accepted, in particular taking into account Principle XIV of the RS Principles.)

Security Limiting Dissemination

If the RS Principles clearly confirm the freedom of States to sense the Earth from space and store, process, and disseminate the data and information thus gathered, the instrument is less outspoken on the possible limitations to this freedom, for reasons of national security or other. The main provision aimed at harmonizing the positions of developed and developing States in this respect can be found in Principle IV, the latter part of which provides that remote sensing activities *"shall be conducted on the basis of respect for the principle of full and permanent sovereignty of all States and peoples over their own wealth and natural resources, with due regard to the rights and interests, in accordance with international law, of other States and entities under their jurisdiction.*" The specific phrasing – "*, which*" instead of "*, and which*" – suggests that the obligation of due regard for the rights and interests of other States and entities under their jurisdiction is limited, in the first place, to the context of States' and peoples' wealth and natural resources. However, the provision complements this by adding that "*[s]uch activities shall not be conducted in a manner detrimental to the legitimate rights and interests of the sensed State,*" thus requiring that the general rights and interests of other States be taken into account in the performance of remote sensing activities as covered by the RS Principles.

As Principle IV applies to "remote sensing activities," all aspects of such activities, including the dissemination of stored and processed data and information, may find limitations in the legitimate rights and interests of the sensed State. Principle XII expands on these rights of the sensed State specifically as regards its rights of access to data and information collected through remote sensing. According to this Principle, the sensed States shall have access to "*primary data and processed data concerning the territory under its jurisdiction,*" as soon as they are produced and "*on a non-discriminatory basis and on reasonable cost terms.*" In addition, the sensed States shall have access to "*available analysed information concerning the territory under its jurisdiction in the possession of any State participating in remote sensing activities on the same basis and terms, taking particularly into account the needs and interests of the developing countries.*" While Principle XII focuses on the rights of the sensed State, its provisions are grounded in assumptions of free access to and dissemination of data and information obtained by the sensing State or its nationals. The requirement to make primary and processed data available to the sensed State on a nondiscriminatory and reasonable cost basis implies that, in general, such data must be freely – though not *gratis* – accessible to everyone.

Principle XII could be interpreted as allowing some exceptions to the free dissemination of remote sensing information, as it explicitly requires sensing States to grant the sensed State access to processed information when "available" only. Though the security implications of remote sensing were undoubtedly well-known at the time of negotiation of the RS Principles, the text does not clarify how the availability of processed information may be restricted, and if the justifications for such restrictions could include reasons of national security. And, as noted, the Principles do contain a similar qualification when regulating the access to primary and processed remote sensing data. All of this should perhaps not be too surprising, since, during the negotiations of the 1986 Resolution, only classified military and intelligence-service satellites specifically developed for the purpose of government security applications were able to produce sufficiently high-quality data so as to have possible security implications. Access to these data was already closely regulated on the basis of specific governmental laws and policies. Hence, the principle of the free exploration and use of outer space could easily be confirmed as applying to all aspects of remote sensing activities in a document that, in any case, only addressed such activities for specific nonmilitary purposes.

This is not to say that the end result is satisfactory, or that the historic circumstances surrounding the conclusion of the Principles Resolution should necessarily have resulted in a text that leaves it to the reader to determine what exactly is the balance between the right of free access to and dissemination of remote sensing data, and limitations for reasons of national security. Compare, for example, the ambiguity of the 1986 Resolution in this respect with the provisions of the 1978 Convention on the Transfer and Use of Data of Remote Sensing of the Earth from Outer Space. (Adapted 19 May 1978, entered into force 21 August 1979. The text of the Convention is annexed to a letter of the USSR representative to the UN Secretary-General dated 28 June 1078, UN. Doc. A/33/162 of 29 June 1978.) Sponsored by the then USSR, the treaty entered into force a year later and was ratified by eight States. (These States are/were Cuba, Czechoslovakia, the German Democratic Republic, Hungary, Mongolia, Poland, Romania, and the USSR.) Even though the Convention is hence of limited legal relevance, it is noteworthy for the limitations it imposes on the distribution of potentially sensitive remote sensing data. In particular, we may refer to Article IV of the Convention, which provides that *"[a] Contracting Party in possession of initial data of the remote sensing of the Earth from outer space, with a better than 50 metres resolution on the terrain, relating to the territory of another Contracting Party, shall not disclose or make them available to anyone except with an explicit consent thereto of the Contracting Party to which the sensed territories belong, nor shall it use them or any other data in any way to the detriment of that Contracting Party."* While the 1978 Convention has, for all intents and purposes, passed into disuse, its approach has been partially replicated in national laws on remote sensing where they explicitly allow for limitations on the access to remote sensing data when its resolution is sufficiently high so as to potentially reveal information that may affect the State's national security. (See, for example, the Annex to EU Regulation 1159/2013).

Security as a National Exception to Principles of International Law

The manifest focus of the 1986 Principles on the free access to and dissemination of data gathered through remote sensing could raise the issue of conformity with this international instrument of such national security exceptions (see Lyall and Larsen 2009, 424–425). To be sure, the question may be of limited importance, since UNGA resolutions are legally nonbinding documents, and the formulation of the 1986 Principles as they have been adopted leaves ample discretion for States to implement its provisions in such a manner as they see fit. Nevertheless, it is commonly accepted that the Principles on Remote Sensing set an important legal precedent since they successfully managed to end years of negotiations in the form of a text that was adopted with consensus at all stages. More importantly, the principles are generally followed in State practice, even to the extent that they are sometimes believed to either declare or have become, at least in part, norms of customary international law. (See Answers from the Chair of the Space Law Committee of the International Law Association (ILA) to questions by the Chair of the Working Group of the LSC, UN Doc. A/AC.105/C.2/2015/CRP.25 of 22 April 2015: *"[u]nlike the OST, the 'Principles Relating to Remote Sensing of the Earth from Outer Space' is a non-binding instrument except when declaring customary international law.*" The Chair opines in this regard that Principle XIV of the 1986 Resolution addressing international responsibility is not of customary nature, for "*[t] his Principle would be confining the scope of Article VI of the OST to 'states operating remote sensing activities'*" (with reference to Cheng 1997, 572–597).) Whether national security exceptions in domestic laws are, then, in conformity with the 1986 Resolution depends as much on the interpretation of the latter document as on the uniformity of State practice that allows for exceptions in the name of national security. (The consistent inclusion of these exceptions in the national policies and laws of some of the most important spacefaring States has even prompted some to argue that the national security exception in turn could be considered a rule of CIL itself. See De Beer, June 2015.)

While the latter issue will be dealt with later on in this chapter, the matter of interpretation of the RS Principles as allowing, or not, restrictions on the access to and dissemination of information for reasons of national security should be answered in the affirmative. As noted, Principle IV requires that remote sensing activities "*shall not be conducted in a manner detrimental to the legitimate rights and interests of the sensed State.*" To the extent that the dissemination of remote sensing data and information is considered an activity subject to authorization under international space law, UNGA Resolution 68/74 of 2013 confirms that national security concerns may be taken into consideration in the implementation of this international principle (Recommendations on National Legislation relevant to the Peaceful Exploration and Use of Outer Space, UN Doc. A/RES/68/74 of 11 December 2013). The resolution confirms that the conditions for such authorization "*should be consistent with the international obligations of States, in particular under the United Nations treaties on outer space, and with other relevant instruments, and may reflect the national security and foreign policy interests of States*" (Paragraph 4 of the Resolution). Principle III, mirroring Art. III OST, also notes that remote sensing activities shall be performed in accordance with international law, in particular the Charter of the United Nations and the relevant instruments of the International Telecommunication Union (ITU). International law commonly accepts national security interests as "legitimate rights and interests," which should thus be taken into account by other States in their remote sensing activities. Moreover, such interests are also generally accepted exceptions to the fundamental freedom of information, as widely codified in both universally and regionally applicable instruments on human rights (see Achilleas 2001). Article 19 of the International Covenant on Civil and Political Rights (adapted 16 December 1966, 999 *UNTS* 171) defines the freedom of expression as the freedom of everyone to "*seek, receive and impart information and ideas of all kinds, regardless of frontiers, either orally, in writing or in print, in the form of art, or through any other media of his choice.*" While this right may hence be invoked to receive and impart remote sensing data and information, it is subject to restrictions when these are "*provided for by law and are necessary [. . .] [f]or the protection of national security or of public order (ordre public), or of public health or morals.*" (See further Art. 19 of the Universal Declaration of Human Rights, UN Doc. A/RES/217A (III) of 10 December 1948. For codifications of this right and security exceptions thereto at the regional level, see Art. 10 of the Convention for the

Protection of Human Rights and Fundamental Freedoms of 4 November 1950, 213 *UNTS* 222; Art. 13 of the American Convention on Human Rights of 22 November 1969, 1141 *UNTS* 123; Art. 9 of the African Charter on Human Rights and Peoples' Rights of 27 June 1981, 1520 *UNTS* 217.) Similar exceptions grounded in concerns for national security are recognized in the provisions of the ITU Constitution on the public's right to correspond through telecommunications as well. (While the ITU Constitution – a binding treaty ratified by 193 States – recognizes the right of the public to correspond by means of the international service of public correspondence, its *"Member States also reserve the right to cut off, in accordance with their national law, any [...] private telecommunications which may appear dangerous to the security of the State or contrary to its laws, to public order or to decency."* See Arts. 33 and 34 (2) of the Constitution and Convention of the International Tele-communication Union, most recently amended in 2018, published in the Basic Texts of the ITU, 2019.)

States may, thus, according to international law, restrict access to data collected through remote sensing for reasons of public security in their national legislation. Even if the RS Principles do not explicitly contain a provision in this regard, their general phrasing leaves room for discretion by States when promulgating domestic regulations on remote sensing, as is indeed confirmed by other, binding sources of international law on human rights and telecommunications. (It has been argued that States that do not provide for checks on the dissemination of remote sensing data would violate their obligations under Article VI OST: see Schmidt-Tedd and Kroymann 2008, 105. As noted, this argument relies on an extension of the definition of "remote sensing activities" of the RS Principles to the Outer Space Treaty.) Still, States are not granted unfettered discretion to determine at will the grounds for and means of restricting access to remote sensing data and informa-tion. International human rights and telecommunications law are adamant that exceptions to the freedom of expression are applied restrictively, and are only allowed when provided for by law, and when they are actually necessary to protect national security. Likewise, the RS Principles insist that the rights and interests of States must be legitimate in order to be taken into account by other States. Whether national laws conform to these requirements will depend on how the national security exception is phrased, and what actions authorities may take to safeguard these interests.

In the end, we shall see that most national laws providing for national security exceptions are clearly in line with the Principles and, indeed, have been inspired by the 1986 instrument (see section National Security Restrictions in the SDSA). (See the discussion of the applicable French and German laws in, respectively, Achilleas 2008, 2; Schmidt-Tedd and Kroymann 2008, 104–105.) Moreover, regulatory frameworks governing the access to remote sensing data at the regional level also confirm the importance of the RS Principles, which, in turn, encourage States to provide for the establishment and operation of data collecting and storage stations and processing and interpretation facilities, *"in particular within the framework of regional agreements or arrangements wherever feasible"* (see section Regional Level) (Principle VI).

National Level

Introduction

Despite a steep rise in recent years in the number of States that have adopted domestic space legislation, many countries active in space are still lacking a general space law framework. (For an overview of national regulatory frameworks for space activities, see the UN national space law collection at http://www.unoosa.org/oosa/en/ourwork/spacelaw/nationalspacelaw/index.html. See also the Space Legaltech platform of the French Sirius Chair, which provides an interactive overview of a much larger cross-section of national space legislation across the world: http://spacelegaltech.chaire-sirius.eu/.) Further, most national space laws that have been adopted focus on the authorization and supervision of national space activities in general, without touching on issues of remote sensing, or the dissemination of sensitive data and information collected by private commercial actors in this sector. To the extent that they are not covered by national rules in related fields such as export control, it follows that, for these countries, the collection and distribution of remote sensing data remain mainly regulated by international space and telecommunication law, if at all (Achilleas 2008, 4).

Those States that have promulgated rules on access to remote sensing data and information typically include provisions allowing for the restriction of access to and dissemination of data collected from remote sensing satellites and/or the information derived therefrom. Canada subjects licenses to operate remote sensing space systems to a sensitivity check designed to safeguard the national security interests and international obligations of Canada. (Act Governing the Operation of Remote Sensing Space Systems, Bill C-25 of 25 November 2005. Art. 8 (1) of the Act provides that "*the Minister may, having regard to national security, the defence of Canada, the safety of Canadian Forces, Canada's conduct of international relations, Canada's international obligations and any prescribed factors, (a) issue a provisional approval of a licence application; (b) issue a licence; or (c) amend or renew a licence.*" Such license may be suspended and subsequently cancelled if the continued operation of the licensed system would be injurious to these same interests or inconsistent with Canada's international obligations (Arts. 11 (1) and 12 of the Act). For the purposes of the Act, a remote sensing space "system" includes "*the facilities used to receive, store, process or distribute raw data from the satellites, even after the satellites themselves are no longer in operation*" (Art. 2).) Likewise, Title 51 of the US Code requires that licensees of private remote sensing space systems shall "*operate the system in such manner as to preserve the national security of the United States and to observe the international obligations of the United States*" (51 US Code Subchapter III – Licensing of Private Remote Sensing Space Systems, § 60122). The French Law on Space Operations requires remote sensing data providers to submit a declaration of their activities to the relevant administrative authorities, who must then verify that these activities do not damage the "fundamental interests of the nation," including in particular the national defense, foreign relations and international obligations of France. (Articles 23, para. 1 and 24, para. 1

of the Loi Relative aux Opérations Spatiales, N° 2008518, 3 June 2008, as most recently amended on 30 May 2013, at https://www.legifrance.gouv.fr/affichTexte.do?cidTexte=LEGITEXT000018939303. For a discussion of these provisions, see Achilleas 2008, 7–9.)

While most of these national laws provide only for a general basis for limiting space remote sensing operations for reasons of national security, foreign relations, or international obligations of the State, the actual regulation is left to the relevant authorities to decide at policy level. (Some States have only adopted remote sensing policies, without any specific legal basis. See, for example, the Indian Remote Sensing Data Policy of 2011, RSDP-01:2001, at https://nrsc.gov.in/Remote_Sens ing_Data_Policy. According to the RSDP, the Government of India reserves the general right to impose control over imaging tasks and the distribution of data from Indian remote sensing satellites when it is of the opinion that national security, the international obligations or the foreign policies of India so require.) Two recent exceptions are the German and Japanese laws dealing specifically with the security implications of remote sensing data. In 2016, Japan adopted the Act on Securing Proper Handling of Satellite Remote Sensing Records, according to which the Prime Minister can prohibit transfer of particular remote sensing data for a limited time if it is necessary for national security (Art. 19 of Act No. 77 of 2016, available with English translation at https://www8.cao.go.jp/space/english/rs/rs_act.pdf). Nine years prior to that, Germany promulgated the 2007 Act to give Protection against the Security Risk to the Federal Republic of Germany by the Dissemination of High-Grade Remote Sensing Data (Satellite Data Security Act, SDSA). Unlike its Japanese counterpart, the Satellite Data Security Act law did not complement a general national space law framework, as it constituted the first foray of the German legislator in space law regulation. (The Japanese Remote Sensing Records Act complements the country's 2008 Basic Space Law, which does not address remote sensing data handling as such: Act No. 43 of 28 May 2008, with unofficial English translation in 34 *J. Space L. 2008*, 471. On this law and its relevance for remote sensing, see Aoki 2010.) As a result, the Act takes a comprehensive approach that touches upon various aspects of the space activity of remote sensing, specifically addressed from a national security perspective.

Scope of the SDSA

In the preparation for the 2007 Satellite Data Security Act, the UN Principles on Remote Sensing were actively taken into consideration as "*the guiding principle for the practice of Earth remote sensing*" (Schmidt-Tedd and Kroymann 2008, 105). As an Act adopted by a key Member State of the European Union, the European Space Agency and NATO, the SDSA provisions are also careful to take into account the regional context. The Act is thus a uniquely useful tool for assessing the interactions between the regulations on remote sensing data dissemination at the national, international, and regional level. Like the Remote Sensing Principles, the SDSA applies both to the operation of high-grade Earth remote sensing systems and to the handling of data generated by such systems, until their moment of dissemination (§ 1, (1), 1 and 2 SDSA). Such data include the signals from satellite sensors as well as *"all products derived from the same, regardless of their degree of processing and their type of storage or representation"* (§ 2, (1), 2 SDSA). While the Act thus does not allow for a distinction between primary data, processed data and analyzed information, its overall scope matches the broad definition of the RS Principles' approach to "remote sensing activities," though for different objectives.

The SDSA also mirrors the UN Principles on Remote Sensing by excluding the application of the Act to military remote sensing activities, though the exception is phrased more narrowly and conditional upon actual protection of the information generated by such military activities. As such, the SDSA does not limit its scope of application to remote sensing operations undertaken for specific types of civilian purposes, but only excludes its application for *"the operation of high-grade earth remote sensing systems by a State agency with military or intelligence duties, provided that the possibility of unauthorized third parties gaining knowledge of the generated data is excluded"* (§ 1, (2) SDSA). Should the applicable legal framework or the specific contractual provisions for the operation of military remote sensing systems not meet this condition, the SDSA will remain applicable, regardless of the non-civilian nature of the activity. This discrepancy between the SDSA and the RS Principles may be explained by the different objectives of the two regulations – general principles on remote sensing versus security restrictions – and the need to fit the German Act into the existing patchwork of domestic laws, including export control regulations.

National Security Restrictions in the SDSA

As a domestic act concerned with the national security implications of remote sensing data for Germany, the Satellite Data Security Act naturally restricts its application to remote sensing activities having a clear link with Germany. For the handling of data, such link is established when the dissemination is done by German nationals or foreign legal persons with their head office in Germany, or when the data are disseminated from within the territory of Germany (§ 1, (1), 1 and 2 SDSA). The Act defines a "data provider" as any person who disseminates data generated by a high-grade Earth remote sensing system but limits "dissemination" to the initial act of *"bringing data into circulation or making data accessible to third parties"* (§ 2, (1), 3 and 6 SDSA). Hence, the SDSA only applies to initial data providers but not to remote sensing service providers or data resellers (Schmidt-Tedd and Kroymann 2008, 108). Moreover, such dissemination may only be subject to restrictions for security reasons if the data has a particularly high information content (§ 2, (1), 4 SDSA).

The determination of what exactly constitutes "particularly high information content" is left to the discretion of the Federal Ministry of Economics and Technology (§ 2, (2) SDSA). However, the conditions under which this determination is to be made are specified in the SDSA, in a provision that combines technical data with security considerations. As such, the information content of remote sensing data shall be determined not only according to the geometric, spectral, radiometric, and temporal resolution but also taking into consideration *"the possible effects of disseminating data with particularly high information content on the vital security interests of the Federal Republic of Germany, the peaceful co-existence of nations and the foreign relations of the Federal Republic of Germany"* (ibid.). For this reason, data providers wishing to distribute high information content data must apply for a dissemination license subject to conditions of reliability, and technical and operational measures to prevent unauthorized persons from gaining access to the installations for receiving, processing, or storing this data (§ 11, (1) and § 12, (1), 1 and 2 SDSA). Such dissemination license is separate from the license that the operator of a high-grade Earth remote sensing system must obtain under the SDSA (§ 3, (1) SDSA).

The SDSA thus takes national security interest into account for the determination of its scope, to the extent that the "particularly high information content" of generated data to which the Act applies must be assessed in light of their potential security sensitivity. However, the initial categorization of data as having particularly high information content is, as such, not a sufficient ground for restricting its dissemination by data providers. The SDSA imposes a double sensitivity check, to be performed by the data provider as well as the relevant authority, before high-grade data is disseminated, in a two-step process. First, data providers that receive a request for dissemination of data that falls under the Act have to perform a sensitivity check, taking into account not only the information content itself but also the represented target area, the ground segment, and the time between the data generation and possible dissemination (§ 17, (1) and (2) SDSA). The actual information content of data is thus but one factor that determines its sensitive nature, which rather follows from a combination of the information about a certain area and the timing of dissemination. (Schmidt-Tedd and Kroymann 2008, 107. See also infra, Art. 14 (1) (b) Regulation 1159/2013, according to which the sensitivity of remote sensing information shall be assessed based, inter alia, on the time between acquisition of inputs and dissemination of the information.)

It is only when the above factors combined reveal the possibility of harm being caused to the security interests of Germany as defined earlier, that a request for the dissemination of data shall be deemed sensitive. If the request is sensitive, the data provider is prohibited from complying, unless a dissemination permit is obtained, separate from the dissemination license noted earlier. Such permit must also be obtained if the data providers wishes to disseminate high information content data without a preceding request (§ 19, (1) SDSA). Such permit will only be granted if the dissemination of data *"in the individual case does not harm"* the vital security interests of Germany, does not disturb the peaceful co-existence of nations and does not substantially impair the foreign relations of Germany (§ 19, (2) SDSA). If this sensitivity check is not passed the authority may deny the request, or grant a permit on the condition that the request is altered, e.g., by ordering that the resolution is lowered or by imposing a time delay on the dissemination (Schmidt-Tedd and Kroymann 2008, 110). Moreover, the authorities may take other, proportionate measures against a data provider *"in the individual case to ensure the due performance of its obligations*," in particular by requiring that the dissemination of data to be adapted to the state of the art, or temporarily prohibiting the dissemination of data (§ 16 SDSA).

Though approaching the issue from a different angle, the actual provisions of the German Satellite Data Security Act confirm the UN Principles on Remote Sensing, both in their restrictive approach to imposing limitations on the free dissemination of data, and in the close circumscription of the justifications for such limitations in considerations of international peace and security. First, the SDSA implicitly confirms the principle of free use of outer space through remote sensing activities, including the free and nondiscriminatory dissemination of data therefrom, by requiring that a comprehensive sensitivity check is performed on a case-by-case basis at various intervals by different actors based on a variety of factors, rather than concluding from the general high information content nature of data gathered by a remote sensing program that its dissemination should always be restricted. In also requiring a dissemination license and a separate permit for the dissemination of individual sensitive data, the SDSA goes further than the RS Principles, by extending the authorization and continuing supervision from the operation of the remote sensing system to all aspects of the remote sensing activity as defined in the latter document. Finally, the restrictive and typically temporary measures that authorities can take when national security interests may be compromised indirectly confirms the basic principle of free data dissemination. (Other rules support this as well: see, for example, § 20 of the SDSA, allowing for the granting of collective permits for data dissemination, even if such permits are still subject to strict limits and conditions.)

Second, the careful manner in which the security interests are phrased is clearly in line with the limits to the free dissemination of data contained in Principles III and IV of the 1986 Resolution. Mindful of their exceptional nature, the security grounds in the SDSA have been closely circumscribed as justifying limits to the dissemination of remote sensing data only when they could damage the *vital* security interests of Germany or *substantially* impair the State's foreign relations. These qualifications ensure that the exceptions are in line with the limits to remote sensing implied by the obligation to have due regard for the legitimate rights and interests of States (see Schmidt-Tedd and Kroymann 2008, 105 ("*fully in line with the inherent limits of remote sensing and the rights resulting therefrom*")). The third and final ground for limiting the free access to remote sensing data under the SDSA, the peaceful co-existence of nations, undoubtedly conforms to the limits of general international law as applicable by virtue of Principle III of the RS Resolution. As noted, this Principle requires that remote sensing activities shall be conducted in accordance with international law, in particular the Charter of the United Nations. Article III OST adds to this that such application of norms of international law must be "*in the interest of maintaining international peace and security and promoting international cooperation and understanding*." The peaceful co-existence of nations is an obvious restatement of this fundamental concern of international law, and national exceptions to the free dissemination of remote sensing data are thus clearly in line with the limits of international law as incorporated in Principle III RS Principles and Article III OST.

Regional Context and the SDSA

The Satellite Data Security Act is overall very conscious of the regional and international security context in which it is to operate. This context is taken into account, in the first place, at the stage of application of the Satellite Data Security Act, which should be excluded when the laws of other countries already apply that are comparable to the provisions and protected interests of the Act. Though mainly concerned with avoiding unnecessary duplication of substantively similar regulations, the grounds for exclusion also take into account the strategic partnerships of Germany, by differentiating between the laws of the Member States of the European Union and third countries. As such, the SDSA provides that it *"may not be applied"* to the operation of a high-grade earth remote sensing system that is permitted under the applicable law of another EU Member State with substantively comparable provisions (§ 1, (2) SDSA). However, the responsible authority *"may waive the application of the Act"* if the legal provisions of a third country satisfy these same requirements and if there is an international treaty between Germany and this third country that affirms the comparability of the provisions and protected interests. (Ibid. This disparity may also be explained by requirements of internal EU law, which aims to eliminate as much as possible any obstructions to the free exchange of information between its Member States. See in this regard also the preamble to EU Regulation 1159/2013, which provides that *"GMES data and information policy should [. . .] respect the rights and principles recognized in the Charter of Fundamental Rights of the EU, in particular [. . .] the freedom to conduct business."*)

The regional and international context is also taken into account in the performance of the sensitivity check for data whose dissemination has been requested. In performing this test, the administrative authority must take into account, inter alia, *"the obligations assumed and agreements entered into by [Germany] with the Member States of the European Union, the parties to the North Atlantic Treaty of April 4, 1949, [. . .] and Australia, Japan, New Zealand and Switzerland, [as well as] the existing rules under which the requesting party could further transmit the data and the availability of comparable data on international markets"* (§ 17, (3) SDSA). Through this provision, the regional interests of allies are integrated in the sensitivity check to be performed by the national authority under the SDSA. Moreover, the Act also requires, in general, that the personal characteristics of the requesting party must be taken into account when performing a sensitivity check (§ 17, (2) SDSA). It follows that requests for information by EU and NATO Member States will typically not be deemed sensitive. (Schmidt-Tedd and Kroymann 2008, 113: *"[a]s a result [of the criteria for the sensitivity check], data requests of NATO member states will not be sensitive in most instances."*) However, the interests of regional partners or allies are still not grounds for a blanket denial *c.q.* approval of requests to disseminate remote sensing data, as they are only one of many factors to be taken into account in an individualized security test.

Finally, it may be added that security interests and regional alliances are not only justifications for restrictions on the free dissemination of remote sensing data but are also taken into account by the German legislator as grounds for priority treatment of requests for such dissemination. As such, a data provider is obliged to give priority to complying with requests for the dissemination of data from the German government, *"in the event of the casus foederis [defensive alliance] in accordance with Article 5 of the North Atlantic Treaty [. . .], in case of defence [. . .], if the requirements for the internal state of emergency [. . .] are satisfied, in the event of tension [. . .] or if there is a current danger to military or civil forces of [Germany] deployed in a foreign country or to employees of the diplomatic service [. . .]"* (§ 21 SDSA). While these grounds are more specific than the grounds for restricting access to remote data, they may also appear slightly broader. However, this should not concern us when considering the conformity of the SDSA with the RS Principles, as the security interests listed may not be invoked as justification for limiting the dissemination of remote sensing data, but rather require the data provider to provide such data on a priority basis (see also § 22 SDSA).

Regional Level

Introduction

The Satellite Data Security Act proved a useful example of the regulation of the dissemination of remote sensing data at the national level, for the comprehensive manner in which it approaches this issue. The Act was also highlighted because it was adopted by a State whose regulations are deeply entwined with those of a key regional player in space, the European Union. The EU manages Copernicus – formerly known as GMES (Global Monitoring for Environment and Security) – a pioneering Earth observation program for both environmental and security purposes, implemented in close partnership with the EU Member States and the European Space Agency. The Copernicus remote sensing space system has a uniquely dual composition, comprised, on the one hand, of dedicated satellites developed by the European Space Agency named Sentinels, and so-called contributing mission satellites provided by the Member States and third parties, on the other (see the Copernicus contributing missions overview page at https://spacedata.copernicus.eu/web/cscda/missions).

As an EU Member State with advanced national remote sensing programs, Germany contributes to Copernicus through its TerraSAR-X satellite and its "twin" TanDEM-X, as well as the RapidEye 5-satellite constellation. (See the Copernicus page for TerraSAR-X at https://spacedata.copernicus.eu/web/cscda/missions/terrasar-x. The satellite has been registered by Germany in the UN Space Objects Register: UN Doc. ST/SG/SER.E/526 of 29 February 2008; See the Copernicus page for TanDEM-X at https://spacedata.copernicus.eu/web/cscda/

missions/tandem-x. The satellite has been registered by Germany in the UN Space Objects Register: UN Doc. ST/SG/SER.E/608 of 30 November 2010; See the Copernicus page for the RapidEye mission at https://spacedata.copernicus.eu/web/cscda/missions/rapideye. The five satellites of the constellation have all been registered by Germany in the UN Space Objects Register: UN Doc. ST/SG/SER.E/569 of 16 July 2009.) The dissemination of data from these satellites is hence subjected, in theory, to a dual national and regional legal regime. (The data disseminated by these satellites will only be subject to the Satellite Data Security Act if it is handled, as noted, *"by German nationals or by legal persons or associations of persons under German law, b) by foreign legal persons or foreign associations of persons with their head office within the territory of [Germany], or c) where the data are disseminated from within the territory of [Germany]"* (§ 1, (2) SDSA). TerraSAR-X and Tan-DEM-X were both financed and implemented as a public-private partnership between DLR German Aerospace Center and Airbus Defence and Space. See the page for these satellites at the DLR website at https://www.dlr.de/dlr/en/desktopdefault.aspx/tabid-10378/566_read-426/#/gallery/345. The German branch Airbus DS Geo GmbH holds the exclusive commercial exploitation rights for both TerraSAR-X and TanDEM-X, while *"DLR is responsible for providing [. . .] data to the scientific community, mission planning and implementation, radar operation and calibration, control of the two satellites"* (ibid.). As for RapidEye, however, the issue is less straightforward. While the constellation of five satellites was registered by Germany and initially owned and operated by the German company RapidEye AG, ownership and operation of the constellation was later transferred following the acquisition of RapidEye AG to Canadian company BlackBridge in 2011 (prompting a name change from RapidEye AG to BlackBridge in 2013), and subsequently to US Planet Labs in 2015. As Planet Labs has its head office in San Francisco, the continued application of the SDSA depends on whether the data collected by the RapidEye satellites is disseminated from within German territory. Despite the transfer in ownership of the constellation, the RapidEye ground segment and spacecraft control centre remain located in Berlin, though the S-band and X-band downlink stations are located in Svalbard, Norway. See the information on the RapidEye constellation at the Earth Observation Portal Directory at https://directory.eoportal.org/web/eoportal/satellite-missions/r/rapideye. Hence, it would appear that the SDSA continues to apply to the dissemination of the data generate by the RapidEye constellation.) It was noted, however, that the German Satellite Data Security Act may not be applied to the operation of high-grade Earth remote sensing systems that are permitted under the applicable law of another Member State of the EU. Presumably this should include programs such as Copernicus that are owned and managed by the EU, and implemented by the Member States, at least as far as the contributing missions of other Member States are concerned. Yet the SDSA will only be set aside if the applicable rules contain comparable provisions and protected interests, which requires taking a closer look at the applicable Copernicus regulations. As we will see, these regulations clearly distinguish, in the elaboration of their data policy, between dedicated and contributing mission data, thus minimizing the potential for overlap between the national and regional regulatory level.

Copernicus Regulations: Scope

The basic instruments governing Copernicus data and information are Regulation 377/2014 establishing the Copernicus Programme (Regulation (EU) No 377/2014 of the European Parliament and of the Council of 3 April 2014 establishing the Copernicus Programme and repealing Regulation (EU) No 911/2010, *OJ L* 122/44 of 24 April 2014) and Regulation 1159/2013 defining, inter alia, criteria for restricting access to such data and information. As is clear from the program's original name – Global Monitoring for Environment and Security – the scope of the remote sensing activities included in Regulation 377/2014 is wider than that of the UN Remote Sensing Principles. The Copernicus service component consists of services related to atmosphere monitoring, marine environment monitoring, land monitoring, climate change, emergency management, and, finally, the security service (Art. 5 (1) Regulation 377/2014). The objective of the Copernicus security component is "*to provide information in support of the civil security challenges of Europe improving crisis prevention, preparedness and response capacities, in particular for border and maritime surveillance, but also support for the Union's external action, without prejudice to cooperation arrangements which may be concluded between the Commission and various Common Foreign and Security Policy bodies, in particular the European Union Satellite Centre*" (Sub (f) of Art. 5 (1) Regulation 377/2014). While it is clear that this service does not, as such, fall under the scope of the RS Principles, the dual-use nature of Copernicus makes it difficult to carve the security component entirely out of the general regulations governing the program. This is clear from the definitions of the terms "Copernicus data" and "Copernicus information" – a distinction reflecting the RS Principles but not found in the SDSA – which do not differentiate between data and information collected for security or other services. (Art. 3, (7) and (8) Regulation 377/2014: "*'Copernicus data' means dedicated mission data, contributing mission data and in situ data; 'Copernicus information' means information from the Copernicus services referred to in Article 5(1) following processing or modelling of Copernicus data [. . .].*")

Open Access Policy and Security Restrictions

The EU regulation of Copernicus data and information is strongly rooted in the general open data policy developed by the Union in a string of instruments. As such, Regulation 1159/2013 notes that GMES (Copernicus) data and information should "*strongly contribute to*" ((2) of the preamble of Regulation 1159/2013) the open data policy promoted by the EU in the 2003 Directive on the Re-use of Public Sector Information (Directive 2003/98/EC of the European Parliament and of the Council of 17 November 2003, *OJ L* 345/90 of 31 December 2003), the 2010 Commission Communication "A Digital Agenda for Europe" (Commission Communication COM(2010) 245 final/2 of 26 August 2010. See also preambular paragraph (9) of Regulation 377/2014: "*Copernicus data and Copernicus information should be available freely and openly to support the Digital Agenda for Europe, as referred* to in the Commission Communication of 26 August 2010 entitled: A Digital Agenda for Europe."), and the 2011 Commission Decision on the Reuse of Commission Documents. (Commission Decision 2011/833/EU on the Reuse of Commission Documents of 12 December 2011, *OJ L* 330/39 of 14 December 2011.) Though the application of these EU policy documents is confirmed for all Copernicus data and information, Regulation 1159/2013 only applies to "GMES dedicated data" and "GMES service information," while the relevant provisions of Regulation 377/2014 on data policy likewise single out "dedicated mission data" and "Copernicus information" (Art. 23 Regulation 377/2014). Dedicated mission data refers to *"spaceborne Earth observation data from dedicated missions for use in Copernicus"* (Art. 3, (7) Regulation 377/2014. Compare the definition for "GMES dedicated data" in Regulation 1159/2013 as referring to *"data collected through the GMES dedicated infrastructure and their metadata"* (Art. 2 (c)).), thus contrasting it with contributing mission data such as data generated by the German TerraSAR-X and TanDEM-X missions. However, the open access policy does extend to all "Copernicus information," which should be understood as all *"information from the Copernicus services referred to in Article 5 (1) following processing or modelling of Copernicus data"* (Art. 3, (8) Regulation 377/2014. See also Art. 2, (a) and (b) Regulation 1159/2013). The latter term in turn refers to a combination of *"dedicated mission data, contributing mission data and in situ data."* (Art. 3, (7) Regulation 377/2014. "In situ data" means *"observation data from ground-, sea- or air-borne sensors as well as reference and ancillary data licensed or provided for use in Copernicus"* (Art. 3, (5) Regulation 377/2014).) Thus, while the access policy and possible restrictions thereto elaborated in Regulations 1159/2013 and 377/2014 will apply to all information produced on the basis of all Copernicus data, the availability of the data from contributing missions that may be used as the basis for such information is still subject to the respective national regulations.

The basic rule governing access to dedicated data and information of Copernicus is laid down in Article 23, (2) of Regulation 377/2014. According to this provision, *"[d]edicated mission data and Copernicus information shall be made available through Copernicus dissemination platforms, under pre-defined technical conditions, on a full, open and free-of-charge basis."* However, such full and open access is subject to a number of predefined restrictions, including the *"security interests and external relations of the Union or its Member States"* (Art. 23, (2), (c) Regulation 377/2014). The precise nature and application of this restriction is elaborated in Regulation 1159/2013. After confirming that the open dissemination principle applies to all users of "GMES dedicated data" and "GMES service information," and noting that such data and information *"may be used worldwide without limitations in time"* (Arts. 3 and 7, (2) Regulation 1159/2013), Articles 11 and 16 of this Regulation cover a limited number of grounds on which the access to this data and information may be restricted.

According to Article 12 of Regulation 1159/2013, the Commission *"shall restrict [the] dissemination [of dedicated mission data and Copernicus information]"* only when their open dissemination *"presents an unacceptable risk to the security interests of the Union or its Member Stats due to the sensitivity of the data and information.*" This sensitivity must be assessed by the Commission according to a predefined set of criteria. Like the Satellite Data Security Act, these criteria combine technical characteristics of the data and the time between the acquisition and dissemination with the existence of closely circumscribed situations affecting national and international security (Arts. 13 and 14 Regulation 1159/2013). These situations specifically refer to "*the existence of armed conflicts, threats to international or regional peace and security, or to critical infrastructures [, and] the existence of security vulnerabilities or the likely use of [dedicated mission data or Copernicus information] for tactical or operational activities harming the security interests of the Union, its Member States or international partners*" (Sub (c) and (d) of Arts. 13 and 14 Regulation 1159/2013). The explicit reference to international peace and security, and the inclusion of qualifiers such as "critical" when describing the security interests make clear that the Regulation is in line with the legitimate rights and interest of States that may limit the open dissemination of Copernicus data and information under international law. (The assessment of the security implications is made, inter alia, against the backdrop of the rules on EU classified information. According to Art. 25, (5) of Regulation 377/2014, "*[w]here EU classified information is generated or handled within Copernicus, all participants shall ensure a degree of protection equivalent to that provided by the rules set out in the Annex to Decision 2001/844/EC and in the Annexes to Decision 2013/488/EU.*" The latter documents define "EU classified information" – which, as a legal term of art, should be distinguished from the technical term "information" in the Copernicus regulations – as "*any information or material designated by a EU security classification, the unauthorised disclosure of which could cause varying degrees of prejudice to the interests of the European Union or of one or more of the Member States*" (Art. 2 (1) of Council Decision 2013/488/EU of 23 September 2013 on the Security Rules for Protecting EU Classified Information, *OJ L* 274/1 of 15 October 2013.)

Unlike the SDSA, which does not distinguish between data and information, Regulation 1159/2013 differentiates between dedicated data collected by the Sentinels and Copernicus information for the purposes of the sensitivity check. While the sensitivity check is always mandatory for Copernicus *information*, the *dedicated data* will only be subject to such test if produced by a space-based observation system capable of generating data of particularly high resolution, as defined in a separate annex to the Regulation. (Art. 13, (1) Regulation 1159/2013. The Regulation's annex on "Characteristics of Space-based Observation System as Referred to in Article 13" defines these systems as being technically capable of generating data of geometric resolution of 2.5 m or less in at least one horizontal direction, of 5 m or less in at least one horizontal direction in the 8–12 μ spectral range (thermal infrared), of 3 m or less in at least one horizontal direction in the spectral range from 1 mm to 1 m (microwave), and of 10 m or less in at least one horizontal direction in at least one spectral channel.) And it is only where dedicated mission data are produced by a space-based observation system which does not meet *any* of the characteristics listed in the Annex, that they shall be presumed not to be sensitive (Art. 14, (2) Regulation 1159/2013). This distinction makes sense, for the sensitivity of remote sensing *information*, is not directly proportionate to the resolution of the remote sensing observation system, unlike raw satellite data before processing and interpretation.

The sensitivity check is performed by the Commission, though it must take into account the national interests of the Member States of the EU in doing so. The preamble of Resolution 1159/2013 adds in this regard that restrictions to the open dissemination of dedicated mission data and Copernicus information *"should respect the obligations of Member States that have adhered to a common defence organisation under international treaties"* (PP (13) Regulation 1159/2013). This provision calls to mind the requirement in the SDSA for the authorities to take into account the obligations of Germany under the North Atlantic Treaty when performing the sensitivity check. Despite this integration of the concerns of Member States, the outcome of the sensitivity check may not be challenged by them, unless the conditions under which the assessment were made have changed. Only in that case may the Commission reassess the situation, either at its own initiative or at the request of Member States with a view to restricting, suspending or allowing the acquisition of dedicated mission data or the dissemination of Copernicus information (Art. 15 Regulation 1159/2013). Nevertheless, the Member States retain a final form of control through the Council, which, despite the overall responsibility of the Commission for the Copernicus security framework, *"shall adopt the measures to be taken whenever the security of the Union or its Member States could be affected by data and information provided by Copernicus."* (Art. 25, (4) Regulation 377/ 2014. See also preambular paragraph (40) of Regulation 377/2014: *"As some Copernicus data and Copernicus information, including high-resolution images, may have an impact on the security of the Union or its Member States, in duly justified cases, the Council should be empowered to adopt the measures in order to deal with risks and threats to the security of the Union or its Member States."*)

In addition to national interests of the Member States, the Commission must also take into consideration the interests of the Copernicus users and the importance of the program for the broader society. As such, Regulation 159/2013 requires that *"security interests shall be balanced against the interests of users and the environmental, societal and economic benefits of the collection, production and open dissemination of the data and information in question."* (Art. 16 (1) Regulation 1159/2013. (2) of the same Article requires the Commission to also consider, when making its security assessment, *"whether restrictions will be effective if similar data are in any event available from other sources."* Similar considerations must inform the outcome of the test under the SDSA.) Such general concern for the overall situation provides for a more comprehensive assessment under EU law than is the case for most national remote sensing regulations. The broader concern than mere security considerations in Regulation 1159/2013 is also apparent from the fact that access to dedicated data and Copernicus information must also be restricted when their open dissemination would conflict with other rights, including when such dissemination would *"affect in a disproportionate manner the rights and principle recognised in the Charter of fundamental Rights of the EU, such as the right for private life or the protection of personal data"* (Art. 11 Regulation 1159/2013).

Sentinel Data

While all Copernicus information is subject to the same rules governing their open dissemination and possible restrictions, the data generated by the satellites in the Copernicus constellation will be subject to a different regime depending on whether they are acquired by Sentinels or contributing missions, or provided by third parties. As noted, Regulations 1159/2013 and 377/2014 apply only to dedicated mission data. In addition, the data generated by the ESA-developed Sentinels are also covered by a separate Sentinel data policy set jointly by the EU and ESA in 2013. (Preambular paragraph (38) of Regulation 377/2014 provides that *"[t]he access rights to Copernicus Sentinel data granted under the GMES Space Component Programme as approved by the ESA Programme Board on Earth Observation on 24 September 2013 should be taken into account."*) This policy is in turn based on the Joint Principles for a Sentinel Data Policy developed by ESA and the GMES Bureau in 2009 (ESA/PB-EO(2009)98, rev. 1 of 23 October 2009). Pursuant to these Joint Principles, the Sentinel Data Policy may not distinguish between access for public, commercial, or scientific users, or between European or non-European users; the licenses for the Sentinel data shall be free of charge; and the Sentinel data will be made available online, free of charge, subject only to a generic registration process ("Joint Principles for a Sentinel Data Policy," Annex to ESA/PB-EO(2009)98, rev. 1 of 23 October 2009, 1). The Joint Principles do foresee possible security restrictions that may affect data availability or timeliness, but limit their reach by requiring that they only be implemented for *"specific Sentinel data,"* and result in the activation of *"specific operational procedures"* (ibid., 1–2).

Following the adoption of the Joint Principles, the European Space Agency consolidated its data policy provisions for its various remote sensing missions (ERS or European Remote Sensing, Envisat and Earth Explorer) in a single document, and adapted its existing Earth Observation Data Policies to the content of the Joint Principles (see European Space Agency, "ESA Data Policy for ERS, Envisat and Earth Explorer missions," October 2012, at https://earth.esa.int/c/document_library/get_file?folderId=296006&name=DLFE-3602.pdf). The consolidated ESA policy is based on the principle that, though ownership of data provided by the ERS, Envisat, and Earth Explorer missions remains with ESA, all these data *"shall be available in an open and non-discriminatory way"* (ibid., 1). To be sure, this principle of free and open access is subject to some restrictions, as the policy distinguishes between free datasets and retrained datasets. However, the restrictions to the latter are explicitly *"due to technical constraints"* only, and are justified by the fact that *"data are not systematically processed and made available on-line"* (ibid., 2). In light of the technical reasons for these restrictions, access requests to restrained datasets are reviewed based on factors of relevance and feasibility rather than considerations of national security. Data from restricted datasets remains free of charge, unless the request concerns very large datasets, in which case a contribution may be asked.

The applicable EU and ESA documents demonstrate a clear consistency of both regional organizations' remote sensing data policies with the UN Remote Sensing

Principles. ESA explicitly declared as much, by noting that the *"distribution of [data provided by the ERS, Envisat and Earth Explorer missions] shall be consistent with the United Nations Resolution A/RES/41/65 dated 3 December 1986 on Principles relating to Remote Sensing of the Earth from Space"* (ibid., 1). In defining the scope of application of its policy, ESA also stipulated that it applies to *"all primary and processed data (up to level 2) as defined according to the UN terminology (UN resolution 41/65)."* This is also in keeping with the Joint Principles elaborated with the EU, which state that the UN Resolution on Remote Sensing should be taken into account as one of *"the instruments and documents [that] provide the legal and programmatic framework for the Sentinel Data Policy"* (Joint Principles for a Sentinel Data Policy, 2). Specifically, the EU-ESA Copernicus programme is also clearly in line with Principle VI of the UN Resolution, which recommends that States provide for the establishment and operation of data collecting and storage stations, and processing and interpretation facilities for remote sensing data, *"in particular within the framework of regional agreements or arrangements whenever feasible."* (See also preambular paragraph 6 of Regulation 377/2014 (*"To promote and facilitate the use of Earth observation technologies both by local authorities and by small and medium-sized enterprises (SMEs), dedicated networks for Copernicus data distribution, including national and regional bodies, should be promoted"*).)

Contributing Mission Data

As noted, Regulation 377/2014 makes a terminological distinction between dedicated mission data and contributing mission data, and subsequently applies its provisions on data security policy to the former only. Likewise, Regulation 1159/2013, as per its title, only defines the criteria for restricting access to "GMES dedicated data and GMES service information." It follows that different rules may govern the access to and distribution of contributing mission data, compared to Sentinel data. The same terminological distinction is retained in Commission Implementing Decision 2018/621 on the technical specifications for the Copernicus space component. (Commission Implementing Decision (EU) 2018/621 of 20 April 2018 on the Technical Specifications for the Copernicus Space Component pursuant to Regulation (EU) No 377/2014 of the European Parliament and of the Council, *IJ L* 102/56 of 23 April 2018.) This Decision, which defines the Copernicus space component as including data dissemination, and in particular the provision, archiving, and dissemination of contributing mission data (see Sections 1 and 2.1 of the Annex to the Decision), distinguishes between Dedicated Copernicus Missions (Sentinels) data and data from Copernicus Contributing Missions (CCMs) (Annex Sections 3 and 4, respectively). While the Decision confirms the open access policy for the former, it notes that *"[l]icensing conditions for data shall be negotiated with contributing mission data providers for data that needs to be procured. These licensing conditions could depart from the open data policy."* (Annex Section 2.5, (a): *"users have a free, full and open access to Copernicus Sentinel Data and Service Information without any express or implied warranty, including as regards quality and suitability for any purpose*"; Annex Section 4.3. Section 4.1 also provides that "*[d]ata from CCMs could be either free of charge or could be procured under specific licensing conditions.*") To what extent the legal regulation for access to contributing missions data differs from the general EU data policy will hence depend, in practice, on the licensing terms agreed with the providers of such CCM data. These terms will depend, in part, on the municipal rules under which the data provider is operating its remote sensing activities, and may hence impose stricter limitations for reasons of national security.

Conscious of the dependency of Copernicus on the contributing missions of Member States, the Commission has committed itself to cooperating with Member States "*in order to improve the exchange of data and information between them [. . .] to ensure that the required data and information are available to Copernicus.*" (Art. 13 Regulation 377/2014. The provision adds that "*[t]he Member States' contributing missions, service and in situ infrastructures are essential contributions to Copernicus.*" Still, CCMs are officially intended only as a complementary part of Copernicus, in that "*[d]ata from CCMs shall be obtained by Copernicus to fulfil the data requirements [. . .], whenever these cannot be met by the Sentinels*" (Section 4.1 of Decision 2018/621).) This is an application of the fundamental principle of sincere cooperation enshrined in Article 4, (3) of the Treaty on European Union (TEU), pursuant to which the Union and its Member States shall, "*in full mutual respect,*" assist each other in carrying out tasks that flow from the EU Treaties. In theory, the mutuality criterion underlying this principle also limits the discretion of Member States to restrict the remote sensing data flow under Copernicus when a more open dissemination would be considered necessary to allow the EU to perform its tasks. However, it is also an accepted principle of EU law that "*[t]he provisions of the Treaties shall not preclude the application of the [rule that] no Member State shall be obliged to supply information the disclosure of which it considers contrary to the essential interests of its security.*" (Art. 346 (1) of the Treaty on the Functioning of the European Union (TFEU). To the extent that national restrictions on the free dissemination of remote sensing data would be considered as limiting the free traffic between EU Member States, Art. 36 TFEU stipulates that the general prohibitions on quantitative restrictions "*shall not preclude prohibitions or restrictions on imports, exports or goods in transit justified on grounds of public morality, public policy or public security.*" See also Achilleas 2008, 9.) This follows from the fundamental notion that national security remains the sole responsibility of each EU Member State. (Art. 4, (2) TEU ("*The Union shall respect the equality of Member States before the Treaties as well as their national identities, inherent in their fundamental structures, political and constitutional, inclusive of regional and local self-government. It shall respect their essential State functions, including ensuring the territorial integrity of the State, maintaining law and order and safeguarding national security. In particular, national security remains the sole responsibility of each Member State*").) For this reason, the Copernicus regulations explicitly note that the means of disseminating all Copernicus data and information – including dedicated mission data – and the criteria and procedures for restricting access thereto may only be regulated by the Commission "*respecting third party data and information policies and without prejudice to rules and procedures applicable to space and in situ infrastructure under national control or under control of international organisations*" (Art. 24 (1) (a) and (c) Regulation 377/2014).

As illustrated by the German Satellite Data Security Act, such national laws may well require that restrictions to the dissemination of data should take into account existing obligations of the State under EU law. However, the explicit deference of both the founding treaties of the EU and the specific Copernicus regulations to the national security of Member States means that Member States will retain a large measure of discretion to withhold remote sensing data that it considers detrimental to its own interest. This does not mean, however, that the same data, or comparable data with similar security implications, could not be acquired by dedicated Copernicus Sentinels. To be sure, like all acts of the EU, the provision of Copernicus services must take into account the general principles of subsidiarity and proportionality (Art. 5 TEU). Regulation 377/2014 therefore provides that "*[p]rocurement of new data that duplicate existing sources shall be avoided, unless the use of existing or upgradable data sets is not technically feasible, cost-effective or possible in a timely manner*" (Art. 5, (2) of the Regulation). Theoretically, however, information withheld by contributing mission partners, for reasons of national security or other, fall under this category, since their use is not "possible in a timely manner." If the Copernicus Sentinels may hence still conceivably acquire these data – and the sweeping nature of remote sensing satellites may well make this unavoidable – their dissemination remains subjected to the general sensitivity check of Copernicus dedicated data.

Conclusion

This chapter reviewed a selection of key instruments adopted to regulate the dissemination of Earth data collected by remote sensing satellites at the international, national, and regional levels. While these instruments naturally differ in terms of regulatory focus, they demonstrate sufficiently broad similarities in scope, concepts, and substantive rules so as to reveal a shared approach to the regulation of remote sensing activities.

At the international level, the 1986 UN Remote Sensing Principles focus on the freedom of States and their private actors to collect, store, and distribute remote sensing data and information, without explicit exceptions for reasons of national security. This is perhaps understandable, considering their elaboration by the Committee on the Peaceful Uses of Outer Space as a means to clarify the application of the Outer Space Treaty, whose key principles serve to safeguard the freedom of States to explore and use outer space. However, the Principles' integration of the legitimate rights and interests of other States, as well as a general concern for international peace and security, reveals a broader context of international law that allows for an implied exception of national security to the free dissemination of remote sensing data.

This implication is rendered explicit at the national level, as illustrated, in this chapter, by the German Satellite Data Security Act of 2007. Though the Act focuses on restrictions to the free dissemination of remote sensing data, the careful formulation of these limitations indirectly confirms the free access thereto as their underlying principle. The SDSA also confirms the importance of regional integration and alliances as an important factor restricting limitations to free remote sensing data dissemination, rooted in concerns for international peace and security. In turn, the regional regulation of remote sensing, as analyzed through the lens of the rules and policies of the Copernicus programme, confirm the fundamental principle of open access to remote sensing data, while incorporating the security concerns of individual Member States of the EU. As such, instruments at both the national and regional levels not only confirm the abiding importance of the nonbinding UN Remote Sensing Principles but also clarify their application by means of a detailed set of rules and policies that aim to strike a balance between public security and public access to information in a free market.

References

International Instruments

Treaties

African Charter on Human Rights and Peoples' Rights of 27 June 1981, 1520 *UNTS* 217
American Convention on Human Rights of 22 November 1969, 1141 *UNTS* 123
Chicago Convention on International Civil Aviation of 7 December 1944, 15 *UNTS* 295
Constitution and Convention of the International Telecommunication Union (2018 version), published in the Basic Texts of the ITU, 2019
Convention for the Protection of Human Rights and Fundamental Freedoms of 4 November 1950, 213 *UNTS* 222
Convention on the Transfer and Use of Data of Remote Sensing of the Earth from Outer Space of 19 May 1978, annexed to UN. Doc. A/33/162 of 29 June 1978
International Covenant on Civil and Political Rights of 16 December 1966, 999 *UNTS* 171
Paris Convention relating to the Regulation of Aerial Navigation of 13 October 1919, *Br. Treaty Series* 1923, No. 14
Treaty on Principles Governing the Activities of States in the Exploration and Use of Outer Space, including the Moon and Other Celestial Bodies of 27 January 1967, 610 *UNTS* 8843

UNGA Resolutions

Concerted Action for Economic Development of Economically Less Developed Countries, UN Doc. A/RES/1515 (XV) of 15 December 1960
Permanent Sovereignty over Natural Resources, UN Doc. A/RES/1803 (XVII) of 14 December 1962

Principles Relating to Remote Sensing of the Earth from Outer Space, UN Doc. A/RES/41/65 of 3 December 1986

Recommendations on National Legislation relevant to the Peaceful Exploration and Use of Outer Space, UN Doc. A/RES/68/74 of 11 December 2013

Universal Declaration of Human Rights, UN Doc. A/RES/217A (III) of 10 December 1948

UN Documents

Answers from the Chair of the Space Law Committee of the International Law Association (ILA) to questions by the Chair of the Working Group of the LSC, UN Doc. A/AC.105/C.2/2015/CRP.25 of 22 April 2015

Information furnished in conformity with the Convention on Registration of Objects Launched into Outer Space, UN Doc. ST/SG/SER.E/526 of 29 February 2008

Information furnished in conformity with the Convention on Registration of Objects Launched into Outer Space, UN Doc. ST/SG/SER.E/569 of 16 July 2009

Information furnished in conformity with the Convention on Registration of Objects Launched into Outer Space, UN Doc. ST/SG/SER.E/608 of 30 November 2010

Report of the Legal Sub-Committee on the Work of its Eleventh Session (10 April–5 May 1972), UN Doc. A/AC.105.101 of 11 May 1972

European Instruments

Commission Communication COM(2010) 245 final/2 of 26 August 2010

Commission Decision 2011/833/EU on the Reuse of Commission Documents of 12 December 2011, *OJ L* 330/39 of 14 December 2011

Commission Delegated Regulation (EU) No. 1159/2013 of 12 July 2013 supplementing Regulation (EU) No. 911/2010 of the European Parliament and of the Council on the European Earth monitoring programme (GMES) by establishing registration and licensing conditions for GMES users and defining criteria for restricting access to GMES dedicated data and GMES service information, *OJ L* 309/1 of 19 November 2013

Commission Implementing Decision (EU) 2018/621 of 20 April 2018 on the Technical Specifications for the Copernicus Space Component pursuant to Regulation (EU) No 377/2014 of the European Parliament and of the Council, *IJ L* 102/56 of 23 April 2018

Council Decision 2013/488/EU of 23 September 2013 on the Security Rules for Protecting EU Classified Information, *OJ L* 274/1 of 15 October 2013

Directive 2003/98/EC of the European Parliament and of the Council of 17 November 2003, *OJ L* 345/90 of 31 December 2003

European Space Agency. ESA Data Policy for ERS, Envisat and Earth Explorer missions, October 2012. https://earth.esa.int/c/document_library/get_file?folderId=296006&name=DLFE-3602.pdf. Accessed 18 Apr 2019

Joint Principles for a Sentinel Data Policy, ESA/PB-EO(2009)98, rev. 1 of 23 October 2009

Regulation (EU) No 377/2014 of the European Parliament and of the Council of 3 April 2014 establishing the Copernicus Programme and repealing Regulation (EU) No 911/2010, *OJ L* 122/44 of 24 April 2014

National Instruments

Canada, Act Governing the Operation of Remote Sensing Space Systems, Bill C-25 of 25 November 2005

France, Loi Relative aux Opérations Spatiales, N° 2008518, 3 June 2008, as most recently amended on 30 May 2013. https://www.legifrance.gouv.fr/affichTexte.do?cidTexte=LEGITEX T000018939303

Germany, Act to give Protection against the Security Risk to the Federal Republic of Germany by the Dissemination of High-Grade Earth Remote Sensing Data of 23 November 2007, 2590 *Federal Gazette (BGBl.)*, Year 2007, Part I, No. 58, 28 November 2007

India, Remote Sensing Data Policy of 2011, RSDP-01:2001. https://nrsc.gov.in/Remote_Sensing_ Data_Policy Accessed 18 Apr 2019

Japan, Act on Securing Proper Handling of Satellite Remote Sensing Records, Act No. 77 of 2016. Available with English translation via https://www8.cao.go.jp/space/english/rs/rs_act.pdf. Accessed 18 Apr 2019

Japan, Basic Space Law, Act No. 43 of 28 May 2008, with unofficial English translation in J Space L 34(2008):471

United States, 51 US Code Subchapter III – Licensing of Private Remote Sensing Space Systems, § 60122

Doctrine

Achilleas P (2001) High-resolution remote sensing imagery and human rights. Proc Colloq Law Outer Space 44:234

Achilleas P (2008) French remote sensing law. J Space Law 34:1–9

Aoki S (2010) Japanese law and regulations concerning remote sensing activities. J Space Law 36:335–364

Bosc PA (2015) Earth observation for security and dual-use. In: Schrogl K-U et al (eds) Handbook of space security: policies, applications and programs. Springer, New York

Cheng B (1997) Studies in international space law. Clarendon, Oxford

De Beer AC (2015) The refusal of access to high resolution remote sensing data for reasons of national security: a (new) rule of customary international law? https://repository.up.ac.za/ bitstream/handle/2263/53120/DeBeer_Refusal_2016.pdf;sequence=1. Accessed 18 Apr 2019

Lee RJ (2001) Military use of commercial remote sensing data. Proc Colloq Law Outer Space 44:248

Lyall F, Larsen PB (2009) Space law – a treatise. Ashgate, Farnham

Mantl L (2015) Principle XIV [of the RS Principles] (International responsibility). In: Hobe S, Schmidt-T B, Schrogl K-U (eds), Cologne Commentary on Space Law, vol 1. Outer Space Treaty, Carl Heymanns Verlag, Köln

Schmidt-Tedd B, Kroymann M (2008) Current status and recent developments in German remote sensing law. J Space Law 34:97–140

Smith LJ, Reynders M (2015) Historical background and context [of the RS Principles]. In: Hobe S, Schmidt-T B, Schrogl K-U (eds), Cologne Commentary on Space Law, vol 1. Outer Space Treaty, Carl Heymanns Verlag, Köln

Part IV

Space Security Programs Worldwide and Space Economy Worldwide

Space Security Programs and Space Economy: An Introduction

58

Christina Giannopapa

Contents

Abstract

This chapter provides an introduction to Part 4 of the second edition of the *Handbook of Space Security* addressing the subject of "Space Security Programs Worldwide." It covers expert views on space security activities and programs of established spacefaring nations: the United States, Russia, China, Europe – including dedicated chapters on France, Italy, the United Kingdom – and India. It also reviews space security policies of emerging space countries – Australia, Iran, and Pakistan – to showcase a wide range of space activities and programs worldwide. Additionally it covers the economy of the space sector and it provides an overview of the UN space security-related organizations.

Introduction

Space activities in the past were only the privilege of the governments of a few space-faring nations. Over the past years there has been an increase in space activities, with the number of nations with space priorities also increasing. Space

C. Giannopapa (✉)
European Space Agency (ESA), Paris, France
e-mail: christina.giannopapa@esa.int

© Springer Nature Switzerland AG 2020
K.-U. Schrogl (ed.), *Handbook of Space Security*,
https://doi.org/10.1007/978-3-030-23210-8_63

programs are designed around two main political arguments: access to independent information to support government interests and access to critical technologies and capabilities. The governments' space programs have traditionally been designed to respond to programmatic objectives-related launchers and satellites, and nowadays also include applications. The launchers and satellite programs, civilian or military oriented, are often developed to reflect security objectives. The dual nature of space technologies and applications makes it difficult to strictly define space security programs in isolation.

In the twentieth century, launcher programs have been closely related to classified military projects demonstrating intercontinental ballistic missile capabilities (ICBM). Over the years they have seen a gradual level out to commercial launches separated from the military activities. Many countries still consider their launch program as a priority in order to gain independent access to space. Today, besides the United States and Russia, the European Space Agency, Japan, China, India, Iran, and Israel also possess launch capabilities. Most of these countries or agencies are offering services also on a commercial basis providing through their launch vehicles access to put satellites to space to many more countries around the world (Euroconsult 2019).

On October 4, 1957, the Soviet Union's space program put in space the first artificial satellite Sputnik 1. Initial satellite programs were set to fulfill science and military objectives. Today, there are more than 50 countries in all continents that have a satellite in orbit. Almost 1,000 operational satellites are now in orbit with diverse Earth observation, telecommunications, navigation, and positioning missions. In parallel to the growing importance of these down-to-earth applications, science and space exploration remain key missions of space agencies, invigorating international scientific cooperation. The United States leads with more than 250 satellites in orbit, followed by Russia and China (Euroconsult 2019). The new landscape of space-faring nations is the result of two trends: the ambition of many countries around the world to develop independent national space programs and the globalization of the aerospace and defense industry.

This chapter of the Handbook introduces different space activities and launch/satellite programs of established spacefaring nations: the United States, Russia, China, Europe – including dedicated chapters on France, Italy, the United Kingdom – and India and also emerging space countries – Australia, Iran, and Pakistan – to showcase a wide range of space activities and programs worldwide. The section covers also the space sector economy, the political economy of outer space security, and the New Space economy consequences as indicated by the space programs worldwide. It concludes with an overview of the UN space security-related organizations, due to the influence that they may have over the development of the space programs across the world.

Space Security Programs

The space programs of the *United States* (US) military and intelligence organizations constitute by far the world's largest space program. The services provided by these programs include telecommunications, surveillance, missile early warning, meteorology, positioning/timing, radio interception, nuclear detonation detection, and data relay. ▶ Chapter 59, "Satellite Programs in the USA" describes how the space systems provide both tactical and strategic services to the US military and intelligence agencies and in some cases to those of its allies. Strategic functions include monitoring international security treaties, analyzing the security forces of current and potential adversaries, and providing information to the President and the Secretary of State. Tactical functions include supporting US military and intelligence forces around the world. Overall, the US military space program continues to dwarf (a) the military space programs of all other countries combined and (b) of US civilian agencies such as NASA. The United States is unique in deploying military satellites of all types and on a global basis, and there is little sign that this will change in the next decade.

Russia can be considered today as having the most complete launch program in the world. Russia currently operates four types of launch vehicles, the Rockot, Soyuz, Zenit, and Proton. ▶ Chapter 60, "Russian Space Launch Program" explains how Russia has been successfully engaged in space activities for more than 60 years, having entered the space age as part of the Soviet Union and striding on as a separate state. On the one hand, after the dissolution of the USSR, Russia inherited the huge scientific and technical potential and technological developments of one of the two most powerful space nations of that time. But on the other hand, Russia was deprived of a large part of technologies and infrastructure put in place earlier. The launch vehicles that used to be Soviet became foreign, and the key launch site turned out to be located outside Russia's national territory. Also, it proved to be difficult for Russia to use remnants of its own technologies. For Russia, space is thus not only a question of national defense and security or its position in the market of commercial launch services but also, and more importantly, a question of the status of Russia as a highly developed nation in terms of science and technology.

In *Europe*, the 2019 Regulation for a Space Programme proposes the development of Governmental Satellite Communications (GOVSATCOM) and Space Situational Awareness (SSA) programs to accompany the satellite navigation program Galileo and the Earth observation program Copernicus. The European family of launchers includes the Ariane 5, Vega, and Soyuz that secure Europe's independent access to space and are launched through the Guiana Space Centre. During the Ministerial Council 2019, the European Space Agency (ESA) has recently adopted a safety and security program and has also secured the transition to the next generation of launchers: Ariane 6 and Vega-C, as well as the Space Rider, ESA's new reusable spaceship. The chapters on institutional and national space and security programs in Europe describe how the aforementioned institutional programs are intertwined with the national and multilateral programs of the European States based on their national budgets and contributions to organizations such as ESA, EU, European Defence Agency (EDA), and NATO. Overall, the space activities and programs of European States address the fields of Earth observation, satellite communication, Global Navigation Satellite System (GNSS), SSA, space transportation, satellite operations, and detection, tracking, and warning. The French, Italian, and British space programs are presented in separate chapters in greater detail.

France is historically the third space-faring nation. After more than 50 years of dedicated vision and investments in space industry and programs, ► Chap. 65, "Future of French Space Security Programs" explains how France has developed innovative and credible capabilities which ensure its independence and constitute a tangible asset contributing to the development of coherent European defense space cooperation. In direct connection with the new Defense Space Strategy sought by the President of the Republic, modernization of current capabilities is the challenge of the coming years for France.

Italy has a very long history and heritage in the space sector. ► Chapter 66, "Italy in Space: Strategic Overview and Security Aspects" describes how beginning in the 1950s, with the development of the first national satellite, the San Marco 1, and its launch, operated by an Italian team, Italy entered into a period of rapid growth for space technologies, applications, and cooperation at international level. Today, Italy is the third contributor of the European Space Agency, and it has, at national level, the knowledge and the experience in the whole chain of supply in space.

Britain stands at something of a crossroads in its development of spacepower, not least in its military and security elements. ► Chapter 67, "British Spacepower: Context, Policies, and Capabilities" explains how spacepower is coming of age within the United Kingdom, not least in the military dimension. The British state is at something of a crossroads when making choices about its future space development as it has to respond to turbulence across the transatlantic region.

China's development of launch vehicles is sticking to the "self-reliance and independent innovation" path. ► Chapter 69, "Chinese Space Launch Program" describes how, with more than 50 years of experience, China has successfully developed more than 10 models of launch vehicles and managed the transition from research test to flight application, and from flight application to industrialization. The Long Mach is its primary expendable launch system family. This chapter provides an overview of China's space launch plan. This chapter mainly presents the development history of China's launch vehicles, launch vehicles in service, the new generation of launch vehicles under development, as well as the efforts made by China in the field of space security. In addition, ► Chap. 68, "Chinese Satellite Program" explains the development history and outlines specific satellite programs: Earth observation, communications and broadcasting, navigation and positioning, and scientific and technological test satellites. In addition, it addresses the satellite program's future perspectives and China's international exchanges and cooperation.

The *Indian* launch program started with the vision to utilize the potential of space technology and its applications for national development in the 1960s. ► Chapter 70, "Indian Space Program: Evolution, Dimensions, and Initiatives" describes the significant progress that India has made in launch vehicle technology. Over the years India has been developing the Satellite Launch Vehicle which retired in 1984: the Augmented Satellite Launch Vehicle; the Polar Satellite Launch Vehicle; and the Geosynchronous Launch Vehicle, and is making efforts for Reusable Launch Vehicles. One of the important missions accomplished in the recent past is the development and operationalization of the new launch vehicle, GSLV MkIII. India's next milestone mission of Human Spaceflight has been initiated and the first crewed flight is expected by 2022. As of September 2019, Indian Space Research Organisation (ISRO) has accomplished 184 missions, including 101 satellite missions, 73 launch vehicle missions, and 10 technology demonstration missions.

The space sector in *Australia* is experiencing an unprecedented level of public interest and government support. National security considerations and the economic benefits of a fast-growing world market for space products and services are inextricably linked as drivers for a range of government and industry initiatives, as explained in ▶ Chap. 71, "Australia's Space Security Program." The government's clear intention is to enhance Australian Defence sovereign space capabilities progressively through access to allied and commercial space-based capabilities. Australia as a nation is exposed to and provide evidence that Australia is now treating space as an integral component of its role in the protection of its national security interests and in the advancement of its international responsibilities.

Despite political, technological, and economic constraints, *Pakistan* is considered an aspiring space power with a relatively modest space program compared to the larger, more successful ones of China and India. ▶ Chapter 72, "Pakistan's Space Activities" describes how the country can utilize available resources to improve its nascent space infrastructure through collaborative efforts to gain eventual self-sufficiency for socioeconomic and strategic purposes in the South Asian region. The concept of multilateral collaboration, utilizing available resources and public-private partnerships to empower its space program, enhances its domestic scientific and technological base and builds an indigenous space industry that can reap dividends at home and abroad.

The Economy of the Space Sector

The *Space Sector Economy* over the years has become more commercial, and different space applications have emerged outside the traditional research and development (R&D), calling for a wider definition of space economy. This wider "space economy" can be defined using different angles. It can be defined by its products (e.g., satellites, launchers), by its services (e.g., broadcasting, imagery/data delivering), by its programmatic objectives (e.g., military, robotic space exploration, human spaceflight, Earth observation, telecommunications), by its actors/value chains (from R&D actors to users), and by its impacts (e.g., direct and indirect benefits). ▶ Chapter 73, "Space Sector Economy and Space Programs World Wide" describes how the worldwide national space budgets have continued to grow in 2017 and 2018. In the area of defense space programs, funding is expected to increase until 2025. In 2018, the United States is dominating with 71.7% of the world total. In relative terms, Asian spending on military space activities has more than doubled over the period between 2008 and 2018, strongly driven by China along with significant Japanese budget allocations, and is expected to increase further from $3.49 billion to $5.19 billion in 2025. Europe is the fourth largest region in terms of space defense spending, totaling $1.30 billion following the United States, Asia, and CIS countries (Russia). Ariane 6 development is approaching its final phase, which may lead to reduced defense space contributions in Europe. The foreseen growth in space defense spending is mainly driven by the United States, China, India, and Japan (Euroconsult 2019).

These figures are justified by *New Space Economy* as a new paradigm able to revolutionize space activities. ▶ Chapter 74, "The New Space Economy: Consequences for Space Security in Europe" states that there is a large consensus of perceptions which tends to define this set of US-led policies and investments as a shift for all space activities. This is the reason why it is triggering effects worldwide and creating a new context for European space activities. In Europe, space policies have often been characterized by a dialectic between plurality and collaborative common frameworks. The fragmented European spacescape has always found it difficult to respond to the critical mass developed by US public and private policies. This is today the case with the "new space economy," where the United States is fundamentally renewing its public policy in order to create a more efficient and business-friendly approach to space, a shift which has helped create a new breed of space entrepreneurs, IT tycoons investing in space with considerable financial capabilities, and tech incubator methodologies. This paradigm has renewed investments, technologies, and a new spirit of space conquest and puts enormous pressure on the European space sector, which has to renew its classic approach. Europe needs to come up with a more strategic space vision that is able to take into account not only the outcomes of new technology but also the inbred characteristics of space-based technology production and data fluxes, with key issues in terms of security.

As human civilizations increasingly explore, utilize, and compete in space, the man-made security challenges are evolving and the strategies and *Political Economic* rationales become increasingly relevant for analysis. As such, ▶ Chap. 75, "Political Economy of Outer Space Security," sustainability and efficiency call for exploitation of static economies of scale and scope in space industries and services, yet the trade-offs in control, governance, and dynamic innovation point toward autonomy and oligopolistic structures with overcapacity. The economic sustainability becomes a key element of the dynamic pursue of space policies and objectives at national and partnership levels. In the latter case, specialization and its implications for the wide economy through externalities and indirect effects receive increasing attention as space becomes contested, congested, and competitive. Notwithstanding the fact that they are largely government controlled, aerospace industries play a crucial role in trading patterns. Hence, they can be considered a fiscal government spending element similar to defense expenditure. The country specializations and their evolution in commercial markets and alliances are focal points in the current global trade policy paradigm shifts, affecting performance

The Role of International Organizations for Space Security

The various initiatives being discussed in the different UN *fora* indicate the increasing pressures facing the international community in addressing all aspects of space security and strengthening the multilateral regime governing the use of space.

However, they also reveal substantial differences among states over priorities, methodologies, mechanisms, and programmatic settings to address and tackle space security issues. ▶ Chapter 76, "Views on Space Security in the United Nations" describes how this proliferation of initiatives which offers distinct approaches that look at both short- and long-term solutions might pose more challenges to the existing discussions. More worrisome, however, may be another challenge: deadlocks in the work of one body can be transferred over other fora, limiting progress even on those aspects on which there has traditionally been consensus. Existing frictions among different UN bodies are also a limiting factor in pushing forward a shared agenda on space security, and so too is the long-standing division between civilian and military uses of space embedded in the UN machinery. While divergent perceptions and priorities among leading space powers still exist (and will remain), the existence of political will that accommodates, rather than eliminates, these differences can prove effective in finding common ground for future action that can be acceptable for the interests of different states. As the nature of risks to space infrastructure and services, and the available responses, are similar (although not entirely) for all space actors – whether civilian, commercial, or military – common threat perceptions may influence how states choose to cooperate and readily serve as a basis for developing common responses and finding future consensus.

Conclusions

Space programs aiming at the development and utilization of space technologies and applications constitute an essential part of capabilities in the security domain related to both civil and defense. The dual nature of space technologies and applications makes it difficult to strictly define space security programs in isolation. As the number of countries involved in space activities is increasing, there is an increasing need for ensuring both space for security and security in space.

References

Euroconsult (2019) Government space programs. Benchmarks, Profiles & Forecasts to 2028

Satellite Programs in the USA

59

Pat Norris

Contents

P. Norris (✉)
VAPN Ltd, West Byfleet, UK
e-mail: pat@vapn2.com

© Springer Nature Switzerland AG 2020
K.-U. Schrogl (ed.), *Handbook of Space Security*,
https://doi.org/10.1007/978-3-030-23210-8_39

Abstract

The space programs of the US military and intelligence organizations are described. The services provided by these programs include telecommunications, surveillance, missile early warning, meteorology, positioning/timing, radio interception, nuclear detonation detection, and data relay. Both unclassified and classified programs are described, with less detail and more speculative information for the latter. Recent trends that indicate the focus of future programs are discussed.

Introduction

In this chapter what is by far the world's largest space program is described. Up until 2011 the space budget of the Department of Defense (DOD) was reported in the media as being between $20 billion and $22 billion each year this century. Since then the published figure is about half that, i.e., $10–12 billion per annum, apparently due to removal of intelligence-related space programs from the reported budget figures. For the period 2011 to 2013, figures reported in *The Washington Post* based on information leaked by Edward Snowden indicated that the National Reconnaissance Office (NRO) and the National Geospatial-Intelligence Agency (NGA) had a combined budget of $15–16 billion per annum, which should probably be considered as part of US Government space security activities. Information about NRO and NGA funding in more recent years has not been published, but it seems likely that the full US space security budget is currently in excess of $20 billion per annum.

The space programs of the US military and intelligence communities are treated in this chapter as being one and the same. The NRO is a joint agency of DOD and the Central Intelligence Agency (CIA) in order to deliver space systems that serve both parent organizations.

The large number of programs is broken down in this chapter as follows:

- Telecommunications (Section "Telecoms")
- Global Positioning System (Navigation) (Section "Global Positioning System (GPS)")
- Surveillance (Section "Surveillance")
- Other (Section "Other Satellites")

The descriptions include the ground elements for control of the relevant missions but exclude the user equipment and other stand-alone ground systems. In general, only operational systems are described, with just a few of the very large number of research, prototype, demonstration, and preoperational space systems mentioned in section "Other Satellites."

Background

The space systems described in this chapter provide both tactical and strategic services to the US military and intelligence agencies and in some cases to those of its allies. Strategic functions include monitoring international security treaties, analyzing the security forces of current and potential adversaries, and providing information to the President and the Secretary of State. Tactical functions include supporting US military and intelligence forces around the world.

The higher cost of many US military space systems compared to commercial systems is due in part to their hardening against nuclear explosions in space, their ability to resist jamming and other forms of interference, and their ability to operate autonomously in the absence of ground control systems.

The launchers used by the DOD are not described in this chapter since they are used for civilian as well as military missions. From 2007 to 2016, United Launch Alliance (ULA), a joint venture between Boeing and Lockheed Martin, was the sole source provider of launch services to DOD and NRO for medium and heavy lift. Despite steadily rising launch costs, the company maintained its position based on the performance and reputation of its two vehicles, Atlas V and Delta IV, and the lack of viable competition. In 2016 SpaceX broke the ULA monopoly by winning a contract to launch a GPS-III satellite on its Falcon 9 rocket at a price of $83 million, about 40% less than ULA had previously charged, according to Lt. Gen. Samuel Greaves, then head of the Air Force's Space and Missile Systems Center. As of March 2019, SpaceX has won ten Air Force launch contracts, eight for Falcon 9, and two for Falcon Heavy: the latter two are the Space Test Program-2 (STP2) demonstration mission that comprises 25 small spacecraft (2019 launch) and the Air Force Space Command (AFSPC)-52 satellite whose mission is classified (2020 launch). The price of the Falcon Heavy for the AFSPC satellite was $130 million. The price differential between SpaceX and ULA was highlighted in February 2019 by the award of three launch contracts to each for 2021–2022 launches: $297 million to SpaceX to launch AFSPC-44, NROL-85, and NROL-87, averaging $99 million per launch, and $441.76 million to ULA to launch SBIRS GEO-5, SBIRS GEO-6, and Silent Barker, a classified space situational awareness mission, averaging 48% more than SpaceX at $147 million per launch.

In October 2018 the Air Force awarded three contracts for the development of Blue Origin's New Glenn ($500 million), Northrop Grumman's OmegA ($792 million), and ULA's Vulcan Centaur ($967 million) launch vehicle prototypes. In 2020 two contractors are to be selected to compete for up to 25 national security launches to be awarded from 2022 to 2027 based on a selection process beginning in 2019 (which SpaceX and any other launch supplier can enter). One of the triggers for this initiative is the Congressional ban on the use of the Russian RD-170 engine in ULA's Atlas V rocket from 2022. Some Members of Congress are questioning the fairness of the award to the three companies because of their concerns that it puts SpaceX at a competitive disadvantage for the 2020 selection.

Congress enacted $2.1 billion for five launches in Fiscal Year (FY) 2019, and $1.7 billion has been requested by DOD for FY2020 for four launches.

Smaller launcher options for DOD and NRO include Northrop Grumman's Minotaur and Pegasus.

The following references have been used in compiling this chapter:

- Information about unclassified programs has mainly been drawn from:
  - The DOD program fact sheets
  - The DOD AU-18 Space Primer
- Information about classified programs has mainly come from:
  - *Watching Earth from Space* by the author of this chapter and references therein, especially:
        The US Intelligence Community by J T Richelson

The information in these references has been updated based mainly on information in the trade press, including *Aviation Week & Space Technology*, *Air & Cosmos* (in French), and *Space News* and budgetary information reported in *The Washington Post* on August 29, 2013, based on documents leaked by Edward Snowden.

Telecoms

Introduction

80% of all communications to US deployed military forces travel through satellite. And 80% of that traffic is carried through commercial communications satellites – DOD spent over $600 million in 2010 leasing services from commercial satellite operators. The remaining 20% of the 20% of traffic to deployed forces is carried by special purpose military communications satellites. Commercial satellites are used wherever possible so that in case of an urgent need, capacity is available on the military satellites.

The use of commercial satellites has led to some embarrassing moments for the US military. On several occasions, the transmissions from remotely controlled aircraft have been intercepted by journalists and made public. There has been no persuasive rationale offered as to why these transmissions are not encrypted other than that the remotely controlled aircraft were of a vintage that did not carry encryption facilities. Given the ubiquitous and low cost of mass market encryption/decryption technology as used, for example, in cell phones and pay-TV, many commentators have wondered why the encryption technology is supposedly too modern or expensive for remotely controlled aircraft that cost more than $1 million each. The proliferation of remotely controlled aircraft since 2003 has called for an increasing amount of satellite communications capacity – particularly the large long-duration drones that are controlled from bases in the USA even when flying in Central Asia.

The US's specialist military communications satellites provide a mix of strategic and tactical services to forces on land, at sea, and in the air. Three parts of the radio spectrum are used for the transmissions, ultrahigh frequency (UHF, below 1 GHz), X-band (6–8 GHz), and extremely high frequency (EHF, 20–30 GHz).

Satellites currently in orbit include a mix of a new generation of systems and those of the previous generation. By 2020 it is expected that the DOD satellite communication services will be carried by 16 of the new generation systems – 4 AEHF, 8 WGS, and 4 MUOS – described in the next three sections. These 14 will be augmented by (a) EHF payloads on polar-orbiting satellites to provide Arctic coverage (Section "Interim and Enhanced Polar System (IPS/EPS)"), (b) data relay satellites (Section "Data Relay"), and (c) commercial satellites.

In the past each service had its own communications satellites, but with the current systems, the user doesn't necessarily know which satellite he is using. Connections are provided by the Global (or Regional) SATCOM Support Center (GSSC) based on the availability of the appropriate bandwidth in the relevant areas. As of December 2018, GSSC is part of the US Air Force Space Command. Previously, commercial satellite services had been the responsibility of a unit of the Defense Information Systems Agency (DISA). Air Force Space Command will absorb the Enhanced Mobile Satellite Services program, also known as the Iridium satellite program office, as well as DISA's commercial satellite leasing team. As of November 8, 2018, DISA had 91 active commercial SATCOM leases that are worth about $4.5 billion, according to Clare Grason, division chief for satellite communications at DISA.

To improve the value for money of the commercial SATCOM services it purchases, the DOD is no longer insisting on lowest price technically acceptable (LPTA) selection of supplier for each purchase. The alternative Blanket Purchase Agreement (BPA) approach has been used, for example, to purchase O3b medium Earth orbit (MEO) satellite services from SES. This BPA allows military users to purchase SES O3b MEO managed services up to $516.7 million over 5 years. By putting the BPA in place, the DoD has been able to drastically cut the amount of time needed to acquire satellite services, as well as negotiating a bulk purchase price as opposed to the high spot prices often associated with the LPTA approach.

The remainder of this section is structured as follows:

- AEHF and Milstar (Section "Advanced Extremely High Frequency (AEHF) and Milstar")
- MUOS and UFO (Section "MUOS and UFO")
- WGS (Section "Wideband Global SATCOM (WGS)")
- Data Relay (Section "Data Relay")
- Interim and Enhanced Polar System (Section "Interim and Enhanced Polar System (IPS/EPS)")

Advanced Extremely High Frequency (AEHF) and Milstar

The White House National Security Council, chaired by the President, uses the AEHF satellites (see Fig. 1) to control tactical and strategic forces on missions around the globe and at all levels of conflict. Four AEHF satellites are in geosynchronous orbit, launched in 2010, 2012, 2013, and 2018, with two more planned in 2019 and 2020. They combine (a) a strategic communications mission, i.e., assured jam-proof connectivity between the President and nuclear forces, and (b) protected tactical communications. These two quite distinct missions share equipment on the satellite such as a digital core processor. The strategic mission requires the satellites to be radiation hardened to survive a nuclear explosion in space and other threats, which critics have argued makes AEHF more expensive than two separate satellites would have been (one for the strategic mission, the other for the tactical).

Each AEHF weighs about 6½ t when launched on an Atlas 5 from Cape Kennedy and is based on the Lockheed Martin A2100 spacecraft bus (also used for commercial satellites) with a Japanese IHI BT-4 apogee thruster and an electric propulsion system. The electric ion propulsion system is intended for maneuvering the satellite in its final orbit but was called into action on AEHF-1 when the apogee thrust motor failed. The slow evolution from the elliptical transfer orbit into which the Atlas 5 placed the satellite on August 14, 2010, and the eventual geostationary orbit took 16 months using the low thrust ion engine. A further 5 months were required for on-orbit testing until AEHF-1 became operational in March 2012. Ironically, in early 2012 Boeing announced the sale to two commercial satellite operators of four "all electric" satellites, which do away entirely with chemical propulsion. These satellites will take 4 to 6 months to reach their operational orbit but are significantly lighter (and therefore cheaper) than the conventional satellites that achieve their operational orbit in days rather than months.

Fig. 1 AEHF satellite

Credit: USAF
Artist's impression of AEHF in orbit

The space segment consists of a cross-linked constellation of two satellites with a planned extension to four, forming a ring at geosynchronous orbit that provides continuous communications without the need for a ground-hop. They provide communications at a variety of data rates from 75 bps to approximately 8 Mbps and broadly provide ten times the throughput of the 1990s-era Milstar satellites with a substantial increase in coverage for users. System uplinks and cross-links operate in the extremely high frequency (EHF) range and downlinks in the superhigh frequency (SHF) range. The communications are channeled through an antenna farm on each satellite comprising:

- Two SHF downlink phased arrays
- Two cross-link antennas
- Two uplink/downlink nulling antennas
- One uplink EHF phased array
- Six uplink/downlink gimbaled dish antennas
- One each uplink/downlink Earth coverage horn

The four-satellite constellation will allow the military to deliver an extended data rate (XDR) service, which in practical terms will deliver video in hours rather than days, and allow data to be transferred in seconds rather than minutes. But the XDR capability will not likely be on offer until more of the ground and airborne terminals are available for use, in about 2020 (more than a decade after the first satellite was launched). The terminals for airborne and ground-based users that have not yet been fielded are known as the Family of Beyond Line-of-Sight Terminals (FAB-T).

In 2015, the ground/air terminal program (prime contractor Raytheon) was split into (a) command post terminals to facilitate communications for nuclear and conventional forces through the E-6 and E-4 aircraft and (b) force element terminals to provide the capability to the B-2 and B-52 bomber and the RC-135 Rivet Joint intelligence, surveillance, and reconnaissance aircraft. The Government Accountability Office has said that all six AEHF satellites are expected to be on-orbit before FAB-T is operational.

Under a contract valued at $6.6 billion, Lockheed Martin supplied four AEHF satellites, and a further two have since been ordered to bring the total order value to about $9 billion. Northrop Grumman is responsible for the EHF payload.

The AEHF constellation is also used by Canada (at a cost of $592 million), the Netherlands, and the UK, while Australia, Japan, and NATO are negotiating to join.

For the future, the Evolved Strategic System (ESS) is expected to provide a next-generation replacement for the AEHF constellation. One option being considered is to disaggregate the satellite into tactical and strategic variants. The planned sequence is for a prototype phase followed by a fielding phase in 2025 with an initial operational capability 5 years later.

Five of the Milstar satellites (see Fig. 2) that AEHF supersedes are in orbit, launched in 1994 through 2003. Each Milstar satellite serves as a smart switchboard in space by directing traffic from terminal to terminal anywhere on the Earth. The satellite actually processes the communications signal and can link with other

Fig. 2 Milstar satellite (artist's impression)

Milstar satellites through cross-links. Milstar was the first US military communications satellite with this smart switchboard capability which provides flexibility for users with a reduction in ground controlled switching. The satellite establishes, maintains, reconfigures, and disassembles required communications circuits as directed by the users. Milstar terminals provide encrypted voice, data, teletype, or facsimile communications. A key goal of Milstar is to provide interoperable communications among the users of Army, Navy, and Air Force Milstar terminals.

Each Milstar satellite weighs 4.5 t and generates 8 kW of electrical power. They are all in inclined geosynchronous orbit. The cost of each satellite was about $800 million. Prime contractor was Lockheed Martin Missiles and Space.

The Protected SATCOM Division of the Space and Missile Systems Center's MILSATCOM Directorate is the program office for the AEHF system and is responsible for acquisition of the space and ground segments as well as the Air Force terminal segments. The Army and Navy acquire their own terminals.

The AEHF satellites are operated by the 4th Space Operations Squadron (4 SOPS) located at Schriever AFB, Colorado. The mission control segment controls satellites on orbit, monitors satellite health, and provides communications system planning and monitoring.

MUOS and UFO

The four Mobile User Objective System (MUOS) satellites (plus one on-orbit spare) are designed to replace the ultrahigh frequency (UHF) follow-on (UFO) constellation that provides narrowband communications (64 kbps or less). MUOS (see Fig. 3) is designed to support users that require mobility, high data rates (up to 384 kbps "on the move"), and improved operational availability. A 2012 US Navy information bulletin described the UHF spectrum as "the military's communications workhorse, as it is the most effective military radio frequency for penetrating jungle foliage, inclement weather, and urban terrain on the move." The current UFO constellation comprises eight operational UFO satellites augmented by two pre-UFO FLTSAT

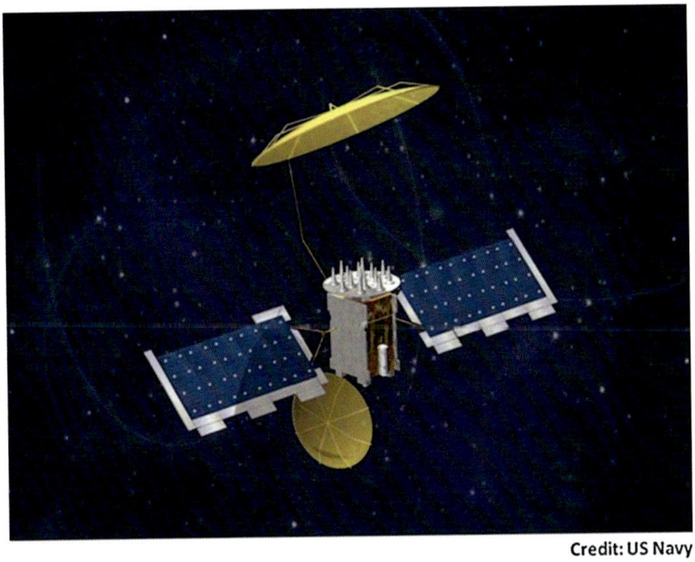

Credit: US Navy

Artist's impression of MUOS in orbit

Fig. 3 MUOS satellite

satellites and leased services on communications satellites. The UFO satellites achieved initial operational capability in November 1993 and full operational capability in February 2000. The final satellite in the series, UFO-F11, was launched in December 2003.

Each MUOS satellite carries a legacy payload that allows terminals compatible with the UFO system to continue in service. In addition, MUOS carries a payload that provides a military variant of the commercial 3-G wideband cellular service using Wideband Code Division Multiple Access (WCDMA) technology. Fully fielded, MUOS provides an aggregate of 40 Gbps for the military user, compared to the legacy UFO system's aggregate of 2.7 Gbps – a 15-fold increase. The increase means users have more than 16,332 simultaneous accesses (voice, video, data) at 2.4 kbps, compared to 1,111 accesses provided by the UFO satellite system at the same data rate.

User information flows to the satellite via UHF WCDMA links, and the satellites relay this by Ka-band feeder link to one of the four Earth stations in Italy, Australia, Hawaii, and Virginia that are interconnected by a fiber-optic terrestrial network. These facilities identify the destination of the communication and route the information to the appropriate ground site for Ka-band uplink to the satellite and then relay via UHF WCDMA downlink to the correct users. MUOS uses Internet Protocol versions 4 and 6 (IPv4/IPv6) to give users connectivity to the military internet.

The four operational MUOS satellites were launched in 2012, 2013, and 2015 (two launches) and are now in geosynchronous orbit at longitudes 15.5°W,

100°W, 177°W, and 75°E each with an inclination of 5°, plus MUOS-5 as an in-orbit spare in a slightly elliptical orbit with an inclination of about 8° and a longitude close to 108°W (a longitude where minimal East-West station-keeping fuel is required).

The fifth MUOS satellite launched in June 2016 suffered a propulsion failure that prevented it reaching its planned orbit. The Navy ground operations staff used the onboard fuel for station keeping to propel the spacecraft into an orbit that was geosynchronous and with enough fuel left to provide operational utility.

MUOS-1 weighed 6.8 t (15,000 lbs) at its February 24, 2012, launch which was the heaviest satellite ever carried by an Atlas 5 rocket. Like AEHF, the satellite is based on prime contractor Lockheed Martin Space Systems' A2100 platform bus. The solar arrays produce 15.4 kW (end-of-life performance). Overall cost of the five satellites plus associated ground infrastructure was $7.4 billion.

A key feature of MUOS (and of commercial satellites such as Inmarsat-4, Thuraya and ACES that offer similar services to users on the move) is a large antenna that unfurls in orbit to a diameter of 14 m (46′ – almost the width of a basketball court). Each satellite has 16 WCDMA beams: four WCDMA beams on four 5 MHz carriers and UHF channel bandwidths of 17.5 kHz and 21.5 kHz.

The wideband CDMA for communications in motion is intended to be compatible with the Joint Tactical Radio System (JTRS) series, a program that has had a troubled history. In August 2018 MUOS was approved for expanded operational use by US Strategic Command. The Marine Corps will be the first service to widely deploy MUOS, largely due to its 6-year investment in MUOS portable radios. The Marine Corps began initial MUOS fielding to its AN/PRC-117G radios in the fourth quarter of 2018, followed by initial operational capability in the first quarter of 2019. In contrast to the Marines, according to an April 2018 Government Accounting Office report, the Army Handheld, Manpack, and Small Form Fit Radios (HMS) program is not scheduled to enter full production until 2021 – HMS continues efforts begun under the Joint Tactical Radio System.

Once its on-orbit testing has been completed, MUOS is operated by the Naval Satellite Operations Command in Point Mugu, California.

The possibility of selling UHF satellite service to the US military tempted commercial satellite operator Intelsat to build a suitable payload into its IS-27 satellite that was destroyed in a launch vehicle failure 40 s after lift-off in January 2013. Similar commercial "hosted payload" deals have been taken up by defense agencies in several countries including Britain, France, Spain, and Australia.

The USA opened MUOS to international partners in 2015. Canada has requested access to the system and as of late 2018 was in negotiation with the USA over the details of the agreement. Canada had considered co-financing a sixth MUOS satellite with Lockheed Martin but has instead decided to pay the US Government for assured access to the MUOS services.

Wideband Global SATCOM (WGS)

The Wideband Global SATCOM (WGS – previously called wideband gap-filler satellite – see Fig. 4) offers wideband broadcast and communications services.

The system supports continuous 24-h per day wideband satellite services at X-band and Ka-band to tactical users and some fixed infrastructure users, and it has the ability to cross-band between the two frequencies onboard the satellite. WGS supplements X-band communications that had been provided by the Defense Satellite Communications System III (DSCS-III) and augments the one-way Global Broadcast Service (GBS) satellites through new two-way Ka-band service.

The first WGS in orbit, WGS-1, with its 2.4 Gbps wideband capacity, provided greater capability and bandwidth than all nine DSCS-III satellites combined that it superseded.

The WGS design includes 19 independent coverage areas that can be positioned throughout the field of view of each satellite. This includes:

- Eight steerable and shapeable X-band beams formed by separate transmit and receive phased arrays
- 10 Ka-band beams served by independently steerable, diplexed antennas, including three with selectable RF polarization
- Transmit/receive X-band Earth coverage beams

WGS supports communications links within the government's allocated 500 MHz of X-band and 1 GHz of Ka-band spectrum. Each WGS satellite can filter and route 4.875 GHz of instantaneous bandwidth. Depending on the mix of ground

Fig. 4 WGS satellite

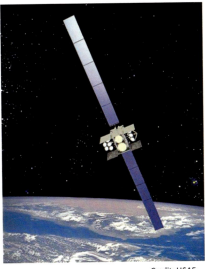

Credit: USAF
Artist's impression of WGS in orbit

terminals, data rates, and modulation schemes employed, each satellite can support data transmission rates ranging from 2.1 Gbps to more than 3.6 Gbps. By comparison, a DSCS-III satellite supported up to 0.25 Gbps.

Prime contractor Boeing has built WGS on its commercial 702 platform, which includes the Boeing xenon ion propulsion system (XIPS) for on-orbit station keeping and station changes. The gallium arsenide solar cells provide about 13 kW of power, and the mass at launch is about 6 t.

Ten WGS satellites have been launched into geostationary orbit in 2007, 2009 (two launches), 2012, 2013 (two launches), 2015, 2016, 2017, and 2019. WGS 4, 5, and 6 comprised the Block 2 version of WGS. WGS-6 launched on August 7, 2013, was funded by the Australian Department of Defense. A total of 12 WGS have now been ordered, and WGS-7 through WGS-10 are the Block 2 follow-on version, with WGS-9 funded by an international consortium comprising Canada, Denmark, New Zealand, Luxembourg, and the Netherlands. The cost of WGS-5 through WGS-7 is about $350 million each, while WGS-8 and WGS-9 cost about $55 million more each due to being outfitted with enhancements that boost their capacity by 30%. WGS-11 and WGS-12 were unexpectedly ordered in 2018 when Congress added $600 million to the DoD budget that had not been requested by the Pentagon. They are likely to be ready to launch in about 2022.

In 2019, Congress approved $49.5 million for a new DoD budget line for commercial satellite communications. Congressional appropriators took the money from the $61 million the Air Force had requested for so-called pathfinder initiatives to supplement or replace WGS services. With the addition of a new program line for commercial SATCOM, Congress is forcing DoD to figure out a strategy for the procurement of commercial services.

Spacecraft platform control is performed by the 3rd Space Operations Squadron (3 SOPS) at Schriever AFB in Colorado Springs, Colorado. Payload commanding and network control is handled by the Army 53rd Signal Battalion headquartered at nearby Peterson AFB, Colorado, with subordinate elements A Co. at Fort Detrick, Maryland; B Co. at Fort Meade, Maryland; E Co. at Fort Buckner, Okinawa, Japan; C Co. Landstuhl, Germany; and D Co. Wahiawa, Hawaii.

The sale of WGS satellites and services to countries such as Australia and Canada has been criticized by some in industry as an encroachment of DOD onto commercial territory. Commercial companies such as Paradigm (European) and Xtar (US and Spanish) offer military satellite communications services outside the USA.

Data Relay

The USA has a number of special relay satellites in high orbits so that US surveillance satellites can transmit data to a US ground station no matter where in the world the surveillance satellite is. The relay satellites are either in a geostationary orbit that is always 36,000 km above the equator or an orbit that takes it further north at a similarly high altitude. For this scheme to work, in addition to the relay satellites themselves, you have to also place special equipment on the surveillance satellites to enable them to communicate with the relay satellites.

The US military and intelligence communities have two sets of relay satellites at their disposal. First is NASA's Tracking and Data Relay Satellite System (TDRSS) which is described in various NASA publications. (The TDRSS page on the NASA website has links to a variety of documents describing the system and satellite: www.nasa.gov/directorates/heo/scan/services/networks/tdrs_main). The NRO's postulated imaging radar satellites (see section "Imaging Radar Satellites") are said by some commentators to use TDRSS to get their images back home quickly, but other US military satellites such as the optical surveillance series (see section "Optical Imaging Satellites") are said to use a military version of TDRSS that civilian commentators usually refer to as Space Data Systems (SDS) but according to information leaked by Edward Snowden to *The Washington Post* is called QUASAR.

According to several commentators, ten QUASAR satellites have been orbited since 1998 into a mix of geostationary and inclined elliptical orbits. Seven, in 2000, 2001, 2011, 2012, 2014, 2016, and 2017, were placed in geosynchronous orbit from where in principle they can serve surveillance satellites as far as 70° north and south, i.e., to the edge of the Arctic and Antarctic regions (the 2012 satellite is said by some civilian analysts to be a replacement for the one launched in 2001). However beyond about 50° north or south, the surveillance satellite in low orbit has increasing difficulty to stay locked on the QUASAR satellite as the line of sight gets closer and closer to the horizon – by the time the surveillance satellite is at 70° north (or south), the QUASAR geosynchronous satellite appears very close to the Earth and is difficult to track. For this reason some of the QUASAR satellites are placed in what is often called an elliptical Molniya orbit named after the Soviet communications satellites that popularized the orbit 40 years ago. A Molniya orbit is high in the sky over the northern hemisphere and low over the southern hemisphere. The Soviets chose this orbit to provide broadcasting and communications to the whole of the Soviet Union many of whose inhabitants live far to the north. The US military chose it for the same reason – to give good coverage of all of the Soviet Union, especially the militarily interesting regions around Murmansk (69° north) and the Bering Strait (67° north) next to Alaska. Out of its 12-h orbit, 8 h will be in the high northern part, so three such satellites can provide continuous 24-h coverage between them. Three QUASAR satellites launched in 1998, 2004, and 2007 are said to be in these Molniya orbits.

Interim and Enhanced Polar System (IPS/EPS)

The Interim Polar System (IPS) provides protected communications (anti-jam, anti-scintillation, and low probability of intercept) for tactical users in the Arctic region. IPS was deployed to meet the demand for protected polar satellite communications to support submarines, aircraft, and other platforms and forces operating in the high northern latitudes that have been steadily increasing over the last 20 years. The existing IPS payloads provide EHF low data rate (75 bps to 256 kbps) communications to users above 65° north latitude by using satellites in high elliptical orbit. Initial operational capability of the IPS was achieved in 1998 with the first payload, which was launched in 1997 or 1998 and is reported to be still providing uninterrupted service. The second IPS payload was announced as being in orbit in 2007, although it may have been launched some time earlier. The third payload will replace the first payload at an unknown date in order to maintain full operational capability of the system. The satellites that host these IPS payloads are not publicly identified. The IPS payloads are manufactured by Boeing Satellite Systems.

The Enhanced Polar System (EPS) is expected to provide an essential adjunct to the MILSATCOM mid-latitude systems. EPS will provide continuous coverage in the polar region for secure, jam-resistant, strategic, and tactical communications to support peacetime, contingency, homeland defense, humanitarian assistance, and wartime operations. EPS characteristics will include:

- Protected communications services and communications services without continuous system command and control
- Integrated capability allowing different levels of planners to manage their resources
- Interconnectivity between EPS satellites and mid-latitude users via an EPS Gateway located at a Global Information Grid Point of Presence
- Data rates between 75 bps and 1.28 Mbps (threshold)
- AEHF Extended Data Rate (XDR)-interoperable waveform

The first two operational EPS payloads were delivered by prime contractor Northrop Grumman to the Air Force in 2013 and were placed in orbit on unidentified satellites probably in 2016 and 2017. The initial EPS system comprises two EHF communications payloads hosted on satellites operating in highly elliptical orbits, modified AEHF communications terminals, a gateway to provide connectivity into other communication systems and the Global Information Grid (GIG), and an extension of the AEHF Mission Control Segment (MCS) hardware and software to accommodate EPS. The antenna farm on the payload includes a spot beam aimed at the gateway, a user spot beam, and a user Earth coverage beam. In 2018 Northrop Grumman was awarded a $429 million contract to provide two Extremely High Frequency eXtended Data Rate (EHF XDR) payloads for the Enhanced Polar System-Recapitalization (EPS-R). As with the EPS payloads, the EPS-R payloads will maintain continuity of protected military satellite communications in the North Polar Region and are designed to enable hosting on a separately procured satellite with other payloads.

Global Positioning System (GPS)

The Global Positioning System (GPS) provides accurate and reliable positioning and timing to military and civil users around the world. At a minimum, the GPS constellation needs 24 satellites in six orbital planes (see Fig. 5) in order to ensure

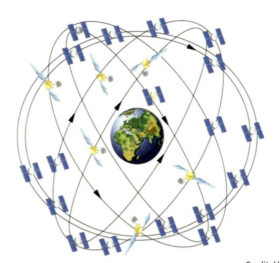

Credit: USAF

Artist's impression of GPS Constellation (not to scale)

Fig. 5 GPS constellation

that at least four satellites are in view by the user at all times. The constellation flies in a semi-synchronous orbit at approximately 20,000 km above the Earth.

GPS satellites have a semi-synchronous orbit, with a period of 11 h and 58 min. The 2-min offset from 12 h accommodates the movement of the Earth around the sun during a single GPS satellite orbit, i.e., the GPS orbital period is based on a sidereal day and not the solar day.

Due to better than specified reliability in orbit, the current constellation consists of about 30 satellites. The added redundancy offers improved accuracies and availability to users. As of late 2018, the GPS operational constellation comprised 1 Block IIA, 11 Block IIR, 7 Block IIR-M, and 12 Block IIF satellites – the various Blocks will be described below and are summarized in the 2 accompanying diagrams: "GPS modernization program" (Fig. 6) and "GPS program evolution" (Fig. 7). The oldest of the 31 satellites was launched in 1993, and in October 2018, their average age was 11.1 years.

GPS offers two types of services to its user base – the standard positioning service (SPS) and the precise positioning service (PPS). The SPS is available for anyone's use – military or civil. SPS offers 3–5-m accuracy. The PPS can only be accessed by authorized personnel – those with the correct decryption keys such as the US military or its allies. PPS accuracy is 2–4 m. There are several signals and codes that make up each of the GPS services.

Today, GPS transmits on the frequencies L1 (1575.42 MHz) and L2 (1227.6 MHz). The codes transmitted on these frequencies are the coarse acquisition (C/A) code and the pseudorandom (P(Y)) code. Currently, the C/A code is transmitted on L1 and P(Y) is on both L1 and L2. The C/A code is what everyone receives

Fig. 6 GPS modernization program

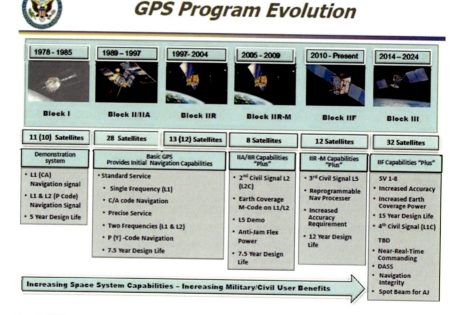

Fig. 7 GPS program evolution

– it is the code within the SPS. The P(Y) is an encrypted code and can only be received by those with the appropriate keys. This is the code that is obtained when a user is subscribed to the PPS.

Block IIA and IIR satellites transmit only two signals and codes – C/A on L1 and P(Y) on both L1 and L2. The Block IIR series are "replenishment" satellites developed by Lockheed Martin. Each IIR satellite weighs 4,480 lb (2,030 kg) at launch and 2,370 lb (1,080 kg) once on orbit. The first attempted launch of a Block IIR satellite failed on January 17, 1997, when the Delta II rocket exploded 12 s into flight. The first successful launch was on July 23, 1997. Twelve satellites in the series were successfully launched.

Within the Block IIR-M, the M-code, a second civil signal (L2C), and flex power have been added. L2C enables civil receivers to correct for the ionosphere. Flex power allows power to be transferred from one signal to another, thus providing some additional anti-jam capability. The first Block IIR-M satellite was launched on September 26, 2005, and the eighth and final launch of a IIR-M was on August 17, 2009. The prime contractor was Lockheed Martin.

The Block IIF series are "follow-on" satellites developed by Boeing and have everything that IIR-M offers but add a third civil signal on L5 (used during "safety of life" applications). Boeing built a total of 12 Block IIF satellites. The first was launched in May 2010 on a Delta IV rocket, and the twelfth on February 5, 2016. The spacecraft has a mass of 1,630 kg (3,600 lb) and a design life of 12 years. Compared to previous generations, GPS-IIF satellites have a longer life expectancy and a higher accuracy requirement. Each spacecraft uses a mix of rubidium and cesium atomic clocks to keep time within eight billionths of a second per day. They provide an improved military signal and variable power for better resistance to jamming in hostile environments.

The average contracted cost of a GPS-IIF satellite is $121 million although it is reported that Boeing has had to spend $306 million on each of the first three satellites. The price of the final nine GPS-IIFs was set in 2000.

The most advanced block in production and being deployed is GPS-III. It has all the same capabilities as IIF and many added capabilities. It has increased power in the form of a spot beam, and it has slightly better accuracy, mostly due to cross-links that greatly reduce the age of data. Each GPS-III also carries three improved rubidium clocks and transmits four civil signals: L1 C/A, L1C, L2C, and L5. Each spacecraft weighs 8,115 lbs (3,681 kg) at launch and 4,764 lbs (2,161 kg) when on-orbit. The first was launched in late 2018 (a slip of 4 years since the program evolution diagram above was published). That was also the first GPS satellite launched on the SpaceX Falcon 9 rocket – as of late 2018, SpaceX had won five of six GPS-III launch contracts so far awarded, the first for a price of $83 million, the others $97 million each (about one quarter the price of the United Launch Alliance rockets used for earlier GPS launches).

Lockheed Martin received a $1.5 billion contract to develop the first two GPS-IIIs, followed by a $2.1 billion extension for satellites three through eight ($350 million each) and a $395 million extension for satellites nine and ten ($200 million each). In September 2018 Lockheed Martin was selected to build 22 GPS-IIIF (follow-on) satellites at a price of $7.6 billion ($345 million per satellite). Additional features include a 100% digital navigation payload, Regional Military Protection, and new search-and-rescue payloads. Further features may be added at three "technology insertion points" in the GPS-IIIF program beginning with satellites 7, 13, and 19, respectively. The first launch of a GPS-IIIF satellite is scheduled for 2026, with the last foreseen in 2034. The GPS-III satellites are all built on the Lockheed Martin A2100 commercial satellite platform with refinements such as additional radiation shielding to ensure the required 15-year lifetime in the medium Earth orbit (MEO) where the radiation environment is less benign than in geostationary orbit where most A2100s are located.

The headquarters for the control segment is the master control station (MCS) at Schriever AFB in Colorado Springs, operated by the 2nd Space Operations Squadron (2 SOPS) of the 50th Space Wing of the US Air Force. In addition to the master control station, there are six dedicated GPS monitoring stations and ground antennas, located at Colorado Springs, Kwajalein Atoll, and Hawaii in the Pacific Ocean, Ascension Island in the Atlantic Ocean, Diego Garcia in the Indian Ocean, and Cape Canaveral in Florida.

GPS satellites are commanded and controlled via the monitor stations, the MCS, and the ground antenna. Monitor stations are essentially high-quality GPS receivers placed at various precise locations throughout the world. These receivers track the satellites just like a normal receiver would – they obtain the satellites' ephemeris and any downlinked data. Information is then transferred to the MCS where it is processed to calculate the orbit and clock offsets of each satellite. The MCS transmits that data back to the satellite via the ground antenna to update the satellite with its true location and time. These upload corrections occur at least once per day, although the system can operate autonomously for an extended period if necessary – albeit with reduced accuracy.

To give a snapshot of the scale of the US military commitment to GPS, the US Congress appropriated $1.4 billion for it in FY2019 which included $451 million for GPS-III satellites. $513 million was for the accompanying ground system, the centerpiece of which is a next-generation ground control system, called OCX, that has been plagued by delays, prompting the Air Force to pump more money into upgrading the existing GPS system as a stopgap. The DOD request to Congress for FY2020 includes $1.7 billion for GPS satellites and its ground system and user terminals.

A $1.5 billion OCX contract was awarded to Raytheon in 2010 to develop and deploy a major new version of the GPS control segment. The initial OCX capability was expected to be online in 2015, but additions and extensions have pushed that date back to the early 2020s, with costs rising to about $6 billion. Because of the delays to OCX, Lockheed Martin received a $96 million contract modification in 2016 to adapt the existing GPS ground system so that it provides a stopgap level of support for the GPS-III satellites that would allow them to be launched. Because of delays to the launch of the first GPS-III, the Block 0 version of OCX (known as the Launch and Checkout System, LCS) was deployed to support that launch.

Surveillance

Introduction

Large elements of the US military surveillance satellite capability are classified, and the information in this chapter is drawn only from unclassified sources with the inevitable risk that the information is incomplete or wrong. The scale of the programs is enormous, including what the Head of the National Reconnaissance Office described as "the world's largest satellite." The following is the breakdown of the topic in this section:

- Optical Imaging Satellites (Section "Optical Imaging Satellites")
- Imaging Radar Satellites (Section "Imaging Radar Satellites")
- Missile Early Warning Satellites (Section "Missile Early Warning Satellites")
- Nuclear Detonation Detection Systems (Section "Nuclear Detonation Detection System")
- Signals Intelligence Satellites – non-Maritime (Section "Signals Intelligence Satellites: Non-maritime")
- Maritime Signals Intelligence Satellites (Section "Maritime Signals Intelligence Satellites")
- Weather Satellites (Section "Weather Satellites")

Optical Imaging Satellites

Since the 1980s the US military has relied on a series of surveillance satellites referred to by the media as Keyhole satellites although the name for the more recent satellites is Enhanced Imagery System (EIS). They are large expensive spacecraft, each weighing 10–15 t and costing about $1.5 billion (of which a third is for the launcher) – one was said by NRO to have cost $2 billion less than originally projected, which suggests a multibillion dollar price tag. They orbit at between 250 and 1,000 km altitude and provide optical images that are said to have a resolution of 8–10 cm (3–4 in.) and less detailed infrared images. There are thought to be seven in orbit in early 2019 launched in 1995, 1996, 2001, 2005, 2011, 2013, and 2019, and there must be some doubt about whether the two older ones are still fully functional. The four launched in 2001 through 2013 are upgraded versions that provide very wide coverage as well as detailed images. They can rotate their camera to dwell on a scene as they pass across the sky. The infrared feature means that they can take nighttime images.

General Bruce Carlson (see Fig. 8) when he was the Director of the National Reconnaissance Office that buys and operates these satellites promoted a cost-saving concept called Next-Generation Electro-Optical reconnaissance satellite or NGEO. He said that it "will require a little bit of up-front investment" but will result in a modular and flexible satellite allowing NRO to insert new technology more rapidly than at present as well as reducing costs. He was determined to halt the "erosion in

Fig. 8 Gen. (retired) Bruce
Carlson

Credit: NRO
NRO Director Gen. Bruce Carlson

the [NRO's] science and technology base" that he described as "the seed corn for the future." Improving the technology to put into its expensive operational satellites is also addressed using tiny low-cost "CubeSats." NRO had 12 CubeSats under construction in 2009 according to NRO official Karyn Hayes-Ryan. She noted that they can be built in less than 6 months and thus enable NRO to "keep pace with Moore's law." Batteries, solar cells, computer processors, gyroscopes, and radios were some of the technologies she listed as being tested on CubeSats. As well as quickly and cheaply testing new ideas in space, this approach helps train people and is "more risk tolerant" according to Ms Hayes-Ryan. Two launches in 2013 carried a total of 40 NRO-related CubeSats into orbit.

General Carlson noted that many of the current satellites have lasted far longer than expected, and they deliver useful intelligence mainly because of "the young people that write software" to adapt the images the satellites produce. He joked that "we have satellites inside our very aging constellation that are old enough to vote and some, that are still operating, are old enough to drink. We don't let them drink, but they are old enough to drink." On another occasion he quipped that "half the constellation is geriatric." He pointed out however that because of the improvements in software "if you look at the product that we got ten years ago and compare it to the product that we have today, in many cases, it's an order of magnitude better in quality whether it's accuracy or clarity or timeliness."

An eighth satellite similar to the seven just mentioned may also be in orbit at about 800 km altitude and is said to be a stealth satellite, impervious to detection from the ground. Launched in 1999 it was part of a program called MISTY that is thought to have to been cancelled in 2006 when projected costs got out of hand. Intelligence specialist Jeffrey Richelson who first broke the story about the MISTY satellites is skeptical about their stealthiness. He notes that civilian observers were able to keep track of the first MISTY satellite launched in 1990 and watch its various maneuvers.

The EIS satellites are thought to resemble the Hubble Space Telescope (see Fig. 9) in shape and size, and are supplied by the same prime contractor, Lockheed Martin. This suggested resemblance was reinforced in 2012 when the NRO donated

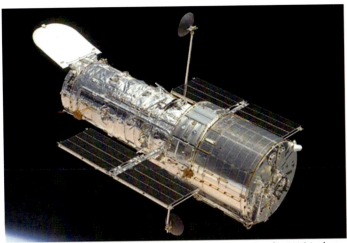

Credit: NASA photo

Hubble Space Telescope in orbit

Fig. 9 Hubble space telescope

two unused optical assemblies (essentially telescopes) to NASA similar in size to those of Hubble. The telescopes are thought to have been left over from the cancelled program to develop a replacement for the EIS satellites, called the Future Imagery Architecture. The telescopes weigh 1.7 t each and have the same sized primary mirror as Hubble, 2.4 m diameter, but lack cameras (detectors, filters, prisms, etc.), stabilization, power, and so on. They have a wider field of view than Hubble, and the optical quality of the mirror surface is not too dissimilar – about 60 nm versus Hubble's 30 nm. Budget figures leaked by Edward Snowden indicate that development of a new series of satellites began in 2012 called the Evolved Enhanced CRYSTAL System (EESC) – NROL-71/USA-290 launched in early 2019 is said by some analysts to be the first of this new series of imaging spy satellites. Unofficial orbital information puts it in a 400-km near-circular orbit inclined at 73.6°. The leaked funding figures also suggest that production of EIS satellites is being run down.

The US military has been a major purchaser of commercial satellite images (such as those of France's SPOT series) for many years. In 2003, the National Geospatial-Intelligence Agency (NGA) placed two contracts guaranteeing to purchase $150 million per annum of imagery from each of two suppliers and offered about half of the $500 million cost of building and launching each of the high-resolution optical satellites. The action by NGA followed on from a $100 million contract for high-resolution imagery in 1999 to a third company. A smaller contract followed in 2004 to that third company, and in 2005 that company merged with one of the 2003 winners leaving DigitalGlobe and GeoEye as NGA's two high-resolution commercial suppliers. In 2010 the NGA reinforced its commitment to using commercial imagery with the award of contracts valued at $7.3 billion in total over 10 years to the two companies, although cutbacks driven by the Federal budget deficit reduced the per year value of these contracts somewhat, partially as a consequence of which the two companies combined in 2013 under the name DigitalGlobe. In 2018, NRO assumed responsibility for the NGA contract and then awarded DigitalGlobe a $900 million extension to provide commercial imagery until August 2023. DigitalGlobe supports the NRO with three satellites – WorldView-1, World-View-2, and WorldView-3. WorldView-4 (launched in 2016) was to have been available should there be any interruptions or degradation in the performance of the earlier satellites (ground resolution of 30 cm is achievable) but stopped producing usable imagery in early 2019 due to failure of its control moment gyros. The company is building a $600 million WorldView Legion constellation, which will comprise at least three satellites, with the first scheduled to launch in 2021.

The formation of the NRO as a joint venture of the DOD and the CIA illustrates that the data from imaging satellites is used by two distinct communities – the military and the intelligence agencies. The friction between the requirements of the two communities surfaces from time to time, e.g., the urgent, tactical needs of the military in today's conflict zone versus the background gathering of information about tomorrow's enemy by the intelligence agencies. Budget pressures are said to have increased this tension and even to have contributed to the relatively early retirement of General Carlson as NRO Director.

The NGA, with a budget of about $5 billion per annum, is the arm of the military and intelligence communities that "develops imagery and map-based intelligence solutions for US national defense, homeland security, and safety of navigation." In other words they process the imagery from spy satellites.

Imaging Radar Satellites

In the summer of 2008, the US Administration declassified the fact that the country has radar imaging satellites. This was hardly Earth-shattering news since the "secret" had been widely known and discussed for more than a decade. It is also common knowledge that US efforts to develop a new generation of imaging radar satellites had become so expensive that the effort was cancelled in 2008. Perhaps this more cost conscious approach explains why the number 2 official at the NRO, Betty Sapp, was able to tell a Congressional Subcommittee in April 2010 that "the NRO received an Unqualified Opinion on its fiscal year 2009 Financial Statement." More surprisingly, she added that "this was the first clean audit for a defense intelligence agency since 2003."

As of late 2018, some analysts consider that there are five of the most recent generation of radar satellites in orbit, produced by Lockheed Martin. These Topaz satellites (often called "Lacrosse" or sometimes "Onyx" by the media) are thought to weigh about 4 t and are extremely large (50 m across with their solar arrays deployed) in order to generate the 10–20 kW of power needed to drive the radar – and correspondingly expensive at $1½ billion each. They provide imagery through cloud and in the dark with a resolution of about 1 m. They orbit at about 1,000–1,100 km altitude with orbital inclination of about 60° (or 120°), while earlier models orbited about 400 km lower and were considerably heavier (14 t). Ground-based imagery taken by amateur observers appears to show a gold-colored circular or ellipsoidal radar dish of about 50 m diameter – probably with a mesh surface rather than a solid one. Leaked budget figures show that procurement of a Block 2 series of the Topaz satellites began in 2013.

The USA has turned to the international commercial marketplace to augment the Topaz radar images. In 2008, the NGA began to buy small quantities of radar imagery from Canadian company MDA which owns and operates the Radarsat satellites. Then in early 2010, the NGA placed contracts to buy up to $85 million worth of data over a 5-year period from each of the three foreign imaging radar satellite families: Germany's TerraSAR-X and Tandem-X, Italy's four COSMO/SkyMed satellites, and Canada's Radarsat-1 and Radarsat-2. The US Southern Command has also purchased some Israeli TecSAR imagery probably to help in the war on drugs in Central and South America.

Missile Early Warning Satellites

For more than 40 years, the US satellites used to spot missile launches (also called early warning satellites) have been labeled as Defense Support Program (DSP) satellites. These have now been replaced by the Space-Based Infrared System (SBIRS).

The first DSP was launched in November 1970, although a series of experimental satellites, with the designations MIDAS and RTS, had been launched throughout the 1960–1966 period to demonstrate the concept of launch detection and try out various techniques – different orbits, sensors, data recovery methods, and so on. The general idea was to watch for the bright flash of light and flare of heat given off by a rocket motor. They proved very successful at spotting Soviet and Chinese missile launches. DSP data allowed the launch site to be pinpointed to within 3–15 km and the launch heading to within 5–25° depending on various factors such as the relative location of the launch and the DSP satellite. There were a few false alarms but only for smaller missiles such as submarine-launched missiles in the northern hemisphere summer (due to the glint of the sun on the sea).

Any bright flash triggered the DSP sensors, but software algorithms sorted out the missile launches from ammunition dump explosions, forest fires, gas pipeline fires, the burnup of satellites reentering the Earth's atmosphere, and military jet aircraft using afterburner – even the July 1996 explosion of TWA Flight 800.

As the capabilities of DSP satellites have grown, so have their weight and power. Unlike the old lightweight, low-power satellites, the newest generation of DSP satellites weighs over 5,000 lb, and the solar arrays generate 1,285 W of power. The most recent DSP satellite is approximately 33 ft long and 22 ft in diameter. The system comprises the satellite vehicle, also referred to as the bus, and the sensor. The satellites are placed in geosynchronous orbit. Global coverage can be efficiently achieved with three satellites. Additional satellites can provide dual or triple coverage, providing for more accurate and timely event reporting.

The DSP satellite spins around its Earth-pointing axis, which allows the infrared (IR) sensor to sweep across each point on the Earth.

The DSP sensor detects sources of IR radiation. A telescope/optical system and a photoelectric cell (PEC) detector array, comprised primarily of lead sulfide detectors and some Mercad-Telluride cells for the MWIR detection capability, are used to detect IR sources.

The SBIRS replacements for the DSP missile warning satellites are now in orbit. The first two elements of the highly elliptical orbit (HEO) part of the SBIRS constellation were launched in 2006 and 2008, piggybacking on other unidentified (for security reasons) US military satellites. The third and fourth SBIRS HEO payloads have also been launched – some analysts have identified NRO payloads USA-259 launched on December 13, 2014, and USA-278 launched on September 24, 2017, as their host spacecraft. Four SBIRS have been launched into geosynchronous orbit in 2011, 2013, 2017, and 2018. Two more (GEO-5 and GEO-6) are in production, due to be launched in 2021 and 2022. Lockheed Martin Space Systems in Sunnyvale, CA, is the SBIRS prime contractor, with Northrop Grumman Electronic Systems in Azusa, CA, as the payload subcontractor, and Lockheed Martin Information Systems and Global Services in Boulder, CO, as the ground system subcontractor.

The technology used in SBIRS is much more powerful than in DSP. Unlike DSP, the output of the sensor is an image not just a location, so further analysis can be performed on it.

The GEO spacecraft bus consists of a militarized, radiation-hardened version of the Lockheed Martin A2100 spacecraft, providing power, attitude control, command and control, and a communications subsystem with five separate mission data downlinks to meet mission requirements. The GEO infrared payload consists of two sensors: a scanner and a step-starer. The scanning sensor continuously scans the Earth to provide 24/7 global strategic missile warning capability. Data from the scanner also contributes to theater and intelligence missions. The step-staring sensor, with its highly agile and highly accurate pointing and control system, provides coverage for theater missions and intelligence areas of interest with its fast revisit rates and high sensitivity. Similar to the GEO scanning sensor, the HEO sensor is a scanning sensor, with sensor pointing performed by slewing the full telescope on a gimbal. Both the GEO and HEO infrared sensors gather raw, unprocessed data that are downlinked to the ground, so that the same radiometric scene observed in space will be available on the ground for processing. The GEO sensors also perform onboard signal processing and transmit detected events to the ground, in addition to the unprocessed raw data.

The GEO satellite weighs 5,525 lbs (2,506 kg) in orbit including the 1,100 lbs (499 kg) two-sensor payload and 430 lbs (195 kg) of fuel. The HEO sensor payload weighs 530 lbs (240 kg).

The sophistication of the facilities on the ground has also been enhanced – more powerful computer processing to analyze the information and faster relaying of results to the troops, ships, or planes that can do something about it. Even before the first SBIRS was launched, the timeliness and quality of DSP information had improved because of this. One industry official claims SBIRS will be ten times better than DSP, able to detect dimmer and more fleeting targets, and presumably the effects of the upgraded ground facilities will improve that even further.

The second SBIRS piggybacking on an unnamed military satellite is also sending back high-quality information. Figure 10 shows the track of a missile above the clouds as seen by SBIRS, indicating that the trajectory can be spotted and analyzed in quite some detail.

Already 7 years behind schedule, the first geosynchronous SBIRS satellite fell foul of a very modern problem in 2009 in that its software was found to be faulty, resulting in a delay and modification that cost $750 million. The faults were so bad that USAF Colonel Robert Teague said "the [software] design and architecture had fundamental flaws. The solution essentially required starting from scratch." General Robert Kehler, the head of the USAF Space Command, sounded resigned to even more delays and cost increases when in the fall of 2009 he said that "I don't know if there will be any other impacts to schedule and cost. I'm not optimistic or pessimistic. We are where we are with SBIRS." What is now considered to be the full constellation of two geosynchronous and two elliptical satellites costing about $15 billion was due to be operational in 2014, but the system won't achieve full use of any GEO satellite staring sensors until the ground systems are modified, around 2018.

SBIRS is an integrated "system of systems" consisting of ground components as well as the satellites described above. The ground component consists of control stations such as the Mission Control Station located at Buckley AFB, Colorado, which is responsible for consolidating event data from dispersed legacy DSP ground systems. Remote ground stations (including mobile and deployable stations) receive missile warning data from the satellites and feed the data via secure communications

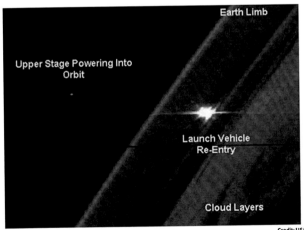

Credit: USAF

Space-Launch, Orbital Insertion Engine and Launcher Re-Entry detected by SBIRS HEO-1

Fig. 10 SBIRS HEO-1 Image

links to ground stations for processing. The ground stations assess system reliability, attempt to identify the type of launch occurring, and generate a launch report. Crews send these reports to the NORAD operations centers at Cheyenne Mountain AFS, Colorado; the Alternate Missile Warning Center at Offutt AFB, Nebraska; and other command centers. Air Force Space Command's 460th Operations Group is responsible for conducting HEO, GEO, and DSP operations at all fixed ground sites.

Because of the lengthy delays and cost overruns in deploying the SBIRS satellites, an alternative approach was begun leading to the launch of a flight demonstrator in September 2011. The Commercially Hosted Infrared Payload (CHIRP) was hosted on the geostationary commercial telecommunications satellite, SES-2, developed by Orbital Sciences. The sensor payload was supplied by SAIC of McLean Virginia. CHIRP is a wide field of view staring sensor designed to detect and track the heat signature of missiles at launch. The budget of $216 million is considerably lower than the >$1 billion budget for the SBIRS satellites, but now that those are finally being deployed, the CHIRP concept seems likely to be discontinued.

In May 2018 the planned purchase of two further geosynchronous SBIRS satellites (GEO-7 and GEO-8) was cancelled and a plan announced to buy three upgraded geosynchronous Overhead Persistent Infrared (OPIR) satellites from Lockheed Martin and two polar-orbiting spacecraft from Northrop Grumman. Northrop was awarded $47 million to begin analysis and risk reduction work for the first two polar-orbiting satellites, while Lockheed Martin was authorized to begin up-front work on a potential $2.9 billion contract. Fiscal Year 2019 funding for OPIR was $806 million, and the FY2020 request to Congress was for $1.6 billion. The satellites are scheduled for launch on a rapid time line, with the first in 2023. The planned constellation is due to be in place by 2025, 4 years earlier than the original plan for an Evolved SBIRS.

Nuclear Detonation Detection System

The GPS satellites described in section "Global Positioning System (GPS)" carry an additional payload suite to support a nuclear detonation detection system (NDS). The NDS payload is also carried by the DSP early warning satellites described above (which are gradually being phased out) and was initially the primary mission of the VELA satellites. The sensor array includes optical, x-ray, dosimeter, and electromagnetic pulse (EMP) sensors which detect and measure light, x-ray, subatomic particle, and EMP phenomenology to pinpoint the location and yield of a surface or airborne nuclear detonation. A nuclear explosion starts off with a short very intense flash and then a short dimming while expanding gas overtakes the fireball and masks it from view until the gas has thinned and cooled enough for the fireball to be visible again – this dual flash enables the NDS to distinguish a nuclear explosion from other events. The information sensed on the GPS NDS system is relayed to the ground-based Integrated Correlation and Display System (ICADS) via a dedicated channel,

L3 (1381.05 MHz). NDS supports several tasks, such as treaty monitoring and nuclear force management.

GPS-III satellites carry a newly designed nuclear detection payload, details of whose performance are classified.

Signals Intelligence Satellites: Non-maritime

The first US communications intelligence (COMINT) satellite was launched in 1968. Called CANYON it was the first of a series that is thought to continue today although the name has changed over the years – to CHALET then VORTEX then MERCURY and currently NEMESIS and ORION. The name changes were to some extent caused by security lapses such as that by British spy (and pedophile) Geoffrey Prime and US spy Christopher Boyce.

The technology of these eavesdropping satellites is impressive. They are located 36,000 km out in space in the geostationary arc. Being so far from the ground, the satellites need enormous radio antennas to pick up radio signals. The MUOS antenna (see section "MUOS and UFO," above) is 14 m in diameter when unfurled in orbit. But the secret eavesdropping satellites are thought to have already been 50 m wide by 1994 and 90 m by 2006 – at least one newspaper report claimed they had dishes 150 m wide (almost 500 ft). The Head of the NRO, General Bruce Carlson, said that the satellite launched on November 21, 2010, from Cape Canaveral was "the largest satellite in the world" – thought to be one of the ORION series mentioned below.

The first satellite to "watch" the Earth was in fact an electronic intelligence (ELINT) satellite – the tiny GRAB satellite. GRAB was launched into orbit in 1960 and provided hitherto inaccessible information about radar systems deep inside the Soviet Union. GRAB worked by receiving radio signals across a very wide range of frequencies and relaying them to a station on the ground. The satellite didn't process the signals in any way, just acted like a "bent pipe" in taking radio signals that it heard at its 1,000 km altitude and beaming them downwards to a friendly station. By having a station suitably located near the border, signals from 5,500 km inside the Soviet Union could be picked up. Signals from two Soviet radar systems, one associated with SAM-1 antiaircraft missiles and the other with missile early warning, were detected and could then be analyzed. The advantage of knowing the characteristics of an adversary's radar was not only that you could in future know what it was when you detected it, but you could design electronic countermeasures to jam it or confuse it. GRAB was the first of a long line of what became known as ELINT satellites.

The success of GRAB encouraged the USA to use satellites extensively for analyzing its enemy's radars and other military signals. The tiny GRAB evolved into massive satellites with enormous antennas as discussed above. Many of these satellites were placed in geostationary orbit, 36,000 km above the Earth, hence the need for their giant antennas. One series already mentioned above, initially targeting communications intercepts, began with the launch of CANYON in 1968 which evolved a decade later into CHALET. A second series that was targeted at data from missile tests began in 1970 with the launch of the first RHYOLITE satellite (later renamed AQUACADE). (The names of US spy satellites are always written in uppercase – the reason for this pretentious habit is not known to the author.) This series evolved to pick up communications transmitted across the Soviet Union via microwave towers. It was succeeded in 1985 by the first of two MAGNUM satellites. In 1994 a new generation of satellites began to appear. As we approach the present day, details about the current US eavesdropping satellites become more and more vague.

In addition to satellites in geostationary orbit, another group of eavesdropping satellites has been picking up radio signals in what is called a Molniya orbit discussed in section "Data Relay" – this is an elliptical orbit that is at its maximum height over the northern hemisphere and at its lowest over the southern. Their maximum height is similar to a geostationary satellite with a correspondingly broad view of the Earth below. The big advantage is that they cover northern latitudes better than from geostationary orbit, but the disadvantage is that they move along their orbit from north to south and thus are over the northern area of interest only for 8 or so hours a day. Three satellites in Molniya orbit are sufficient to provide continuous coverage over a given area. Initially called JUMPSEAT satellites, these evolved into the TRUMPET satellites that were launched in 1994, 1995, and 1997. The first in what is said to be a TRUMPET replacement series was launched in 2006. Launches on December 13, 2014, and September 24, 2017, carried payloads that may be part of this series.

There are thought to be about a dozen operational US COMINT satellites of this type currently in orbit. Leaked 2011–2013 budget documents list three series called RAVEN (unfunded since before 2011), NEMESIS (unfunded in 2012 and 2013), and ORION. Recent launches of satellites that fall into one or other of the COMINT categories are said to include NRO-26 in January 2009, NRO-32 in November 2010, NRO-15 in June 2012, and NROL-37 in June 2016.

The cost of these satellites is astronomical. Press reports of the January 2009 launch of the 6 t NRO-26 referred to it as a "highly upgraded ORION electronic eavesdropping satellite" and said that "the combined cost of the spacecraft plus booster is roughly $2 billion." Leaked budget documents show that a "SIGINT High Altitude Replenishment Program (SHARP)" is underway costing about $800 million per annum. This suggests that the satellites that emerge from this program will continue to cost in excess of $1 billion each. The NROL-67 launch on April 10, 2014, may have been one of the satellites from this program. Recent cost overruns of several types of US spy satellites have become public knowledge, and ELINT satellites are no exception. An attempt to move to a more cost effective approach was discussed in the trade press for several years and would have involved a larger number of cheaper satellites – but it seems to have been a failure, since the large expensive type is still being launched every 2 or 3 years.

The high cost of these monster satellites means that they are designed to last for 15 or more years – it would be too expensive to replace them more frequently. But communication is an area which changes dramatically every few years, so the expensive long-lasting satellites risk becoming obsolete before they have served their term.

Maritime Signals Intelligence Satellites

The USA has for many years deployed constellations of satellites in orbit groups of two, three, or four satellites so that they could triangulate the signals from a ship using the different time and direction of the ship's radio signals as seen by each satellite in the group. The various series of US satellites originally had friendly names (at least for use publicly) starting with Poppy in the 1960s and then White Cloud in the 1970s and 1980s. More recently these satellite groups are referred to as NOSS (Naval Ocean Surveillance Satellite) followed by a −1, −2, or −3 to signify which generation (White Cloud being NOSS-1), and the program has the unfriendly official name, INTRUDER.

Pairs of satellites (the USAF designates one of them as "debris" presumably in an attempt to disguise their mission) launched as NRO-34 in April 2011, NROL-36 in September 2012, NROL-55 in October 2015, and NROL-79 in March 2017 are said to be members of the NOSS-3 series.

A pair of NOSS satellites, each weighing about 3 t, plus the rocket on which they are launched into orbit costs about $600 million – there are thought to be 20 or so in orbit altogether, spaced so as to give as near as possible continuous coverage of the ocean areas of interest. A database of all ship movements is maintained based on NOSS information plus data from Navy and Coast Guard air and sea surveillance. Civilian as well as military ships are tracked reflecting the threat from terrorists and the interest in the navies of countries such as China and Iran.

In addition to detecting ships and working out their exact location by triangulation, these satellites allow the ships' radar signals and radio communications to be collected for later analysis. In recent years the US military have acknowledged that specific launches are in the NOSS category but not that they are a group – civilian observers have detected additional objects close to the acknowledged one, but the military refer to those objects as "debris." The current INTRUDER (NOSS-3) satellites come in pairs as has been recorded by many ground-based observers just by watching them with binoculars. The satellites are 1,000–1,200 km high with an orbital inclination of about 63° and one of the pair trails the other by 50–250 km. The height is a compromise between being as close to the Earth as possible in order to detect faint emissions and being as high as possible to get a broad view. The relatively small separation of the pair ensures that they both see the same target, thus allowing the signals to be compared and the target's location and direction to be triangulated.

There is occasional press and web speculation that INTRUDER satellites possess radars of their own and thus detect ships that are observing radio and radar silence. Other reports say that INTRUDER satellites can detect the faint radio emissions from a ship's engines and the magnetic effects of a ship's hull.

Weather Satellites

The Defense Meteorological Satellite Program (DMSP) has been providing weather information for the Department of Defense for half a century (initially it was called Program 35).

Two primary operational DMSP satellites are always in near-circular, sun-synchronous, 835 km altitude, near-polar orbits with a period of 101.6 min and an inclination of 98.75°. The primary weather sensor on DMSP is the Operational Linescan System, which provides continuous visual and infrared imagery of cloud cover over an area 1,600 nautical miles wide (see Fig. 11). Global coverage of weather features is accomplished every 14 h providing essential data over data-sparse or data-denied areas. Additional satellite sensors measure atmospheric vertical profiles of moisture and temperature. Military weather forecasters can detect developing patterns of weather and track existing weather systems over remote areas, including the presence of severe thunderstorms, hurricanes, and typhoons.

The DMSP satellites also measure space environmental parameters such as local charged particles and electromagnetic fields to assess the impact of the ionosphere on ballistic-missile early warning radar systems and long-range communications. Additionally, these data are used to monitor global auroral activity and to predict the effects of the space environment on satellite operations.

DMSP-5D3-F18 was launched in 2009 on an Atlas V 401 and DMSP-5D3-F19 in 2014. Compared with earlier versions, the current Block 5D-3 satellites include upgraded instruments, increased power capability, improved on-orbit autonomy

Credit: USAF

DMSP's Operational Linescan System produces day & night visual & infrared imagery, such as this night-time view of East Asia showing Japan (right), S Korea (centre) with N Korea almost entirely dark above it, and China (left)

Fig. 11 DMSP image of East Asia

(60 days), enhanced design-life duration of 5 years, solid-state data recorders, and a UHF downlink which enables data to be sent directly to tactical users. DMSP-5D3-19 suffered a failure of its onboard command and control system in February 2016 and was declared a complete loss in July that year. DMSP-5D3-17, launched in 2006, was reinstated as the second operational unit together with DMSP-18. The only remaining unlaunched DMSP is now in a museum, Congress having cancelled funding to store it in flight-ready condition. The fragile condition of old DMSP satellites was illustrated in February 2015 when the defunct DMSP-5D2-F13 launched in 1995 spun off a large number of pieces of debris, probably due to a battery explosion.

The next-generation DMSP military weather satellites and civilian National Oceanic and Atmospheric Administration (NOAA) satellites were to have been merged into the National Polar-orbiting Operational Environmental Satellite System (NPOESS). In 2010, with $6 billion having been spent on NPOESS already and the total budget still rising, the program was cancelled. The DOD initiated the Defense Weather Satellite System (DWSS) to replace it, but in 2012 that too was cancelled due to budget constraints.

The DMSP 5D-3 satellites weigh 1¼ ton of which 350 kg is the sensor payload, and the solar panels produce 2.2 kW of power. The satellites are over 7 m long with solar panels deployed.

As a near-term alternative to DMSP, in September 2018 the Air Force awarded a $119 million contract for the Operationally Responsive Space-8 (ORS-8) satellite to Sierra Nevada Corp. ORS-8 will be an experimental satellite providing cloud characterization and theater weather imagery and will be launched in 2022. As of early 2019, this contract was on hold following a protest by Space Systems Loral (a Maxar Technologies company).

Longer term, two series of Weather System Follow-on (WSF) satellites are envisaged – one carrying mainly microwave instruments and the other carrying mainly optical and infrared instruments. Ball Aerospace received a $94 million contract in November 2017 to design and develop the WSF Microwave satellite, with options to construct two flight models for a total contract value of $458 million, with the first to be launched in the 2022–2024 period. A request for information to industry for the WSF electro-optical infrared (WSF-E) satellite was issued in November 2017, envisaging launch of the satellite in 2023 or 2024.

As part of the NPOESS convergence plan, DMSP operations were transferred from the Defense Department to the Commerce Department in 1998, with funding responsibility remaining with the Air Force. Satellite operations were moved to Suitland, MD, where NOAA's Office of Satellite Operations provides the command, control, and communications for both DMSP and NOAA's Polar-orbiting Operational Environmental Satellite System. Although NPOESS has been cancelled, NOAA continues to operate the DMSP satellites using a common civil/military ground system.

Tracking stations at New Boston Air Force Station, NH; Thule Air Base, Greenland; Fairbanks, Alaska; and Kaena Point, Hawaii, receive DMSP data and electronically transfer them to the Air Force Weather Agency at Offutt Air Force Base,

Nebraska. Tactical units with special equipment can also receive data directly from the satellites.

The Space and Missile Systems Center at Los Angeles AFB, California, is responsible for development and acquisition of DMSP systems. The ORS-8 spacecraft procurement is managed by the Space Rapid Capabilities Office (part of Air Force Space Command) at Kirtland Air Force Base, NM.

Other Satellites

Space-Based Space Surveillance (SBSS)

The first Space-Based Space Surveillance (SBSS) satellite was launched in 2010 able to detect debris, spacecraft, or other distant space objects without interference from weather, atmosphere, or time of day. SBSS is especially useful for monitoring small objects in geostationary orbit – ground-based optical telescopes are the alternative solution for this function. Under a $189 million contract, Boeing has overall responsibility for the SBSS system, including the ground system and initial mission operations and including Ball Aerospace which is providing the satellite and sensor. The 1 t satellite has a sensor sensitive to visible wavelengths with a large aperture that provides a wide field of view. The sensor is mounted on a two-axis gimbal so that it can be quickly and efficiently moved between targets. It has twice the sensitivity and triple the probability of detecting threats of alternative techniques such as the MSX Space-Based Visible Sensor (SBV), which ceased operation in December 2008, and can monitor objects as small as a 1-meter cube all the way out to the geosynchronous belt. The SBSS satellite is in a 390-mile high circular sunsynchronous orbit (inclination 97.8°). It is operated from the Satellite Operations Center located at Schriever AFB, Colorado Springs. Initial operational capability was delayed until August 2012 due to faulty electronics which reset when passing through the South Atlantic Anomaly (a low hanging spur of the ionosphere) requiring a software upload in April 2012.

The 113 kg ORS-5 satellite was launched in August 2017 to act as a bridge between the SBSS satellite and the yet to be confirmed SBSS follow-on program. Also referred to as SensorSat, ORS-5 is in an equatorial orbit at 600 km altitude. The onboard sensor is a CCD imager that provides continuous imaging of the geosynchronous belt. The primary ground station is at the Air Force Satellite Control Network facility in Guam, with backup in Diego Garcia plus limited backup in Hawaii. Prime contractor for the $87 million development contract was the Massachusetts Institute of Technology's (MIT) Lincoln Laboratory. The Northrop Grumman (Orbital ATK at the time) Minotaur-IV launcher costs $24 million. What is now the Space Rapid Capabilities Office at the Kirtland Air Force Base managed the program. In orbit the satellite is operated by the Air Force Space Command's 50th Space Wing at Schriever AFB.

The USAF is reviewing the requirements for follow-on satellites that would make up a constellation in space taking into account the results from the SBSS-1 and ORS-

5 satellites and from improvements in ground-based technology such as the $110 million Space Surveillance Telescope at White Sands Missile Range, New Mexico. The SBSS follow-on is intended to continue the expansion of data collection of the geosynchronous belt and to complement the up close inspection capabilities of the GSSAP system (see section "GSSAP").

GSSAP

Four Geosynchronous Space Situational Awareness Program (GSSAP) satellites are in orbit. They are able to perform rendezvous and proximity maneuvers to allow close-up looks at spacecraft in geosynchronous orbits with their electro-optic sensors and can provide the location, orbit, and size of satellites and space objects. With a price tag thought to be about $700 million, GSSAP operates in a "near-geosynchronous orbit regime" to provide accurate tracking and characterization of man-made orbiting objects (Fig. 12).

In 2015, Gen. John Hyten, the head of Air Force Space Command, said that the Defense Department had used the satellites to capture "truly eye-watering" images for unspecified users while they were in test mode. The system improves the US capability to rapidly detect, warn, characterize, and attribute disturbances to space systems in the geosynchronous environment. One unclassified use of the system was to check on the MUOS-5 communications satellite (see section "MUOS and UFO") that ran into propulsion problems about halfway to geosynchronous orbit.

The satellites were built by Northrop Grumman (formerly Orbital ATK) of Dulles, Virginia. The first two were launched in July 2014, and the program achieved initial operational capability in September 2015. Two replenishment satellites were launched in August 2016 and accepted into operation in September 2017. The Air

Fig. 12 MiTEx mission badge

Force ordered two more GSSAP satellites from Northrop Grumman for an undisclosed price in late 2016.

Operated by the Space & Missile Systems Center at Los Angeles AFB during its in-orbit test phase, GSSAP was turned over to 1st Space Operations Squadron of the 50th Space Wing of the 14th Air Force at Schriever AFB in Colorado CO for the operational phase.

GSSAP is one element of the US infrastructure aimed at addressing Space Situational Awareness. The other main elements are on ground – a worldwide network of radars and telescopes that provide the main inputs to a detailed catalogue of space-based objects. A recent upgrade to that ground infrastructure called the Space Fence is scheduled to begin operation in 2019. In June 2014 Lockheed Martin won the $914 million contract to build the Space Fence, an S-band phased array radar that the Air Force says will improve its ability to detect and track Earth orbiting objects. Defense Department officials said they are optimistic that on the best days, the Space Fence may be able to track objects as small as 1 cm. That's a marked improvement over the current network of radars and sensors, which tracks objects 10 cm and larger. That additional precision means the Air Force will have tracking data for 200,000 objects, up from the approximately 20,000 objects it tracks today. In turn, the new data could lead to an orders of magnitude increase in collision warnings.

The initial Space Fence facility on Kwajalein Island in the west-central Pacific Ocean will deliver 80 percent of the envisioned overall capability. After that enters operation, work is planned to start on a second site in Western Australia which would become operational around 2022 and improve the timeliness and accuracy of observations.

Two MiTEx satellites launched in 2006 were precursors to GSSAP, demonstrating a new way to monitor objects in geostationary orbit – go there and look around. The 225 kg satellites are small and thus difficult to detect from the ground. Their ability to inspect geostationary orbiting objects was demonstrated in December 2008 and January 2009 when they were commanded to approach the malfunctioning 2½ ton DSP-23 early warning satellite (see section "Missile Early Warning Satellites") that was drifting eastward across the geostationary arc at about 1° per week from its 8.5°E longitude starting point – posing a threat to other satellites as it drifted. The MiTEx microsatellites tested a range of technologies for the Defense Advanced Research Projects Agency (DARPA), the US Air Force, and the US Navy. These include avionics, advanced communications, fuels that spontaneously ignite on contact, solar cells, and new software. One was built by Northrop Grumman (Orbital Sciences at the time), and the other was by Lockheed Martin.

Research Satellites

US military technology proving and mission demonstration satellites are typically the subject of two or three launches each year. The satellites are often members of a series with names such as TacSat, Operationally Responsive Space (ORS), STPSat,

FalconSat, etc. Increasingly, CubeSats and other forms of micro- and nano-satellites are part of the mix. Given their ad hoc and nonoperational nature, they will not be described here – with the following exception.

The most widely reported USAF research space mission in recent years has been that of the X-37B series so that will now be briefly described. The X-37 began as a NASA project in 1999, before being transferred to the Department of Defense in 2004.

Two X-37B robotic spaceplanes (mini, unmanned versions of the space shuttle) have been into orbit and reentered successfully and then landed horizontally and autonomously. The schedule of the flights of Orbital Test Vehicle (OTV) -1 and OTV-2 is shown in Table 1. Each vehicle weighs about 5 t (barely one-twentieth that of the space shuttle), and they have been launched from Cape Canaveral, Florida, on an Atlas V 501 or Falcon 9 rocket and landed at Vandenberg Air Force Base, California, or at Kennedy Space Center. Their orbits have not been publicly announced, but amateur enthusiasts have documented variations of orbit during each flight, generally in the 300–400 km altitude region with an inclination of about 40°. For example, OTV-2 was in a 330 × 340 km × 42.8° orbit until a series of burns lowered it to 280 km 3 weeks before its reentry.

X-37B program manager Lt. Col. Tom McIntyre noted that "with the retirement of the space shuttle fleet, the X-37B OTV program brings a singular capability to space technology development. The return capability allows the Air Force to test new technologies without the same risk commitment faced by other programs."

The details of the X-37B's mission are classified, as is its payload. The US Air Force's deputy undersecretary for space programs, Richard McKinney, described it in 2010 as "a test vehicle to prove the materials and capabilities, to put experiments in space and bring them back and check out the technologies. This is, pure and simple, a test vehicle so we can prove technologies and capabilities."

The X-37B project is managed by the USAF's Rapid Capabilities Office located at Joint Base Anacostia-Bolling in Washington, DC, with mission control for orbital flights based at the 3rd Space Experimentation Squadron at Schriever Air Force Base in Colorado. The vehicle was developed by Air Force Research Laboratory offices AFRL/RV at Kirtland AFB, New Mexico, and AFRL/RB (Air Vehicles) at Wright-

Table 1 OTV flight schedules

	Flight 1	Flight 2	Flight 3	Flight 4	Flight 5
Vehicle	OTV-1	OTV-2	OTV-1	OTV-2?	OTV-1?
Launcher	Atlas V 501	Atlas V 501	Atlas V 501	Atlas V 501	Falcon 9
Launch	April 22, 2010	March 5, 2011	Dec. 11, 2012	May 20, 2015	Sept. 7, 2017
Landing	Dec. 3, 2010	June 16, 2012	Oct. 17, 2014	May 7, 2017	TBD
Landing site	Vandenberg AFB	Vandenberg AFB	Vandenberg AFB	Kennedy Space Center	TBD
Mission length	224 days	469 days	674 days	717 days	>500 days

Patterson AFB, Ohio, with Boeing Phantom Works. This sets it apart from other USAF space projects which are mostly managed out of Los Angeles and indicates a closer connection with the "air" side of the Air Force.

Future Prospects

The short-term upgrades and replacements for the various satellites described in this chapter have been covered alongside the description of the current systems. There are some general trends across all of the programs including:

- There is an increased, but still not universal, willingness within the US military and intelligence communities to use commercial satellite services.
  - In telecommunications, the military is a major purchaser of ad hoc commercial services but has avoided the long-term relationships with commercial suppliers of the type adopted in, for example, the UK and Spain.
  - In the surveillance world, the US military dominates the world marketplace for optical imagery. Recognizing its ability to influence the commercial market, it has underpinned the financing of the combined DigitalGlobe/GeoEye company, by entering long-term data and service purchase agreements with it. In parallel, it continues to build and operate very sophisticated systems itself without any sign of a letup.
  - The GPS navigation system remains a DOD program despite the fact that the service is the mainstay of an enormous global commercial business. With the entry of Russian, Chinese, and European systems with performance similar to GPS, and all providing the navigation signal free of charge, DOD will almost certainly have to continue funding the system.
- DOD space systems have with rare exceptions been late and over-budget. Many new systems in telecommunications and surveillance have been cancelled in recent years because of this, some of which have been mentioned in the previous sections.
  - The NRO Director from 2009 to 2012, General (Ret.) Bruce Carlson, worked wonders in overseeing the successful deployment of several much-delayed and much-needed surveillance systems in 2010–2012. It remains to be seen if NRO can continue this successful record after his departure.
  - Systems such as MUOS (Section "MUOS and UFO") and SBIRS (Section "Missile Early Warning Satellites") have been placed in orbit despite the fact that their support systems will not be fully available for some years to come. GPS-III (Section "Global Positioning System (GPS)") is likely to follow suit.
- There is some recognition in the US military that large sophisticated satellites are sitting ducks for anti-satellite weapons and that more and simpler satellites might provide a more robust service. However, this intellectual analysis has largely failed to influence the latest US military space systems, which are generally larger than the previous generation. This situation is broadly at odds with the trend in the commercial space marketplace where small satellites have established an

important niche for customers willing to seek a compromise of price versus performance versus schedule.

- For the USA, "interoperability" usually means that the partners have to be compatible with US systems and often are required to purchase US equipment. The dominance of US forces in NATO means that this tendency is likely to continue.

Conclusions

The US military space program continues to dwarf (a) the military space programs of all other countries combined and (b) of US civilian agencies such as NASA. The USA is unique in deploying military satellites of all types and on a global basis, and there is little sign that this will change in the next decade.

References

Anon (2009) AU-18 space primer. Air University Press, Maxwell Air Force Base, Alabama
Norris P (2010) Watching earth from space. Springer Praxis, London/New York
Richelson JT (2008) The US intelligence community, 5th edn. Westview Press, Boulder

Russian Space Launch Program

60

Elina Morozova

Contents

Abstract

This chapter provides an overview of the current status of the Russian space launch program, including its main launch vehicle families and operational cosmodromes. A historical background, in particular on the achievements of the Soviet age and

E. Morozova (✉)
International Legal Service, Intersputnik International Organization of Space Communications, Moscow, Russian Federation

International Institute of Space Law (IISL), Paris, France
e-mail: morozova@intersputnik.com

© Springer Nature Switzerland AG 2020
K.-U. Schrogl (ed.), *Handbook of Space Security*,
https://doi.org/10.1007/978-3-030-23210-8_102

their application in the post-Soviet times, is given, and future developments are described, including those which are guided by government policy in the field of space activities, the federal space program, and specific dedicated projects, which are adopted to secure efficient implementation of the relevant tasks (see Federal Dedicated Program. Development of Spaceports to Support Space Activities, 2017–2025 (approved by the Government of the Russian Federation on 13 Sept 2017) and Federal Space Program of Russia for 2016–2025 (approved by the Government of the Russian Federation on 23 March 2016).

Introduction

In early October 1957, the world's first artificial Earth satellite launched into orbit by the Soviet R-7 rocket opened the way to outer space. A few pioneering accomplishments in the exploration and use of outer space achieved by the USSR followed shortly. They include the first space object that reached another celestial body and hard-landed on the Moon, the first human spaceflight and the first woman in space and the first spacewalk and the first docking and crew exchange between two manned spacecraft.

In the ensuing long-term space race of the USSR and the USA, considerable progress was made in space industry, including the development of launch vehicles of various classes and the construction of space launch facilities. Today, Russia, the USSR's successor since the very end of 1991, is sometimes called a "space carrier." Indeed, the existing fleet of space rockets helps to solve complex tasks of delivering all types of payloads to different Earth orbits. More so, in particular launch areas such as the delivery of crews to the International Space Station, Russian technology enjoys a monopoly.

During the entire history of post-Soviet space activities from 1992 to 2018, Russia remained the world leader in terms of the number of annual launches most often for a total of 19 years, 6 years second to the USA (1996–1999, 2003, and 2017), and 2 years second to the USA and China (2016 and 2018). For 13 years where the largest number of launches took place (1992–1995, 2000–2002, 2005–2006, and 2008–2010), Russia nipped and tucked or even outpaced the USA and China combined. In 2019, Russia is planning to carry out about 45 launches and regain the lead (Opening remarks by Dmitry Medvedev 2019). Along with that, one of the priority tasks remains to commercialize Russia's space industry and increase its share in the international market of launch services (Briefing by Yuri Borisov 2019).

Russia has all the trumps in hand to accomplish this task.

Main Launch Vehicles

The R-7 Intercontinental Ballistic Missile (ICBM), used for the first time as a carrier of space objects, was named Sputnik. Having made two successful launches only, it was sent for further modernization. Sputnik-1, the first artificial satellite, weighed just about 80 kilograms, while it was required to put much heavier payloads into various Earth orbits. The mass of the third artificial satellite exceeded 1.3 tons, a convincing indication that significant progress could be made in improving launchers' characteristics within a very short period of time.

Indeed, the 1960s competition between the Soviet Union and the USA to be the first on the Moon made them contend at the limit of scientific and technical capabilities. The USSR developed the Luna and the Molniya launch vehicles, which succeeded in performing the first hard and the first soft landings on the Moon of the Luna-2 and Luna-9 missions in 1959 and 1966, respectively. The Molniya was also successful in launching automatic interplanetary stations to another nearest planets of the solar system – Venus in 1961 and Mars in 1962. In 1963–1964, the Polyot launching vehicle flew Soviet maneuvering satellites.

On April 12, 1961, the first ever cosmonaut of the planet, Yuri Gagarin, was successfully launched on the Soviet Vostok rocket carrier. In order to provide the ability of launching manned space vehicles of a greater mass, the Vostok vehicle had to be modified. In turn, the Voskhod was capable of launching manned spacecraft with a crew of two or three people. Further modernization of the Voskhod, which was named Soyuz, made it possible to launch new manned spacecraft with the same name, Soyuz, capable of maneuvering and docking in orbit. Subsequently, the Soyuz was repeatedly refined and improved and significantly expanded the launch capabilities of the USSR and later of Russia.

Then the two spacefaring rivals were challenged to try strength against each other in the superheavy class to send a human to the Moon. In 1960, the development of Soviet high-power launch vehicles began, and the project was named N1-L3. In 1964, the USSR Government decree stated, for the first time, that the most important challenge to be met in space exploration by the N1 launch vehicle was to explore the Moon including an expedition landing on its surface and its subsequent return to the Earth. Besides the vehicle itself, the complex included a lunar system called L3 intended for the launching to the Moon and returning back to the Earth of a crew of two people with only one of them to be actually put down on the lunar surface. Four trial launches of the N1 were in varying degrees a failure. Just before the fifth one, after a considerable modification, the program was shut down in 1974. The most extensive experience gained in designing and manufacturing the N1 was later employed in the development of the Energia heavy-lift launch vehicle capable of injecting payloads of up to 100 tons as part of the Energia-Buran reusable space system. The first and the second successful launches in 1987 and 1988 proved that the Energia was a unique achievement of the world's rocket building industry with no analogues. However, a total recession in space industry coinciding in time with the collapse of the USSR had an unfavorable effect on the Energia-Buran project. In 1992, all the works were terminated.

In the Soviet epoque, hundreds of ministries and thousands of organizations were involved in the creation of launch vehicles of various classes and types, and the largest scientific and production centers were located not only in Russia but also in Ukraine which is famous for its Cyclone and Zenit launch vehicles, Belarus, and other USSR republics. A number of successful developments of that age continue to be successfully employed in post-Soviet countries, while Russia itself is currently using quite a few launch vehicle families.

Angara Launch Vehicle Family

Angara is the youngest family of Russian launch vehicles, a large part of which is still being developed and tested. Its commissioning is considered a national task of special importance as it will allow Russia to launch all types of spacecraft from its own territory and, thus, ensure its independent-guaranteed access to outer space.

A production order for Angara was placed by Roscosmos State Corporation and the Ministry of Defence of the Russian Federation meaning that the new launch vehicles are expected to satisfy the whole range of domestic needs, including civil, scientific, and military. The Angara space rocket complex uses exclusively domestic hardware elements by Khrunichev State Research and Production Space Center, one of the core companies in Russia's space industry, in cooperation with other Russian industrial and construction entities.

Another distinctive feature of the new complex is that the Angara launch vehicles are environmentally friendly. They do not use aggressive and toxic heptyl-based rocket fuels. This can significantly improve the environmental safety of the complex both in the regions adjacent to launch sites and in areas where spent stages land.

The Angara family includes launch vehicles of various classes. The basis for the development of the different classes is oxygen-kerosene all-purpose space rocket modules, the total number of which in the first stage determines the carrying capacity of the launch vehicle. The modules of the first and second stages are equipped with the liquid jet engine RD-191, the upper stages' modules – with the RD-0124A engine. When the design and construction stage is completed, Angara will provide a full range of lift services – light, medium, and heavy, putting up to 37.5 tons of payload into low Earth orbits (For a launch to the low reference Earth orbit by the Angara-A5V modification.). The development of a heavy modification for manned spacecraft launches, the Angara-5P, is also being discussed.

Unique technical solutions and extensive use of unification allow the whole Angara family to be launched from a single site. The ground infrastructure of the Russian Plesetsk Cosmodrome is able to provide launches of Angara light, medium, and heavy launch vehicles. The second launch complex is planned to be built at the new Cosmodrome Vostochny (The Angara created for launches from the Vostochny Cosmodrome is also called the Amur.).

The prototype of the first stage of the light Angara-1.2 launch vehicle passed flight tests three times – in 2009, 2010, and 2013 – as part of the first Korea Space Launch Vehicle (KSLV-1). A contract for KSLV-1 had been signed by Khrunichev Space Center in 2004. Flight testing of the Angara space rocket complex at Plesetsk began in 2014 – first, of the light-class rocket Angara-1.2.PP and, then, of the heavy-class Angara-A5.1L. Each performed one successful launch. The next flight of the heavy Angara-5 is scheduled for 2019, and in 2020 the whole scope of works on its development and manufacture is to be commissioned.

Proton-M

The Proton-series launch vehicles, with more than 400 launches behind, are well known around the globe and enjoy a high commercial potential on the world market. They are the basis of the Russian space transport system as well, being frequently used under various Russian state and commercial programs. Khrunichev State Research and Production Space Center is the vehicle's designer and manufacturer.

The currently utilized Proton-M is the advancement of the Proton-K launch vehicle with improved energy and mass, operational and ecological characteristics, and reliability. A new control system based on an onboard digital computing complex has been installed on the upgraded Proton-M. It helps burn up all fuel, thus improving the energy characteristics and environmental performance of the launch vehicle, and in spatial maneuvering during the active part of the vehicle's flight, thus extending the range of possible inclinations of the reference orbits; simplifies onboard electronic systems; and promptly assigns and changes a flight task. Proton-M has also received a lighter and a more voluminous head fairing which significantly increases space for payload accommodation, including the accommodation of several payloads of various types for carrying out group launches.

Proton-M is a heavy-class launch vehicle consisting of three stages with RD-series engines – the RD-276 in the first stage, RD-0210 and RD-0211 in the second stage, and RD-0213 and RD-0214 in the third stage. The first Proton-M launch with the Breeze-M upper stage took place in 2001. Today, such launches can inject payloads weighing more than 6 tons into geosynchronous transfer orbit or up to 3.3 tons directly into geostationary orbit. The Proton launch complex at Baikonur Cosmodrome consists of two launch sites united by a shared communications network and servicing facilities and operated autonomously. Kazakhstan's permission to fly the Proton-M from Baikonur is valid till 2025.

In 2021–2022, Proton-M can be replaced with the new Angara or the new Soyuz-5 launch vehicle. Both variants have been discussed; however, the most recent public announcements favor Angara (Telesputnik 2019). A while ago, there was discussed the Proton-Medium, a light version of the Proton-M without its third stage. An agreement of intent on the delivery of ten Proton-Medium was signed with OneWeb; however, no solid contracts were reached. It was therefore decided to offer OneWeb ten Proton-M in the existing version at the cost of a lighter Proton-Medium. Such a substantial discount would be beneficial to Khrunichev Space Center as well, as it did not need to invest heavily in the refurbishment of the launch pad for the sake of ten launches. With no economic grounds and a confirmed demand, there is no reason to expect that the Proton-Medium will be discussed again.

Soyuz-2 Launch Vehicle Family

Soyuz-2 is a new-generation launch vehicle with significantly improved technical and operational parameters, developed by Progress Rocket Space Center, a Russian joint stock company, on the basis of the legendary Soyuz-U which completed almost 800 launches in 1973–2017.

The launch vehicle was upgraded in two phases. During phase 1A, control and propulsion systems of the first and second stages were improved, and the Soyuz-2-1a modification was thereby provided with an enhanced injection accuracy and an increased payload mass for launches to low Earth orbits. During phase 1B, the launch vehicle's third stage was equipped with a modern engine, which further improved the energy capabilities of the Soyuz-2-1b modification. Both modifications consume environmentally friendly fuel components – kerosene and liquid oxygen.

Due to the growing demand for launches of small satellites, further rework of the Soyuz-2-1b modification was done. As a result of an additional modernization phase, phase 1 V, the two-stage light-class Soyuz-2.1v vehicle was developed, which can dramatically reduce manufacturing, launch, and operational costs for spacecraft in low circular and elliptical Earth orbits. After the flight design tests are completed, Soyuz-2.1v will expand the capabilities of the Soyuz-2 launch vehicle family.

Russia's Soyuz-2-1a, Soyuz-2-1b, and Soyuz-2.1v launches are carried out from Baikonur, Plesetsk, and Vostochny cosmodromes. For launches from the Guiana Space Centre which is a French and European spaceport to the northwest of Kourou in French Guiana, there were designed the Soyuz-ST-A and the Soyuz-ST-B modifications of the Soyuz-2 launch vehicle. For that purpose, the Soyuz-ST-series was adapted to the Guiana Space Centre specific requirements in terms of security as to receive the flight termination telecommand from the Earth, fitted with transmitters operating in the ultrahigh frequency band with a European frame structure telemetry, and adjusted to a higher level of humidity and a prior shipment by sea. The launch vehicles are equipped with a head fairing of the ST-type that meets international requirements and makes it possible to deliver the widest range of payloads into orbit.

The Soyuz-ST-A and the Soyuz-ST-B launches from the Kourou spaceport in French Guiana are serviced by Arianespace. Both modifications have been successfully launched since 2011, including for the European Galileo project. At the end of February 2019, Soyuz-ST-B launched the first six satellites of the low-orbit group OneWeb.

Soyuz-FG

Soyuz-FG is another Russian launch vehicle developed on the basis of Soyuz-U and built by Progress Rocket Space Center. It is a middle-class three-stage launch vehicle using environmentally friendly kerosene and liquid oxygen as fuel. In order to increase the specific impulse of propulsion systems and the load capacity of the vehicle, upgraded engines with new nozzle heads are used in the blocks of the first and second stages. (The Soyuz launch vehicle modification FG was named after the English transliteration of "ФГ" in Russian, which stands for "форсуночные головки" translated as "nozzle heads.")

It is designed to launch into near-Earth orbit automatic spacecraft for socioeconomic, research, and special purposes, as well as Soyuz-type manned spacecraft and Progress-type cargo spacecraft under the International Space Station (ISS) project. Since 2011, after the completion of the US Space Shuttle program, Soyuz-FG has been the world's only means of delivering crews to the ISS. (The situation may change in the second half of 2019, when SpaceX's Crew Dragon can start to carry astronauts to and from ISS. SpaceX successfully launched its first Crew Dragon capsule to ISS in March 2019. The Demo-1 flight was uncrewed, however proved its operability. Sierra Nevada Corporation is also working on a reusable multi-mission space utility vehicle Dream Chaser which is awaiting its first flight in 2021.) Soyuz-FG is capable of launching up to 7.2 tons of payload with a manned space mission, while the payload weight of a cargo mission amounts to 7.4 tons.

With the confirmed indicator of operational reliability of 0.985 (Progress Rocket Space Center 2014–2018), Soyuz-FG is one of the safest launch vehicles in the world. The only failure happened in October 2018 is when a Soyuz-FG rocket carrying a Soyuz MS-10 spacecraft with a crew on board crashed due to the abnormal separation of one of the side blocks, which hit the central block near the fuel tank with the bow, what led to its depressurization and, as a result, to the loss of stabilization of the launch vehicle. Russian cosmonaut Alexey Ovchinin and NASA astronaut Nick Hague, who were onboard Soyuz MS-10, were not injured, and their landing was secured by an emergency rescue system. Both crew members accompanied by Christina Koch of NASA arrived safely at ISS on March 2019, following a successful Soyuz-FG launch and docking of their Soyuz MS-12 spacecraft.

Soyuz-FG is also used for commercial and scientific launches, among them the launch in 2003 of the first planetary mission currently being conducted by the European Space Agency (ESA) for the study of the Red Planet titled Mars Express and the launch in 2005 of the first Venus exploration mission of ESA named Venus Express.

All Soyuz-FG launches are performed from the Baikonur Cosmodrome, which is the only spaceport with a Soyuz-FG launch site. In 2020, the use of Soyuz-FG is planned to be ceased as the existing stock of the Soyuz-FG rockets is going to be exhausted by the end of 2019. Manned launches will be transferred to Soyuz-2.1a. An order for the first three launch vehicles of the new type has already been placed with Progress Rocket Space Centre, and the first launch is scheduled for spring 2020.

Strela

In 1980, the silo-based ICBM RS-18 (SS-19 modification 2 Stiletto under the NATO classification), which had been developed by the Soviet Central Design Office of Machine Building, was accepted for service. Capable of carrying up to 6 warheads, it was the most numerous strategic weapons in the Soviet nuclear fleet with a launch heritage of over 150 flights. A total of 360 missiles were placed on high alert. The mass production of the missile by the Moscow Khrunichev Machine-Building Plant stopped in 1985. With 35 years' service time, it is going to meet its end of life in 2020, unless extended.

In accordance with the 1993 Strategic Arms Reduction Treaty, a bilateral agreement between Russia and the USA on the reduction and limitation of strategic offensive arms also known as START II, the Russian strategic nuclear forces had to remain armed with not more than 105 RS-18, converted into single-block missiles. There arose the question of how to reduce the arsenal of these missiles. One of the options was to convert them to launch vehicles and use them for commercial launches.

Thus the RS-18 missile became the basis for the Strela and the Rockot launch vehicles, developed respectively by NPO Mashinostroyeniya, the successor to the developer currently known as Joint Stock Company Military Industrial Corporation NPO Mashinostroyeniya, and Khrunichev State Research and Production Space Center, the successor to the batch production plant manufacturer.

The Strela is a more recent conversion version of the decommissioned Russian ICBM. Unlike the Rockot, it requires only minor modifications to the original missile. It is a liquid-propellant two-stage light-class launch vehicle made according to a tandem scheme. The Strela launch vehicle can be equipped with one of two types of space head parts, which differ in fairings and, as a consequence, in the size of the payload placement space. The use of a specific type of space head depends on the features of the payload. It is capable of launching payloads of up to 2 tons in near-Earth orbits or in the upper atmosphere of the Earth.

The Strela's launches are carried out from the Baikonur Cosmodrome. The launch vehicle is located in a transport and launch container; the launch is carried out from a launch silo. Only three launches were performed in 2003, 2013, and 2014 each, all being successful. Besides the first test run, the Strela launched two Condor-series remote sensing satellites developed by NPO Mashinostroyeniya, the latter launched in the interests of a foreign customer (NPO Mashinostroyeniya 2019).

Rockot

As well as the Strela, the Rockot was also developed based on stocks of the decommissioned Russian RS-18 ICBM, however, by another space industry entity – Khrunichev State Research and Production Space Center. The first Rockot test suborbital launches were performed in 1990 and 1991 from the Baikonur Cosmodrome. The first demonstration commercial space launch took place in 2000.

The Rockot is a light-class launch vehicle designed for launching small- and medium-size spacecraft of up to 1.95 tons into low Earth orbits (Khrunichev Space Center 2019). It is particularly suitable for launches into Sun-synchronous, near polar, and highly inclined orbits. The Rockot's head fairing allows placement of one or more spacecraft and is, therefore, suitable for co-launches.

The vehicle is made according to a three-stage scheme with a sequential arrangement of stages. (This is the main difference from the Strela launch vehicle which is a later version of the RS-18 ICBM conversion with a minimized change in the design of the rocket and the launch complex.) The first two are the booster stack of the converted missile, and the third is the Breeze-KM upper stage. High-efficiency liquid rocket engines operate on nitrogen tetroxide and asymmetric dimethyl hydrazine. Experts from Khrunichev Space Center worked hard to ensure the environmental safety of launches, including the selection of routes of injection and the arrangement of the spare stages landing fields.

The Rockot's high performance largely stems from the use of the Breeze-KM upper stage, which provides a wide range of options for spacecraft injection into orbits. The upper stage equipment is capable of controlling the attitude of the spacecraft with high accuracy, as well as providing its power supply during the launch and during the orbital flight for up to 7 hours. The ability to repeatedly, up to eight times (Roscosmos State Corporation 2019a), turn on the propulsion system of the upper stage makes complex co-launches possible. A special transition system allows the spacecraft to be separated from the upper stage with minimal disturbances.

Both state and commercial Rockot launches are performed from dedicated launch facilities, a transport and launch container, at the Plesetsk Cosmodrome, which is currently the only "home" spaceport for the Rockot. Five silo launchers for the RS-18 ICBM were earlier available at the Svobodny Cosmodrome; however no launches of Rockot vehicles have been carried out from that facility till its closure in 2007 (Lenta.ru 2007).

In 2000–2018, Rockot performed 29 single, dual, and multiple payload launches, 25 of which were fully successful. The most recent mission occurred in November 2018, when three Strela-3 M communications satellites for the Russian Ministry of Defence were launched, bringing the number of government payload launches to 15. Fourteen missions were purely commercial, with the latest launch of the Sentinel-3B remote sensing satellite in April 2018. The Sentinel-3 mission is jointly operated by ESA and EUMETSAT to deliver operational ocean and land observation services to support ocean forecasting systems, environmental monitoring, and climate monitoring.

Commercial launches are operated by Eurockot Launch Services GmbH, the Bremen, Germany, based joint venture of ArianeGroup and Khrunichev Space Center, which invested financially in the development of modern satellite preparation and launch facilities dedicated to the Rockot at the Plesetsk Cosmodrome.

In February 2015, the Ukrainian government banned the supply of components for the Rockot launch vehicle to be used for Russia's military space missions; however, it did not withdraw from their shipment for civil space programs (Lenta. ru 2015). In March 2018, Khrunichev Space Center reported that, in order to ensure import substitution and full Russian production management, the design documentation was being worked out in order to give a second wind to the Rockot 2 vehicle (Zvezda 2018). In addition, in 2020, the lifetime of the RS-18 ICBM produced before 1985 will expire. In this regard, the Russian Ministry of Defence intends to abandon the conversion of Rockot systems, replacing them with the Russian Angara.

Dnepr

Dnepr is another example of a converted launch vehicle developed on the basis of ICBM technology. The RS-20 (SS-18 Satan under the NATO classification), one of the heaviest missiles (Roscosmos State Corporation 2019b), was developed in Soviet times by the Yuzhnoye Design Office based in Dnepropetrovsk, Ukraine, and remained in operational service for at least 15 years (RS-20 service life was subsequently extended.). When the period of its operation came to an end, it was decided to use the carrier of the nuclear warhead in the conversion program called Dnepr, that is, to adapt the rocket for peaceful space launches. This decision was preceded by scientific research and preliminary design efforts, which confirmed that minimum modifications would guarantee a cost-effective solution. First, there is no need to scrap the system. Secondly, the missile performs an important peaceful mission. Finally, it helps the military to make sure that the RS-20 technology, despite an advanced age, is still reliable.

During 1992–2003, a team of Russian and Ukrainian companies together with the Russian Ministry of Defence were involved in developing a space launch system based on the RS-20 being withdrawn from service. International Space Company Kosmotras, a Russian joint stock company, was tasked to commercially operate this RS-20 ICBM-based space launch system.

The Dnepr is a liquid-fueled three-stage launch vehicle. The first and the second stages remain original with no modifications, while the third stage received an upgraded control system that implements the required flight program of all three stages. The launch vehicle is designed for rapid high-precision launches into near-Earth orbits with altitudes of 300–900 km of single or multi spacecraft weighing up to 3.7 tons. Environmental monitoring after the Dnepr launches confirms their complete safety for life and the environment. Environmentalists have never recorded a negative impact from the components of rocket fuel, its combustion products, or an excessive acoustic impact (Rossiyskaya Gazeta 2014).

The first commercial Dnepr mission lifted off in 1999. Over 20 years of operations, the launch vehicle has performed 22 launches, only 1 of which was a failure, thereby providing the mission reliability factor 0.97 (International Space Company Kosmotras 2019a), with a total of 128 payloads injected into Earth orbits in the interests of renowned universities, major space industry companies, and space agencies from France, Germany, Italy, Japan, Saudi Arabia, the UK, the USA, Thailand, South Korea, and other countries (International Space Company Kosmotras 2019b). Dnepr launches are currently performed from the Baikonur Cosmodrome and the Yasny launch base.

Operating Cosmodromes

Guaranteed access to outer space is one of the cornerstones of Russia's state policy in the field of space activities and a significant factor influencing geopolitical interests of modern Russia. Today, it operates five cosmodromes – Baikonur, Plesetsk,

Vostochny, Kapustin Yar, and Yasniy – and exports Russian launch vehicles abroad to be employed at the Guiana Space Centre. Baikonur, even though located in the territory of Kazakhstan, still remains the main spaceport of Russia; however, all future developments of the exploration and use of outer space are mainly associated with the newest and the first purely civil spaceport in Russia called Vostochny. After the purchase deal of Sea Launch by Russian S7 Group in 2016, a sixth Russian spaceport can be added, though not governmental but private.

Baikonur Cosmodrome: Kazakhstan

The territory of Baikonur amounts to 6717 square kilometers. It is not only the largest, it is also the oldest spaceport in the world with a great history. The construction project was approved as long ago as 1955. Within an unusually short time, by 1957, there was erected a launch complex for the R-7 ICBM, the prototype of the modern Soyuz launch vehicle. In the same year, the first artificial Earth satellite, Sputnik-1, was launched from Baikonur, followed by the first ever space-flight in the history of humankind performed by Yuri Gagarin in 1961 and the first spacecraft's launches to the Moon, Mars, and Venus. The cosmodrome performed almost 5000 space launches in total. Today, Baikonur remains the only spaceport that delivers manned spacecraft to the ISS. (See also the section on the Soyuz-FG which is the launch vehicle used for manned spacecraft missions.) Manned missions used to be carried from the John F. Kennedy Space Center in Florida, USA, until the NASA Space Shuttle program was closed in 2011; however, they can be resumed in the second half of 2019 by SpaceX's Crew Dragon followed by Sierra Nevada Corporation's Dream Chaser which is awaiting its first flight in 2021. Cargo space-craft to the ISS are also launched from Baikonur.

After the dissolution of the USSR, Baikonur Cosmodrome found itself in a foreign state. It was therefore agreed that Baikonur Cosmodrome and the city of the same name, which make up the Baikonur complex, were rented by Russia from Kazakhstan. A contract between the Government of Kazakhstan and the Government of Russia was executed in 1994 (see the Rent Contract for the Baikonur Complex Between the Government of the Republic of Kazakhstan and the Government of the Russian Federation dated December 10, 1994, as later amended by the protocols on amendments of 2008, 2017, and 2018 (the latter two are still pending entry into force)) and provided for a 20-year rental term which could be extended, in the absence of objections, for another 10 years, meaning that the initial rent term might expire as soon as in 2024. In November 2018, a protocol on amendments to the contract was signed by both states to prolong the rent of the Baikonur complex till 2050. The protocol, and the extension itself, will come into force upon imple-mentation by both parties of the necessary domestic procedures. The rental of the Baikonur complex costs Russia 115 million US dollars annually. The basic princi-ples of Russia's use of Baikonur, including settling of international space law matters and the launching state issue in particular (see the Agreement between the Russian Federation and the Republic of Kazakhstan on the Basic Principles and Conditions of the Use of the Baikonur Cosmodrome dated March 28, 1994.), are determined in a bilateral international agreement. (In the event of damage caused by the activities of Baikonur Cosmodrome in the course of Russian space programs, Russia is liable as a launching state in accordance with the Convention on International Liability for Damage Caused by Space Objects of March 29, 1972. In this case, Kazakhstan is not considered as a party to the joint launch or a launching state. In the case where the launch of a space object is carried out by Russia together with Kazakhstan, liability for damage is determined by Article V of the 1972 Liability Convention. In the event that the launch of a space object is carried out by Russia together with other states, these states shall be jointly and severally liable for any damage caused in accordance with the 1972 Liability Convention, and Kazakhstan is not considered as a participant in the joint launch or a launching state.)

In order to reduce its dependency on Baikonur, Russia has started to upgrade its own space facilities.

Plesetsk Cosmodrome

The Plesetsk Cosmodrome is the one that has been upgraded and is currently the main government test cosmodrome of the Ministry of Defence of the Russian Federation. In mid-2001, the cosmodrome was included in the Russian Space Forces. It is also highly involved in Russian and international space programs related to scientific and commercial launches of unmanned spacecraft. Light-class Angara-1.2, Soyuz-2.1v, and Rockot, as well as middle-class Soyuz-2.1a and Soyuz-2.1b and heavy-class Angara-A5 launch vehicles, are operated at the Plesetsk Cosmodrome.

With a total area of 1762 square kilometers, Plesetsk is one of the world's largest spaceports. It is located in the Plesetsk district of the Arkhangelsk region in the Northwest of Russia and the northernmost location for a spaceport on the globe. Though the technical facilities of Plesetsk make it possible to launch various spacecraft into all types of orbits, including geostationary orbit, such a long distance from the Earth equator imposes certain limitations. (The linear speed of rotation of the Earth at the latitude of Plesetsk is 212 meters per second; at the latitude of Baikonur, 316 meters per second; and at the latitude of the Guiana Space Centre at Korou, French Guiana, 460 meters per second. Plesetsk is particularly convenient for launches of spacecraft into high inclination and polar orbits.)

However, the choice of the cosmodrome location was largely determined by the tactical and technical characteristics of the R-7 ICBM, as Plesetsk had been initially deployed as a military unit of missile regiments armed with the R-7. First, it was essential that the territories of potential adversaries were within missile range, and the need for special secrecy was also taken into account. Secondly, the possibility of conducting and controlling missile test launches to the Kura missile range located in the Kamchatka region was important.

The history of the Plesetsk Cosmodrome begins in 1957, with the USSR Government's decree on the creation of a military facility with the conventional name

Angara. Since the launch of the first spacecraft from Plesetsk in 1966, more than 1600 space launches have been performed and about 2100 spacecraft for various purposes have been put into Earth orbits. Since 1968, the cosmodrome has been involved in the implementation of international space programs. In April 1972, for the first time in the USSR, a small French spacecraft, MAS-1, was launched from Plesetsk. In 1970–1990, the Plesetsk Cosmodrome held the world leadership in the number of space launches (From 1957 to 1993, 1372 launches were carried out from Plesetsk, while Baikonur took the second place with a total of 917 launches for the same period.) and is historically one of the sites with the world's highest launch frequency.

From Svobodny to Vostochny Cosmodrome

The question of deploying a new Russian cosmodrome was raised before the leadership of the Russian Ministry of Defence in 1992. The main reason was that as a result of the collapse of the USSR, Baikonur was geographically outside the territory of Russia. Though the launches of spacecraft by light and medium-class launch vehicles could be carried out from the Plesetsk Cosmodrome, the issue of launches of heavy-class launch vehicles was particularly acute. Heavy Proton launch complexes were only available at the Baikonur Cosmodrome.

To choose a location for the new cosmodrome, the map of Russia was analyzed. Despite the vast territory of the country, only southern regions of Russia's Far East and Sakhalin Island were considered potentially suitable, and no places closer to the European part of Russia were appropriate for the placement of the cosmodrome. A significant distance from the main capacities of the Russian space industry was a great disadvantage. More so, Sakhalin Island turned out to be one of the most remote parts of Russia having no suitable railroad, resource, and production capacities. The final assessment of the several selected areas resulted in the choice of the district of the city of Svobodny in the Amur region as the location of the new Russian cosmodrome.

The history of the new Russian cosmodrome began in 1997. The Svobodny Cosmodrome, which can be translated into the English language as the Free Space-port, is officially named the Second State Test Cosmodrome of the Ministry of Defence of the Russian Federation. During the whole time of its existence, only five space launches were performed by the Start 1.2 launch vehicle with the latest one in 2006. Four of them were made in the interests of foreign customers – the USA, Israel, and Sweden. According to some reports, in 2005, the Security Council of Russia decided to close the cosmodrome as part of the reduction of the armed forces due to the low intensity of launches and insufficient funding. In 2007, the closure of the cosmodrome was publicly announced, and since then it has not functioned.

Very soon, the question of a new Russian full-fledged multitask cosmodrome, as an alternative to Baikonur, was once again posed. The decree on the construction of the Vostochny Cosmodrome was signed by the President of Russia in 2007, and the memorial sign was laid in honor of the start of works on the first Russian civil cosmodrome in 2010. There were some other tasks behind such a significant project – the cosmodrome was built not only to provide independent access to outer space, to guarantee the implementation of international and commercial space programs, to reduce the cost of the Baikonur Cosmodrome, but also to improve the socioeconomic situation in the Amur region of Russia.

Vostochny means "Eastern" in the Russian language. Indeed, it occupies an area of about 700 square kilometers in the Far East of Russia, rather close to the Earth equator which makes launches from there more efficient, if compared with Plesetsk. The Far East area is underpopulated meaning that space safety is not a big issue for this location, and the trajectory of launch vehicles does not pass over the territories of foreign States. Vostochny is located 45 kilometers north of the Svobodny Cosmodrome which means the proximity of an already well-developed railway, highways, and airfields.

The construction of the launch complex for the light- and medium-lift launches, being the major part of the first construction stage, began in 2012 and was completed in 2016. Today, the Soyuz launch pad is fully operational, and three of four launches have been a success. By the end of 2018, the first stage was expected to be fully completed. According to Roscosmos State Corporation, which is tasked with the Vostochny construction, all the first stage works have been actually done in time, while some papers required for a formal acceptance are being close to finalization (Ria Novosti 2019).

The second construction stage involves the deployment of a launch pad suitable for heavy launches in 2019–2025. A backup complex should be built for the start of the heavy Angara, which already exists in Plesetsk. The first Angara-A5 launch is scheduled for 2023. The existing infrastructure of Vostochny technically allows for a transfer of the manned spacecraft program from Baikonur after minor modifications, which is also planned to be done in the coming years by launching a manned spacecraft, the Federation, in an unmanned version first and with a crew shortly afterward. Recently, the President of the Russian Federation gave instructions to deploy a launch complex for the superheavy class at Vostochny no later than 2023, which constitutes the third stage of construction (Briefing by Yuri Borisov 2019).

Kapustin Yar

The Kapustin Yar missile range occupies an area of about 650 square kilometers in the Northwestern part of the Astrakhan region of Russia, near the border with Kazakhstan. Officially, it is named the Fourth State Central Multipurpose Firing Range of the Russian Federation (Declassified photos of the Kapustin Yar missile range are available on the web-site of the Ministry of Defence of the Russian Federation at http://mil.ru/files/files/kapyar/photos/index.html).

Kapustin Yar was deployed in 1946 as the test site for the first Soviet missile weapons, both ballistic and surface-to-air. In October 1947, the first ever launch of a Soviet ballistic missile was conducted from Kapustin Yar, which remained the only place to test Soviet missiles till 1957. At the same time, the missile range was also used for the exploration of outer space.

In 1951, two dogs, Dezik and Tsygan, were launched and safely returned to the Earth being the first higher mammals in outer space. To deliver the animals to the near-Earth space, a liquid-fuel single-stage R-1-type rocket was used, which is one of the first Soviet geophysical missiles employed in scientific research and experimental works. The rocket rose to a height of about 101 km reaching the Karman Line. In a state of weightlessness, the dogs spent about 4 minutes. Fifteen minutes after the launch, the container with the dogs landed safely a few kilometers from the launch pad. The first flight of Dezik and Tsygan was a significant step forward in outer space exploration. Post-flight examinations showed no serious changes in the animals' physiological state and no harmful effects on their organisms. They perfectly suffered overloads and weightlessness. This gave confidence to Soviet researchers in their plans to launch a human to outer space, what actually happened 10 years later.

In March 1962, with the launch of the Cosmos-1 satellite, Kapustin Yar became a cosmodrome. In 1969, the Intercosmos-1 satellite manufactured by a group of socialist countries took place, and this made Kapustin Yar open for international cooperation, which shortly continued with flying Indian and French satellites. Later it was mainly used for small research satellites' launches by the light-class Cosmos series launch vehicle. Today, Kapustin Yar serves as a site for launches of ballistic, geophysical, and meteorological missiles, as well as low-mass space objects.

Yasniy Cosmodrome

Yasniy is the name of the fifth cosmodrome in Russia after Baikonur, Plesetsk, Vostochny, and Kapustin Yar. It is located in the Yasnenskaya district of Russia's Orenburg region and used for space launches on the Dnepr rocket. International Space Company Kosmotras, a Russian joint stock company, operates the cosmodrome and executes commercial contracts with customers. It orders and pays for launches carried out by the Ministry of Defence of the Russian Federation.

The first space launch from Yasniy dates back to mid-July 2006, when the US Genesis I satellite was launched into near-Earth orbit. A total of ten launches have been made so far, both single and multiple, of up to 33 space objects per launch, mainly for foreign customers. All launches were successful. The latest launch took place in 2015, when South Korea's KOMPSAT-3A remote sensing satellite was inserted into its target orbit.

Guiana Space Centre

Russia and Kazakhstan, former USSR republics, are not the only countries which employ Russian launch vehicles' sites. The launch site of the Soyuz is also located at a distance of almost 10,000 kilometers from Russia – near the city of Korou in

French Guiana, an overseas department and region of France on the North Atlantic coast of South America.

The Guiana Space Centre is jointly utilized by the European Space Agency and the French National Centre for Space Studies (CNES). The spaceport became operational in April 1968. In 2003, the Russian-European project began when a decision was made at the government level to operate Soyuz launch vehicles at the Guiana Space Centre. A number of agreements, including intergovernmental (See the Agreement between the Government of the Russian Federation and the Government of the French Republic on Long-Term Cooperation in the Field of Development, Creation and Use of Rockets and the Placement of the Carrier Rocket Soyuz-ST at the Guiana Space Centre dated November 7, 2003.), were concluded, which laid the legal basis for signing contracts for the creation of the ground infrastructure of the Soyuz launch complex, the manufacture and shipment to French Guiana of the Soyuz-ST-series launch vehicles and the Fregat upper stages.

In May 2011, a ceremony was held to officially turn over the new launch complex to the European Space Agency and Arianespace, the operator of the spaceport. Two versions of the Soyuz launch vehicle, which was modified specifically for the Guiana Space Center, supplement the spaceport capabilities with medium-class launches. Soyuz entered service in October 2011 with an on-target maiden flight that orbited the first two European Galileo navigation satellites. Since then, two to three launches are carried out every year both in the interests of ESA and in the interests of commercial companies.

Sea Launch

Sea Launch Company, an international consortium to develop and operate a sea-based space launch system, was established in 1995. The US Boeing; Russian S.P. Korolev Rocket and Space Corporation Energia; Norwegian ship-building enterprise Aker Solutions, previously known as Kvaerner; and Ukrainian space industry entities Yuzhnoye and Yuzhmash were the founders of this purely commercial international venture. By 1998, the buoyant complex in the equatorial waters of the Pacific Ocean near Christmas Island was constructed. The complex included a "home" seaport in Long Beach, California, a mobile sea-based launch platform named Odyssey and an assembly and command ship dubbed Sea Launch Commander. Such a unique location provided a significant increase in mass of the payloads to be launched due to the proximity to the equator, while the sea offered favorable conditions in terms of spacecraft's delivery.

Sea Launch was tailored for the Zenit-3SL, a three-stage middle-class launch vehicle operating on environmentally friendly liquid oxygen and kerosene. It was based on the two-stage Zenit-2 launch vehicle developed by Yuzhnoye State Design Office, Ukraine, which performed 37 launches in 1985–2009 with the vast majority being successful. The Zenit-3SL is capable of putting payloads weighing more than

6 tons into medium and high Earth orbits, both circular and elliptical, as well as to geostationary transfer orbits.

Marketing research carried out during the implementation of the Sea Launch project revealed that it was advisable to carry out a number of launches of spacecraft weighing up to 4 tons not from the sea-based platform, but from the Zenit-M launch site of the Baikonur Cosmodrome. Such a diversification could also make the use of the available stock of the Zenit-series launch vehicles more efficient. An alternative commercial project was named Land Launch and opened another niche for Sea Launch Company.

The first commercial sea launch in 1999 was followed by another 35 launches, most of which were successful. However, in 2009, Sea Launch Company announced its bankruptcy, and the launch activity was suspended in 2014. According to some reports, losses could be caused by the fact that the company failed to ensure the planned intensity of launches: it was originally scheduled to carry out 2–3 consecutive launches in one exit to the launch pad (Kommersant 2010). Until 2016, negotiations were held on the sale of the project, as well as on measures to be taken to resolve financial and legal issues. In September 2016, Russian S7 Group became the owner of the Sea Launch complex. The launch activity was assigned to S7 Space, a dedicated branch of the group, which planned to resume launches at the very end of 2019 to be followed by another three launches in 2020 and four launches per year in 2021 and 2022.

In 2017, Yuzhmash, the Production Association Yuzhny Machine-Building Plant named after A.M. Makarov, announced the conclusion of a contract for the production and supply of 12 Zenit launch vehicles, but in March 2019, it became known that S7 Space canceled the order. The cancelation may be associated with the decision to replace the Ukrainian Zenit with the new Russian-made launch vehicles Soyuz-5 or Angara. Both alternative options were discussed; however, no final decision has been announced so far.

Conclusions

Russia has successfully engaged in space activities for more than 60 years, having entered the space age as part of the Soviet Union and striding on as a separate state. On the one hand, after the dissolution of the USSR, Russia inherited the huge scientific and technical potential and technological developments of one of the two most powerful space nations of that time. But on the other hand, Russia was deprived of a large part of technologies and infrastructure put in place earlier. The launch vehicles that used to be Soviet became foreign, and the key launch site turned out to be located outside Russia's national territory. Also, it proved to be difficult for Russia to use remnants of its own technologies because suppliers of components had been scattered all over the enormous Soviet Union, and now Russia had to either import them or replace them with similar domestic parts. Political and economic turbulences during the late 1980s and the early 1990s dramatically affected Russia's space activities and had a crucial negative influence on its space capabilities. A number of promising projects were abandoned. Many years have passed for new ones to come to life.

Nevertheless, Russia rightfully remains one of the leading space-faring nations and confidently occupies its well-deserved place among the world space industry leaders. Russian launchers rank among the best on the globe owing to their fine quality, high reliability, and cost-efficiency. Russia's capabilities are employed in launching scientific, civil, and military space missions, both domestic and in the interests of foreign partners. Russia remains open to international cooperation, including for sending crews and cargos to the ISS and for the purpose of deep space exploration and human space programs across the solar system. Ongoing technological advancement and capacity building is traditionally a key element of Russia's space strategy. The challenge ahead is to sustain, secure, and strengthen its position for the decades to come.

To keep Russia's guaranteed access to and presence in outer space, all major cosmodromes need to be retrofitted, upgraded, and kept functioning. In the first place, these are Vostochny, Plesetsk, and Baikonur that have to support federal programs and long-range dedicated tasks for the benefit of science and national economy as well as international and commercial space projects.

The main trend in the development of launch vehicle families in the period until 2025 is to reduce their lineup to two series based on the Soyuz and the Angara. The gamut of operating rocket carriers is also going to be reduced to six types to secure the whole variety of launches – light, medium, and heavy (Vzglyad 2016). Until 2030 there has to be developed a system with a superheavy booster capable of lifting no less than 50 tons, including for missions to the Moon, Mars, Jupiter, or other celestial bodies in the solar system (Main Provisions of the Principles of State Policy 2013).

The superheavy launch vehicle may be named Yenisei and will consist of flight-proven components – the first stage will be based on the first stage of the middle-class Soyuz-5, the second stage will get the famous Russian RD-180 engine that is exported to the USA, and the third stage will be borrowed from the heavy Angara-5 V (Lenta.ru 2019). The new Russian superheavy launch vehicle will use the heritage of the Soviet N1 and Energia. Thus, the Yenisei will be capable of carrying payloads of more than 100 tons to low-earth Earth orbits. The first vehicle is expected to be manufactured in 2027 with the first launch slated for 2028 from the Vostochny Cosmodrome. (This modification is also referred to as the Irtysh.) After 2030 it is planned to increase the lift capacity of the superheavy launcher to 130–180 tons and develop a reusable rocket system with a reusable first stage.

Today, the leadership in space industry remains of strategic importance. For Russia, this is not only a question of national defense and security or its position in the market of commercial launch services but also, and more importantly, a question of the status of Russia as a highly developed nation in terms of science and technology.

References

Briefing by Yuri Borisov, Deputy Prime Minister of Russia responsible for defence industrial complex, at the end of the meeting on the financial and economic condition of Roscosmos State Corporation and its subordinate organizations, 23 Jan 2019. http://government.ru/news/35460/. Accessed 29 March 2019

Federal Dedicated Program. Development of Spaceports to Support Space Activities, 2017–2025 (approved by the Government of the Russian Federation on 13 Sept 2017)

Federal Space Program of Russia for 2016–2025 (approved by the Government of the Russian Federation on 23 March 2016)

International Space Company Kosmotras (2019a) Reliability. http://www.kosmotras.ru/en/reliability/. Accessed 29 March 2019

International Space Company Kosmotras (2019b) Dnepr launch record. http://www.kosmotras.ru/en/zapuski/. Accessed 29 March 2019

Khrunichev State Research, Production Space Center (2019) Rockot launch vehicle. http://www.khrunichev.ru/main.php?id=43. Accessed 29 March 2019

Kommersant (2010) Sea launch to finish in Russia. https://www.kommersant.ru/doc/1364744. Accessed 29 March 2019

Lenta.ru (2007) The Governor of the Amur region agreed to close the Svobodny Cosmodrome. https://lenta.ru/news/2007/03/14/cosmodrome/. Accessed 29 March 2019

Lenta.ru (2015) The Ukraine bans the supply of components for the Rockot launch vehicle to Russia. https://lenta.ru/news/2015/02/12/rokot/. Accessed 29 March 2019

Lenta.ru (2019) The "minimum" cost of the Russian super heavy rocket is revealed. https://lenta.ru/news/2019/03/25/iss/. Accessed 29 March 2019

Main Provisions of the Principles of State Policy in the Field of Space Activities, till 2030 and Further Prospect (approved by the President of the Russian Federation on 19 April 2013)

NPO Mashinostroyeniya (2019) Strela rocket space complex. http://www.npomash.ru/activities/en/krk.htm. Accessed 29 March 2019

Opening remarks by Dmitry Medvedev, Prime Minister of Russia, at the meeting on the financial and economic condition of Roscosmos State Corporation and its subordinate organizations January 23, 2019. http://government.ru/news/35460/. Accessed 29 March 2019

Progress Rocket Space Center (2014–2018) Soyuz-FG launch vehicle. https://www.samspace.ru/products/launch_vehicles/rn_soyuz_fg/. Accessed 29 March 2019

Ria Novosti (2019) Dmitriy Rogozin has commented on the first stage works on construction of Vostochny cosmodrome. https://ria.ru/20190322/1552036765.html. Accessed 29 March 2019

Roscosmos State Corporation (2019a) The Breeze-series upper stages. https://www.roscosmos.ru/450/. Accessed 29 March 2019

Roscosmos State Corporation (2019b) RS-20. https://www.roscosmos.ru/470/. Accessed 29 March 2019

Rossiyskaya Gazeta (2014) 33 satellites of the "Satan". Nuclear missiles work for peaceful space. https://rg.ru/2014/06/20/zapusk-site.html. Accessed 29 March 2019

Telesputnik (2019) Proton-M will compete with the falcon rocket by saving on pre-launch training. https://telesputnik.ru/materials/tekhnika-i-tekhnologii/news/proton-m-budet-konkurirovat-s-rn-falcon-za-schet-ekonomii-na-predstartovoy-podgotovke/?utm_source=newsletter&utm_medium=email. Accessed 29 March 2019

Vzglyad Business Newspaper (2016) The government approved the Federal Space Program for 2016–2025. https://vz.ru/news/2016/3/17/800054.html. Accessed 29 March 2019

Zvezda (2018) Specialists from Khrunichev space Center have reported when the reusable version of Angara launch vehicle will go into production. https://tvzvezda.ru/news/opk/content/201801281917-qhyv.htm. Accessed 29 March 2019

Institutional Space Security Programs in Europe

61

Ntorina Antoni, Maarten Adriaensen, and Christina Giannopapa

Contents

Abstract

European institutional actors are increasingly involved in space and security programs. This chapter addresses the programs of institutional actors engaging in space and security activities in Europe, namely, the European Union (EU), the European Space Agency (ESA), and the North Atlantic Treaty Organization (NATO). The recent policy developments in these organizations and European

N. Antoni (✉)
Eindhoven University of Technology, Eindhoven, The Netherlands
e-mail: ntorina.antoni@gmail.com

M. Adriaensen · C. Giannopapa
European Space Agency (ESA), Paris, France
e-mail: maarten.cm.adriaensen@gmail.com; christina.giannopapa@esa.int

© Springer Nature Switzerland AG 2020
K.-U. Schrogl (ed.), *Handbook of Space Security*,
https://doi.org/10.1007/978-3-030-23210-8_112

countries as described in ▶ Chap. 23, "Strategic Overview of European Space and Security Governance" of this Handbook are inextricably intertwined with the national programs of the European Member States that are presented in the corresponding chapters of Space and Security Programs in the Largest, Medium-Sized and Smaller European Countries. This chapter includes space activities and programs in the fields of Earth observation (EO); Satellite Communication (SATCOM); Positioning, Navigation and Timing (PNT); and Space Situational Awareness (SSA).

Introduction

The increased need for security in Europe and for Europe's space activities had led to several activities and programs developed by institutional actors in the space and defense sectors, towards the strengthening of European security and defense.

In this context, the European Union (EU), the European Space Agency (ESA), and their respective Member States have strengthened their existing partnerships and have established new ones with security stakeholders in Europe. In addition, the North Atlantic Treaty Organization (NATO) has recently adopted a space policy for security and defense purposes. Hence, the main institutional actors addressed in this chapter are the European Union (EU) and the European Space Agency (ESA), while it is worth addressing the activities and programs that NATO has in place.

The EU is a supranational organization composed of 28 Member States. Article 189 of the Treaty on the Functioning of the European Union is the legal basis for EU space affairs. The Council of the EU, the European Parliament, and the European Commission (EC) play an important role in the EU space program. The EU Global Strategy, adopted by the European Council in June 2016, the European Commission Space Strategy for Europe, launched in October 2016, and the European Defence Action Plan 2016 all stress the importance of having access to European Governmental Satellite Communications (GOVSATCOM) capabilities. More recently, in June 2018 the EU has presented the proposal for a regulation for a Space Programme for the EU.

ESA is an intergovernmental organization with 22 Member States (Canada and Slovenia are Associated States) with its mandate enshrined in the 1975 ESA Convention. ESA has increasingly contributed to space security in and from space as reflected in the Council Document "Elements of ESA's Policy on Space and Security" issued in June 2017 (Giannopapa et al. 2018). The document acknowledged ESA's involvement in space and security activities and provided for the guiding principles of ESA's future involvement in the domain of space and security in view of strengthening the dialogue with Member States and supporting the current and future activities of the Agency. In December 2019, the Council at the Ministerial level adopted the safety and security pillar and associated programs.

NATO is an international organization with 29 Member States. The 2010 NATO Strategic Concept identifies collective defense, crisis management, and cooperative security as the essential core tasks that NATO must continue to fulfil to assure the security of its members. In June 2019, NATO adopted a new overarching Space Policy. While both the NATO space strategy and the political designation of space as a potential warfighting domain, in November 2019 NATO recognized space as an operational domain.

Accordingly, this chapter addresses space activities and programs in the fields of Earth observation, Satellite Communication (SATCOM), Global Navigation Satellite Systems (GNSS), Space Situational Awareness (SSA), space transportation, satellite operations, and detection, tracking, and warning. Cross-domain applications, general research and technology development, and business development and technology transfer are not structurally analyzed.

The European Union

The EU is a supranational organization composed of 28 Member States (27 now at the time of the writing because of Brexit). The EU legal basis for defense affairs is situated in Articles 42 and 43 of the Treaty on European Union, while the EU legal basis for space is situated in Article 189 of the Treaty on the Functioning of the European Union. Both the Council of the EU, the European Parliament, and the European Commission play an important role in the EU space program. The EU has been increasingly investing in European space programs. The EU implements the space component development phase of its two flagship programs Copernicus and the European Geostationary Navigation Overlay Service (EGNOS)/Galileo via delegation agreement with ESA. The 2021–2027 proposed by the European Commission space budget of €16 billion for the EU Space Programme Components and the EU Agency for Space (not including space research under Horizon Europe) consists of (European Commission 2018a):

- Galileo and EGNOS: €9.7 billion
- Copernicus: €5.8 billion
- Space Surveillance and Tracking (SST) and Governmental Satellite Communication (GOVSATCOM): €500 million

Additionally, on 9 April 2019, the Directorate-General for Communications Networks, Content and Technology (DG Connect) of the European Commission and the European Space Agency (ESA), signed a technical agreement to collaborate in designing a Quantum Communication Infrastructure (QCI) from Europe. The Euro QCI aims to represent the next generation of ultra-secure communications in Europe, allowing information to be transmitted and stored ultra-securely, linking critical data and communication assets all over the continent. The Euro QCI is expected to consist of components on earth and in space and is expected to boost Europe's capabilities in cybersecurity and communications. The European Space Agency would be the system architect on behalf of the Commission on the space component.

The European Union Space and Security Institutions

European Commission

The European Commission has released its regulatory proposal for the EU space program for the period 2021–2027 on 6 June 2018 (European Commission 2018b). This is in line with the Communication that the European Commission released in October 2016 on its Space Strategy for Europe. At the time of the writing, the proposal is under discussions with the Council of the EU. The European Commission has released its regulatory proposal for the EU Research Framework Programme in the period 2021–2027 (Horizon Europe) on 7 June 2018. Space research is included in the Cluster Digital, Industry, and Space (15 billion EUR in total – space amount to be determined). The proposal refers to Galileo/EGNOS, Copernicus, secure satellite communications, SSA, 5G and Next Generation Internet (NGI), non-dependence and sustainability of the supply chain, new innovative technologies, and space science (European Commission 2019a). On 30 September 2019, the European Commission and the European Space Agency (ESA) organized a joint industry day on "quantum communication infrastructure (QCI) for Europe – space segment" to share and discuss various options for potential space infrastructure for different use cases with QCI stakeholders. This was followed by a declaration that until early 2020 has been signed by 20 EU Member States (European Commission 2019b). This Declaration builds on the technological advances made by the EU space programs and by ESA to significantly boost Europe's capabilities in optical communications and cybersecurity. One of their objectives is to use satellite technologies for delivering Quantum Key Distribution (QKD) services (a technology that uses the principles of quantum mechanics to perform cryptographic tasks), which are not achievable by terrestrial solutions alone (European Commission 2019b).

Space is further relevant in the areas of Open Science and Open Innovation (Space Equity Pilot in Horizon 2020, SME-instrument). The European Commission is in the process of elaborating a Strategic Research and Innovation Agenda (SRIA) for Space Technology under Horizon Europe, taking benefit from work performed in the Space Policy Expert Group (SPEG), Steering Group of the ESA-Commission-EDA Joint Task Force on Critical Space Technologies and Harmonisation activities and the Working Group of the ASD-Eurospace Space Technologies: European Partnership PLATFORM (STEPP) Project.

Within the European Commission, the Directorate-General (DG) for Internal Market, Industry, Entrepreneurship and SMEs (DG GROW) held the main responsibility for space activities of the EU. In addition to GROW, also the DG for Communications Networks, Content and Technology (DG CONNECT), the Joint Research Centre (JRC) in the Directorate E – Space, Security and Migration, the DG For European Civil Protection and Humanitarian Aid Operations (DG ECHO), the DG for Migration and Home Affairs (DG HOME), the DG for Maritime Affairs and Fisheries (DG MARE), the DG for Mobility and Transport (DG MOVE), the European Political Strategy Centre (EPSC), and the DG for Research and Innovation (RTD), are concerned.

At the end of 2019, a new DG for Defence, Industry and Space (DEFIS) has been realized with responsibilities that were previously covered by DG GROW. DEFIS is created under the responsibility of the Commissioner for Internal Market. The Commissioner is also responsible for DG CONNECT and DG GROW (European Commission 2019a). The new DG will most likely (OSW 2019):

- Implement the European Defence Fund
- Ensure an open and competitive European defense equipment market and enforcing EU procurement rules on defense
- Implement the Action Plan on Military Mobility (in collaboration with the DG Mobility and Transport)
- Foster an innovative space industry in the EU
- Implement the EU Space Programme

Horizon2020 is the EU Framework Programme for R&D and innovation for the period 2014–2020. It builds further on FP7 and brings together the innovation parts of the Competitiveness and Innovation Framework Programme (CIP) and the EU Institute of Innovation and Technology (EIT). The objective if the maximize the contribution of the EU funded research and innovation to sustainable growth and jobs.

In 2017, the Commission established a SPEG subgroup in view of an improved coordination and policy-based approach regarding space technologies and associated actions to be supported by the Union. The SPEG subgroup is composed of representatives from the EU Member States and nominated observers from the European Space Agency and the European Defence Agency (ESA 2017a).

Political and Security Committee

The Political and Security Committee (PSC) meets at the ambassadorial level as a preparatory body for the Council of the EU. Its main functions are keeping track of the international situation and helping to define policies within the Common Foreign and Security Policy (CFSP) including the Common Security and Defence Policy (CSDP). It prepares a coherent EU response to a crisis and exercises its political control and strategic direction. The European Union Military Committee (EUMC) is the highest military body set up within the Council. It is composed of the Chiefs of Defence of the Member States, who are regularly represented by their permanent military representatives. The EUMC provides the PSC with advice and recommendations on all military matters within the EU. In line with relevant Council Conclusions, the European Commission restated its determination to closely collaborate with the European External Action Service (EEAS), the European Defence Agency (EDA), and the EU SatCen, together with Member States and the European Space Agency (ESA) "to explore possible dual-use synergies in the space programs."

European External Action Service

Established in 2011, the European External Action Service (EEAS) is the EU diplomatic service and manages the EU's diplomatic relations with other countries outside the bloc and conducts EU foreign and security policy. Within the Security Policy Office (SECPOL) under the Deputy Secretary General for CSDP and Crisis Response, the Security Policy and Space Policy Unit (SECPOL 3) is charged with space affairs in the EEAS, working side by side with the EEAS Space Task Force led by the Special Envoy for Space. Complementing SECPOL, space is relevant in the frame of the INTCEN (EU Intelligence and Situation Centre), the CMPD (Crisis Management and Planning), and the CPCC (Civilian Planning and Conduct Capabilities). The EU Military Staff (EUMS) is also involved in space with stakeholders in the EUMS MPCC (Military Planning and Conduct Capability). As the EU diplomatic service, the EEAS focuses on space security, safety, and sustainability in a global multilateral context (the post Code of Conduct) and in the emerging domain of economic diplomacy including space. The EU Intelligence Analysis Centre (INTCEN) (formerly known as SITCEN, and its origins go back to the WEU) is a Directorate at the EEAS. Its mission is to provide intelligence analyses, early warning and situational awareness to the High Representative and to the European External Action Service, to the various EU decision-making bodies in the fields of the CFSP and CSDP, and counterterrorism, as well as to the EU Member States.

European Defence Agency

EDA is an EU Agency that promotes and facilitates integration between member states within the EU's Common Security and Defence Policy (CSDP). The EDA is headed by the High Representative (HR), and reports to the Council of the EU. EDA is Europe's defense capability development actor. EDA and EEAS together form the Secretariat for Permanent Structured Cooperation (PESCO). EDA performs the Coordinated Annual Review on Defence (CARD). The EDA Capability Development Plan (CDP) was approved by EDA Steering Board in June 2018 (EDA 2018a). EDA is contributing to the preparation of the European defense research program (EDRP) and is involved in the preparation of the European defense industrial development program (EDIDP). The PESCO participating Member States have committed to use EDA as the European forum for joint capability development and consider OCCAR as preferred collaborative program management organization. An initial list of 17 projects to be developed under PESCO was adopted by the Council on 6 March 2018. A second batch of 17 projects to be developed under PESCO was adopted by the Council on 19 November 2018. And finally, a third batch of 13 additional projects to be developed under PESCO was adopted by the Council on 12 November 2019 (EDA PESCO 2018). EDA activity in the domain space is indispensable as space-based communication, situational awareness, and navigation and earth-observation capabilities play an increasingly critical role in security and defense. Satellite reconnaissance is one of the key functions allowing countries to gather information about military build-up or movement of troops worldwide. Precision-guided munitions, missile warning and launch detection systems, and space-based missile defense systems are other examples of space enabled defense capabilities (EDA 2017a).

The importance of space assets and applications for defense capabilities is reflected in the revised CDP approved by the EDA Steering Board in June 2018 (EDA 2018c). Space-related priorities are included notably in the so-called priority areas of Space Based Information Services (Earth Observation (EO); Positioning, Navigation and Timing (PNT); Space Situational Awareness (SSA); Satellite Communication); Information superiority (Radio Spectrum Management; Tactical Communication and Information System (CIS); Information management; Intelligence, Surveillance and Reconnaissance (ISR) capabilities); Air Superiority (i.a. Ballistic Missile Defence) and; cyber defense in space. (EDA 2018a). With regard to synergies between civil and defense research and technology (R&T), the R&T mandate of EDA includes the promotion of collaboration with other stakeholders. EDA, ESA, and the European Commission have cooperated in the frame of "Critical Space Technologies for European Strategic Non-Dependence" (ESA 2018c). ESA and the EDA are also cooperating in the development of new AI-based capabilities in the field of guidance, navigation, and control (GNC) – knowing where an asset is and steering where it is going. Advanced, autonomous GNC is set to become an indispensable element of ambitious future space missions such as rendezvousing with asteroids and comets or the active removal of hazardous space debris from orbit (EDA 2020b). EDA is the defense community facilitator towards the European Commission and EU Agencies and acts as the interface upon Member States' request, exploiting wider EU policies to the benefit of defense and acting as a central operator regarding EU funded defense-related activities. In that regard, EDA has an indispensable role, in support of its Member States, in harmonizing military user requirements for all European space programs.

European Union Satellite Centre

The European Union Satellite Centre (Satcen), previously EUSC and initially founded as the Western European Union Satellite Centre, is an EU Agency that supports the EU's decision-making in the field of the Common Foreign and Security Policy (CFSP), including crisis management missions and operations, by providing products and services resulting from the exploitation of relevant space assets and collateral data, including satellite and aerial imagery, and related services. SatCen's director reports to a governing board chaired by the EU's High Representative for Foreign Affairs and Security Policy (HR/VP). The EU Satcen provides high-level geospatial analyses based on satellite imagery to the EEAS, the EU Member States and to some international organizations such as NATO, the International Atomic Energy Agency (IAEA), the Organization for Security and Co-operation in Europe (OSCE), and the Organisation for the Prohibition of Chemical Weapons (OPCW). Based on the 2014 Council Decision, SATCEN shall support the decision-making and actions of the Union in the field of the CFSP and the CSDP, including European Union crisis management missions and operations, by providing, at the request of the Council or the HR, products and services resulting from the exploitation of relevant space assets and collateral data, including satellite and aerial imagery, and related services (Council of the EU 2014). Its main users are the European External Action Service (EEAS), Member States and international organizations.

Satcen's mission is aimed at providing the exploitation of relevant space assets and collateral data and services. Satcen uses open source data, commercial satellite imagery, and governmental imagery sensors; Helios II, Cosmo-Skymed, SAR-Lupe. In 2016, the classified COSMO-SkyMed link between SatCen and the Italian authorities was installed and declared operational. SatCen has since the ability to order and download classified COSMO-SkyMed products in an easy, secure, and fast manner. The SAR-Lupe classified link between SatCen and the German ground segment is fully operational and is used to place requests and download SAR-Lupe classified imagery (Satcen 2016). Satcen performs delegated operation on behalf of the European Commission: Copernicus in Support to EU External Action (SEA), Copernicus Border Surveillance (in support to FRONTEX), Copernicus Service Evolution (SEA and border surveillance), and Initial Space Surveillance and Tracking (SST) services. Satcen bridges the gap between research and applications in the field of space and security, operating as a dual-use agency serving civil and military users. Satcen priorities for space research include requirements for next-generation Intelligence-Surveillance-Reconnaissance (ISR), geoportals for CSDP missions and EEAS users, cross-cutting dual-use applications using EU space program assets, the use of artificial intelligence for remote sensing applications and deployment of cloud-based solutions for secure data access and distribution, and use of satcom solutions for dissemination of geospatial data and products.

European Union Space Security Activities/Programs

Earth Observation- EO

Copernicus Operational Services

Copernicus is the European Flagship Programme for monitoring the Earth. It is coordinated and managed by the European Commission. The services address six thematic areas: land, marine, atmosphere, climate change, emergency management, and security. They support a wide range of applications, including environment protection, management of urban areas, regional and local planning, agriculture, forestry, fisheries, health, transport, climate change, sustainable development, civil protection, and tourism. The Copernicus legal structure between the EU and ESA is composed of the following elements: the Copernicus EU Regulation, the European Commission Delegation Decision, the EU-ESA Agreement, and ESA contracts within the ESA name. As part of the Copernicus Programme, the European Commission and ESA have awarded in July 2018 an Airbus-led consortium a contract for the provision of satellite-based seamless coverage of the whole of Europe at very high-resolution. The consortium includes Airbus Defence & Space, Planet, Deimos Imaging, IGN-France, and space4environment.

The missions in the Copernicus Space Component are: Sentinel-1; Sentinel-2; Sentinel-3; Precursor Sentinel-4; Sentinel-5; and Sentinel-6. The European Commission DG GROW delegates Copernicus activities to the European Environment Agency (EEA), the European Maritime Safety Agency (EMSA), the European

Border and Coast Guard Agency (Frontex), Satcen, ESA, the European Organisation for the Exploitation of Meteorological Satellites (EUMETSAT), the European Centre for Medium-Range Weather Forecasts (ECMWF), and Mercator Ocean (European Commission DG GROW Website). Frontex and EMSA are European Commission decentralized agencies. The operational services of Copernicus are situated in three major axes: Border surveillance (Frontex); Maritime surveillance (EMSA); Support to External Action (Satcen).

Frontex supports, coordinates and develops European border management in line with the Treaties including the Charter of Fundamental Rights of the EU as well as other international obligations. Frontex coordinates operational and EU measures to jointly respond to exceptional situations at the external borders. Frontex develops capacities at Member States and European level as combined instruments to tackle challenges focusing of migration flows, but also contributing to fight cross-border crime and terrorism at the external borders. Frontex builds the capacities and capabilities in the Member States aiming at develops a functioning European Border and Coast Guard. In November 2015 Frontex signed a Delegation Agreement with DG GROW amounting to €47.5 million for the period 2015–2020 to implement the Border Surveillance component of the Copernicus Security Services. The objective of this component is to provide increased situational awareness when responding to security challenges at the external border through detection and monitoring of cross-border security threats, risk assessment and early warning systems, and mapping and monitoring (Council of the EU 2019a).

EMSA supports, coordinates and develops European maritime safety and security, including prevention of and response to pollution. EMSA Services are offered to all EU and EFTA Member States in accordance with existing access rights and provide enhanced features for, among others, environmental monitoring, search and rescue, and traffic monitoring purposes. It allows Member States to make full use of the integrated vessel reporting information from terrestrial and satellite Automatic Identification System (AIS), Long-Range Identification and Tracking (LRIT), Vessel Monitoring System (VMS), as well as national vessel position data such as coastal radar, patrol assets, and leisure craft. EMSA provides maritime border control support to Frontex under the auspices of the European Border Surveillance System (Eurosur). Vessel information originates from both terrestrial and satellite-based systems as well as other available positioning data and are correlated against satellite aperture radar and optical imagery derived vessel detections. EMSA further provides operational support for the European Fisheries Control Agency (EFCA) coordinated Joint Deployment PlaN operations (JDP) for fisheries activities in the Mediterranean, North & Eastern Atlantic and the North Sea waters. It includes a real time maritime awareness operational picture fusing and correlating VMS, terrestrial AIS, satellite AIS, and LRIT position reports together with visual sightings, as well as establishing a common fishery vessel registry (EMSA 2014).

EDA Activities and Programs

In July 2016, the EDA and Satcen formalized their close and fruitful cooperation, already in place since 2004, with an exchange of letters. With the exchange of letters, the EDA and SATCEN establish a more structured cooperation meaning that they will even more focus on activities of mutual interest, such as studies, workshops, projects, and programs. EDA and SATCEN have also identified specific cooperation areas such as imagery exploitation, geospatial analysis and applications, future space-based earth observation systems, cyber defense, Big Data exploitation in the space and security domain, space situational awareness, or maritime surveillance. The two Agencies will also develop a joint roadmap for cooperation detailing the activities of common interest as included in the respective work programs. The roadmap will be updated annually (EDA 2017a).

In addition, the SULTAN (Persistent Surveillance Long Term Analysis) study was a 2014–2015 cooperative study from EDA and ESA executed by Airbus Defence & Space, aimed at identifying potential options to enhance collection capabilities in Imagery Intelligence (IMINT) through innovative and technologically feasible solutions, able to deliver an operational capability at 2025–2030 timeframe, to meet the need of persistent surveillance of wide areas in defense and security operations. To this extent, the project has provided analysis and description of operational scenarios, technology roadmaps, rough estimation of cost, time schedule and respective merits of assets/systems based on geostationary satellite systems, constellations of optical and radar small/mini satellites in low earth orbit, High Altitude Pseudo-Satellite Systems (HAPS), and Remotely Piloted Aircraft Systems (RPAS). The study represented a significant step in the EDA Intelligence, Surveillance and Reconnaissance capability development process (analysis, identification and selection of IMINT capacities). The dual-use dimension of the surveillance systems and the joint requirement analysis performed on security and defense scenarios could also be used in the framework of future activities in cooperation with the European Commission (EDA 2016a).

Moreover, EDA has a Project Team Space-Based Earth Observation (PT SBEO). The PT regroups all the EDA activities concerning Space Based Earth Observation and is currently elaborating a Common Staff Requirement (CSR) for Earth Observation Requirements in the time frame 2025–2030, following a Common Staff Target (CST) approved in July 2017. This task is executed in cooperation with Member States representatives, European Military Staff and the European Satellite Centre, as well as taking benefit of the experience and expertise at ESA and the European Commission. EDA is investigating the use of Copernicus for defense users, in collaboration with the European Commission DG GROW (ESA 2017a).

The EDA-Satcen cooperative project Radar Imagery Applications Supporting Actionable Intelligence (REACT) has enhanced imagery operators' abilities to manage the complexity of working with radar imagery, especially by providing practical information on the various steps of the workflows to be followed and established credible working procedures for radar imagery exploitation. The work undertaken also allowed us to evaluate new tools and sophisticated algorithms for radar imagery exploitation, such as Automatic Target Detection & Recognition. After the initial REACT Study, EDA has a follow-up activity REACT 2 study on imagery exploitation activities. Working again with the EU SatCen, the focus will be on making SAR IMINT workflows more efficient through defining operating procedures and the use of business process workflow tools. One key objective is to increase the speed of analysis within the radar IMINT tasking cycle (EDA 2017b).

EDA and ESA have signed an Implementing Arrangement on cooperation on Earth observation requirements in the time frame 2025–2030. EDA and ESA will implement a joint study to be contracted in 2018, elaborating mission concepts and a roadmap of technologies needed for future security Earth observation missions. This study will feed into the EDA CSR process (EDA 2018b).

Satellite Communications: SATCOM and Cross-Domain Applications

Governmental Satellite Communications: GOVSATCOM
The European Commission has proposed the establishment of a new EU Space program component aimed at pooling and sharing of secure governmental satellite communications (European Commission 2018b). The EU Global Strategy, adopted by the European Council in June 2016, the European Commission Space Strategy for Europe, launched in October 2016, and the European Defence Action Plan which followed the subsequent month all stress the importance of having enough access to European GOVSATCOM capabilities. Furthermore, in March 2017, the Council's Political and Security Committee has endorsed the document of High Level Civil-Military User Needs for GOVSATCOM, thus further consolidating civil-military synergies in the field (EDA 2017c).

The November 2013 Steering Board at Ministerial level endorsed EDA's proposal and roadmap on developing a future GOVSATCOM capability in support of national defense efforts. On this basis, the 19 December 2013 European Council, "committed to delivering key capabilities and addressing critical shortfalls through concrete projects by Member States, supported by the European Defence Agency," welcomed "preparations for the next generation of Governmental Satellite Communication through close cooperation between the Member States, the Commission and the European Space Agency." Following this mandate delivered by EU Heads of States and Governments, EDA has developed in 2014 the defense needs related to the use of governmental satellite communications assets (ESA 2017a).

In the frame of the 2014 EDA CDP, provision of satellite communication capabilities was agreed as a prioritized action. A landscape of SATCOM capabilities in Europe demonstrated that almost all current national assets will have to be renewed or completed before 2020/2025. In the European Council of December 2013, EDA has been mandated to support participating MS in the capability development process of the next generation of governmental satellite communications (European Council 2013). This step triggered the launch of the project preparation phase which has led to the adoption of a Common Staff Target, describing military user needs, on 4 November 2015, and through Common Staff Requirements and Business Case in March 2017. EDA Steering Board approved in February 2018 a mandate for EDA to act as facilitator in support of the Ministries of Defence within the EU GOVSATCOM Programme (ESA 2017a).

EU SatCom Market

The EU Satcom Market comprises 31 contributing members. The EU SatCom Market has, since the project started in 2012, provided an efficient option to source commercially available SatCom services and from 2017, Communication and Information System (CIS) services are also added to the whole portfolio of services offered to its members. The project goals are to provide a cost effective commercial SatCom and CIS solutions for members and to reduce costs, ease access, and improve operational efficiency for members. In January 2016, the second Framework Contract was awarded to Airbus Defence & Space. In July 2017, the first Framework Contracts for Communication and Information System services were awarded to Airbus Defence & Space and Thales Communications & Security. In January 2020, the third Framework Contract for SatCom services was awarded to an Airbus Defence & Space and Marlink consortium. From the first request in 2013 up to January 2020, €35.4 million in total ordering value for SATCOM services has been placed through the framework contracts (EDA 2020a).

EDA GOVSATCOM

The EDA Project Team on Satellite Communications is part of the Ad Hoc working Group preparing the establishment of the EDA GOVSATCOM Demonstration Project (Pooling and sharing) to start activities in 2018. The Working Group elaborates a Project Arrangement (PA) for the GSC demo supported by a Concept of Operations (CONOPS) document that will serve as the framework for the PA. The program objectives are to demonstrate the benefits of a European dual-use approach for the development of such capability and to provide EDA Member States and European CSDP actors with access to a GOVSATCOM capability based on existing, pooled, governmental SATCOM resources. EDA is in the process of finalizing and signing with its participating Member States the Project Arrangement on the EDA GOVSATCOM demonstration project with operations to commence before the end of 2018 (EDA 2017c; ESA 2017a).

EDA-ESA

EDA and ESA signed in March 2017 an Implementing Arrangement on exploiting space-based assets to advance capabilities in Chemical, Biological, Radiological, Nuclear, and explosive (CBRNe) threats. ESA launched a feasibility study in 2017 in order to assess the technical and economic viability of deploying services based on Satellite Communications, Satellite Navigation, Earth Observation data, and/or other space assets to support CBRNe operations (e.g., catering for detection and situational awareness, prediction, early warning, and response planning) for the benefit of both institutional and commercial users. The study will eventually propose a roadmap for service(s) implementation and demonstration. The feasibility study also included an Option for ESA-EDA cooperation in order to demonstrate the benefits that the National Stakeholder Communities (NSC) in charge of combating CBRNe threats in EU Member States can get from the utilization of services integrating space and terrestrial technologies. The NSC includes different bodies engaged with the planning and execution of actions in response to CBRNe threats (ESA 2017c).

The Administrative Arrangement between EDA and ESA signed in June 2011 provides the legal framework for cooperation projects between EDA and ESA.

Positioning, Navigation, and Timing: PNT

Galileo

Galileo is a fully European program. The European Commission has overall responsibility for the program, managing, and overseeing the implementation of all activities on behalf of the EU. Galileo's deployment, its design, and the development of the new generation of systems and the technical development of infrastructure is entrusted to the European Space Agency (ESA). ESA oversees the R&D phase and IOV phase and co-finances the Galileo program of about 50%. The EU remains the sole owner of the infrastructures and the services (Schrogl et al. 2015).

The four services of Galileo are: Open Service (OS), Commercial Service (CS), Public Regulated Service (PRS), and Search and Rescue Service (SAR). The Commission has delegated the operational management of the program to the GSA, which oversees how Galileo infrastructure is used and ensures that Galileo services are delivered as planned and without interruption (European Commission Galileo Website). The European Commission has set up a specific directorate for the EU Satellite Navigation Programme, under the authority of the DG GROW and two expert groups: Security Board for European GNSS (6 Working Groups) and the Search and Rescue Galileo Operations Advisory Board. The European GNSS Programme Committee (4 Working Groups) monitors and supports the European Commission activities and forms the space for dialogue for Member States. The Galileo Security Board (with 10 Working groups) with Member States representatives primarily from or on behalf of the Ministries of Transport is and advisory body supporting DG GROW in the execution of the Galileo Programme (Sitruk and Plattard 2017).

ESA is charged with the deployment of the European satellite navigation program Galileo through the so-called Galileo Deployment Delegation Agreement. The first 22 satellites of the Galileo "Full Operational Capability" phase are built by OHB (Germany) and Surrey Satellite Technology Ltd. (UK), which is producing the payloads. ESA is now being tasked with the preparation and procurement of the Transition Satellites and supporting equipment qualification activities (ESA 2018e).

Galileo has two control centers: Fucino in Italy generates the accurate navigation messages that are then broadcast through the navigation payloads, and Oberpfaffenhofen in Germany controls the constellation of satellites. A new telemetry, tracking, and command station last year was constructed in Papeete on Tahiti, in the South Pacific in 2017. Establishing Galileo's ground segment was among the most complex developments ever undertaken by ESA, having to fulfil strict levels of performance, security, and safety. Formal responsibility for the operations of this Galileo ground segment was last passed to the GSA in 2017 but ESA continues to oversee its maintenance and growth. Galileo telemetry, tracking, and command stations are situated in the Kiruna, Redu, Kourou, Reunion, Noumea (in New Caledonia), and the Papeete sites. The ground segment also comprises a set of four Medium-Earth Orbit Local User Terminals serving Galileo's search and rescue service, at the corners of Europe and facilities for testing Galileo service quality and security – the Timing and Geodetic Validation Facility and two Galileo Security Monitoring Centres. The Launch and Early Operations Control Centers have the task of bringing new satellites to life, to be handed over to the main Satellite Control Centre in Oberpfaffenhofen within typically a week after launch. The Redu Centre in Belgium set up as Galileo's In-Orbit Test Centre, then puts these satellites through a complex set of testing and checkouts ahead of them joining the working constellation (GPS World 2018).

EGNOS

EGNOS is a Satellite Based Augmentation system (SBAS), created to complement GNSS. The development of EGNOS was managed by the European Space Agency (ESA) under a tripartite agreement with the European Commission and the European Organisation for the Safety of Air Navigation (Eurocontrol). The ownership of the EGNOS assets was transferred from the ESA to the Commission in April 2009 and EGNOS officially entered service on 1 October 2009. Through a contract with the European GNSS Agency (GSA), the service is delivered by the European Satellite Services Provider, ESSP SaS, which was founded by seven air navigation service providers. The GSA has been the EGNOS Programme Manager under delegation from the Commission since 2014 and the ESA is the design and procurement agent working on behalf of the Commission. Efforts are made to extend the geographical coverage for EGNOS services in the European High North, North Africa, and the Middle East (GSA 2017).

Airbus has been selected as the main contractor to develop EGNOS V3. The contract was awarded by ESA, which manages EGNOS development under a working arrangement signed with the European GNSS Agency (GSA), in January 2018. Two EGNOS upgrade versions. EGNOS V3.1 will ensure continuity of EGNOS augmentation of GPS L1, but with a more resilient performance, while EGNOS V3.2 will support a new SBAS service, transmitting on the L5 frequency, which will augment Galileo L1/E1 – L5/E5 along with GPS (Sitruk and Plattard 2017).

GSA

Unlike Copernicus, where the EU still strongly relies on ESA, EUMETSAT and national capabilities for operations, the EU has fully taken over program management for, and exploitation of EGNOS and Galileo through its executive arm, the European Commission. To deal efficiently with those new kinds of missions, it has also created a dedicated entity, the European GNSS Agency (GSA). According to the EU Proposal for a Regulation for a Space Programme for the EU, GSA will be rebranded as the new European Union Agency for the Space Program (EUSPA). EUSPA is ready to succeed and expand the current European GNSS Agency (GSA), which has been managing the EU's Galileo satellite system for 15 years (European

Commission 2018b). In addition to the development of Galileo operations, the EUSPA will manage the use of the Copernicus Earth observation satellite system, prepare the Governmental Satellite Communications (GOVSATCOM) program, and concentrate EU capacities to monitor the Earth's near surroundings (European Commission 2018b; GSA 2020).

With regard to the GSA, in its current form, it is a decentralized (or regulatory) agency of the EU and a key element of European GNSS governance. The GSA was initially set up by the Parliament and the Council through Regulation (EU) N° 912/2010, amended by the Regulation (EU) N° 512/2014 (GSA Regulation). The role of the GSA, its prerogatives, and its internal organization has changed several times since the beginning of the Galileo program. The Agency was created in 2004 to protect the public interest within the former Galileo precise point positioning (PPP) model and given with an important core of related prerogatives (concession contract implementation, security management, PRS-related activities, Member States coordination, etc.). After a drastic reduction of its role, the last regulation gave an important role back to the GSA as the entity that should progressively take charge of system exploitation. The current prerogatives of the GSA are now mainly defined by the GNSS Regulation, completed by the GSA Regulation. They are of two kinds: some prerogatives are fully owned by the GSA ("core prerogatives"), whereas others are delegated by the Commission through a delegation agreement (Sitruk and Plattard 2017).

Space Situational Awareness-SSA/Space Surveillance and Tracking – SST

EU SST Program

The European Commission has proposed the establishment of a new EU Space Programme component for Space Situational Awareness (SSA), meaning a holistic approach to enhance SST capabilities to monitor, track, and identify space objects; to monitor space weather; and to map and network Member States NEO capacities. The SSA Programme will build on and extend the current EU Space Surveillance and Tracking (SST) Support Framework to space weather and near-Earth objects (European Commission 2018b).

After the adoption of the decision establishing a framework on Space Surveillance and Tracking (SST) support in March 2014, the European Commission adopted, in September 2014, a first implementing act on the procedure for participation of Member States in the SST Support Framework. By mid-2015, a Consortium agreement on SST was signed between France (CNES), Germany (DLR), Spain (CDTI), Italy (ASI), and the United Kingdom (UKSA) represented by their respective space agencies (SST Consortium). They became the first Member States participating in the SST Support Framework (European Parliament 2014).

Implementing arrangements have also been concluded between the members of the SST Consortium and the EU Satcen for the delivery of SST services. Since January 2016, the Galileo, Copernicus, and Horizon 2020 programs have been supporting the SST activities aimed at the establishment and operation of the network of sensors, the establishment of the capacity to process and analyze SST data, and the establishment and operation of SST services. The upgrade and renewal of SST sensors is also supported as part of the Horizon 2020 Space Research Programme (European Parliament 2018). The SST Decision identifies as SST Services: Collision Avoidance, In-Orbit Fragmentation and Uncontrolled Re-entry of space objects. To facilitate service provision, the Consortium and EU Satellite Centre (EU SatCen) form the so-called EUSST Cooperation. The work of the Cooperation is governed by the Implementing Arrangement signed in September 2015. Since July 2016, SatCen is providing initial SST services – generated by the SST Consortium – via the first instance of the EU SST Service Provision Portal (EUSST Website).

With a second Implementing Decision in 2016, the Commission launched the second round of Member States applications to join the SST Consortium. Three Member States (Poland, Romania, and Portugal) submitted formal applications to join the SST Consortium by the 19 August 2017 deadline and the procedure should be completed in 2018. The participation of new Member States can help to increase the performance of EU SST. Eight other Member States (Austria, Croatia, Finland, the Czech Republic, Greece, Latvia, Slovakia, and Sweden) expressed their intention to collaborate with the SST Consortium as participating entities in the implementation of the future grants. Private sector contributes to the EU SST, mainly as technology and data provider, and does not participate in the EU SST governance (Council of the EU 2018).

The SST Decision recognizes the sensitive nature of SST and leaves the implementation and management of the EU SST capability to the participating Member States, with assets owned at national level. The Commission's involvement in 2014–2017 was mostly related to monitoring the procedure for Member States' participation, executing grants, interacting informally with the SST Consortium, and drawing up the 2017–2020 coordination plan. The SST Consortium governance structure involves work in steering, technical, and security committees and project and financial coordination, with decisions being taken by unanimity. Most decisions, including those concerning the program management, are taken in the steering committee, where the Commission is an observer since 2017. The coordination committee – SST Consortium and SATCEN – is responsible for the governance of the SST Cooperation (Council of the EU 2018).

EU grants have financed EU SST activities in three main areas: EU SST service provision (1SST); the networking of assets and coordination of actions (2SST); and the upgrade of existing, and development of new, SST assets (3SST). A total of €167.5 million has been allocated for 2015–2020 through various grants under the Copernicus, Galileo, and Horizon 2020 programs, out of which around €70.5 million to implement the actions of the SST Decision (1SST and 2SST grants) and €97 million for the sensors' upgrades (3SST). The activities described in this report were co-financed by the 2015 grants, which were subject to administrative closure in December 2017. The 2016–2017 grants signed in December 2017 should ensure the continuity of the activities and the transition to more comprehensive and effective EU SST services. The topics of Horizon 2020 calls for grants in 2018–2020 were published in 2017. Delivery of three initial SST services started on 1 July 2016: – conjunction analysis and warning (CA), re-entry analysis and information (RE), fragmentation analysis (FG) (Council of the EU 2018).

Other Activities

EDA has concluded in 2016 a study on the Recognised Space Picture (RSP) in cooperation with Satcen. This study aimed at developing a common understanding for an RSP Display focusing on visualization aspects, proposing a draft Operational Concept and analyzing external interfaces and interoperability aspects for the benefit of EU Member States.

The Political and Security Committee of the EU (PSC) adopted in October 2011 a unified set of European Space Situational Awareness (SSA) high-level civil-military user requirements. ESA in the framework of its SSA preparatory program had defined the civil SSA mission requirements (ESA SSA Mission Requirements (MRD) – SSA-GEN-RS-MRD-1000, issue 3.0, dated 29/04/2011), which were concurred by its Member States and approved by its Executive while the EDA had defined military SSA requirements (EDA Common Staff Target (CST) 25 March 2010 – EDA SBD 2010–07), which were approved by the European ministries of defense who compose its Steering Board (Council of the EU 2019b).

After proactive USA positioning in the frame of the International Civil Aviation Organization (ICAO), the European Commission is investigating the so-called new entrants including balloons, solar planes, supersonic planes, suborbital vehicles, and High Altitude Pseudo Satellites (HAPS) HAPS. These are all situated in altitudes above flight level (FL600) and below space (100 km or the Karman Line). Higher Airspace Traffic management (HATM) is considered on top of the existing Air Traffic Management. There have been several interactions among the European Commission, Eurocontrol, the European Aviation Safety Agency (EASA), ESA, EDA, and GSA (Eurocontrol 2019).

The European Space Agency: ESA

ESA Space Security Status

ESA is an international organization with 22 Member States with its mandate enshrined in the ESA Convention. ESA is headed by its Director-General who reports to the ESA Council. ESA's mission is to shape the development of Europe's space capability and ensure that investment in space continues to deliver benefits to the citizens of Europe and the world. ESA programs are funded through mandatory and optional programs, funded by GDP-scale-based contributions to the Level of Resources and voluntary Member States' subscription to programs, respectively. The 2019 ESA budget amounts to €6.68 billion (ESA 2019a).

ESA is increasingly contributing to space security in and from space. The Council Document "Elements of ESA's Policy on Space and Security" acknowledged ESA's involvement in space and security activities and provided for the guiding principles of ESA's future involvement in the domain of space and security in view of strengthening the dialogue with Member States and supporting the current and future activities of the Agency (ESA 2017b). It is the ambition of the ESA to exploit the potential of space in supporting safety and security matters. This ambition is reflected in the four programmatic pillars of the Agency:

- Science and Exploration (scientific program and human and robotic exploration and science support)
- Applications (Earth observation, telecommunications, navigation, and integrated applications)
- Enabling and Support (space transportation, technology development, operations)
- Safety and Security (safety and security in and from space) (ESA 2018a)

Resulting from a mandate given to ESA DG at the Ministerial Council 2016, and the decisions of Member States at the Ministerial Council 2019, ESA has recently adopted a safety and security programmatic pillar with associated program proposals (ESA 2019b).

The EU and ESA signed a Joint Statement on Shared Vision and Goals for the Future of European Space on 26 October 2016. EU and ESA emphasized their intention to reinforce their cooperation in the future as foreseen in the ESA/EU Framework Agreement of 2004 (ESA 2016).

ESA has an Administrative Arrangement in place with EDA since June 2011. Formal cooperation agreements (including exchange of letters) has been implemented for Unmanned Aircraft System (UAS); Remotely Piloted Aircraft Systems (RPAS); Chemical, Biological, Radiological, Nuclear, and high yield Explosives (CBRNe); Cyber Defence for Space; Cyber Defence Training and Exercise; GOVSATCOM; Unmanned Maritime Systems (UMS); Earth Observation Requirements (EDA 2016b).

Furthermore, the cyber defense for space cooperation provides that EDA and ESA will perform a joint study, in two phases: A first phase to identify and priorities critical space technologies which present vulnerabilities with respect to cyber threats, and a second phase to jointly develop and implement initial solutions to address such vulnerabilities. This cooperative effort seeking to maximize dual-use synergies builds upon work already carried out respectively by both ESA and EDA. ESA has also and Administrative Arrangement in place with Satcen since January 2018.

ESA Space Security Activities/Programs

Earth Observation: EO

ESA's Earth Observation program consists of: Earth Observation Envelope Programme (EOEP), the Earth Watch, the GMES Space Component, the Meteosat Third Generation (MTG), and the MetOp Second Generation (MetOp-SG).

At the ESA Ministerial Council 2019, it was agreed that substantial reorientation of ESA's Earth observation (EO) programs into three groups (Future EO, Operational EO, and Customised EO) to reinforce Europe's global leadership in this fast-evolving domain through the implementation of an innovative and competitive set of activities in order to serve European science and business, and fulfil societal objectives. Accordingly, the Future EO program, formerly the Earth Observation Envelope Programme (EOEP), is the backbone program that develops Earth Explorer missions as well as science and technology for all Earth observation missions conducted by ESA and its partners. The Ministers agreed to support the creation of new Scout and Φ-sat missions to develop opportunities in the NewSpace domain. In addition, it was considered important to make full use of new technologies and methods such as artificial intelligence, big data analytics, and High-Altitude Platform Stations (HAPS) (ESA 2019b).

Operational EO comprises meteorological missions and Copernicus, the latter conducted through the successful partnership between the European Union and ESA that will focus, in the framework of ESA's Copernicus Space Component (CSC) Programme, on the development of six new Sentinel missions and the related ground segment, with a view to providing critical data and information to support European policy priorities and EU user requirements related, for example, to climate change, agriculture, biodiversity, the Arctic, Atlantic and Africa (ESA 2019b).

Copernicus

Regarding EU space programs, the fundamental Agreement between the EU represented by the EC and ESA on the implementation of the Copernicus Programme including the ownership of Sentinels of 28 was signed in October 2014 (ESA 2018b). An amendment to the 2014 Copernicus Agreement was by the EU and ESA in January 2019, adding €96 million to ESA's space component budget for Copernicus. This additional contribution covers ESA's additional tasks such as the development of the Copernicus Sentinel-6 mission and the new European Copernicus Data Access and Information Services. In total, the budget for the Copernicus space component amounts to €3.24 billion for the 2014–2021 period. While the European Union leads the Copernicus program, ESA develops and builds the dedicated Copernicus Sentinel satellites. It also operates some of the missions and ensures the availability of data from other partners (ESA 2019d).

The Copernicus Space Component comprises two types of satellite missions, ESA's families of dedicated Sentinels and missions from other space agencies, called Contributing Missions. A unified ground segment, through which the data are streamed and made freely available for Copernicus services, completes the Space Component. ESA is establishing a mechanism to integrate, harmonize, and coordinate access to all the relevant data from the multitude of different satellite missions. This is being carried out in close cooperation with national space agencies Eumetsat and where relevant with owners of non-European missions contributing to the Copernicus objectives (ESA Copernicus Website). While the deployment of the first generation of Sentinels was completed, the ESA Member States at Ministerial Council 2019 agreed to contribute €1.8 billion for the expansion of the Copernicus

Space Component (CSC Expansion program). The six new potential Copernicus Sentinel missions, so-called High Priority Candidate Missions (HPCM), are as follows: Anthropogenic CO2 monitoring mission (CO2M), High Spatio-Temporal Resolution Land Surface Temperature Monitoring Mission (LSTM), Passive Microwave Imaging Mission (CIMR), Hyper Spectral ImagingMission (CHIME), L-Band SAR Mission (ROSE-L), and the Polar Ice and Snow Topographic Mission (CRISTAL) (ESA Copernicus Website).

High-Altitude Pseudo Satellites: HAPS

ESA is additionally investigating services enabled by High-Altitude Pseudo Satellites (HAPS) complemented by satellites, most recently in a feasibility study completed in September 2019. The latter is supported by EMSA, Satcen, and Frontex, with potential follow-up demonstration projects. Maritime, security, and emergency response services are considered as domains where HAPS can play a key role in the short-term. It is expected that soon users in the civil domain engaged in monitoring and surveillance operations will consider utilization of these assets as a complement to the current data sources, being it space, airborne, or in situ. In this context, the European Maritime Safety Agency (EMSA), Frontex, and the European Union Satellite Centre (SatCen) have expressed their interest to support the current study and will provide requirements and guidance (ESA 2019c).

Satellite Communications: SATCOM

ESA's SATCOM programs consist of the Advanced Research in Telecommunications Systems (ARTES) programs: Future Preparation Phase 6 and Phase 7, Core Competitiveness, ScyLight, ARTES C&G Phase 2, Advanced Technology Phase 2, European Data Relay Satellite System (EDRS), Large Platform, IRIS Phase 2.1 and 2.2, Small Geostationary Satellite (SGEO), Next Generation Platform (NEOSAT), Integrated Applications Promotion (IAP), Satellite Automatic Identification System (AIS) SAT-AIS, Electra, Quantum, ICE, Indigo, ECO, Pioneer, Governmental Satellite Communications (GOVSATCOM) Precursor and Aidan.

Secure Satcom for Safety & Security (4S)

At the ESA Ministerial Council 2019, it was agreed that the Programme of Advanced Research in Telecommunications Systems, ARTES 4.0, as a successor to the ARTES Programme which remains in force, and comprising coordinated but financially autonomous program lines either of a strategic nature or for general purpose, shall be conducted within the framework of the Agency pursuant to the corresponding Declaration entering into force on this day following its subscription by the participating Member States concerned; and NOTES that ARTES 4.0 will provide a streamlined and more responsive programmatic toolset to support industry in the increasingly dynamic and fierce competitive market environment and, reaching beyond telecommunications, will add capabilities and value for other ESA activities and programs (ESA 2019b).

GOVSATCOM Precursor Program

In December 2017, ESA and EDA formalized their cooperation on GOVSATCOM through an Implementing Arrangement. The purpose of this cooperation is to maximize the synergies between the ESA GOVSATCOM Precursor and EDA GOVSATCOM Pooling & Sharing Demonstration. The ESA ARTES program proposal for GOVSATCOM Precursor(s) has been supported at ESA's Ministerial Council in 2016 (CM16) for implementation and in-orbit demonstration during 2017–2020, in preparation of an operational GOVSATCOM program, which is anticipated to be under EC lead (ESA 2017b).

Satellite operators and service providers are developing the first GOVSATCOM precursor public–private partnership projects called PACIS: Pacis 1, Pacis 3, Pacis 5, and Pacis 6. These projects aim to develop secure mission control systems and operations centers and demonstrate the benefits of pooling and sharing to users in the field. The PACIS Projects have been defined as part of a future federation of demonstration projects in the 2017–2020 timeframe. The ESA Govsatcom Precursor program will be implemented in synergy with the Govsatcom Pooling & Sharing Demonstrator that is implemented by the European Defence Agency (EDA). (ESA 2018a).

Additionally, preparations have been initiated for a coherent ARTES programmatic framework for Secure Satcom for Safety & Security (4S) in consistency with the Safety and Security program of ESA for the 2019 ESA Council at ministerial level (ESA 2018a).

Remotely Piloted Aircraft Systems: RPAS

Several ARTES studies relevant to security have been conducted, on certification requirements and performance standards of satellite communication links for RPAS and Air Traffic Services (ATS) as well as in support to emerging System Concepts for Unmanned Aircraft System (UAS) Command and Control via Satellite. Future studies have been prepared to analyze cyber-threats on satellite communication networks and impacts on the global society and to study UHF innovative waveform and satellite communication ground segments (ESA 2017b).

ScyLight

The ARTES ScyLight (Secure and Laser Communication Technology) Element is dedicated to the development and early demonstration of optical communication technology and includes a dedicated programmatic line addressing quantum cryptography technology and Quantum Key Distribution (QKD) missions. The proposals on an ITT for a study on space-based QKD systems to protect Critical European Infrastructures have been evaluated in close coordination with the European Commission's DG GROW and DG CONNECT. Under ScyLight, a first contract on QKD Services was signed on 2 May 2018 about QUARTZ – Quantum Cryptography Telecommunication System between ESA and SES. The project will start with the technology development of a Quantum Key Distribution (ESA 2018a).

Future Air Traffic Management (ATM) will increasingly rely on information sharing based on datalink, requiring a higher level of safety and cyber-security. The ARTES element Iris aims at supplying a validated satellite-based ATM communication solution for Europe. Iris is conducted in close coordination with the Single European Sky ATM Research (SESAR) initiative of the SESAR Joint Undertaking, established as a public-private partnership founded by the European Union and Eurocontrol (ESA 2017b).

Positioning, Navigation, and Timing: PNT

ESA's PNT program consists of the European GNSS Evolution Programme (EGEP) and the Navigation Innovation and Support Programme (NAVISP). Third Party contribution accounts for Galileo Full Operational Capability (FOC), Galileo Deployment Phase, Galileo GSA WA, European Geostationary Navigation Overlay Service (EGNOS) GSA WA, and Horizon 2020.

NAVISP has demonstrated in its first 3 years of implementation its capacity to generate innovative ideas on matters ranging from weather monitoring based on GNSS and sensor data crowdsourcing to the use of pulsars as an independent timescale and of quantum sensing to complement space-based techniques for PNT. During its Phase 2, NAVISP will continue to act along the entire PNT value chain so as to maintain and develop a technological edge beyond the scope of H2020 and future Horizon Europe activities with a view to continuing to support the development of competitive products and services, and thereby generating tangible economic benefits (ESA 2019b).

ESA oversees the implementation of the space infrastructure and R&T preparatory activities for the European GNSS Programme (Galileo and EGNOS) funded by the European Union. The specific high-level security objectives for Galileo are to:

- Ensure the ability of the EU to control the system.
- Ensure the ability for the EU to develop, build, and maintain an independent and autonomous GNSS infrastructure.
- Ensure the ability for the EU to prevent the security issues specific to GNSS infrastructure.

The specific high-level security objectives for EGNOS are to:

- Ensure the ability for the EU to protect the system from malicious or hostile attack that could impact the safe use of the system.
- Ensure the ability of EU to protect any data whose disclosure could adversely affect the ability to provide the EGNOS service (ESA 2017b).

ESA is charged with implementing the security requirements for the program and supports security-related accreditation activities necessary for the correct and successful accreditation by the GNSS Security Accreditation Board (GSAB). The PRS Arrangement between ESA and the European Commission and ESA was under negotiation in 2017. The Arrangement covers:

- The deployment phase of the Galileo program under the EC-ESA Galileo Deployment Delegation Agreement (the agreement lasts until 31st December 2021).
- The exploitation phase of the Galileo program under the European GNSS Agency-ESA Working Arrangement (the agreement is foreseen to last until 31st December 2021).
- The early phases of the evolution of the Galileo program under the EC-ESA Delegation Agreement on the implementation of H2020 Framework Programme.

Parallel to the Galileo FOC deployment completion tasks, potential new or enhanced PRS capabilities are being investigated by ESA as part EC/ESA H2020 Delegation Agreement covering GNSS evolution activities (ESA 2017b).

Space Situational Awareness: SSA/Space Surveillance and Tracking – SST

Space Safety Programme (S2P)

At the ESA Ministerial Council 2019, the Ministers agreed that the Space Safety Programme (S2P), as a successor to the Space Situational Awareness (SSA) Programme, aims to contribute to the protection of Earth, humanity, and assets from hazards originating in space, shall be conducted within the framework of the Agency. S2P will contribute significantly by establishing overall complementarity and maximizing synergies among space safety initiatives in Europe, providing opportunities for cooperation among the Agency, Member States, the European Union and other partners to those ends (ESA 2019b).

The establishment of the Space Safety Programme, in response to increased awareness worldwide of the underlying issues, aims to provide ESA and its Member States with the necessary tools to conduct effective risk management, addressing hazards originating in space through the identification of their different types, the analysis of their status, severity, and magnitude; the prevention of such hazards materializing in the form of genuine threats of damage being caused to Earth or ESA's space infrastructure by activating mitigation measures and providing appropriate information to support Member States activities aimed at ensuring efficient crisis management, thus giving Europe a lead role in the field (ESA 2019b).

The innovative and promising approach offered by the program activities aim at establishing a long-term framework for in-orbit servicing starting with active debris removal performed on an ESA space object, thus offering ESA the opportunity to apply more efficiently in the coming years its policy and guidelines aimed at reducing the harmful effects of space debris and opening up new market opportunities to industry;

Activities related to innovative security and safety applications on Earth in support of European and national policies are also conducted within different optional programs under the Applications pillar (ESA 2019b).

SSA Programme

ESA's SSA Programme so far has been is focusing on three main areas:

- Space Weather (SWE): monitoring and predicting the state of the Sun and the interplanetary and planetary environments, including Earth's magnetosphere, ionosphere, and thermosphere, which can affect spaceborne and ground-based infrastructure thereby endangering human health and safety.
- Near-Earth Objects (NEO): detecting natural objects such as asteroids that can potentially impact Earth and cause damage.
- Space Surveillance and Tracking (SST): watching for active and inactive satellites, discarded launch stages, and fragmentation debris orbiting Earth (ESA 2018d).

ESA's activities in the SST segment focus on the research and development (R&D) for SST data processing techniques, applications, and related standardization of data formats. An effective and efficient exchange between an SST system and external sensors is ensured through researching, developing the technologies, and demonstrating the capabilities of the SST Expert Centers. During Period 2 (2013–2016) of ESA's SSA Programme, these activities integrated the available SST sensors, the simulators within the data processing chain, and the applications, preparing for the SST Core software. The development of an Expert Centre system supporting Laser and Optical tracking technologies and expertise is continuing (ESA 2018d).

Concerning Space Weather (SWE), the SSA Programme has received the mandate to start the development of a SWE mission to L5 (phases A/B), as well as to continue with the implementation of the European SWE system based on a growing number of SWE services. It is also aiming at developing the required SWE instrumentation including magnetographs, radiation monitors, extreme ultraviolet lithography (EUV) imagers, and coronagraph. In the Near-Earth Object (NEO) domain, both the sensors – with the development of the NEO Fly-Eye telescope – and the data and services delivery are addressed, through the deployment of the Near-Earth Objects Dynamic Site (NEODYS) system in the NEO Coordination Centre at ESRIN (ESA 2017b).

Within the evolving SST landscape in Europe, the SSA Programme is continuing its R&D activities in SST in Period 3, namely, in radar, telescope, and laser system technologies and including providing support to the development of national SST capabilities. It will initiate the development of a space-based optical component for statistical sampling of small-sized debris, to be flown as a hosted payload. It will also develop a community approach to SST core software that will be made available to Participating States for SSA data processing in line with the established SST architecture. The program is also in the process of performing preoperational endurance test and validation activities of the available components of the SST system, including radar and optical sensors, data processing and fusion, correlation, and cataloguing (ESA 2017b).

North Atlantic Treaty Organization: NATO

NATO Space and Security Status

NATO was founded in 1949 and has 29 members. Common-funding arrangements principally include the NATO civil and military budgets, as well as the NATO Security Investment Programme (NSIP). NATO's 2019 civil budget was €250.5 million and the military budget €1.395 billion (NATO 2018a).

Space has always been regarded as essential by NATO. NATO relies on space-based services provided by Alliance Nations on a voluntary basis. NATO has a dedicated space policy as of June 2019 and has recognized space as an operational domain alongside air, land, sea, and cyberspace (NATO 2019a, Defense News 2019).

The principal political decision-making body is the North Atlantic Council, which exchanges intelligence, information and other data, compares different perceptions and approaches, harmonizes its views, and takes decisions by consensus, as do all NATO committees. The 2010 NATO Strategic Concept identifies collective defense, crisis management, and cooperative security as the essential core tasks that NATO must continue to fulfil to assure the security of its members. Deterrence based on an appropriate mix of nuclear, conventional, and ballistic missile defense capabilities remains a core element of NATO's overall strategy. The NATO Defence Planning Process is the primary means to identify and prioritize the capabilities required for full-spectrum operations and to promote their development and delivery (NATO Crisis Management).

The NATO Communications and Information (NCI) Agency was established in 2012 as a result of the merger of the NATO Consultation, Command and Control Agency (NC3A), the NATO ACCS Management Agency (NACMA), the NATO Communication and Information Systems Services Agency (NCSA), and the ALTBMD program office and elements of NATO HQ. The NCI Agency delivers secure, coherent, cost effective, and interoperable communications and information systems and services in support of consultation, command & control, and enabling intelligence, surveillance, and reconnaissance capabilities. Within the NCI Agency structure, the Joint Intelligence, Surveillance and Reconnaissance Service Line (JISR SL) is a subelement of the Directorate of Application Services. The NCIA operates the Network Control Centre (Mons and Brunssum) which provides service allocation and network monitoring (NCIA Website). The NCI Agency has a vital role in NATO's current space capabilities, supporting in SATCOM, JISR, Navigation Warfare (NAVWAR), and Ballistic Missile Defense (BMD). The BMD Programme, a multi-billion Euro effort, enables NATO nations to act as a single unit when responding to a ballistic missile threat or attack, which requires very quick and coordinated action. The BMD Programme consists of multiple projects that when combined, constitute a capability that connects the entire Alliance to share information and be effective in space, on the ground and at sea (NATO 2019b). Furthermore, The Agency has also participated in the NATO Bi-Strategic Commands' Space Working Group since 2012. The Bi-Strategic Command Space Working Group reports directly to the NATO Military Committee and is committed to the evolution of space support for alliance military operations. The Space Working Group has been essential for the preparation of the NATO Space Policy and the declaration of space as an operational domain (NITECH 2019).

The Joint Air Power Competence Centre (JAPCC) in Kalkar, Germany, is the NATO Department Head for Space Support to Operations. JAPCC was formed in January 2005 to provide the strategic level proponent for Joint Air and Space (A&S) Power that was missing in NATO. Based on a Memorandum of Understanding, the JAPCC is sponsored by 16 NATO nations (including the UK) who provide a variety of experienced Subject Matter Experts (SME) that come from all three services (JAPCC 2016). Within the JAPCC, a writing team consisting of national experts has been advancing in January 2018 on an update of the 2016 NATO Standard AJP 3.3 Allied Joint Doctrine for Air and Space Operations. JAPCC promotes the full recognition of space as a domain, like land, air, maritime, and cyber domains. JAPCC promotes the complete integration of space in the NATO organization. The areas of NATO space support include: PNT, ISR, SATCOM, METOC (meteorology and oceanography), SEW (Shared Early Warning), and SSA (JAPCC 2016). Within the frame of the actual mandate "NATO's approach to space – follow-on work," and as part of the Bi-Strategic Command Space Working Group, JAPCC contributed to the production of a "NATO Overarching Space Policy (OSP)" that was published in June 2019. JAPCC contributed to an advice paper for the International Military Staff (IMS) that led to a declaration in December 2019 that NATO recognizes Space as an operational domain (JAPCC 2019b). Additionally, JAPCC is involved in the 2-year-long Science and Technology Organization (STO) project as an active contributor. Today, space-related data and products from different NATO nations include variations in data protocols and sensor attributes (JAPCC 2019b).

The NATO Support and Procurement Agency (NSPA) provides follow-on logistic support for NATO-funded and non-NATO-funded communication and information systems. These include SATCOM Satellite Ground Terminals (SGT), Transportable SGTs (TSGT), and Man-Packs. The services provided by NSPA cover: Supply from central and distributed stocks; Centralized/consolidated procurement of parts; Direct Exchange (DX); Depot Level Maintenance (DLM); Technical/engineering services; and Transportation (NSPA Website).

NATO Space and Security Activities/Programs

Earth Observation: EO/Intelligence, Surveillance, and Reconnaissance – ISR

Under the NATO Civilian Structure, in the NATO Air Force Armaments Group and within the Defence Investment Division, there is a dedicated Joint Capability Group Intelligence, Surveillance, and Reconnaissance (NATO Organization).

With the frame of NATO Capabilities, Joint Intelligence, Surveillance and Reconnaissance (JISR) is considered crucial. JISR brings together data and information gathered through projects such as NATO's Alliance Ground Surveillance (AGS) system or NATO AWACS aircraft as well as a wide variety of national JISR assets from the space, air, land, and maritime domains. NATO's use of space-based

ISR is centered on the following elements: Intelligence: the final product derived from surveillance and reconnaissance, fused with other information; Surveillance: the persistent monitoring of a target; and Reconnaissance: information-gathering conducted to answer a specific military question (NATO 2018b).

NATO has a German-led smart defense initiative called Multinational Geospatial Support Group (GSG). The GSG provides enhanced standardized geospatial information, such as mapping and terrain imaging to NATO operations and planning (geospatial operational support, data management, terrain analysis, reproduction capability through maps and digital products, in theatre data collection). The following members are included in the GSG: Canada, Czech Republic, Germany, Denmark, Spain, France, UK, Greece, The Netherlands, Norway, Poland, Romania, and Turkey. Italy, Austria, Finland, Croatia, and the USA are considering membership (NATO 2014a).

JISR brings together data and information gathered through projects such as NATO's Alliance Ground Surveillance (AGS) system or NATO Airborne Warning & Control System (AWACS) surveillance aircraft as well as a wide variety of national JISR assets from the space, air, land, and maritime domains (NATO 2018b). The NATO AGS is acquired by 15 Allies (Bulgaria, Czech Republic, Denmark, Estonia, Germany, Italy, Latvia, Lithuania, Luxembourg, Norway, Poland, Romania, Slovakia, Slovenia, and the United States). All NATO Allies contribute to the development of the AGS capability through financial contributions covering the establishment of the AGS main operating base, as well as to communications and life-cycle support of the AGS fleet. Some Allies replace part of their financial contribution through interoperable contributions in kind (national surveillance systems that will be made available to NATO). The NATO-owned and NATO-operated AGS Core capability will enable the Alliance to perform persistent surveillance over wide areas from high-altitude, long-endurance (HALE) aircraft, operating at considerable stand-off distances and in any weather or light condition. Using advanced radar sensors, these systems will continuously detect and track moving objects throughout observed areas and will provide radar imagery of areas of interest and stationary objects. The Main Operating Base for AGS is located at Sigonella Air Base in Italy, which will serve a dual purpose as a NATO JISR deployment base and data exploitation and training center (NATO 2017).

The air segment consists of five RQ-4B Global Hawk Block 40 aircraft and remotely piloted aircraft (RPA) flight control element. The aircraft will be equipped with a state-of-the-art, multi-platform radar technology insertion program (MP-RTIP) ground surveillance radar sensor, as well as an extensive suite of line-of-sight and beyond-line-of-sight, long-range, wideband data links. The ground segment will provide an interface between the AGS Core system and a wide range of command, control, intelligence, surveillance, and reconnaissance (C2ISR) systems to interconnect with and provide data to multiple deployed and non-deployed operational users, including reach-back facilities remote from the surveillance area. The ground segment consists of several ground stations in various configurations, such as mobile and transportable, which will provide data-link connectivity, data-processing, and exploitation capabilities and interfaces for interoperability with C2ISR systems (NATO 2017).

Satellite Communications: SATCOM

NATO use of Satellite Communications (SATCOM) is centered on the following uses:

- Command, Control and Communications (C3) Systems
- Remotely Piloted Aircraft Systems (RPAS) operations
- Beyond Line-Of-Sight (BLOS) communications (NATO 2017)

On February 12, 2020, NATO concluded a Memorandum of Understanding between four nations for the provision of critical satellite communications services to NATO for the next 15 years. The memorandum between France, Italy, the United Kingdom, and the United States enables the four Allies to provide space capacity from their military satellite communications (SATCOM) programmers to NATO. Nations began delivering the capability on 1 January 2020 (NATO 2020). In 2019, NATO authorized 1 €billion for satellite SATCOM services for the next 15 years. The NCI Agency is responsible for operating the satellite communications capability to deliver services to NATO.

NATO has owned a series of military communication satellites, encrypted for use by NATO and designed to link the capital cities of NATO countries. NATO-1 and -2 were launched in 1970 and 1971, respectively, and four of the much larger NATO-3 satellites between 1976 and 1984. Each NATO-3 could support hundreds of users and provide voice and facsimile services in UHF- (ultra-high frequency), X-, and C-bands (see frequency bands). Two NATO-4 satellites, which operate in the same bands but have still more channels, were launched in 1991 and 1993. NATO does not today own any on orbit spacecraft but does own and operate several terrestrial elements such as SATCOM anchor stations and terminals (NATO 2016).

Since 2005, NATO has been making use of Member States' capacities, dealing with core strategic communications and services from the private sector. The C2 for SATCOM is managed by NATO Communication and Information Agency (NCIA) and operated by NATO CIS Group (NATO 2016). Concerning effective C2, the concept of Multi-Domain Operations (MDO) is used to "converge air, space, and cyber capabilities to meet the challenges of these contested domains." Accordingly, "space superiority is the number one priority. NATO won't necessarily win the war with it but losing in space virtually guarantees that NATO will lose it" (JAPCC 2019a).

Under the current NATO SATCOM Post-2000 (NSP2K) program – SATCOM Capability Package CP5A0030 – a consortium formed by the British, French, and Italian governments provide NATO with advanced SATCOM capabilities for a 15-year period from January 2005 until the end of 2019. Under a memorandum of understanding, the program provided NATO with access to the military segments of three national satellite communications systems – the French SYRACUSE 3, the Italian SICRAL 1 and 1Bis, and the British Skynet 4 and 5. This satellite capability has replaced the two NATO-owned and NATO-operated NATO IV communications satellites, which stopped their operational services in 2007 and 2010, respectively, after a combined operational life of 19 years. The SYRACUSE, SICRAL, and Skynet 4/5 satellites provided Super High Frequency (SHF) communications with military hardening features, while Ultra High Frequency (UHF) communications were provided by the SICRAL and Skynet satellites (NATO 2011). The NSP2K agreement was terminated at the end 2019.

As, the NSP2K came to an end, NATO has addressed the Follow-up Capability Package (CP9A0130) or CP130 which begins in 2020. CP9A0130 provides: SHF, UHF, and Extremely high frequency (EHF) space segment capabilities for 2019–2034, and ground and control segment capabilities such as new and/or upgrades for ground terminals, modems, and network control (NATO 2014b). With the funding approval pending, the planned 15-year €1.5billion CP130 will be devoted to upgrading the space segment and improve the ground segment. The objective is that the third-generation transportable satellite ground terminals will be upgraded, and the new fourth-generation terminals will be purchased. The current consortium of nations will be expanded to include the US, while contingency arrangements will be put in place with commercial providers in order to meet urgent requirements (NITECH 2019).

Positioning, Navigation, and Timing: PNT

NATO operations rely significantly on space support services given by the member nations. One of the most essential is the Positioning, Navigation, and Timing (PNT) service, provided by the United States'.

Global Positioning System (GPS) constellation (NATO 2017). NATO's use of space-based PNT addresses the following uses (NATO 2017):

- Precision strike
- Force navigation movements
- Logistic support
- Network timing
- Navigation Warfare (NAVWAR)
- Land and coastlines accurate survey
- Countering Improvised Explosive Devices (C-IED)
- Ballistic missile warning and defeat
- Cryptographic support (key generation)

According to JAPCC, military receivers rely on protected signals. In the case of GPS, the military uses the Precise Positioning Service (PPS), based on the so-called P(Y) code, which can use two different frequencies. Galileo is protected by the Public Regulated Service (PRS) and uses two frequencies as well, different from the GPS frequencies. NATO's threat assessment concludes that a combination of GPS and Galileo will increase the jamming resistance. An opponent will either opt for more jammers to affect the additional frequencies or concentrate on specific geographic areas or sectors to jam PNT signals (JAPCC 2019d).

NATO's ongoing work in the field of PNT is carried out within the NATO Command, Control and Consultation Board (NC3B). This board is organized into capability panels (CaP). CaP "Identification and Navigation" includes various Capability Teams such as NAVWAR, Interface Standardization between receivers, and PNT relevance for air space procedures (NATO 2017).

Space Situational Awareness: SSA/Space Surveillance and Tracking (SST)

NATO reliance on Space Situational Awareness (SSA) services is based on the voluntary contribution of allied nations (JAPCC 2016). NATO use of SSA addresses (NATO 2017):

- Space assets defense, including notification of potential collisions with Space debris
- Surveillance of Space
- Prevention of an adversary's ability to use Space systems and services for purposes hostile to a Nation's security interests

JAPCC 2019 White Paper "Command and Control of a Multinational Space Surveillance and Tracking Network" provides a NATO perspective on emerging multinational SST endeavors. It emphasizes the advantages of an EU SST network for the alliance and promotes the importance of Space Situational Awareness and SST in support of NATO military operations. The White Paper mentions that "since national SST resources often include a significant percentage of military assets, a serious multinational governmental SST endeavor cannot exclude the national Ministries of Defence." It adds: "from a military perspective, this situation is a source of both concerns and opportunities. In one sense, it raises concerns about the opportunity of sharing classified data regarding military spacecraft (own or allied). Additionally, since SST is often a secondary task for most military sensors, nations should carefully consider their level of commitment to SST because it drains resources from their primary task, usually airspace control or missile warning" (JAPCC 2019c).

Conclusion

Security and defense policies are progressively and gradually being supported by space programs of institutions in Europe. In Europe, the European Union and the European Space Agency are the main players together with their Member States. The various space subdomains they get involved in are notably, Earth observation; SATCOM; and Positioning, Navigation, and Timing. The European Defense Agency and the European Union Satellite Center follow suit in space programs for security and defense purposes, with an increasing interest by NATO. Yet, the Ministries of Defence in the Member States have a central role in the decision-making related to space and security in these organizations. The different nature and purpose of each organization leads to diversified development of activities and programs. In addition, whereas civil and defense space activities mainly comprise separate programs, more and more new programs follow a dual-use approach, right from their conception. This means that both civil as well as military funding may contribute in funding the same program. Additionally, both civil and military user communities are envisaged for specific space assets and applications. These developments in space programs frame to an increasing extent the necessity of Europe to take its security and defense at the next level by incorporating existing assets and jointly building new ones.

Disclaimer The contents of this Chapter and any contributions to the Handbook reflect personal opinions and do not necessarily reflect the opinion of the European Space Agency (ESA).

References

Copernicus Website, About Copernicus. Available at: https://www.copernicus.eu/en/about-coperni cus/copernicus-brief

Council of the EU (2014) Decision 2014/401/CFSP of 26 June 2014 on the European Union Satellite Centre and repealing Joint Action 2001/555/CFSP on the establishment of a European Union Satellite Centre, 27 June 2014. Available at: http://eur-lex.europa.eu/legal-content/EN/ TXT/PDF/?uri=CELEX:32014D0401&from=GA

Council of the EU (2018) Council of the European Union, Report from the Commission to the European Parliament and the Council on the implementation of the Space Surveillance and Tracking (SST) support framework (2014–2017), 8891/18, 18 May 2018, Brussels. Available at: http://data.consilium.europa.eu/doc/document/ST-8891-2018-INIT/en/pdf

Council of the EU (2019a) Frontex draft Programming Document 2020–2022, 20 November 2019, Brussels. Available at: https://data.consilium.europa.eu/doc/document/ST-14362-2019-INIT/ en/pdf

Council of the EU (2019b) Promoting synergies between the EU civil and military capability development – European space situational awareness high level civil-military user requirements document, 15715/11, 24 October 2011. Brussels

Defense News (2019) NATO names space as an 'operational domain,' but without plans to weaponize it, 20 November 2019. Available at: https://www.defensenews.com/smr/nato-2020-defined/2019/11/20/nato-names-space-as-an-operational-domain-but-without-plans-to-weapon ize-it/

EDA (2016a) European Defence agency, persistent surveillance long term analysis (SULTAN), 19 January 2016. Available at: https://www.eda.europa.eu/what-we-do/activities/activities-search/ persistent-surveillance-long-term-analysis-(sultan)

EDA (2016b) European Defence Agency, Towards Enhanced European Future Military Capabili-ties, European Defence Agency Role in Research & Technology:2016. Available at: https:// www.eda.europa.eu/docs/default-source/eda-publications/eda-r-t-2016-a4%2D%2D-v09

EDA (2017a) European Defence agency, three questions to Pascal Legai – director of the EU satellite Centre (SatCen). EDM 13:2017. Available at: https://eda.europa.eu/docs/default-source/eda-magazine/edmissue13lowresweb.pdf

EDA (2017b) European Defence agency, REACT: EDA study to boost imagery intelligence. EDM 13:2017. Available at: https://eda.europa.eu/docs/default-source/eda-magazine/edmissue13 lowresweb.pdf

EDA (2017c) European Defence agency, governmental satellite communications, fact sheet, 16 June 2017. Available at: https://www.eda.europa.eu/docs/default-source/eda-factsheets/2017-06-16-factsheet_govsatcom

EDA (2018a) European Defence agency, capability development plan, June 2018. Available at: https://www.eda.europa.eu/what-we-do/our-current-priorities/capability-development-plan

EDA (2018b) European Defence agency, chief executive Domecq at ESA, 5 February 2018, https:// eda.europa.eu/info-hub/press-centre/latest-news/2018/02/05/chief-executive-domecq-at-esa

EDA (2018c) European Defence agency, information sheet on space, 21 September 2018, https:// www.eda.europa.eu/docs/default-source/documents/eda-information-sheet-on-space.pdf

EDA (2020a) European Defence agency, factsheet: EU SatCom market, 4 February 2020. Available at: https://www.eda.europa.eu/info-hub/publications/publication-details/pub/factsheet-eu-satcom-market

EDA (2020b) European Defence Agency, ESA and EDA joint research: advancing into the unknown, 9 January 2020. Available at: https://www.eda.europa.eu/info-hub/publications/publication-details/pub/factsheet-eu-satcom-market

EDA PESCO (2018) European Defence Agency, Current list of PESCO projects. Available at: https://www.eda.europa.eu/what-we-do/our-current-priorities/permanent-structured-cooperation-(PESCO)/current-list-of-pesco-projects

ESA (2016) Joint Statement on Shared Vision and Goals for the Future of Europe in Space by the EU and ESA, 26 October 2016. Available at: http://www.esa.int/About_Us/Welcome_to_ESA/Shared_vision_and_goals_for_the_future_of_Europe_in_space

ESA (2017a) European Space Agency, Europe's master plan for space technology by ESA and the EU, 20 December 2017

ESA (2017b) European Space Agency, Status Report on ESA's security related developments and activities, ESA/C (2017)62, June 2017, Paris

ESA (2017c) European Space Agency, Space-based Services in Support of CBRNe Operations, Feasibility Study, May–August 2017. Available at: https://business.esa.int/funding/invitation-to-tender/space-based-services-support-cbrne-operations

ESA (2018a) European Space Agency, Status Report on ESA's Safety and Security-Related Developments and Activities, ESA/C(2018)58, 29 May 2018, Paris

ESA (2018b) European Space Agency, The Contribution of ESA to Copernicus 2.0, ESA/PB-EO (2018)8, 2 February 2018, Paris

ESA (2018c) European Space Agency, Copernicus evolution – Copernicus Space Component (CSC) Long Term Scenario (LTS), 9 May 2018, ESA/PB-EO (2018)12. France, Paris

ESA (2018d) European Space Agency, laser-tracking and optical expert Centre status and first results, ESA/PB-SSA(2018)16, 16 May 2018. France, Paris

ESA (2018e) European Space Agency, draft amendment no 5 to the Galileo deployment delegation agreement between ESA and the EU, ESA/AF(2018)11, 7 February 2018, Paris

ESA (2019a) European Space Agency, N° 22–2019: ESA ministers commit to biggest ever budget, 28 November 2019. Available at: https://www.esa.int/Newsroom/Press_Releases/ESA_ministers_commit_to_biggest_ever_budget

ESA (2019b) European Space Agency, Resolution on ESA programs: addressing the challenges ahead, adopted on 28 November 2019, ESA/C-M/CCLXXXVI/Res.3. Available at: https://www.esa.int/Newsroom/Press_Releases/ESA_ministers_commit_to_biggest_ever_budget

ESA (2019c) European Space Agency, Services enabled by HAPS – Services enabled by High Altitude Pseudo Satellites (HAPS) complemented by satellites. Available at: https://business.esa.int/projects/services-enabled-haps

ESA (2019d) European space agency, new financial resources for Copernicus space component, 22 January 2019. Available at: https://www.esa.int/Applications/Observing_the_Earth/Copernicus/New_financial_resources_for_Copernicus_space_component

ESA Ariane 6, European Space Agency, Ariane 6. Available at: https://www.esa.int/Enabling_Support/Space_Transportation/Launch_vehicles/Ariane_6

ESA Copernicus Website, Space Component overview. Available at: https://www.esa.int/Applications/Observing_the_Earth/Copernicus/Space_Component

European Commission (2018a) Press release, EU budget: a €16 billion space Programme to boost EU space leadership beyond 2020, Brussels, 6 June 2018

European Commission (2018b) Proposal for a Regulation of the European Parliament and of the Council Establishing the Space Policy Programme of the European Union, Relating to the European Union Agency for Space and Repealing Regulations (EU). No 1285/2013, No 377/2014 and No 912/2010 and Decision 541/2014/EU, COM (2018) 447 final 2018/0236 (COD), 6 June 2018, Brussels. Available at: https://ec.europa.eu/commission/sites/beta-political/files/budget-june2018-space-programme-regulation_en.pdf

European Commission (2019a) Orientations towards the first strategic plan implementing the research and innovation framework Programme horizon Europe, Co-design via web open consultation, summer 2019. Available at: https://ec.europa.eu/research/pdf/horizon-europe/ec_rtd_orientations-towards-the-strategic-planning.pdf

European Commission (2019b) The future is quantum: EU countries plan ultra-secure communication network, 13 June 2019. Available at: https://ec.europa.eu/digital-single-market/en/news/future-quantum-eu-countries-plan-ultra-secure-communication-network

European Commission, DG GROW Website, internal market, industry, Entrepreneurship and SMEs Available at: https://ec.europa.eu/growth/about-us_en

European Commission, Galileo Website. Available at: https://ec.europa.eu/growth/sectors/space/galileo_en

European Council (2013) Conclusions, EUCO 217/13, 20 December 2013, Brussels. Available at: http://data.consilium.europa.eu/doc/document/ST-217-2013-INIT/en/pdf

European Parliament (2014) Decision no 541/2014/EU of the European Parliament and of the council of 16 April 2014 establishing a framework for space surveillance and tracking support OJ L 158, 27 May 2014

European Parliament (2018) Parliamentary questions, E-006244/2016, answer given by Ms Bieńkowska on behalf of the commission, 18 January 2018. Available at: http://www.europarl.europa.eu/sides/getAllAnswers.do?reference=E-2016-006244&language=EN

EUSST Website, EU SST Support Framework, EU SST Governance. Available at: https://www.eusst.eu/project/who-we-are/

Giannopapa C, Adriaensen M, Antoni N, Schrogl K-U (2018) Elements of ESA's policy on space and security. Acta Astronautica 147:346–349

GPS World (2018) Galileo ground segment keeps constellation on track, 28 March 2018. Available at: https://www.gpsworld.com/galileo-ground-segment-keeps-constellation-on-track/?utm_source=gps_navigate&utm_medium=email&utm_campaign=gps_navigate_04032018&eid=376813635&bid=2055906;?utm_source=gps_navigate&utm_medium=email&utm_campaign=gps_navigate_04032018&eid=376813635&bid=2055906

GSA (2017) EGNOS Open Service (OS) Service Definition Document, 2017. Available at: https://www.gsa.europa.eu/sites/default/files/brochure_os_2017_v6.pdf

GSA (2020) Space is an enabler of security and defense, 7 February 2020. Available at: https://www.gsa.europa.eu/newsroom/news/space-enabler-security-and-defence

JAPCC (2016) Joint Air Power Competence Centre, Multinational Space Surveillance and Tracking Initiatives from a NATO Perspective. Available at: https://www.japcc.org/looking-up-together-multinational-space-surveillance-tracking-initiatives-nato-perspective/

JAPCC (2019a) Joint air power competence Centre, Joint air & Space Power Conference 2019, Shaping NATO for multi-domain Operations of the future. Multi-Domain Operations and Counter-Space, By Major Richard W. Gibson. Available at: https://www.japcc.org/what-is-a-multi-domain-operation/

JAPCC (2019b) Joint air power competence Centre, Annual Report 2019. Available at: https://www.japcc.org/wp-content/uploads/JAPCC_AR_2019_screen.pdf/

JAPCC (2019c) Joint air power competence Centre, command and control of a multinational space surveillance and tracking network, June 2019. Available at: https://www.japcc.org/wp-content/uploads/JAPCC_C2SST_2019_screen.pdf

JAPCC (2019d) Joint air power competence Centre, is NATO ready for Galileo? How the combination of GPS and Galileo could increase NATO's resiliency in PNT, June 2019. Available at: https://www.japcc.org/is-nato-ready-for-galileo/

NATO (2011) SATCOM Post-2000 – Improved satellite communications for NATO, last updated 15 December 2011. Available at: https://www.nato.int/cps/en/natohq/topics_50092.htm

NATO (2014a) NATO, Multinational Geospatial Support Group, information flyer, JUNE 2014

NATO (2014b) NCIA Exploring the Future for NATO Communications Capabilities NATO, Communications Services (NCS) Roadmaps. Available at: http://citeseerx.ist.psu.edu/

viewdoc/download;jsessionid=9F8558DA0CB4D5C0A9043E7189DFDD17?doi=10.1.1.663.
3478&rep=rep1&type=pdf

NATO (2016) NATO standard AJP-3.3 – Allied Joint Doctrine for Air and Space Operations,
Edition B version 1, April 2016. Available at: https://www.gov.uk/government/uploads/system/
uploads/attachment_data/file/624137/doctrine_nato_air_space_ops_ajp_3_3.pdf

NATO (2017) The key role of space support in NATO Operations, Three Swords Magazine, 32/
2017, July 2017. Available at: http://www.jwc.nato.int/images/stories/_news_items_/2017/
SPACESUPPORT_NATO_ThreeSwordsJuly17.pdf

NATO (2018a) NATO agrees 2019 civil and military budgets for further adaptation, December
2018. Available at: https://www.nato.int/cps/en/natohq/news_161633.htm?selectedLocale=en

NATO (2018b) Joint Intelligence, Surveillance and Reconnaissance, June 2018. Available at:
https://www.nato.int/cps/en/natohq/topics_111830.htm

NATO (2019a) Foreign ministers take decisions to adapt NATO, recognize space as an operational
domain, 20 November 2019. Available at: https://www.nato.int/cps/en/natohq/news_171028.
htm

NATO (2019b) NATO's ballistic missile Defence Programme gets a makeover, 25 November 2019.
Available at: https://www.nato.int/cps/en/natohq/news_171028.htm

NATO (2020) NATO begins using enhanced satellite services, 12 February 2020. Available at:
https://www.ncia.nato.int/NewsRoom/Pages/20200212-NATO-begins-using-enhanced-satellite
-services.aspx

NATO Crisis Management. Available at: https://www.nato.int/cps/en/natohq/topics_49192.htm

NATO, NATO Organization. Available at: https://www.nato.int/cps/su/natohq/structure.htm

NCIA Website, NATO Communications and Information Agency, Joint Intelligence, Surveillance
and Reconnaissance. Available at: https://www.ncia.nato.int/Our-Work/Pages/Joint-Intelligence
-Surveillance-and-Reconnaissance.aspx

NITECH (2019) NATO innovation and technology, issue 2, October 2019. Available at: https://
issuu.com/globalmediapartners/docs/nitech_issue_02_oct_2019?fr=sNjc5YTMyNjE3NQ

NSPA Website, NATO Support and Procurement Agency, Satellite Communication Systems.
Available at: https://www.nspa.nato.int/en/organization/Logistics/WSES/satellite.htm

Satcen (2016) EU Satellite Centre Annual Report:2016. Available at: https://www.satcen.europ
a.eu/key_documents/EU%20SatCen%20Annual%20Report%20201658e24cb1f9d7202538bed
52b.pdf

Schrogl K-U et al (eds) (2015) Handbook of space security: policies, applications and programs.
Springer, New York

Sitruk A, Plattard S (2017) The Governance of Galileo, ESPI Report 62, January 2017. Available at:
https://www.espi.or.at/images/Rep62_online_170203_1142.pdf

Space and Security Programs in the Largest European Countries

62

Ntorina Antoni, Christina Giannopapa, and Maarten Adriaensen

Contents

N. Antoni (✉)
Eindhoven University of Technology, Eindhoven, The Netherlands
e-mail: ntorina.antoni@gmail.com

C. Giannopapa · M. Adriaensen
European Space Agency (ESA), Paris, France
e-mail: christina.giannopapa@esa.int; maarten.cm.adriaensen@gmail.com

© Springer Nature Switzerland AG 2020
K.-U. Schrogl (ed.), *Handbook of Space Security*,
https://doi.org/10.1007/978-3-030-23210-8_147

Abstract

This chapter presents space and security programs in the five largest European countries and indicates their main priorities and trends. Recent policy developments in European countries, this have influenced national and institutional space security programs. The current chapter addresses different types of space activities and programs in the fields of Earth observation (EO), Intelligence-Surveillance-Reconnaissance (ISR), Satellite communication (SATCOM), positioning, navigation, and timing (PNT), and Space Situational Awareness (SSA). The countries presented are France, Germany, Italy, Spain, and the United Kingdom (UK).

Introduction

The increased need for security in Europe and for Europe's space activities has led to several activities and programs developed in aerospace and defense. The main actors in Europe engaging in space and security activities are the European countries, the European Union (EU), and the European Space Agency (ESA). ▶ Chap. 61, "Institutional Space Security Programs in Europe" describes how security and defense policies are progressively and gradually being supported by space programs of the EU and ESA, while the North Atlantic Treaty Organization (NATO) is also gradually involved. The European countries' space security policies are to a large extent determined by national needs and priorities, as explained in ▶ Chap. 25, "Space and Security Policy in Selected European Countries." In addition, the policies are influenced by the overall space and security governance and their participation in relevant organizations as explained in ▶ Chap. 23, "Strategic Overview of European Space and Security Governance."

The current chapter addresses space and security budgets, activities, and programs of selected European countries in the fields of Earth observation (EO), Intelligence-Surveillance-Reconnaissance (ISR), Satellite communication (SATCOM), positioning, navigation, and timing (PNT), and Space Situational Awareness (SSA). European countries may be distinguished not only on the basis of their membership to space and security related organizations but also their space budget. In the absence of an official grouping, their ESA annual budget and their defense expenditure as a share of their Growth Domestic Product (GDP) are used for their classification (Sagath et al. 2018).

The group presented in this chapter includes ESA Member States with a GDP above €1.2 trillion and an annual ESA space budget (2018 ESA budget) above €200 million: France, Germany, Italy, Spain, and the UK (Sagath et al. 2018). This chapter complements the chapters on medium-size and smaller European countries presented in this section of the handbook. The content is up to date until January 2019.

France

Space and Security Budget

France had an estimated defense expenditure of €42.748 billion or 1.84% of GDP in 2018, below the NATO guideline of 2% of GDP. 24.4% of the 2018 defense budget was allocated to equipment expenditure (NATO 2019). The Bill on Military Planning 2019–2025 foresees an expenditure of €295 billion on defense for the period 2019–2025 (French Republic 2018b; RTL INFO 2018). The French Senate adopted on 29 May 2018 the Military Program Law (*Loi de Programmation Militaire* - LPM) 2019–2025 on military funding program. The LPM 2019–2025 aims to renew the French armed forces and end decades of falling defense spending to reach NATO's guideline of 2% of GDP by 2025, according to the French Ministry of the Armed Forces. The final text of the LMP 2019–2025 was agreed by a joint committee of the National Assembly and Senate (Le Figaro 2018). President Macron formally signed the LPM on 13 July 2018. The French Senate and lower house National Assembly had previously examined the draft law, and both had voted in favor of the bill. One of the amendments added to the bill was an annual parliamentary review of spending (Defence News 2018a).

French government expenditure for space-related activities is the largest in Europe. It amounted to €2.8 billion in 2018, of which €2.23 billion covered civil activities and €569 million defense activities (Euroconsult 2019). French Space Agency - CNES's budget for the national program amounted to €1.2 billion in 2018, financed by the Ministry of Research to support innovation and competitiveness. The civil space budget is divided between the national program (€1.187 million in 2018) and the contributions to ESA (€963 million) and to EUMETSAT (€84 million). The defense budget fluctuates according to procurement cycles. Telecommunication was the primary application in 2018 for the development of Syracuse IV. The Military Program Law plans €3.6 billion investment over 2019–2025 to replace the current systems and acquire electronic intelligence capacity (Euroconsult 2019). Defense space expenditure was estimated at €569 million in 2018, primarily channeled toward telecommunications, Earth observation, technology, and space security. In 2018 the Ministry of the Armed Forces allocated €60 million for the CERES (*Capacité de Renseignement Électromagnétique Spatiale*) military electronic intelligence mission (Euroconsult 2019).

Space and Security Activities and Programs

Earth Observation (EO): Intelligence-Surveillance-Reconnaissance (ISR)

The civilian satellite *SPOT 1- Systeme Probatoire d'Observation de la Terre* (Probative System for Earth Observation) was developed and launched under the responsibility of scientific staff from CNES (Schrogl et al. 2015). The dissemination of information was allocated to the company Spot Image, which was created to distribute the images of SPOT. SPOT 4 was launched in March 1998 from Kourou. It featured two imaging instruments offering panchromatic and multispectral acquisition mode. SPOT 4 stopped functioning in July 2013. SPOT 5 was launched on 4 May 2002 from Kourou and stopped functioning in March 2015 (Schrogl et al. 2015). SPOT 6, launched in December 2012, and SPOT 7, launched in June 2014, share orbits with Pléiades. The SPOT satellites are dual-use civilian satellites, related to military programs. The United States (US) military has been a major purchaser of SPOT series commercial imagery. SPOT has been transferred to Airbus Defence and Space for operations and development of replacement capabilities (Schrogl et al. 2015).

Hélios is a French optical satellite imaging constellation for military reconnaissance. The Hélios-1 program was the result of a cooperation. of Francewith Italy, and Spain. The first-generation Hélios satellites, Hélios 1A and 1B, were launched in 1995 and 1999. Featuring daily revisit capability, the satellites had a resolution of about 1 meter (IAI 2003). The Hélios program was expected to reach a new phase with the joint development of Hélios-2by France and Horus by Germany: Hélios in the optical field and Horus in that of radar. Due to political and budgetary difficulties in Germany, the project remained on a national level. France consequently decided to pursue the Hélios-2program alone with fitting in bilateral cooperation along the way (IAI 2003). The second-generation of the program began with the launch of Hélios 2A on 18 December 2004. Hélios 2B was launched on 18 December 2009. Hélios 2B is managed by the French procurement agency *Direction générale de l'armement* (*DGA*). DGA has delegated the responsibility for the space segment to CNES. The Hélios-2 satellites – H-2A and H-2B – provide defense users with high-resolution imagery from low Earth orbit (LEO). Hélios-2 is a cooperation program of France with Belgium, Spain, Greece as co-owners, and Italy and Germany via exchange of data. Hélios-2 is the only fully military optical Earth observation system currently in operation in Europe. The Hélios system is operated by CMOS in collaboration with the CNES Toulouse Station Retention Centre (CMP). Belgium and Spain participate in the program since 2001, Italy since 2005 and Greece since 2007. France's stake remains at 90% (French Senate 2018; Schrogl et al. 2015).

Pléiades is the dual-use successor to France's SPOT optical Earth observation satellite constellation and was funded through the Defense Budget Program 191, a dual-use research program (French National Assembly 2016; French Republic 2018a). Access to data from the two Pléiades satellites is prioritized to military users. The Pléiades satellites are in a sun-synchronous orbit. The first Pléiades satellite was launched on 17 December 2011 on a Europeanized Soyuz rocket from Kourou. The second one was launched on 2 December 2012. Austria, Belgium, Spain, and Sweden are participating to the costs of the system in exchange for access to data. The Pléiades satellites contribute also to the Optical and Radar Federated Earth Observation system (ORFEO) (Schrogl et al. 2015). Pléiades imagery is today commercially available through Airbus GeoStore. Pléiades High Resolution (PHR) provides imagery for dual-use with priority for defense users. Imagery to civil users is provided, complementary with Hélios-2, via public service delegation to the commercial provider Airbus Defence and Space. Pléiades has been led and run by France in cooperation with Austria, Belgium, Spain, and Sweden. There is also cooperation with Germany via capacity exchange – programming rights – on SAR-Lupe and cooperation with Italy on Synthetic Aperture Radar (SAR) via exchange of data on the dual-use Cosmo-SkyMed (French Senate 2018). In 2019, Airbus and CNES agreed to co-finance a constellation of four Earth observation satellites with a dual purpose for the next 12 years aiming at "providing imagery to the French government, including scientific and military users, and imaging capacity for Airbus to commercialize" (Space News 2019). The first two Airbus-built Pléiades Neo imaging satellites have started comprehensive environmental testing (Airbus 2020).

The *Segment sol d'observation PHAROS* (Portail d'Acces au Renseignement de l'Observation Spatiale) application is a virtual federation of the military Hélios-2, SAR-Lupe, COSMO-SkyMed, and Pléiades systems. PHAROS is enabled through an access portal for imagery intelligence – equally in stationary and deployable configurations – and has been operational since June 2012. PHAROS provides multisensor unified for end users in France and on theatre of operations to all systems, including German and Italian assets. In return of the programming rights negotiated with Italy and Germany on their assets, France has ceded programming rights on Hélios-2 (French Senate 2018).

The *multinational space-based imaging system* (*MUSIS*) was originally a European surveillance program intended to supply visible, radar and infrared imagery and replace previous national platforms (Bird & Bird 2016). MUSIS today revolves around bilateral cooperation established between France and partnering countries. The *Composante Spatiale Optique* (CSO) is the French space component. France awarded in 2011 a contract to Airbus Defence and Space for CSO-1 and CSO-2 for €795 million. CSO applies to defense users only, with three identical optical satellites. CSO-1 will perform traditional surveillance tasks where CSO-2 will permit target identification. CSO-3 will improve revisiting time. France provides co-funding opportunities to other members of CSO in return for optical imagery. Germany agreed in 2015 to fund 2/3 of CSO-3 for an amount of €210 million. Belgium and Sweden are also cooperating. MUSIS has a similar functionality foreseen as in PHAROS with access to SARah system (Germany), COSMO-SkyMed Second Generation system (Italy), Ingenio system (Spain) and the French legacy components (Hélios). The first CSO high-resolution optical satellite fleet was launched in December 2018, while it is expected to be fully deployed in 2021. The German-French agreement from March 2015 establishing cooperation between radar and optical Earth observation facilitated the implementation of MUSIS (French Senate 2018).

The *OTOS* program (Super-Resolved Optical Earth Observation) is a preparatory program for future generation space satellites for Earth observation and defense running since 2011 – under the name CXCI – until 2020. OTOS is a technological demonstrator that aims to prepare the necessary technologies on the one hand for a future high-resolution post-Pléiades Earth observation program and on the other hand for the rest of the CSO Defence Programme (French Republic 2018a). The *CO3D* program – optical constellation in 3D – is a constellation of optical mini satellites answering the needs of a of digital terrain model mission and a 3D global model mission for civilian and military needs. These two objectives are based on the same concept of a small satellite with a competitive recurrent cost and a system architecture designed. The launch is foreseen in 2022. The constellation aims to provide continuity for the stereo imagery currently provided by Pléiades (French Republic 2018a).

Regarding big data, the French Ministry of the Armed Forces has defined the requirements for a digital platform to optimize the use of big data capabilities. In April 2018, and following a competitive tender, the French procurement agency DGA selected three companies or groups of companies to take part in the first phase of the *ARTEMIS* (Architecture de Traitement et d'Exploitation Massive de l'Information multi-Sources) innovation partnership. ARTEMIS is a 15-year framework agreement aiming to provide sovereign big data solutions to stay abreast exponential growth in data volumes, transmissions speeds, and formats in the design of information systems (Thales 2018). Six use cases of the project will enable the Ministry of the Armed Forces to take advantage of new capabilities resulting from digital technologies for knowledge sharing, better monitoring of soldiers' health, predictive maintenance of equipment, treatment and visualization of strategic and tactical information (DGA 2019a).

In the context of maritime security, the 2014/2015 project *Trimaran* concerns a concept of a one-stop shop for access, including a range of satellite services for maritime surveillance. In 2016, the French Navy awarded a 4-year contract to a consortium of Telespazio France and Airbus Defence and Space to provide a satellite-based maritime surveillance service. Under Trimaran 2, a follow-up to Trimaran 1, maritime zone commanders will have access to a portal for surveillance services using optical and radar imaging and AIS (Automatic Identification System) data to enhance the effectiveness of their national maritime missions. The radar and optical satellite images will be interpreted by Telespazio France and Airbus Defence and Space. Reports will then be produced and sent to the users indicating the types of vessels identified and their position, speed, and direction (Airbus 2016).

Satellite Communications (SATCOM)

France's first steps in SATCOM systems were the experimental satellites Symphonie and Telecom 1. The current French military communications programs are implemented through the French Syracuse program and the French-Italian program *Athena-Fidus* (for the latter see more under the section for *Italy*).

The *Syracuse III System* (Système de Radiocommunication Utilisant un Satellite) is the first exclusive French military satellite communications system, with extremely high frequency (EHF) capacity and a global coverage from Mexico to Indonesia, for at least three theatres of operations simultaneously. Syracuse IIIA military telecommunications satellite was launched in October 2005 and was scheduled to operate for at least 12 years. Syracuse IIIB was launched in 2006. A third satellite, SICRAL 2 was realized in cooperation with Italy and deployed in 2015 with the aim to create redundancy and increase coverage. Syracuse III is maintained to assure transition with Syracuse IV. The SATCOM capability responds to the need identified in the 2013 White Paper on Security and Defence and assures interoperability with NATO (French Senate 2018). In 2004, the French procurement agency DGA appointed Thales as prime contractor for the Syracuse III ground segment. The contract included the development, acquisition, and through-life support services of the full complement of Syracuse III communication systems and equipment. Thales provided service availability under operational level agreements until 2020 and delivered a total of 600 SATCOM terminals. The contract value amounted to €1.3 billion. Under the Syracuse III program, Thales developed the M21 modem, which successfully completed testing under real jamming conditions in 2003. The modem is compliant with NATO standardization agreement standards. Aristote is one of the central systems in the Syracuse III ground system. Aristote provides communications services between France and units deployed in the theatre of operations and optimizes the transmission capacity of the Syracuse III satellites. The Aristote system's capacity provides the armed forces with the interoperability they require during out-of-area deployments of joint multinational forces on several operational exercises (Defence Aerospace 2004).

Syracuse IV was formerly known as COMSAT-NG. In September 2012, the French procurement agency DGA awarded contracts to Thales Alenia Space and Airbus Defence and Space to conduct design studies on the next-generation military satellite communications system as per the military requirements. In December 2015 the consortium agreed to construct and supply the COMSAT-NG telecommunications system. The contract included construction of a ground control segment and Ka-band anchor stations, modernization of ground stations in France, and options for additional satellites. The consortium provided also operations and maintenance support. Thales Alenia Space owns 65% stake in the consortium, while the remaining is owned by Airbus Defence and Space. (French Republic 2018a).

In June 2019, the Ministry of the Armed Forces decided to proceed with the realization of the ground segment of the satellite communications program Syracuse IV. Accordingly, DGA signed a contract with Thales to design and build this capacity. This system brings together space assets (two satellites) and ground assets for users (terminals) and the operator of military networks (land network connection stations, management centers). The Syracuse IVA and Syracuse IVB satellites, ordered at the end of 2015, will be put into service before the end of 2022 to take over from the Syracuse IIIA and IIIB satellites launched in 2005 and 2006. They will be joined by 2030 by a third satellite optimized for use by aeronautical platforms such as increased connectivity, drones, etc. Current ground resources are compatible with these new satellites but must gradually be replaced, modernized, and completed in order to be able to fully exploit the new capacities of these satellites. This will allow the French forces to enhance communications capacities in terms of speed, availability, and resistance to threats, in particular for the equipment that will be delivered during the 2019–2025 Military Programme Law as part of the Scorpion programs; Rafale F4, Defense and Intervention Frigate and Force Supply Building (DGA 2019b).

In support of the Syracuse IV program, Telemak was funded through the *Programme 191 Dual Research*. Telemak activities consist of improving the performance of Ka-band covers while developing protection against interference and aggression, and to secure ongoing technological developments on the charges useful with a transparent digital processor (French Republic 2018a). Also, FAST is a dual-use project aimed at removing technological risks of the next-generation commercial telecommunication satellites and Syracuse IV, including the development of new generation chips common to civil space and defense programs as well as the development of a positioning, navigation, and timing 3G transparent digital processor (French Republic 2018a).

Overall, the French approach regarding national and multinational SATCOM capabilities development is based on a vision of different layers contributing to the overall capability. The *sovereign core* is composed of Military Satellite Communications Systems (MILSATCOM) with hardened capacity and military use only (Syracuse and SICRAL). The *extended core* is composed of governmental SATCOM with assured access and use by military and public authorities (Athena-Fidus and GOVSATCOM). The *augmented core* complements the picture by adding Commercial Satellite Communications (COMSATCOM) with access upon availability and use by military, public authorities, and civil private customers (EU SATCOM market). France has a long-term position to keep MILSATCOM out of Governmental Satellite Communications (GOVSATCOM) in line with this layered approach. The GOVSATCOM program should give added value to the European Defence Technological and Industrial Base (DGA 2015).

Positioning, Navigation, and Timing

The Ministry of the Armed Forces has access to the civil and military signals of the American Global Positioning System (GPS) system through military receivers developed in France under US license or acquired via Foreign Military Sales (FMS). GPS is evolving, with the introduction of GPS III satellites and the modernization of the ground segment. These changes will require a renewal of the receivers so that they are compatible with the new military signal "Code M," which will be operational by 2020. OMEGA operation aims to equip the armed forces with jam-resistant receivers and access to several constellations, typically GPS and Galileo (PRS). The realization of the receivers and their integration will be in its majority done after 2019 (French National Assembly 2016).

In the frame of its multiannual program for Research and Technology CNES is looking to prepare the next-generation orbital infrastructures for navigation systems, location, and data collection, improving the performance of technologies, measurement, and systems until 2030/2040, and to prepare, for the short term, technologies and good use of the downstream sector of current generation systems. In 2018, CNES supported time-frequency research aimed at atomic clocks, oscillators and advanced timing and frequency transfer techniques, and performance increase of civil and defense services based on current and future systems, including technologies and relevant signal processing for the user segment. The 2018 call for ideas aimed to meet the future challenges of autonomous vehicles such as drones, transport, agriculture, and miniaturization which requires increase of accuracy, robustness, and integrity of positioning measurements. CNES is preparing for the evolutions in space infrastructure including GPS system, the deployment of Galileo, multi-constellation mode operations (Galileo, GPS, BEIDOU, GLONASS), as well as the use of satellite-based augmentation system (SBAS) services such as European Geostationary Navigation Overlay Service (EGNOS) and the Wide Area Augmentation System (WAAS) (CNES 2018).

Space Situational Awareness (SSA)

The French dual-use operational Space Situational Awareness (SSA) activities are organized on three levels. At the programmatic and decision-making level, CNES and the French Ministry of the Armed Forces closely cooperate to define policy, capabilities, and priorities. At the sensors level, the Ministry of the Armed Forces operates the *GRAVES* (Grand Réseau Adapté à la Veille Spatiale) survey radar and several tracking radars that CNES uses on a regular basis. CNES together with the National Centre for Scientific Research (CNRS) also operates three TAROT telescopes located at Calern, in Chili, and in La Réunion for survey and tracking purposes. At the operational level, there are two operations centers in France. On the one hand, the military COSMOS Ops Centre of the French Air Force is responsible for Air and Space Defence and reports to the Office of the Prime Minister, as provided for by the French Code of Defence. On the other hand, the CNES Ops Centre consists of a 24/7 on-call team of ten specialists dedicated to conjunction assessment, alerts, and recommendations of collision avoidance maneuvers to spacecraft operators and owners (UNCOPUOS 2017).

After approaches and inspections of French satellites by foreign governments, Air Force Gen. Jean- Pascal Breton of the French Joint Space Command stated in January 2018 that the capability to detect and identify the suspect of an unfriendly or aggressive act an "essential condition for protection." The capability to track exo-atmospheric space activity will be gradually "strengthened to allow identification and classification of objects in orbits that are of interest to France," he added. France first introduced a space surveillance asset called Stradivarius located in Celar, in Brittany. France was obliged to dismantle its space surveillance asset recognizing the "space-dominant" role of the USA in NATO (Defence News 2018b).

The current GRAVES system is operated by the French Air Force. It is the space component of France's SCCOA (*Système de Commandement et de Conduite des Opérations Aérospatiales*, System for Command and Control of aerospace operations) (French National Assembly 2016). GRAVES was developed by ONERA (*Office National d'Etudes et de Recherches Aérospatiale*, French national aerospace research center), which was delegated by DGA. It took 15 years for the installation to become operational in November to December 2005. GRAVES is composed of a bi-static radar installation – emission site near Dijon and reception site near Albion – and exploitation server located at the Air Force base of Lyon Mont Verdun. Operations are performed by COSMOS (*Centre Opérationnel de Surveillance Militaire des Objets Spatiaux*) (ONERA 2016).

The GRAVES functionality ranges from modelling the passage of foreign observation or listening satellites to the detection of new objects such as spy satellites. GRAVES had detected the orbital presence of "anomalies," in other words satellite browser spies, especially US ones. Expanded surveillance French space allowed by GRAVES led to a strategic deal with the US, with both partners keeping the information on each military asset in space classified. Information on France's military satellites is therefore no longer available on open US sources. At the national level, the strategic benefit of the investment represented by GRAVES in terms of spatial surveillance is perceived as very high, even more considering the relatively minor cost of the system at €30 million. In this regard, GRAVES aims to strengthen the status of France as a strategic partner of the US (French National Assembly 2016).

To continue to give France leverage in terms of influence and to keep a central place in the future European Space Surveillance and Tracking (SST) architecture, improvements to GRAVES are required in addition to sharing the capabilities with partners. A 2012 report submitted to the French Senate identified the following actions to be taken: GRAVES obsolescence treatment activities, installation of additional sensors, and inching identification capabilities through optical imagery. Identification is currently achieved through the assistance of the German Tracking and Imaging Radar (TIRA). France and Germany have concluded an agreement for sharing GRAVES and TIRA data. France is considering how it can achieve this capability at national level without dependency on partner assets and capabilities (French National Assembly 2016). At the end of 2016, a maintenance in operational condition – renovation – contract was concluded between DGA and ONERA/Degreane Horizon (Groupe Vinci Energies) for the maintenance activities allowing the GRAVES infrastructure to remain operational until 2030 for €40 million. The contract was awarded for 5 years with a further option for 3 years of operations. Apart from that initial modernization, ONERA is studying options for future upgrades which would allow the system to detect mini satellites weighing less than 500 kg and micro-satellites weighing less than 150 kg (ONERA 2016). The European Union Space Surveillance and Tracking (SST) Support Framework contributes to these activities. In the current condition, GRAVES does not detect all sizes of satellites. The ongoing renovations will enhance capabilities, but detection of nanosatellites is expected to be achieved only by 2025 (ONERA 2016).

Commissioned in 1992, *Monge A601* is a missile-range instrumentation vessel, flagship of the Trials Squadron, and belongs to the DGA. The vessel has DRBV 15C air search and two navigation radars; its mission equipment includes the Stratus Gascogne, Armor (two), Savoie and Antares (two) missile tracking radars, a laser radar, an optronic tracking unit, and 14 telemetry antennae (FRS 2015). Monge could be partially used for SST functions like acquisition of orbital parameters in LEO. Monge can monitor missile launches including the Ariane rocket family. The Monge SATCOM system was made compatible with Syracuse III in 2015 after a major refit (FRS 2015).

SATAM is composed of four radars belonging to the French Air Force and is used for air defense purposes. SATAM may perform additional tracking in low Earth orbit. The tasks are implemented by the Air Force Command based on requests by COSMOS (FRS 2015). The radars are used for monitoring debris for management of collision risk determination and atmospheric reentry analysis. OSCEGEANE (Observation Spectrale et Caraterisation des Satellites Geostationnaire) is an experimental project to determine spectral signature of GEO objects. Operations are undertaken by COSMOS (French Joint Space Command 2016).

The LPM 2019–2025 foresees a modernization of GRAVES and SATAM, also using the opportunities presented by a future EU SST/SSA program (French Republic 2018b).

Electronic Intelligence (ELINT)

Since 1995 France has been placing experimental ELINT payloads and satellites in orbit. The objective is to eventually have satellites that allow France to figure out the location of and, where possible, identify individual transmitters and radars. This information would help French forces avoid detection in any military conflict, but initially the objective has been to understand how much information can be obtained to justify the funding of a future operational system (Schrogl et al. 2015).

France operates *ELISA*, an experimental electronic intelligence satellite, due to be replaced with the operational CERES system in 2020. ELISA (Electronic Intelligence Satellites) is a demonstration project, composed of four small satellites that were launched in December 2011, for spotting radar and other transmitter positions. The project is run and operated by DGA and CNES, who tasked EADS Astrium as a prime contractor with developing the space segment. ELISA could lead to an operational program of space-based electronic intelligence. Like Essaim, the ELISA satellites used the Myriade platform with a mass of roughly 130 kg each (Schrogl et al. 2015). Essaim (French for "swarm") was a family of four small satellites for electronic intelligence. They were commissioned by DGA, and they were launched on 18 December 2004 (together with, among others, Helios-2A) from Kourou. The Essaim constellation deorbited in 2010 (Schrogl et al. 2015).

The *CERES* (Capacité de Renseignement Électromagnétique Spatiale) military electronic intelligence mission will use three formation-flying satellites detecting and locating radio communications and radars (Bird & Bird 2016). Preparatory phases of the project started in June 2007. DGA signed a contract with Airbus Defence and Space in December 2015 for the three satellites in low Earth orbit with an estimated program cost of €400 million. The objective of CERES is to provide the ability to intercept and locate electromagnetic emissions – radio communication and radar – from space. The CERES system forms part of the national joint force intelligence chain by contributing to permanent monitoring, surveillance, and support for operations. The CERES program aims at obtaining an operational listening capacity allowing the interception and localization of electromagnetic emissions from space (French Senate 2018). CERES will be using ISIS. The ISIS project (initiative for space innovative standards) is the production of an interoperability repository based on a line of ground segment products for generic control of next-generation satellites (French Republic 2018a).

According to the LPM 2019–2025, early warning contributes to surveillance of proliferation and ballistic activity; the identification of potential aggressors, with a view to the implementation of deterrence or conventional counterforce actions; alert populations based on the estimated areas targeted; and meet NATO commitments. NATO planned in November 2010 to establish a full anti-missile defense capability, the NATO Integrated Air and Missile Defence System. France aims to contribute to this system by 2020–2021. System architecture studies conducted in 2011–2012 confirmed the interest in developing two types of complementary sensors: space-based optical sensors with infrared detectors and very long-range UHF radars installed on land or at sea (French Senate 2018).

The first ability was tested by the SPIRALE program. *SPIRALE* was a French space-borne early warning capacity demonstrator (*Systeme Preparatoire Infra-Rouge pour l'Alerte*) consisting of two satellites launched in 2009 and remaining in orbit until 2011 (French Ministry of the Armed Forces 2012). Between 2002 and 2011, for a total cost €137 million, an Earth database essential to understanding natural and physical phenomena likely to generate false alarms was created. For budgetary reasons, the continuation of the program has been postponed beyond military programming for 2014–2019.

Regarding the radar component of the system, a demonstrator of a very long-range radar was ordered in 2011 and completed in 2016. The radar is currently in testing and validation phase. The originally envisaged timetable, which provided for a delivery of an early warning system in 2021, was considered unrealistic given the resources provided for in the LPM 2014–2019. However, the 2017 Strategic Defence Review and National Security Council expressly confirm the interest of the project. The acquisition of an early warning system is open to cooperation. In this respect, it should be noted that France, Italy, and Turkey signed a letter of intent in November 2017 to strengthen their cooperation in the field of armaments, including missile defense (French Senate 2018).

In the frame of NATO, France contributes to the ALTBMD Program (Active Layered Theatre Ballistic Missile Defence). Missile launch detection, tracking, and warning activities are implemented through cooperation with the US and the Space-Based Infrared System (SBIRS). Regarding anti-missile capabilities, France has a ground-air system called SAMP/T, the ballistic anti-missile capability of which has been demonstrated in 2011. In addition to the 10 ground-based road mobile SAMP/T batteries, France holds the Principal Anti Air Missile System (PAAMS), the S1850M Radar, and the European Multifunction Phased Array Radar (EMPAR), all implemented on two Horizon-class frigates, and 12 ground-based road mobile Crotale Next Generation short-range air defense batteries (French Ministry of the Armed Forces 2012).

Fibally, ONERA is conducting a technology demonstrator, dubbed DRTLP, for an over-the-horizon radar to detect and track ballistic missile launches. ONERA took delivery in 2017 of the demonstrator built on a reduced scale of one-eighth for a very long-range radar, and tests have begun in 2018. The demonstrator studies a capability to detect and track a missile launch and forecast the point of impact and studies detection and tracking of satellites. ONERA aims to optimize the technology with upstream research allowing the DGA to decide how to pursue the project. (Defence News 2018c).

Germany

Space and Security Budget

Germany's estimated defense expenditure was €42.12 billion or 1.38% of GDP in 2018, below the NATO guideline of 2% of GDP. 16.6% of the 2018 estimated budget was allocated to equipment expenditure. To reach NATO's 2% of GDP target by 2024, defense expenditures would have to more than double within 7 years (NATO 2019). German government space expenditure was estimated at €1.9 billion in 2018 including €1.74 billion for civil activities and €166 million for defense (Euroconsult 2019). The 2018 budget increased by 1% compared to 2017. The breakdown of civil space expenditures between the national and ESA programs has remained constant since the beginning of the decade, with national programs receiving between 43% and 45% of the budget and the rest going to ESA. In 2018, Germany was the second largest contributor to ESA, almost on par with France. The German contribution grew 7% in 2018, reaching €920million. Defense space expenditure is not publicly released (Euroconsult 2019).

Space and Security Activities and Programs

Earth Observation (EO): Intelligence Surveillance-Reconnaissance (ISR)

In November 2017, OHB System AG was awarded a contract for the development of a global electro-optical satellite system, nicknamed "Georg. This was intended for reconnaissance and consisted of two satellites to be placed under the authority of the German Secret Service (BND). The rationale for the system was to guarantee assured access and direct access to electro-optical imagery for German intelligence. The contract amounted to €400 million and the satellite is expected to be launched in 2022 with full operational capacity in 2022–2023 (La Tribune 2017). The budget was approved in June 2017. When the German government signed the contract with OHB, some French observers criticised Berlin of walking back on Franco-German agreements – with France focusing on electro-optical sensors and Germany on radar and partners exchanging data – adding that Germany had yielded to influence from its national space industry and refusing French industrial expertise (International Institute for Strategic Studies 2017).

SAR-Lupe is Germany's first satellite-based reconnaissance system. SAR-Lupe is a SAR (Synthetic Aperture Radar) reconnaissance satellite imaging project of the German government, implemented by the German Ministry of Defense (BMVg) and the former Federal Office of Defence Technology and Procurement (BWB). The SAR-Lupe program consists of five identical (770 kg) satellites, launched between December 2006 and July 2008 and developed by OHB System AG as prime contractor. The satellites are controlled by a ground station operated by the Bundeswehr Geoinformation Centre (BGIC) under the Strategic Reconnaissance Command (KSA). Within the KSA, the Zentrale Abbildende Aufklärung (ZentrAbbAufkl) operates the user segment for SAR-Lupe. Delivery of the overall system was officially accepted by the Federal Office of Equipment, Information Technology and Utilization (BAAINBw), in September 2008. A bilateral agreement (Schwerin Agreement) with the French government was signed in 2002 for data from SAR-Lupe to be provided in exchange for data from Helios (OHB Website). The SAR-Lupe original program cost was €350 million (Satellite Observation 2016). Operating in X-band, the radar satellites have two modes. The first mode "stripmap," in which the satellite maintains a fixed orientation regarding Earth, provides extended time imaging with a fixed direction of the antenna. The second one "spotlight," in which the satellite or the sensor direction rotates to keep pointing at a specific target area, is used for high-resolution imagery. The actual resolution values of SAR-Lupe are classified. The only official statement is that the spatial resolution is much better than 1 m (Schrogl et al. 2015). Germany's development of this program was directly related to its experiences during the NATO action in Kosovo, particularly to difficulties in getting the US to share satellite intelligence of direct relevance to the protection and security of non-US allied forces. These experiences convinced Germany of the need for its own space-based intelligence-gathering assets (European Parliament 2006).

In July 2013, OHB System AG signed a contract with the Federal Office of Equipment, Information Technology and Utilization (BAAINBw), within the German Armed Forces, to develop the *SARah* (Satellite-based Radar Reconnaissance) system consisting of three second-generation satellites aimed to replace the SAR-Lupe constellation. OHB agreed to build two passive-antenna Synthetic Aperture Radar (SAR) satellites at 500 km non-SSO orbits, and Astrium GmbH would build a larger, phased-array-antenna satellite at 750 km SSO dawn-dusk orbit under contract for OHB (SpaceX 2013). In 2019 BAAINBw signed an additional contract with OHB adjusting the initial requirements to implement the SARah system in response to threats in the area of IT security and satellite communications (Bloomberg 2019). The system with one ground station is planned to be delivered and to become operational in 2020 after launch on SpaceX Falcon 9. The ground segment for SARah is already operating the SAR-Lupe satellites as of February 2018. Germany is in the process of preparing the SARah next-generation system with decisions to be taken before the end of 2020 to ensure seamless transition between systems (EO Portal Website).

Germany had considered in the past establishing *HIROS* (High-Resolution Optical System) in close collaboration with the US. HIROS was to be a triplet optical satellite expected to be built by OHB System AG. After funding issues at the German Secret Service (BND) side, efforts were dropped (Satellite Observation 2016). The German Ministry of Defense, instead, decided to invest €210 million in the French CSO system (*Composante Spatiale Optique*) in 2015, the follow-up project of Helios-2 (La Tribune 2017).

TerraSAR-X/TanDEM-X is a high-resolution interferometric SAR (Synthetic Aperture Radar) mission of the German Aerospace Center, DLR, together with the partners EADS Astrium GmbH and Infoterra GmbH in a Public Private Partnership consortium (Schrogl et al. 2015). The mission concept is based on a second TerraSAR-X radar satellite flying in close formation to achieve the desired interferometric baselines in a highly reconfigurable constellation (Schrogl et al. 2015). The elevation model yields important information to military planners preparing for tasks such as special forces operations, target designation for bombings, and surveillance missions. Following an announcement from DLR in October 2016, the TanDEM-X global elevation model was completed, exceeding the 10-meter accuracy. Between January 2010 and December 2015, the two radar satellites transmitted more than 500 terabytes of data to Earth via the worldwide reception network. In parallel, systematic creation of elevation models began in 2014. TerraSAR-X and TanDEM-X have long exceeded their specified service lives and continue operating (DLR 2016). Airbus Defence and Space holds the commercial marketing rights for TanDEM-X data, worth €359 million for the data licenses, processing software, and running costs. The US had urged Germany to grant access to the TanDEM-X global elevation model (Spiegel 2015). Airbus Defence and Space and the German Ministry of Defense subsequently signed a contract for the utilization of TanDEM-X mission data. On 14 December 2015, the US National Geospatial-Intelligence Agency and the German Bundeswehr Geoinformation Centre signed an agreement to strengthen worldwide geospatial data-sharing partnerships and increase the accuracy and quality of NGA products and services (NGA 2015).

Building on the success of TanDEM-X, *Tandem-L* is a DLR proposal for a highly innovative radar satellite mission to monitor dynamic processes on the Earth's surface with hitherto unknown quality and resolution. Important mission goals include the global measurement of forest biomass, the systematic monitoring of deformations of the Earth's surface, the quantification of glacier motion and melting processes in the polar regions, and observations of the dynamics of ocean surfaces and ice drift. The implementation of Tandem-L means a worldwide unique remote sensing system exceeding the performance of existing systems. According to current planning, the Tandem-L satellites could be launched in 2023 (DLR Tandem-L). In addition, the *TerraSAR-X NG* is intended to succeed the current TanDEM-X and TerraSAR-X. The TerraSAR-X NG mission is intended to take the data and service continuity well beyond 2025 taking benefit of a 9.5 years satellite lifetime. The space segment, initially a single spacecraft, will be launched into the reference orbit, while the first-generation systems will still be operational. This is a project of Airbus Defence and Space Geo-Intelligence/Infoterra GmbH, Friedrichshafen, Germany (EO Portal Website).

EnMAP (Environmental Mapping and Analysis Program) is a hyperspectral satellite mission that aims at monitoring and characterizing the Earth's environment on a global scale. EnMAP serves to measure and model key dynamic processes of the Earth's ecosystems by extracting geochemical, biochemical, and biophysical parameters, which provide information on the status and evolution of various terrestrial and aquatic ecosystems. The EnMAP long-term program is based on a cooperative approach involving various German institutions including DLR and the German Research Center for Geosciences. In November 2006, DLR awarded a design contract to Kayser-Threde GmbH (OHB owned) as prime contractor of EnMAP. In 2008, a contract for the realization phase was signed for a total contract value of €90 million. Launch is expected for 2021 (OHB 2008).

The *RapidEye* mission was a commercial remote sensing mission by the German Company RapidEye AG. It was supported by German Aerospace Center (DLR) with funds from the Ministry of Economics and the Brandenburg state government. The total sum invested in the project amounted to around €160 million, of which about 10% is funded by DLR. The RapidEye's complete constellation of five satellites was successfully launched on a single DNEPR-1 rocket (a refurbished ICBM missile) on 29 August 2008 from the Baikonur Cosmodrome in Kazakhstan and became commercially operational in February 2009. In November 2013 RapidEye officially changed its name to BlackBridge (Schrogl et al. 2015). The RapidEye constellation was acquired by Planet Labs, a US private company, in 2015.

The DLR *FireBIRD* (Fire Bispectral InfraRed Detector) mission consists of a pair of satellites – TET-1 (Technology Experiment Carrier) and BIROS (Bispectral Infrared Optical System) – that detect high-temperature events from space. Both satellites are based on the small satellite BIRD – operational from 2001 to 2004 – which was developed by the DLR Institute of Optical Sensor Systems. TET-1 has been orbiting Earth since 2012. BIROS has also been in orbit since 2016, adopting an open constellation to support TET-1. The satellite data is mainly received at the DLR ground station in Neustrelitz and then processed, archived, and made available worldwide for scientific purposes by the German Remote Sensing Data Center. FireBIRD has the capability to detect smaller fires. This enables more precise mapping and therefore analysis of their impact on the climate (DLR 2017a).

The German Remote Sensing Data Center (DFD) is an institute of DLR. DFD and DLR's Remote Sensing Technology Institute (IMF) together comprise the Earth Observation Centre (EOC). With its national and international receiving stations, DFD offers direct access to data from Earth observation missions, derives information products from the raw data, disseminates these products to users, and safeguards all data in the National Remote Sensing Data Library for long-term use. The DFD Ground Station Network consists of stations located in Germany, Canada, and Antarctica (DLR 2018). DLR signed a bilateral cooperation agreement with the Canada Centre for Remote Sensing (CCRS) for receiving satellite data in Canada. The CCRS makes land available in the context of this contract. The commercial partner, PrioraNet Canada (PNC) – a joint venture between the Canadian company Iunctus Geomatics and the Swedish Space Corporation – has been responsible for the maintenance and development of the site. The Antarctica station has been important in the frame of data reception for the TanDEM-X mission (DLR 2018).

The Fraunhofer Institute is developing a *nanosatellite ERNST* (experimental spacecraft based on nanosatellite technology) to reduce development costs and time to orbit. When carried into orbit in 2021, ERNST will be equipped with an infrared camera for Earth observation. ERNST will be put in a 700 km SSO orbit. The main purpose of the ERNST is to evaluate the utility of a 12 U nanosatellite mission for scientific and military purposes (Fraunhofer 2018).

In the field of *HAPS* (high-altitude platform station), a start-up company called Alphalink, in cooperation with the Technical University of Berlin, is developing a high-altitude platform for delivering Internet connection to remote locations. Interest in the applications has been shown by disaster management institutions, the Federal

Office of Disaster Management and Civil Protection in Germany, as well as the International Disaster Management Association (Space News 2017a).

Satellite Communications (SATCOM)

Germany currently holds two communication satellites, COMSAT 1 and COMSAT 2, and ground stations (*SATCOMBw* system). SATCOMBw system is the first owned, dedicated communications system of the German Armed Forces. DLR's German Space Operations Centre (GSOC) is responsible for monitoring and controlling the spacecraft. The satellites were launched in October 2009 and May 2010 to geostationary orbit with an orbital lifetime foreseen of 15 years (Air Force Website). The German Armed Forces awarded a contract to a team led by MilSat Services in July 2006, to carry out SATCOMBw stage 2 military communications program for 10 years. The contractual scope included in-orbit delivery and operation of two communications satellites, development of anchor station and ground user terminal segment, as well as modernization of the central command, control, and network management centers. Germany had also awarded a subcontract to Thales Alenia Space to design, build, integrate, test, and deliver COMSATBw-1 and COMSATBw-2 satellites. Tesat, a subsidiary of Airbus Defence and Space, supplied the payloads (Air Force Website).

In March 2016, Airbus Defence and Space GmbH was awarded a contract from the German Armed Forces procurement agency – BAAINBw for continued operations of the SATCOMBw satellite communications system until 2022 (Shephard 2016). Germany is currently defining requirements for the future systems replacing the existing SATCOMBw stage 2 capacities. COMSAT 1 and COMSAT 2 will reach their end of life by 2027. To fill this potential gap, prospective solutions are investigated including acquisition and in-service aspects also in the frame of future EU and NATO programs (IAI 2018).

DLR signed in July 2017 a contract with OHB Systems for an experimental telecommunications satellite, *Heinrich Hertz*, that will be used partly by the Federal Armed Forces. The launch of the satellite is expected in 2021. The mission aims to explore and test new communications technologies in space at a technical and scientific level. Heinrich Hertz will carry approximately 20 technology experiments as well as a fully functioning Ku- and Ka-band military communications payload. Heinrich Hertz will use the SmallGEO satellite platform designed under the European Space Agency's ARTES program. DLR is responsible for Heinrich Hertz's project planning and implementation, while the Ministry for Economic Affairs and Energy is financing the program (OHB Website).

Furthermore, within the project OSIRIS, optical communication systems are optimized especially for small satellites that are developed. To enable robust communication links, DLR also deals with the development and implementation of forward error correction algorithms which are optimized for free-space optical communication links (DLR 2017b).

Germany has developed a satellite-based Modular Warning System (MoWaS), for defense and crisis situations, and has also made it available to the federal states as a central warning and information system (BBK 2017). In order to cope with crises, the federally owned satellite-supported warning system (SatWaS) was developed since 2001. The further development of SatWaS into the Modular Warning System (MoWaS) was completed in 2013. This is geographic information system (GIS)-based. Through a single transmission protocol, MoWaS can control all the devices and applications imaginable today (such as smoke detectors, mobile devices, apps). This includes already existing but also future warning channels. This is possible by using the Common Alerting Protocol as the open data format of the alerts (BBK 2017).

Positioning, Navigation, and Timing (PNT)

In September 2017, the German Ministry of Defense selected Rockwell Collins' NavHub navigation system to provide GNSS availability to a variety of its military vehicles. The NavHub system serves as a next-generation GNSS- and military-code (M-code)-enabled solution for the German Armed Forces. Customizable for ground and maritime platforms, NavHub provides a variety of vehicle interfaces, meets the standards required by military vehicle operators, and allows users to receive data from multiple secure and open-service GNSS constellations to simultaneously confirm the navigational solution (Selective Availability/Anti-spoofing Module (SAASM) GPS receivers). Access to multi-constellation GNSS and GPS M-code provides a significantly enhanced navigational solution over the GPS-only solution (GPS World 2017). Since 2014, the German government has been sponsoring a special prize for Public Regulated Service (PRS) applications as part of the European Satellite Navigation Competition (ESNC), with the aim to perform joint testing activity with Belgium and other Member States and advance technology on further miniaturization and simplification of PRS receiver technology (GSA 2017).

The Institute of Communications and Navigation of DLR is involved in development of many advanced signal processing algorithms for GNSS applications with stringent requirements toward service performance and reliability. DLR's Multi-output Advanced Signal Test Environment for Receivers (MASTER) is a unique and powerful hardware simulation tool for testing and quality assessment of Global Navigation Satellite System (GNSS) Receivers (DLR Website). In addition, the *German Satellite Positioning Service* (SAPOS) has set up and permanently operated a multifunctional differential GNSS service. The system is based on a network of approximately 270 GNSS reference stations which are operated by the Surveying Authorities of the States of the Federal Republic of Germany (AdV). This service is widely available with high reliability and comprises three service areas with different properties and accuracies (AdV Website).

Space Situational Awareness (SSA)

The Space Situational Awareness Centre (GSSAC), run jointly by the German air Force and DLR, operates the *Tracking and Imaging Radar* (TIRA) system, which is owned by the Fraunhofer Institute (French National Assembly 2016). TIRA is an adapted radar to track objects in low Earth orbit through characterization and localization. TIRA performs the "characterization and track" phase but needs input from other kinds of radar, like the French GRAVES, to survey the space zone and identify the object that needs to be observed and tracked (FRS 2015). France and Germany have an agreement for sharing GRAVES and TIRA data (French National Assembly 2016). TIRA's typical tasks, apart from orbit determination and damage analysis, include the identification and technical analysis of satellites. This is possible due to its radar images which are characterized by high radiometric and spatial resolution. All phases of the space mission, extending from the launch and operational phases to the reentry phase, can be supported with the radar data from TIRA. TIRA has monitored the reentry of the Chinese Space Station Tiangong-1 in 2018 (Fraunhofer Institute 2018).

In 2016, Germany decided to expend its SSA center originally established in 2009, in the aftermath of the 2007 Chinese Fengyun-1C ASAT test. The Centre is run jointly by DLR and the German Air Force. The new German Experimental Space Surveillance and Tracking Radar, or GESTRA, became operational in 2019. The new space surveillance radar GESTRA (German Experimental Space Surveillance and Tracking Radar), which is currently being developed by Fraunhofer Institute for DLR, is equipped with an electronically steerable antenna that can scan large areas of the sky based on semiconductor technology. GESTRA is designed to operate continuously to create a catalogue of the debris in near-Earth space (Fraunhofer Institute 2018). GESTRA received its first signals reflected by objects in space on 27 November 2019 (DLR 2019).

Italy

Space and Security Budget

Italy's defense expenditure was €21.18 billion or 1.22% of GDP in 2018, below the NATO guideline of 2% of GDP (NATO 2019). The Italian government expenditure for space-related activities amounted to €1 billion in 2018 of which €950 million covered civil activities and €50 million defense (Euroconsult 2019). ESA represented 61% of Italy's civil space budget in 2018 at €578 million. Earth observation is the largest area of investment of the Italian space program representing 43% of the national civil expenditures in 2018, which is 24% of the civil budget. The budget for defense space programs is based on the procurement cycles of the national Earth observation and satellite communications systems. The defense space budget reached its lowest level across the decade at €50 million in 2018, following the launches of Athena-Fidus (2014), SICRAL 2/Syracuse IIIC (2015) and the optical satellite SHALOM (2017) (Euroconsult 2019).

Space and Security Activities and Programs

Earth Observation (EO): Intelligence Surveillance-Reconnaissance (ISR)

COSMO-SkyMed is the first space-borne Earth observation system implementing a dual-use architecture for civilian and defense needs in both national and international contexts. The system has been commissioned and funded by the Italian Space Agency (ASI) and the Italian Ministry of Defence in 2003 with Thales Alenia Space Italy (TASI) as prime contractor. The system design and development has been led by TASI in collaboration with a large industrial team comprising many other small- and medium-sized Italian companies mainly belonging to the Finmeccanica Group. The dual nature of the system can be retrieved already in its mission objectives relaying mainly on the capability to provide information and services useful for many activities and applications for both civilian and defense users (Schrogl et al. 2015).

The COSMO-SkyMed system consists of a space segment composed by a constellation of four low Earth orbit midsized very high-resolution satellites, each carrying a multi-mode high-resolution Synthetic Aperture Radar (SAR) instrument. Both ground and space segments are conceived to support dual-use, the space segment thanks to an antenna granting a wide spread of resolutions and swaths, whereas the ground segment thanks to the duplication of its key elements and a series of rules and procedures granting a secure data circulation. The success achieved by this space-borne mission is testified by the interest that it has generated in other countries. As such, ASI operates as a procurement agency for the French defense administration. The COSMO-SkyMed French defense user ground segment is fully operative, and according to Italian and French government partnerships, a bilateral image-trading protocol has been established: COSMO-SkyMed SAR image products are exchanged with Helios-2 optical image products to support institutional defense applications (Schrogl et al. 2015). Thales Alenia Space and Arianespace signed in September 2017 a launch contract for two COSMO-SkyMed Second Generation (CSG) satellites manufactured for ASI and the Italian Ministry of Defence (Arianespace 2017a). The CSG COSMO-SkyMed constellation is a satellite system designed to ensure operational continuity of SAR observation services provided by the four first-generation COSMO-SkyMed satellites, launched between 2007 and 2010 (Telespazio 2019). Built by Thales Alenia Space, the CSG COSMO-SkyMed satellites will each weigh approximately 2.3 tons at launch and will be positioned in SSO dawn-dusk orbit at an altitude of 619 km (Defense Aerospace 2017). The first CSG COSMO-SkyMed (CSG-1) was launched from the Kourou European Space Centre in French Guyana on December 18, 2019.

In June 2017, OHB Italia and Arianespace announced the signature of a Vega launch contract in 2018 for *PRISMA* (*Precursore iperspettrale della missione applicative*). PRISMA is an Earth observation satellite fitted with an innovative electro-optical instrument, combining a hyperspectral sensor with a medium-resolution panchromatic camera (Arianespace 2017b). The PRISMA mission main objectives are the in-orbit demonstration and qualification of an Italian state-of-the-art hyperspectral imager, the implementation of a preoperative mission and the validation of the end-to-end data-processing chain for new applications based on high spectral resolution images. This allows to provide products for environmental observation and support to risk management. The PRISMA Space Segment consists of a single satellite placed on a Sun-synchronous low Earth orbit and an expected operational lifetime of 5 years. PRISMA is civilian system under civil control, but the Italian Ministry of Defence has demonstrated interest in its applications including agriculture, environment, climate change, and costal monitoring (Loizzo 2016).

OPTSAT-3000 consists of a satellite in a Sun-synchronous low Earth orbit and of a ground segment for in-orbit control and for data acquisition and processing. OPTSAT-3000 aims to provide high-resolution images of any part of the globe, providing Italy with an autonomous national capability of Earth observation from space with a high-resolution optical sensor. The system is supplied by Leonardo through its joint venture Telespazio. As prime contractor, Telespazio leads an international group of companies including, among others, Israel Aerospace Industries, which built the satellite within an international cooperation agreement between Italy and Israel as well as OHB Italia that is responsible for the launch and will make use of the VEGA European launcher. OPTSAT-3000 will jointly operate with the second-generation COSMO-SkyMed system of radar satellites – which has also been developed by Italian industry. Thales Alenia Space and Telespazio integrate optical and radar data to provide the Italian Ministry of Defence with accurate information and state-of-the-art analysis (Leonardo 2017). OPTSAT was launched with Vega from Kourou on 2 August 2017. The OPTSAT-3000 program was developed to provide an independent, national high-resolution optical space-based Earth observation capability, integrated with the first- and future second-generation COSMO-SkyMed SAR satellite constellation in support of the requirements of the Italian Ministry of Defence, as well as national agencies operating in the field of safety and security and international customers (Janes.com 2017).

The *SHALOM* (*Spaceborne Hyperspectral Applicative Land and Ocean Mission*), which is in collaboration with the Israel Space Agency (ISA), may be considered as an upgraded OPTSAT-3000 satellite regarding hyperspectral imagery). SHALOM and OPTSAT-3000 give Italy key space intelligence capability, based on defense cooperation between Italy and Israel (Janes.com 2017). The mission objectives are to provide high-resolution (spectral, spatial, temporal) data of geochemical, geo-physical, and geo-biological variables; provide thematic digital maps of the above parameters such as environmental quality, crisis monitoring, search for mineral and natural resources, monitoring water bodies, and assisting precision agriculture activity; enable quantitative measurement of currently immeasurable (space) parameters that are required by a wide range of end users; and provide high-quality calibrated data as input for generating thematic maps and models for monitoring those parameters (Janes.com 2017). ASI and Israel Space Agency signed in 2015 an agreement for the development of the SHALOM mission (EARSC 2015).

Satellite Communications (SATCOM)

SICRAL (*Sistema Italiano per Comunicazioni Riservate e Allarmi*) is Italy's first dedicated military telecommunications satellite and was the product of the industrial consortium comprising Alenia Spazio (70%), FiatAvio (20%), and Telespazio (10%) (European Parliament 2006). The program is divided into three phases. The first began in 2001 with the launch of the SICRAL 1, a satellite that is still in operation. The second phase began in 2009 with the launch of SICRAL 1B, which has an estimated operational life span of 13 years (Telespazio 2020a). The third phase, in cooperation with France (see above in the section of France), became operational after April 2015 with the launch of SICRAL 2, with an estimated operational life span of 15 years. SICRAL 2 is a geostationary satellite, able to enhance the capability of military satellite communications already offered by SICRAL 1 and SICRAL 1B and by France's Syracuse System. SICRAL 2 supports satellite communications for the Italian and French Armed Forces, anticipating the needs of growth and development in the next few years. The satellite has an additional backup function to the current capacity of the French Syracuse III system and that of SICRAL 1B allocated to NATO communications (Telespazio 2020b).

Positioning, Navigation, and Timing (PNT)

The Italian Space Agency (ASI) Strategic Document 2016–2025 has a dedicated section on Galileo including a Galileo Public Regulated Service (PRS) national program, considered an area of strategic importance. For Italy, the real challenge and opportunity for the future come from the integration of downstream services of navigation with services of telecommunications and Earth observation. To take advantage of these opportunities, a national support program, the Mirror Galileo, is being studied and envisages the development of MEO (Medium Earth Orbit) platforms for navigation payloads, to facilitate the competitiveness of our national industry in the upstream sector, technological developments for Galileo components and subsystems, and also in a developmental perspective (ASI 2016).

The *PRESAGO* project, with the involvement of potential institutional domestic users, has defined the preliminary design of the baseline PRS, that is of the infrastructures, systems, and services necessary for supporting and making the use of the PRS efficient, both inside and outside of the national borders. The domestic program for the Galileo PRS infrastructure comprises national capability for manufacturing PRS receivers, including relative security modules, and the development of domestic activities. The latter consist of the development of the PRS security center, the interference monitoring system, the PRS terminals, the network and interfaces for domestic users, etc. The main benefits expected are the possibility for Italy to use its own infrastructure to support other European countries in using the PRS services; the access to the potential market relative to the implementation of PRS structures by other European countries; the development of value-added service solutions, based on the concept of PRS servers; and the development of value-added service solutions that integrate other satellite technology solutions (ASI 2016).

A private network of more than 150 Global Navigation Satellite Systems (GNSS) permanent sites, named ItalPoS (Italian Positioning Service) and uniformly covering the entire Italian territory, was established in April 2006 by the Italian Division of Leica Geosystems S.p.A. This network also involves several GPS stations of the INGV (Italian National Institute of Geophysics and Volcanology) RING (real-time

Integrated National GPS) network and GPS stations from other public and private bodies (Castagnetti et al. 2010).

Space Situational Awareness (SSA)

The ASI Strategic Vision document outlines the Italian plans for contributing to EU SST. ASI is expected to make available the sensors located at the SGC (Space Geodesy Centre). The role of SGC is that of Centre of Expertise for Civil Protection in monitoring the uncontrolled reentry of space debris. The Ministry of Defence will contribute, with optical telescopes and radar used for detection and tracking, together with the National Operations Centre. INAF (the National Institute for Astrophysics) will contribute with the Sardinia Radio Telescope and the "Northern Cross" Radio Telescope located in Bologna (ASI 2016).

The Matera Laser Ranging Observatory (MLRO) is owned by ASI and operated by e-GEOS. The observatory provides very precise laser ranging for satellites, suitable for satellite tracking applications. The observatory has not yet been tested for space debris tracking. The SPADE optical telescope at Matera is owned by ASI and operated by e-GEOS. The telescope contributed in a LEO and GEO space debris campaign supporting the work of the Inter-Agency Space Debris Coordination Committee (IADC) (Telespazio 2017). The Sardinia Radio Telescope (SRT) is the result of a scientific and technical collaboration among three Structures of the Italian National Institute for Astrophysics (INAF): the Institute of Radio Astronomy of Bologna, the Cagliari Astronomy Observatory, and the Arcetri Astrophysical Observatory in Florence. Funding agencies are the Italian Ministry of Education, Universities and Research, the Sardinia Regional Government, and ASI. The manufacturing of the SRT mechanical parts and their assembly on-site was commissioned in 2003 to the company MTM (Germany). The final tests and acceptance of the instrument were performed in August 2012 (Tofanie 2008).

Space debris monitoring is part of the Italian Institute for Astrophysics (INAF) and Cagliari Astronomical Observatory research activity in the framework convention ASI/INAF "Space Debris—IADC activities support and SST preoperative validation." In this framework, the INAF participation concerns the testing of the SRT's operative capacities in the detection of signals scattered by space debris. ASI, INAF, and the Ministry of Defence have signed a framework agreement for the SST program. The agreement – which runs from June 2015 until end 2020 – foresees a Steering Committee for Space Surveillance and Tracking Activities (OCIS), responsible for coordinating the national activities in the European SST initiative and directing ASI as national entity representing Italy within the European SST Consortium (Telespazio 2017).

In January 2015, thanks to the cooperation with the Air Force's IV Brigade, Telecommunication and Systems for Air Defence and Flight Assistance, a test was carried out in which the Selex ES (now Leonardo) RAT-31/DL FADR (Fixed Air Defence Radar) played the leading role. The system has become the backbone for the air surveillance of NATO countries, over the years. The test that successfully recorded the trajectory of several small satellites showed the advantages of using radar in the search of space debris. In comparison to the benefits of optical telescopes, the radar has significant advantages. The testing campaign has given results that confirm the great potential of the FADR in this field, even opening the possibility of exploiting other Air Defence radar system networks across Europe (including numerous RAT-31/DLs installed in Austria, Poland, Hungary, the Czech Republic, Germany, and other countries) to provide satellite monitoring and surveillance services (Leonardo 2015).

Spain

Space and Security Budget

Spain had an estimated defense expenditure of €11.172 billion or 0.92% of GDP in 2018, below the NATO guideline of 2% of GDP. 19.3% of the 2018 estimated budget was allocated to equipment expenditure (NATO 2019). Spain has committed itself to "regularly increase" defense budgets to achieve the agreed objectives. In this sense, the government aims to increase its defense investment spending in the medium term "successively" so that it reaches 20% of the total expenditure on this matter and will increase the expenses dedicated to defense research and technology for approximate them to 2% of the total expenditure (The Diplomat 2017). Spain's total space budget was estimated €354 million in 2018, including €293 million for civil activities and €61 million for defense. The budget driven by the contribution to ESA started to recover in 2016 after experiencing a decrease in 2013 followed by a 3-year stagnation. The 2018 budget grew by 18% compared to 2017 (Euroconsult 2019).

Space and Security Activities and Programs

Earth Observation (EO): Intelligence Surveillance-Reconnaissance (ISR)

Spain has developed its own dedicated EO system. The Spanish *PNOTS (National Earth Observation Programme)* is a complete system, based on SEOSat/Ingenio and PAZ (Schrogl et al. 2015). The development of both satellites was established under the Space Strategic Plan 2007–2011 on flagship missions (CDTI 2008). With PNOTS Spain acquired a fully independent operational satellite remote sensing capability (Schrogl et al. 2015). PNOTS is funded and owned by the government of Spain. The project development of SEOSat/Ingenio is overseen by the European Space Agency (ESA) as a national contribution within the framework of Europe under a procurement assistance agreement signed between ESA and the Centre for the Development of Industrial Technology (CDTI) in 2007 (Space News 2018a). The National Institute of Aerospace Technology, INTA, is managing the ground segment of the two missions. The ground segment is being developed by an industrial consortium including Deimos, GMV, and Isdefe. The System will be operated from the ground station of Torrejon de Ardoz (Spain), which will be the primary control center of the mission and using Maspalomas (Canary Islands) as backup station. The dual-use nature of the PAZ mission implies security constraints within its ground segment (SPIE 2017).

HISDESAT, together with INTA, is responsible for in-orbit operations and commercial operations of both satellites. Airbus Defence and Space Spain (formerly EADS CASA Espacio) is the prime contractor leading the industrial consortia of both missions. A major objective of PNOTS is to maximize the common developments and services and to share the infrastructure between both missions (whenever possible). Both missions will also contribute to the European Copernicus program. According to the contract, the ESA SEOSat/Ingenio project team must ensure that the European ground segment will allow the SEOSat/Ingenio system to become a candidate national mission contributing to Copernicus and to participate to the ESA third-party mission scheme within the EO multi-mission environment and therefore to support HMA (Heterogeneous Mission Access) services. SEOSat/Ingenio is the first Spanish Earth observation satellite financed by the Ministry of Economic Affairs, Industry and Competitiveness (formerly the Spanish Ministry of Industry) and built by a consortium of industries of the Spanish space sector with Astrium España (now Airbus Defence and Space Spain) as the prime contractor (with SENER, TASE, INDRA, GMV, and INTA) (Schrogl et al. 2015).

Ingenio is a satellite system with a foreseen 7-year operational lifetime in Sun-synchronous orbit (SSO) at 670 km. The payload consists of a multispectral imager (MS), a panchromatic imager (PAN), and an Ultraviolet and Visible Atmospheric Sounder (UVAS). SENER is responsible for the design, manufacturing, integration, alignment, and verification of the primary payload of the mission. The launch is scheduled for the first half of 2020 onboard a Vega launcher from Kourou (ESA 2014; SpaceWatch global 2019). PAZ will be one of the first satellites to combine Earth observation data with a sophisticated Automatic Identification System (AIS), which will allow to make the best possible monitoring of the world maritime environment. The expected operational lifetime is 5 and a half years. PAZ was launched in February 2018 by SpaceX from Vandenberg military base (SpaceX 2018).

HISDESAT is responsible for the launch and commercial operations of both satellites of the observation system in cooperation with the INTA, which is to provide ground control. Both satellites allow for Earth observation for multiple purposes: border control, intelligence, environmental monitoring, protection of natural resources, military operations, enforcement of international treaties, surface monitoring, city and infrastructure planning, monitoring of natural catastrophes, and high-resolution mapping, among many other (IDS 2014). The main center of the ground segment, located in the INTA, is completely installed with the integrated systems and in phase of interoperability tests. Spain's participation in the Pléiades program has allowed the national industry to reach industrial capabilities through the development of the ground segment for the Spanish Ministry of Defence (Spanish Ministry of Defence 2015a).

In the medium term, the amount, type, and cost of the PAZ satellite images necessary to cover the needs of the Ministry of Defence are expected to be assessed. In relation to the observation capacity, optics should analyze the options available once the operational life of Helios is over. High-resolution optical images could be obtained through the kickoff of a program with industry participation without ruling out a possible cooperation of Spain in the optical component (CSO) of the MUSIS as a substitute for Helios. There are possibilities for a reorientation of employment of the facilities of the ground segment of the Pléiades, for a future national program or within the framework of the MUSIS program (Spanish Ministry of Defence 2015a).

Satellite Communications (SATCOM)

HISPASAT was incorporated in 1989 to design, develop, manage, and deliver a commercial network capability as fleet operator in Ku band and Governmental services in X-band Satellite Communications System (SATCOM) for the Spanish government (Spanish Ministry of Defence 2015b). HISPASAT is owned by Abertis, a Spanish toll road company; SEPI, an industrial holding company; and Centre for the Development of Industrial Technology (CDTI) (Space News 2018b).

HISDESAT was incorporated in 2001 to define, develop, and operate new governmental SATCOM (Spanish Ministry of Defence 2015b). HISDESAT has been providing secure satellite communications services to the Spanish Ministry of Defence in support of all international missions of the Armed Forces, among others. It has also extended these services to other governments. HISDESAT services now include Earth observation, satellite communications, and AIS services through exactEarth (HISDESAT 2016). HISDESAT customers include the Spanish Ministry of Defence, Spanish Ministry of the Interior (CNI), Spanish Ministry of Economic Affairs, Industry and Competitiveness, the Spanish Regional Authorities, the Danish Ministry of Defence, Belgian Ministry of Defence, Norwegian Ministry of Defence, and US defense and intelligence agencies. HISDESAT operates SPAINSAT, Xtar-Eur, Paz and Ingenio (CDTI 2011; HISDESAT 2016). In 2007 an agreement was signed between the Ministries of Defence and Foreign Affairs to develop secure communications for the government's foreign activities. The project is based on two central hubs located at the Ministry of Foreign Affairs, with the entire network being connected through the geostationary satellites SPAINSAT and Xtar-Eur (IDS 2014). HISDESAT-operated secure communications services are provided by Ministry of the Interior to control borders between Spain, Portugal, and different African countries through the "Seahorse" program, strengthening security and illegal immigration control operations (IDS 2014).

The capacity of military satellite communications is covered by the service offered by the satellites *SPAINSAT*, as the main satellite, and Xtar-Eur as a redundant satellite. This set of satellites has a nominal life of 15 years, so, with its entry into operation in 2005 and 2006, the operational requirements of the Ministry of Defence are met until the year 2021. To meet this need, the Ministry of Defence and the companies HISDESAT and HISPASAT signed a framework agreement for the implementation of a satellite military communications system on 31 July 2001 Spanish Ministry of Defence (2015a). In May 2019, HISDESAT appointed Thales Alenia Space and Airbus to build two SPAINSAT Next Generation (NG) satellites for governmental communications that will replace the existing SPAINSAT and Xtar-Eur satellites. They will be launched in 2023, with an operational lifetime

15 years, aiming to guarantee the continuity of the secure communications services to the Spanish Ministry of Defence and Governmental Agencies using the current fleet (Space News 2019).

Positioning, Navigation, and Timing (PNT)

Permanent GNSS network and associated services are organized at the level of autonomous communities/regions. HISDESAT holds a stake of 27% in exactEarth (HISDESAT 2016). Based in Canada, exactEarth is a leading organization in the field of global automatic identification system (AIS) vessel tracking, collecting the most comprehensive ship monitoring data and delivering the highest quality information to customers.

Space Situational Awareness (SSA)

The Spanish Space Surveillance and Tracking (S3T) system provides services with two main objectives: to ensure the long-term availability of space infrastructures which are essential for the safety and security of worldwide citizens and to provide the best available information to governmental and civil protection services in the event of uncontrolled reentries of entire spacecraft or space debris thereof into the Earth's atmosphere. The SST services comprise collision risk assessment; the generation of conjunction data messages between objects in space; the detection and characterization of in-orbit fragmentations and collisions; and characterization and surveillance of uncontrolled reentries of space objects into the Earth's atmosphere (CDTI 2017).

The S3T system is currently contributing to the provision of these services by means of a national SST Operations Centre (S3TOC) and a set of ground-based sensors (S3TSN) which include optical surveillance and tracking telescopes and a surveillance radar. From a functional point of view, the S3TOC consists of a data-processing function, a sensor planning and tasking function, and a service provision function. The data-processing function is devoted to sensors' observation data processing, including correlation, orbit determination, and maintenance of a catalogue of space objects observed by the S3T sensors. The routine operations were initiated in July 2016. ESA is supporting CDTI in the development and procurement of the S3T system (CDTI 2017). Centu-1 is owned by Deimos, and it has been operationally contributing to the S3T system. It is used for surveillance, contributing to build up and maintain the S3T catalogue (CDTI 2017).

The Monostatic Space Surveillance Radar (MSSR) is a close-monostatic L-band radar, owned by the European Space Agency (ESA). It is located at the Santorcaz military naval base, about 30 km from Madrid (Spain). Through an agreement between the Spanish Ministry of Defence and ESA, the radar has been operational within the S3TSN since the end of the year 2016. The Ministry of Defence has participated in the selection of the site of the future advanced surveillance radar SST, as well as data policy and information security. In June 2015, it carried out the Transfer of the Operative Control of the Radar SSA of Santorcaz from ESA to the Air Force (CDTI 2017). The monostatic breadboard surveillance radar was developed within ESA's SSA Program. Following the completion of its development and upon the Spanish government's request, the operation of the Radar was transferred on a loan basis to the Spanish Ministry of Defence for use in Spain's national SST activities. In relation to surveillance systems and spatial tracking, the Ministry of Defence supports the CDTI in the participation of Spain to the EU SST program (Spanish Ministry of Defence 2015a).

The United Kingdom

Space and Security Budget

Following the British vote to leave the European Union in 2016, the UK's economy has been experiencing political and economic uncertainty (Euroconsult 2019). UK covered an estimated defense expenditure of GBP 45.2 billion or 2.14% of GDP in 2018, above the NATO guideline of 2% of GDP. 22.4% of the 2018 estimated budget was allocated to equipment expenditure (NATO 2019). The UK government expenditure on space amounted to GBP694 million in 2018. 65% of the space budget was dedicated to civil space activities with top three expenditures in space science and exploration, Earth observation, and telecommunications. The UK's GBP 455 million civil space budget comprised of GBP 100 million invested in national programs, GBP 293 million contribution to ESA, and GBP 62 million contribution to EUMETSAT (Euroconsult 2019). The defense space expenditures equaled to GBP 240 million in 2018 primarily channeled toward military satellite communications (Skynet 5), while it is expected to grow for the future generation of the Skynet system and the British GNSS as an alternative to Galileo (Euroconsult 2019).

Space and Security Activities and Programs

Earth Observation (EO): Intelligence Surveillance-Reconnaissance (ISR)

Traditionally, the UK has not put much emphasis on developing its own Earth observation satellites, because it has been relying on privileged access to relevant US assets (Schrogl et al. 2015). The UK Air and Space Doctrine explicitly mentions the relevance of environmental monitoring from space for security objectives (UK Ministry of Defence 2017).

UK Space Agency published in October 2013 a Strategy for Earth Observation from Space (2013–2016) in the context of the National Space Policy and the National Space Security Policy (April 2014). The strategy concentrates on civil EO requirements but recognizes that some civilian space systems could be dual-use in nature and be capable of supporting national security requirements (UKSA 2013).

In response to a recent parliamentary question, the Ministry of Defence elaborated on UK activities in the field of intelligence, surveillance, targeting, acquisition, and reconnaissance capability (ISTAR), stating that the activities are in line with the

Strategic Defence and Security Review "A Secure and Prosperous UK" of 23 November 2015. Intelligence, Surveillance, Target Acquisition and Reconnaissance and Information Systems and Services (ISS) are managed by the Ministry of Defence under a combined portfolio approach (UK Ministry of Defence 2012).

The Royal Air Force is strengthened by its ISTAR fleet in aerial reconnaissance. In the field of combat-aircraft ISTAR, the tactical imagery intelligence wing is an independent group force element, based at RAF Marham and covering a wide span of imagery intelligence missions. Its tasks include the exploitation of EO and infrared (IR) imagery, producing intelligence products in direct support of deployed operations (RAF 2012).

Resulting from the 2017 Defence Equipment Plan, the UK intends to spend GBP 5 billion through the ISTAR Operating Centre in the period 2017–2027. This investment includes spending on chemical, biological, radiological, and nuclear (CBRN) detection and countermeasures; electronic countermeasures; a range of equipment including communications, intelligence, surveillance, and reconnaissance; air defense; air traffic management; and tactical data links (UK Ministry of Defence 2018).

Civil spending in UK space activities is split between ESA, EUMETSAT, and national programs. In 2011, the UK invested GBP 21 million into NovaSAR-S, a low-cost SAR satellite developed by SSTL with maritime, forestry, flooding, and agriculture applications. The UK government provided GBP 21 million to assist in the development and launch of NovaSAR-S and will also benefit from access to the SAR data, significantly boosting the UK's sovereign Earth observation capabilities for applications such as ship detection and identification, oil spill detection, forestry monitoring, and disaster monitoring, particularly flood detection and assessment (SSTL 2017). NovaSAR-S was launched in September 2018.

The Zephyr S next-generation High-Altitude Pseudo Satellite (HAPS) is a new variant of the Zephyr family of unmanned aerial vehicles (UAVs) owned by Airbus Defence and Space. Zephyr S is a production variant of Zephyr 8 demonstrator. The new variant is intended for a variety of military, security, and civil missions, including maritime surveillance, border patrol, intelligence, reconnaissance, navigation, satellite-like communications, missile detection, environmental surveillance, signals intelligence (SIGINT), continuous photo capturing, and humanitarian and disaster relief. Development on the Zephyr 8 HAPS program began in April 2014. The UK Ministry of Defence awarded a contract to Airbus Defence and Space for the production and operation of two Zephyr S solar-powered unmanned aircraft systems in February 2016. Airbus has partnered with four British small and medium enterprises, and two universities to help develop key technologies in aerostructures, energy storage, and propulsion for what it described as "the next generation of Zephyr" (AIN 2018). Combining solar power plus rechargeable batteries, the Zephyr S reached a world record for longest duration flight for an unmanned aircraft without refueling in January 2019. Just under 26 days, Zephyr S demonstrated that a long-term mission is feasible for a solely solar-electric, stratospheric-level unmanned aerial vehicle (UAV) flying above the weather and conventional air traffic (PowerElectronics 2019).

Satellite Communications (SATCOM)

The UK's secure SATCOM capability is provided through a Private Finance Initiative (PFI) with Airbus, which is managed by Joint Forces Command. It provides a secure and resilient communications capability through the Skynet series of satellites and other SATCOM resources from other providers. UK partners and allies also use Skynet bandwidth, which bolsters collaborative ties, and, similarly, lost or degraded capabilities can be replaced by negotiating access to their space services. Commercial bandwidth can provide redundancy for military systems, but there are potential security risks if military communications are enabled by commercial satellites, which could also host foreign payloads. There are also risks in using commercial bandwidth because the terms of service provision could be significantly less than that provided through a dedicated military system (UK Ministry of Defence 2017).

Astrium Services (ASV) (now Airbus) is the service provider that has developed the widest array of SATCOMs for military and security purposes in a market-oriented pattern. With its Paradigm subsidiary, Airbus has built a commercial capacity under a long-term PFI contract with the UK Ministry of Defence, for the provision of military satellite communications services to 2022. Since 2003, Paradigm operates the five satellites of UK Skynet (Schrogl et al. 2015). In 2012, the UK government announced it would retake ownership of the Skynet system including the four spacecraft and ground segment at the end of the supply contract in 2022. The excess capacity is leased to other military customers in the USA, Canada, France, and NATO. Though the PFI with Airbus has permitted military requirements being met, the outsourcing may have decreased Ministry of Defence technical know-how and resources, hampering program management of the follow-up scheme which could be valued at 6 billion GBP (Schrogl et al. 2015).

A first element of a new British military satellite communications capability to replace the current *Skynet* 5 network has been awarded to Airbus Defence and Space without a competition. Negotiations to complete the deal to supply the Skynet 6A satellite are ongoing. The Skynet 6 program is packaged into three elements: the stopgap spacecraft to be built by Airbus, a service delivery package to manage ground operations from 2022, and an enduring capability program to provide future communication system capacity beyond the end of the next decade. Ministry of Defence officials said that the default position on the two future, and larger, parts of the Skynet 6 program would be competitive. A third, smaller competition to appoint an acquisition partner to act as the customer's friend in the Skynet 6 procurement is also expected to move forward. Officials said they expect a competition to appoint a new service delivery partner to take over the running of the ground operations from 2022. The enduring capability is also in line to be competed as things stand. The final part of the Skynet 6 requirement will be the introduction of a future enduring communications capability, which will partly be provided by satellites (Defense News 2017). The Skynet 6A geostationary military communications satellite is scheduled to be operational by mid-2025 (Space News 2017b).

Resulting from the UK Defence Equipment Plan 2017, published on 31 January 2018, the UK plans to invest GBP 22.9 billion on Information Systems and Services

(ISS) in the period 2017–2027, including satellite communications. The UK foresees a change in its procurement strategy for the Future Beyond Line of Sight Strategic Communications program based on a PFI, the acquisition of satellites, and cost reduction for the existing Skynet 5 network (UK Ministry of Defence 2017).

The advanced multi-mode secure SATCOM modem called Proteus was initially developed by Airbus UK for the Ministry of Defence, specifically for use on the Skynet 5 system. Airbus is developing a stripped-down version that could be used outside the UK for partner countries. The Proteus modem offers robust protection against jamming and interception and is suitable for installation in fixed and mobile platforms through super high-frequency and extremely high-frequency SATCOM bands (Airbus 2018).

Almost all civil spending on satellite communications technology development is channeled via ESA.

Positioning, Navigation, and Timing (PNT)

The UK has announced its plans to develop a national alternative to the EU's Galileo system (The Times 2018). The Government Office for Science procured a study into GNSS dependency rendered public in January 2018. The report sets out the findings of a review exploring the UK's dependency on GNSS covering threats and vulnerabilities, sector dependencies, mitigation strategy, and standards and testing. The Blackett report recommends measures to make UK critical services more resilient to disruption or loss of GNSS. Implementation of the Blackett recommendations is being overseen by a UK Cabinet Office Blackett Review Implementation Team (BRIG). The technical aspects of implementing the recommendations are being led by a (PNTTG), reporting to the BRIG (UK Government 2018).

Announced in May 2017, QinetiQ and Rockwell Collins aimed to develop under a new partnership next-generation multi-constellation Global Navigation Satellite System (GNSS) receivers. The focus would lie on developing a family of multi-constellation "open-service" GNSS receivers. Based on the high-level overview they have provided, the two companies are collectively looking to provide military and government aircraft operators with the ability to use and switch between use of existing and future GNSS constellations. Outside of aircraft GNSS development, QinetiQ and Rockwell Collins will also look to develop GNSS receivers designed to "reduce operational costs for ground troops, vehicles and high-dynamics GNSS-guided weapons" (Aviation Today 2017).

The British Isles continuous GNSS Facility (BIGF) supports research scientists with archived RINEX format GNSS data, metadata, and derivative products. This unique facility in the UK is hosted at the Nottingham Geospatial Institute – a center for related postgraduate teaching and research, at the University of Nottingham. BIGF was funded by the Natural Environment Research Council (NERC) from 2002 to 2018 and is now funded by United Kingdom Research and Innovation (UKRI) since 2018. The archive comprises data from GPS and GLONASS satellites, from a high-density network of around 160 continuously recording stations, located throughout mainland Britain, Northern Ireland, and Ireland (BIGF 2012).

Brexit does not prevent the UK as a third country from using the encrypted signal of Galileo provided that the relevant agreements between the EU and the UK are in place. Merely having access to PRS at some future date (as requested by the US and Norway) will not be good enough for London. The UK wants British companies to continue to participate in all aspects of the development and build of Galileo. Major contributions have been made by Surrey Satellite Technology Limited (SSTL), which has prepared the navigation payloads on every operational satellite in the sky; the UK arm of Airbus, which controls the satellites at its center in Portsmouth; and CGI (formerly Logica), which has been instrumental in designing the PRS itself (BBC 2018).

UKSA has written in May 2018 to all UK companies currently authorized to work on the secure elements of Galileo (including for instance SSTL, QinetiQ, and CGI), asking them to consult UKSA before taking on future contracts relating to the design and development of the program and its encrypted service. By reminding British companies that they need the express security clearance from ministers to engage in new contracts, London was essentially saying to Brussels that it has the power to stop those companies from handing over technical knowledge on PRS to firms in the EU-27 (BBC 2018). Yet, a UK GNSS will not add significantly new capabilities for defense as the UK will continue to have access to the precise signals of the American GPS and may access Galileo PRS after conclusion of a specific agreement for PRS access for third countries (as requested by the US and Norway). UK GNSS would only address *one* of the major losses in space power to the UK because of Brexit. It does not address being shut out of GOVSATCOM, EU SSA, and Copernicus (Defence in Depth 2018).

Space Situational Awareness (SSA)

The UK holds three space surveillance systems, two radar systems and one optical system. The latter is a telescope managed by the British National Space Centre (BNSC) and named Starbrook. It is located in Cyprus and has an added experimental survey sensor since 2006. The two radar systems are the Fylingdales complex and the Chilbolton facility. The first is part of the US Space Surveillance Network and is operated by British armed forces. The second, CAMRa (Advanced Meteorological Radar), is managed by the Rutherford Appleton Laboratory and is mainly used for atmospheric and ionospheric research (European Parliament 2015).

The Daedalus experiment – part of the Space Situational Awareness Project in DSTL's Space Programme – is exploring the effect on satellites of so-called Icarus "deorbit sails." When deployed, the sail increases drag, causing a controlled descent into the Earth's atmosphere where the satellite will burn up (UK Government 2017).

Concluding Remarks

This chapter provides evidence of the trend toward increasing relevance for security and defense in national space and security programs. "Space and security," both in its security from space and security in space aspects, are progressively contributing to further integration of space activities. Traditionally, only civilian space activities including for instance Earth observation, telecommunications, human spaceflight, space transportation, and technology development were subject of cooperation projects at intergovernmental and supranational levels. Security or defense related space programs were kept at the national level or dealt with bilaterally or multilaterally in *ad hoc* cooperative programs. These trends demonstrate an evolution of the largest European countries' priorities from strictly civil-oriented applications to also encompassing security and defense ones, facilitating synergies based on the dual nature of space. National space programs with security or defense dimensions, in combination with the EU and ESA programs, demonstrate alignment toward the use of space for security and defense. To conclude, the increasing relevance of security and defense in Europe, to some extent, could be framed as the necessity for Europe to take its security and defense into its own hands *vis-à-vis* other global space powers. Hence, this appears to be a strong driver in the current geopolitical context.

Disclaimer   The contents of this chapter and any contributions to the handbook reflect personal opinions and do not necessarily reflect the opinion of the European Space Agency (ESA).

References

AdV Website, Surveying authorities of the States of the Federal Republic of Germany, SAPOS – the German satellite positioning service. http://www.adv-online.de/Products/SAPOS/
AIN (2018) Online, Airbus Zephyr Pseudo-satellite gets new funding, 19 January 2018. https://www.ainonline.com/aviation-news/defense/2018-01-19/airbus-zephyr-pseudo-satellite-gets-new-funding
Air Force Technology Website, Military communications satellite systems, SATCOMBw Military Communications Satellite System. https://www.airforce-technology.com/projects/satcombw-military-communications-satellite-system/
Airbus (2016) Airbus Defense & Space and Telespazio selected to provide satellite-based maritime surveillance service for the French Navy, 2 June 2016. https://www.airbus.com/newsroom/news/en/2016/06/airbus-defence-and-space-and-telespazio-france-selected-to-provide-satellite-based-maritime-surveillance-service-for-the-french-navy.html
Airbus (2018) Network for the Sky – Secure networked airborne communications, July 2018. https://www.nfts.airbus.com/wp-content/uploads/2018/07/A4_Brochure_V5_Final.pdf
Airbus (2020) Pléiades Neo well on track for launch mid-2020, 24 February 2020. https://www.airbus.com/newsroom/press-releases/en/2020/02/pleiades-neo-well-on-track-for-launch-mid2020.html
Arianespace (2017a) Italian Space Agency and Italian Ministry of Defense choose Arianespace to launch COSMO-SkyMed Second-Generation (CSG) satellites manufactured by Thales Alenia Space on the occasion of the 34th French-Italian summit, 27 September 2017. https://www.arianespace.com/press-release/italian-space-agency-and-italian-ministry-of-defense-choose-arianespace-to-launch-cosmo-skymed-second-generation-csg-satellites-manufactured-by-tha les-alenia-space-on-the-occasion-of-the-34th-french/
Arianespace (2017b) OHB Italia, on behalf of the Italian Space Agency, and Arianespace sign contract to launch PRISMA Italian satellite, 19 June 2017. https://www.arianespace.com/press-release/ohb-italia-on-behalf-of-the-italian-space-agency-and-arianespace-sign-contract-to-launch-prisma-italian-satellite/
ASI (2016) Strategic Vision Document 2016–2025. https://www.researchitaly.it/en/projects/space-strategic-vision-2016-2025-document-published-by-asi/

Aviation Today (2017) QinetiQ, Rockwell Collins to develop new GNSS receivers, 26 May 2017. https://www.aviationtoday.com/2017/05/26/qinetiq-rockwell-collins-develop-new-gnss-receivers/

BBC (2018) UK ups the ante on Galileo sat-nav project, 14 May 2018. https://www.bbc.com/news/science-environment-44116085

BBK (2017) Federal Office for Civil Protection and Disaster Assistance, services for modern civil protection, September 2017. https://www.bbk.bund.de/SharedDocs/Downloads/BBK/EN/Services_for_modern_civil_protection.pdf?__blob=publicationFile

BIGF (2012) British Isles continuous GNSS Facility Folder, 2012

Bird & Bird (2016) Satellite bulletin – February 2016: significant increase in French spending on military space programs, 26 February 2016. https://www.twobirds.com/en/search?r=595be8b8-fac8-4244-a958-e39a6ef5dc14&s=51897736-c9a6-4562-a8ba-f54b70a0532d&nt=IndustryNews

Bloomberg (2019) OHB SE: satellite-based radar reconnaissance for Germany – OHB awarded additional contract for SARah project, 7 August 2019. https://www.bloomberg.com/press-releases/2019-08-07/dgap-news-ohb-se-satellite-based-radar-reconnaissance-for-germany-ohb-awarded-additional-contract-for-sarah-project

Castagnetti C., Casula G., Capra A., Bianchi M. G., Dubbini M. (2010), Adjustment andtransformation strategies of ItalPoS Permanent GNSS Network, Bulletin of Geodesy and Geomatics, vol. LXIX n. 2-3, p. 299-317, ISSN 0006-6710

CDTI (2008) Centre for the Development of Industrial Technology, Spanish Earth Observation Programme, Presentation, 17 November 2008. https://www.cdti.es/recursos/doc/Programas/Aeronautica_espacio_retornos_industriales/Agencia_Espacial_Europea/1245_251125112008103422.pdf

CDTI (2011) Centre for the Development of Industrial Technology, Spanish-Norwegian Space Industry Event, Presentation, May 2011. https://www.cdti.es/recursos/doc/Programas/Aeronautica_espacio_retornos_industriales/Espacio/25260_1011101120111405.pdf

CDTI (2017) Centre for the Development of Industrial Technology, Architecture of the Spanish Surveillance and Tracking System, ESA 7th European conference on space debris, 18–21 April 2017. https://conference.sdo.esoc.esa.int/sites/default/files/Final%20programme%207th%20European%20Conference%20on%20Space%20Debris.pdf

CNES (2018) Appel a Idees Externe R&T Systemes Orbitaux, 26 June 2018. https://rt-theses.cnes.fr/sites/default/files/users/2018-10946_APPEL%20A%20IDEES%20EXTERNE%20RT_2019-VF.pdf

Defense Aerospace (2004) Thales Appointed Prime Contractor for EUR 1.3 Billion Syracuse III Ground Segment Contract, 2 December 2004. http://www.defense-aerospace.com/article-view/release/49797/thales,-alcatel-share-syracuse-iii-contract-(dec.-3).html

Defence in Depth (2018) Better the devil you know? Galileo, Brexit, and British Defence Space Strategy, 23 May 2018. https://defenceindepth.co/2018/05/23/better-the-devil-you-know-galileo-brexit-and-british-defence-space-strategy/

Defense News (2017) Airbus scores British military satellite deal without competition, 31 July 2017. https://www.defensenews.com/space/2017/07/31/airbus-scores-uk-militarys-satellite-deal-without-competition/

Defense News (2018a) Macron signs French military budget into law. Here's what the armed forces are getting, 16 July 2018. https://www.defensenews.com/global/europe/2018/07/16/macron-signs-french-military-budget-into-law-heres-what-the-armed-forces-are-getting/

Defense News (2018b) Foreign governments are approaching French satellites in orbit, says space commander, 26 January 2018. https://www.defensenews.com/space/2018/01/26/foreign-governments-are-approaching-french-satellites-in-orbit-says-space-commander/

Defense News (2018c) France tests radar to detect and track ballistic missiles, satellites, 23 March 2018. https://www.defensenews.com/intel-geoint/sensors/2018/03/23/france-tests-radar-to-detect-and-track-ballistic-missiles-satellites/

Defense Aerospace (2017) Italian Space Agency and Italian Ministry of Defense Choose Arianespace to Launch COSMO-SkyMed Second-Generation (CSG) Satellites, 27 September

2017. https://www.arianespace.com/press-release/italian-space-agency-and-italian-ministry-of-defense-choose-arianespace-to-launch-cosmo-skymed-second-generation-csg-satellites-manufactured-by-thales-alenia-space-on-the-occasion-of-the-34th-french/

DGA (2015) Direction générale de l'armement, Communication Satellites for European Defence and Security: French Perspectives, Presentation by P.-E. Schoumacher, 25 November 2015, Betzdorf, Luxembourg. http://www.eu2015lu.eu/en/agenda/2015/11/25-seminaire-SatCom/index.html

DGA (2019a) Direction générale de l'armement, Notification de la deuxième étape du projet ARTEMIS de big data du ministère des Armées, 6 June 2019. https://www.defense.gouv.fr/dga/actualite/notification-de-la-deuxieme-etape-du-projet-artemis-de-big-data-du-ministere-des-armees

DGA (2019b) Direction générale de l'armement, Syracuse IV, le ministère des armées commande la réalisation du segment sol, 17 June 2019. https://www.defense.gouv.fr/english/dga/actualite-dga/2019/syracuse-iv-le-ministere-des-armees-commande-la-realisation-du-segment-sol

DLR (2016) New 3D world map – TanDEM-X global elevation model completed, 4 October 2016. https://www.dlr.de/content/en/articles/news/2016/20161004_new-3d-world-map-tandem-x-global-elevation-model-completed_19509.html

DLR (2017a) FireBIRD monitors forest fires in California, 19 December 2017. https://www.dlr.de/content/en/articles/news/2017/20171219_firebird-monitors-forest-fires-in-california_25507.html

DLR (2017b) Optical satellite downlinks at DLR – OSIRIS, Presentation, 12 July 2017. https://artes.esa.int/sites/default/files/AM%2003_20170712_Fuchs_OpticalSatelliteDownlinksatDLR_final.pdf

DLR (2018) Earth observation for humanity – German Remote Sensing Data Center (DFD) at the DLR Earth Observation Center (EOC), Presentation, 6 March 2018

DLR (2019) GESTRA space radar passes its first test, 29 November 2019. https://www.dlr.de/content/en/articles/news/2019/04/20191129_latest-radar-technology.html

DLR, Tandem-L – A Satellite Mission for Monitoring Dynamic Processes on the Earth's Surface. https://www.dlr.de/content/en/downloads/publications/brochures/tandem-l-brochure_1663.pdf?__blob=publicationFile&v=11

DLR Website, Institute of Communications and Navigation, GNSS Software Receiver. https://www.dlr.de/kn/en/desktopdefault.aspx/tabid-2066/3255_read-9556/; GNSS HW-Simulator. https://www.dlr.de/kn/en/desktopdefault.aspx/tabid-2066/3255_read-9429/

EARSC (2015) ASI and ISA strengthen the cooperation in the Earth Observation Field, 20 October 2015. http://earsc.org/news/asi-and-isa-strengthen-the-cooperation-in-the-earth-observation-field

Earth Observation (EO) Portal Website, Sharing Earth Observation Resources, Satellite Missions, SAR-Lupe Mission. https://directory.eoportal.org/web/eoportal/satellite-missions/s/sar-lupe. Terra Mission. https://directory.eoportal.org/web/eoportal/satellite-missions/t/terra

ESA (2014) SEOSAT/INGENIO, A Spanish High-Spatial-Resolution Optical Mission, Presentation, 2014

Euroconsult (2019) Government space programs – benchmarks, profiles & forecasts to 2028, July 2019

European Parliament (2006) Europe's space policies and their relevance to ESDP, June 2006. https://www.files.ethz.ch/isn/26583/06_06_Space_Policy.pdf

European Parliament (2015) Space, sovereignty and European security – building European capabilities in an advanced institutional framework, 2014, 39. https://www.europarl.europa.eu/RegData/etudes/etudes/join/2014/433750/EXPO-SEDE_ET(2014)433750_EN.pdf

Fraunhofer (2018) Traveling into space – safely, quickly and cost-effectively, Press Release, 10 April 2018. https://www.fraunhofer.de/en/press/research-news/2018/april/traveling-into-space-safely-quickly-and-cost-effectively.html

Fraunhofer Institute (2018) Space observation with radar to secure Germany's space infrastructure, Press Release, 23 March 2018. https://www.fhr.fraunhofer.de/en/press-media/press-releases/2018/space-observation-with-radar-to-secure-germanys-space-infrastructure.html

French Joint Space Command (2016) SSA: first priority of French military space policy 2025, Presentation, Brigadier General Jean-Daniel TESTÉ, Commander, Joint Space Command, Tokyo, March 2016. http://www.jsforum.or.jp/stableuse/2016/pdf/15.%20Teste.pdf

French Ministry of the Armed Forces (2012) Détecter un tir de missile balistique le plus tôt possible, 26 March 2012. https://www.defense.gouv.fr/actualites/dossiers/l-espace-au-profit-des-opera tions-militaires/les-fonctionsspatiales-au-service-des-operations/detecter-un-tir-de-missile-balistique-le-plus-tot-possible

French National Assembly (2016) Commission de la défense nationale et des forces armées, 17 May 2016, Compte rendu n° 48. http://www.assemblee-nationale.fr/14/cr-cdef/15-16/c1516048.asp

French Republic (2018a) PLF 2018 – Extrait du Bleu Budgétaire de la mission: Recherche et Enseignement Supérieur – Programme 191: Recherche Duale (civile et militaire), 2 October 2018. https://www.performance-publique.budget.gouv.fr/sites/performance_publique/files/far andole/ressources/2019/pap/pdf/DBGPGMPGM191.pdf

French Republic (2018b) President of the French Republic, Bill on Military Planning 2019/2025, 19 January 2018. https://www.defensenews.com/global/europe/2018/07/16/macron-signs-french-military-budget-into-law-heres-what-the-armed-forces-are-getting/

French Senate (2018) Projet de loi de finances pour 2018: Défense: Équipement des forces. https://www.senat.fr/rap/a17-110-8/a17-110-8_mono.html

FRS (2015) Fondation pour la Recherche Stratégique, The European Space Surveillance and Tracking service at the crossroad, October 2015. https://www.frstrategie.org/en/publications/defense-et-industries/european-space-surveillance-and-tracking-service-crossraoad-2015

GPS World (2017) German Defense chooses Rockwell Collins NavHub system for GNSS, 14 September 2017. https://www.gpsworld.com/german-defense-chooses-rockwell-collins-navhub-system-for-gnss/

GSA (2017) Galileo Public Regulated Service ready for action, 4 April 2017. https://www.gsa.europa.eu/newsroom/news/galileo-public-regulated-service-ready-action

HISDESAT (2016) Spatial Resources in Ports of the Future, Presentation, April 2016. https://www.aeit.es/sites/default/files/migrate/content/downloads/sebastiancatolfi.pdf

IAI (2003) Istituto Affari Internazionali, Space and Security Policy in Europe, November 2003, ESA Study. https://www.esa.int/About_Us/Corporate_news/Space_and_security_policy_in_Europe

IAI (2018) Istituto Affari Internazionali Boosting Defence Cooperation in Europe: an analysis of key military capabilities, June 2018. https://www.iai.it/en/pubblicazioni/boosting-defence-coop eration-europe

IDS (2014) Spain Defence & Security Industry. https://www.infodefensa.com/servicios/publicaciones/spain-defence-security-industry-2014.html

International Institute for Strategic Studies (2017) France and Germany: drivers of European defence-industrial cooperation, 18 December 2017. https://www.iiss.org/blogs/military-bal ance/2017/12/european-defence

Janes.com (2017) OPTSAT-3000 gives Italy key space intelligence capability, 7 August 2017. https://www.leonardocompany.com/en/press-release-detail/-/detail/optsat-3000-satellite-launch-earth-observation

La Tribune (2017) Observation spatiale: quand Allemagne se joue de la France, 8 December 2017, https://www.latribune.fr/entreprises-finance/industrie/aeronautique-defense/observation-spatiale-quand-l-allemagne-se-joue-de-la-france-760854.html

Le Figaro (2018) Florence Parly veut développer des capacités militaires spatiales, 21 June 2018. https://www.lefigaro.fr/flash-actu/2018/06/21/97001-20180621FILWWW00047-florence-parly-veut-developper-des-capacites-militaires-spatiales.php

Leonardo (2015) Being protected from space RAT/31DL. https://www.leonardocompany.com/en/news-and-stories-detail/-/detail/proteggersi-spazio-protected-from-space

Leonardo (2017) OPTSAT-3000 Earth observation satellite for Italy's Ministry of Defence is ready to be launched, 31 July 2017. https://www.leonardocompany.com/en/press-release-detail/-/detail/optsat-3000-satellite-launch-earth-observation

Loizzo R (2016) The Prisma Hyperspectral Mission, living planet symposium 2016, 9–13 May 2016, Prague. http://www.prisma-i.it/images/Articoli/2016_Loizzo-et-al.pdf

NATO (2019) Defence Expenditure of NATO Countries (2013–2019) Press Release, PR/CP (2019) 123, November 2019. https://www.nato.int/nato_static_fl2014/assets/pdf/pdf_2019_11/20191129_pr-2019-123-en.pdf

NGA (2015) NGA, Germany's Bundeswehr Geoinformation Centre sign geospatial data sharing agreement, 18 December 2015. https://www.nga.mil/MediaRoom/News/Pages/NGA,-Germany%E2%80%99s-Bundeswehr-Geoinformation-Centre-sign-geospatial-data-sharing-agreement-.aspx

OHB (2008) OHB subsidiary Kayser-Threde awarded contract for implementing the national earth observation mission EnMAP by the German Aerospace Center, 11 November 2008. https://www.dgap.de/dgap/News/corporate/ohb-technology-ohb-subsidiary-kayserthrede-awarded-enmap-contract/?newsID=369111

OHB (2013) OHB System AG awarded contract for the development and integration of the SARah radar satellite reconnaissance system for the German federal armed forces, 2 July 2013. https://www.ohb-system.de/press-releases-details/ohb-system-ag-awarded-contract-for-the-development-and-integration-of-the-sarah-radar-satellite-reconnaissance-system-for-the-ge.html

OHB Website, Current Programs, Security, SAR-Lupe. https://www.ohb-system.de/sar-lupe-english.html; Current Programs, Communication, Heinrich Hertz. https://www.ohb-system.de/heinrich-hertz-english.html

ONERA (2016) GRAVES: vers une surveillance spatiale française plus performante, Press Release, 12 December 2016. https://www.onera.fr/en/node/95

PowerElectronics (2019) Solar-powered "Pseudo-satellite" aircraft logs first flight of 26 days, 10 January 2019. https://www.powerelectronics.com/markets/automotive/article/21864274/solarpowered-pseudosatellite-aircraft-logs-first-flight-of-26-days

RAF (2012) Royal Air Force, the view from above – RAF ISTAR, 14 November 2012

RTL INFO (2018) Les principales mesures de la loi de programmation militaire 2019–2025, 8 February 2018. https://www.rtl.be/info/monde/france/les-principales-mesures-de-la-loi-de-programmation-militaire-2019-2025-994090.aspx

Sagath D, Papadimitriou A, Adriaensen M, Giannopapa C (2018) Space strategy and governance of ESA small member states. Acta Astronaut 142:112–120

Satellite Observation (2016) A new German Space Policy? 28 December 2016. https://satelliteobservation.net/2016/12/28/a-new-german-space-policy/

Schrogl K-U et al (eds) (2015) Handbook of space security: policies, applications and programs. Springer, New York

Shephard (2016) Media, Airbus DS awarded COMSATBw contract, 10 March 2016. https://www.shephardmedia.com/news/digital-battlespace/airbus-ds-keep-operating-comsatbw-satellites/

Space News (2017a) How close are high-altitude platforms to competing with satellites? 26 October 2017. https://spacenews.com/how-close-are-high-altitude-platforms-to-competing-with-satellites/

Space News (2017b) U.K. military seeks to ride wave if commercial space innovation, 7 November 2017. https://spacenews.com/u-k-military-seeks-to-ride-wave-of-commercial-space-innovation/

Space News (2018a) Morocco satellite launch could accelerate Spanish space efforts, 5 January 2018. https://spacenews.com/morocco-satellite-launch-could-accelerate-spanish-space-efforts/

Space News (2018b) EUTELSAT completes 302 million EUR HISPASAT divestiture, 19 April 2018. https://spacenews.com/eutelsat-completes-302-million-euro-hispasat-divestiture/

Space News (2019a) Airbus to build four imaging satellites for French space agency, mulls 20-plus constellation, 12 July 2019. https://spacenews.com/airbus-to-build-four-imaging-satellites-for-french-space-agency-mulls-20-plus-constellation/

Space News (2019b) Airbus inks two-satellite deal with Spain's satellite operator Hisdesat, 6 May 2019. https://spacenews.com/airbus-inks-two-satellite-deal-with-spains-satellite-operator-hisdesat/

SpaceWatch global (2019) Spain's SEOSAT/Ingenio Earth Observation Satellite Built by Airbus Ready for Pre-Launch Testing, June 2019. https://spacewatch.global/2019/06/spains-seosat-ingenio-earth-observation-satellite-built-by-airbus-ready-for-pre-launch-testing/

SpaceX (2013) OHB signs contract for Germany's next-gen radar satellites, 2 July 2013. https://spacenews.com/36091ohb-signs-contract-for-germanys-next-gen-radar-satellites/

SpaceX (2018) PAZ Mission, 22 February 2018. https://www.spacex.com/news/2018/02/22/paz-mission

Spanish Ministry of Defence (2015a) Ministerio de Defensa, DGAM, Plan Director de Sistemas Espaciales, July 2015. https://www.defensa.gob.es/Galerias/dgamdocs/plan-director-sistemas-espaciales.pdf

Spanish Ministry of Defence (2015b) Presentation on SECOMSAT – Spanish approach – present and future satcom for defence purposes, 25 November 2015. http://www.eu2015lu.eu/en/agenda/2015/11/25-seminaire-SatCom/7_Roberto-PELAEZ-HERRERO.pdf

SPIE (2017) Digital Library, SEOSAT/INGENIO: a Spanish high-spatial-resolution optical mission, 17 November 2017. https://www.spiedigitallibrary.org/conference-proceedings-of-spie/10563/105632M/SEOSATINGENIO-a-Spanish-high-spatial-resolution-optical-mission/10.1117/12.2304151.full?SSO=1

Spiegel (2015) Spiegel Online, USA drängen auf Daten-Deal mit der Bundeswehr, 3 November 2015. https://www.spiegel.de/politik/deutschland/bundeswehr-usa-draengen-auf-daten-deal-a-1060912.html

SSTL (2017) SSTL announces NovaSAR-S data deal with Australia's CSIRO, 26 September 2017. https://www.sstl.co.uk/media-hub/latest-news/2017/sstl-announces-novasar-s-data-deal-with-australia-

Telespazio (2017) Telespazio and TAS, Space Situation Awareness & Space Surveillance Tracking, 25 May 2017, Presentation. http://www.cesmamil.org/wordpress/wp-content/uploads/2017/05/9-_-Matarazzo-Brancati-_-Thales-Telespazio.pdf

Telespazio (2019) The first COSMO-SkyMed Second Generation satellite has been launched successfully, 18 December 2019. https://www.telespazio.com/en/news-and-stories-detail/-/detail/171219-the-first-second-generation-cosmo-skymed-satellite-has-been-launched-successfully

Telespazio (2020a) Exploring the SICRAL programme. https://www.telespazio.com/en/news-and-stories-detail/-/detail/171219-the-first-second-generation-cosmo-skymed-satellite-has-been-launched-successfully

Telespazio (2020b) The first COSMO-SkyMed Second Generation satellite has been launched successfully, 18 December 2019. https://www.telespazio.com/en/news-and-stories-detail/-/detail/sicral-telespazio

Thales (2018) French Defence Procurement Agency Selects Thales and Sopra Steria to compete for Artemis an innovation partnership for the Armed Forces Ministry's Big Data Platform, Press Release, 11 April 2018. https://www.thalesgroup.com/en/worldwide/security/press-release/french-defence-procurement-agency-selects-thales-and-sopra-steria

The Diplomat (2017) Spain adheres to EU's common framework of defence and security, 11 November 2017. https://thediplomatinspain.com/en/spain-adheres-to-eus-common-framework-of-defense-and-security/

The Times (2018) UK will launch satellite system to rival EU's Galileo, 2 May 2018. https://www.thetimes.co.uk/article/uk-will-launch-satellite-system-to-rival-eu-s-galileo-8127sl8fc

Tofanie G (2008) Status of the Sardinia Radio Telescope project, proceedings of the international society for optical engineering, January 2008

UK Government (2017) DSTL Scientists tackle growing problem of space junk, press release, 24 July 2017. https://www.gov.uk/government/news/dstl-scientists-tackle-growing-problem-of-space-junk

UK Government (2018) Satellite-derived time and position: Blackett review – Review exploring our dependency on global navigation satellite systems (GNSS), 30 January 2018. https://www.gov.uk/government/publications/satellite-derived-time-and-position-blackett-review

UK Ministry of Defence (2012) The Defence Equipment Plan 2012. https://www.gov.uk/government/publications/the-defence-equipment-plan-2012

UK Ministry of Defence (2017) Joint Doctrine Publication 0-30 (JDP 0-30) – UK Air and Space Power, second edition, December 2017. https://assets.publishing.service.gov.uk/government/uploads/system/uploads/attachment_data/file/668710/doctrine_uk_air_space_power_jdp_0_30.pdf

UK Ministry of Defence (2018) The Defence Equipment Plan 2017, 31 January 2018. https://www.gov.uk/government/publications/the-defence-equipment-plan-2017

UKSA (2013) UKSA, Strategy for Earth Observation from Space 2013–16, October 2013. https://assets.publishing.service.gov.uk/government/uploads/system/uploads/attachment_data/file/350655/EO_Strategy_-_Finalv2.pdf

UNCOPUOS (2017) General presentation of French activities and views for the long-term sustainability of outer space, in relation with the implementation of the first set of guidelines (A/71/20, Annex), A/AC.105/C.1/2017/CRP.26, 3 February 2017. https://www.unoosa.org/res/oosadoc/data/documents/2017/aac_105c_12017crp/aac_105c_12017crp_26_0_html/AC105_C1_2017_CRP26E.pdf/

Space and Security Programs in Medium-Sized European Countries

63

Ntorina Antoni, Christina Giannopapa, and Maarten Adriaensen

Contents

N. Antoni (✉)
Eindhoven University of Technology, Eindhoven, The Netherlands
e-mail: ntorina.antoni@gmail.com

C. Giannopapa · M. Adriaensen
European Space Agency (ESA), Paris, France
e-mail: christina.giannopapa@esa.int; maarten.cm.adriaensen@gmail.com

© Springer Nature Switzerland AG 2020
K.-U. Schrogl (ed.), *Handbook of Space Security*,
https://doi.org/10.1007/978-3-030-23210-8_148

Abstract

This chapter presents the space and security programs of medium-sized European countries and indicates their main priorities and trends. In particular, the current chapter addresses the different types of space activities and programs in the fields of Earth observation (EO), Intelligence-Surveillance-Reconnaissance (ISR), Satellite Communication (SATCOM), Positioning, Navigation, and Timing (PNT), and Space Situational Awareness (SSA). The countries presented are Austria, Belgium, the Netherlands, Norway, Sweden, and Switzerland.

Introduction

The increased need for security in Europe and for Europe's space activities has led to several activities and programs developed by European countries in aerospace and defense. The main actors in Europe engaging in space and security activities are the European countries, the European Union (EU), and the European Space Agency (ESA). ▶ Chap. 61, "Institutional Space Security Programs in Europe" in this handbook describes how security and defense policies are progressively and gradually being supported by space programs of the EU, ESA, and the European Defence Agency (EDA), with the North Atlantic Treaty Organization (NATO) is also getting gradually involved. The European countries' space security policies are to a large extent determined by national needs and priorities, as explained in ▶ Chap. 25, "Space and Security Policy in Selected European Countries." In addition, policies are influenced by the space and security governance and their participation in relevant organizations as explained in ▶ Chap. 23, "Strategic Overview of European Space and Security Governance."

As such, the various space security activities and programs are to a large extent determined by national needs and priorities as well as participation to space and security-relevant organizations, including the EU and ESA. The current chapter addresses the space and security budgets, activities, and programs of selected European countries in Earth observation (EO), Intelligence-Surveillance-Reconnaissance (ISR), Satellite Communication (SATCOM), Positioning, Navigation, and Timing (PNT), and Space Situational Awareness (SSA). The European countries may be distinguished not only based on their membership to the space and security-related organizations but also based on their space budget. In the absence of an official grouping of these countries, their ESA annual budget and their defense expenditure as share of their growth domestic product (GDP) are used to classify them into three groups (Sagath et al. 2018).

The group presented in this chapter includes Member States with a GDP between €350 billion and €1.2 trillion and an annual ESA space budget (2018 ESA budget) typically between €50 and €200 million: Austria, Belgium, the Netherlands, Norway, Sweden, and Switzerland (Sagath et al. 2018). This chapter complements the chapters on the largest and smaller European countries presented in the corresponding chapters of the Handbook. The content is up to date until January 2019.

Austria

Space and Security Budget

Austria's defense budget was estimated at €2.26 billion in 2018 (0.58% of GDP) and €2.29 billion in 2019 (0.57% of GDP), and is expected to rise to €2.42 billion in 2020 (0.58% of GDP). The figures of the budgetary framework 2018/2019–2022 will increase, and budgetary regulations will be adapted to support the National Defence Plan 21.1 (OSCE 2018). The Austrian government expenditure for space-related activities amounted to €67.9 million in 2018 (Euroconsult 2019). The Austrian contribution to ESA in 2018 was €47 million. Austria is a member of the European Organization for the Exploitation of Meteorological Satellites (EUMETSAT), with a contribution in 2018 of €12 million. The main pillars of the Austrian space activities are the Austrian contributions to the programs of ESA and EUMETSAT, the contributions to EU space activities and the national space program (Euroconsult 2019). No military space budget has been identified by Euroconsult.

Space and Security Activities and Programs

The Austrian Space program is in principle based on two national programs that have been carried out since 2002: the Austrian Space Applications Programme (ASAP) and the Austrian Radio Navigation Technology and Integrated Satnav services and products Testbed (ARTIST). Through ASAP, Austrian research institutions as well as commercial enterprises have been supported in their efforts in conducting space science and exploration and in developing space technologies, products, and services. ARTIST is intended as an Austrian testbed for navigation applications. Possible future applications and services of the European satellite system GALILEO have been tested and evaluated through demonstration projects with respect to their innovative character and technical and economic benefits (BMVIT 2017).

Earth Observation (EO) – Intelligence – Surveillance-Reconnaissance (ISR)

The Technical University of Vienna, the *Zentralanstalt für Meteorologie und Geodynamik* (ZAMG), and private sector actors established in 2014 the Earth Observation Data Centre for Water Resources Monitoring (UN SPIDER 2014). Austria participates to the French Pléiades program. Moreover, Austria is part of the Copernicus program in the frame of the EU and ESA. Austria's contribution to ESA's optional Earth observation programs amounted to €9.63 million in 2018 (Euroconsult 2019). Further, Austria is a member of the following international organizations: the European Centre for Medium-Range Weather Forecasts (ECMWF); the Global Earth Observation System of Systems (GEOSS); the CEOS (Committee on Earth Observation Satellites); and the WMO (World Meteorological Organization).

Satellite Communications (SATCOM)

The Technical University in Graz has a dedicated Institute of Communication Networks and Satellite Communications. The Research Priorities on Satellite and terrestrial broadband communication are Internet via satellite; digital TV and interactive Internet applications; 40 GHz fixed-broadband wireless access; tele-education, telemedicine, and emergency communications as applications; communications systems for air platforms (TU Graz Website). Additionally, sharing of national SATCOM assets is facilitated through the EDA EU SATCOM Market and the EDA Project Team on Satellite Communications, while technology development is facilitated through ESA. Austria participates in satellite communication activities through ESA's optional ARTES programs, with a budget of €8.12 million for 2018 (Euroconsult 2019). Further, Austria is a member of the International Maritime Organization (IMO) Global Maritime Distress and Safety System.

Positioning, Navigation, and Timing (PNT)

Austria carrie out space research and development at national level: TACTIC (Creating Awareness of Galileo Public Regulated Service-PRS at Critical Infrastructures) and PRS-Austria. PRS Austria studied the impact and countermeasures of Austrian PRS application scenarios in GNSS denied environments (TU Graz Website). Moreover, Austria participates in the optional ESA navigation programs in conjunction with the EU EGNOS/Galileo flagship program. Additionally, Austria is part of the COSPAS-SARSAT international satellite system for search and rescue.

Space Situational Awareness (SSA)

The Satellite Laser Ranging Station at the Lustbühel Observatory in Graz is a world leading research institute in space surveillance and tracking (BMVIT 2017). Austria's contribution to the optional SSA program at ESA was €0.55 million in 2018 (Euroconsult 2019).

Belgium

Space and Security Budget

Belgium has an estimated defense expenditure of €4.1 billion or 0.93% of GDP in 2018, below the NATO guideline of 2% of GDP. From 2019 on, the budgetary path followed a progressive and linear growth expected to reach a defense effort of 1.3% of the GDP by 2030 (NATO 2019). The Belgian Federal Parliament approved in May 2017 a law authorizing €9.2 billion of investments for the military until 2030, including remotely piloted aircraft systems (RPAS), airplanes, and frigates (Belgian Parliament 2017). The Belgian government expenditure for civil space-related activities amounted to €219 million in 2018, with the largest contribution of €47 million going to ESA (Euroconsult 2019). An amount of €14.5 million was allocated to EUMETSAT in 2018 (EUMETSAT 2018). The Law from May 2017 approving

€9.2 billion in military investments includes satellite communications (SATCOM), Imagery intelligence (IMINT), and intelligence, surveillance, target acquisition, and reconnaissance (ISTAR) (Belgian Parliament 2017). Accordingly, the decision of the Ministry of Defense to join the CSO French reconnaissance program has boosted the defense space expenditures (Belgian Ministry of Defence 2016).

Space and Security Activities and Programs

Earth Observation (EO) – Intelligence-Surveillance-Reconnaissance (ISR)
STEREO III (Support to the Exploitation and Research in Earth Observation) is the ongoing Belgian program for remote sensing research. The thematic research priorities include global monitoring of vegetation and evolution of terrestrial ecosystems; management of the environment on a local and regional scale; interaction between change in land cover and climate change; epidemiology and humanitarian aid; security and risk management (BEOP Website). Furthermore, the Flemish Institute for Technological Research (VITO) is involved in the processing of remote sensing imagery. VITO is charged with the payload data ground segment of the PROBA-V mission. VITO develops and operates Earth observation systems that are used to collect data about population growth, urban development, agriculture, the forestry industry, and natural disasters (VITO Website). In addition, the VEGETATION program is the result of a collaboration among Belgium, France, Italy, Sweden, and the European Commission. The system consists of two observation instruments, SPOT 4 and SPOT 5 satellites that study the state of vegetation at a global level and track its spatial and temporal evolution. One of the most important contributions of Belgium is the funding and the hosting of the Image Processing Centre at VITO Mol (BELSPO Website).

Belgium takes part in the NAOS (National Advanced Optical System) Program with Luxembourg. Luxembourg is benefiting from Belgian expertise in the frame of its Earth observation program NAOS, which will provide Luxembourg along with EU and NATO partners with additional ISR capabilities. LUXGOVSAT will become the operator and the Belgian Ministry of Defense will support payload operations for LUXGOVSAT (SpaceNews 2017). Belgium also participates in the French Pléiades program, the Multinational Space-based Imaging System (MUSIS) program, and the Hélios-II program. According to the strategic vision for Defence (2016–2030), Belgium aims to invest €45.7 million in the French *Composante Spatiale Optique* (CSO) program (Belgian Ministry of Defense 2016). Moreover, Belgium is part of the Copernicus program in the frame of the EU and ESA. Belgium participates in ESA's optional Earth observation programs, with a budget of €31 million for 2018 (Euroconsult 2019). Further, Belgium is a member of the following international organizations: ECMWF, GEOSS, CEOS, and the WMO.

Satellite Communications (SATCOM)
Since 2006, Belgium leases X-band satellite communications capability through a multiyear contract for indeterminate duration from Hisdesat for X-band capacity of

XTAR-EUR & SPAINSAT as C-band communications were insufficient to meet current and future needs. The satellite communications provide the required link between the operational center of the defense staff with military detachments overseas through the mobile stations and the BEMILSATCOM ground station in Marche-en-Famenne. Belgium leases communications from EUTELSAT and INTELSAT since 1999 (Skywin 2015). In addition, NEWTEC offers state-of-the-art commercial off-the-shelf satellite communications technology and solutions for the government and defense market. The Newtec Dialog VSAT platform is a scalable, flexible, and efficient multiservice satellite communications platform that allows network operators and satellite service providers to build and adapt their networks easily and in an affordable way as government and defense operations evolve (NEWTEC 2018).

In addition, Thales is a long-standing partner of the Belgian Armed Forces, supplying tactical communication systems and a variety of onboard sensors for armored vehicles, ships, helicopters, and unmanned aerial vehicles (UAV). Thales group, based in Belgium, provides encryption systems and secures satellite communications for NATO. Belgium is a partner of the Wideband Global SATCOM-9 (WGS), through the sharing of Luxembourg's allocation in the WGS system (SpaceNews 2017). Moreover, Belgium is supporting Luxembourg on governmental satellite communications. LuxGovSat's first satellite, GovSat-1 was launched in January 2018 and operates in X-band and military Ka-band to provide secure and assured Government services for EU and NATO. Belgium and Luxembourg have also cooperated in the frame of PACIS-1 (ESA 2017). Belgium is a member of the IMO Global Maritime Distress and Safety System.

Additionally, sharing of national SATCOM assets is facilitated through EDA EU SATCOM Market and the EDA Project Team on Satellite Communications, while technology development is facilitated through ESA. Belgium participates in satellite communication activities through ESA's optional ARTES programs, with a budget of €20.8 million for 2018 (Euroconsult 2019).

Positioning, Navigation, and Timing (PNT)

The first continuously operating GPS reference station in Belgium was installed in 1992 by the Royal Observatory of Belgium (ROB). Over the years, several additional tracking stations have been installed in Belgium. The ROB network was extended, but also each of the regions (Flanders, Wallonie, and Brussels) installed their own network of permanent GPS stations. The ROB has installed several continuously operating reference stations in Belgium. The primary goal of these stations is the integration of Belgium to international reference systems (ROB Website).

Moreover, the Redu Centre in Belgium has expanded to host Galileo's in-orbit test centre and a Galileo sensor station. The main purpose is to verify the performance of the payloads on the four Galileo In-Orbit Validation (IOV) satellites, including radio frequency and baseband parameters, easurements (GPS World 2018). Most of Belgium's military capabilities will continue to rely on GPS in the near term, although alternative Global Navigation Space System (GNSS) capabilities are being pursued to improve resilience. Further, Belgium participates in the optional ESA navigation programs in conjunction with the EU EGNOS/Galileo flagship program. Additionally, Belgium is part of the COSPAS-SARSAT international satellite system for search and rescue.

Space Situational Awareness (SSA)

The PROBA2 Science Centre (P2SC), located at the Royal Observatory of Belgium in Brussels, oversees scientific operations and data processing for ESA's PROBA2 spacecraft. The P2SC has a science operations center, where instrument observing plans are devised with support also by ESA's Spacecraft Operations Centre in Redu. Finally, the P2SC serves as the main site for coordination of the PROBA2 Science Working Team (Proba2 Science Center Website).

In February 2017, BELSPO entered into an agreement with USSTRATCOM to share SSA services and information. The arrangement enhances awareness within the space domain and increases the safety of spaceflight operations. Belgium joined other nations (the UK, the Republic of Korea, France, Canada, Italy, Japan, Israel, Spain, Germany, Australia, and the United Arab Emirates), intergovernmental organizations (ESA and EUMETSAT), and more than 50 commercial satellite owner/operator/launchers already participating in SSA data-sharing agreements with USSTRATCOM (US Strategic Command 2017).

In addition, Belgium hosts the ESA SSA Space Weather Coordination Centre (SSCC) at the Space Pole. The SSCC coordinates the provision of space weather services available either at the SWE Data Centre located in Redu or at federated sites. The SSCC monitors the service network and provides the first-level user support with the help of scientific experts in the topics of solar weather, space radiation, ionospheric weather and geomagnetic conditions (ESA SSCC Website). The SSCC is operated by a Belgian consortium on behalf of ESA's SSA program Office. Belgium's budget for the optional ESA SSA program was €1.6 million in 2018 (Euroconsult 2019).

The Netherlands

Space and Security Budget

The Dutch economy continues to grow faster than the average in the eurozone, with a GDP growth at 2.7% in 2018 compared to the previous year, mostly driven by domestic demand (Euroconsult 2019). The Netherlands defense expenditure equaled to €9.45 billion or 1.36% of GDP in 2018, below the NATO guideline of 2% of the GDP (NATO 2019). The Netherlands had an estimated space budget of €150 million in 2018 covering all public civil and defense expenditure at national and international level. This amount comprises of €91 million contribution to ESA (26% increase compared to 2017), €24.8 million contribution to EUMETSAT, and €34 million to the Dutch National Program. The space defense expenditure for the field of telecommunications was €2 million in 2018 (Euroconsult 2019).

Space and Security Activities and Programs

Earth Observation (EO) – Intelligence-Surveillance-Reconnaissance (ISR)

The Netherlands has provided as in-kind contribution to Sentinel-5P a measurement instrument called *Tropomi*, which conducts research/monitoring of the atmosphere. Sentinel-5P was launched on 13 October 2017 from Plesetsk. The Dutch industry and public research organizations worked together for 7 years on the Tropomi development, spending an estimate of €133 million on the project. The project was funded and supported by the Ministry of Economic Affairs and Climate, the Ministry of Infrastructure and Water Management, and the Ministry of Education, Culture and Science. The Dutch government invested €98 million with the remainder financed by ESA. NSO worked together with Royal Dutch Meteorological Institute (KNMI), the Netherlands Organization for Applied Scientific Research (TNO), and the Dutch national expertise institute for scientific space research (SRON). Airbus Defence and Space Netherlands was the prime contractor (Dutch Government 2017a).

The *Scanning Imaging Absorption Spectrometer for Atmospheric Cartography (SCIAMACHY)* instrument on board ESA's Envisat mission (2002–2012) was a spectrometer that could measure a wide range of the electromagnetic spectrum. It was able to support research into the composition of the atmosphere. Absorptions in this spectrum provide information about the concentration of trace gases in the atmosphere, which in turn are a measure for air quality and climate change. The SCIAMACHY instrument was jointly realized by the Netherlands, Germany, and Belgium (NSO Website).

The NSO provides no-cost access to raw, unprocessed satellite data from several optical and SAR sensors through its Satellitedataportaal (satellite data portal). NSO purchased such data, financed by the Ministry of Economic Affairs and the Directorate of Infrastructure and Water Management, and made it publicly available to everybody in the Netherlands. Since 2015, a part of the satellite data has been made available free of charge via the European Copernicus program (NSO Website).

In 2012, the NSO entered into a 3-year agreement with MDA to license *RADARSAT-2* data with coverage over the Netherlands. The standard mode, dual-polarized data is processed and delivered to the Satellietdataportaal where Dutch users can access the data. During the contract term, the NSO expects to acquire close to 600 Standard, dual-polarized RADARSAT-2 images of the Netherlands and its surrounding waterways. The focus of the work by the NSO and the Dutch geospatial community is on agricultural and interferometric synthetic aperture radar (InSAR) applications (MDA 2015).

Additionally, the Dutch Geodata for Agriculture and Water (G4AW) program stimulates sustainable food production, a more efficient use of water in developing countries, and aims to alleviate poverty by enhancement of sustainable economic growth and self-reliance in the G4AW partner countries. G4AW provides a platform for partnerships of private and public organizations. Together they provide food producers with relevant information, advice, or financial products. G4AW is a program of the Dutch Ministry of Foreign Affairs with policy priorities for food security and water, which is executed by the NSO (NSO Website).

In September 2015, the Ministry of Justice and Security launched the innovation program Satellite applications in a so-called triple helix collaboration with external partners such as NSO, TNO, and The Hague Security Delta. Implementing organizations within the Ministries of Defence; Finance, Infrastructure, and the Environment; and Social Affairs and Employment have joined components of this program. The aim is to develop innovative products and services based on public-private partnerships to develop and implement satellite technology in security and justice. In addition, the Ministry of Justice and Security wants to gain experience with the use of new technology. In this program furthermore, attention is also paid to the risks of dependence on satellite technology and the protection of privacy (NSO 2016).

Moreover, *Ozone Monitoring Instrument (OMI)* is an Earth observation instrument, commissioned by the Netherlands and Finland, which measures the concentration of ozone and other trace gases in our atmosphere. OMI maps the worldwide ozone concentration daily. This information is useful for the prediction of the intensity of ultraviolet radiation and smog formation. The OMI instrument is part of the EOS-AURA satellite, built under the responsibility of NASA and launched in 2004. OMI was developed by Airbus Defence and Space Netherlands (then called Dutch Space) and TNO. The KNMI was scientific leader of the project and SRON was also closely involved in the project. OMI derives its heritage from NASA's Total Ozone Mapping Spectrometer (TOMS) instrument and the ESA Global Ozone Monitoring Experiment (GOME) instrument (on the ERS-2 satellite). OMI measures criteria pollutants such as O3, NO2, SO2, and aerosols (Airbus Defence & Space OMI Website).

Further, the Netherlands is part of the Copernicus program in the frame of the EU and ESA. The Netherlands contribution to ESA's optional Earth observation programs amounted €12.4 million for 2018 (Euroconsult 2019). The Netherlands is a member of the following international organizations: ECMWF, GEOSS, CEOS, and the WMO.

Satellite Communications (SATCOM)

The Dutch Ministry of Defense has awarded a contract to Thales to supply naval satellite communications (SATCOM) terminals on the four ocean patrol vessels that are being built for the Royal Netherlands Navy. Under the contract, the company will deliver a complete SATCOM solution including state-of-the-art dual-band stabilized antenna, with civilian SATCOM Ku-band and military X-band. The SATCOM solution incorporates leading-edge multiband and stabilization capabilities, using multiple transmission systems, and ensuring perfect onboard integration. With the naval satellite communications terminals, the ocean patrol vessels will be able to communicate over secure, high-bandwidth satellite links with other national and allied force elements. The SATCOM solution is already in service in several NATO countries, including France, Belgium, and Norway (Naval Technology 2009).

The Ministry of Defense has a program called *Militaire Satelliet Communicatie lange termijn* (Defensiebreed) – MILSATCOM. The project covered a total cost of €132.3 million between 2007 and 2019 and is aimed at guaranteeing high-frequency satellite capacity, establishing ground stations in the Netherlands and Curacao as well as procuring mobile ground segment. The Netherlands intends to replace their MILSATCOM ground terminals by 2020 (Dutch Ministry of Defense 2017).

In November 2017, the Royal Netherlands Air Force (RNAF) and the Netherlands Aerospace Centre (NLR) signed two agreements. These agreements contribute to RNAF's strategic ambition to use information-driven functionalities to improve and accelerate its response to potential military threats. The first agreement concerns the development and launch of a so-called nanosatellite for demonstration of technological possibilities relevant for staging military operations. The second agreement concerns the renewal of the existing covenant between RNAF and NLR aimed at joint knowledge building in the field of military air and space operations (NLR 2017a).

The nanosatellite, named "Brik-II" is a 6 U CubeSat, is expected to be launched in 2020. The main tasks of Brik-II are navigation, communication, and Earth observation. NLR is responsible for the instrument detecting radio waves from sources including radar installations. This will enable RNAF to better map its surroundings. NLR not only developed the idea for the instrument but designed and constructed it as well. NLR is also responsible for "ground processing," which involves the collected raw data being converted into data with military relevance. The nanosatellite's foremost purpose is to acquire knowledge about the space domain and to prove military-relevant applications. The nanosatellite will be a precursor to future satellites, for which both the acquired technological knowledge and operational experience will be used to develop the next generation of Dutch military satellites. The second agreement concerns the renewal of the covenant between RNAF and NLR relating to military air and space operations. Its purpose is to jointly develop this field of knowledge and align the strategies of the two parties (NLR 2017a).

The RNAF's strategy is to develop capabilities to enable it to carry out military operations from the air and from space. Parallel to this, RNAF will undergo a transformation to become an information-driven and agile air force organization with advanced operational capabilities. Obtaining the right volume of intelligence at the right time plays a central role in this. Getting hold of the right data before an enemy does will enable RNAF to increase its operational pace and improve its effectiveness and thereby to respond faster and better than its opponent (NLR 2017a).

The Netherlands is a founding member along with Canada, Denmark, Luxembourg, and New Zealand of the WGS-9 (part of the US-led WGS system). WGS-9 was launched in March 2018. The contribution gives the countries access to the WGS military communications system (Dutch Air Force 2017). The Netherlands is a member of the IMO Global Maritime Distress and Safety System. Additionally, sharing of national SATCOM assets is facilitated through NATO, while technology development is facilitated through ESA's optional ARTES programs with a budget of €8.4 million for 2018 (Euroconsult 2019).

Positioning, Navigation, and Timing (PNT)

The Netherlands Positioning Service (NETPOS) is the GNSS reference network of the Land Registry and the Directorate of Infrastructure and Water Management. NETPOS makes it possible for surveyors with one GNSS receiver to determine their position in the terrain within a few seconds and within a few centimeters. Companies that work for Ministry of Infrastructure and Water Management or the Land Registry may also make use of this. NETPOS has several GNSS reference stations spread over the Netherlands, with an average distance of 40 kilometers. This provides good coverage for the whole of the Netherlands. The data from the GNSS reference stations are used for obtaining geo-information, navigation, weather forecasts, education, and scientific research. NETPOS uses navigation satellites from both GPS (American) and GLONASS (Russian). In the future, the European Galileo will be added to this (NETPOS Website).

Funded by the Ministry of Infrastructure and Water Management, the IKUS Project (*Inventarisatie Kwetsbaarheden Uitval Satellietnavigatie*) addresses the issue of GNSS dependency and backup PNT solutions (Dutch Ministry of Justice and Security 2015). In August 2017, the Dutch Government released the report describing the resilience of critical GNSS infrastructure, focusing on the effects of space weather (Dutch Government 2017b).

The NLR undertook GPS and Galileo PRS interference field tests in 2017 (NLR 2017b). Further, the Netherlands participates to the optional ESA navigation programs in conjunction with the EU EGNOS/Galileo flagship program. In the frame of Galileo, the Galileo Reference Centre (GRC), the performance monitoring hub for the EU's global satellite navigation system, was officially opened in Noordwijk, in May 2018. The Centre will be operated by GSA with as main objective to ensure continuous Galileo accuracy below 20 cm. The GRC missions are as follows (GSA 2018): performing independent monitoring and assessment of Galileo service provision; assessing, when feasible, the compatibility and interoperability of Galileo vis-a-vis other GNSS; providing service performance expertise to the program; supporting investigations of service performance and service degradations; providing an archiving service for performance data over the nominal operational lifetime of the system; and integrating data and products from EU Member States and Norway and Switzerland.

Space Situational Awareness (SSA)

The Ministry of Defense is, in collaboration with the TNO, NLR, and KNMI, researching space weather (Ministry of Justice and Security 2015). The Joint Meteorologiche Group (JMG) located at the *Vliegbasis Woensdrecht* foresees the Armed Forces with space weather reports. The Netherlands' contribution to the optional SSA program at ESA amounted to €0.5 million in 2018 (Euroconsult 2019).

Norway

Space and Security Budget

Norway had an estimated defense expenditure of NOK 61.349 billion (€5.88 billion) or 1.8% of GDP in 2018, below the NATO guidelines of 2% of GDP (NATO 2019). In 2018, Norway's total space budget amounted to just above NOK 1 billion, remaining stable compared with 2017. Norwegian investments for civil space activities have been steadily increasing in the past years, growing from NOK 569 million in 2014 to NOK 958 million in 2018. Defense space expenditure is estimated at NOK 80 million annually in the period 2016–2018 (for military satellite communications). The 2018 ESA contributions covered NOK 644 million. Although Norway gives preference to ESA programs over the national civil program, which represented 18% of the total civil expenditures in 2018, both budgets have increased in the past 5 years. The contribution to EUMETSAT for 2018 amounted to NOK 142 million (Euroconsult 2019). Norway is not a member of the EU but does work closely with the EU and its member states, as part of the European Economic Area (EEA) and the European Free Trade Association (EFTA).

Space and Security Activities and Programs

Earth Observation (EO) – Intelligence-Surveillance-Reconnaissance (ISR)

Norway relies on imagery procured from commercial providers, such as Inmarsat, Iridium, and Eutelsat, and international partners to support its tasks (Satellite Today 2015). The Norwegian CryoClim project is developing services for monitoring the cryosphere, i.e., the sea ice, snow cover, and glaciers in Norway and Svalbard. CryoClim will combine data from several different optical, meteorological, and radar satellites, including Radarsat-1 and Radarsat-2 (Norwegian Space Center 2018).

The NOFO (Norwegian Clean Seas Association for Operating Companies) and KSAT (Kongsberg Satellite Services) have signed an extended agreement for satellite-based remote sensing, on the Norwegian Continental Shelf, for detection of acute pollution from petroleum activity. This agreement represents a solid boost for the detection of oil spills on the Norwegian shelf, enabling rapid deployment of an oil recovery operation by having all fields in production monitored daily by satellite. KSAT has delivered its satellite-based oil spill detection service to the industry, through NOFO, since 2005 through KSAT ground stations (KSAT 2016).

Moreover, Norway is part of the Copernicus program in the frame of the EU and ESA. Norway participates in ESA's optional Earth observation programs, with a budget of €10 million for 2018 (Euroconsult 2019). Further, Norway is a member of the following international organizations: ECMWF, GEOSS, CEOS, and the WMO.

Satellite Communications (SATCOM)

The Norwegian Space Centre's focus on infrastructure development has been of vital importance to the High North. The agency owns 50 percent of KSAT, with Kongsberg Defence & Aerospace owning the other 50 percent. KSAT operates satellite stations in Svalbard, Tromsø, and Antarctica. Due to the unique location of its polar ground stations, KSAT can provide maritime monitoring services and data from several radar and optical sources, delivered straight to the end user. The company also owns the fiber-optic cable linking Harstad on the Norwegian mainland to Longyearbyen, Svalbard. The Norwegian Space Centre has also leased capacity from Telenor for broadband communications to Antarctica via the Thor 7 satellite (Norwegian Ministry of Industry and Trade 2012).

The Norwegian government announced in March 2018 the investment of NOK 1 billion for a new Arctic communications project implemented by Space Norway AS. The project is based on a system of two satellites providing coverage 24 h a day in the area north of 65 degrees N latitude. The expected life span of the satellites is 15 years and the satellites are expected to be launched in 2022. Space Norway AS owns the fiber-optic cable between Svalbard and mainland Norway, a key element of Norway's electronic infrastructure in the Arctic. Since 2015, the company has been working to establish satellite-based broadband communications capacity in the region (Barents Observer 2018).

Thales in Norway has supplied military and enterprise networks since the mid-1980s and has gained a broad experience in defense communications over the years. Thales has been selected by the Norwegian Armed Forces as the supplier for satellite communication systems for both the Nansen-class frigates and the Skjold-class corvettes. Additionally, Thales has been selected as sole supplier to NATO for securing IP networks with high-grade encryption for all classifications, including Cosmic Top Secret (CTS) (Thales Website).

Further, Norway is a member of the US-led WGS program and, also, a member of the IMO Global Maritime Distress and Safety System. Additionally, sharing of national SATCOM assets is facilitated through NATO, while technology development is facilitated through ESA's optional ARTES programs with a budget of €71 million for 2018 (Euroconsult 2019).

Positioning, Navigation, and Timing (PNT)

The Norwegian satellites AISSat-1 and AISSat-2 monitor maritime traffic in Norwegian and international waters by detecting AIS (Automatic Identification Signals) from ships to determine their position, speed, and direction. The satellite project was realized as a collaboration between three Norwegian governmental institutions: the Norwegian Space Centre, the Norwegian Coastal Administration (NCA), FFI, Kongsberg Seatex, and KSAT. The satellite platform itself was purchased from Canada. The first AIS CubeSat was launched on the 12th of July 2010 from India to a polar orbit. AISSat-1 was primarily intended to demonstrate space-based AIS. Building on the experience and success with AISSat-1, a second identical satellite AISSat-2 was launched from Kazakhstan in July 2014. The third satellite in the AIS series, AISSat-3 was lost due to a launch failure from Russia in November 2017. The collaborating institutions use the data from AISSat-1 and AISSat-2 for a variety of purposes, including monitoring fisheries, oil spills, and maritime traffic, to support anti-piracy operations along the coast of Africa and other areas of interest to Norway (Norwegian Space Center Website).

Additionally, the Norwegian satellites NorSat-1 and NorSat-2 were launched in July 2017. The satellites including the satellite-AIS payloads are owned and commissioned by the Norwegian Space Centre, which will also be responsible for the operation of the two microsatellites. The receivers were developed by Kongsberg Seatex of Norway, with support provided by ESA through the SAT-AIS element of

ESA's ARTES program. The two satellites monitor maritime traffic and test science and technology payloads. In the Norwegian government's High North Strategy, further development of space-related infrastructure is a stated target. The satellite-based ship AIS is an important part of the integrated monitoring and notification system for the northern sea areas. The satellites are primarily financed by the Norwegian Coastal Administration and the Norwegian Space Centre (Norwegian Space Center Website).

In January 2018, the Norwegian Space Centre procured the final development NorSat-3 from Space Flight Laboratory in Canada. Combining a navigation radar detector and AIS receiver aims to enhance maritime awareness for the Norwegian Coastal Administration, Armed Forces, and other maritime authorities. The satellite is funded by the Norwegian Coastal Administration and managed by the Norwegian Space Centre. The Norwegian Defence Research Establishment is leading the development of the radar detector payload, which is funded by the Ministry of Defense. NorSat-3 is designed to capture signals from frequencies that the International Maritime Organization has allocated for civil navigational radars (Nordic Space 2018).

In August 2017, NASA signed an agreement with the Norwegian Mapping Authority to develop a Satellite Laser Ranging Station 1046 km from the North Pole that will produce high-precision locations of orbiting satellites, help track changes in the ice sheets, and improve the efficiency of marine transportation and agriculture. The Arctic station will be the latest addition to a global network of space geodetic stations. Under the partnership, the station will be built and installed in Ny-Ålesund, Svalbard. NASA will also provide expert consultation on how to operate the instruments. The Norwegian Mapping Authority started construction work on the new scientific base in 2014. The goal is to have all systems in operation by 2022 (NASA 2017).

Additionally, Norway is part of the COSPAS-SARSAT international satellite system for search and rescue. Further, Norway participates in the optional ESA navigation programs in conjunction with the EU EGNOS/Galileo flagship program.

Space Situational Awareness (SSA)

The Norwegian Armed Forces Globus II radar at Vardø is used for space debris monitoring. The Globus II is a large X-band dish radar located at Vardo in Northern Norway. It is a dedicated sensor in the SSN and is used for tracking deep space objects, including objects in geosynchronous orbits, and for wide-band imaging of space objects. Originally known as HAVE STARE (AN/FPS-129), the radar became operational in 1995 at Vandenberg Air Force base in California. While there, it observed several intercontinental ballistic missile (ICBM) flight tests a as well as two non-intercept tests of the kill vehicle for US national missile defense interceptor then under development (IFT-1A and IFT-2). Beginning in late 1998, HAVE STARE (AN/FPS-129) was dismantled and moved to Vardo, and then renamed Globus II. A previous radar named Globus was operational since the 1960s by Norway at Vardo, in cooperation with the US Air Force, used to monitor Soviet and Russian ballistic missile flight tests (Mostlymissiledefense 2012).

In 2013, the Birkeland Centre for Space Science was established as a Centre of Excellence in research. Its mission has been to increase knowledge of electrical current flows around the Earth, particle showers from space, auroras, gamma-ray bursts, and other links between the Earth and space (Norwegian Ministry of Industry and Trade 2012).

In April 2017, USSTRATCOM entered into an agreement with the Norwegian Ministry of Defense and Norwegian Ministry of Trade, Industry and Fisheries to share SSA services and information. The arrangement will enhance awareness within the space domain and increase the safety of spaceflight operations (US Strategic Command 2017). Norwegian SSA operations are also conducted within the ESA framework, with activities performed by the NMA and TGO. Norway's contribution to the optional SSA program at ESA was €1 million in 2018 (Euroconsult 2019).

Sweden

Space and Security Budget

In line with the Swedish Defence Policy 2016–2020, an increased defense budget is fundamental, particularly considering the deteriorating security situation, but also to address the need to increase warfighting capabilities of the Swedish Armed Forces. To that extent, the political agreement means that this defense bill adds approximately SEK 10 billion extra to the Armed Forces for the period of 2016–2020. Total defense spending in the period 2016 to 2020 amounts to SEK 224 billion for the Armed Forces, including defense intelligence (Swedish Government 2015).

In August 2017, the Swedish government reached an agreement on additional measures to increase the operational capabilities of military units and ensure overall total defense capability. In addition to the investments of SEK 500 million announced in the 2017 Spring Amending Budget, military capabilities and total defense capability will be enhanced with an additional SEK 2.7 billion per year from 2018. Top priority will be given to measures ensuring the implementation of the 2015 Defence Resolution during the period until the end of 2020. Priority will also be given to measures aimed at further enhancing capabilities during the current period and ensuring that military capabilities and total defense capability can increase after 2020. In accordance with the defense agreement, the allocated an additional SEK 80 million in 2018 to the National Defence Radio Establishment and the Military Intelligence and Security Service for efforts targeting counterterrorism, information and cyber security, and security services (Swedish Government 2017).

Swedish government expenditure on space amounted to SEK 1.1 billion in 2018, distributed between ESA programs (64%), national activities (21%), and EUMETSAT (15%). The 2018 ESA contributions covered SEK 696.8 million. The contribution to EUMETSAT for 2018 amounted to SEK 164 million (Euroconsult 2019).

Space and Security Activities and Programs

Earth Observation (EO) – Intelligence-Surveillance-Reconnaissance (ISR)

The Swedish National Space Agency (SNSA – Rymdstyrelsen) ran a National Earth observation program (Technopolis 2012). During the period 2001–2012, the program provided public funding amounting to SEK 219 million, of which SEK 133 million were related to projects in the research part and SEK 86 million to projects in the user's part. Until 2005, the two parts were of roughly equal size, but in recent years, the research part has come to significantly dominate the user's part. The beneficiaries in the research part have been dominated by a few universities and a government agency (Technopolis 2012).

Sweden participates in the French Pléiades program. Sweden is part of the Copernicus program in the frame of the EU and ESA. Sweden's contribution in ESA's optional Earth observation programs amounted €12.9 million for 2018 (Euroconsult 2019). Further, the Netherlands is a member of the following international organizations: ECMWF, GEOSS, CEOS, and the WMO.

Satellite Communications (SATCOM)

Sharing of national SATCOM assets is facilitated through EDA and NATO, while technology development is facilitated through ESA's optional ARTES programs with a budget of €71 million for 2018 (Euroconsult 2019). Further, Sweden is a member of the US-led WGS program and, also, a member of the IMO Global Maritime Distress and Safety System. Additionally, sharing of national SATCOM assets is facilitated through NATO, while technology development is facilitated through ESA's optional ARTES programs with a budget of €6.18 million for 2018 (Euroconsult 2019). Sweden is also a member of the IMO Global Maritime Distress and Safety System.

Positioning, Navigation and Timing (PNT)

A GNSS analysis center project has been started within the Nordic Geodetic Commission, chaired by Lantmäteriet (the Swedish mapping, cadastral, and land registration authority). The EGNOS Ranging and Integrity Monitoring Station that was inaugurated at Lantmäteriet in Gävle has been successfully supported by Lantmäteriet since 2003. SWEPOS is the Swedish national network of permanent GNSS stations and is operated from the headquarters of Lantmäteriet in Gävle (Lantmäteririet 2015).

SWEPOS provides real-time services on both meter level and centimeter level, as well as data processing. The SWEPOS Network RTK Service reached national coverage during 2010. Since data from permanent GNSS stations are exchanged between the Nordic countries, good coverage of the service in border areas and along the coasts has been obtained by the inclusion of 20 Norwegian SATREF stations, 4 Norwegian Leica SmartNet stations, 5 Finnish Geotrim stations, 1 Finnish Leica SmartNet station, 3 Danish Leica SmartNet stations, and 2 Danish Geodatastyrelsen (Danish Geodata Agency) stations (Lantmäteririet 2015).

Sweden is part of the COSPAS-SARSAT international satellite system for search and rescue. Further, Sweden participates in the optional ESA navigation programs in conjunction with the EU EGNOS/Galileo flagship program.

Space Situational Awareness (SSA)

The Onsala Space Observatory (OSO), the Swedish National Infrastructure for Radio Astronomy, provides scientists with equipment to study the Earth and the rest of the Universe. OSO takes part in international radio astronomical projects. The observatory is a geodetic fundamental station. The equipment for Earth sciences includes navigation satellite receivers, a superconducting gravimeter, tide gauges, and radiometers (Chalmers Website).

The Swedish Institute of Space Physics (IRF) is a governmental research institute. Its primary task is to carry out basic research, education, and associated observatory activities in space physics, space technology, and atmospheric physics. IRF has employees in Kiruna, Umeå, Uppsala, and Lund. The research programs portfolio includes Solar Terrestrial and Atmospheric Research (STAR), Solar System Physics and Space Technology, and Space Plasma Physics (IRF Website).

Norwegian SSA operations are also conducted within the ESA framework, with activities performed by the NMA and TGO. Norway's contribution in the optional SSA program at ESA was €0.213 million in 2018 (Euroconsult 2019).

Switzerland

Space and Security Budget

Switzerland's defense spending is estimated to reach USD 6.5 billion in 2023, registering a compound annual growth rate (CAGR) of 4.67% between 2019 and 2023, according to a report by Strategic Defence Intelligence (SDI). Swiss defense budget spending in 2018 reported an increase of 6.7% in comparison with 2017, driven by the country's need to invest in new technologically advanced military equipment and devices to replace outdated equipment. Allocation of capital expenditure is anticipated to increase to an average of 24.7% of the total defense budget over the forecast period, compared with the average of 22.7% recorded during 2014–2018. Over the forecast period, Switzerland is expected to invest in command, control, communications, computers, intelligence, surveillance and reconnaissance (C4ISR), and cybersecurity. The country intends to acquire unmanned aerial vehicles, critical infrastructure protection, and military radar, as well as upgrading its existing communications network and incorporating enhanced levels of Internet security (Army Technology 2018).

The 2018 space budget covered CHF 200 million, with CHF 166 million contribution to ESA and CHF 24 million to EUMETSAT (Euroconsult 2019). The Euroconsult identified budget does not include military space budget.

Space and Security Activities and Programs

Although Switzerland invests most of its direct space funding through ESA, Switzerland has a so-called Programme for National Complementary Activities for Space (*Measures de Positionnement*), to encourage the emergence of space technology projects. It aims to develop niche sectors and to better position Swiss industrial and academic entities, particularly in the frame of ESA activities and other international programs such as the EU Research Framework Programmes. The Swiss Space Office (SSO) mandated the Swiss Space Centre (SSC) to implement the call for proposals 2018 (SSO 2018).

Earth Observation (EO) – Intelligence-Surveillance-Reconnaissance (ISR)

In Switzerland, Earth observation data are being used in many fields, such as weather prediction, climate monitoring (e.g., clouds, air pollution, glaciers), landslide detection, or topographic mapping (SCNAT 2008).

Switzerland has no dedicated Earth observation satellite program. Switzerland's technology development at national level, it takes place namely at the Earth observation and Remote Sensing Group of the ETH Zurich Institute of Environmental Engineering (ETH Zutich Website). As an example of the use of space-based data for food security, the overall aim of the project Remote sensing-based Information and Insurance for Crops in Emerging economies ("RIICE") is to reduce the vulnerability of rice smallholder farmers in low-income countries in Asia and beyond. RIICE is implemented through a public-private partnership between the Deutsche Gesellschaft für Internationale Zusammenarbeit, the Swiss Agency for Development and Cooperation, the International Rice Research Institute, sarmap, and SwissRe (RIICE Website).

The objective of the Swiss Data Cube (SDC) is to support the Swiss government for environmental monitoring and reporting and enable Swiss scientific institutions (e.g., Universities), to facilitate new insights and research using the SDC, and to improve the knowledge on the Swiss environment using EO data. UN Environment/GRID-Geneva and the University of Geneva are currently building the Swiss Data Cube (SDC). Following the work done by Geoscience Australia the "Data Cube" is a new way for organizing Earth observations data. UN Environment/GRID-Geneva and the University of Geneva have developed a strong and fruitful collaboration around the SDC with the Committee on Earth Observations Satellites (CEOS), the Group on Earth Observations (GEO), and Geoscience Australia (Swiss Data Cube Website). The concept of the Data Cube is a series of structures and tools that calibrate and standardize datasets, enabling the application of time series and the rapid development of quantitative information products. The Data Cube approach calibrates this information to make it more accessible, easier to analyze and reduces the overall cost for users (Swiss Data Cube Website).

Moreover, Switzerland is part of the Copernicus program in the frame of the EU and ESA. Switzerland participates in ESA's optional Earth observation programs with a budget of €805.6 million for 2018 (Euroconsult 2019). Further, Switzerland is a member of the following international organizations: ECMWF, GEOSS, CEOS, and the WMO.

Satellite Communications (SATCOM)

Switzerland has no dedicated SATCOM program. Switzerland has technology development at national level, for instance, at EPFL. The State Secretariat for Education, Research, and Innovation (SERI) cooperated with DLR on the development of satellite-based laser communications (DLR 2008).

RUAG and Elbit Systems (Israel) announced in November 2018 the signature of a Memorandum of Understanding to form a Joint Venture Company. The JVC will enable the companies to create synergies and leverage their respective competences and serve as a national Communication and System Competence Centre of Excellence. The Competence Centre will cater to the needs and requirements of the Swiss Federal Department of Defence, Civil Protection and Sport (DDPS). The Centre will act as a knowledge center to support the companies' joint efforts regarding a DDPS communication program of the Swiss Ministry of Defense and other joint endeavors in the future (Israel Defense 2018).

In September 2018, RUAG and SWISSto12 signed a partnership agreement committed to supply future satellites missions, including constellations of telecommunication satellites with crucial antenna solutions. The collaboration will focus on the development, design, and manufacturing of phased array antenna products that leverage the unique and complimentary offerings of both companies. SWISSto12, with their patented 3D printing technology, will provide innovative antenna system designs along with a manufacturing solution to 3D print the antenna in one highly integrated part combining, RF, mechanical and thermal functions. This approach aims at a high performance – at minimal cost and lead time for manufacturing and integration (RUAG 2018).

Further, Switzerland is a member of the IMO Global Maritime Distress and Safety System. Additionally, technology development is facilitated through ESA's optional ARTES programs with a budget of €9.5 million for 2018 (Euroconsult 2019).

Positioning, Navigation, and Timing (PNT)

Swisstopo, the Swiss Federal Office of Topography, is responsible for the development, operation, and maintenance of a fundamental station and permanent stations of Switzerland's automated global navigation satellite systems. Swisstopo contributes, as one of several European processing centers of the European Permanent Network EPN. The Automated GNSS Network of Switzerland (AGNES) is a multipurpose reference network for national first order surveying, scientific research such as geodynamics, and GNSS meteorology and serves as a base for the Swiss positioning service (swipos). Swisstopo maintains the swipos positioning service for real-time applications. Switzerland is integrated into the international (IGS) and European (EPN) permanent networks via its Zimmerwald station. In the framework of the International GPS Service (IGS), the Federal Office of Topography set up a permanent GPS station in Zimmerwald in 1992. The data of this station are used by the Centre for Orbit Determination in Europe (CODE) at the Astronomical Institute of the University of Bern for the determination of highly precise GPS satellite orbits. The IGS provides the data from Zimmerwald geostation to interested parties for scientific research (Swisstopo Website).

The Swiss Federal Office of Civil Aviation, Zurich Airport, Geneva Airport, Switzerland's regional airports, SWISS, EasyJet, the Swiss Air Force, and skyguide have jointly established an innovation program called "CHIPS" for introducing new satellite-based flight procedures (Skyguide Website).

Further, Switzerland participates in the optional ESA navigation programs in conjunction with the EU EGNOS/Galileo flagship program.

Space Situational Awareness (SSA)

The Zimmerwald Laser and Astrometric Telescope (ZIMLAT) has been designed for both satellite laser ranging and optical tracking with charge-coupled device (CCD) cameras, the latter mainly for orbit determination of space debris by means of astrometric positions. For the first time, a titanium-sapphire laser was introduced into a satellite laser ranging (SLR) tracking system. On 20th July 2009, Zimmerwald was the first European station which successfully sent laser pulses to the Lunar Reconnaissance Orbiter (LRO). In March 2012, the Graz and Zimmerwald SLR stations successfully conducted the first ever so-called "bistatic" laser ranging to a noncooperative target; the Zimmerwald SLR station for the very first time successfully detected and time – tagged photons sent by a powerful laser at the Graz SLR station and diffusely reflected by the body of the European ENVISAT spacecraft. In June 2013, photons reflected by a space debris object with a considerably smaller cross section than ENVISAT could be detected (Gutner and Ploner 2006).

The Swiss startup ClearSpace signed a debris-removal contract with ESA tasking the company with deorbiting a substantial piece of a Vega rocket left in orbit in 2013. For the pillar Space safety and security, Switzerland takes the lead in the removal of space debris thanks to ClearSpace, a spin-off from the EPFL Space Center focusing on developing technologies and services to remove unresponsive satellites from space. The startup has received funding for its ClearSpace-1 ADR mission under the ESA ADRIOS program. ESA has also selected ClearSpace to be the leader of the industrial consortium for the project (SpaceNews 2019).

Skyguide announced in March 2018 a partnership with AirMap to develop and deploy a national drone traffic management system for Switzerland. Swiss U-space, as it is known, will power Switzerland's thriving community of drone companies with UTM (Unmanned Traffic Management) services for advanced commercial drone operations like Beyond Visual Line of Sight (BVLOS) (Airmap 2018).

Switzerland's contribution in the optional ESA SSA programs was €1.12 million in 2018 (Euroconsult 2019).

Concluding Remarks

This chapter provides evidence of the trend toward increasing relevance for security and defense in national space and security programs, together with the chapters on the largest and smaller European countries. "Space and security," both in its security from space and security in space aspects, are progressively contributing to further integration of space activities. Traditionally, only civilian space activities including, for instance, Earth observation, telecommunications, human spaceflight, space trans-portation, and technology development were subject of cooperation projects at intergovernmental and supranational level. Security- or defense-related space pro-grams were kept at the national level or dealt with bilaterally or multilaterally in *ad hoc* cooperative programs. These trends demonstrate an evolution of European countries priorities from strictly civil-oriented applications to also encompassing security and defense ones. National space programs with security or defense dimen-sion, in combination with the EU and ESA programs, demonstrate alignment towards the use of space for security and defense. In this group of countries, the level of national space programs engagement varies, and often EU and ESA engage-ment is prime. To conclude, the increasing relevance of security and defense in Europe, to some extent, it could be framed as the necessity for Europe to take its security and defense into its own hands vis-à-vis other global space powers. Hence, this appears to be a stronger driver in the current geopolitical context.

Disclaimer The contents of this chapter and any contributions to the Handbook reflect personal opinions and do not necessarily reflect the opinion of the European Space Agency (ESA).

References

Airbus Defence & Space, Space Netherlands, OMI, Website. Available at: http://www.airbusdefen ceandspacenetherlands.nl/project/omi/

Airmap (2018) Swiss U-space powers life-saving drone pilot program in Switzerland, 6 June 2018

Army-technology (2018) Report: Swiss defence budget to reach $6.5bn by 2023, 4 May 2018

Barents Observer (2018) Two new satellites to boost Norway's Arctic internet, 27 March 2018

Belgian Ministry of Defence (2016) The Strategic vision for Defence, 29 June 2016, 29 June 2016

Belgian Parliament (2017) Wetsontwerp Projet De Loi houdende de militaire programmering van investeringen voor de periode 2016- 2030, 7 November 2016, DOC 54 2137/001

BELSPO Website. Belgian Federal Science Policy Office Earth Observation. Available at: http://www.belspo.be/belspo/index_en.stm

BEOP Website. Belgian Earth Observation, STEREO III. Available at: https://eo.belspo.be/en/ stereo-in-action

BMVIT (2017) Federal Ministry for Transport, Innovation, and Technology, Austrian Technology in Space – an overview of Austrian Space Industry and Research, August 2017

Chalmers Website. Onsala Space Observatory. Available at: https://www.chalmers.se/en/ researchinfrastructure/oso/Pages/default.aspx

DLR (2008) Germany and Switzerland affirm their leading role in satellite communications, 28 November 2008

Dutch Air Force (2017) Air force space command, wideband global SATCOM satellite, 22 March 2017

Dutch Government (2017a) Nederlandse ruimtevaartinnovatie Tropomi gelanceerd, 13 October 2017

Dutch Government (2017b) Rapport 'Inventarisatie Kwetsbaarheid Uitval Satellietnavigatie', 23 August 2017

Dutch Ministry of Defense (2017) Materieelprojectenoverzicht – Prinsesjesdag 2017

Dutch Ministry of Justice and Security (2015) Luchtmacht lanceert Space security center, Nationale Veiligheid en Crisisbeheersing, XIII-5, 2015

ESA (2017) Pooling and sharing for government Satcoms, 21 November 2017

ESA SSCC Website. SSA Space Weather Coordination Centre (SSCC). Available at: http://swe.ssa.
    esa.int/web/guest/service-centre
ETH Zurich Website. Chair of Earth Observation and Remote Sensing – Current Projects. Available
    at: https://eo.ifu.ethz.ch/
EUMETSAT (2018) Annual Report 2018. Available at: https://www.eumetsat.int/website/home/
    AboutUs/Publications/AnnualReport/index.html
Euroconsult (2019) Government Space programs – benchmarks, Profiles & Forecasts to 2028, July
    2019
GPS World (2018) Galileo ground segment keeps constellation on track, 28 March 2018
GSA (2018) Galileo reference Centre now officially open, 18 May 2018
Gutner W, Ploner M (2006) CCD and SLR dual-use of the Zimmerwald tracking system, October
    2006
IRF Website. Swedish Institute of Space Physics (IRF), Research Programmes. Available at: https://
    www.irf.se/en/
Israel Defense (2018) Elbit systems to form joint venture in Switzerland, 15 November 2018
KSAT (2016) Kongsberg satellite services, increased use of satellite monitoring on the Norwegian
    Shelf, 5 October 2016
Lantmäteririet (2015) The Swedish mapping, cadastral and land registration authority, Reports in
    Geodesy and Geographical Information Systems, Geodetic activities in Sweden 2010–2014
MDA (2015) Case study NSO – developing the Netherlands geospatial community, May 2015
Mostlymissiledefense (2012) Space Surveillance Sensors: Globus II Radar June 1, 2012
NASA (2017) Norway to develop Arctic laser-Ranging Station, 07 August 2017
NATO (2019) Defence expenditure of NATO countries (2013–2019) press release, PR/CP(2019)
    123, November 2019
Naval Technology (2009) Royal Netherlands Navy Chooses Thales-Built Satcom Solution,
    12 October 2009
NETPOS Website. Kadaster, The Netherlands Positioning Service NETPOS Available at:
    https://www.kadaster.nl/netpos
NEWTEC (2018) Government and defense COTS SATCOM, March 2018
NLR (2017a) Netherlands Aerospace Centre, Royal Netherlands air Force and NLR strengthen
    partnership for information-driven air force operations, 8 December 2017
NLR (2017b) Netherlands aerospace Centre, GPS P(Y) vs Galileo PRS interference field test,
    presentation, 6 October 2017
Nordic Space (2018) Developing of the Norwegian NorSat-3, 7 March 2018
Norwegian Ministry of Industry and Trade (2012) Between heaven and earth: Norwegian Space
    policy for business and public benefit – Meld. St. 32 (2012–2013) Report to the Storting (White
    Paper)
Norwegian Space Center Website, Norway's satellites. Available at: https://www.romsenter.no/eng/
    Norway-in-Space/Norway-s-Satellites
NSO (2016) Space Policy – Policy Memorandum Space 2016 and NSO advisory report 2016
NSO Website. Netherlands Space Office, National Programs. Sciamachy. Available at: https://www.
    spaceoffice.nl/en/activities/national-programme/national-projects/sciamachy/; Satellite Appli-
    cations, Satellite Data Portal. Available at: https://www.spaceoffice.nl/en/activities/satellite-
    applications/satellite-data-portal/; Geodata for Agriculture and Water (G4AW). Available at:
    https://g4aw.spaceoffice.nl/en/
Organization for Security and Cooperation in Europe – OSCE (2018) Republic of Austria,
    permanent Mission of Austria to the OSCE, exchange of information on the OSCE Code of
    conduct on politico-military aspects of security, 15 April 2015 and 16 April 2018
Proba2 Science Centre Website, About the PROBA2 Science Center. Available at: http://proba2.
    sidc.be/
RIICE Website. Remote sensing-based Information and Insurance for Crops in Emerging econo-
    mies ("RIICE"), About RIICE. Available at: http://www.riice.org/about-riice/about-riice/

ROB Website. Royal Observatory of Belgium, GNSS Research Group, GNSS Networks. Available at: http://www.gps.oma.be/networks_tutorial.php

RUAG (2018) RUAG Space and SWISSto12 partner to supply advanced active antenna solutions for high throughput satellites, 14 September 2018

Sagath D, Papadimitriou A, Adriaensen M, Giannopapa C (2018) Space strategy and governance of ESA small member states. Acta Astronaut 142:112–120

Satellite Today (2015) Denmark Expects MilSatCom capacity Surge to continue, 9 April 2015

SCNAT (2008) Significance of Earth Observation for Switzerland, Report of the Swiss Commission for Remote Sensing SCRS (Schweizerische Kommission für Fernerkundung SKF) Member of the Swiss Academy of Sciences SCNAT 2008

Skyguide Website. Satellite-based navigation. Available at: https://www.skyguide.ch/en/company/innovation/satellite-based-navigation/

Skywin (2015) Wallonie Espace Infos, no. 81, July–August 2015

SpaceNews (2017) Luxembourg eyes earth-observation satellite for military and government, 7 November 2017

SpaceNews (2019) Swiss startup ClearSpace wins ESA contract to deorbit Vega rocket debris, 9 December 2019

SSO (2018) Swiss space office, call for proposals 2018 – to foster and promote Swiss scientific and technological competences related to space activities

Swedish Government (2015) Sweden's defence policy 2016 to 2020, 1 June 2015

Swedish Government (2017) Budget 2018: increased military capabilities and enhanced total defence, 20 September 2017

Swiss Data Cube Website. About Swiss Data Cube Website (SDC). Available at: https://www.swissdatacube.org/

Swisstopo Website. Federal Office of Topography swisstopo, Permanent GNSS Networks. Available at: https://www.swisstopo.admin.ch/en/knowledge-facts/surveying-geodesy/permanent-networks.html

Technopolis (2012) Technopolis Group, Faugert & Co Utvärding AB, Impact evaluation of the Swedish National Space Board's National Earth Observation Programme

Thales Website. Thales Group, Europe, Security in Norway. Available at: https://www.thalesgroup.com/en/countries/europe/norway/security-norway

TU Graz Website. Institute of Communication Networks and satellite Communications – Research Priorities. Available at: https://www.tugraz.at/en/institutes/iks/home/; TU Graz Q/V-Band, TUG, Q/V-band Optimised Transmission system. Available at: https://www.tugraz.at/en/institutes/iks/nachrichtentechnik/aktuelle-projekte/qv-band-optimized-transmission-system/

UN-SPIDER (2014) Austria: earth observation data Centre for water resources monitoring, 7 October 2014

US Strategic Command (2017) Belgium signs agreement to share space services data, 7 February 2017

VITO Website. Flemish Institute for Technological Research (VITO), Remote Sensing technology; Poduct Distribution Portal; VITO, EO data made easily accessible. Available at: https://www.vito-eodata.be/PDF/portal/Application.html#Home

Space and Security Programs in Smaller European Countries

64

Ntorina Antoni, Christina Giannopapa, and Maarten Adriaensen

Contents

N. Antoni (✉)
Eindhoven University of Technology, Eindhoven, The Netherlands
e-mail: ntorina.antoni@gmail.com

C. Giannopapa · M. Adriaensen
European Space Agency (ESA), Paris, France
e-mail: christina.giannopapa@esa.int; maarten.cm.adriaensen@gmail.com

© Springer Nature Switzerland AG 2020
K.-U. Schrogl (ed.), *Handbook of Space Security*,
https://doi.org/10.1007/978-3-030-23210-8_149

Abstract

This chapter presents space and security programs of smaller European countries and indicates their main priorities and trends. In particular, it addresses space activities and programs in the fields of Earth observation (EO), Intelligence-Surveillance-Reconnaissance (ISR), Satellite Communication (SATCOM), Positioning, Navigation, and Timing (PNT), and Space Situational Awareness (SSA). The countries presented are Czech Republic, Denmark, Estonia, Finland, Greece, Hungary, Ireland, Luxemburg, Poland, Portugal, and Romania.

Introduction

The increased need for security in Europe and for Europe's space activities has led to several activities and programs developed in aerospace and defense. The main actors in Europe engaging in space and security activities are the European countries, the European Union (EU), and the European Space Agency (ESA). ► Chap. 61, "Institutional Space Security Programs in Europe" in this handbook describes how security and defense policies are progressively and gradually being supported by space programs of the EU and ESA, while the North Atlantic Treaty Organization (NATO) is also gradually involved. Space security policies of the various European countries are to a large extent determined by national needs and priorities, as explained in ► Chap. 25, "Space and Security Policy in Selected European Countries." In addition, the policies are influenced by the overall space and security governance and their participation in relevant organizations as explained in ► Chap. 23, "Strategic Overview of European Space and Security Governance."

Hence, various space security activities and programs are to a large extent determined by national needs and priorities as well as participation to space and security relevant organizations, including the EU and ESA. The current chapter addresses space and security budgets, activities, and programs of selected European countries in the fields of Earth observation (EO), Intelligence-Surveillance-Reconnaissance (ISR), Satellite Communication (SATCOM), Positioning, Navigation, and Timing (PNT), and Space Situational Awareness (SSA). European countries may be distinguished not only based on their membership to space and security related organizations but also based on their space budget. In the absence of an official grouping of these countries, their ESA annual budget and their defense expenditure as a share of their growth domestic product (GDP) are used in this chapter, to classify them into three groups (Sagath et al. 2018).

The group presented in this chapter includes ESA member states with a GDP up to €350 billion and an annual ESA space budget (2018 ESA budget) typically up to €35 million: Czech Republic, Denmark, Estonia, Finland, Greece, Hungary, Ireland, Luxemburg, Poland, Portugal, and Romania (Sagath et al. 2018). This chapter complements the chapters on the largest and medium-sized European countries presented in this handbook. The content is up to date until January 2019.

Czech Republic

Space and Security Budget

In 2018, the country's defense budget was valued at CZK 59.75 billion (€2.35 billion) or 1.19% of GDP in 2018, below the NATO guideline of 2% of GDP. The national defense budget for 2018 was 10.8% higher compared with 2017 (NATO 2019). The Czech government expenditure for space-related activities amounted to €53.06 million in 2018, resulting from parallel contributions to ESA (€857 million) that doubled compared to 2014 (Euroconsult 2019). The space budget is expected to rise more in the future (Czech Ministry of Transport 2014). No military space budget has been identified by Euroconsult.

Space and Security Program

Czech Republic's space activities include (Ministry of Transport 2016): contributions to the ESA mandatory and optional programs and to the EUMETSAT program; space applications development activities in several sectors, e.g., transport, industry and environment, and resource management; and space-related scientific research at universities and institutes of the Czech Academy of Sciences.

The 2003 launched MIMOSA (Microaccelerometric Measurements of Satellite Accelerations) was a microsatellite of the Czech Republic, designed and developed at the Astronomical Institute/Academy of Sciences (ASU/CAS) of Ondrejov. The project was funded by the Grant Agency of the Czech Republic. The overall mission objective was to obtain total density distributions in space and time of the upper ionosphere by sensitive measurements of the nongravitational orbital perturbations including atmospheric drag and solar radiation pressure (EO Portal Website). Furthermore, VZLUSat-1, a 2 U CubeSat Czech technology nanosatellite of VZLU, developed in cooperation with Czech companies (RITE, HVP Plasma, 5 M, TTS, IST) and universities (CVUT, University of West Bohemia). VZLUSat-1 is still operational (EO Portal Website).

Earth Observation (EO)-Intelligence-Surveillance-Reconnaissance (ISR)

The Czech priorities in Earth observation include development of new geoinformatics products from EO data; application of SAR data for monitoring of infrastructure statics, multispectral, and hyperspectral data for environment applications, land use, land cover, monitoring, natural disasters etc.; cooperation in development of services for downstream GMES/Copernicus market; development of integrated applications using EO data; and development of high-tech optical systems including active and adaptive optics (Czech Ministry of Transport 2016).

The Czech Republic has in place bilateral cooperation with France. In November 2018, the French Space Agency – CNES President underlined the success of bilateral space cooperation between France and the Czech Republic, exemplified by the Taranis mission. The mission aims to study storm phenomena in the atmosphere and plans to launch in 2019. Charles University in Prague has been working closely with France's LPC2E and IRAP research laboratories to deliver the data processing unit and the instrument for detection of energetic electrons for this program. Similarly, Charles University's space physics laboratory is working with IRAP to develop an instrument for Europe's Solar Orbiter mission, which is set to launch in 2020 (CNES 2018).

The Czech Republic is participating to the Copernicus program in the frame of the EU and ESA. The Czech contribution to ESA Earth observation programs amounted to €2.23 million in 2018 (Euroconsult 2019).

Satellite Communications (SATCOM)

The Czech Republic contributes to satellite communication activities through ESA's optional ARTES programs, with a budget of €6.91 million for 2018 (Euroconsult 2019).

The Czech priorities in SATCOM include design of communication terminals; design and manufacture of satellite platform mechanisms; analysis of application of electromagnetic waves propagation models; and development and manufacturing of control and monitoring systems for ground segment (Czech Ministry of Transport 2016).

Positioning, Navigation, and Timing (PNT)

The Czech Republic participates in the optional navigation programs in conjunction with the EU EGNOS/Galileo flagship program. The European GNSS Agency (GSA) seat is in Prague.

The Czech priorities in PNT include monitoring and educational tools for EGNOS/Galileo; design of GNSS receivers and other sensors; GNSS signal processing; Smart GNSS antennas; GNSS only and integrated applications combining EO satnav, satcom, and other technology/data; and analysis of GNSS reliability and security (Czech Ministry of Transport 2016).

The Czech Republic has four complementing GNSS permanent networks: CZEPOS, commercial network of permanent GNSS stations in the Czech Republic operated by the COSMC; GEONAS, geodynamic GNSS network in the Czech Republic operated by the Institute of Rock Structure and Mechanics of the Czech Academy of Sciences; PPGNET, network of permanent GNSS Stations in Greece operated by the Research Institute of Geodesy, Topography, and Cartography, Geodetic Observatory Pecny, and the Charles University in Prague, Faculty of Mathematics and Physics, Department of Geophysics in close cooperation with the

Seismological Laboratory of University of Patras; and VESOG, research and experimental network of permanent GNSS stations in the Czech Republic operated by the Research Institute of Geodesy, Topography, and Cartography, Geodetic Observatory Pecny (Czech GEO Data Portal Website).

The Geodynamic Network of the Academy of Sciences (GEONAS) of the Czech Republic was established in order to make regular geophysical and geodetic observations for current geodynamic studies of the Bohemian Massif and adjacent Central European geological structures. The permanent GNSS stations have been set up, since 2001, by the Institute of Rock Structure and Mechanics, Acad. Sci. (IRSM) within the framework of the national center of Earth Dynamics Research activities. All stations register both NAVSTAR and GLONASS satellite signals. Some of these stations are located inside the regional geodynamic networks for epoch GPS measurements (Schenk et al. 2010).

Space Situational Awareness (SSA)

The Czech Republic contribution to the optional SSA program at ESA amounted to €455 thousand in 2018 (Euroconsult 2019).

The Czech priorities in SSA include (Ministry of Transport 2016): tracking of near-Earth objects; studies, modeling, and monitoring (photometry, follow-up) of asteroids (including NEOs), together with characterization of rotational periods, surfaces of small meteoroidal bodies, and their interactions with the Earth's atmosphere; follow-up astrometry of newly discovered NEOs, recoveries of NEOs, and comets; control and tracking systems for telescopes and ground station antennae; large volume data processing, computing capacities; European Fireball Network (using mostly optical instruments but also including specialized radio antenna for monitoring of daytime meteors) – many stations located in the Czech Republic and abroad; and Ionospheric, Radiation and Interplanetary Space Weather dealing with space plasma physics aimed at the ionosphere and magnetosphere of the Earth (ionospheres and magnetospheres of planets of the solar system) and at the study of solar wind.

The potential use of the high-tech laser center (ELI Beams) in Dolni Brezany has been considered in the frame of A2/AD satellite dazzling and asteroid defense (ESJ News 2018).

Denmark

Space and Security Budget

In 2018, Denmark recorded GDP growth of 1.2%, one of the highest GDPs per capita in Europe (Euroconsult 2019). Denmark had an estimated defense expenditure of DKK 28.78 billion (€3.85 billion) or 1.32% of GDP in 2018, below the NATO guideline of 2% of GDP. 18.1% of the 2018 estimated budget was allocated to equipment expenditure (NATO 2019). The new Defence Agreement for 2018–2023 represents a substantial investment. It aims to increase defense spending with DKK 800 million in 2018 and an increased trend to DKK 4.8 billion more expenditure in 2023. This is an increase in defense spending by more than 20 per cent as well as a significant increase in equipment investment (Danish Ministry of Defence 2017). Denmark does not have a national space budget. Instead it funds its space activities via ESA and the European Organisation for the Exploitation of Meteorological Satellites (EUMETSAT). In 2018 its contribution to ESA was €30.6 million and to EUMETSAT €7.7 million. The space defense budget in the field of telecommunications was €8 million in 2018 (Euroconsult 2019).

Earth Observation (EO)-Intelligence-Surveillance-Reconnaissance (ISR)

The Danish armed forces use Earth observation satellite imagery to provide reconnaissance in support of: detecting oil spills, and enforcing sanctions on environmental polluters, ice breaking and monitoring, and sea search and rescue operations (Danish Ministry of Defence 2016).

The Danish policy paper "Strategy for the Arctic 2011–2020" underlines that security and sovereignty in the arctic space is exercised through the presence of the Danish armed forces. To bolster the reconnaissance capability, this presence provides the Danish Defence Acquisition and Logistics Organization signed in 2016 a public-private partnership with DTU Space and satellite company GomSpace to develop a test surveillance nanosatellite dubbed "Ulloriaq" (means star in Greenlandic) and "GOMX-4A." Launched in February 2018 from the Jiuquan Satellite Launch Center in China (together with GOMX-4B), the project is used to evaluate the usefulness and cost-effectiveness of nanosatellites for monitoring naval vessels and aircraft in the arctic, and for assisting with its search and rescue responsibilities. When the project ends in 2020, it will be clearer whether this type of space equipment can contribute to the task performance of the Danish armed forces. The armed forces are to evaluate the usefulness of such a satellite system in relation to the possibility of obtaining a better "situational picture" of the Arctic region in the future. The idea is that the satellite will contribute to the monitoring of the Danish Ministry of Defence's area of responsibility in the Arctic and form part of the civilian tasks performed by the armed forces in the area including sea rescue (DTU 2018).

GOMX-4 are twin siblings following the successful GOMX-3 mission and represent the new generation of nanosatellite platforms. GOMX-4B is funded by ESA, and it has been designed to be the most advanced CubeSat for IOD. It is based on the innovative and flexible 6 U platform from GomSpace, and it shall demonstrate the operations of six payloads on board. The main payloads are 6 U propulsion modules from NanoSpace, the innovative S-band Inter-Satellite Link (ISL) from GomSpace and the High-Speed Link from GomSpace with high data rate capacities. Additionally, this satellite accommodates the Radiation Hardness Assurance Board, called Chimera, developed by ESA to evaluate the behavior in space of different ceramic memories and two new optical devices, the HyperScout hyperspectral camera from Cosine and a star tracker developed by ISIS. Even with different payloads and mission goals, the two satellites will work together using an Inter-Satellite Link to optimize their capabilities to share data and to transmit it to ground. The life expectancy of the GOMX-4 nanosatellites is 3–5 years, and they are expected to be fully operated by GomSpace during that period (DTU 2018).

GomSpace is a Danish-Swedish nanosatellite manufacturer. The company's business operations are mainly conducted through the wholly-owned Danish sub-sidiary, GomSpace A/S, with operational office in Aalborg, Denmark (Satnews 2017). Beyond GOMX-4A, Denmark possesses no independent Earth observation capabilities. It has to rely on imagery procured from commercial providers, such as Inmarsat, Iridium, and Eutelsat, and international partners to support its tasks (SatelliteToday 2015). Nonetheless, in addition to the arctic strategy, the country's "Defence Agreement 2018-2023," "Space Strategy 2016," and "Agreement on the future missions of the Danish Ministry of Defence," all recognize that heightened economic and military activity in arctic space will necessitate the increased presence of the Danish armed forces in the region and consequently further investments in EO capabilities. As such, there is a likelihood that helicopter and drone-based surveil-lance will gradually be replaced with high-resolution satellite imagery (DTU 2018).

Denmark operates three unarmed maritime patrol aircraft over the Baltic Sea and off Greenland. The 2012 Defence Agreement includes substantial funds for testing different additional surveillance options for the Arctic, including UAVs and the use of existing satellites (SIPRI 2016).

In addition to ESA, EUMETSAT, and the EU, Danish public authorities also take part in other international cooperation collaborations. Through the Arctic Council, Danish authorities are involved in various projects concerning infrastructure in space and the use of satellite data. The Ministry of Defence works together with Canada on the exchange of surveillance data (Danish Government 2016). Denmark's position for sharing and pooling of EO is mainly centered around the bilateral approach, implemented in cooperation agreements with direct partners (mainly Scandinavian Countries, Canada, France, and the US in addition to others).

Denmark is participating to the Copernicus program in the frame of the EU and ESA. Denmark's contribution to ESA's optional Earth observation programs amounted to €4.2 million for 2018 (Euroconsult 2019).

Satellite Communications (SATCOM)

Sharing of national satcom assets is facilitated through NATO, while technology development is facilitated through ESA. Denmark took part in satellite communica-tion activities through ESA's optional ARTES Programs, with a budget of €1.19 million for 2018 (Euroconsult 2019).

The Danish company DataPath Inc. was awarded a €6.5 million contract in 2015 to provide five Wideband Global Satcom (WGS) terminals with related equipment and training for the Danish Air Force and 29 Wideband Global Satcom (WGS) terminals with related equipment and training for the Danish Army (SIGNAL 2015).

Moreover, the Wideband Global Satcom-9 (WGS), part of the US-led WGS system is funded jointly by Canada, Denmark, Luxembourg, the Netherlands, and New Zealand. The system was launched in March 2018. The contribution gives the countries access to the US-controlled WGS military communications system (ViaSatellite 2016). In this regard in May of 2016, the SES S.A. announced that it will provide two anchor stations for the Danish Defence Acquisition and Logistics Organization (DALO). Under the agreement, SES Techcom Services, a wholly-owned subsidiary of SES, will provide and maintain two WGS system anchor stations – one in X-band and one in Ka-band. This will enable the Danish armed forces to communicate through the system, which provides flexible, high-capacity communications for defense operations through the associated satellite constellation and control systems. SES was awarded the contracts based on its experience in providing satellite communication anchor stations, the associated WGS certification process and overall life-cycle cost criteria. The Danish forces will join other nations partnering with the US in the WGS program and thus offer the US State Department satellite-based communication services to users, including marines, soldiers, sailors, airmen, and the White House Communications Agency (SES 2016).

Positioning, Navigation, and Timing (PNT)

The Danish Ministry of Defence has access to the civil and military signals of the US GPS system. GPS is evolving, with the introduction of GPS III satellites and the modernization of the ground segment.

The Division for Geodesy at DTU Space researches the maintenance and development of geodetic infrastructure and development of new techniques for surveying and mapping. Research in the field of geodetic infrastructure is carried out as a basis for spatial infrastructure, including surveying and mapping. Research in the field of positioning is being carried out with the aim of developing new techniques for surveying and navigation purposes. Furthermore, it also makes it possible to monitor the integrity of satellite systems and detect problems. A major effort is also being put into the detection of ice load changes in Greenland based on permanent GPS combined with campaign measurements (DTU Space Website).

Denmark participates in the optional navigation programs in conjunction with the EU EGNOS/Galileo flagship program.

Space Situational Awareness (SSA)

Activities related to the Earth's electromagnetic environment are carried out in Denmark mainly at DMI and DTU Space. DTU Space has been involved in space weather research FP7 Coronal Mass Ejections and Solar Energetic Particles (COMESEP). Additionally, SSA activities are pursued within the ESA framework. Moreover, The Danish Ministry of Defence and US Strategic Command

(USSTRATCOM) signed a data sharing agreement in April 2018 to share space situational awareness (SSA) services and information (SpaceNews 2018). Further, Denmark contributes to the optional SSA program at ESA, with the amount of €0.24 million in 2018 (Euroconsult 2019).

Estonia

Space and Security Budget

Estonia's 2018 defense budget remains one of the most balanced in Europe. According to the NATO defense budget methodology, which differs somewhat from the national methodology, personnel expenditures account for 30%, procurement and investments 27%, and other costs 43% of defense expenditures. An additional €15 million is added to Estonia's 2018 defense budget from the NATO Security Investment Program (NSIP), the majority of which is intended for the training of the NATO Very High Readiness Joint Task Force (VJTF) at the central training area and the hosting of units during a period of crisis. Therefore, a total of more than €586 million, representing 2.11% of the forecast GDP, is available for spending within the area of government of the Ministry of Defence. It is divided in 21% of defense expenditures in 2019, with infrastructure investments and procurement costs accounting for 43% and all other spending 36% (NATO 2019, Estonian Ministry of Defence 2019). The Estonian government expenditures for space-related activities amounted to €6.7 million in 2018, which has doubled compared to 2014 (Euroconsult 2019). No military space budget has been identified by Euroconsult.

Space and Security Program

Estonia sent its first nanosatellite ESTCube-1 into space (€100 thousand cost) in May 2013 after 6 years of development by over a hundred students and scientists. The aim of ESTCube-1 was to popularize science and engineering among students. It served as the basis for 48 research projects, 5 doctoral theses and has generated 6 spin-off companies to date. The first satellite is followed by ESTCube-2 and TTU100 projects. In 2014, Tallinn University of Technology (TUT) established the Mektory Space Centre, where more than 15 academic supervisors and 40 students from various disciplines are involved in the nanosatellite program (Investinestonia 2017).

Earth Observation (EO)-Intelligence-Surveillance-Reconnaissance (ISR)

The Estonian Land Board is responsible for the national mirror site (ESTHub) for providing free access to Sentinel satellite data. ESTHub will collect and archive datasets that cover the Estonian territory and a buffer zone of about 200 km. These datasets will be freely available and downloadable for anyone, but the required notices on the data source must be provided (Estonian Land Board Website).

The Mektory Space Centre is working on the TTÜ100 satellite project. The main mission of the TTÜ100 satellite is Earth observation and demonstration of Earth observation technologies. The satellite includes cameras, image processing, and communication with ground station. The cameras include the RGB sensor for visual light image and the NIR sensor for near-infrared image that can be used for assessing vegetation growth, climate, geology, and sea conditions (TTU Website).

The Estonian Marine Institute Department of Remote Sensing and Marine Optics mainly focuses on developing optically complex remote sensing methods, suitable for coastal sea and inland waters, and applying those methods. The department also studies the relationship between underwater light field and wildlife. Specific research fields are, for example, identifying the flowering of potential toxic phytoplankton and its quantitative sensing with remote sensing methods, mapping of phytobenthos, and the depth of shallow waters using satellites and sensors, which are located on the planes, and researching the global carbon cycle. Furthermore, the department concentrates on developing primary production models and remote sensing methods for evaluating the quality of the water in Baltic Sea and in lakes (Estonian Marine Institute Website).

Additionally, Estonia is involved in the HYPERNETS Project. The HYPERNETS project is coordinated by the Royal Belgian Institute of Natural Sciences. Besides Estonia, there are project partners from six countries: France, Argentina, Italy, Germany, and the UK. In the project which includes Tartu Observatory as a partner, an instrument for the support measurements of optical remote sensing satellites is created. The support measurement instruments will form a worldwide remote sensing network with unified data processing and management (University of Tartu 2018).

Technology development is facilitated through ESA. Estonia participates in the Copernicus program in the frame of the EU and ESA. Estonia's contribution to ESA Earth observation programs covered a budget of €590 thousand in 2018 (Euroconsult 2019).

Satellite Communications (SATCOM)

Estonia is also a member of EDA. Estonia joined the EDA EU Satcom Market in January 2017. The EDA EU Satcom Market since its start in 2012 under the name EU SatCom Procurement Cell (ESCPC). The EDA EU Satcom Market today comprises 28 contributing members including Estonia (EDA Website). In addition, Estonia participates in the EDA Project Team on Satellite Communications and is part of the Ad Hoc Working Group preparing the establishment of the EDA GOVSATCOM Demonstration Project (pooling and sharing). Further, technology development is also facilitated via NATO. Estonia is a member of NATO since 2004. Estonia's primary defense capability forms a military force that supports the activation of NATO's collective defense mechanism. Estonia does not develop all its military capabilities independently, instead as a NATO member, the Estonian Defence Forces can develop capabilities in cooperation with NATO allies (Estonian Ministry of Defence).

Positioning, Navigation, and Timing (PNT)

ESTPOS is the Estonian GNSS-RTK permanent stations network consisting of continuously operating reference stations. The first GNSS (Global Navigation Satellite System) CORS (Continuously Operating Reference Station) in Estonia became operational in 1996 in Suurupi, established by ELB in cooperation with the Finnish Geodetic Institute (FGI). Starting from 2008, four Estonian CORS are incorporated into the EPN (EUREF Permanent GNSS Network). A few years later more Estonian GNSS CORS were established. In 2008 new CORS were set up in Kärdla (KARD), Mustvee (MVEE), and Võru (VOR2, initially denoted VORU) and outdated equipment in MUS2 was replaced. Currently the total number of the resulting ESTPOS reference stations is 28. These ESTPOS stations were interconnected to the national geodetic network by a special GNSS campaign in 2017. Modernized ESTPOS is an important part of Estonian geodetic reference system (Metsar et al. 2018).

The Ministry of Defence makes use of the US Global Positioning System (GPS), which provides position, velocity, and time information to an unlimited number of users, through civilian and encrypted military modes. The Ministry of Defence relies on the GPS system until the full operational capacity of Galileo (including PRS). The dependency on GPS is a logical consequence of NATO membership and the fact that GPS is the only globally available GNSS network. In the future, in all EU and NATO framed operations, military GNSS receivers will simultaneously use Galileo and GPS, increasing accuracy and resilience (Muls 2016).

Estonia cooperated with Switzerland on the construction of a GNSS-RTK permanent station network ESTPOS from May 2014 until September 2015. In addition, cooperation with Finland and Latvia is envisaged (Estonian Land Board 2015).

Space Situational Awareness (SSA)

In the context of SSA, the event-timing devices being developed by Eventech are regarded of good quality and have a promising market both for SSA and in-orbit applications. According to a recent study, on Latvia's participation in the European Space Agency since 2015, the event timers could potentially be considered for developing further the SSA program but also under the GSTP programs of ESA (Invent Baltics OÜ 2019). The Institute of Astronomy of the University of Latvia operates the station RIGL 1884, part of the International Laser Ranging Service105 managed by NASA. This expertise, along with the leading Latvian technology in event-timing devices, could be further explored (Invent Baltics OÜ 2019).

Finland

Space and Security Budget

Finland had an estimated defense expenditure of €2.8 million or 1.23% of GDP in 2017. 19,9% of total defense budget goes to procurement of materiel and 10.8% on real estate expenditure. Finland's Ministry of Defence proposed a €3.2 billion defense budget for 2019, a €326 million, or 11%, increase from 2018. The ministry said on 9 August 2018, that this would increase the country's defense budget's share of GDP from 1.23% in 2018 to 1.32% in 2019. The Ministry of Defence attributed much of the increase –€260 million – to spending on the Laivue (Squadron) 2020 program in order to build a new class of four multipurpose offshore patrol vessels. Nearly €1.2 billion, or 43.6%, of the budget will be allocated to equipment readiness. Defense procurement contracts account for €771 million (Finnish Ministry of Defence 2018).

During 2013–2020, Finland set to invest €400 million in ESA's and European Commission's space programs. Finland's contribution to the ESA budget in 2018 was €20 million (0.6%) which is less than half of the share corresponding to the Finnish GDP (1.40%). The Finnish public sector funding to space activities is €50 million including the payments to ESA and EUMETSAT. The Ministry of Economic Affairs and Employment proposed budget authorizations for the grants of Business Finland, the Finnish Funding Agency for Innovation, of slightly more than €344 million, which is almost €74 million more than in the approved 2018 budget figure. Slightly more than €102 million is proposed as Business Finland's operating costs, which is an increase of more than €18 million compared with the Budget for 2018. The space budget identified by Euroconsult does not include space-related military expenses (Finnish Ministry of Economic Affairs 2018; Euroconsult 2019).

Space and Security Program

The main space activities in Finland include space science, Earth observation, satellite positioning and satellite telecommunications, and the satellite equipment industry. Each area has basic research, applied research, and business and applications exploiting space technology for the needs of the citizens, for the public sector, and for commercial purposes. The foundation of Finnish space activities is the participation to European organizations. ESA's programs are the backbone of the Finnish space activities, which are also influenced by Finland's membership in EUMETSAT, ESO and European Commission's space activities such as Galileo, Copernicus, and the space activities in Horizon 2020 (Finnish Ministry of Economic Affairs and Employment 2013).

There are about 90 operators in Finland that are engaged in the development of space technology applications. There are dozens of satellite navigation companies in Finland including: u-Blox (positioning electronic circuits), HERE (maps), Reaktor (small satellites), Iceye (small satellites), Mobisoft, Aplicom and Paetronics (vehicle management), Beaconsim (simulators), Alpha, Positron, and SSF (pseudolites, which can replace satellite signals in such places as container ports and open-cast mines), Suunto and Sports Tracking Technologies (sports), Tracker (tracking of hunting dogs), and Vaisala (measurement equipment and systems). The combined turnover of the Finnish companies producing satellite navigation systems and equipment is at least €300 million (Finnish Ministry of Transport and Communications 2018). Additionally, Business Finland's New Space Economy program offers funding, networks, and export services for developing international space-related business. The New Space Economy program funds startup companies that are reforming the sector, growth-seeking manufacturing companies, and businesses focused on data utilization. The ESA Business Incubation Centre Finland supports this objective (Business Finland Website).

On 18 April 18 2017, Aalto-2 became the first Finnish-built satellite to be launched into space from Cape Canaveral, Florida. Then on June 23, Finland's Aalto-1 carried the world's smallest hyperspectral imager into space on the Polar Satellite Launch Vehicle sent up by the Indian Space Research Organization. The Aalto-1 and Aalto-2 missions have ignited the rise astropreneurship and the establishment of a NewSpace sector in Finland. Independent space companies, the first space law, and a Finnish space program are set to reshape traditional technologies, develop faster and cheaper access to space than ever before, and advance earth observations far beyond today's satellite capabilities. The first satellite, Aalto-1, had two initial goals: (1) a technology demonstration of the state-of-the-art payload and (2) serve as a learning curve in the operation and management of space missions for future launches (Science Business 2017).

Earth Observation (EO)-Intelligence-Surveillance-Reconnaissance (ISR)

The Finnish startup ICEYE plans an 18-satellite constellation of SAR (synthetic-aperture radar) equipped microsatellites. The ICEYE imaging radar instrument can image through clouds, obscuring weather, and darkness. In 2017, ICEYE secured USD 2.8 million in R&D funding from SME Instrument within EU Horizon 2020 as well as USD 2.8 million financing from private entities. Additional funding was received from the Finnish Funding Agency for Innovations. ICEYE-X1 is ICEYE's first proof-of-concept microsatellite mission with a SAR sensor as its payload. It was launched on January 12, 2018, on ISRO's PSLV-C40 rocket from Satish Dhawan Space Center in India. ICEYE-X1 is also the world's first SAR satellite in this size (under 100 kg), enabling radar imaging of the Earth through clouds and even in total darkness. The satellite is the very first Finnish commercial satellite. Following the January 2018 launch of ICEYE-X1, ICEYE, the two newest SAR satellites were launched in July 2019 (ICEYE 2019, ICEYE Website).

Additionally, the Finnish Meteorological Institute (FMI) Earth Observation Research unit focuses on remote sensing of the atmospheric composition and the cryosphere as well as the development of retrieval methods. The cryospheric processes group investigates the use of Earth observation data in understanding the water and carbon cycles in the Arctic, northern tundra, and boreal forest regions. The group has been involved in research projects funded by the European Space Agency focusing on Polar regions and cryosphere products (FMI Website). In April 2018, Finland and China signed an agreement to establish a joint research center for Arctic space observation and data sharing services. The China CAS and Finnish FMI will jointly build a research center in Sodankyla, north Finland's Lapland. The center will enhance cooperation on cryosphere research with satellites, which will provide information from the Arctic region for use in climate research, environmental monitoring, and operational activities, such as navigation in the Arctic Ocean. The two countries agreed to build the center as a platform for international cooperation in research on the Arctic region, and a model of Sino-European space-based Earth observation application cooperation. In 2016, China Remote Sensing Satellite North Polar Ground Station (CNPGS) was built in Sweden. The agreement is the latest move of China's Digital Belt and Road Program, which was initiated in 2016 to improve environmental monitoring and promote data sharing (CAS 2018).

During the last India-Nordic Summit in April 2018, the Finnish Prime Minister highlighted the opportunities for cooperation between our two countries that exist especially in sectors such as energy, satellites, and education. Finland and India have agreed to deepen cooperation in these fields and will hold further negotiations on concrete projects (Finnish Government 2018). In addition, in the frame of the Finnish Presidency of the Arctic Council and the meteorological cooperation program, FMI considers themes that would include monitoring of the Arctic, especially the increased utilization of satellite date in operational activities and services; supporting research that aims to increase an understanding of the Arctic environment; and launching of new service concepts to support Arctic functions and increase safety in the Arctic. Moreover, in July 2017, the Finnish and Italian Ministries of Defence signed a framework agreement on cooperation regarding the COSMO-SkyMed Second Generation radar imaging satellite system (Finnish Ministry of Defence 2017).

Finland participates in the Copernicus program in the frame of the EU and ESA. Finland's contribution to ESA Earth observation programs covered a budget of €5.1 million in 2018 (Euroconsult 2019).

Satellite Communications (SATCOM)

Norsat International Inc. has provided satellite communication equipment and services for the Finnish Defence Forces since in 2009. Following the success of the network used during an European peacekeeping mission in Chad, the armed forces requested that Norsat assist them in expanding their communication network to support an ongoing mission in Afghanistan. In this field, Norsat's rugged, portable satellite terminals consistently provide communication links in the extreme environmental conditions of desert deployments, wet conditions, and cold Northern winters (Norsat 2012).

Moreover, the Finnish 5GKIRI project aims to support the construction of 5G networks through more agile and nationwide processes. The participating cities are

Espoo, Helsinki, Jyväskylä, Kuopio, Lahti, Oulu, Turku, and Vantaa, and the project is being coordinated by Sitowise Oy. Commercial operations on 5G networks can begin at the start of 2019 (FICORA 2018).

Further, Finland's contribution to satellite communication activities through ESA's optional ARTES programs amounted to €1.4 million for 2018 (Euroconsult 2019).

Positioning, Navigation, and Timing (PNT)

Finland's objectives and priorities regarding PNT and GNSS relate to the specific geographic position of the country. Satellite navigation programs and their associated support systems do not perform optimally at high latitudes and especially in the Arctic region in particular in determining vertical positioning. This is due to the satellite constellation, i.e., the way satellites are situated on their orbits. However, Galileo provides a somewhat better coverage in higher latitudes than GPS, whereas GLONASS provides the best possible coverage due to its highest inclination angle. There are problems with geostationary navigation systems (EGNOS) in Northern Europe, because the satellites that send support signals are poorly visible in northern areas, and the good reception is not possible. Additionally, resilience against potential inference and jamming of signals is a priority (Ministry of Transport and Communications 2018).

The Ministry of Transport and Communications has set out a draft operational program on satellite navigation. The program describes the current state of satellite navigation systems and how they are deployed in different sectors, especially in automated transport. Finland aims to develop automated maritime transport in line with their maritime strategy and policy (Finnish Ministry of Transport and Communications 2018).

The National Land Survey of Finland (NLS) operates a nationwide GNSS network of 20 stations. The NLS Finnish Geospatial Research Institute performs research on Arctic navigation including ice-aware maritime route optimization using satellite imagery (VORIC, STORMWINDS, and ESABALT projects), enhanced situational awareness for maritime operation, and developing methods for autonomous driving in arctic conditions and contributing to Aurora Snowbox, a test area for autonomous driving in Northern Finland (NLS 2017).

Finland participates in the optional navigation programs in conjunction with the EU EGNOS/Galileo flagship program.

Space Situational Awareness (SSA)

The Finnish institutes involved in space weather activities include universities (Helsinki, Oulu, Turku, and Aalto), research institutes (FMI, Finnish National Land Survey), and companies (e.g., Isaware, ASRO, KNL Networks, and RF Shamans). Finnish activities include projects in operational services and research, and missions for new measurement and space transportation technologies. Besides its national service, Finland is active partner also in the operational services by the

Space Situational Awareness (SSA) program of ESA. Finnish contributions are included in the Expert Service Centers on Space Radiation, Geomagnetic activity, and Ionospheric weather. Finland participates in the optional ESA SSA program with an amount of €324 thousand in 2018 (Euroconsult 2019).

Greece

Space and Security Budget

Greece had an estimated defense expenditure of €4.5 billion or 2.28% of GDP in 2018, well above the NATO guideline of 2% of GDP (NATO 2019). Greek government expenditure on space amounted to €19.8 million in 2018. Greece contributed 10.3 million to ESA and €7.5 million to EUMETSAT in 2018 (Euroconsult 2019). Greece announced in 2019 the increase in its contribution to the ESA optional programs from 8.5 million to 33 million for the next 3 years (Infocom.gr 2020). No military space budget has been identified by Euroconsult.

Earth Observation (EO)-Intelligence-Surveillance-Reconnaissance (ISR)

The National Observatory of Athens has an Institute for Astronomy, Astrophysics, Space Applications, and Remote Sensing working on a variety of nationally, EU and ESA funded projects, including Earth observation applications (National Observatory Website). Greece participates in collaborative programs at regional (i.e. Mediterranean, Southeast Europe), ESA, and EU level. There is a strong interest and potential of the Greek Earth Observation community toward downstream services and applications development including security aspects. In the context of ISR, the Greek Ministry of National Defence has recently expressed its intent to lease from Israel seven Heron medium-altitude, long-endurance (MALE) unmanned aerial vehicles (UAVs) for 3 years. The systems will be used to monitor refugee flows in the country (C4Defense 2018).

Since 2007 Greece has participated in the French Helios 2 program. In addition, Greece is a member of the French program Composante Spatiale Optique (CSO). Since 1999 the governments of Belgium, France, Germany, Greece, Italy, and Spain have been working on an agreement called the Common Operational Requirements for Global European Earth Observation System by Satellites – more commonly known by its French acronym BOC. The aim was to define common requirements for military or dual-use Earth observation systems in the visible infrared and radar domains (Schrogl et al. 2015).

Greece is participating to the Copernicus program in the frame of the EU and ESA. Greece participates in ESA's optional Earth observation programs. Additionally, in March 2019 Greece and ESA agreed on the development of a national program on Greek Earth Observation Microsatellites for 5 years (Mononews 2019).

Satellite Communications (SATCOM)

Since 2019, Greece together with Luxembourg are the only countries, apart from the largest five countries in Europe, that they have their own operational governmental satellite communications system. The Greek system is referenced as GreeCom or KLEIDDI (key in Greek) operated by the Ministry of National Defence in cooperation with the General Secretariat of Telecommunications and Post of the Ministry of Digital Governance responsible for civilian space matters (TaNea 2019). The civilian part is installed in all high-level governmental offices, parliament, civil protection, police, and embassies outside Greece. It is based on the exploitation of Greek frequencies and orbital position of the Hellas Sat satellites. These are operated by Hellas Sat which is a satellite communications solutions provider offering coverage over Europe, Middle East, and Southern Africa. It has operated since 2001 upon licenses provided by Greece and Cyprus, and the Greek license was renewed in 2017 (Euractiv 2019). Hellas Sat 2 was launched in May 2003; Hellas Sat 3 was launched in June 2017; and Hellas Sat 4 was launched in February 2019. Further, Greece engages with the EU, EDA, and ESA ARTES in view of the European Governmental Satellite Communications (GOVSATCOM).

Positioning, Navigation, and Timing (PNT)

Greece participates in the EGNOS Ground Segment, which comprises a network of Ranging Integrity Monitoring Stations (RIMS), two Mission Control Centers (MCC), and two Navigation Land Earth Stations (NLES) per GEO satellite. The main function of the RIMS is to collect measurements from GPS satellites and to transmit these raw data every second to the Central Processing Facilities (CPF) of each MCC. The initial configuration included 34 RIMS sites located over a wide geographical area. In order to improve the performance of the EGNOS system and enlarge the area where the EGNOS services can be used, additional RIMS were deployed in Athens, Greece, as well as in Spain and Egypt (ESA Navipedia).

Space Situational Awareness (SSA)

The National Observatory of Athens has an Institute for Astronomy, Astrophysics, Space Applications and Remote Sensing working on a variety of national, EU and ESA funded projects, including space weather (National Observatory Website).

The Arichtarchos telescope has been used in the frame of near-Earth object (NEO) activities analyzing lunar impact. NELIOTA is an activity initiated by ESA, which was launched in February 2015 at the National Observatory of Athens and was running until November 2018. The project determined the frequency and distribution of NEOs by monitoring the non-illuminated side of the Moon for flashes caused by NEO impacts, which result in the formation of a crater on the surface of the Moon. NELIOTA will help assess the threat of small NEO collisions to orbiting spacecraft and to future ESA Moon missions (National Observatory Website).

Greece participates in the ESA SSA program.

Hungary

Space and Security Budget

The 2018 defense budget of Hungary was valued at HUF 484 billion. The country had an estimated defense expenditure of 1.21% of GDP in 2018 below the NATO guideline of 2% of GDP (NATO 2019). Euroconsult estimates the Hungarian 2018 space budget at €10.5 million, with €6.2 million contribution to ESA and €3.5 million to EUMETSAT (Euroconsult 2019). No military space budget has been identified by Euroconsult.

Space and Security Program

Masat-1, the first Hungarian satellite, captured the first satellite space photographs on March 8, 2012 over the southern section of the African continent. Masat-1 reentered the atmosphere in January 2015 (Innoteka 2018). The 3-year-long operations of MaSat-1, the Hungarian contribution to the Rosetta mission, and the accession to the European Space Agency have laid down the foundations for the future of Hungarian space activities (Hungarian Space Office – HSO 2016).

In June 2016, Hungary announced the work on two additional satellites, RadCube and Smog-1, to be launched in 2020. The new space laboratory established at the Hungarian Academy of Sciences Research Centre (MTA EK) is instrumental for the testing and validation of the new satellites (Innoteka 2018).

Earth Observation (EO)-Intelligence-Surveillance-Reconnaissance (ISR)

Within the Government Office of Budapest, the Earth Surveying and Remote Sensing and Land Office in the Cosmic Geodesy Department is the central surveying and mapping organization in Hungary. The activities of the Institute include the maintenance of national data bases, professional information systems, data processing, research and development in the fields of surveying, mapping, and remote sensing. The Department's Satellite Geodetic Observatory (SGO) is an internationally recognized center of satellite geodesy. It is active in basic research (geodynamics, meteorology), applications, development and services related to Global Navigation Satellite Systems (GNSS), as well as interferometric synthetic-aperture radar (InSAR) research and its applications (HSO Department of Geodesy).

A combined national multi-technique (SAR, GNSS, leveling) geodetic network was established with the support of ESA in 2015. They have been studying various scientific and practical applications of SAR data from the Sentinel–1 satellites of the EU-ESA Copernicus program. The Department is playing a key role in national and European land monitoring activities (HSO Department of Geodesy).

Satellite Communications (SATCOM)

The BME Department of Broadband Infocommunications and Electromagnetic Theory has participated to the ESA Alphasat programme with a propagation and a communication experiment. It developed the power subsystem and Langmuir-probe experiment for the ESEO satellite. The coordination of the Masat-1 project, furthermore to the development of the satellite's onboard communications system, and the establishment and operation of the ground control station were major achievements of the Department. The Department also participates to the radar image processing of the Sentinel–1A satellite (HSO BME).

Hungary became a founding member of the Intersputnik International Organization of Space Communications in 1971. Intersputnik has 26 Member States. The organization serves the purpose of the co-development of space-based telecommunications capabilities.

Technology development in Hungary is facilitated through ESA.

Positioning, Navigation, and Timing (PNT)

The current national permanent GNSS network "GNSSnet.hu" consists of 54 permanent stations: 35 in Hungary and 19 in adjacent countries through cross-border data exchange. The Cosmic Geodesy Observatory has created and maintains this GNSS network for GNSSnet.hu, capable of performing authentic geodetic tasks. The stations' data are continuously transmitted to the GNSS service provider in real time, where a central software balances the measurements together to produce the corrections necessary to make field measurements more accurate (GNSSnet.hu Website).

Space Situational Awareness (SSA)

The research group at the Eötvös Loránd University, focuses on satellite remote sensing, research related to Earth's magnetosphere and involvement in international scientific projects. In the field of space physics, and in particular space weather research, the group focuses mainly on the theory of electromagnetic wave propagation in magnetized plasmas and its practical applications (HSO, Eötvös).

Ireland

Space and Security Budget

Defense spending increased in the 2019 Irish budget by €47.5 million to almost €1 billion, though most of the €994 million was absorbed by the pay bill. Capital spending jumped from €29 million to €106 million (Irish Examiner 2018a). Euroconsult estimates the Irish 2018 national space budget at €17.4 million ESA contribution and €6.5 million EUMETSAT contribution (Euroconsult 2019). No military space budget has been identified by Euroconsult.

Space and Security Program

Ireland's first satellite, the EIRSAT-1 completed the Critical Design Review phase of the European Space Agency's program in September 2018. The successful completion of this first stage of the Fly Your Satellite (FYS) program is regarded as an important milestone in the project, with the satellite currently on track to be completed within the next 2 years. EIRSAT-1 is led by a team of students from UCD and funded by the Irish Research Council, Science Foundation Ireland and the European Space Agency. The CubeSat will be launched from the International Space Station, where it will join a multitude of other objects in Earth's orbit. As it stands the EIRSAT-1, team expect the satellite to be delivered to the ESA by mid-2020. After launch the satellite will operate for between 6 and 12 months in orbit, communicating data from space to the mission control in UCD (Irish Examiner 2018b).

The ESA Space Solutions Centre Ireland and its ESA Business Incubation Centres (BIC) are one of 16 ESA BICs developed to create viable businesses and new jobs (Tyndall Website).

Earth Observation (EO)-Intelligence-Surveillance-Reconnaissance (ISR)

Engagement with space-based EO has primarily been within specific interest areas, such as the work of Met Éireann with ESA and EUMETSAT and research links with the international satellite observations community. Ireland's strategic location on the western boundary of Europe and its relatively clean environment have been factors in attracting research interest and links to the international satellite analysis community. Long-term operational use of satellites has been confined to the area of meteorology, with only a limited appreciation of how they can be used in other areas. Consequently, focused investment in promoting education, research, development, and operational use of EO has not occurred, while there has been a major investment in other sensor development areas and associated information and communications technology (ICT). This has resulted in inertia among the potential user communities, which prefer to rely on conventional methodologies and data streams for core activities, and a dearth of a significant well-trained and informed research community in this area (EPA 2011).

Moreover, the National Centre for Geocomputation (NCG) was established in 2004 at the Maynooth University. The center covers a diverse range of research areas including Earth observation; mobile mapping systems, including unmanned aircraft systems (UAS); geospatial modeling; spatial statistics; geovisualization; interoperability; and cloud-based architectures (Maynooth University 2014).

Additionally, Ireland is participating to the Copernicus program in the frame of the EU and ESA. Ireland participates in ESA Earth observation programs with a budget of €1.06 million in 2018 (Euroconsult 2019). Ireland signed a Technical Collaborative Agreement with ESA in October 2017 regarding the access to Copernicus data. This arrangement provides Ireland with unprecedented access to the Copernicus program's near-real-time Earth observation information. The data will be stored locally for research, commercial development and policy informing purposes (Irish Government 2017).

Satellite Communications (SATCOM)

The US company Viasat announced in April 2018 the opening of its expanded Dublin office, with nearly 100 team members located in Dublin primarily developing innovative software solutions for the commercial aviation industry. Viasat expects to more than double its headcount in Dublin over the next few years and extend development beyond connected aircraft software to include broader software and mobile application support for international maritime customers, European residential broadband and Wi-Fi markets, government systems as well as support for Viasat's next-generation ultra-high capacity satellite platform known as ViaSat-3 (DBEI 2018).

Moreover, Ireland participates in satellite communication activities through ESA's optional ARTES programs, with a budget of €1.35 million in 2018.

Positioning, Navigation, and Timing (PNT)

The active GNSS network for Ireland is the result of collaboration between Ordnance Survey Ireland (OSi) and the Ordnance Survey of Northern Ireland (OSNI). The Irish National Grid (ING) is realized through a continuously operating reference system (CORS) network of active GNSS systems consisting of the 17 OSi network stations and 6 OSNI Stations (FIG 2013).

Further, Ireland participates in the optional navigation programs in conjunction with the EU EGNOS/Galileo flagship program.

Space Situational Awareness (SSA)

The Rosse Solar-Terrestrial Observatory (RSTO) at the Dublin Institute for Advanced Studies (DIAS) has three eCallisto solar radio burst monitors operating at 10–400 MHz, a 4-element LOFAR/LBA test array, ionospheric monitors, and a magnetometer operated by DIAS Geophysics (DIAS Website).

Luxembourg

Space and Security Budget

Luxembourg defense expenditure totaled €301 million or 0.56% of GDP in 2018, below the NATO guideline of 2% of GDP. 45.1% of the 2018 estimated budget was allocated to equipment expenditure (NATO 2019). The Luxembourg total space expenditure amounted to €78 million in 2018 of which €49 million covered civil activities and €29 million defense. The amount has significantly increased from €53 million in 2017 (Euroconsult 2019). Since Luxembourg's adhesion to ESA in 2005, its annual contribution has increased €3.3million to €26.6 million in 2018. Initially, Luxembourg's military space expenditure covered entirely its contribution to the US WGS military communication program, but later it increased due to the funding of the SES16/GovSat with investment by both SES and the government (Euroconsult 2019). The Euroconsult numbers may not fully reflect the expenditure by Luxembourg on Govsat-1, the National Advanced Optical System (NAOS) and WGS.

Earth Observation (EO)-Intelligence-Surveillance-Reconnaissance (ISR)

After deciding on the establishment of a national GOVSATCOM system, the next priority for Luxembourg was the establishment of a national Earth observation system, in line with the ambitious defense strategy implementation by 2025. Luxembourg wanted governmental system (not military) with first goal to have national capacity for imagery and at the same time be provider for EU, NATO, and partner countries (role of image provider at governmental level (Security Luxembourg 2018).

The National Advanced Optical System – NAOS – aims to be a single satellite system operating in LEO SSO Polar orbit, equipped with a panchromatic and hyperspectral camera payload. Two ground stations will provide data upload and download. Two TT&C stations in Luxembourg will be deployed. Luxembourg is making an approved client list (including EU, NATO, and other organizations and nations) and partners will be able to program tasks. Expected operational lifetime is 10 years (Paperjam 2018). OHB Italia was responsible for the preliminary study completed in April 2018. The Luxembourgian Parliament approved the Government Proposal for NAOS and corresponding draft law (€170 million excluding VAT in the military equipment fund until 2028) in July 2018, with 2 votes against out of 60 votes. The launch and start of operations are expected in 2022. The NAOS satellite will increase Luxembourg's military spending from 0.4% to 0.6% of GDP (Paperjam 2018).

Luxembourg is benefiting from Belgian expertise in the NAOS Programme. LUXGOVSAT will become the operator and the Belgian Ministry of Defence is supporting payload operations for LUXGOVSAT. Luxembourg will be the sole owner of the system, but Belgium will acquire programming rights in cooperation with Luxembourg (Belgian Ministry of Defence 2017).

Presented by the Ministry of Economy in April 2018, Luxembourg has invested through ESA (EOEP, InCuBed, IAP, Copernicus) and national programs (collaborative ground segment and high-power computing) €45 million in support of EO infrastructure and service development (Ministry of Economy 2018).

Satellite Communications (SATCOM)

Luxembourg entered global space activities in 1985 through the creation of the Société Européenne des Satellites (SES), a landmark for satellite telecommunications and a major player in this sector. This initiative led to the development of an entire space industry in Luxembourg (The Luxembourg Government 2016). Luxembourg's involvement in satellite communications is built on the heritage and capabilities that revolve around SES, created in 1985, and builds on the legacy of ESA's ARTES program for technology R&D activities. The involvement of SES in European projects provides opportunities for the Luxembourg space industry. For example, LuxSpace (an OHB subsidiary) will provide a microsatellite platform to support the demonstration mission of the European space-based Automatic Dependent Surveillance Broadcast (ADS-B) air traffic control project (Euroconsult 2019).

Luxembourg's MilSatcom needs are covered by participation to WGS and NATO. Luxembourg's GOVSATCOM needs are now covered by the national GOVSAT system.

The 2015 procurement of GOVSAT-1 contributes to Luxembourg's strategy of becoming more involved with NATO (Euroconsult 2019). On 31 January 31 2018, Luxembourg launched the GovSat-1 communication satellite. GOVAT-1, which possesses reliable satellite communication capacity for military and civilian purposes, meets growing demand from institutions and the defense sector. While part of the capacity of the GOVAST-1 satellite will be used to meet the Grand Duchy's needs in terms of satellite communication using military frequencies, it is intended that its remaining communication capacity will be sold on to allied and partner countries and to international organizations (including NATO and the EU). The satellite is being operated by LUXGOVSAT SA, a public-private joint venture created in 2015 by the Luxembourg State and SES (Security Luxembourg 2018).

GovSat-1 is a multi-mission satellite that offers X-band and Military Ka-band capacity over Europe, Africa, the Middle East, and substantial maritime coverage over the Mediterranean and Baltic Seas, as well as over the Atlantic and Indian Oceans. GovSat-1 is a highly secure satellite with encrypted command and control, and anti-jamming capabilities (SES 2018). The GOVSAT applications include Institutional Security Applications: Civil-military interagency collaboration, Strategic and tactical networks, Emergency response, Disaster recovery, Protection of natural resources, and Remote government offices; and Defense Applications: Remote army operations (Mission to HQ communications), Communications on the move (COTM), Communications on the pause (COTP), Maritime operations, and Aero operations (ISR) (GovSat 2018).

In November 2016, GovSat (brand under LUXGOVSAT SA) announced that it has been granted a long-term commercial SATCOM contract to support the operational phase of NATO Alliance Ground Surveillance (AGS). The contract for an end-to-end service includes the delivery of satellite capacity in commercial Ku-band as well as associated capacity management support to provide the required command and control as well as sensor data communications between the NATO Global Hawk UAVs and ground segment over the AGS operational area. GovSat will be ensuring the provision of the Satcom services out of its security cleared facilities with dedicated security cleared personnel, also deployed within NATO premises. With this agreement, Luxembourg Authorities and the NCI Agency as procurement executive agent respectively acquire and manage the services provided by GovSat (GovSat 2016). The Luxembourgian program also fits in the EU GOVSATCOM initiative (Space News 2017).

Positioning, Navigation, and Timing (PNT)

In May 2018, SES signed a long-term agreement with Spaceopal. The contract is part of the Galileo Service Operator (GSOp) framework agreement between Spaceopal – a joint venture between Telespazio and DLR GfR mbH – and the GSA. Under the agreement, SES will provide Spaceopal with services to support the maintenance and seamless operations of the Galileo Global Navigation Satellite System (GNSS). SES will be responsible for in-orbit measurements for the Galileo satellite constellation and provide VSAT managed services to Telespazio for the Galileo Data Dissemination Network (GDDN) (SES 2018).

Space Situational Awareness (SSA)

SSA is currently mainly pursued through cooperation in the frame of ESA.

Poland

Space and Security Budget

Poland had an estimated defense expenditure of PLN 42.824 billion or 2.00% of GDP in 2018, matching the NATO guideline of 2% of GDP (NATO 2019). The 2016 Euroconsult numbers do not account for any military spending on space in Poland. The Polish government expenditure for space-related activities amounted to PLN

339.5 million in 2018, distributed between contributions to ESA (43%), national activities (39%), and EUMETSAT (18%) (Euroconsult 2019).

Space and Security Program

The National Space Program of Poland includes support for the development of satellite systems, technological development as well as integration and development of infrastructure for the purposes of data processing and sharing center, support for administration, education, and higher education (Science in Poland 2017). Under this program, the funds provided by the Ministry of Investment and Economic Development are to be allocated to support the development of an astronomical observation satellite, a SAR microsatellite, and a number of other R&D projects in Poland's space sector, among others (Space R&D 2018).

Earth Observation (EO)-Intelligence-Surveillance-Reconnaissance (ISR)

The SAT-AIS-PL project is implemented by a consortium of Polish scientific, research, and business organizations. The aim is to create a satellite-based automated maritime traffic identification system (SAT-AIS) for the defense and security of Poland. The project is implemented in close cooperation with ESA and the European Maritime Safety Agency (EMSA). The first Polish satellite built for the use in the SAT-AIS-PL system will be a part of a constellation of satellites of other EU countries with a view to increasing the efficiency of the automatic maritime traffic identification system by collaborating with those countries for the exchange of satellite data (Polish Government 2017).

With support from EU regional development funding, the National Institute of Meteorology and Water Management (IMGW) in consortium with the Space Research Centre of the Polish Academy of Sciences, the Academic Computer Centre CYFRONET AGH, and POLSA is conducting since the end of 2017 the project "Operating system for gathering, sharing and promotion of digital information about the environment" (Sat4Envi). The aim of the Sat4Envi project is to create infrastructure for receiving, storing, processing, and distributing data from Sentinel-1, Sentinel-2, Sentinel-3 satellites, and satellite products based on existing IMGW resources. Thanks to its implementation, data from, among others, Copernicus program missions available for Poland will be enabled to public administration to be applied in its activities related to environmental protection, area development and planning, development of urbanization, and transport networks, and to private entities to create commercial services. The project has received over PLN 17.9 million funding under the Operational Programme Digital Poland (Science in Poland 2018).

On June 8, 2017, the Polish subsidiary of Thales Alenia Space, Poland's state-run defense group PGZ, and Warsaw University of Technology publicly announced their intention to jointly found a concurrent design facility. The facility is to create a satellite design center, also known under the name of Concurrent Design Facility under the responsibility of the Ministry of National Defence. It shows the country's effort to construct its first Earth observation satellite development of satellite technology by 2030 (Defence24 2017).

On December 18, 2017, two Polish companies, Creotech Instruments (SA) and Wroclaw Institute of Spatial Information and Artificial Intelligence (WIZIPISI) signed an alliance with Planetek Italia. The cooperation aims to further increase research in the field of Earth observation along with the development and commercialization of joint solutions (Planetek Italia 2017).

To reach the objectives of the Polish Space Strategy and to meet the requirements of the Ministry of Defence, Poland is investing in optoelectronic satellites. The tender pertaining the creation of reconnaissance satellites was planned by 2017 with a program value estimated at 700 million PLN (Defence24 2016a). The first high-resolution satellite could be used both for military, as well as for civilian applications, a dual-use purpose system, delivering the information both for the Army, as well as for the national services. The very-high resolution satellite would be used solely for military purposes. The main goal of the whole program would be to create a national visual reconnaissance system while stimulating the Polish aerospace and space industry for technology development. In accordance with the 2017 Polish Space Strategy, the satellites should be launched and fully operational by 2024 (Defence24 2016b).

Experience gained during the work on the optoelectronic satellite will facilitate the construction of the radar reconnaissance satellite. Also, national know-how from the construction of ground-based radars will be used, with a view to completing the project by the end of 2025 (Polish Government 2017). The initial feasibility study on the SAR system was ordered by POLSA at the end of 2015. A consortium led by Airbus Defence and Space PZL "Warszawa-Okęcie" S.A. branch was the contractor. The Polish Army uses the COSMO-SkyMed Second Generation system in accordance with the agreement between the Polish Ministry of Defence and its Italian counterpart. The Polish Ministry of Defence is participating in the COSMO-SkyMed program with the role of "international defense partner" (Telespazio 2015). Moreover, a specialized satellite reconnaissance center is being created in Poland. The Polish Defence User Ground Segment station is going to be created, within the territory of Poland, in Bialobrzegi, north of Warsaw with full capability expected in 2020 (Defence24 2016a).

Furthermore, Poland participates in the Copernicus program in the frame of the EU and ESA. Poland participates to ESA Earth observation programs with a budget of €3.72 million in 2018 (Euroconsult 2019).

Satellite Communications (SATCOM)

Poland has begun the initial stage of the conceptual works related to a Polish telecommunications satellite which, above all, would be tasked with a defense-related mission. Acquisition of a satellite of this class would secure the needs of our country within the domain of obtaining independent transfer capabilities pertaining to the military satellite data. System of this type would also facilitate communications between the military units, including the units deployed abroad. The above program is especially significant when one considers the NATO commitments made by Poland, along with the involvement of the Polish armed forces in foreign deployments (Defence24 2016b).

The recent experiences suggest that permanent access to a broadband data connection needs to be provided for the units of the Polish Military Contingent, to maintain communications with command. At the same time, it shall be expected that, as high quantity and volume of data is becoming more and more accessible, the demand for expanding the capabilities within that scope would also be an area of a continuous growth. Up until now, access to the telecommunications-transferred data for the Polish forces has been provided through provision of access to the US Army and NATO satellite assets, as well as through application of satellite connectivity assets provided by commercial entities (Defence24 2016b).

In the establishment of Polish satcom capabilities, it is agreed at national level that civil-military cooperation in the development of the Polish communication satellite system is needed. It is also important for Poland to participate in international space programs that will improve the qualifications of Polish scientists and to allocate more funds for research and development activities (Polish Institute of Aviation 2018).

Poland participates to satellite communication activities through ESA's optional ARTES programs, with a budget of €0.94 million for 2018 (Euroconsult 2019).

Positioning, Navigation, and Timing (PNT)

The Armament Inspectorate of the Polish Ministry of Defence announced a tender in 2016 to acquire 1244 military-grade GPS units featuring the SAASM (Selective Availability Anti-Spoofing Module) system (Defence24 2016c). Also, Poland has requested to buy to the US eight universal position navigation units, 34 low-cost reduced-range practice rockets and 1,642 guidance and control section assemblies for GMLRS along with other test sets, devices, and GPS receivers (Defence News 2017).

The European GNSS Agency GSA-supported POSITION (POlish Support to Innovation and Technology IncubatiON) project has worked to increase E-GNSS market penetration and general awareness within the country. Specifically, the project focuses its efforts on startups and early stage investment opportunities for Polish companies looking to utilize E-GNSS technology (Inside GNSS 2017).

Further, Poland participates to the optional navigation programs in conjunction with the EU EGNOS/Galileo flagship program.

Space Situational Awareness (SSA)

SSA is a priority for Poland, identified in the Polish Space Strategy. Before the adoption of the Strategy in 2017, Poland had been active in the development of its national SSA capabilities. Poland has an objective to establish an operations center for the surveillance and tracking of space objects (Space Surveillance and Tracking – SST) enabling the acquisition and processing of information about the current and predicted situation in space (Polish Government 2017).

In July 2015, the National Centre for Research and Development had announced a competition, the aim of which is to create an automated system for optical monitoring and tracking, under the name of SST PL. SST PL was considered a necessary condition and stepping stone to join the EU SST Consortium. Within the framework of the SST-PL project realized by the National Centre for Research and Development, two telescopes are built, with wide and narrow fields of view, making it possible to detect and assess the status of LEO space debris (Defence24 2016d).

The telescopes will be part of the so-called Polish sky observation system (Polish SSA System), used for defense and civil purposes. Creotech is involved in several initiatives related to space surveillance, including mapping of the space debris trajectories and NEO monitoring. A 2016 project concerned the construction of an automated space surveillance and tracking system. The program was realized within the frame of the State Defence and Security program, managed by the National Centre for Research and Development. The consortium responsible for implementing the initiative included the Air Force Institute of Technology, along with Creotech Instruments S.A. and the Military University of Technology (Defence24 2016d).

Poland has been also working on active debris removal. The PW-Sat2 project's purpose is to design a satellite to test an innovative technology of the deorbiting system. The PW-Sat2 satellite will be launched into a Sun-synchronous orbit of approx. 575 km altitude, from the Vandenberg Air Force Base in the US. Thanks to the financing obtained at the beginning of 2016 from the Ministry of Science and Higher Education, it was possible to issue a call for proposals and select a launch provider for the satellite. Innovative Space Logistics B.V. (ILS) was chosen to do this task. PW-Sat1 was deployed in 2012 and decayed from orbit in October 2014. The PW-Sat1 project cost was 500 thousand EU and was financed through ESA PECS program (Kosmonauta 2016).

The 2018 ESA budget for SSA amounts to €22.9 million. Poland participates in the optional ESA SSA program with a budget of €1.16 million in 2018 (Euroconsult 2019).

Portugal

Space and Security Budget

The country had an estimated defense expenditure of €2.874 billion or 1.52% of GDP in 2018, below the NATO guideline of 2% of GDP (NATO 2019). The Portuguese Government proposed in October 2018 that its 2019 budget would include €2.34 billion for defense, a 17.5% increase compared with the €1.99 billion spent in 2018. However, the earmarked budget for defense in 2018 was €2.151

billion. The proposed budget includes €275 million for defense equipment and €6.9 million for military cooperation. €60 million will support Portuguese participation in peacekeeping missions, with the UN providing an additional €five million (Jane's 360 2018). Euroconsult estimates Portugal's space budget at €25 million including contributions to ESA (€16.1 million) and to EUMETSAT (€6.5 million) (Euroconsult 2019). No military space budget has been identified by Euroconsult.

Earth Observation (EO)-Intelligence-Surveillance-Reconnaissance (ISR)

INFANTE is an R&D project for the development and in-orbit demonstration of technology for a small satellite, precursor for Earth observation constellations. The project's Space segment includes a low-cost and modular bus; a communications system based on software-defined radio equipped with advanced functions; a small propulsion system; deployable panels with solar arrays and antennas; SAR and multispectral camera and a payload bay for scientific experiments; and equipment for validation.The Ground segment includes the development of an innovative toolkit to fast-track assembly, integration, and testing of small satellites and a "data hub" to aggregate, process, and disseminate data. INFANTE is led by TEKEVER ASDS and puts together renowned Portuguese companies in the field of Space, such as Active Space Technologies, GMV, HPS, Omnidea, and SpinWorks, among others; internationally recognized R&D centers such as CEIIA, FCT-UNL, FEUP, INL, IPN, ISEP, ISQ, ISR Lisboa, IT Aveiro, and UBI; national partners like Deimos Engenharia, Edisoft, and Optimal; and end users such as IPMA, INIAV, or the Portuguese Maritime Authority and international organizations (Space Today 2018).

Portugal has made use of the EU civil protection mechanism, following requests for assistance to battle forest fires in the country. The Copernicus EMS Rapid Mapping has been activated by the Portuguese National Authority for Civil Protection to produce delineation (showing the extent of damage) and grading (showing the magnitude of damage) maps for Areas of Interest (Copernicus 2017).

Satellite imagery is used indirectly by the three branches of the Portuguese armed forces (AAFF) in their use of cartography and geointelligence in operational planning. Geointelligence is obtained from organizations to which Portugal belongs or through agreements or protocols with third countries or entities, including for instance SatCen. The Center for Military Intelligence and Security of the Armed Forces General Staff is the body responsible for disseminating geospatial intelligence to support the planning and conduct of military operations (IUM 2016). The Army, specifically the CIGeoE, uses satellite Earth surface imagery in the international project Multinational Geospatial Co-Production Program. This project aims to develop a global geographic information system. For that purpose, it uses high-resolution satellite imagery, from which cartographic intelligence is produced. The intelligence acquired is essentially used for support to weapons systems, military operations, humanitarian aid, and disaster situations. The images are obtained from a protocol with the Portuguese Sea and Atmosphere Institute, which in turn obtains them from EUMETSAT (IUM 2016).

The Ministry of Environment and the Portuguese Institute of Sea and Atmosphere initiated a project to build an infrastructure for the Portuguese Sentinel data – IPSentinel. This infrastructure will allow access to the data of the Sentinel satellites of the Portuguese territory and the search and rescue area in the Atlantic Ocean under the responsibility of Portugal. The project is promoted by the General Directorate of Marine Policy through the Financial Mechanism of the European Economic Area (ICACI 2015).

Portugal is participating in the Copernicus program in the frame of the EU and ESA. Portugal contributes to ESA Earth observation programs with a budget of €1.6 million for 2018 (Euroconsult 2019).

Satellite Communications (SATCOM)

The PoSAT-1 was launched into orbit in September 1993, on board the Ariane 4 rocket. The satellite was developed by several public, private, and military entities and was partly financed by the government. The program aimed to prepare the Portuguese industrial fabric to gain presence in the space industry and to promote Portugal's technological and scientific development. The PoSAT-1 was used by both military and civilian organizations until 2006. The PoSAT-1 allowed the AAFF to carry out tactical and strategic communications, data transmission of encrypted and unencrypted messages, reception of images, and meteorological information. Data from the PoSAT-1 is no longer used by the armed forces, which now relies on private contractors and agreements with third countries or institutions.

Both the supreme military body of Portugal and the branches of the Portuguese armed forces contract private service providers, but only the former contracts communications to military satellites (X-band). The Navy uses military satellite communications (X-band) in the Frigates, the Submarines, and in the Replenishment Ship. Nearly all naval assets are equipped with the Global Maritime Distress and Safety System, which provides INMARSAT coverage beyond the 300-mile line. Satellite communications are particularly important to the Army and to the National Deployed Forces (IUM 2016).

Both the Navy and the Army consider satellite communications especially important for operational contexts. However, it should be noted that the Navy still relies on radio communication to ensure redundancy, even if it is only partial. The Navy also relies on the Metocmil system to ensure redundancy in satellite communications. This dependence can be seen in the several uses for data from satellite communications in Portuguese military operations outside national territory and in the use of these images for cartography and meteorological forecasts (IUM 2016).

Portugal participates in satellite communication activities through ESA's optional ARTES Programs, with a budget of €2 million for 2018 (Euroconsult 2019).

Positioning, Navigation, and Timing (PNT)

The Portuguese Armed Forces (AAFF) use both civilian and military GPS signals. The Navy uses the civilian GPS signal for navigation, except for frigates, which use the P(Y) GPS signal. However, the use of PNT data is important not only to the Navy, but to maritime navigation in general. Portugal has differential GPS stations to improve the precision of the GPS signal by comparing the information received with a known location. The use of PNT data is also deemed essential to the cartographic process of the CIGeoE (AAFF Geospatial Information Centre). This process uses data from GNSS, which combines signals received from the GPS, Glonass, and, in the future, the Galileo systems. The Armed Forces use civilian and military GPS for several weapons systems, mainly for aeronautical navigation (IUM 2016). Aircraft navigation systems can be exclusively inertial or mixed. With a mixed navigation system, more technologically advanced aircraft combine GPS signal with inertial signal, thereby reducing the latter's intrinsic error and increasing precision. The civilian GPS signal is currently used for navigation in the field of Research, Development and Innovation, in the research line of unmanned aerial vehicles (IUM 2016).

Since 2006, the Ministry of Environment has been working on the densification and the upgrade of the GNSS CORS (Continuously Operating Reference Stations) network – ReNEP – with two main goals: (i) the maintenance of the national reference frames; and (ii) to provide a real-time precise point positioning service. In 2015, the Directorate-General of Territory (DGT) finalized the network, which consists of 48 CORS: 42 in the mainland and 7 on the islands (Azores and Madeira Archipelagos). All the stations collect both GPS and GLONASS data, six are part of the EPN (EUREF Permanent Network) and three of these also belong to the IGS (International GNSS Service) network (ICACI 2015).

Further, Portugal participates in the optional Navigation programs in conjunction with the EU EGNOS/Galileo flagship program.

Space Situational Awareness (SSA)

At national level, SST fits into the Research and Innovation Strategy for Intelligent Specialization (EI&I) 2014–2020. It contributes to the Industrial Development Strategy for Growth and Employment (EFICE) 2014–2020, promoting better exploitation of national and European funds, and allows optimizing the fulfillment of the objectives defined in the Strategic Concept of National Defence. At the end of 2018, Portugal joined the EU SST Consortium. Portugal's participation in the EU SST program aims to build a national SST capacity, which is properly articulated with other national programs in space, integrating sensors, processing capacity and services, currently dispersed by several entities (IUM 2016).

The Portuguese initial SST capacity was prepared in the frame of the Space Surveillance and Tracking Project Group, abbreviated GPSST. It is financed through funds entered in the Military Programming Law, in "Central Services-Joint Capabilities," through the project "Support Defence Technological and Industrial Base (BTID)." Up to a maximum amount of €1.4 million is to be increased through the budgets of other participating entities (Portuguese Government 2017).

The Portuguese initial SST capacity provides for the installation of the National Operating Centre in the Science and Technology Park of Terceira Island (TERINOV) and also an optical observatory with two telescopes in the Pico do Areeiro – Madeira. There are additional ideas to establish a radar component on Flores and Corvo. The GPSST was tasked to provide to the Council of Ministers by the end of October 2018 the future governance model of the national SST program after its functions have been terminated, proposing its structure, identifying sources of financing for the sustainability of the national SST infrastructure, considering the possible engagement to SSA, and ensuring the necessary compliance with the 2018 National Strategy for Space (Portuguese Government 2018).

Portugal participates in the optional ESA SSA program for €133 thousand in 2018 (Euroconsult 2019).

Romania

Space and Security Budget

Romania had an estimated defense expenditure of RON 17.181 or 2.04% of GDP in 2018, matching the NATO guideline of 2% of GDP (NATO 2019). Romanian government expenditures for space-related activities amounted to RON 252.3 million in 2018, resulting from parallel contributions to ESA (RON 190.3 million) and EUMETSAT (RON 24.4 million) (Euroconsult 2019). No military space budget has been identified by Euroconsult.

Space and Security Program

The Ministry of Education and Research funds the Research, Development and Innovation STAR Program – Space Technology and Advanced Research for the 2012–2019 period, through the Romanian Space Agency (ROSA). The program includes three subprograms: Research, Infrastructure, and Support (ROSA Website). The Science and Technology pillar of the S3 strategy of ROSA focuses on participation of relevant institutions and relevant programs. The Services pillar of the S3 strategy of ROSA focuses on participation of relevant institutions and relevant programs. The Security pillar of the S3 strategy of ROSA focuses on participation of relevant institutions and relevant programs (ROSA Website).

Earth Observation (EO)-Intelligence-Surveillance-Reconnaissance (ISR)

Romania has an IMINT and GEOINT training facility in Bucharest, led by ROSA and the Military Technical Academy (MTA Website). The Geospatial Intelligence Center provides the spatial foundation essential for the analysis of information from all sources, providing the baseline starting point for all geographically referenced analytic efforts. It brings together cartography, imagery analysis, geospatial analysis, geodesy, aeronautical analysis, maritime analysis, and regional analysis in Romania, providing unique intelligence capability (MTA Website).

ROSA is implementing studies for a national EO ground and space system, including optical and radar observation facilities (EISC 2016). Romania has a Centre for Remote Sensing Applications in Agriculture, called CRUTA. Important contributors to national remote sensing activities include the ROSA Research Center, the Remote Sensing Laboratory; the Institute of Geodesy, Photogrammetry, Cartography, and Land Management; the Institute of Soil Sciences and Agrochemistry; the National Institute of Meteorology; the Geological Institute of Romania; the Military Surveying Service; the Romanian Forest Research and Management Institute; the Institute of Studies and Design for Land Reclamations; and the Institute of Geology and Geophysics (ESA 2015).

Romanian ministries are increasingly using space-based service in support of a variety of policy areas. As an example, the Ministry of Environment contracted TerraSigna, a GIS service provider, and a team of volunteering IT specialists to create "Inspectorul Padurii" (The Forest Inspector) in the frame of combatting illegal deforestation in Romania. "The Forest Inspector" is a geographic information system that can ingest radar and high-resolution satellite images from Sentinels 1 and 2, Landsat, OpenStreet, and Google Maps. The software then scans successive images of the same areas to identify changes in the forest that indicate where logging has taken place. The data is refreshed every 2 to 7 days, which enables authorities to become aware of any illegal logging in almost real time. The satellite-based map is further enriched with information coming from the government's digital database and tracking system. Such information can include who has permits to cut what and where, what the truck's license plate number is, and when the logging took place. Protected areas are monitored through the platform. In addition, the map can be used to identify fake GPS loading points, faster (Eurisy 2017).

Through the European Maritime Spatial Planning (MSP) Platform, financed under the European Maritime and Fisheries Fund (EMFF), the National Institute of R&D for Optoelectronics implements RO-CEO, the Romanian Cluster for Earth Observation, a project running from November 2016 until October 2019 (MSP Platform Website). The RO-CEO project aims to increase the capacity of Romanian organizations to contribute to ESA's Earth observation programs and projects, by setting-up the Romanian Cluster for Earth Observation, a formal association of organizations, with its own statute and agenda. RO-CEO project unites key players at national level, addressing several important aspects of the Earth observation activities, to support geophysical algorithm development, calibration/validation and the simulation of future spaceborne earth observation missions. The main roles of the cluster are: to promote the specific interests and relevant capacities at national level; to attract more investments and contracts for the Romanian institutions; to improve the provision of services to end users by joining complementary skills and expertise, and; to ensure the sustainability of the EO sector in Romania by enabling EO market development (MSP Platform Website).

Romania is advancing on the establishment of a satellite EO data reception and processing center. The National Institute of Research and Development for Optoelectronics has a Center for Atmospheric REmote Sensing and Space Earth observation. The National Institute for Marine Research and Development "Grigore Antipa" has a Space Technologies Competence Centre Dedicated to the Romanian Marine and Coastal Regions Sustainable Development. The Polytechnic University Research Center for Space Information in Bucharest has a competence center for smart sensors and big data technology for space applications (EISC 2016).

The Romanian government has depended on the International Charter on Space and Major Disasters and the Copernicus Emergency Management Service, with space-based imagery delivered to ROSA and ministerial stakeholders in the frame of national disaster management, notably triggered by floods (Eurekalert 2005).

In December 2017, ESA and ROSA signed an Understanding for the Sentinel Collaborative Ground Segment Cooperation to facilitate Sentinel data exploitation in Romania. Under the agreement, ROSA will coordinate ground segment activities in Romania – such as hosting, distributing, ensuring access, and archiving Sentinel data – and act as an interface between ESA and national initiatives. ROSA also plans to cooperate with different European partners and institutes. The Sentinel Collaborative Ground Segment aims to provide complementary access to Sentinel data and/or to specific data products or distribution channels. The collaborative elements bring specialized solutions to further enhance the Sentinel missions' exploitation in various areas, such as data acquisition and (quasi-) real-time production, complementary products and algorithms definitions, data dissemination and access, development of innovative tools, and applications as well as complementary support to calibration and validation activities (ESA 2017).

Romania participates in ESA Earth observation programs with a budget of €4.16 million in 2018 (Euroconsult 2019).

Satellite Communications (SATCOM)

The Romanian Ministry of Defence wants to launch its own telecommunication satellites into geostationary orbit as part of a strategic project, according to a press released following a meeting between the Defense Minister Gabriel Les, the head of Romanian Space Agency (ROSA), and ESA experts (Business Review 2018).

Romania participates in satellite communication activities through ESA's optional ARTES programs, with a budget of €2.53 million in 2018 (Euroconsult 2019).

Positioning, Navigation, and Timing (PNT)

Romania has a national permanent GNSS network of 74 reference stations (in 2016). The system provides GNSS augmentation services under ROMPOS – Romanian Position Determination System. ROMPOS is part of the Central and East Europe ground station augmentation system named EUPOS. ROMPOS services include DGNSS service (dm accuracy), RTK service (cm accuracy), and GEO (geodetic service – cm/mm accuracy). ROMPOS services are provided through the National Center of ROMPOS Services (NCRS), which is intended to monitor and control the GNSS stations activity for the automatic transfer of data to central data server, which are used for post-processing position determination and positioning services and products for real-time applications (EUREF 2016).

Further, cooperation is facilitated through ESA. Romania participates in the optional navigation programs in conjunction with the EU EGNOS/Galileo flagship program.

Space Situational Awareness (SSA)

Romania is participating to ESA's Space Situational Awareness optional program since 2012, through the Romanian Space Agency. The main areas where national priorities are placed contribute to achieving ESA security objectives as stated in the SSA Program Declaration. The national SSA overall objectives are being fulfilled at the same time by rising national industry potential in SSA, especially in the fields of software development, optical, and radar technology and developing space weather science potential.

Romania's priority for SSA is the retrofitting of the Cheia Ground Station for SST purposes (ROSA 2018). The Cheia SST 32 m antenna may be refurbished and modified for mono- and bistatic radar tracking of NEO and space debris (EISC 2016).

Moreover, cooperation is facilitated through ESA and EU.

ROSA represents Romania in the Space Surveillance and Tracking Committee and organized by European Commission. In August 2017 Romania applied to be part of the EU SST Consortium. On May 31, 2018, Romania was admitted in this consortium as full member.

Romania participates in the optional ESA SSA program with a budget of 1.6 million in 2018 (Euroconsult 2019).

Concluding Remarks

This chapter provides evidence of the trend toward increasing relevance for security and defense in national space and security programs in smaller European countries. In particular, this is reflected through dual-use utilization of "Space and security," both in its security from space and security in space aspects which are progressively contributing to further integration of space activities. Traditionally, only civilian space activities including for instance Earth observation, telecommunications, human spaceflight, space transportation, and technology development were subject of cooperation projects at intergovernmental and supranational level. Security- or defense-related space programs were kept at the national level or dealt with bilaterally or multilaterally in *ad hoc* cooperative programs. These trends demonstrate an evolution of European countries priorities from strictly civil-oriented applications to also encompassing security and defense ones. Dual-use allows the possibility to accommodate various needs with limited resources. The national space programs with security or defense dimension, in combination with the EU and ESA programs, demonstrate alignment toward the use of space for security and defense in smaller European countries. To conclude, the increasing relevance of security and defense in Europe, including the smaller European countries, to some extent could be framed as the necessity for Europe to advance civil-defense synergies and join resources in order to achieve autonomy as a global player.

Disclaimer   The contents of this chapter and any contributions to the Handbook reflect personal opinions and do not necessarily reflect the opinion of the European Space Agency (ESA).

References

Belgian Ministry of Defense (2017) Coopération belgo-luxembourgeoise dans le domaine de la Défense, 23 November 2017

Business Review (2018) Romania's Ministry of Defense aims to launch its own satellites into orbit, 22 November 2018

C4Defense (2018) Greece to lease Heron UAVs, 8 February 2018

CAS (2018) China, Finland to enhance arctic research cooperation, 31 October 2018

CNES (2018) Space cooperation with the Czech Republic - CNES celebrates Czech Republic's 10 years of success in ESA, 13 November 2018

Copernicus (2017) Copernicus EMS rapid mapping activation for forest fires in Portugal, 18 October 2017

Czech GEO Data Portal Website, GNSS Data Portal, GNSS data portal. Available at: http://www.czechgeo.cz/en/article/default/gnss-data-portal

Czech Ministry of Transport (2014) National Space Plan 2014–2019, 27 October 2014

Czech Ministry of Transport (2016) CZ Space Catalogue, 2016

Danish Government (2016) National Space Strategy, Denmark Government

Danish Ministry of Defence (2016) Armed Forces, National Role

Danish Ministry of Defence (2017) A strong defence of Denmark, Proposal for New Defence Agreement 2018–2023, October 2017

DBEI (2018) Focus on Aerospace and Aviation – December 2018DIAS Website, Astro – Research

Defence 24 (2015) Optoelectronic data for Poland: Polish satellite programme and expansion of the agreement with Italy, 17 December 2015

Defence 24 (2016a) Space plans made by the Polish Ministry of Defence. "Defined Needs" vs. Great Challenges, 16 April 2016

Defence 24 (2016b) Poland Looks Towards Space. Satellite Programme Begins, 18 February 2016

Defence 24 (2016c) Polish army looks to acquire rugged GPS units, May 2016

Defence 24 (2016d) Polish space situation awareness system is being developed, 25 May 2016

Defence 24 (2017) Satellites: concurrent design facility to be established in Poland, May 2017

Defence News (2017) Poland to buy Lockheed-made rocket launcher, November 2017

DTU (2018) Unique Danish satellite in orbit around the Arctic, 5 February 2018

DTU Space Website, Geodesy and Earth Observation. Available at: https://www.space.dtu.dk/english/research/research_divisions/geodesy_and_earthobservation

EISC (2016) Romanian space activities, presentation, October 2016

EO Portal Website, Satellite Missions, MIMOSA and CatoSat-2E. Available at: https://directory.eoportal.org/web/eoportal/satellite-missions/c-missions/cartosat-2e

EPA (2011) Climate Change Research Programme (CCRP) 2007–2013 Report Series No. 8, 2011

ESA (2015) Land remote sensing in Romania, Presentation, 2015

ESA Navipedia, EGNOS Ground Segment. Available at: https://gssc.esa.int/navipedia/index.php/EGNOS_Architecture

ESJ News (2018) Lasers and Space as Czech Contributions to NATO, EU, 17 October 2018

Estonian Land Board (2015) GNSS Services in Estonia: ESTPOS as a case study, Presentation, December 2015

Estonian Land Board Website, National Mirror site (ESTHub). Available at: https://www.maaamet.ee/en/objectives-activities/geoinformatics/national-mirror-site-esthub

Estonian Marine Institute Website, Department of Remote Sensing and Marine Optics, Available at: https://mereinstituut.ut.ee/en/institute/department-remote-sensing-and-marine-optics

Estonian Ministry of Defence (2019) Estonian defence budget 2019, September 2019

Euractive (2019) Today the launch of HellasSat 4 – the advantages and Kleiddi, 5 February 2019

EUREF (2016) National Report Romanian GNSS permanent network, presentation, 2016

Eurekalert (2005) Disaster Charter brings satellites to bear on Romanian flooding, August 2005

Eurisy (2017) Romania turns to satellites to crackdown on illegal deforestation; MSP-Platform Website, RO-CEO – Romanian Cluster for Earth Observation

Euroconsult (2019) Government Space Programs – Benchmarks, Profiles & Forecasts to 2028, July 2019

FICORA (2018) Finland to become a leader in 5G: the 5GKIRI cooperation project between Finland's largest cities supports construction of future communications networks, 1 November 2018

FIG (2013) The performance of network RTK GNSS Services in Ireland, February 2013

Finnish Government (2018) Prime Minister Sipilä: Finland and India have good opportunities for cooperation, 17 April 2018

Finnish Ministry of Defence (2018) Division of Defence Spending, Defence Budget 2018, February 2018

Finnish Ministry of Economic Affairs (2018) The 2019 draft budget of the Ministry of Economic Affairs and Employment: Further investments in innovation funding

Finnish Ministry of Economic Affairs and Employment (2013) The national strategy for Finland's space activities in 2013–2020 – to space through Europe, global benefits and prosperity to Finland from space activities

Finnish Ministry of Transport and Communications, Efficient deployment of satellite navigation systems in Finland – Action Plan 2017–2020, June 2018

FMI Website, Finnish Meteorological Institute, Earth Observation Research. Available at: https://en.ilmatieteenlaitos.fi/earth-observation

GNSSnet.hu Website. Available at: https://www.gnssnet.hu/index.php?r=site%2Frealtime

GovSat (2016) GovSat Press Release: NATO AGS Contract Awarded to GovSat, 8 November 2016

GovSat (2018) GovSat-1 Brocure, A new concept in secure communications, 29 January 2018

HSO (2016) A new era of Hungary in space, 14 April 2016

HSO Eötvös, Loránd University Space Research Group. Available at: http://sas2.elte.hu/urfiz_hullamkiserletek_1EN.html

HSO BME, Department of Broadband Infocommunications and Electromagnetic Theory. Available at: https://www.vik.bme.hu/en/departments/

HSO Department of Geodesy, Remote Sensing and Land Offices. Available at: https://eurogeographics.org/member/department-of-geodesy-remote-sensing-and-land-offices/

ICACI (2015) Portugal National Report – ICA 2015, 23 August 2015

ICEYE (2019) Two recently launched ICEYE SAR Satellites Commissioned – added to commercial constellation, Press Release, 12 September 2019

ICEYE Website, History Timeline. Available at: https://www.iceye.com/company/history

Infocom.gr (2020) Increased contributions of Greece to ESA, 3 February 2020

Innoteka (2018) Magyar űrstratégia kell!, 4 May 2018

Inside GNSS (2017) Poland in position for a big role in GNSS, 7 April 2017

Invent Baltics OÜ, Study on Latvia's Participation in European Space Agency Since 2015, September 2019

Investinestonia (2017) Turning theory into practice: the short story of the Estonian space race, March 2017

Irish Examiner (2018a) Ireland's defence spending is lowest in Europe at 0.3% of GDP, 8 June 2018

Irish Examiner (2018b) First Irish satellite one giant leap closer to space odyssey, 25 September 2018

Irish Government (2017) News Service, Ireland signs Earth Observation agreement with the European Space Agency, Press Release, 13 October 2017

IUM (2016) Space in the Portuguese Defence Sector, April 2016

Jane's 360 (2018) Portugal proposes defence spending increase to Parliament, 18 October 2018

Kosmonauta (2016) Polish PW-Sat2 will be launched into space at the end of 2017 on-board Falcon 9, December 2016

Luxembourg Government (2016). Innovation.public.lu; SES (2018a) GovSat-1 Brochure, 29 January 2018

Maynooth University (2014) 8th Irish Earth Observation Symposium, October 2014

Metsar J, Kollo K, Ellmann A (2018) Modernization of the Estonian national GNSS reference station network. Geodesy Cartography 44(2):55–62

Mononews (2019) Agreement Greece – ESA for the development of microsatellites, 26 March 2019

MTA Website, Geospatial intelligence centre – about. Available at: https://www.mta.ro/geoint/about.html

Muls (2016) De opbouw van het Galileo Navigatiesysteem, Revue Militaire Belge, December 2016

NATO (2019) Defence Expenditure of NATO Countries (2013–2019) Press Release, PR/CP(2019) 123, November 2019

NLS (2017) Finnish Geospatial Research Institute and its Department of Navigation and Positioning, Presentation, August 2017

NORSAT (2012) International, case study: Finnish Defence Forces, 17 September 2012

Paperjam (2018) Lancement d'un nouveau satellite en «triple win», 26 July 2018

Planetek Italia (2017) Planetek strengthens its alliance with Polish space companies, December 2017

Polish Government (2017) The Polish Space Strategy, 2017

Polish Institute of Aviation (2018) Program of the Polish communication satellite – a debate at the Institute of Aviation, March 2018

Portuguese Government (2017) Resolução do Conselho de Ministros 116/2017, 2017

Portuguese Government (2018) Resolução do Conselho de Ministros 113/2018, August 2018

ROSA (2018) The 3S – new ROSA strategy, presentation, September 2018

Sagath D, Papadimitriou A, Adriaensen M, Giannopapa C (2018) Space strategy and governance of ESA small member states. Acta Astronaut 142:112–120

SatelliteToday (2015) Denmark Expects MilSatCom capacity Surge to continue, 9 April 2015

Satnews (2017) Four Additional Satellites Ordered from GomSpace A/S and Aerial & Maritime, 19 December 2017

Schenk et al (2010) GEONAS - geodynamic network of permanent GNSS stations within the Czech republic. Acta Geodyn Geomater 7(1). (157):99–111

Schrogl K-U (et al.) (Eds.) (2015) Handbook of Space Security: Policies, Applications and Programs. Springer, New York

Science Business (2017) Aalto-2 satellite launched into space, 20 April 2017

Science in Poland (2017) Polish Space Agency: Draft National Space Program is ready, 27 December 2017

Science in Poland (2018) FP Space launches the Polish commercial satellite Intuition-1 project, 26 February 2018

Security Luxembourg (2018) Defense – Luxembourg to buy an observation satellite, 19 January 2018

SES (2016) SES enables Danish Defence wideband global satcom system connectivity, 12 May 2016

SES (2018) SES Provides Managed Services for Galileo, 30 May 2018

SIGNAL (2015) Danish Military Procures SATCOM Terminals, 7 April 2015

SIPRI (2016) Military capabilities in the Arctic: a New Cold War In The High North?, SIPRI Background Paper, October 2016

Space News (2017) Luxembourg eyes Earth-observation satellite for military and government, 7 November 2017

Space R&D (2018) Polish Space Agency eyes $420M program to develop satellites, 6 March 2018

Space Today (2018) Microsatellite constellations from Portugal, 16 October 2018

SpaceNews (2018) International SSA agreements could pave the way for further space cooperation, panelists said, 18 April 2018

TaNea (2019) The confidential plan "KLEI.D.DI." for the phones of Pavlopoulos, Tsipras and ministers, 9 March 2019

Telespazio (2015) COSMO-SKYMED: the collaboration between Italy and Poland is strengthened, 2 September 2015

TTU Website, TTÜ100 Satellite. Available at: https://www.ttu.ee/projects/mektory-eng/satellite-programme-3/satellite-programme/

University of Tartu (2018) Tartu observatory develops new instruments for global remote sensing network, 17 April 2018

ViaSatellite (2016) Denmark gains access to WGS constellation via SES anchor stations, Satellite today, 17 May 2016

ESA (2017) Pooling and Sharing for Government Satcoms, 21 November 2017

Future of French Space Security Programs

Jean-Daniel Testé

Contents

J.-D. Testé (✉)
OTA (l'Observation de la Terre Appliquée), Peujard, France
e-mail: jd.teste@wanadoo.fr

© Springer Nature Switzerland AG 2020
K.-U. Schrogl (ed.), *Handbook of Space Security*,
https://doi.org/10.1007/978-3-030-23210-8_122

Abstract

Our Western societies are increasingly dependent on space systems. These have become essential to many daily activities as well as to strategic activities of prime importance. In an international context that is currently particularly troubled, their security becomes a major issue. France has been interested in this subject for several years and has developed unique space surveillance capabilities in Europe. These means, precursors 30 years ago, must today be modernized and adapted to current risks and threats. This technical evolution will have to be carried out in coherence with a doctrinal evolution based on simple and global principles. In regard to the costs of such technical changes, the evolution shall be considered within a wide European cooperation considering that finally EU sovereignty will be largely reinforced by these systems.

Introduction

At a time when rivalry among the world great powers is spilling over to space, the safety of French satellites could be threatened, directly or indirectly. Directly, because some powers consider France as the indefectible ally of a rival power or as an autonomous entity with the ability to monitor space and therefore with the ability to reveal potential suspicious actions in space. Indirectly, because space is a particular medium where Kepler's laws dictate that an activity on a precise orbital position may have consequences to other orbits. The Indian March 27 missile shooting down a satellite in low Earth orbit is the most recent example.

Given the importance of own space capabilities for society, for efficient diplomacy, and for military operations, France cannot risk losing control. However, one realizes that we do not have today the necessary means to deter a potential adversary from conducting an action in space that could harm one or more satellites. Nevertheless, in the short term, the application of some basic principles should help to limit the consequences of this deficiency.

Dependency

For many years France has assessed the ability of the space domain supporting autonomous situational awareness and decision-making while simultaneously being a unique force multiplier in support of military operations. In this regard space is considered strategic. Today, more than 30 years after the launch of the first French military payload in space (Syracuse I Telecom payload in 1982), space capabilities have become increasingly important in the whole range of defense and security operations. We are at the turning point where military space and space infrastructures at large have evolved from necessity to dependency driven. Entering into the second decade of the twenty-first century compels us to review our space heritage from the Cold War. Almost nine billion people on planet Earth, with a growing number of major space-faring nations and nuclear powers, pose challenges to the military who push their limits through an unprecedented series of operations. This is a new paradigm illustrated by fights in Afghanistan, Libya, Mali, Central African Republic, and now Iraq and Syria.

France is historically the third space-faring nation. After more than 50 years of dedicated vision and investments in space industry and programs, France has developed innovative and credible capabilities which ensure its independence and constitute a tangible asset contributing to the development of coherent European defense space cooperation. The main capacities the French armed forces will have to fulfill in the following years are stipulated in the 2013 French White Paper on Defense and National Security (2013 Livre blanc de la Défense et sécurité nationales): entry at first, precision strikes, special forces, and work as a framework nation within an international coalition.

Accordingly, before a "go" decision is taken, the following conditions are required:

- Prior knowledge of the theater of operations and its environment (geography)
- Precise knowledge of the enemy and forces engaged (Intelligence)
- Precise knowledge of intended targets (Targeting)
- Capacity to perform missions in complete safety and without collateral damage (precise Navigaton)
- Ability to dispatch orders and assign responsibilities (Command & control)

In addition to that, modern soldiers must face three new emerging game changers:

1. The first one is *time*. Time is critical, the time of decision-makers, the time of the media, and the time of social network and globally the new cyber dimension.
2. The second is *tempo*. Operations have to be conducted fast, with visible results and limited footprint in the shrinking military capabilities and operating context.
3. And the third is *adversary*. We are facing a new kind of enemy. Indeed, its nature covers now a wide range from small terrorist groups to organizations using conventional weapons and cyberspace capabilities, easily switching from one side of the scope to the other such as Daesh or Al-Qaida.

The great news is that we are now able to fulfill all these challenges. This is only possible because we have completed a full spectrum of space capabilities:

- Observation and localization are required for geography.
- Observation and eavesdropping are necessary for intelligence.
- Observation and accurate localization enable accurate targeting.
- Navigation and SATCOM are a unique support for operations.
- SATCOM are mandatory for command and control.

The availability of all these space capabilities for all warfighters allows French armed forces to perform military operations at the right *time*, with the right *tempo*, and whatever the kind of *adversary* they have to fight against.

Space provides the fantastic leverage one needs, but on the other hand, we need to cope with the dependency on our own space capabilities without any efficient alternatives. Hence, these capabilities need to be protected and preserved at a time where more countries are experimenting anti-satellite weapons.

The Threat: Increasing Suspicious Developments

Today we are witnessing a resurgence in the rivalry among the great powers of our world. Each of them intensifies their efforts in the space domain in order to demonstrate a mastery or even a supremacy in this strategic field. To this end, their research activities are endowed with much larger budgets than in the past, which enables them to experiment, on the ground or in orbit, with new technologies that may ultimately constitute threats or even weapons against operational orbital systems. In September 2018, the French Minister of the Armed Forces (Ministre des armées) Mme Parly gave a very specific illustration of this type of threat (Discours publics, Les discours dans l'actualité, Déclaration de Mme Florence Parly, ministre des armées, sur la défense spatiale, à Toulouse le 7 septembre 2018, http://discours.vie-publique.fr/notices/183001732.html).

Yet, in order to understand and anticipate potential future conflicts in space, it is essential to define the threats and the potential aggressors. There are two types of risks in space: artificial risks (collision with debris) and natural risks (disruptive phenomena related to solar activity). Although risks constitute an important concern today, we will mainly address threats here. The main difference between risks and threats lies in the intent of the latter.

Weaponization of space refers to the deployment of real weapons into orbit and no longer simply supporting systems for armed ground operations (militarization). Weaponization has been experimented since the 1960s. While the activities in this area were dominant during the Cold War, the latest significant activities date back to 2007–2009, when the Chinese and the Americans each destroyed one of their deactivated satellites to demonstrate that they had a real operational capability. Weaponization, therefore, seems to have become a credible risk with multiples shapes as presented below.

Attack Directed Against the Satellite Itself from the Ground

The attack by a customized ballistic or antiaircraft missile is feasible only against low orbit satellites. This is a precision attack that requires the aggressor to get accurate data on the orbit and missile technology with devastating consequences. The Chinese experience of firing a ballistic missile against one of its own satellites has caused an unprecedented number of debris (about 30,000). On 27 March 2019, India as well managed to successfully fire a ground-based anti-satellite weapon against a satellite in low Earth orbit (The Wall Street Journal, India Successfully

Tests Satellite-Killer Missile, 27 March 2019, https://www.wsj.com/articles/india-successfully-tests-satellite-killer-missile-11553679166).

High-Altitude Nuclear Weapons

A nuclear explosion in space via missile, although experimented by the Russians and the Americans, is far too destructive without any selectivity and therefore a very unlikely threat.

Directed-Energy Weapons (AED): Lasers and Microwaves

Low-energy lasers are a very likely threat as well because the technology is mature. They can dazzle (deny the ability to take a picture) the observation satellite and damage the components of the optical chain (degradation the quality of the image). High-energy lasers can damage a satellite's structures, including its solar panels or optical parts. This technology that requires high power is not yet fully achieved, but it is estimated that the three main space powers have experimented and are developing such weapons. High-powered microwave (MFP) weapons emitted by radiation from the ground can destroy the satellite's electronic components by creating a magnetic field. This technology is also under research and development among big powers.

Hard Attack Against the Ground Segment

This attack can be targeted and is already technologically and operationally feasible. The consequences of a hard attack on the user ground segment (mission center, receiving antenna, and computer system) could disable the satellite programming and data receiving. The consequences against the control/control segment could be the disruption of communication with the satellites (in both directions). The satellites could then go out of order. An attack on the antennas (2Ghz) could prevent sending commands to the satellite.

Jamming

The jamming of telecommunications and navigation signals represents maybe the most important and probable threat against space services. The technology is already well mastered, and operational concepts are very similar to those used against ground devices. The direct consequence could be to prevent or disrupt all communications and navigation. For an observation satellite, the consequence could be to render it temporarily out of service. This threat category includes also spy satellites sometimes known as "browsers." Spy satellites manage to get close to the target satellites in order to spy and sometimes jam their communications.

Cyberattack

Cyberattacks are quite feasible and probable. Particularly complex to detect and assign, they are likely to be used extensively against ground segments of space systems even in peacetime.

Satellites Able of Making Co-orbital Spatial Rendezvous

Many spacefaring nations are today interested in the ability of a satellite to maneuver on its orbit. They are conducting studies and experimenting under the umbrella of active debris removal. But these technologies will certainly allow a satellite to capture a debris and bring it out of orbit. In addition, they will enable to approach, capture, and deorbit another satellite. This threat is already part of the capabilities of the three main space powers. This inventory is certainly not exhaustive but revealing the different research activities and experiments conducted by the major space powers. Their goal is to have the ability, if necessary, to significantly reduce the advantage that satellites provide to an adversary in conflict, be it diplomatic, economic, or military.

Fully aware of the strategic capabilities of space assets, France has very early developed capabilities for space surveillance.

French Space Surveillance Capabilities

Origins

French military capabilities for space surveillance have their origin in national nuclear deterrence. Indeed, given the very high level of availability and security required, training of ground operations and armed forces at times had to take place out of the sight of countries potentially affected by these capabilities. That is to say, undetected by their means of gathering information and, in particular, by observation and eavesdropping satellites.

The easiest way to avoid being detected by an intelligence satellite is certainly to operate out of range of it at the time when they are far from the area of operations. It is therefore essential to know their orbits with great precision. Although initially the orbital data of the enemy satellites were provided to the French Defense by North American Aerospace Defense Command (NORAD), it quickly became clear that such dependence was not consistent with the need for total sovereignty of France's deterrent capability.

Thus, in the 1980s, the Direction générale de l'armement (DGA) and Office National d'Etudes et de Recherches Aérospatiales (ONERA) developed capacity demonstrators in various fields, radar surveillance, optical surveillance, and electromagnetic listening:

- The Grand Réseau Adapté à la VEille Spatiale – Grand Network Adapted to Space Surveillance (GRAVES) space surveillance radar which allows the creation of a catalog of space objects and their orbital characteristics in complete independence
- The Systeme Probatoire d'Observation du Ciel (SPOC) which could complete and specify the catalog of GRAVES
- An eavesdropping antenna to determine the activity of space objects

Although the GRAVES radar was put into operational service in the armies in the early 2000s, the other two systems remained at the demonstrator stage.

This brief historical reminder of the origins of the surveillance of the military space in France makes it possible to explain the current limits. In fact, the performances specified in the GRAVES system were to be able to detect and track the foreign observation and listen to satellites (at the time of spacecraft weighing more than 1 ton, measuring more than 1 meter, and orbiting between 300 and 1200 km altitude).

Organization and Capacities

In addition to its performances, which are still unique in Europe, the GRAVES radar also had motivated the French Defense to organize itself in order to implement this new system and get the best benefit from it.

The system was entrusted to the Air Force and more particularly to the Command of Air Defense and Air Operations (CDAOA) which operates from the air base of Lyon-Mont Verdun. The unit in charge of this activity is COSMOS (Operational Center for Military Surveillance of Space Objects).

For the great benefit of the armed forces, the COSMOS mission is therefore to implement space surveillance sensors, to manage the national catalog of space objects, to detect abnormal behavior, to provide space support for military operations, and to manage the space meteorology. In the event of abnormal behavior, security risk, or spatial danger, COSMOS is also responsible for disseminating the alert and informing the authorities.

In addition to the GRAVES radar, the COSMOS also implements SATAM radars which are very accurate and can specify the orbitographic data of the catalog when it is particularly necessary (e.g., calculation of avoidance maneuver). Finally to monitor distant orbits (median or geostationary), the COSMOS has access to the GEOTRACKER system and can program CNES TAROT telescopes.

In 2008, the White Paper on Defense and National Security recommended that "the doctrine of space operations and programs be placed under the responsibility of an identified and dedicated joint command under the authority of the Chief of the Joint Staff." As a result, the Joint Space Command (CIE) was created in 2010 to develop and implement the French military space policy.

In its current form, the CIE exercises only a simple coordination of national space activities. With the exception of the field of international relations applied to the military space in which he is the leader, under the control, however, of the international relationships division of the Joint Staff, the CIE has no command. Its responsibilities are at best shared at worst, attributed to other military organizations.

The preparation of the future and the space programs are conducted by the CIE under the kind control of the Deputy Chief of Staff Plans of the Armed Forces. Military Intelligence Directorate (DRM) and Information Systems Directorate (DIRISI) have full authority, respectively, over the observation and telecommunications space systems. Authority over space surveillance activities is shared with DRM and CDAOA. The DGA and the CNES exercise all their authority over prospective studies and R&D activities. Faced with these Defense heavyweights, the CIE's room for maneuver is very narrow.

Nevertheless, following space events in recent years concerning French military satellites, the Ministry of the Armed Forces became aware, on the one hand, of the dependence of military operations on space support and the vulnerability of the space component of Defense on the other hand.

On 13 July 2019, the President of the Republic affirmed the position of France by recalling that space is "essential for our operations" and that "by the incredible potential it offers but also by the conflicts that 'it raises', it 'is [. . .] a real stake of national security.'" During this speech he asked for the drafting of "a defense space strategy" which "will also have a vocation to be applied, in all relevant aspects, at European level" (France 24, Macron announces creation of French space force, 13 July 2019, https://www.france24.com/en/20190713-macron-france-space-force).

As a result, a new national Defense space strategy is being developed. It should be published this year and restore the military space necessary dynamism to meet the challenges of security in space and give the CIE a real role of commander of French military space capabilities.

As detailed above, the majority of French space surveillance capabilities date as early as the 2000s and therefore require urgent modernization in order to be adapted to the new international space context.

The latter is characterized by a drastic reduction in the size of space systems in orbit and the development of threats against operational satellites by the majority of space powers.

It is no longer a matter of simply monitoring the space; it is also necessary to be able to guarantee the safety of our satellites.

The effort will be tremendous both financially and in terms of human resources. The Ministry of the Armed Forces will have to make choices, and they will be dramatic since they are not anticipated: more space will certainly mean less of one (or more) other capacity. The new capacities will have to be developed in respect of the principles detailed below.

Basic Principles for Enhanced Security in Space

First Principle: Resilience

Resilience is the ability of a system to react to a disturbance, to reorganize itself, and to continue to operate in the same way as before the disturbance. In terms of military space, resilience makes any attack on the resilient system particularly difficult or even ineffective. In addition to hardening the space systems involved, which makes them more expensive and heavier, resilience can be achieved by dispersing the single satellite function across multiple systems, a solution called distributed architecture. For example, a telecommunication satellite constellation is a distributed architecture solution, which makes this telecommunication system particularly difficult to disrupt since several satellites in the constellation would have to be destroyed in order to significantly degrade the system's capacity. Furthermore, a constellation of small satellites with spare satellites has a greater capacity for easy renewal than a single large satellite, which further increases this resilience.

The USA has perfectly integrated this distributed architecture concept since they have found on the one hand their dependence on space systems and on the other hand the fact that the geostationary orbit is no longer invulnerable. By focusing on the very design of the system, they provide a relevant answer, reliable, efficient, and less expensive than other solutions.

The distributed architecture concept can also apply to other aspects. One of them is to disseminate a same capacity on several systems between several allied countries (observation, GNSS, communication). For example, the Galileo and GPS systems both provide a GNSS signal, so it is possible to subscribe to both systems and thus have continuity of service if one of them is unavailable because of disruption by unfriendly country or enemy.

The principle of resilience is thus a first step toward what could be called "spatial deterrence." Indeed any act of destruction of any spatial capacity would de facto create a "deterrent" cost both financially and diplomatically. The next step to be fully deterrent is the ability to identify the aggressors without any doubt.

Second Principle: Knowledge

"To know how to defend oneself, one must first know against what to defend." As such, two deadlines need to be considered.

In the short term, the regular monitoring of the enemy's behavior will allow to understand his tactical approach and to decode his strategy. The intelligence work is essential to decode the enemy's behavior and to anticipate it. It is primarily founded on the surveillance function that is to say the regular monitoring of activity in space. The major space powers of the world perform this watch using both terrestrial means (radar, telescopes) and space means. Then the spatial situation has to be elaborated and constructed from the analysis of the movements of space objects to detect any potential threat; this is what is called the SSA (Space Situational Awareness).

Although France is one of the few nations to have developed this capacity, it remains very modest compared to the major space powers and must now be modernized and improved through a real ambitious SSA capacity procurement plan.

Later, the technologies changes will bring out new capacities that will have to be taken into consideration. Thus the electrical propulsion applied to space (of low thrust but of very high yield) makes it possible to board masses of propellant 6 to 20 times lower, thus allowing greater maneuvering in space. This technology, combined with the rapid development of robotics and coupled with artificial intelligence, will enable friendly or unfriendly automatic space appointments. These developments make it even more necessary to update our surveillance capabilities so as not to be threatened and defenseless.

Above all, it is first and foremost a question of technological foresight and operational innovation by developing a clear vision of the strategic objectives in this field. With regard to the latter, its horizon of deployment, and our own scientific and technological capabilities, our political ambitions, and the importance given to our autonomy of appreciation and action, it will then be possible to define areal space strategy and to develop the technical, financial, programmatic, and operational means to implement it.

Third Principle: Protection

On the one hand, protection is performed by the detection of the hostile activity then on the other hand by the security of the threatened satellite (self-protection or maneuver). If anticipation makes it possible to know what the possible aggressions are, it is necessary to detect the attack, which is complex to realize in orbit. That is why the capacity of self-protection of a satellite becomes essential regarding the emergence of the nano-satellites particularly difficult to detect from the ground. Today self-protection systems on board or fly-by are being tested in the USA; they are an effective means of detecting the threat, without having to monitor all the orbits. Such systems deserve to be developed and should be fitted at least on any satellite.

In the event of detected hostile activity, the priority will be to put the targets in security, out of reach of the attacking systems. Maneuvers can be calculated and commanded from the ground stations based on the information collected by our space surveillance systems and taking into account the knowledge gained about enemy modes of action. Subsequently, self-protection systems will enable our satellites to respond autonomously to the threat.

Fourth Principle: Action

The best way to deter an enemy from attacking our space systems is to let him know that we have the means to deal with his own satellites in comparable or even greater losses. These means of action must make it possible to conduct offensive operations similar and proportionate to the attack. This is why it is essential to have a complete panoply to act on the ground, from the ground, or completely in space.

As deterrence is concerned, a proper communication campaign should be designed to demonstrate capabilities to potential adversaries but also reassure allies and neutral countries that EU will fulfill its international commitments in terms of regulation and space law (no weaponization of space, no production of space debris).

Starting today, without developing any additional specific means, it may be possible to respond to an attack in space by targeting the ground infrastructures of the enemy's space systems. He must feel that we know the precise location of these vital centers for the implementation of its space component (space communication networks, satellite control centers, space command center, etc.), just as we have conventional weapons to destroy them by many means (operations in cyberspace, electromagnetic jamming systems, air-to-air or naval cruise missiles).

In order to broaden our field of capacities and to be able to better adapt our response, we need to have, in the medium term, systems of action from the ground to the space such as lasers, directional jammers, and microwaves weapons.

Finally, in the long term, it will be necessary to develop new capacities such as active systems in space, using new technologies and with disruptive courses of action.

Once we will master all these new kinds of operations, we will be able to deter an enemy from attacking our space systems and thereby continue to use them for the benefit of our society and the effectiveness of our diplomacy and military operations.

Next Step

Modernization

In direct connection with the new Defense Space Strategy sought by the President of the Republic, modernization of current capabilities is the challenge of the coming years.

As it has already been written, it will have to go far beyond the simple replacement of the current obsolete systems. It will, first of all, be a question of extending the knowledge of space activity to the totality of the exo-atmospheric space: from low orbits (200 km) to geostationary orbit (34,000 km).

National sovereignty in this strategic area is obviously sought, so it will be necessary to be able to unambiguously identify each of the space objects in orbit. This ability is certainly the most complex to acquire and therefore the most expensive. Two different paths are usually taken to achieve this:

- The first, the least complicated but also the least accurate, is to be able to detect each new space launch and then to follow the new object since its separation from the launcher until it goes into operational orbit. The space object is thus registered in the catalog and assigned to a country, but its mission and its performances remain however to be determined by other means of intelligence. This method requires a surveillance system capable of detecting and tracking space launches, which is then supplemented by a space monitoring device.

- The second most accurate but more complex is to make very high-resolution images of the object and then determine the nationality, mission, and performance by image analysis. This method requires accurate, high-performance, ground-based, or deployed in-orbit imaging systems.

Combination of both methods is possible, extremely powerful, but also extremely expensive.

In accordance with the principles set out in the previous chapter, knowing the activity in space is not enough to guarantee the safety of satellites; it is now necessary to be able to protect them. The first step in this direction will be to equip them with self-protection devices, in order to detect the threat, to characterize it, to signal it, and to plan an emergency evasion maneuver. Next-generation satellites should have such devices.

A first action capability in space or to space should also be programmed. The latter is essential to the credibility of the overall posture. Indeed, how to deter an adversary from attacking our space systems if we do not have the ability to make him fear a self-defense as powerful as his attack?

The form of this component remains to be determined, certainly according to the available financial means but especially in respect of the international commitments of our country. From the ground up, the first capability that comes to mind is laser illumination to disrupt or even destroy the electronic devices of an opposing satellite. It is already technically accessible. In space, we have been mastering for several years the technique of orbital rendezvous; the conception of a space system of interception or close investigation is conceivable in the medium term and should be decided if the evolution of the nearby threats of our satellites is precise.

Finally, all these new and renovated capabilities require a command system dedicated to space. It will have to be designed in order to connect the different actors and systems of the SSA (Space Situational Awareness) chain of the Armed Forces, to provide the essential elements for the decision-making of the higher national authorities.

Sensors on the ground and in space, self-protection of the satellites, capacities of action in or toward space, and the effective realization of all these components are certainly necessary but out of reach of a budget of the Defense already under-equipped to answer the needs of deterrence and traditional components. It is therefore only at the European level that these capacities can be developed.

Cooperation Within the EU

However, since its origins and regardless of the field concerned, the European Union has never accepted the leadership of a single country being an expert in the field. Thus, the development of all the necessary capabilities for European sovereignty in space cannot be properly initiated and carried out relying solely on a coalition of ad hoc political wills. Several methods and modes of governance can lead to the result; those that seem to the author the most relevant are described below. In any case it will be necessary to convince to share an already late awareness and to mobilize competences, budgets, and efforts to reach a satisfactory result within operationally acceptable but realistic deadlines.

The first method is that of the Space Surveillance and Tracking (SST) program, based on a limited number of participants, who organize themselves to share governance and thus the allocated budget and responsibilities. The top five countries were Germany, Spain, Great Britain, Italy, and France. The initiative seemed ambitious and seductive, but it proved itself quickly unproductive. Its main weakness was to disregard the existing capacities demonstrated by each of the participating countries which led to an equal share of budgets between each and finally to a general disappointment. Indeed, countries that need to modernize their capacities (France and Germany) have not received enough European budget to achieve this. The countries without capacity (Italy and Spain) and wishing to invest to acquire one were on the other hand overbudgeted but unable of using usefully the allocated amount. As for Britain, as usual, it tried to place itself, unsuccessfully, as a referee above the fray. Failure is proven and this method of governance has now to be avoided.

A second possible method would be to attribute the realization of a first European capacity to the only two countries with developed means and experience in that field, Germany and France. Then it will be up to these two nations to agree to modernize their capacities and bring them into coherence with one another in order to achieve the complementarity required to satisfy the most urgent needs. However, it will be necessary to obtain prior approval from all other member countries and for this to give guarantees of service availability to all, knowledge and know-how transfer to those who would later be interested, and finally plan in the medium term, the creation of a true European space security center, different from the European Union's satellite center, which will have to remain focused on image analysis.

The third method that the author envisions and which he advocates will begin with an objective inventory of what exists in Europe (capabilities, services, know-how, and experience). Countries and companies offering real guarantees will then be audited to contribute to the creation of the European space security capability. Complementarity should be sought and redundancies eliminated. The European External Action Service could be responsible for this work in order to eventually create a European Space Operations Center (EOC) or EUSpOC (EU Space Operations Center).

Here is a possible agenda:

1. Inventories of existing capacities in the states of the European Union (radars, telescopes, catalogs, databases, operations centers)
2. Inventories of the commercial services of the space domain
3. Choice of capacities/services to be selected for the first European component
4. Establishment of the European Space Operations Center
5. Constitution of European capacity (connection of selected systems/services)

6. Constitution of the European catalogue of space objects/transfer of know-how to voluntary member countries
7. Review of necessary upgrades/additions
8. Achievement of selected improvements involving new actors

At the end, the European Union could be endowed with space security capabilities at the height of the protection of its space assets.

As already said the last method has the author's preference because it is the one that will allow the best taking into account the existing capabilities in each of the European countries while allowing a transfer of know-how to new players wishing to contribute to the European capacity. Governance, provided by the EEAS (European External Action Service), will overcome the risk of hegemony of large countries on the project and facilitate the involvement of new ones. Finally, the support and then the adhesion of all will be much easier to obtain.

Conclusion

Like other countries in the European Union, France has developed extremely powerful satellites to ensure its sovereignty, its independence, and its autonomy of appreciation of situation and decision. In a desire to strengthen its political component, the European Union has done the same. As a result nowadays, the European society, like the other Western societies, is dependent on space for its daily functioning, security, and defense. Current space systems, although essential, are fragile and almost unprotected in space. At the same time, the main space powers are experimenting anti-satellite systems increasingly efficient and elaborate.

We must take into consideration this trend and deploy a real capacity of spatial deterrence. Since the advent of nuclear weapons, the term deterrence has been reserved exclusively for the corresponding doctrine. However, if we want to maintain the control of the use of our space systems regardless of their application, it is now necessary to deter the countries that equip themselves with systems to attack our satellites either by spying, scrambling, deception, or even more coercive actions. If it is already possible to operate in space at a low level, the acquisition of this deterrent capacity will take time and a lot of money. Acquisition of a full and independent capacity will be possible only at the European level, distributing services to all member states. This is the price to pay for ensuring the security of EU Space component, essential for European independence and sovereignty.

References

République Française, The French White Paper on Defence and National Security 2008 (Le Livre blanc sur la défense et la sécurité nationale), 17 June 2008
République Française, French White Paper on Defence and National Security 2013 (Le Livre blanc sur la défense et la sécurité nationale), 29 April 2013

Italy in Space: Strategic Overview and Security Aspects

66

Francesco Pagnotta and Marco Reali

Contents

F. Pagnotta (✉) · M. Reali
Presidency of the Council of Ministers – Office of the Military Advisor, Rome, Italy
e-mail: F.Pagnotta@governo.it; M.Reali@governo.it

© Springer Nature Switzerland AG 2020
K.-U. Schrogl (ed.), *Handbook of Space Security*,
https://doi.org/10.1007/978-3-030-23210-8_111

Abstract

Space is relevant for humankind. This is more than a concept to be agreed upon or not; it is a fact. People on Earth rely on space capabilities, applications, and services more than in the past, and it is quite impossible to think of "a day without space." Italy has a very long history and heritage in the space sector. Starting from the period of the "space race," Italy researched and developed technology, systems, operations, and applications over the years, and today, Italy has national telecommunication and Earth observation systems, participates in the main space-related international organizations, and has very strong bilateral collaborations in the space sector. Considering the aforementioned, Italy recognizes that the space domain is becoming more and more congested, contested, and competitive. Therefore the need for a space security program is one of the pillars in the national strategy.

Introduction

Space in Italy begins at the end of the 1950s with a team of researchers and operators from both academic and military sectors. Space activities came out of the pioneering phase faster than expected. Just after the last century, the space sector entered into a period of rapid growth.

Today, space applications and services are an integral part of daily life. If we think about the weather, communication, or navigation, each smartphone can integrate them. Earth observation, telecommunication, and navigation systems are essential in order to provide support to people on Earth, for both civil and military activities.

Key factors of this increasing relevance of space activities and capabilities are high technology, miniaturization, big data, and digitalization. These are just some of the many aspects supporting the transition to a new era in the space domain. In 2012, the USA proposed a scenario for the space domain in 2025: congested, contested, and competitive. This scenario seems to have arrived about 5 years earlier than expected. Some international initiatives, at both American and European level, have been proposed to foster the cooperation and the exchange of data and information among countries. Safety and security of "access to" and "operations in" space seem to be the most relevant topics in decision-making at the higher levels.

Today, space capabilities are essential, and their unavailability is considered more and more as a distress. It represents a vulnerability, and any similar ground-based capability cannot replace it. Therefore, in the last years, it is necessary to have a clear and shared space security strategy in order to mitigate the natural and human-made threats in and from space. Space Situational Awareness, Space Traffic Management, Cyber Threats, and Cyber Security are some of the critical topics, and they are currently under discussion in most of the international space fora.

We can easily affirm that space and aerospace are interdisciplinary sectors where scientific and technical aspects merge with socioeconomic, political, and regulatory aspects, and we need to integrate them adequately. This leads to the adoption of a comprehensive approach in space, which takes into account all aspects and optimizes the use of capabilities according to the availability of funds at both national and international level. Finally, we have to consider that any applications and operational services support both civil and military operations. This underlines the relevance and opportunity to apply a comprehensive approach as well in the research and development of new technologies and capabilities, in order to build systems based on needs from both civilian and military field. The coordination between all stakeholders and actors – at both national and international level – becomes a key factor in the space and aerospace sectors, in order to find synergies, to encourage the development of new initiatives for international collaboration, and to support new financial means.

In this spirit, talking about new trends and markets, the commercialization of space deserves a particular mention. Initiated in the USA to facilitate and support funding from private entities in the space sector, this new trend is growing faster and faster all over the world. In Italy, in 2016, a space economy strategic program was approved with the objective to merge old and new funding lines, based on public and private support. This is a really new trend in all Europe, trying to move from the old business model, based on institutional investments to the new one based mainly on the private investments for the procurement and public investments for the services.

During the 11th Conference on European Space Policy, held in January 2019 in Brussels, the European Parliament and the European Space Agency mentioned the new "European space economy." The objective is to establish a new space market not only limited to institutional customers but also open to private and commercial sectors. This is something similar to the Italian public-private partnership that has been used during the last 10 years, with reference to the procurement of national space systems (e.g., SICRAL 1B satellite).

Space in Italy

The Origin

In the beginning of the 1950s, Italy recognized space as a strategic domain. Space assets, satellites, and orbital platforms were equipped with the latest technology.

In 1964, Italy was the third country to conduct the launch of a satellite (United Kingdom, with the launch of Ariel 1 in April 1962, was the third county to operate a satellite, and Canada, with the launch of Alouette 1 in September 1962, was the third country to build a satellite.), the San Marco 1, with the aim to study the lower atmosphere within the framework of collaboration between the Aerospace Research Center of the University of Rome and NASA. This was just the first of a series of satellites since 1967, taking place at the United States Wallops Island Flight Facility base in Virginia (USA) and launched from the base at the Italian center of Malindi in Kenya.

History

After more than 50 years of space history, today, Italy is at the frontier of research and technology in space.

The knowledge and experience in space has been increasing in most of the areas of the space domain: satellites manufacturing, launch and operation, launchers, enabling technologies and equipment (antennas, rockets, etc.), telecommunication and earth observation, human spaceflight and exploration, and ground segment.

Academic research and support to military and civilian operations have mainly been the triggers the rapid growth in space technologies. These two factors enabled the most of the technologies and lead to the miniaturization of electronics and structures in order to fit the stronger requirements of both military and civilian assets.

Starting with the San Marco 1 launch, Italy has developed a scientific and industrial background leading to over 100 space missions. Among others we can mention the participation to the International Space Station with astronauts, experiments, and the Italian Multi-Purposes Logistic Modules and the scientific missions like Cassini-Huygens, Rosetta, Mars Express, Hipparcos, and BeppoSAX.

International cooperation played a key role in this "Italian space adventure." Italy is a member of the European Space Agency and collaborates with most of the space agencies all around the world: the National Aeronautics and Space Administration (NASA), the State Space Corporation ROSCOSMOS, the China Manned Space Agency and China National Space Administration, and the Japan Aerospace Exploration Agency (JAXA) are just some of the collaborations in place. In this context, space diplomacy is considered vital in order to face global challenges and research in space in a collaborative approach.

After the launch of the first national satellite, San Marco 1, the Italian scientific and industrial community worked together in order to develop other and more complex space systems. Today we have in Italy a complete chain of products in the space sector. The first Italian launch site was located near the coast of Malindi (Kenya), a floating launch site composed of three platforms where an Italian team launched 27 rockets and 9 satellites (with an American launcher, the Scout B) between 1967 and 1988. The launch platform was named San Marco, and the other two platforms – power support and control room – Santa Rita 1 and Santa Rita 2.

In 1977, the first Italian telecommunication satellite (Sirio) was launched from Cape Canaveral. In 1988, the Italian Space Agency together with the Italian manufacturing and scientific community started to realize more and more national missions. In 1991, the second Italian telecommunication satellite was launched (Italsat 1), and just 1 year after, in 1992, the first Italian astronaut (Franco Malerba) operated the Italian mission named Tethered from space.

In 2007, Italy joined the challenge of building a European navigation system: Galileo. In 2011, the first two satellites of the constellation were launched (Giove A and Giove B), and today, the constellation is operational, and two more satellites are planned to be launched in 2020.

In 2007, Agile, an Italian satellite for gamma ray evaluation, was launched, and after 5 years, in 2012, the first launch of the European VEGA launcher was conducted. The studies and the development of the launcher were carried out by the European Space Agency, and the prime contractor was an Italian company. Within 20 years, the Italian Space Agency became a significant player in the world in developing satellite technologies and mobile systems for exploring the universe and space science.

Recent Past

In the recent past, the main players concerned in space were the Ministry of Education, University, and Research and the Ministry of Defence. The promotion, development, and dissemination of scientific and technological research applied in the space sector, including coordination and management of national projects and Italian participation in the European and International programs, were in charge of the Italian Space Agency, within the Ministry of Education, University, and Research. The Joint Defence Staff, within the Ministry of Defence, was in charge of military exploitation of space, including related research and development activities. The Defence was in charge of space operations and responsible for planning, coordinating, and supervising technical and operational space-related activities in support to military operations, including the homeland security aspects.

The political coordination of space-related topics was in charge of the Ministry of Foreign Affairs and International Cooperation, with the supervision of the Presidency of the Council of Ministers, through the Office of the Military Advisor to the Prime Minister.

New Governance in Italian Space and Aerospace Sector

The space and aerospace sectors have been recently reorganized with the Law 11 January 2018, n.7, which reshapes the governance of national space policies. This Law, at the forefront of the European context, gives to the President of the Council of Ministers top management, general political responsibility, and coordination of the Ministries' policies relating to space programs. The Law establishes an "Inter-Ministerial Committee for Space and Aerospace policies," in which 12 Ministers (Defence; Interior; Cultural Heritage and Activities and Tourism; Agricultural, Food and Forestry Policies; Education, University and Research; Economic Development; Infrastructure and Transportation; Environment and Protection of Land and Sea; Foreign Affairs and International Cooperation, Economy and Finances; South; European Affairs.) and the President of the Conference of the Regions and the Autonomous Provinces participate.

The Law reconfirms the strategic importance of the space sector to support and enhance the economic, social, and industrial development in Italy. In particular, the new Law has conferred to the Prime Minister the general political responsibility and the coordination of the Ministries' space programs.

Upon invitation, the President of Italian Space Agency participates in order to provide technical and scientific support. The Military Advisor to the Prime Minister was appointed as Secretary, in charge of the activities of support, coordination, and secretariat of the Inter-Ministerial Committee.

The Committee was created in order to ensure the guidance and coordination in space and aerospace sectors with reference to related operational services.

The Committee, with the modalities defined by its own internal regulation adopted in its first session and taking into account the guidelines of the national foreign policy and the European Union's space and aerospace policy, has many tasks; some of those are:

- Defines the government's guidelines on space and aerospace with reference also to research, innovation, technology, and impact on the production sector
- Directs and supports the Italian Space Agency in the definition of international agreements and in relation with international space organizations
- Ensures the coordination of programs and activities of the Italian Space Agency with the programs and activities of central and peripheral administrations
- Identifies the main guidelines for participation in the European programs of the European Space Agency and for the development of bilateral and multilateral agreements
- Defines the guidelines for the development of synergies and cooperation in the space sector between research bodies, public administration, universities, and companies, with particular reference to small and medium-sized companies in the sector
- Defines the framework of the financial resources available for the implementation of space and aerospace policies, according to criteria for the promotion and development of innovative satellite services of public interest, pursuing objectives of synergy between public and private resources, destined for the creation of space and aerospace infrastructures.

The Secretary is also the Chairman of a Coordination Board, established in order to analyze the dossiers, propose solutions to the political forum, and verify that the activities are in accordance with the political guidelines, provided by the Inter-Ministerial Committee. The Coordination Board is a Technical forum where the representatives of the 12 Ministries, the Conference of the regions and the Italian Space Agency coordinate and define the national positions for the space and aerospace programs and policies, for the approval of the Inter-Ministerial Committee. The main tasks of the Coordination Board are as follows:

- Coordinate and define the national positions, presented by Italian delegates in the European Union, European Space Agency, and in bilateral cooperation
- Implement the Inter-Ministerial Committee's decisions

- Elaborate studies and evaluations for the space and aerospace sector
- Finalize proposals and documents for the approval of the Inter-Ministerial Committee

The representatives of the 12 Ministries and the Conference of the regions (as delegates of the Inter-Ministerial Committee) take part, together with representatives of the Italian Space Agency, in the Coordination board, as permanent members with voting rights.

The Departments for Civil Protection, Regional Affairs, Cohesion Policies, Intelligence, and the Italian Institute for Environmental Protection participate, without voting rights, in the meetings of the Coordination Board.

In addition, representatives of industrial associations, industries, academic institutions, and research entities may be invited as technical consultants, based on opportunities and depending on the agenda of the meeting.

From Operations to Applications

Launchers

From past experience starting with the Scout B launches in 1988, Italy conceived the VEGA project; a launcher based on the Zefiro motor for the European Ariane program and developed by the Italian industry Bombrini Parodi Delfino.

VEGA (Italian acronym for Advanced Generation European carrier Rocket) is an expendable launch vehicle, developed by the European Space Agency in 1998. The first launch took place in 2012 from the French Guyana Space Center.

Vega is a single-body launcher with three solid rocket stages (named P80, Z23, and Z9, respectively, the first, the second, and the third stage) and an upper liquid rocket stage (named AVUM). The payload range is from 300 to 2,500 kg satellites to polar and Low Earth Orbit (LEO). The reference mission is a polar orbit with 1,500 kg satellite at 700 km.

Italy is the leading contributor to the VEGA program, with 65%, followed by France with 13%, and other participants with 22%, including Spain, Belgium, the Netherlands, Switzerland, and Sweden.

The development costs for the launcher were about €710 million, with spending an additional €400 million to sponsor five development flights between 2012 and 2014. On 14 February 2012, VEGA carried out the first launch, and, today, it placed in orbit more than 20 payloads into orbit, including small satellites (Cubesat), the Intermediate eXperimental Vehicle, a part of the SkySat constellation by Terra Bella of Google, the experiments on gravitational waves Lares, and Lisa Pathfinder.

VEGA can be considered as a great success for both Avio and the over 40 European companies that contributed to the project (Fig. 1).

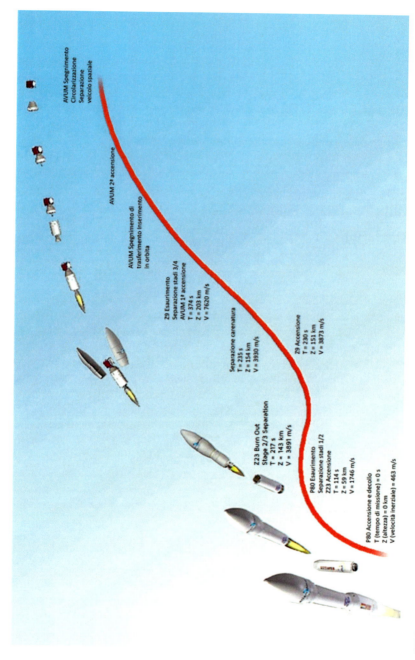

Fig. 1 VEGA launch mission profile scheme – Credit: AVIO S.p.a

Satellites

Telecommunication

After the launch of the first Italian telecommunication satellite, Sirio, Italy begun the development and launch of a series of national satellites: Italsat 1 (1991), Italsat 2 (1996), managed by the Italian Space Agency and not operational so far, the military SICRAL 1 (2001), SICRAL 1b (2009), and, in cooperation with France, SICRAL 2 (2015) and the dual-use ATHENA-FIDUS (2014) (Fig. 2).

The SICRAL (Italian acronym for Italian System for Secure Communications and Alerts) is the national system (three satellites are operational) for military communications, providing interoperability between the networks of defence, law enforcement, and civil emergency support to the management of strategic infrastructure. The program is divided into three phases, and the third one started in 2015 in cooperation with France. The system, starting from 2001, provides communication in UHF, SHF, and EHS/ka bands, anticipating the needs of growth and development in the next few years.

The SICRAL control team operates from the SICRAL Joint Control Center, located in Vigna di Valle (Roma), with a backup center in the civilian "Piero Fanti" Space Centre in Fucino (L'Aquila).

Fig. 2   SICRAL 2 artistic view – Credit: Telespazio S.p.a

The ATHENA-FIDUS satellite (Access on THeatres and European Nations for Allied forces – French Italian Dual-Use Satellite) is a French-Italian telecommunications system, developed as a part of a collaboration between Italian and French space agencies and the Ministries of Defence. The satellite complements some capabilities already offered by the Syracuse and SICRAL systems in EHF and ka bands. The ATHENA-FIDUS is operated from a French control center.

Earth Observation

Italy has a strong tradition in research technology and development activities, together with robust synergies among scientific community, governmental institutions, and national industries. A high maturity level was reached in different application domains where Italy today plays a key role: one of them is the Earth observation sector.

COSMO-SkyMed is an Earth observation program, funded by the Ministry of Defence and the Ministry of Education, Universities, and Scientific Research. It consists of a constellation of four satellites equipped with high-resolution X-band radar sensors. The system allows operations under any weather or visibility conditions and with very high revisit frequency. The program is dual-use and provides support in the context of international agreements, including Copernicus, the European program for Earth observation.

The constellation was implemented in stages: the first satellite was launched in June 2007, the second in December 2007, the third in October 2008, and the fourth in November 2010. The ground segment is located at the "Piero Fanti" Space Centre, in Fucino (L'Aquila), for the control of the constellation. There are other locations around Italy for the planning and the acquisition, processing, and distribution of satellite data (Fig. 3).

OPTSAT-3000 is a high-resolution optical satellite for imagery, owned by the Ministry of Defence, with a mission life estimated in 7 years. A team from the

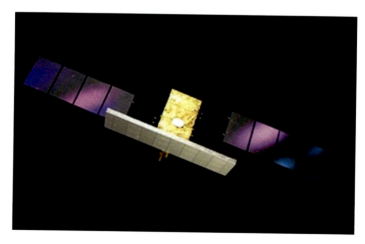

Fig. 3 Cosmo SkyMed artistic view – Credit ASI

SICRAL Defence Control Center, located in Vigna di Valle (Rome), operates and controls the system, launched in 2017 with VEGA launcher from the French Guyana Space Center. The mission planning, acquisition, and processing of data is made in other sites in Italy. The Israel Aerospace Industries, based on inter-governmental agreements, built the satellite. The system is able to interoperate with the second-generation of radar satellites COSMO-SkyMed, planned to launch in the near future. It provides to the Ministry of Defence the capability to integrate optical and radar data generated by the two systems.

European Space Infrastructure and Italian Participation

Italy, as part of the European Union and the European Space Agency, participates in the Copernicus and Sentinel programs. The European Space Agency is developing six families of Sentinel satellite missions, providing high-resolution radar and optical images and according to the needs of Copernicus. The European Commission coordinates and manages the Copernicus program, in order to provide accurate, timely, and easily accessible information of the Earth, including the marine and atmospheric environments.

Italy participates also in the European Galileo program, a collaboration of the European Union with the European Space Agency to ensure technological independence of Europe and to provide international standards for Global Navigation Satellite Systems. At full operational capability, Galileo consists of 30 Medium Earth Orbit satellites and a ground infrastructure consisting of two Galileo Control Center, one in Germany, in Oberpfaffenhofen (Munich), and the other one in Italy, at the "Piero Fanti" Space Centre, in Fucino (L'Aquila).

In 2014, the validation campaign for the early provision of four types of Galileo services was initiated: open service (OS); commercial (CS); public regulated service (PRS); and search and rescue support service (S&RSS). The types of services offered by the Galileo system are distinguished in open or encrypted based on the needs of the end users.

The Italian government ensures political and industrial support for these European programs, in order to reinforce international cooperation and to guarantee capabilities and applications from and in space.

Ground Segment

"Piero Fanti" Space Centre – Fucino (L'Aquila)

The Fucino Space Centre is the largest civil space center in the world with approximately 170 antennas and 250 employees (engineers, specialist technicians, and operational staff). Owned by Telespazio S.p.a., the center is active since 1963.

The main activities carried out are:

- In-orbit satellite control
- Launch and early orbit phase control
- Telemetry, tracking, and command services

- Ground station network services
- Flight dynamics for all types of satellites and orbits
- Telecommunication, television, and multimedia services
- Operational logistics, field, and hosting services (Fig. 4)

SICRAL Joint Control Center – Vigna di Valle (Roma)

The SICRAL Joint Control Center is located in Vigna di Valle, within the Air Force Base "Luigi Bourlot," near Roma. It is the only control center in Europe operated by military personnel, and the main missions are, according to the North Atlantic Treaty Organization applicable documentation (Air Joint Publication 3.3 B, regarding the Allied joint doctrine for air and space operations):

- C2 Satellite/Space Control (SICRAL e OPTSAT-3000)
- Space Force Enhancement (satellite communication)

Fig. 4 "Piero Fanti" Space Center, Fucino (L'Aquila) – Credit: Telespazio S.p.a

• Space Situational Awareness (space situational awareness) (Fig. 5)

Matera Space Center – Matera

The Matera Space Center is active since 1994 and is located within the Space Geodesy Center of the Italian space agency (opened in 1983).

It is part of ground station network operated by Telespazio S.p.a. and dedicated to the Earth observation activities. The center acquires, processes, stores, and distributes data from the main remote sensing satellites, and, in addition, produces images, products, and services in near real time for maritime surveillance. The Matera Space Center is one of the three stations of the Core Ground Segment of the European Space Agency, within the Copernicus program.

From Upstream to Downstream, Through Midstream

Italy has a complete chain of products in the space sector. The space economy is the whole chain of value, starting from research, development, and realization of systems (upstream) to the furniture of products and services (downstream). All systems that collect, process, and provide the data and information to the user are called "midstream."

Since 2016, the Italian Government began to promote a new policy in the space sector based on the signals and activities initiated at European level:

Fig. 5   SICRAL Joint Control Center, Vigna di Valle (Roma) – Credit: Ministry of Defence

- Scientific and technological programs within the European Space Agency, in collaboration with European Union
- Space program directly funded by the European Union
- The European research and innovation program (Horizon 2020)
  The main programmatic lines were:
- Institutional funding for research, based on the Italian Space Agency budget, and for Earth observation and telecommunication considering also the Defence funds
- Technological programs based on funding lines from the Ministry of Economic Development

The space economy is based on the needs provided by the big, medium, and small enterprises in the space and aerospace sectors, in order to provide new services and products based on big data collection and processing capability. At the same time, the Italian Government started to put together all funding lines in space and aerospace sectors, in order to provide a sort of joint funding line (merging the existing ones).

A "strategic plan for space economy" was published in order to provide guidelines and create a National Strategy for Smart Specialization. Six programs started at national level:

- Telecommunication program
- Galileo support program
- Galileo Public Regulated Service infrastructure support program
- Copernicus support program
- Space Surveillance and Tracking support program
- Exploration and space technology development program

Space Security

Space Security

Space can be considered as the fourth environment of operation, and, therefore, it affects directly national security. Space services and applications – once used only for research purpose – are increasingly entering the everyday life. The services offered by all infrastructures in space will be used more and more by national, regional, and local institutions, as well as by private companies for their business and by citizens to support their daily activities. Furthermore, the use of telecommunication, navigation, and earth observation space capabilities is becoming synergistic with robotics, city management, and drones programs.

The topic of space security is discussed in the world among many national and international for, yet, currently, without a defined and shared definition so far. Usually security is something related to military aspects, not only limited to the services, applications, and exchanged data or information. The demand for space applications is rapidly increasing, due to the awareness of the relevance and flexibility of space services (like telecommunication, Earth observation, and navigation) in support of everyday life, to operations, and to emergency. At the same time, the quality of services requested by the user is becoming more and more challenging for the industry; therefore, space systems become more complex and harder to manage and control. This leads to the new era of digitalization, artificial intelligence, internet of things, and 5G. Concurrently more countries start to enter in the space and aerospace sectors and markets.

Considering the aforementioned, the number of satellites is increasing, and the technology is rapidly changing. These two aspects are important with respect to the safety and security of space infrastructures and applications.

Therefore, in the last years, the importance of understanding and defining space security is becoming relevant and critical. At the same time, it is strategic to consider the safety of operations in space. Both security and safety in space affect military and civil applications, at national and international level. This means that it is not only a fact of national security. It is also important to take into account the collaboration and cooperation with other nations, interested in ensuring the sustainability of operations in and access to space.

Security, safety, resilience, and sustainability seem to increase their relevance today. How is the relationship between these words? In order to ensure the resilience of space infrastructures and to guarantee the sustainability of space operations, we need to strengthen both the security and the safety of space. The first step is to identify and define what the threats from and in space are.

Space Threats

It is possible to categorize the space threats in many different ways. One of them takes into account two main groups: intentional and not intentional. The first group is usually related to the security aspects and the second one is more related to the safety. If we consider the intentional threats, they can vary significantly depending on the type of effects they create, the kind of technology they use, and the level of resources needed to develop them. With this in mind, a first classification of threats is possible as below:

- Kinetic physical (counter-space weapons to strike or detonate a warhead)
- Non-kinetic physical (laser, high-powered microwaves, and electromagnetic pulse weapons)
- Electronic (jamming or spoofing radio frequency signals)
- Cyber (attacking the data and not the radio frequency signals)

With respect to the not intentional threats, they are usually related to the natural threats (asteroids, meteors, etc.) or to debris and/or not operational satellites or rocket bodies. We are talking about the so-called environmental safety and monitoring.

At the same time, we can introduce a transversal classification, if we look at the origin of the threats. Therefore, looking at the not intentional threats, we have ground-based threats (e.g., natural occurrences and power outages), space-based threats (e.g., solar, cosmic radiation, temperature variation, or space objects, like debris or rocket bodies), and interference-based threats (e.g., unintentional human interference). Moving on the intentional threats, we can have sabotage or physical destruction for ground-based, interceptors, or directed-energy weapons for space-based and cyber-attacks or jamming for interference-based.

Space Surveillance

Italy is one of the main contributors to space in Europe, both in the European Space Agency and in the European Union. The Italian Space Agency has participated in the Space Surveillance and Tracking activities within the European Space Agency program and Italy joined from the beginning the European Commission initiative named Space Surveillance and Tracking Support Framework. The European Union recognized the relevance of space infrastructures and investments, due to the Galileo and Copernicus programs, and decided to begin the Space Surveillance and Tracking initiative with the objective to bring together the Member States' surveillance and processing capacities, in view of setting up an initial European service, consisting of conjunction, reentry, and fragmentation analysis. Since June 2015, Italy is part of the European Union Space Surveillance and Tracking Consortium, together with France, Germany, Spain, and United Kingdom. In 2019, Poland, Portugal, and Romania joined the Consortium.

In particular, Italy has dual-use national space assets both for scientific and operational purposes. Italy recognized the strategic importance of the European Union initiative and decided to join the Consortium, in order to protect national and European access to space and to ensure the safety of operations in space. The Italian Space Agency, the Ministry of Defence, and the National Institute of Astrophysics signed a framework agreement at national level to participate in the European Union Space Surveillance and Tracking Consortium jointly, bringing together resources, knowledge, competences, and capacities in the space surveillance sector.

The Italian Air Force operates the Italian Space surveillance and tracking Operations Center, located at Pratica di Mare Air Force Base, near Roma. The Center uses highly standardized and reliable tools and qualified and skilled personnel to provide:

• Planning and schedule of national sensors
• Processing and evaluation of data and information collected from the sensors
• Exchange of information with the other operation centers (both European and international, like the Combined Space Operations Center in Vandenberg, United States of America)
• Services according to the European Union Space Surveillance and Tracking Service Portfolio (including conjunction, reentry, and fragmentation analysis)

The national sensor's architecture consists of three telescopes (named SPADE, CAS, and PdM-MITE), three radars (named MFDR, BIRALES, and BIRALET), and one laser (named MLRO). In the near future, the architecture could be extended to other existing or new sensors (either telescopes or radars).

Cyber Threats

The increased interest of governments, companies, and new actors in space leads to the need for a clear understanding of security and cyber issues. It seems that space policy is not yet prepared for the expansion and transformation of space activities, and this leads to growing security risks at national level. It is relevant to understand in a better way what the threats are, also considering the new era of digitalization, strictly connected to the cyber and space domain.

There are simply recommendations for the security and safety of space activities, yet not a regulation or a strategy neither at national nor at international level. We have sensors and processing centers to monitor the situation in space, in order to provide a clear picture of the threats, and, at international level, several initiatives are already in place. Now, the list of threats seems to be clear and updated, but the kind and the type of actions to put in place in order to secure activities and services are probably not so concrete and ready to act.

The number of actors in space is increasing, and the level of threat is rapidly increasing at the same. Therefore, to some extent space is becoming a warfighting domain. Cyber security is facing an increasing level of relevance, and most of the spacefaring nations are discussing, nationally and internationally, how to face the related issues.

Looking at the new digital domain, cyber threats will increase their relevance, and the issue could probably raise its importance, becoming critical for all space infrastructures, both space-based and ground-based.

5G, internet of things, and artificial intelligence are increasing their importance within most of the national and international programs. With this in mind, we need to understand clearly if we are ready to face this new "digitalization era."

Considering the aforementioned, Italy is currently preparing a National strategy for space security, in view of the relevance of the topic and the need of an international and collaborative approach in this issue.

Policy, Priorities, and Strategy

New Policy

According to the Law, the Inter-Ministerial Committee has to define a policy document named "Government Guidelines for the Italian Space and Aerospace Policy." The President of the Council of Ministers approved the document March 25, 2019.

This document intends to give a general introduction related to the importance of the space and aerospace sectors in Italy and, then, providing the way ahead at national level in terms of strategy (main pillars as programs and strategy to implement) and activities (prioritized).

This document provides the guidelines for the preparation of two other documents:

- The "Strategic Document on National Space Policy," to define the political strategy and the lines of financial intervention for the development of innovative industrial technologies and space applications and services in favor of national economic growth
- The "Strategic Vision Document for Space," to identify strategic goals and objectives for the development of the national space sector

Furthermore, the Coordination Board created a working group with the task to define a National Space Security Strategy, to focus on the protection of space infrastructures from not intentional (e.g., debris or space weather) or intentional threats (e.g., direct actions on ground or in-orbit, electromagnetic interference, and cyber-attacks). Although still being developed, the main contents are related to:

- *Adequate level of continuity of services*
  Reliability, resilience, robustness, and, therefore, continuity of services, as for the Regulated Public Services (Galileo PRS model), in communications (e.g., Govsatcom), and Earth observation (e.g., Copernicus Security Services). In addition, the European Union seems to be moving in the same direction. Italy is supporting the internalization of institutional services (avoiding outsourcing the management to Entrusted Entities), within the EU Space Program Regulation (2021–2027). Security in space cannot be delegated or outsourced, considering its relation to national security.
- *Space safety – natural events/not intentional threats*
  The increased number of satellites corresponds to the increasing in vulnerabilities for all space infrastructures. Low Earth Orbit is starting to be more and more congested. If we look at the new mega-constellation of satellites the number of orbital object could increase drastically and lead to a collapse of all orbits (Kessler Syndrome). At the same time, we can start envisaging the same trend also for Geostationary Orbit. Therefore, it is essential to reduce the risk of in-orbit collisions with an adequate national Space Surveillance and Tracking, integrated with Space Weather. In this regard, a working group, led by the Prime Minister Office, is working to define and implement a shared "programmatic-financial planning," synergistic with the international initiative for Space Situational Awareness activities at international level.
- *Space security – intentional threats*
  New technologies to enable the so-called in-orbit servicing – "maneuvering satellites," "co-located satellites," and "anti-satellite satellites" – lead to the need of an increased space security. Without regulations and agreements in place, it is essential to create a robust infrastructure for monitoring the situation

in space and safeguarding national satellites. In addition to space operations, access to and return from space must be ensured. This will probably lead to the consolidation of Space Traffic Management.

- *Cyber security*
Space security is also related to the intentional threats, including the cyber-attacks or disturbing actions on ground segment, so-called electromagnetic interference. All national space infrastructures shall meet the user needs related to cyber security and electromagnetic interference.

In addition, the Government Guidelines provide indication for the definition of upstream, midstream, and downstream development plans.

The space and aerospace sectors and their capabilities and technologies, through the development and upgrading of applications and operational services will improve the daily life of citizens. Therefore, the institutional user will define the requirements for new systems, applications and space programs focusing on the needs of the user communities.

Pillars

The Inter-Ministerial Committee, with the Government Guidelines for Space and Aerospace Policy, has provided a list of programs and related strategy to implement at national level. The pillars of the Italian strategy are, as per the Government Guidelines document:

- To support of industrial policy and new technological chains, in order to increase the competitiveness at international level
- To strengthen the international cooperation with all international organization in the space and aerospace sectors
- To provide a multiannual planning from both financial and programmatic perspective, in order to provide a clear view of the short-, mid-, and long-term plan for development and realization of space programs
- To reinforce the value of space applications and technologies
- To consolidate a space economy plan in order to support the public-private partnership
- To define a National Space Security Strategy, including the monitoring of objects in space and the mitigation of cyber threats
- To facilitate the access to funds for the small and medium enterprises
- To valorize the national resources
- To support the development of all chains of the space domain. With this in mind, it will define the following:
  - Downstream development plan
  - Midstream development plan
  - Upstream development plan

Priorities

According to the Law, "The Committee defines the policy directive of the Government in space and aerospace matters, regarding research, innovation, technology and return for the industrial sector, as guidelines for editing the Strategic Document on National Space Policy." The document is the "Government Guidelines for the Italian Space and Aerospace Policy," where the national strategic areas for space and aerospace sectors are listed and prioritized:

• Telecommunication, Earth observation and navigation, and services and applications will be used by the community of users, and they need to be promoted by the institutions.
• The study of the universe, through the participation to the international cooperation programs and activities, in particular with European Space Agency and National Aeronautics and Space Administration, also considering the relevance of scientific and technological leadership of Italy at international level.
• Access to space, with reference to the traditional and future concepts, as strategic capability and part of the national industrial sector, in order to foster the commercial exploitation at national and international level.
• Suborbital flights and stratospheric platforms, also considering the development of national spaceport (the first one is Taranto-Grottaglie spaceport).
• In-orbit servicing and de-orbiting, in order to ensure the resilience and the continuity of the operations in-orbit and also the technological applications.
• Robotic exploration of space, related to the moon, the asteroids and other planets, considering the competence of the national scientific and industrial communities, through the development and operation of equipment and experiments onboard the satellites and probes developed within the framework of international collaboration at bilateral and multilateral level.
• Human exploration of space, ensuring and promoting the national excellence in the sector developed in the past years, consolidating the participation in the International Space Station and the new initiatives at international level, such as the exploration of the moon and the developing of new space stations.

Needs for a Strategy

A National Strategy for Space Security is one of the pillars of the Government Guidelines. Italy recognizes the need of ensuring institutional and commercial users of space an adequate level of resilience from space threats and continuity of services. Intentional or not intentional threats to space-based and ground-based assets are needed to be adequately secured and made safe. Space Surveillance and Tracking, Space Situational Awareness, and, in the future, Space Traffic Management are enablers for the safety and sustainability of space activities. Cyber security is enabler for security and resilience of space infrastructures, both space-based and ground-based.

With this in mind, space diplomacy and international collaboration are two pillars in order to foster a safe and secure space. Bilateral and multilateral approaches to these issues are the best way to increase the resilience of space infrastructure and the sustainability of space activities. Some initiatives at international level have already started, and, in the near future, they need to be promoted and enhanced.

Conclusions

Italy has a very long history and heritage in the space sector. Beginning in the 1950s, with the development of the first national satellite, the San Marco 1, and its launch, operated by an Italian team, Italy entered into a period of rapid growth for space technologies, applications, and cooperation at international level. Today, Italy is the third contributor of the European Space Agency, and it has, at national level, the knowledge and the experience in the whole chain of supply in space. Earth observation, communication, and navigation are key areas of exploitation and development for all user communities, in order to provide continuity and resilience of services. The increasing numbers of the threats in space, due to the increasing interest of the traditional and the new spacefaring countries, lead to considering strategic the security and the safety of the space domain. With this in mind, the National Law has established an Inter-Ministerial Committee in order to coordinate the space and aerospace activities and policies. The Committee has approved a document named "Government Guidelines for the Italian Space and Aerospace Policy" with the pillars and the priorities in the space and aerospace sectors for Italy. The document provides the guidelines for the preparation of two other documents to be issued within this year: the "Strategic Document on National Space Policy" and the "Strategic Vision Document for Space." In addition, the Inter-Ministerial Committee will approve a document named "National Strategy for Space Security," in order to provide the national guidelines for space security and safety.

References

Defense Intelligence Agency (USA) (2019) Challenges to security in space. Defense Intelligence Agency, Washington, DC
Delibera del Presidente del Consiglio (25 marzo 2019) Government guidelines on space and aerospace. http://presidenza.governo.it/AmministrazioneTrasparente/Organizzazione/Artic olazioneUffici/UfficiDirettaPresidente/UfficiDiretta_CONTE/COMINT/DEL_20190325_aeros pazio-EN.pdf
https://www.cfr.org/report/cybersecurity-and-new-era-space-activities
http://www.telespazio.com/programmes-programmi
Legge (7/2018) Misure per il coordinamento della politica spaziale e aerospaziale e disposizioni concernenti l'organizzazione e il funzionamento dell'Agenzia spaziale italiana. https://www.normattiva.it/uri-res/N2Ls?urn:nir:stato:legge:2018-01-11;7!vig=
NATO Standard AJP-3.3 (Ed. B v.1) (2016, April). NATO Standardization Office (NSO) © NATO/OTAN
Piano Strategico Space Economy (2016) Quadro di posizionamento nazionale Ver. 1.0. https://www.mise.gov.it/images/stories/documenti/all_6_Piano_Strategico_Space_Economy_master_13052016_regioni_final.pdf

British Spacepower: Context, Policies, and Capabilities

67

Bleddyn Bowen

Contents

Abstract

This chapter examines ongoing British military and security space activities in the context of persistent grand strategic problems for the UK. This chapter shows how spacepower is coming of age within the UK, not least in the military dimension. The British state is at something of a crossroads when making choices about its future space development as it has to respond to turbulence across the transatlantic region.

Introduction

We've got to have this... I don't mind for myself, but I don't want any other Foreign Secretary of this country to be talked at, or to, by the Secretary of State in the United States as I just have with Mr Byrnes. We've got to have this thing over here, whatever it costs. We've

B. Bowen (✉)
University of Leicester, Leicester, UK
e-mail: bb215@leicester.ac.uk

© Springer Nature Switzerland AG 2020
K.-U. Schrogl (ed.), *Handbook of Space Security*,
https://doi.org/10.1007/978-3-030-23210-8_75

got to have the bloody Union Jack on top of it. – Ernest Bevin, UK Foreign Secretary, 1946, in reference to atomic bomb technology (Hennessy 2010: 50–51)

In the mid-1940s, Britain had to adapt to its new role as a weakened Empire, dependent on the economic and scientific prowess of the United States, encapsulated by the abrupt exclusion of British scientists and engineers from the American nuclear weapons program shortly after the Second World War. For much of the time since the Second World War, space had been seen as something of an "expensive jolly," an area where the Americans could expend their wealth and Britain could subsequently capitalize upon through the Five Eyes partnership with minimal financial investment in sovereign British military space capabilities. That old consensus has fractured.

Today, Britain stands at something of a crossroads in its development of spacepower, not least in its military and security elements. Britain is on the cusp of breaking from its past where successive governments never saw much significance to space, and supporters of a greater indigenous military spacepower capability were branded derisively as "space cadets" (SpaceWatch Global 2018). While the British space economy is reaping the benefits of miniaturization of information technology and global data infrastructure, Whitehall continues to grow more interested in space as a wealth generator and source of security and military capabilities in the context of the geopolitical ruptures caused by Brexit and the behavior of the new American president. Indeed, transatlantic security politics continue have repercussions in orbit; space systems and industry do not exist in a geopolitical vacuum. Spacepower is increasingly seen as a form of power any significant actor in the international system must possess, akin to navies, air forces, and nuclear weapons and not just the preserve of the two or three most powerful states in the system. The British state has historically been behind the curve in grasping this geostrategic truth. Since the abandoned attempt at a British missile and rocket capability, Britain has conducted scientific and commercial space activities through the European Space Agency (ESA), the European Union (EU), and private companies, which "removed the need for an autonomous British space program. It took the British government more than thirty years to reach a different conclusion" (Paikowsky 2017: 99). This is typified by the fact that Britain must rely on other states and their resident companies to access this geostrategic environment, as other "middle-powers" have developed their own means of space launch. To use a naval analogy, it would be as if Britain could build ships but would have to contract the Indian state to launch them into the sea. For the first time in its history, the British state has no sovereign means of accessing a vital common and geostrategic environment to its security, prosperity, and influence, in stark contrast to its history with air and sea power. This is not a clarion call for launcher development, rather, it is to show the relative impotence of the UK as a sovereign strategic actor in terms of spacepower – a reality that is often at odds with a British strategic culture more accustomed to a history of imperial dominance through the global commons.

Today, there is a keen effort to put the "bloody Union [Flag]" in space, beyond the highly visible exploits of the astronaut Tim Peake. In 2010, the UK Space Agency

(UKSA) was formed, the *UK Military Space Primer* was published by the UK Ministry of Defence (MoD), and in 2013 and 2017, the Joint Doctrine Publication 0-30 *UK Air and Space Doctrine* was also revised and updated. The year 2014 saw the release of the UK's first National Space Security Policy (NSSP), and in 2015, the first National Space Policy (NSP) was published. At the time of writing, the MoD is in the process of creating the first Defence Space Strategy which is meant to outline space capability gaps across the defense sector that the MoD wishes to fill (MoD 2018a). Spacepower has never enjoyed such sustained attention as a tool of grand strategy and national security in Whitehall as it has in recent years. For the purposes of this chapter, grand strategy is defined as "a purposeful and coherent set of ideas about what a [state] seeks to accomplish in the world, and how it should go about doing so" (Milevski 2016: 140). Grand strategy refers to the way to meet the overarching objectives of a state as articulated by supreme political decision-makers, such as the maintenance of British influence, wealth creation and exploitation, and security within the current form of the international system and its web of alliances. This chapter provides a brief overview of British spacepower from a defense and security perspective. First, it grounds the overview by placing spacepower – which is a cardinal concept for defense and military planners in the twenty-first century – in its British grand strategic context. Second, it provides an overview of the military aspects of British space policy. Third and finally, activities and potential investments in communications, space tracking, remote sensing, navigation, and launchers are summarized.

Spacepower: Its Time Has Come

Spacepower is a concept whose time has come. The 1991 Gulf War was described in some circles as the "First Space War" (Anson and Cummings 1991: 45), because military space systems were instrumental in carrying out battlefield operations and surprise attacks. Since the twilight years of the Cold War, many states have invested in military space applications and have underpinned and enabled their own force modernization programs, resulting the precision-warfare and network-enabled strike forces across Europe and Asia complete with a proliferation in satellite-based command and control, reconnaissance, and navigation systems. With the proliferation of satellites and the growth in the number of space actors throughout the post-Cold War era, spacepower is not only more important for military and economic power, but Earth orbit is also a more target-rich environment. Two decades into the twenty-first century, then, it should be no surprise that three of Earth's most capable military powers continue to develop or have resuscitated weapons program that range from land, sea, and air-launched kinetic-kill satellite interceptors or sophisticated electronic warfare and laser dazzling anti-satellite capabilities (Weeden and Samson 2018).

Spacepower is "the ability of a [state] to exploit the space environment in pursuit of [state] goals and purposes and includes the entire Astronautical capabilities of the [state]. A [state] with such capabilities is termed a space power" (Lupton 1988: 7).

In other words, a space power is an adjective to describe an actor (usually a state) which possesses some form and degree of spacepower, which is a noun or collection of things denoting an ability to use outer space and project influence there. In that sense, spacepower is "just another" tool of grand strategy as one geographic form of power among many (e.g., land, sea, air) made up of several different kinds of thematic forms of power (e.g., military, economic, political, symbolic, scientific) that a state can attempt to marshal for defense and security objectives. Although this chapter focuses on the military and security aspects of British spacepower, they exist in a larger context; military and economic power in their broadest senses are never disconnected in practice as "in the long run one is helpless without the other" (Carr 1974: 132). Alongside sea and air power, spacepower is now a major geostrategic asset and threat for states in the twenty-first century and cannot be ignored. Spacepower produces acute and chronic threats for states, but also short-term and long-term boons through the multitude of military, economic, and industrial applications it enables. After the land, the sea, and the air, space – and Earth orbit in particular – is the fourth geostrategic environment.

British Space Policy and Strategy

The past and present of British spacepower is one of integration and dependence. Britain has integrated with the United States for military and intelligence spacepower, while it has become a significant commercial and scientific spacepower within the European Union (EU) and European Space Agency (ESA) (Bowen 2018a). However, with the incrementally decreasing costs of accessing space, the increasing capabilities afforded to miniaturized computers and more efficient on-board systems, and the growing demand for terrestrial support services and downstream applications, smaller and specialized economies can tap into space development and the exploitation of Earth orbit to a degree that was not possible decades ago. This presents several opportunities for developed economies and middle-power states such as the UK. For decades, the British state has not been pressured to increase its presence in space due to the high level of integration the MoD and intelligence services have enjoyed with the United States under the Five Eyes partnership and the wider "special relationship" which includes the nuclear, space, and missile sectors. However, with American spacepower becoming increasingly taxed in its resources and ability to meet purely American security needs alone, coupled with the increasing benefits of a strong space industrial base, there is an increasing desire for more sovereign UK capabilities in military and intelligence space systems. At present, it is reasonable to portray Britain as a significant second tier space power, sharing its rank with Japan, France, and Germany (Bowen 2018a: 324–330). Britain has clear dependencies in space in many areas – in particular in launch and imagery satellites – but it also has strengths in the commercial sector, research and development, and secure communications.

The publication of the UK's first space policies, as well as fresh revisions to its spacepower doctrine, reveal a consistent statement of intent on tapping into spacepower in a more concerted effort than what many in the sector familiar with given the British state's track record in space (Table 1).

Space is arguably not so much of a missing link in British grand strategy and national security policy as it was in the past, if the flurry of official publication activity is any indication. The top-level documents seen here are more educational in their nature, rather than concrete roadmaps for capability acquisition or a shift in de facto policy behavior. Any detail on investments is notoriously hard to pin down, with public statements remaining vague. The documents of most interest here are the 2014 NSSP, the 2015 NSP, the 2017 JDP 0-30, and the embryonic DSS. In the summer of 2018, the DSS was revealed as a document that was still being worked on, and some headline themes for the document were mentioned in conferences and in a leaflet, which are discussed below. The NSSP and NSP recognized space systems as critical national infrastructure and detailed the functions of such infrastructure for the range of government capacities, both civil and military, for the needs of national security (Bowen 2018a: 330–332). This developing ecosystem of documents demonstrates how the British state's organs increasingly recognize spacepower as "just another" source of capability and factor in national security and military power, and that the environment of Earth orbit is just another strategic environment, like the sea and the air. The MoD's space doctrine crystallizes this further by stating in no uncertain terms that:

the military element is only one component of the UK's space power capability, since space is also used to maximise the influence created through diplomatic and economic instruments of national power. A strong commercial space sector, allied space capabilities and civil and scientific expertise are all vital contributors to UK space power... UK space power, along with maritime, land, air and cyber power, form the interdependent levers of the military instrument of national power.... The Space domain should therefore be considered as

Table 1 Select UK documents relevant to space policy after 2010. Space-specific documents highlighted in gray

Select UK documents relevant to space policy after 2010		
Year	Document	Lead Department/Organization
2010	UK Military Space Primer	Ministry of Defence
2010	National Security Strategy (NSS)	Cabinet Office
2012	Civil Space Strategy, 2012–2016	UK Space Agency
2013	Joint Doctrine Publication (JDP) 0-30, 1st edition	Ministry of Defence
2014	National Space Security Policy (NSSP)	Cabinet Office
2015	National Security Strategy (NSS)	Cabinet Office
2015	National Space Policy (NSP)	UK Space Agency
2017	Joint Doctrine Publication (JDP) 0-30, 2nd edition	Ministry of Defence
2018	Prosperity From Space	Space Growth Partnership
2018	Industrial Strategy	Business, Energy, & Industrial Strategy
2018	Towards a Space Defence Strategy (DSS, leaflet)	Ministry of Defence

routinely as the other operating domains, and must be included in military planning processes. (MoD 2017: 74–75, 117)

Spacepower is being established within the British state's institutions. This ecosystem of documentation and their relations to more generic capstone documents – such as the 2015 NSS – "challenges prevailing notions in the field of space security that 'space deterrence' can be considered as a matter separate from and different to deterring war in general" (Bowen 2018a: 330–331). Spacepower is finally emerging from the shadow of nuclear weapons, missile defense, and the "Revolution in Military Affairs" (Gray and Sheldon 1999: 23–25), which has prevented the comprehension of Earth orbit as a co-equal geostrategic environment rather than as a technical component of other terrestrial weapons systems for decades. A hostile act in space will be interpreted and responded to based on political context, strategic effects, and economic costs. The fact that something happens in space does not make it politically unique. The fact that such an act may be in orbit does not condition responses to be more or less violent or escalatory, it depends on the overall strategic context which must take effects, events, and objectives on Earth into account. Space warfare is very much the continuation of Terran politics by other means (Bowen 2019: 556). These are important declaratory moves that challenge persistent and erroneous views throughout society, government, and academia of space activity that space is a realm for scientists and engineers and their experiments alone, or that space is somehow above or beyond terrestrial politics.

This series of space documentation also acknowledges how British freedom of action, sovereignty, or the ability to carry out operations is reliant on allies and the commercial sector. The MoD's space doctrine notes its reliance on American and European allies for key services, not least position, navigation, and timing systems, as well as Earth observation and space-based intelligence capabilities (Bowen 2018a, b). But they remain very general in their description. At the time of writing, the DSS is no more than ministerial statements and a leaflet. At a conference in May 2018, Guto Bebb MP, a former junior minister at the MoD, outlined four general objectives for the DSS: "to enhance the resilience of space systems, to improve operational effectiveness, to enhance space support to frontline troops, and to support wider government activities" (Bowen 2018c). As well as these general themes, the DSS and remarks at the May 2018 conference highlight a desire to develop battlefield-relevant space capabilities, particularly in intelligence, surveillance, and reconnaissance (ISR) support for terrestrial military operations, and increasing the number of space personnel at the MoD from 500 to 600 over the next 5 years (MoD 2018b).

Top-level policy documents such as these often leave "wiggle-room" because the practice of grand strategy and the maintenance of national security is often more intuitive, or "an instinctive understanding as to how to handle the instruments of power" (Strachan 2013: 257). When considering British space policy and strategy, it is worth remembering that politics and policy can upend forecasts, long-term commitments, or political priorities and sensitivities at a stroke and change direction faster than any bureaucracy can adapt to on paper. The 2014 NSSP and 2015 NSP did not foresee the geopolitical rupture of Brexit, which threatens to upend the British space sector's full integration in flagship EU space projects and the EU's space industry (Bowen 2018a: 335–336; Bowen 2018c; Besch 2018; UK House of Commons 2018). It is not the task of such documents to predict the future, but such a massive shift in major geopolitical and macroeconomic position necessitates a radical rethink of the British state's place in the international system.

Yet there are recurring themes that will plague British grand strategy, security policy, and decision-makers as they continue to evolve Britain's defense capabilities and nurture a domestic space industry, no matter how the dust settles on Brexit and President Trump. Richard Aldrich captures this general strategic context well when he argues that:

> any venture into space meant tough political as well as technical choices. In the mid-1960s Harold Wilson's cabinet had opted for the British-made Skynet military communication satellite, instead of GCHQ's (Government Communications Headquarters) preferred option of an American-made model. Whether to buy cheap and reliable from the Americans, or to invest in expensive British national capacity (and jobs), or indeed even to join with the Europeans, was a perennial issue. (2010: 438)

These tensions remain, though the EU route for security capabilities will be cut off due to withdrawal from Europe, leaving bilateral options outside EU structure as the only European route for industrial-scale space security asset development. None of the documents above give a clear and consistent idea of which sort of capabilities – beyond encouraging British industry in small satellites – that the MoD and the other organs of state may be interested in developing a sovereign British system or continuing to buy into American system, or integrate with European projects. However, it is not necessarily for those documents to outline a capability program.

Capabilities

When considering existing capabilities and any development and acquisition priorities, it is worth remembering that defense and security cannot be separated from economics, and vice versa. On the international stage, military and economic capabilities and dependencies in space create different forms of influence that shape the governance of outer space and any evolution of legal and security regimes in orbit. Britain is lacking in many government and military satellites with exception to the Skynet military communications satellite system (Bowen 2018a: 325). The UK has no wholly government-owned or military operated Earth observation or ISR satellites, unlike many European states, with exception to some commercial testbed satellites such as the Carbonite-2 live video capture satellite and the NovaSAR small radar satellite. The British commercial space sector has been booming, and is a familiar story to space professionals. At present, the UK space sector employs around 40,000 people, and turnover in the sector has grown from £4.1bn to £13.7bn between 2000 and 2015 (House of Commons 2017: 8–10). The bulk of the British space sector's income is via space applications (74%), with operations (15%) and manufacturing (8%) trailing far behind in 2015 (UK Space Agency 2016: 6). The success of the British space sector – in particular in commercial communications, small satellite designs and payload manufacturing, and scientific research and innovation – provides a firm footing for British and allied military-industrial complexes to tap into for security and military platforms and personnel, if the political decision was made to expand the UK military and security space sector. Indeed, British-based investors and the UK are a significant source of activity and presence within the start-up space business scene in the global space economy (BryceTech 2018).

If the small satellite research, design, and manufacturing base can survive and maintain its competitive advantages into the 2020s, it places Britain in a decent position to capitalize on the global space economy into the twenty-first century before other states and economic blocs are able to successfully play catch-up and erode Britain's leadership in a niche capability in small satellites and downstream applications. This places Britain in an advantageous position to develop sovereign space security, defense, and intelligence capabilities based on the strengths of its domestic space industrial base. Not unlike the British Empire's maritime past (Kennedy 1976: xv–xvi), Britain's "rise" in military spacepower will only last in the long term if it takes advantage of, and supports in a grand strategic sense, the British state's and society's economic exploitation of outer space. There are many "low-hanging fruit" of military space capabilities that the UK can acquire given its currently small space personnel and hardware footprint, and bring it into parity with its individual European neighbors in terms of sovereign space assets.

Although there is no single British space security program, it is possible to outline British space security and military activities. Britain's difficult position of balancing dependencies is exacerbated on the procurement level by the competing principles of Technology Advantage and Open Procurement. This means that even should Britain decide to acquire, develop, or maintain a capability, there are many technical and practical questions and uncertainties over which specific systems, providers, and operators – domestic or international – are to be pursued to fulfil that capability. Technology Advantage is the MoD's term for the pursuit of defense capabilities on a sovereign basis, those that are to be developed within the UK and to minimize dependence on allies and partners as well as any security liabilities induced by the private sector's modus operandi, regardless of the additional costs of pursuing such capabilities on a unilateral basis and "in-house" at the MoD. Cryptographic and electronic warfare technologies are an example. Open Procurement is the acquisition of defense capabilities through an international competitive tendering process, open to commercial providers from approved international partners (Bowen 2018a: 334–335).

Communications

The *Skynet* constellation of communications satellites, placed in geostationary orbit, is the longest-lived and arguably most entrenched British military and intelligence space capability. The system dates back to the 1970s and has enabled secure communications for the British state for decades. Although often seen as an MoD asset, the biggest historical customer of *Skynet* has traditionally been the UK's GCHQ intelligence agency, based in Cheltenham, through the sheer volume of signals intelligence (SIGINT) information they share with the US National Security Agency (NSA) and the limitations and security concerns of pre-fiber transatlantic cables. *Skynet* represented Britain's "first significant step into space" and a concerted effort to maintain some encrypted satellite voice communications capability within Britain by the political leadership, rather than purchasing American equivalents "off the shelf" (Aldrich 2010: 347–348). Today, the *Skynet* system consists of seven satellites, including four *Skynet* 5 satellites, and three *Skynet* 4 satellites. These satellites provide coverage for most of the globe, apart from the Pacific Ocean. The *Skynet* system provides encrypted high-frequency flexible spot-beam voice communications and data transmission between any terminal at fixed sites and mobile users on land and at sea.

Procurement for the sixth generation of *Skynet* is currently being deliberated in Whitehall, as the new generation is meant to be ready by the early 2020s. There are some concerns over how to build a long-lasting communications satellite given the rapid pace of technological change in communications technology, and may be delaying the decision (Henry 2018). However, the MoD has been consistent in stating that it wishes to bring the *operation* of *Skynet* 6 back in-house, as opposed to the civilian operators of Airbus which has overseen the operation of *Skynet* in recent years. Britain does contain some civilian and commercial expertise in satellite communications, as there are many communications companies registered and based within the UK, not least Inmarsat – a major provider of global maritime satellite communications.

Through the Five Eyes partnership with the United States, the MoD gains significant additional bandwidth for communications. Britain is integrated with the American Advanced Extremely High Frequency (AEHF) and Wideband Global Satcom (WGS) constellations, and *Skynet* is envisioned as part of a system of systems that complements allied communications needs as well as providing a certain degree of sovereign British communications capacity, without having to completely rely on others (Erwin 2018a). Britain's future participation in the EU's GOVSATCOM program is subject to considerable doubt given Brexit and the security protocols in such programs, as well as the cessation of British companies' rights to bid for contracts to develop the future security-relevant hardware and software of *Galileo* due to Britain's decision to rescind its membership of the EU. Given the fact that Britain has continued to invest in this secure communications capability on a sovereign basis and the MoD and Government are increasing their interesting the value of space systems for national security, wealth generation and exploitation, as well as direct British military power, it is not surprising that there is a desire to regain lost skills of satellite operations within the MoD. "Buying British" for *Skynet*, as well as bringing operations in-house to the MoD would help meet the Government's desire to stimulate and support the UK space industry. If the MoD once again oversees the daily operations of its primary satellite program, it will create a larger corps of space operators and spacepower literacy, as well as a greater demand for the services of the Space Operations Centre (SpOC) at RAF High Wycombe. As far as communications are concerned, British space interests seem to be renewing Technology Advantage and the expense of Open Procurement options.

SSA

Space Situational Awareness (SSA) is another space capability where Britain contributes to allied efforts in building an operational picture of outer space and missile launch detection. Although ostensibly a missile-launch detection and tracking radar, the RAF's Fylingdales radar station in the North Yorkshire Moors can also be used to detect and track space objects, providing more raw data for British and American SSA analysts to identify. Fylingdales also provides atmospheric reentry information relevant to UK territory. However, its primary mission is still for nuclear attack early warning and missile defense for the American system. As such, Fylingdales is not optimized for SSA tasks, but it is still a significant asset in this endeavor. The data from Fylingdales is sent to the United States' Combined Space Operations Centre (CSpOC), where it is compiled with their Space Data Association and the Space Surveillance Network (SSN) and the data from which is then shared back with Britain and provides the data for Joint Forces Command (JFC) and SpOC at RAF High Wycombe, which provides SSA intelligence products to the organs of the British state.

Beyond stating a desire to continue to improve collaboration with America and Europe on SSA data-sharing, as well as signing new memoranda of cooperation with Australia and New Zealand, and SSA-specific multilateral information sharing agreements within the Five Eyes structure (Annett and Dennis 2018: 18), SSA is somewhat muted as a capability for investment from the UK Government. Some argue that it may be best for the British state to provide more freedom for its commercial entities to develop SSA capacities and products (Annett and Dennis 2018: 19, 23). However, as spacepower and Earth orbit becomes more critical for military security and power projection, the need for more sovereign SSA to monitor hostile activities in space may one day encourage Whitehall to support sovereign SSA assets. At present, JFC at the MoD provides the British state with a focal point for SSA assessments and inputs for terrestrial military operations, and is at least an institutional and intellectual basis for the continued development of space operations skills and space intelligence in the UK.

ISR/EO

An area of apparently greater priority is in ISR and EO satellites. Being able to conduct reconnaissance for deployed forces at sea, on land, and in the air would reduce the UK's currently total dependence on allies and commercial providers for such space-based capabilities. Crucially, such abilities would allow the UK to target imagery assets and acquire sensitive multispectral intelligence in a more focused and secure manner than through using the open market and allies. Specifically, it would help the RAF partially fill a glaring hole in one of its four roles – providing space-based ISR assets and analysis to the "joint" battlefield. At present, neither the RAF nor any other organ of the British state has space-based ISR to call its own. There is a risk that allied space systems may be overwhelmed or otherwise engaged in a crisis or conflict, and such assets may not always be available to the UK on a priority basis from other states. Notably, the MoD and the Defence Science and Technology Laboratory (DSTL) have been a partner in encouraging and developing the Carbonite-2 live video imagery satellite and the NovaSAR small synthetic aperture radar satellite. With some sums of money already invested in these two satellites, it shows how the MoD can contribute to wider industrial strategy goals of nurturing the strengths of British-based space industry, and how the government can become a significant client for the already-successful and sustainable British small satellite sector. Recently, DSTL has advertised a tender worth £750,000 for a feasibility study for a defense-orientated SAR small satellite program, called Project Oberon, with a suggestion that funding for the next phase could be as high as £4 m (Government Online 2018).

Though lacking facilities to rapidly mass produce small satellites at present, stimulating the sector through MoD purchases and requisitions of military variants of small commercial satellites for short-lived or "pop-up" space-based ISR assets for specific battlefield operations could induce greater manufacturing capabilities in Britain. This would also demonstrate a further interest by the MoD of bringing space-based ISR and EO technologies under the "Technology Advantage" umbrella, though the components and supply chains of off-the-shelf technologies will remain a liability. Joint Forces Command (JFC) is the body in the MoD that tasks, requests, and processes ISR support from allies and other providers in space. An MoD equipment plan for 2018–2028 mentions no specific capability investment. However, the "placeholder" is there with a reference to future satellite capabilities (MoD 2018c: 27).

In the American Schriever space wargame of 2018, the UK delegation was given a stronger lead role in commanding the Special Capabilities Integration Cell which "gamed" how the UK, the United States, Canada, and Australia would combine space assets with existing capabilities and potential future capabilities as well (Erwin 2018b). Pop-up small satellite ISR from the British could be a new development to consider in the conduct of future military space operations, particularly as established American ISR infrastructure can be accounted for by potential adversaries in advance of a conflict. Some kind of ISR capabilities for the British state is a feasible goal given that it not only taps into existing strengths in British space industry and would stimulate it further, but it is also relatively affordable. The UK Government invested £21 m in NovaSAR and £4 m in Carbonite-2, with SSTL marketing its range of small satellites in the tens of millions of US dollars. These figures are dwarfed by the UK's proposed £3-5bn spend on a Galileo replacement UK global navigation satellite system (GNSS).

Navigation

Despite having supported American attempts to prevent the development of the Galileo satellite navigation system in 2003, in subsequent years, the British space industry became the lead in technology development and manufacturing for Galileo's navigation payload. However, Brexit has upended Britain's privileged position in the development of the Public Regulated Service (PRS) element of the Galileo system. EU regulations, themselves drafted with British input and approval, maintain that only EU member states' industries may win contracts for and develop the security hardware, software, and coding technologies for the PRS. In early spring 2018, the relatively new Secretary of State for Defence, Gavin Williamson, protested at this loss of access to develop PRS components of Galileo for British-based industry as ESA reached decision time on a new round of Galileo procurement. Airbus since 2016 had been planning to move such operations "back into" EU member states as it was cognizant of the EU's space industrial policy and its principle of georeturn to EU member states and security restrictions to non-EU member states.

With much media coverage and ministerial hyperbole, the Galileo system – one few people outside the space sector had previously heard of – had become a household name. The British government has seemingly accepted the figure of £3-5bn for building a replacement GNSS system to sustain the UK's GNSS industrial capabilities and has allocated £92 m for a "feasibility study." Given the costs of SSTL satellites and the UK's investments in technology development ranging in the tens of millions, and UKSA's budget which rests at approximately £375 m per year, 75% of which goes into the common pool at the European Space Agency, these figures for both the proposed British GNSS and the feasibility study are quite large given the supposed capability achieved. Spending approximately one-quarter of the value of UKSA's annual budget on a "feasibility study" for a project that will cost billions over its lifetime to procure and deploy, and many hundreds of millions per annum to operate once deployed, does strike one as more of political theatre rather than serious space capability planning given the daunting opportunity costs imposed by large project in light of other, cheaper, and more acute capability gaps in British military and security space capabilities. This is even more the case when one considers that a British GNSS will triplicate a capability for British MoD. The British will still have access to the secure signals of GPS – of which British industry has never had a part in developing. Meanwhile, the EU has confirmed it would seek to allow the UK passive access to the PRS signal and has already begun considering the application of the United States and Norway to be passive PRS users. Given Britain's value to European security, and that EU rules and statements have only concerned the development of and automatic access to PRS, it is not unrealistic to foresee negotiations in future for Britain to receive passive access to the PRS signal and an ability to manufacture receivers, on the same basis that it does with America's GPS (UK House of Commons 2018). Nevertheless, the future for GNSS expertise in Britain is not certain, as it relied on an EU-funded space program, and the British taxpayer will face a hefty fee if it wishes to keep its Technology Advantage in GNSS capabilities. To maintain it will be an opportunity cost for

British military capability development rivalled perhaps only by the aborted Blue Streak program itself.

Launch

The opportunity costs of a Galileo replacement system is put into sharper relief when recent British activities in the launch sector are considered. Though the small satellite launch market is rather bleak (Niederstrasser 2018) and not directly militarily relevant as of yet, it is an area of targeted investment from the UK Government. The UK Government is continuing its investment in the British small satellite sector, as well as attempting to develop a UK spaceport for small polar launch vehicles. In 2018, the Space Industry Bill was passed, updating the legislation of 1986 and making it easier to create future space legislation to streamline the oversight and regulation of "NewSpace" or "Space 4.0" actors within the UK. The year 2018 also saw the government attempt to induce commercial small satellite launch providers to develop spaceports in the UK, with London declaring a site on the northern coast of Cataibh (Sutherland) in northern Scotland as its preferred site for vertical polar launches. £50 m in total has been allocated towards developing British spaceport capabilities, with the bulk of the funding allocated towards Lockheed Martin – which owns Rocket Lab, the operators of the Electron launch vehicle – and Orbex, a European small launcher company. What little is left of the £50 m may go to alternative sites such as Tewynblustri (New Quay) in Cornwall which may be a site for Virgin Orbit horizontal launches, and Llanbedr in Gwynedd, near the Eryri (Snowdonia) mountains in Wales, which borders restricted maritime airspace.

A UK-based launch capability based on Open Procurement between American or European owned small launch providers could provide the MoD and the British state with a timelier space launch capability for its pop-up ISR and EO space systems, as well as providing a small launch site closer to home and for European small satellite companies. Such a joined-up capability in terms of sovereign UK small satellite space assets, plus a UK-based space capability, would be transformative for UK space power and the extent to which it could take unilateral actions in high-intensity combat operations on Earth. A sovereign space capability in this sense would go a long way to meeting the UK's stated grand strategic objectives of being able to conduct and lead military operations where it must. It remains to be seen however whether this future will come about as major investments are needed on both tracks for it to occur, especially in the context of the economic and security tradeoffs between Technology Advantage and Open Procurement.

Conclusion

The UK is at an interesting but turbulent juncture. The UK Government is far more interested than ever before in the military and commercial potential of the so-called "New Space" economy and the UK's niche strengths in it, while the industry must deal with the possible disruptions of Brexit, epitomized by the UK's self-imposed exclusion from the continued development of the PRS elements of Galileo as a result of larger political forces. However, Britain's successes in the research and commercial sectors of space place a significant industrial base for the Government to draw upon if it chooses to invest in more capabilities on the ground and in space, and has already crafted the bulk of the necessary top-level documentation that ensures that Whitehall and the Devolved Governments take space security seriously. Battlefield ISR capabilities, hoisted onto small satellites, provide an experimental and relatively affordable pathway for Britain to develop its sovereign ISR space capabilities and reduce its current total dependence on allies.

Whether or not Britain decides to build a replacement GNSS for Galileo and absorb the opportunity costs it presents, the episode has shown that space policy is very much the continuation of Terran politics by other means, and has entrenched the reality of everyday space infrastructure and the military applications of space technology in the minds of a previously ignorant political and media elite, and the general public. However, no matter the scandal or political hubris of the day, British spacepower will always remain caught between the United States and the European Union; the two giants of spacepower on its doorstep. Spacepower has intellectually come of age within Whitehall and is recognized as a tool of grand strategy for any great power in the international system. Britain is starting at a relatively low level of sovereign military and security space capabilities and Britain will have to seek an unfamiliar role in space as it invests more in sovereign assets. This is epitomized by the relationship in access to outer space between Britain and India, where the former imperial possession and postcolonial economy now can provide its former imperial master with the access to outer space that it lacks, for a fee. Given how small the British military space enterprise has been to date, it appears the only way is up for the "bloody Union [Flag]" in space.

References

Aldrich R (2010) GCHQ: the uncensored story of Britain's most secret intelligence agency. Harper Press, London

Annett I, Dennis R (2018) Increasing resilience in space-based capabilities for the UK through improved space situational awareness and regulatory control. RUSI J 163(2):16–26

Anson P, Cummings D (1991) The first space war: the contribution of satellites to the Gulf War. RUSI J 136(4):45–53

Besch S (2018) A hitchhiker's guide to Galileo and Brexit. Centre for European Reform, 3rd May 2018. https://www.cer.eu/insights/hitchhikers-guide-galileo-and-brexit. Accessed 24 Nov 2018

Bowen BE (2018a) British strategy and outer space: a missing link? Br J Polit Int Rel 20(2):323–340

Bowen BE (2018b) The RAF and space doctrine: a second century, a second space age. RUSI J 163(3):58–65

Bowen BE (2018c) Better the devil you know? Galileo, Brexit, and British defence space strategy. Defence in Depth, 23 May 2018. https://defenceindepth.co/2018/05/23/better-the-devil-you-know-galileo-brexit-and-british-defence-space-strategy/. Accessed 24 Nov 2018

Bowen BE (2019) From the sea to outer space: the command of space as the foundation of spacepower theory. J Strateg Stud, 42(3–4):532–556

Bryce Space and Technology (2018) Start-up space: update on investment in commercial space ventures, Alexandria. https://brycetech.com/downloads/Bryce_Start_Up_Space_2018.pdf. Accessed 02 Dec 2018

Carr EH (1974) The twenty years' crisis 1919–1939. Macmillan, Basingstoke

Erwin S (2018a) UK MoD still uncertain on how to procure satellite communications. Space News, 6th Nov 2018. https://spacenews.com/u-k-mod-still-undecided-on-how-to-procure-satellite-communications/. Accessed 24 Nov 2018

Erwin S (2018b) As satellites become targets, UK military seeks closer ties with space industry. Space News, 6th Nov 2018. https://spacenews.com/as-satellites-become-targets-u-k-military-seeks-closer-ties-with-space-industry/. Accessed 27 Nov 2018

Government Online (2018) DSTL Project Oberon contract, 22 November. http://www.government-online.net/dstl-project-oberon-contract/. Accessed 22 Dec 2018

Gray CS, Sheldon JB (1999) Space power and the revolution in military affairs: a glass half full? Airpower J 13:23

Hennessy P (2010) The secret state: preparing for the worst 1945–2010. Penguin, London

Henry C (2018) HTS, megaconstellations feed UK indecisiveness about Skynet 6 program. Space News, 7th Nov 2018. https://spacenews.com/hts-megaconstellations-feed-uk-indecisiveness-about-skynet-6-program/. Accessed 24 Nov 2018

House of Commons (2017) Space sector report, London. https://www.parliament.uk/documents/commons-committees/Exiting-the-European-Union/17-19/Sectoral%20Analyses/34-Space-Report.pdf. Accessed 2 Nov 2018

Kennedy P (1976) The rise and fall of British naval mastery. Allen Lane, London

Lupton D (1988) On space warfare. Air University Press, Montgomery

Milevski L (2016) The evolution of modern grand strategic thought. Oxford University Press, Oxford

Niederstrasser C (2018) Small launch vehicles – a 2018 state of the industry survey. In: Paper presented at the 32nd AIAA/USU conference on small satellites. Paper ID: SSC18-IX-01

Paikowsky D (2017) The Power of the Space Club, Cambridge, Cambridge University Press

SpaceWatch Global (2018) British military space: warfare in space domain part of UK modernising defence programme review. https://spacewatch.global/2018/12/british-military-space-warfare-in-space-domain-part-of-uk-modernising-defence-programme-review/. Accessed 22 Dec 2018

Strachan H (2013) The direction of war: contemporary strategy in historical perspective. Cambridge University Press, Cambridge

UK House of Commons (2018) Exiting the EU select committee, 'Oral evidence: the progress of the UK's negotiations on Eu withdrawal', HC 372, 9th May 2018. http://data.parliament.uk/WrittenEvidence/CommitteeEvidence.svc/EvidenceDocument/Exiting%20the%20European%20Union/The%20progress%20of%20the%20UK%E2%80%99s%20negotiations%20on%20EU%20withdrawal/oral/82783.pdf. Accessed 24 Nov 2018

UK Ministry of Defence MoD (2017) Joint Doctrine Publication 0–30, Shrivenham, Doctrine Development and Concepts Centre

UK Ministry of Defence (MoD) (2018a) Towards a defence space strategy. https://assets.publishing.service.gov.uk/government/uploads/system/uploads/attachment_data/file/712376/MOD_Pocket_Tri-Fold_-_Defence_Space_Strategy_Headlines.pdf. Accessed 4 Nov 2018

UK Ministry of Defence (MoD) (2018b) UK poised for takeoff on ambitious Defence Space Strategy with personnel boost, 21 May 2018. https://www.gov.uk/government/news/uk-poised-for-take-off-on-ambitious-defence-space-strategy-with-personnel-boost. Accessed 24 Nov 2018

UK Ministry of Defence (MoD) (2018c) The Defence Equipment Plan 2018, 5th Nov 2018. https://assets.publishing.service.gov.uk/government/uploads/system/uploads/attachment_data/file/753785/20181102-MOD_EquipmentPlan2018_FINAL-v1.pdf. Accessed 27 Nov 2018

UK Space Agency (2016) Summary report: the size and health of the UK Space industry, London, December 2016

Weeden B, Samson V (2018) Global counterspace capabilities: an open source assessment. Secure World Foundation, Washington, DC

Chinese Satellite Program

68

Xiaoxi Guo

Contents

Abstract

The Chinese government considers the space industry as an important part of the nation's overall development strategy. After generations of hard work, China's space industry has created a unique approach of development that conforms to its national conditions, made brilliant achievements symbolized by the successful implementation of several major projects, and obtained a lot of innovative results in space science. Centering on China's satellite programs, this chapter briefly introduces the development history, then outlines specific satellite programs:

X. Guo (✉)
China Academy of Space Technology (CAST), Beijing, China
e-mail: vivian_gxx@163.com

© Springer Nature Switzerland AG 2020
K.-U. Schrogl (ed.), *Handbook of Space Security*,
https://doi.org/10.1007/978-3-030-23210-8_118

Earth observation, communications and broadcasting, navigation and positioning, and scientific and technological test satellites. Finally, it addresses the satellite program's future perspectives and China's international exchanges and cooperation.

Introduction

Since its establishment in 1956, China's space industry has gradually built up a complete industrial base and has steadily increased its research and development capacity. Currently (as of March 2019), China has the second largest number of spacecrafts in orbit. (Data from the Union of Concerned Scientists, https://www.ucsusa.org/nuclear-weapons/space-weapons/satellite-database.)

The purposes of China's satellite programs are to expand the understanding of the Earth and the universe; to promote human civilization and social progress; to meet the economic demand, technological development, and national security; to raise the scientific and cultural qualities; and to safeguard national interests and enhance comprehensive national power. China has always adhered to the exploration and utilization of outer space for peaceful purposes.

China's satellite engineering was developed on the basis of weak infrastructure industries, relatively underdeveloped scientific and technological level, and limited national funding. Satellite development in China can be divided into three phases.

- *Technology Preparation Phase* (1958–1970). China's satellite development began in the late 1950s. In February 1968, the China Academy of Space Technology (CAST) was established. In April 1970, China launched the first man-made satellite DFH-1, which indicated that China became the world's fifth country to independently develop and launch man-made satellite.
- *Technology Test Phase* (1971–1984). In 1975, China successfully launched and recovered a remote-sensing satellite for the first time. In 1984, China launched the first GEO communications satellite DFH-2.
- *Satellite Application Phase* (1985-Now). On the basis of several successful tests, recoverable satellites and communications satellites were put into practical application. After that, China successfully developed and launched meteorological satellites, communications and broadcasting satellites, navigation and positioning satellites, resources satellites, ocean observation satellites, scientific and technology test satellites.

China unveiled the Belt and Road Initiative (BRI) in 2013, and introduced the Space Silk Road concept in 2014, aiming at creating an entire range of space capabilities including satellites, launch services, and ground infrastructure and at supporting related industries and service providers going global. In this process, China has followed the ancient "silk road spirit" and worked to build a "community with shared future for mankind."

Satellite Programs Development

China's space industry has witnessed rapid progress manifested by significantly enhanced capacity in independent innovation and access to outer space, constant improvement in space infrastructure, smooth implementation of major projects, and substantial achievements in space science, technology, and applications.

Through continuous efforts, China gradually formed a full range of satellite series, including Earth observation series, communications and broadcasting series, navigation and positioning series, scientific and technology test series. Various satellites have been widely used in many fields, such as society, economy, science and technology, and culture and education.

In recent years, China's commercial space industry has got off to a fast start with the help of microsatellite technologies, and applications such as communications, Earth observation, and technology tests that have been carried out, providing a favorable supplement to China's space system architecture.

Earth Observation Satellites

China has established a comprehensive system of satellite observation. The performance of the Fengyun (Wind and Cloud), Haiyang (Ocean), Ziyuan (Resources), Gaofen (High Resolution), Yaogan (Remote-Sensing), and Tianhui (Space Mapping) satellite series and the Small Satellite Constellation for Environment and Disaster Monitoring and Forecasting has improved. China's commercial remote sensing satellites are being launched intensively in the last few years.

Fengyun

The Fengyun polar orbit meteorological satellites have succeeded in networking observation by morning and afternoon satellites, while its geostationary Earth orbit (GEO) meteorological satellites have formed a business mode of "multi-satellites in orbit, coordinated operation, mutual backup and encryption at the appropriate time." So far, China has launched 17 meteorological satellites, eight of which are in orbit, making it the third country after the United States and Russia to have both polar orbit and geostationary meteorological satellites. The Fengyun-4A, the first of China's second-generation GEO meteorological satellites launched in December 2016, has begun serving users in China and across the Asia-Pacific region since May 2018. The Fengyun-4A is enabled with vertical atmospheric sounding and microwave detection capabilities to address 3D remote sensing at geostationary altitudes (Fig. 1).

Apart from the traditional Fengyun series satellites, China also launched a Carbon Dioxide Observation Satellite (TanSat) in December 2016, making China the third country after Japan and the United States to monitor greenhouse gases through its own satellite. The satellite helps understanding climate change and provides China's policy makers with independent data. On a three-year mission, the satellite will thoroughly examine global CO_2 levels every 16 days, accurate to at least 4 ppm (parts per million).

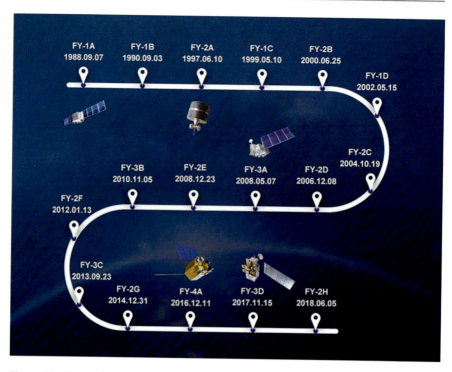

Fig. 1 Timeline of the Fengyun Series Satellites. (Courtesy of the National Satellite Meteorological Center)

Haiyang

China operates two families of Haiyang satellites – the Haiyang-1 series and Haiyang-2 series – that carry different sets of oceanography instruments. By the end of 2018, Haiyang-1 A/B/C satellites and Haiyang-2 A/B satellites had been launched, with the capability to observe the ocean dynamic environment. China has basically formed an ocean monitoring capability covering the whole world. As a high-tech means of ocean observation, ocean satellite remote sensing has been widely used in many fields. The latest members of the Haiyang series, Haiyang-1C and Haiyang-2B, were launched in September and October 2018, one after another. Haiyang-2B will replace Haiyang-2A launched in 2011 and will be joined by Haiyang-2C and 2D satellites to form a network around 2020.

Ziyuan

The Ziyuan-1 02C satellite was launched, the Ziyuan-3 01 and 02 stereo-mapping satellites have achieved double-satellite networking and operating. As China's first civil high-resolution transmission type stereo-mapping satellites, the Ziyuan-3 integrate functions of surveying, mapping, and resources investigation, and acquire high-resolution stereopsis and multispectral images with coverage of the whole nation continuously, steadily, and rapidly for a long time.

CHEOS

In order to improve the comprehensive capabilities of China's Earth observation system, the Chinese government approved to implement China High-resolution Earth Observation System (CHEOS) in 2010, which is established as one of the Major National Science and Technology Projects. The program consists of a fleet of multispectrum imaging and synthetic aperture radar satellites, intended to provide spectral and spatial monitoring that will aid in disaster prevention and reduction, climate, global change monitoring, hydrology, meteorology, and environmental management.

Gaofen

Since the launch of the Gaofen-1 01 satellite in 2013 as part of the country's high-resolution Earth observation project, twelve Gaofen satellites have been launched, six of which were launched in 2018. The Gaofen-5 can detect the state of air pollution in China through observing pollutants, greenhouse gases, and aerosols. It will help reduce the heavy reliance on data on air pollutants and greenhouse gases generated by foreign satellites. The Gaofen-6 satellite has a similar function to the Gaofen-1 satellite but with better cameras, and its high-resolution images can cover a large area of the Earth. Its data will also be applied in monitoring agricultural disasters such as droughts and floods, evaluation of agricultural projects and surveying of forest and wetlands (Table 1).

The system has been combined with other observation means to form all-weather and all-day Earth observation capabilities. By the end of 2018, the Gaofen-1, Gaofen-3 and Gaofen-4 satellites have achieved 100% effective national coverage. Through the implementation of this major project, China has been able to independently acquire all kinds of high-resolution data.

Yaogan

Yaogan series satellites use remote sensing technology and equipment to observe Earth's land cover and natural phenomena, mainly in the fields of territorial resources surveys, environmental monitoring and protection, urban planning, crop yield estimation, disaster prevention and reduction, and space science experiments. Since the Yaogan-1 satellite launched in 2006, China has sent more than 50 Yaogan series satellites into space. The first group of Yaogan-31 and the first group of Yaogan-32 remote sensing satellites were sent into space in April and October 2018 separately.

Small Satellite Constellation for Environment and Disaster Monitoring and Forecasting

The Huanjing-1 A/B/C satellites, members of the Small Satellite Constellation for Environment and Disaster Monitoring and Forecasting have come into service. The final objective of the constellation will consist of four optical satellites and four Synthetic Aperture Radar (SAR) satellites in orbit, to achieve the capability of quantitative, all-weather, all-time disaster forecasting, monitoring, and assessment.

Table 1 List of On-orbit Gaofen Series Satellites

Satellite	Year Launched	Purpose	Data Distribution (as of the end of 2018)
Gaofen-1 01	2013	China's first Gaofen satellite, a multispectral high-resolution wide-field-imaging Earth observation satellite	9,018,139 views
Gaofen-1 02, 03, 04	2018	Similar to Gaofen-1, but features a smaller bus. China's first civil operational constellation of high-resolution optical satellites, used for land resources investigation, monitoring, supervision, and emergency, etc.	
Gaofen-2	2014	A sub-meter optical land observation satellite	6,869,751 views
Gaofen-3	2016	China's first C-band multipolarized synthetic aperture radar (SAR) imaging satellite with a resolution of 1 m	830,144 views
Gaofen-4	2015	World's first geosynchronous orbit high-resolution Earth observation satellite	463,184 views
Gaofen-5	2018	China's first full-spectrum hyperspectral satellite	
Gaofen-6	2018	China's first satellite capable of performing high-precision agricultural monitoring	
Gaofen-8	2015	Optical remote sensing satellites, mainly used for land survey, urban planning, road network design, agriculture, and disaster relief	
Gaofen-9	2015		
Gaofen-11	2018		

Commercial Remote Sensing Satellites

Commercial remote sensing satellites, represented by the Jilin-1, Zhuhai-1, and SuperView-1 satellite constellations, are booming. The Jilin-1 satellites are China's first domestically developed commercial Earth imaging satellites. From October 2015 to December 2018, 10 satellites of the Jilin-1 constellation were launched into orbit to provide remote sensing data and product services for government departments and industrial users. Jilin-1 was developed and operated by Chang Guang Satellite Technology Co. Ltd. (also known as CGSTL), belonging to the Changchun Institute of Optics, Fine Mechanics and Physics (CIOMP) under the Chinese Academy of Sciences (CAS). Imagery from the Jilin-1 constellation is offered on the commercial market and is hoped to find application in a number of areas such as forecasting and mitigation of natural disasters, resource exploration and various monitoring tasks (Fig. 2, Table 2).

As of December 2018, seven satellites of the Zhuhai-1 constellation, launched in June 2017 and April 2018, are in orbit to cover the globe every 5 days on average. The Zhuhai-1 is a commercial remote-sensing micro-nano satellite constellation invested in by Zhuhai Orbita Aerospace Science and Technology Co. Upon completion, it will provide satellite big data services for global agriculture, forestry,

拍摄日期: 2018年3月12日
Date: Mar.12th,2018
成像地点: 中国·高雄市·义守大学
Location: Kaohsiung, China
成像卫星: 吉林一号光学A星
Satellite: JLCG-1
侧 摆 角: -11.7284°
Off-nadir Angle: -11.7284°
长光卫星技术有限公司
Chang Guang Satellite Technology Co.,Ltd.

Fig. 2 Gaoxiong, Taiwan Province, China, as seen by the Jilin-1A optical satellite. (Courtesy of Chang Guang Satellite Technology Co. Ltd)

animal husbandry and fishery, water and soil resources, environmental protection, transportation, smart city, modern finance and other industries.

The first phase of the China Aerospace Science and Technology Corporation (CASC) self-developed commercial remote sensing satellite system has been completed. Four SuperView-1 optical satellites, respectively launched in December 2016 and January 2018, are networked as a constellation in orbit with a resolution of 0.5 m, and the revisit period is shortened to 1 day. The SuperView-1 is the first multimeans high-resolution optical remote sensing satellite constellation for commercial use in China. As of December 2018, SuperView-1 had completed global imaging of more than 225,000 views and an area of 24.56 million square kilometers, and its data is being distributed to over 20 countries and regions.

Communications and Broadcasting Satellites

China has comprehensively advanced the construction of fixed, mobile, and data relay satellite systems. The successful launch of communications satellites such as Yatai (APStar) and Zhongxing (Chinasat) represent the completion of a fixed communications satellite support system whose communications services cover all of China's territory as well as major areas of the world. The Tiantong-1 01 satellite, China's first mobile communications satellite, has been successfully launched. With the Tianlian-1 04 satellite launched in November 2016, the first-generation data relay satellite system composed of four Tianlian-1 satellites has been completed, to provide data relay and control services for China's Shenzhou manned spacecraft, space laboratory and space station missions, as well as other satellites in the low- and medium-Earth orbits.

Table 2 List of the Jilin-1 Satellites

Satellites	Launch	Imaging modes	Mass	Resolution	Width
Jilin-1 Optical A Satellite	2015/10/7	Push-broom; Large-angle sway	420 kg	0.72 m (panchromatic);2.88 m (multispectral)	≥11.6 km
Jilin-1 Smart Video 01/02 Satellite		Staring video	95 kg	1.13 m (color)	4.6 km × 3.4 km
Jilin-1 Smart Verification Satellite		Push-broom; Staring video; Smart imaging; Stereo imaging	57 kg	4.7 m	9.6 km × 9.6 km
Jilin-1 Smart Video 03 Satellite	2017/1/9	Staring video; Nighttime light imaging; Stereo imaging; Space object imaging	165 kg	≤0.92 m (color)	11 km × 4.5 km
Jilin-1 Video 04/ 05/06 Satellite	2017/11/21	Staring video; Push-broom; Nighttime light imaging; Stereo imaging; Space object imaging	208 kg	1 m (color)	19 km × 4.5 km
Jilin-1 Video 07/08 Satellite	2018/1/19	Staring video; Push-broom; Nighttime light imaging; Stereo imaging; Space object imaging	208 kg	≤0.92 m (color)	19 km × 4.5 km

Shijian-13 (Chinasat-16) satellite, China's first high-throughput communications satellite that applies electric propulsion technology, was launched in April 2017 and delivered to the customer in-orbit in January 2018. The satellite, built upon the DFH-3B satellite bus, features a Ka-band broadband communications system capable of transmitting 20 gigabytes of data per second, making it the most powerful communications satellite China has developed to date. Shijian-13 represents the development direction of China's communications satellite technology in the next 5–10 years. The integrated electric propulsion, laser communications, advanced electronic system and other technologies will be standard equipment for future communications satellites.

Up to now, the development of China's DFH satellite buses mainly goes through four generations, namely, DFH-2, DFH-3, DFH-4, and the latest DFH-5 satellite buses. The DFH-5 bus incorporates a number of new technologies like high-thrust ion propulsion, a large trussed structure and a much increased payload capacity. The DFH-5 bus is expected to be launched in 2019, leading the upgrade of communications satellites.

Commercial communications satellites developed by China have successfully entered the international market with increasing competitiveness of the associated satellite products. China has exported whole satellites and provided in-orbit delivery of communications satellites to Nigeria, Venezuela, Pakistan, Bolivia, Laos, Belarus, Algeria and other countries. With the successful launch of the APStar-6C satellite in May 2018, China's total number of whole-satellite exported communications satellites reached 10.

The Chang'e-4 relay satellite, named Queqiao ('Magpie Bridge'), is a satellite launched in advance of and in support of China's Chang'e-4 soft landing mission on the Lunar far side. With its launch in May 2018 and insert into orbit in June 2018, Queqiao became the first communication satellite operating in the Halo orbit around the second Earth-Moon Lagrangian (L2) point, to set up a communication link between the Earth and the Chang'e 4 lunar lander.

The two largest Chinese state-owned enterprises and main contractors for the Chinese space program, the China Aerospace Science and Technology Corporation (CASC) and the China Aerospace Science and Industry Corporation (CASIC), are building global low-Earth orbit (LEO) satellite mobile communications and space internet systems, namely the Hongyan Constellation ("鸿雁星座") program and the Hongyun Project ("虹云工程"). In December 2018, the first experimental satellites in the two constellations were launched into space one after another.

The commercialization of high-orbit communication satellites especially high-Earth-orbit high-throughput satellites is moving forward at the same time.

Navigation and Positioning Satellites

The BeiDou Navigation Satellite System (hereinafter referred to as the BDS) has been independently constructed and operated by China with an eye on the needs of the country's national security and economic and social development. As a space infrastructure of national significance, the BDS provides all-time, all-weather and high-accuracy positioning, navigation and timing services to global users. China has been laying store by the construction and development of the BDS. It started to explore a path to develop a navigation satellite system suitable for its national conditions since 1980s and gradually formulated a three-step strategy of development.

The BDS is mainly comprised of three segments: space segment, ground segment and user segment. The BDS space segment is a hybrid navigation constellation consisting of Geostationary Earth Orbit (GEO) satellites, Inclined Geosynchronous Satellite Orbit (IGSO) satellites and Medium Earth Orbit (MEO) satellites.

The second-generation BeiDou system (BDS-2) has been offering services to customers in the Asia-Pacific region since December 2012. In 2015, China started building up the third-generation BeiDou system (BDS-3) for global coverage constellation. The first BDS-3 satellite was launched on 30 March 2015. By the end of 2018, the BDS-3 primary system had been completed to provide global services, including countries and regions participating in the Belt and Road Initiative (Silk Road Economic Belt and the twenty-first Century Maritime Silk Road).

Currently (as of the end of 2018), the BDS-1 is already retired. 15 satellites of the BDS-2 are in continuous and stable operation. Before formal deployment of the BDS-3 constellation, 5 BDS-3 test satellites had been launched, to carry out in-orbit test and verification. The BDS-3 satellites equip with the higher-performance rubidium atomic clocks with stability of 10^{-14} and hydrogen atomic clocks with stability of 10^{-15}, which has further improved the performance and lifetime of the satellites. 19 networking satellites (including 18 MEO satellites in operation and 1 GEO satellite under in-orbit test) have been successfully launched, stable and reliable inter-satellite links have been established, and deployment of the core constellation for the BDS-3 has been successfully completed (Fig. 3).

Current basic navigation service performance standards of the BDS are as follows:

- System service coverage: global;
- Positioning accuracy: 10 m horizontally, 10 m vertically (95%);
- Velocity measurement accuracy: 0.2 m/s (95%);
- Timing accuracy: 20 ns (95%);
- System service availability: better than 95%.

In the Asia-Pacific region, the positioning accuracies are 5 m horizontally and 5 m vertically (95%).

The mass applications of the BDS enjoy broad prospects. BDS-based navigation services have been widely adopted by e-commerce enterprises, manufacturers of intelligent mobile terminals and location-based services providers. The services have extensively entered into the fields of mass consumption, share economies, and those related to people's livelihood, which are profoundly changing people's production and livings.

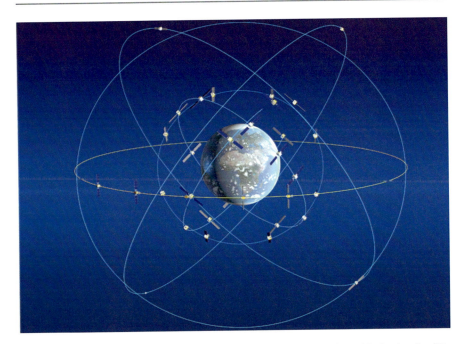

Fig. 3 China BeiDou Navigation Satellite System. (Image by the BeiDou Navigation Satellite System Website)

Scientific and Technology Test Satellites

China has developed and launched a series of scientific and technological test satellites, conducted numerous new technology validation tests and space environment exploration missions, and acquired valuable data of space environment, dark matter particles, microgravity, and stereoscopic seismic observation. During the past few years, several "firsts" have been made in the areas of space science and technological test.

Shijian Series

China's Shijian (Practice) satellites test new technologies. Shijian-1 was China's second satellite and was a platform to test satellite technologies. Shijian-8 was the world's first satellite devoted to crop breeding. Seeds were placed in the satellite and then exposed to the higher radiation levels of space in the hopes that genetic mutations may occur. The seeds were then removed from the satellite after it returned to Earth and grown. The Shijian-9A satellite, launched in 2012, is a remote sensing satellite with a multispectral imager that took part in the search for Malaysian Airlines flight 370. In 2014, the Shijian-11 program received the first prize in the National Science and Technology Progress Award.

Double Star Program

China has implemented the Double Star Program to explore the Earth's magneto-sphere in concert with the Cluster Program of the European Space Agency (ESA), obtaining much new data and making important progress in space physics. Through lunar exploration projects, China has studied the morphology, structure, surface matter composition, microwave properties, and near-moon space environment, further enhancing its knowledge of the moon.

Strategic Priority Program on Space Science

In January 2011, the Chinese Academy of Sciences (CAS) deliberated and approved the implementation plan of Strategic Priority Program on Space Science, signaling its official start. The main goal of the Program is dedicated to deepen the under-standing of the universe and planet Earth, seeking new discoveries and new break-throughs in space science via the implementation of both independent and cooperational space science missions. In November 2017, CAS declared that the four missions making up the Program have been successful. The missions, launched between December 2015 and June 2017, are the dark matter probe "Wukong," the Shijian-10 retrievable satellite, the quantum science satellite "Mozi," and the hard X-ray modulation telescope (HXMT).

Dark Matter Particle Explorer (DAMPE)

China's first space observatory, the Dark Matter Particle Explorer, or DAMPE (also known as Wukong, '悟空' in Chinese), was launched in December 2015. The scientific objectives of DAMPE are as follows: 1) Search for and study of dark matter particles by conducting high-resolution observation of high-energy electron and gamma rays; 2) Study the origin of cosmic rays by observing the high-energy electron and heavy nuclei above TeV; 3) Study the propagation and acceleration mechanism of cosmic rays by observing high-energy gamma rays. There are five payloads onboard the satellite: Si-Pin array, plastic scintillation hodoscope array, BGO calorimeter, neutron detector, and payload data manage-ment system. The first results, accurate measurement of electron cosmic-ray energy spectra in space, were published in the international authoritative aca-demic journal *Nature* in 2017. Up to December 2018, DAMPE had been operated for 3 years and reached its expected life span, and 5.5 billion high-energy particles have been detected (Fig. 4).

Shijian-10

China's first microgravity scientific experiment satellite and the 24th recover-able satellite of China, Shijian-10, was launched and then returned in April 2016. The major scientific objectives of Shijian-10 are to get innovative achievements in kinetic properties of matter and rhythm of life by carrying out various scientific experiments in space. Nineteen scientific experiments were carried out onboard the satellite. Among them ten experiments are for microgravity science research, and the other nine experiments are for space life science research.

Fig. 4 Concept figure of the DAMPE satellite. (Image by the National Space Science Center)

Quantum Experiments at Space Scale (QUESS)

The world's first quantum science satellite, the Quantum Experiments at Space Scale, or QUESS, was launched in August 2016. The satellite is named after the ancient Chinese scientist and philosopher Micius (also known as Mozi, "墨子" in Chinese). QUESS is a proof-of-concept mission designed to facilitate quantum optics experiments over long distances to allow the development of quantum encryption and quantum teleportation technology. As of August 2017, QUESS had accomplished all of its objectives. The QUESS team won the prize delivered by the American Association for the Advancement of Science as it laid the groundwork for ultra-secure communication networks of the future (Fig. 5).

Hard X-Ray Modulation Telescope (HXMT)

Major breakthroughs have been made in space astronomical observation. China's first X-ray astronomy satellite, the Hard X-ray Modulation Telescope or HXMT (also known as Insight, "慧眼" in Chinese), launched in June 2017 to observe black holes, neutron stars, active galactic nuclei, and other phenomena based on their X-ray and gamma-ray emissions. The payloads onboard HXMT include the High Energy X-ray Telescope, the Medium Energy X-ray Telescope, the Low Energy X-ray Telescope, as well as a Space Environment Monitor. Using the direct demodulation method and scanning observations, HXMT can obtain X-ray images with high spatial resolution, while the large detection areas of these telescopes also allow pointed observations with high statistics and high signal to noise ratio. HXMT completed its five-month period of in-orbit calibration and test observations in January 2018, and then officially began science operation.

China Seismo-electromagnetic Satellite

China's first space-based platform for stereoscopic seismic observation system, China Seismo-Electromagnetic Satellite or CSES (also known as Zhangheng-1,

Fig. 5 This photo, taken on December 10, 2016, shows a satellite-to-Earth link established between QUESS and the quantum teleportation experiment platform in Ali, China's Tibet Autonomous Region. (Image credit: Xinhua)

"张衡一号" in Chinese), was launched in September 2017 to help scientists monitor the electromagnetic field, ionospheric plasma and high-energy particles for an expected mission life of 5 years. Covering the latitude area between 65° north and 65° south, it will focus on Chinese mainland – areas within 1000 km to China's land borders and two major global earthquake belts. CSES has studied several 6–7-magnitude earthquakes of 2018 and found that some disturbances of geomagnetic elements might occur 1–5 days before the earthquake, proving the satellite's reliability and detection capability.

Tianyuan-1

Tianyuan-1, China's first in-orbit refueling system for satellites, was lifted into space by a Long March 7 carrier rocket on June 25, 2016, during the rocket's maiden flight. Tianyuan-1 has conducted nine in-orbit tests including the control and refilling of liquid in microgravity and accurate measurement of propellant, according to the National University of Defense Technology in Changsha, in Hunan province, which developed the system. The spacecraft recorded video and data when it filled three types of propellant tanks. The results of these tests showed Tianyuan-1 has met designers' requirements, the university said, adding that the system features a high level of automation and stability. The recent tests performed by Tianyuan-1 will pave the way for large-scale resupply and refueling for China's future manned space station.

Future Prospects

In the next period of time China will uphold the concepts of innovative, balanced, green, open and shared development, and promote the comprehensive development of space science, space technology and space applications, so as to contribute more to both serving national development and improving the well-being of mankind.

China plans to expedite the development of its space endeavors by continuing to enhance the basic capacities of its space industry, strengthen research into key and cutting-edge technologies, and implement the BeiDou Navigation Satellite System, high-resolution Earth observation system, and other important projects. Furthermore, the country is launching new key scientific and technological programs, and further conduct research into space science, promoting the integrated development of space science, technology and applications.

Earth Observation Satellites

In accordance with the policy guideline for developing multifunctional satellites, and creating networks of satellites and integrating them, China will focus on three series of satellites for observing the land, ocean and atmosphere, respectively. China is to develop and launch satellites capable of high-resolution multimode optical observation, L-band differential interferometric synthetic aperture radar imaging, carbon monitoring of the territorial ecosystem, atmospheric Lidar detection, ocean salinity detection, and new-type ocean color observation. China will take steps to build its own capabilities of highly efficient, comprehensive global observation and data acquisition with a rational allocation of low-, medium- and high-spatial resolution technologies, and an optimized combination of multiple observation methods. China will make overall construction and improvement on remote-sensing satellite receiving station networks, calibration and validation fields, data centers, data-sharing platforms and common application supporting platforms to provide remote-sensing satellite data receiving services across the world.

China plans to launch additional nine Fengyun meteorological satellites by 2025, including a number of Fengyun-3 and Fengyun-4 satellites as well as a Sun-synchronous orbit climate-focused satellite and a greenhouse gas detection spacecraft. A total of seven Fengyun-4 satellites are planned to be launched to remain in service through 2037 when a successor program will be inaugurated.

Considering the needs of atmospheric observation by industries and the public on the matter of weather forecasting, atmospheric environmental monitoring, meteorological disaster monitoring and global climate observation, China will continue to improve its large-scale active and passive optical and microwave detection abilities, construct 2 satellite constellations for weather and climate observation, and develop atmosphere detection satellites and hyper-spectrum, laser, polarization observation technologies to detect atmospheric particles, air pollution and greenhouse gas. What's more, China will also intensify satellite data sharing with world meteorological organizations, to form a complete system of atmospheric observation.

In order to meet the great demand in the field of marine resources development and environmental protection, China will develop a variety of optical and microwave observation technologies, construct satellite constellation on marine dynamics and ocean color, develop ocean surveillance satellites, to continuously improve the comprehensive ability of ocean observation. Afterward, China will continue to develop satellites on ocean salinity and marine environmental monitoring, to meet the great demand of marine environmental protection.

Apart from the conventional land observation satellites, China will also develop electromagnetic monitoring satellites, terrestrial carbon monitoring satellites, water cycle observation satellites, and gravity gradient measurement satellites, to monitor the geophysical environment change required by industries specialized on earthquake, disaster prevention and mitigation, and climate change. China aims to foster the monitoring capabilities in geophysics and terrestrial carbon sink to support earthquake prediction and climate change research.

In near future, CASC will develop high-end optical satellites, agile SAR satellites and other commercial satellites to increase the diversity of commercial remote sensing satellites.

Chang Guang Satellite Technology Co., Ltd. plans to complete networking of 60 satellites by 2020 and realize a 30-min revisit to any location on the Earth; complete networking of 138 satellites by 2030 and realize a 10-min revisit to any location on the Earth.

Communications and Broadcasting Satellites

As for the satellite communications broadcasting system, China will continue to operate in a commercial mode and satisfy public welfare needs in the meantime. The main concerns are the development of fixed and mobile communications and broadcasting satellites, and their ground facilities, such as control stations, gateway stations. Through all these efforts, it is expected to provide services of broadband communication, fixed communication, television live broadcasting, mobile communication and mobile multimedia broadcasting, and gradually build satellite communication and broadcasting systems, to be integrated with ground communication networks and covering the world's major regions to facilitate Broadband China Strategy, the globalization strategy and international communication abilities.

As planned, three or four of high-throughput communications satellites will be put into orbit by the latter period of the 13th Five-Year Plan; the Chinasat-18 high-throughput communications satellite will be sent into space in 2019 and its coverage will extend to the entire China in combination with the Chinasat-16; and, two GEO high-throughput satellites with ultra-large capacity will be launched into space roughly in 2023. High-orbit communications satellite can be applied to many areas such as education, marine communications, emergency communications, and outdoor security, and caters for multilayer demands regarding to the Belt and Road Initiative, military–civilian integration, frontier defense, and targeted poverty alleviation.

According to planning, nine Hongyan satellites will be launched by 2020 to test the system, which ultimately will comprise 320 satellites; the Hongyun project will place 156 satellites in orbit. Through these systems, six application services, namely, mobile terminal communication, broadband Internet access, Internet of things, hot spot information push, navigation enhancement, and aviation navigation monitoring, can be provided worldwide, and China's satellite communication coverage will be extended to the ocean and even polar regions for the first time.

Navigation and Positioning Satellites

In the future, the BDS will continue to improve service performance, expand service functions and enhance continuous and stable operation capability. Before the end of 2020, the BDS-2 will launch 1 backup GEO satellite, the BDS-3 will launch another 6 MEO, 3 IGSO and 2 GEO satellites, to further improve global basic navigation and regional short message communication service capabilities, and to achieve the global short message communication, satellite-based augmentation, international search and rescue, and precise point positioning service capabilities. The BDS is planned to provide the following services through various types of satellites in 2020. With the BDS as corn, a comprehensive positioning, navigation, and timing (PNT) system will be established and improved before 2035, which will be more ubiquitous, integrated, and intelligent.

Scientific and Technology Test Satellites

In July 2018, the Chinese Academy of Sciences (CAS) officially launched the Strategic Priority Program on Space Science (Phase II). Upon great scientific achievements by Dark Matter Particle Explorer (DAMPE), Quantum Experiments at Space Scale (QUESS), Hard X-Ray Modulation Telescope (HXMT), ShiJian-10 Recoverable Satellite, and so on from Phase I, the program will launch four more space science satellites in the next 5 years. The new patch of space science satellites, includes Einstein Probe (EP), Advanced Space-based Solar Observatory (ASO-S), ESA-CAS Solar Wind Magnetosphere Ionosphere Link Explorer (SMILE), all of which have officially entered its engineering phase. Gravitational Wave High-energy Electromagnetic Counterpart All-sky Monitor (GECAM) is carrying out Phase A study and will enter engineering phase soon.

International Exchanges and Cooperation

The Chinese government holds that all countries in the world have equal rights to peacefully explore, develop and utilize outer space and its celestial bodies, and that all countries' outer space activities should be beneficial to their economic development and social progress, and to the peace, security, survival and development of mankind.

International space cooperation should adhere to the fundamental principles stated in the Treaty on Principles Governing the Activities of States in the Exploration and Use of Outer Space. China maintains that international exchanges and cooperation should be strengthened on the basis of equality and mutual benefit, peaceful utilization and inclusive development.

Over the years, China has exported communications satellites and remote sensing satellites to several countries and regions. At the same time, China has vigorously promoted international cooperation by sharing satellite resources and cooperating in building space and ground infrastructure. As one of the four major GNSS providers, the BDS persists in open cooperation and resource sharing, actively carries out international exchanges and cooperation, and promotes the development of global satellite navigation. BDS-enabled products have been exported to more than 90 countries, providing users with a variety of choices and better application experience.

China and Brazil, through the mechanism of the Space Cooperation Sub-committee of the Sino-Brazilian High-level Coordination Commission, have conducted constant cooperation in the China-Brazil Earth Resources Satellite (CBERS) program. On the basis of the CBERS-4 launched in December 2014 and other satellites, China and the space agencies of Brazil, Russia, India, and South Africa cosponsored and actively promoted cooperation in the BRICS remote-sensing satellite constellation.

Within the mechanism of the Sino-French Joint Commission on Space Cooperation, the China-France Oceanography Satellite (CFOSat) was successfully launched in October 2018. As the first satellite-related cooperation between China and France, the CFOSat is equipped with the world's most advanced technologies. The CFOSat, complementing other existing oceanography satellites, will study the dynamics of waves and how they interact with surface winds, and deepen our understanding of their formation and physical mechanism.

The mission of China Seismo-Electromagnetic Satellite is part of a collaboration program between the China National Space Administration (CNSA) and the Italian Space Agency (ASI), and developed by China Earthquake Administration (CEA) and Italian National Institute for Nuclear Physics (INFN).

China has strengthened bilateral and multilateral cooperation through the Belt and Road Initiative (Silk Road Economic Belt and the twenty-first Century Maritime Silk Road, also known as BRI). Within the BRI, China actively conducts cooperation with other B&R countries in space technology and makes its applications satellite systems and technologies available to them. In terms of satellite navigation, so far, more than 30 B&R countries, including Pakistan, Saudi Arabia, Myanmar and Indonesia, have signed agreements to embed BDS domestically. By the end of 2018, its services had covered all countries and regions participating in the BRI. In terms of satellite remote sensing, the Gaofen-1 and Gaofen-2 have fulfilled their duties as on-duty satellites under the International Charter on Space and Major Disasters in proving important data support to the monitoring and evaluation of floods in Sri Lanka and Bangladesh. Besides, the Fengyun meteorological satellites have gradually covered about 40 B&R countries.

China encourages and endorses the efforts of domestic scientific research institutes, industrial enterprises, institutions of higher learning and social organizations to develop international space exchanges and cooperation in diverse forms and at various levels under the guidance of relevant state policies, laws and regulations.

Conclusions

China is determined to quicken the pace of developing its space industry and actively carry out international space exchanges and cooperation so that achievements in space activities will serve and improve the well-being of mankind in a wider scope, at a deeper level, and with higher standards. Through independent development efforts in developing application satellites and satellite applications, China has made positive contributions to human space exploration. China has also innovation capability in space. It focuses on implementing important strengthening science and technology space projects to realize leapfrog development in space science and technology by way of making new breakthroughs in core technologies and resource integration. China is actively building a space technology innovative system featuring integration of the space industry, academia, and the research community, with space science and technology enterprises and research institutions as the main participants. It has strengthened in the space field and multiple basic research advanced frontier technologies to increase sustainable innovative capacity in space science and technology. No matter the past, present, and future, China advocates peaceful use of space around the world and is willing to cooperate with other countries to develop satellite programs and realize a win-win situation based on the principles of mutual respect, mutual benefit, and equality.

Further Reading

BeiDou Navigation Satellite System website. http://en.beidou.gov.cn/

China Aerospace Science and Technology Corporation website. http://english.spacechina.com/n16421/index.html

China Daily website. http://www.chinadaily.com.cn/

China National Space Administration website. http://www.cnsa.gov.cn/english/index.html

Chinese Academy of Sciences website. http://english.cas.cn/

Ministry of Foreign Affairs of the People's Republic of China website. https://www.fmprc.gov.cn/mfa_eng/

National Satellite Meteorological Center website. http://www.nsmc.org.cn/en/NSMC/Home/Index.html

Report on the Development of BeiDou Navigation Satellite System (2018) (Version 3.0), China Satellite Navigation Office, December 2018

University of Science and Technology of China website. http://en.ustc.edu.cn/

White Paper – China's Space Activities in 2016 The State Council Information Office of the People's Republic of China, December 2016

Chinese Space Launch Program

69

Lehao Long, Lin Shen, Dan Li, Hongbo Li, Dong Zeng, and
Shengjun Zhang

Contents

L. Long (✉)
Chinese Academy of Engineering, Beijing, China
e-mail: longlh@spacechina.com

L. Shen · D. Li · H. Li · D. Zeng · S. Zhang
China Academy of Launch Vehicle Technology (CALT), Beijing, China
e-mail: tolinsh@sina.com; lidan7814@sohu.com; lhbspace@sina.com; est_zeng@163.com;
zhangsj98@sina.com

© Springer Nature Switzerland AG 2020
K.-U. Schrogl (ed.), *Handbook of Space Security*,
https://doi.org/10.1007/978-3-030-23210-8_121

1401

Abstract

China's development of launch vehicles is sticking to the "self-reliance and independent innovation" path. With more than 50 years' experience, China has successfully developed more than ten models of launch vehicles and managed the transition from research test to flight application and from flight application to industrialization. This chapter provides an overview of China's space launch plan. This chapter mainly presents the development history of China's launch vehicles, launch vehicles in service, the new generation of launch vehicles under development, as well as the efforts made by China in the field of space security.

Introduction

With an increasing capacity of access to space and driven by launch missions, China's development of launch vehicles is sticking to a path of "self-reliance and independent innovation." With more than 50 years' experience, China has successfully developed more than ten models of launch vehicles and managed the transition from research test to flight application and from flight application to industrialization. It has promoted the development of satellites, satellite applications and technology, as well as manned space technology. China has strongly supported the successful implementation of major projects, namely, the "manned space project" and "lunar exploration project" as representative ones.

China aims for safe, reliable, fast, economic, environment-friendly access to space. China's launch vehicles' development goal in the past, present, and future is to promote space exploration technology development and enhance the progress of human civilization.

This chapter mainly presents the development history of China's launch vehicles, launch vehicles in service, the new generation launch vehicles under development, as well as the efforts made by China in the field of space security.

Development Background of China's Launch Vehicles

As early as the Song Dynasty (in the eleventh century), China invented black powder rocket, which was in line with the principle of rocket propulsion. It spread to the Arab World and the West in the thirteenth century.

The development of China's present launch vehicles began in the mid-1960s. With long exploration and thorough efforts, China has now successfully developed the Long March 1 (LM-1) series, the Long March 2 (LM-2) series, the Long March 3 (LM-3) series, and the Long March 4 (LM-4) series of launch vehicles and is developing a new generation of launch vehicles. It formed a family of Long March launch vehicles of more than ten kinds of models. With 40 years development, China's space launch technology has made remarkable achievements. The LM launch vehicles experienced many technological leaps, such as from using conventional propellants to cryogenic propellants, from one-time start of the last stage to multi-start, from tandem configuration to parallel configuration, from one vehicle launching one satellite to launching multi-satellites, and from launching cargos to launching astronauts. Now, they can launch different kinds of satellites and manned spacecraft to different types of low, medium, and high Earth orbits. The launch capacity for low Earth orbit (LEO) and geosynchronous transfer orbit (GTO) is 25 t, and 14 t, respectively. The orbit injection precision reaches leading international level. It can meet the diverse needs of different users. The existing LM launch vehicles have the ability to launch spacecraft to the moon and in the deep space.

On April 24, 1970, China's LM-1 rocket successfully launched the Dongfanghong-1 satellite into low Earth orbit, making China the fifth country in the world that successfully launched its own satellite with its homemade rocket.

In 1999, LM-2F launch vehicle successfully launched the experimental Shenzhou spacecraft and laid a solid foundation for the realization of the strategic goals of China's manned space flight, making China the world's third country of independently developing manned space technology, and further enhanced China's aerospace industry status in the international arena. In October 2003, China's first manned space mission was a success. In June 2012, China successfully accomplished the first manned rendezvous and docking.

In October 2007, LM-3A launch vehicle successfully sent China's first lunar probe satellite "Chang'e-1" into preset orbit, marking China's space industry successfully entered the new field of deep space exploration, Chinese nation's thousands of years' dream of flying to the moon started to become a reality.

Launching artificial Earth satellite, manned spacecraft, and lunar probe are three milestones in the development of China's space industry.

Launching Plan and Development of China's Launch Vehicles

Up to June 5, 2019, LM launch vehicles carried out 306 launches, and the launch success rate reached 95%. China finished the first 100 launches in 37 years, the second 100 launches in 7 years, and the third 100 launches in 5 years. With one single model, more than 100 launches of the LM-3A series have been realized. It fully verified the reliability of LM launch vehicles and promoted the industrialization of China's launch vehicles. The launch plan of China's launch vehicles will continue to maintain a high density. It is said that China will complete 110 launches with an annual average about 20 times during "the 13th Five-Year" period.

Table 1 International commercial launch record of LM launch vehicles (1987–2012)

No.	Payload/SC	Launch vehicle	Customer	Launch date	Remarks
1.	Microgravity test instrument	LM-2C F09	Matra Marconi, France	1987/ 08/05	Piggyback
2.	Microgravity test instrument	LM-2C F11	Intospace, Germany	1988/ 08/05	Piggyback
3.	AsiaSat-1	LM-3F07	AsiaSat, HK	1990/ 04/07	Dedicated
4.	BADR-A/Aussat Dummy payload	LM-2E F01	SUPARCO, Pakistan	1990/ 07/16	Piggyback
5.	Aussat-B1 Aussat	LM-2E F02	Australia	1992/ 08/14	Dedicated
6.	Freja	LM-2C F13	SSC, Sweden	1992/ 10/06	Piggyback
7.	Optus-B2	LM-2E F03	Aussat, Australia	1992/ 12/21	Dedicated
8.	APSTAR-1	LM-3F09	APT, HK	1994/ 07/21	Dedicated
9.	Optus-B3	LM-2E F04	Optus, Australia	1994/ 08/28	Dedicated
10.	APSTAR-II	LM-2E F05	APT, HK	1995/ 01/26	Dedicated
11.	AsiaSat-2	LM-2E F06 (EPKM)	AsiaSat, HK	1995/ 11/28	Dedicated
12.	Echo Star-1	LM-2E F07 (EPKM)	EchoStar, USA	1995/ 12/28	Dedicated
13.	INTELSAT-7A	LM-3B F01	INTELSAT	1996/ 02/15	Dedicated
14.	APSTAR-1A	LM-3F10	APT, HK	1996/ 07/03	Dedicated
15.	ChinaSat-7	LM-3F11	ChinaSat, China	1996/ 08/18	Dedicated
16.	Microgravity test Instrument	LM-2D F03	Marubeni Corp., Japan	1996/ 10/20	Piggyback
17.	Mabuhay sat	LM-3B F02	Mabuhay, Philippines	1997/ 08/20	Dedicate
18.	APSTAR-IIR	LM-3B F03	APT, HK	1997/ 10/17	Dedicated
19.	Iridium	LM-2C/ SD F02	Motorola, USA	1997/ 12/08	Dual
20.	Iridium	LM-2C/ SD F03	Motorola, USA	1998/ 03/26	Dual

(continued)

Table 1 (continued)

No.	Payload/SC	Launch vehicle	Customer	Launch date	Remarks
21.	Iridium	LM-2C/ SD F04	Motorola, USA	1998/ 05/02	Dual
22.	ChinaStar-1 China	LM-3B F04	Orient, China	1998/ 05/30	Dedicated
23.	SinoSat-1	LM-3B F05	SinoSat, China	1998/ 07/18	Dedicated
24.	Iridium	LM-2C/ SD F05	Motorola, USA	1998/ 08/20	Dual
25.	Iridium	LM-2C/ SD F06	Motorola, USA	1998/ 12/19	Dual
26.	Iridium	LM-2C/ SD F07	Motorola, USA	1999/ 06/12	Dual
27.	CBERS-01	LM-4F04	INPE, Brazil	1999/ 10/14	Dedicated
28.	SACI	LM-4F04	INPE, Brazil	1999/ 10/14	Piggyback
29.	CBERS-02	LM-4F08	INPE, Brazil	2003/ 10/21	Dual
30.	APSTAR-VI	LM-3B F06	APT, HK	2005/ 04/12	Dedicated
31.	NigComSat-1	LM-3B F07	NSRDA, Nigeria	2007/ 05/14	Dedicated
32.	ChinaSat-6B	LM-3B F08	ChinaSat, China	2007/ 07/05	Dedicated
33.	CBERS-02B	LM-3B F09	INPE, Brazil	2007/ 09/19	Dedicated
34.	ChinaSat-9	LM-3B F10	China Sat, China	2008/ 06/09	Dedicated
35.	VeneSat-1	LM-3B F11	Venezuelan Ministry of Science and Technology	2008/ 10/30	Dedicated
36.	PALAPA-D	LM-3B F12	PT Indonesia Tbk	2009/ 08/31	Dedicated
37.	Paksat-1R	LM-3B F15	SUPARCO	2011/ 08/12	Dedicated
38.	W3C	LM-3B F17	Eutelsat	2011/ 10/07	Dedicated
39.	NigComSat-1R	LM-3B F18	NSRDA, Nigeria	2011/ 12/20	Dedicated
40.	APSTAR-VII	LM-3B F19	APT, HK	2012/ 03/31	Dedicated

International Commercial Launch of Long March Launch Vehicles

Since China officially announced that the LM launch vehicles entered the international commercial launch market in 1985, it has successfully launched various foreign-made satellites into orbit and occupied a place in the international market of commercial satellite launch services.

By the end of 2017, China accomplished 60 commercial launches for foreign customers and domestic customers, including 15 piggyback launch services. China's rising aerospace industry attracts international counterparts' attention with its good market reputation and first-class brand image. In the future, China's aerospace industry will greatly expand internationally; make comprehensive and multilevel cooperation with many foreign clients in the fields of product development, system construction, satellite application, resource sharing, personnel exchanges, and manned space flight; and will actively realize the goal of using space technology to benefit human beings.

Launch Vehicles in Service

LM-2

The LM-2 series developed in 1970 are mainly used for LEO missions. Currently, the LM-2 series rockets consist of 7 different launchers (LM-2, LM-2C, LM-2C/CTS1, LM-2C/CTS2, LM-2D, LM-2E, LM-2F). But only five of them are showed in Fig 1. Among them, LM-2 and LM-2E are no longer in use.

LM-2C/CTS-1/CTS-2

The LM-2C series have two-stage state and three-stage state, mainly used for launching satellites into LEO, SSO, extremely elliptical orbit (EEO), and GTO. It possesses the capability of launching multi-satellites with one vehicle.

LM-2C in Two-Stage State

With total length of 43 m, diameter of 3.35 m, LM-2C launch vehicle is mainly used for launching LEO and SSO satellites. Its launch capacity for 200 km LEO and 600 km SSO is 4.1 t and 1.5 t, respectively.

In August 1987, the LM-2C successfully provided piggyback launch of microgravity test instrument for French Matra Marconi Company, marking the beginning of China's international cooperation in aerospace industry.

LM-2C in Three-Stage State

LM-2C in three-stage state, namely, LM-2C/CTS, is formed by adding a solid upper stage to the LM-2C. It includes LM-2C/CTS-1 and LM-2C/CTS-2. Its total length is 43 m, and the diameter is 3.35 m. LM-2C/CTS-1 mainly used for launching multi-satellites, and SSO satellites, with 1.9 t launch capacity for 600 km SSO. LM-2C/

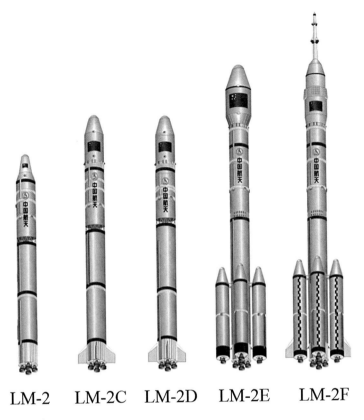

LM-2   LM-2C   LM-2D   LM-2E   LM-2F

Fig. 1 LM-2 series

CTS-2 mainly used for launching EEO satellites and GTO satellites. The GTO launch capacity (inclination 28_) is 1.25 t.

Between 1997 and 1999, LM-2C/CTS-1 successfully completed "one launcher two satellites" launch for seven times, sending 14 Iridium satellites (2 dummy satellites and 12 communications satellites) into orbit.

Specifications of LM-2C series			
	First stage	Second stage	Third stage CTS-1/ CTS-2
Maximum diameter of core stage (m)	3.35	3.35	/
Propellant mass (t)	172	54.6	2.62/0.125
Propellant	UDMH/ N2O4	UDMH/N2O4	Solid
Engine	YF-21C	YF-24E	Solid
Engine thrust (kN)	2961.6	741.4 (main) 11.8 × 4 (vernier)	107

(continued)

Specifications of LM-2C series

	First stage	Second stage	Third stage CTS-1/ CTS-2
Engine-specific impulse (N·s/kg)	2556.6	2922.37 (main) 2834.11 (vernier) (vacuum)	10.78
Booster number	/		
Booster diameter (m)	/		
Lift-off mass (t)	242	.	
Total length (m)	43.027		
Fairing diameter (m)	3.35		
Launch capacity (kg) LEO SSO GTO	4100(two stages)		
	1,500 (two stages), 1,900(600 km SSO)		
	1,250		
Current main mission (orbit)	SSO satellite		
Main launch site	Jiuquan, Taiyuan, Xichang satellite launch center		
Research and development entity	China aerospace science and technology corporation		

LM-2D

LM-2D is a two-stage conventional liquid launch vehicle, mainly used for LEO and SSO missions. It has the capabilities of launching two satellites in parallel and launching multi-satellite with one vehicle.

With total length of 41.056 m and diameter of 3.35 m, LM-2D launch vehicle is mainly used for launching satellites to LEO and SSO. Its launch capacity for 260 km LEO and 600 km SSO is 3.6 t and 1.5 t, respectively.

Specifications of LM-2C series

	First stage	Second stage
Maximum diameter of core stage (m)	3.35	3.35
Propellant mass (t)	124	59
Propellant	UDMH/N2O4	UDMH/N2O4
Engine	YF-21C	YF-24C
Engine thrust (kN)	2961.6	742.04 (main) 47.1 (vernier)
Engine-specific impulse (N·s/kg)	2556.6	2,942 (main) 2,834 (vernier) (vacuum)
Booster number	/	
Booster diameter (m)	/	
Lift-off mass (t)	250	
Total length (m)	41.056	
Fairing diameter (m)	3.35	

(continued)

Specifications of LM-2C series

	First stage	Second stage
Launch capacity (kg) LEO	3,600 (260 km LEO)	
SSO	1,500(600 km SSO)	
Current main mission (orbit)	SSO satellite	
Main launch site	Jiuquan satellite launch center	
Research and development entity	China aerospace science and technology corporation	

LM-2F

LM-2F is a highly reliable and safe launch vehicle developed to meet the demands of China's manned space special project. LM-2F has two variants of launching Shenzhou spacecraft and target spacecraft. As for launching Shenzhou spacecraft, its diameter is 3.35 m, and total length is 58.3 m. It consists of four liquid boosters, first stage, second stage, fairing, and the escape tower. Its launch capacity for LEO is 8.1 t.

On October 15, 2003, LM-2F successfully sent China's astronaut Yang Liwei into space, making China the third country in the world having manned space capability and also marking China's Manned Space Project entering into a substantive application stage.

Specifications of LM-2F

	First stage	Second stage	Booster
Maximum diameter of core stage (m)	3.35	3.35	2.25
Propellant mass (t)	184	83.727	45.277
Propellant	UDMH/N2O4	UDMH/N2O4	UDMH/N2O4
Engine	YF-20 K	YF-24 K	YF-25 K
Engine thrust (kN)	2961.6	741.4 (main) 11.8 × 4 (vernier)	740.4
Engine-specific impulse (N·s/kg)	2556.6	2922.37 (main) 2834.11 (vernier) (vacuum)	2550
Booster number	4		
Booster diameter (m)	2.25		
Lift-off mass (t)	493		
Total length (m)	58.3		
Fairing diameter (m)	3.35		
Launch capacity (kg) LEO	8,100		
Current main mission (orbit)	LEO		
Main launch site	Jiuquan satellite launch center		
Research and development entity	China aerospace science and technology corporation		

LM-3

The LM-3 series are made up of four launch vehicles (see Fig. 2), i.e., LM-3, LM-3A, LM-3B, and LM-3C. As LM-3 has been retired, the rest of the three kinds of rockets are known as the LM-3A series of launch vehicles.

LM-3A

LM-3A is a three-stage liquid launch vehicle, consisting of the first stage, second stage, third stage, and fairing. Its total length is 52.52 m, the diameter of the first stage and the second stage is 3.35 m, and the diameter of the third stage is 3 m. Its standard launch capacity to GTO is 2.6 t.

Fig. 2 LM-3 series of launch vehicles

Specifications of LM-3A

	First stage	Second stage	Third stage
Maximum diameter of core stage (m)	3.35	3.35	3
Propellant mass (t)	171.843	33.207	18.518
Propellant	UDMH/N2O4	UDMH/N2O4	LH2/LOX
Engine	YF-21C	YF-24E	YF-75
Engine thrust (kN)	2961.6	741.4 (main) 11.8 × 4 (vernier)	82.76
Engine-specific impulse (N·s/kg)	2556.6	2922.37 (main) 2834.11 (vernier) (vacuum)	4,300
Booster number	0		
Booster diameter (m)	/		
Lift-off mass (t)	243		
Total length (m)	52.52		
Fairing diameter (m)	3.35		
Launch capacity (kg) LEO SSO GTO	– – 2600		
Current main mission (orbit)	GTO		
Main launch site	Xichang satellite launch center		
Research and development entity	China aerospace science and technology corporation		

LM-3B

LM-3B is a three-stage liquid launch vehicle employing enhanced LM-3A as the core stage and strapped with four liquid boosters. Its total length is 54.84 m. The diameter of booster is 2.25 m. The diameter of the first stage and the second stage is 3.35 m, and the diameter of the third stage is 3.0 m. At present, LM-3B rocket consists of three variants. Its standard launch capacity to GTO is between 5.1 and 5.6 t.

LM-3B is the main launch vehicle to undertake the high orbit international commercial satellite launching services. It is also the launcher of lunar probe "Chang'e III" and "Chang'e IV."

Specifications of LM-3B

	First stage	Second stage	Third stage	Booster
Maximum diameter of core stage (m)	3.35	3.35	3	2.25
Propellant mass (t)	171.843	49.876	18.324	37.756
Propellant	UDMH/N2O4	UDMH/N2O4	LH2/LOX	UDMH/N2O4

(continued)

Specifications of LM-3B				
	First stage	Second stage	Third stage	Booster
Engine	YF-21C	YF-24E	YF-75	YF-25
Engine thrust (kN)	2961.6	741.4 (main) 11.8 × 4 (vernier)	82.76	740.4
Engine-specific impulse (N·s/ kg)	2556.6	2922.37 (main) 2834.11 (vernier) (vacuum)	4,300	2556.6
Booster number	4			
Booster diameter (m)	2.25			
Lift-off mass (t)	427			
Total length (m)	54.84			
Fairing diameter (m)	4.2			
Launch capacity (kg) LEO SSO GTO	– – 5600			
Current main mission (orbit)	GTO			
Main launch site	Xichang satellite launch center			
Research and development entity	China aerospace science and technology corporation			

LM-3C

LM-3C takes LM-3A as the core stage and is strapped with two liquid boosters at the first stage. Its total length is 54.84 m. The diameter of the booster is 2.25 m, the diameter of the first stage and second stage is 3.35 m, and the diameter of the third stage is 3.0 m. The GTO launch capacity of LM-3C reaches 3.8 t.

Specifications of LM-3C				
	First stage	Second stage	Third stage	Booster
Maximum diameter of core stage (m)	3.35	3.35	3	2.25
Propellant mass (t)	171.843	49.876	18.324	37.756
Propellant	UDMH/ N2O4	UDMH/N2O4	LH2/ LOX	UDMH/ N2O4
Engine	YF-21C	YF-24E	YF-75	YF-25
Engine thrust (kN)	2961.6	741.4 (main) 11.8 × 4 (vernier)	82.76	740.4
Engine-specific impulse (N·s/ kg)	2556.6	2922.37 (main) 2834.11 (vernier) (vacuum)	4,300	2556.6
Booster number	2			
Booster diameter (m)	2.25			
Lift-off mass (t)	343			
Total length (m)	54.84			

(continued)

Specifications of LM-3C				
	First stage	Second stage	Third stage	Booster
Fairing diameter (m)	4			
Launch capacity (kg) LEO	–			
SSO	–			
GTO	3800			
Current main mission (orbit)	GTO			
Main launch site	Xichang satellite launch center			
Research and development entity	China aerospace science and technology corporation			

LM-4

The LM-4 series are made up of three launch vehicles (see Fig. 3), i.e., LM-4A, LM-4B, and LM-4C. As LM-4A has been retired, the rest of the two kinds of rockets are known as the LM-4 series.

LM-4B

LM-4B is a three-stage launch vehicle using room-temperature liquid propellants. It is made up of the first stage, second stage, third stage, and fairing. It consists of the vehicle structure, engine, pressurized feeding system, control, telemetry, and external measuring safety subsystems. Its total length is 45.776 m, the diameter of the first and second stage is 3.35 m, and the diameter of the third stage is 2.9 m. LM-4B can implement a variety of orbit missions (SSO, LEO, GTO) and different types of satellite launches. Its launch capacity for 200 km 60_ inclination circular orbit is about 4.6 t, and for 400 km sun-synchronous orbit is about 3.2 t. LM-4B is mainly used to send satellites to SSO.

Specifications of LM-4B			
	First stage	Second stage	Third stage
Maximum diameter of core stage (m)	3.35	3.35	2.9
Propellant mass (t)	181.89	35.408	14.34
Propellant	UDMH/N2O4	UDMH/N2O4	UDMH/N2O4
Engine	YF-21C	YF-24H	YF-40B
Engine thrust (kN)	2961.6	742.04(main)46.09 (vernier)	100.848
Engine-specific impulse (N·s/kg)	2556.6	2942.4(main) 2761.6(vernier)(vacuum)	2,971
Lift-off mass (t)	250		
Total length (m)	45.776		
Fairing diameter (m)	3.35		

(continued)

Fig. 3 LM-4 series of launch
vehicles

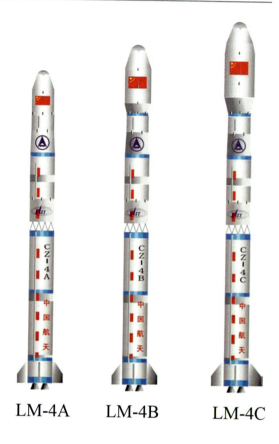

LM-4A   LM-4B   LM-4C

Specifications of LM-4B			
	First stage	Second stage	Third stage
Launch capacity (kg) LEO SSO/400 km GTO	4600		
	3200		
	/		
Current main mission (orbit)	SSO		
Main launch site	Taiyuan satellite launch center		
Research and development entity	China aerospace science and technology corporation		

LM-4C

LM-4C is an enhanced three-stage liquid launch vehicle by adding new technologies
such as restart of the third-stage engine to LM-4B. Its total length is 45.776 m, the
diameter of the first and second stage is 3.35 m, and the diameter of the third stage is
2.9 m. Its launch capacity for 600 km trajectory tilted by 60 degree to the equator is
about 3.7 t, for 800 km SSO is 2.7 t, and for GTO is about 1.3 t. LM-4C is mainly
used to send satellites to SSO.

Specifications of LM-4C

	First stage	Second stage	Third stage
Maximum diameter of core stage (m)	3.35	3.35	2.9
Propellant mass (t)	189.841	34.449	13.971
Propellant	UDMH/ N2O4	UDMH/N2O4	UDMH/ N2O4
Engine	YF-21C	YF-24H	YF-40A
Engine thrust (kN)	2961.6	742.04 (main) 46.09 (vernier)	100.848
Engine-specific impulse (N s/kg)	2556.6	2942.4 (main) 2761.6 (vernier)(vacuum)	2,971
Lift-off mass (t)	250		
Total length (m)	47.977		
Fairing diameter (m)	3.8		
Launch capacity (kg) LEO	/		
SSO/800 km	2700		
GTO	1300		
Current main mission (orbit)	SSO		
Main launch site	Taiyuan satellite launch center		
Research and development entity	China aerospace science and technology corporation		

New Generation Launch Vehicles

China is continuously strengthening the construction of space transportation systems, further perfecting the completeness of LM launch vehicles, enhancing the ability of access to space, and developing a new generation of launch vehicles and upper stages with the LM-5, LM-6, LM-7, LM-11 as the representatives. LM-5 uses nontoxic environment-friendly propellants with LEO launch capacity of 25 t and GTO capability of 14 t. LM-6 is a new and fast responsive rocket, with launch capacity for 700 km SSO not less than 1 t. Launch capacity of LM-7 for LEO and 700 km SSO is 13.5 t and 5.5 t, respectively.

LM-5

LM-5 is a new large rocket developed under the guideline of "one series, two engines, and three modules." The "three modules" refers to the 5 m diameter module using liquid hydrogen and liquid oxygen as propellants, the 3.35 m diameter and 2.25 m diameter modules using liquid oxygen and kerosene propellants. The "two engines" refers to the newly developed liquid oxygen and liquid hydrogen engine of 50 t class thrust and the liquid oxygen and kerosene engine of 120 t class thrust. Based on the design concept of "generalization, serialization, and combination," six configurations with 5 m-diameter core stage will be formed based on the three newly developed modules. Its GTO launch capacity covers from 6 t to 14 t, and LEO launch capacity from 10 t to 25 t. LM-5 configuration for the first flight is the one strapped with four 3.35 m-diameter boosters.

LM-5 uses brand new power system, large vehicle structural design and manufacturing technology, advanced control, and digital technology, which significantly improve the overall level of China's launch vehicles and the capacity of utilizing space resources. Its total length is 57 m. The diameter of the first stage and second stage is 5 m. The diameter of four strap-on booster is 3.35 m. Its maximum GTO launch capacity is 14 t. The maiden flight of LM-5 took place in November 2016.

LM-6

As a member of the new generation rockets, LM-6 is a light and fast responsive liquid launch vehicle. To meet easy and quick launch requirements, it is transported to the simplified launch pad (without tower) by the self-moving vertical car, then is erected, filled up, and launched. It is 29.9 m long and 3.35 m in diameter. Its lift-off mass is about 102 t, and lift-off thrust reaches about 1200 kN. It adopts three-stage configuration and uses nontoxic propellants such as liquid oxygen and kerosene. With monitoring and control limits, its launch capacity for 700 km SSO is approximately 500 kg. Without considering limits of monitoring and control conditions, its launch capacity for 700 km SSO is about 1,000 kg by means of gliding and restarting of the third stage. The maiden flight took place in September 2015.

LM-7

LM-7 is a new medium-sized launch vehicle, mainly used for launching cargo ship to the space station. Its total length is 53.1 m. Lift-off mass is about 595 t. Lift-off thrust is 735 t. The diameter of the core stage is 3.35 m, and the diameter of the four strap-on boosters is 2.25 m. The maximum launch capacity for LEO is 13.5 t. The maiden flight took place in June 2016.

LM-11

LM-11 is a four-stage solid launcher developed by china aerospace science and technology corporation. It is a quick response and low-cost launcher. The total length of LM-11 is 20.8 m, the maximum diameter is 2 m, the lift-off mass is 58 t, and the lift-off thrust is 120 t. LM-11's capacity of SSO at the altitude of 700 km is 400 kg, the capacity of LEO is 700 kg. The maiden flight of LM-11 took place in Sept 2015. The first sea launch of LM-11 was completed in June 2019.

Prospect of Security Policy

China's Space Security Policy

Outer space is the common wealth of human being, and the exploration of outer space is the unremitting pursuit of mankind. At present, the world space activities are booming. Major space nations successively develop or adjust the space development strategy, development plan and development objectives, status and role of the aerospace industry in the country's overall development strategy have become increasingly important. The impact of space activities on human civilization and social progress has been enhanced.

China puts the development of space industry as an important part of the country's overall development strategy and always adheres to the policy of exploring and using outer space for peaceful purposes. In recent years, China's space industry develops rapidly with some important technology areas having reached the world's leading level. Space activities play an increasingly important role in China's economic construction and social development.

In the future, China will focus on the national strategic goals, strengthen independent innovation, expand opening up and cooperation, and promote sound and rapid development of space industry. Meanwhile, China is willing to work together with the international community to jointly safeguard a peaceful and clean outer space and make new contributions to promote the human peace and development.

China's aims of developing space activities are as follows: to explore the outer space and enhance the understanding of the Earth and the universe; to peacefully use the outer space and promote human civilization and social progress for the benefit of all human beings; to meet the needs of economic construction, scientific and technological development, national security, and social progress; to improve the scientific and cultural quality of human beings; to safeguard national interests; and to enhance overall national strength.

The principle of China's space security policy is subordinate to and serves the overall national development strategy and adheres to the scientific planning, self-development, the peaceful use, and open cooperation principles.

Scientific planning is defined as respecting science and law and comprehensively balancing and scientifically developing space technology, space application, space science, and other space activities based on the development reality of the aerospace industry, to maintain the comprehensive, coordinated, and sustainable development of aerospace industry.

The meaning of self-development always adheres to the independent and self-reliant path of development, which mainly relies on our own power to self-develop the space industry according to national conditions and national strength, to meet the basic needs of the country's modernization drive.

Peaceful use means always adhere to the peaceful use of outer space, opposes weaponization of outer space and arms race in outer space, rationally develops and utilizes space resources, and effectively protects the space environment, letting space activities bring benefits to the people.

   Open cooperation refers to adhere to the combination of being independent and cooperating openly, on the basis of equality and mutual benefit, peaceful utilization and common development, actively carrying out international exchanges and cooperation in the space industry, and committed to the common progress of human space industry. International cooperation in the space field should follow the basic principles of United Nations' *Declaration on International Cooperation in the Exploration and Use of Outer Space for the Benefit and in the Interest of All States, Taking into Particular Account the Needs of Developing Countries.*

   In the next 5 years, China will strengthen the basic capacity building of aerospace industry, advance the deployment of cutting-edge research, and continue to implement major scientific and technological projects and priority projects in key areas, such as the manned spaceflight, lunar exploration, high-resolution Earth observation system, satellite navigation and positioning system, and the new generation of launch vehicles. It will comprehensively improve space infrastructure, promote the development of satellite industry and its application industry, conduct space science research in depth, and enhance the comprehensive, coordinated, and sustainable development of the aerospace industry.

Space Debris Mitigation of Long March Launch Vehicles

Established in 1993, the Inter-Agency Space Debris Coordination Committee (IADC) aims to strengthen the research and coordination of member states in the field of space debris. China is also a member of the committee. As a space power, China actively participates in the related anti-space debris activities and conscientiously fulfills the duties and obligations that it commits to IADC, in order to protect the image and status of China's aerospace industry in the international arena.

   China organizes experts in the fields of aerospace and space policy to study the feasibility of the design and management of space debris mitigation and formulated the Guideline for Orbital Debris Mitigation in 2006. From technical aspects, the standard puts forward basic requirements on the design of orbital debris in each step of space activities. Those basic requirements are consistent with IADC's guidelines for space debris mitigation. Based on the Guideline for Orbital Debris Mitigation, China is gradually setting design and management standards for space debris mitigation.

   In accordance with international conventions, China will greatly promote the space debris mitigation design of Long March launch vehicles. For LEO missions, life of the last stage in orbit is less than 25 years, and passivation measures are generally adopted; for SSO missions, active de-orbit measures or passivation measures are taken.

   In the future, China will continue to strengthen the space debris monitoring, mitigation and spacecraft protection in the field of space debris. It will develop space debris monitoring and collision warning technology and carry out the monitoring and collision warning of space debris and small near-Earth celestial bodies. It will establish evaluation system of space debris mitigation design and actively take space debris mitigation measures on the after-mission spacecraft and launch vehicles. It will test digital simulation technology of space debris impact and promote the construction of space debris protection system.

Conclusion

Free exploration, development and utilization of outer space and celestial bodies, is the equal right shared by countries in the world. Each country's outer space activities should contribute to its national economic development and social progress and should benefit human security, survival, and development. China's launch vehicle technology development will contribute to the technological progress of world's space exploration. China advocates strengthening international exchange and cooperation and promoting inclusive development in the space industry on the basis of equality and mutual benefit, peaceful use, and common development.

References

China Academy of Launch Vehicle Technology, website: http://calt.spacechina.com/n482/n498/index.html

China Aerospace Science and Technology Corporation, website: http://www.spacechina.com/n25/n146/n238/n12985/index.html

China's Space Activities (2011), China National space Administration website, http://www.cnsa.gov.cn/n6758824/index.html

China's Space Activities (2016), China government website: http://www.gov.cn/xinwen/2016-12/27/content_5153378.htm; China National space Administration website: http://www.cnsa.gov.cn/n6758824/index.html

Indian Space Program: Evolution, Dimensions, and Initiatives

70

Hanamantray Baluragi and Byrana Nagappa Suresh

Contents

Abstract

With a modest start in the 1960s by Dr. Vikram A. Sarabhai – known as the father of Indian space program – the latter has matured in the last six decades through the use of space technologies and applications for national development. Undoubtedly there has been a major evolution since its inception. In this direction, programs and missions developed by the Indian Space Research Organisation (ISRO) consist of launch vehicle development, Earth observation, satellite communications, satellite navigation and space science and planetary

H. Baluragi (✉) · B. N. Suresh
Indian Space Research Organization (ISRO), Bangalore, India, Bangalore, India
e-mail: baluragi@isro.gov.in; byranasuresh@gmail.com; bnsuresh@isro.gov.in

© Springer Nature Switzerland AG 2020
K.-U. Schrogl (ed.), *Handbook of Space Security*,
https://doi.org/10.1007/978-3-030-23210-8_38

exploration, and satellite applications. One of the important missions accomplished in the recent past is the development and operationalization of the new launch vehicle, GSLV MkIII. India's next milestone mission of Human Spaceflight has been initiated, and the first crewed flight is expected by 2022. As of September 2019, ISRO has accomplished 184 missions, including 101 satellite missions, 73 launch vehicle missions, and 10 technology demonstration missions. This technical chapter describes the evolution of the Indian space program, its dimensions, and new initiatives.

Introduction

The primary objective of the Indian space program is to build indigenous capabilities in space technology, develop applications to meet the developmental needs of the country, and harness the benefits of space applications for socioeconomic development (Suresh 2015; Pant 1986). In this direction, programs and missions drawn up by the Indian Space Research Organisation (ISRO) consist of launch vehicle development, Earth observation, satellite communications, satellite navigation, and space science and planetary exploration. As of September 2019, a total of 184 missions have been accomplished. This includes 101 satellite missions, 73 launch vehicle missions, and 10 technology demonstration missions including Space Capsule Recovery Experiment (SRE), Crew Module Atmospheric Re-entry Experiment (CARE), Reusable Launch Vehicle-Technology Demonstrator, scramjet engine, pad abort test, Indian nano-satellite missions, and microsatellite missions. In addition, 297 foreign satellites from 33 countries have been successfully launched from Sriharikota location using ISRO's Polar Satellite Launch Vehicle.

The Earth observation program consists of state-of-the-art remote sensing satellites such as Resourcesat, Cartosat, Oceansat, radar imaging satellite and weather/meteorological satellites. These satellites observe and monitor Earth's resources and provide systematic information at wide-ranging spatial resolutions available at regular intervals. The satellite communications program includes state-of-the-art INSAT (Indian National Satellite) and the GSAT (Geosynchronous SATellites) communications satellites. These satellites render telecommunication, television broadcasting, meteorological data dissemination, and emergency and strategic communication. They also provide a variety of societal applications for tele-education, tele-medicine, and the Village Resource Centers. The satellite navigation program comprises of a constellation of 8 Indian Regional Navigation Satellite System (IRNSS) with associated ground segment infrastructure. They are intended to provide accurate positioning and timing service supported by the GPS-aided GEO augmented navigation (GAGAN). The space science and planetary exploration program includes satellites designed for studying outer space and planetary exploration; these are the Chandrayaan-1, Mars Orbiter Mission, AstroSat, and Chandrayaan-2.

India has started its launch vehicle development to orbit satellites in a self-reliant manner. Starting with the development of the Satellite Launch Vehicle (SLV-3) during the 1970s, it has progressed through the Augmented Satellite Launch Vehicle

(ASLV), Polar Satellite Launch Vehicle (PSLV), and Geosynchronous Satellite Launch Vehicle (GSLV). Recently, the development of the next-generation launch vehicle, the Geosynchronous Satellite Launch Vehicle Mark III (GSLV MkIII), has been completed and has become operational.

Several other advanced technologies such as the semi-cryogenic boosters and air-breathing propulsion and the development of reusable launch vehicles have been undertaken to meet the long-term demands of the country. Recently, ISRO has initiated the development of a Small Satellite Launch Vehicle toward achieving quick turnaround, on-demand launch services in a cost-effective manner. This vehicle is expected to undergo flight test in the second quarter of 2020. After successful development and demonstration of critical technologies for human space-flight, India has committed to develop end-to-end systems for launch and safe recovery of Indian astronauts by 2022.

Evolution of the Indian Space Program

With a modest start in the early 1960s, the Indian space program has now matured and reached operational stage. From its very inception, the emphasis of the Indian space program has been on the application and utilization of space technology for national development (Suresh 2015; Pant 1986) with gradual transformation in pursuit of self-reliance. Harnessing space applications (Pant 1986) for national development involves:

(i) The use of space for mass communication including communication broad-casted through television and radio, etc.
(ii) The use of space for navigation
(iii) The use of space for remote sensing including meteorology to sense, collect, process, and disseminate the resource and weather data for effective use in planning, monitoring, development, and exploitation of natural resources
(iv) The use of satellites for space science

To realize these goals, the Indian space program has developed capability to (i) carry out efficient utilization of communication and remote sensing satellite data and services; (ii) design, build, and operate communication and remote sensing satellites; and (iii) develop space transportation systems for orbiting satellites. ISRO has made significant progress over the past six decades and has gone through four distinct phases which are shown in Fig. 1.

Initiation Phase (1960–1970s)

The first "initiation phase" was a learning phase. During this period, ISRO established laboratories, trained manpower by undertaking R&D tasks and took up initial development of sounding rockets and associated scientific payloads. Major

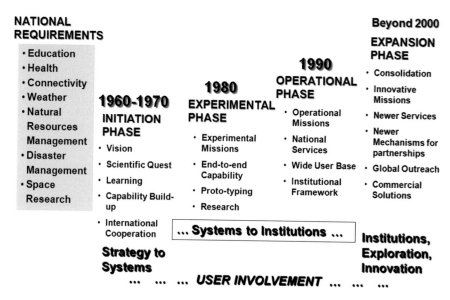

Fig. 1 Evolution of the Indian Space Program

experimental projects were undertaken for the development of space technology oriented toward applications and appropriate hardware realization. Experiments such as the Satellite Instructional Television Experiment (SITE) and the Satellite Tele-communication Experimental Project (STEP) in the field of space-based communi-cations and broadcasting demonstrated the importance of satellites for national development. The SITE project (during 1975) connected 2400 villages and demon-strated the largest mass communication experiments ever carried out in the world. The STEP project (1977–1979) helped ISRO to understand the problems associated with space and ground systems for telecommunications satellites. To summarize, during this phase, ISRO defined its vision for the space program, strategy for its systems, plan for capability building within the country, and the ability to effectively cooperate with other nations.

Experimental Phase (1980s)

In the second "experimental phase," ISRO developed a series of satellites for different applications such as Aryabhata, Bhaskara-I, Bhaskara-II, and Ariane Pas-senger Payload Experiment (APPLE). Following the successful demonstration of Bhaskara-I and Bhaskara-II satellites launched in 1979 and 1981, respectively, India began to develop the indigenous Indian Remote Sensing (IRS) satellites to support national development in the areas of agriculture, water resources, forestry and ecology, geology, water sheds, marine fisheries, and coastal management. Toward developing expertise in satellite imagery, ISRO designed and built the first remote sensing missions, i.e., IRS-1A, and launched onboard the Vostok-2M launcher.

Similarly, for technology development of communications satellites, the Indian National Satellite (INSAT) system was commissioned with the launch of the INSAT-1B in August 1983. The INSAT system ushered a revolution in India's television and radio broadcasting, telecommunication, and meteorological sectors. It enabled the rapid expansion of television and modern telecommunication facilities even to remote areas and offshore islands. During this period, ISRO also developed the first-generation launch vehicle SLV-3, which placed the 40 kg Rohini satellite into low Earth orbit, thereby making India the sixth member of an exclusive club of spacefaring nations. The successful culmination of the SLV-3 project showed the way to advanced launch vehicle projects. The Augmented Satellite Launch Vehicle (ASLV) was designed to enhance the payload capacity to 150 kg and enable further technology developments in the areas of strap-on booster, bulbous payload fairing, canted nozzles, closed-loop guidance, and overall mission management. The ASLV proved to be a low-cost intermediate vehicle demonstrating and validating critical technologies needed for future launch vehicles. At the same time, ISRO developed the Polar Satellite Launch Vehicle (PSLV) for the launching of operational remote sensing satellites into the Sun-synchronous polar orbit. For this purpose, it developed state-of-the-art materials and manufacturing methods, and it established a launch pad and major facilities like propellant casting and assembly, integration, and testing.

Operational Phase (1990s)

During the "operational phase," ISRO entered into the operation and quasi-operation of space applications. The INSAT-1 series of the first-generation multipurpose satellites for communication and meteorology and the Indian Remote Sensing (IRS) were launched and made operational. The PSLV provided a quantum jump in the development of critical technologies like large solid motor, Earth-storable liquid engines, composite motor case, strap-down navigation system, etc. This vehicle became a versatile platform for the host of missions such as low Earth orbit (LEO), Sun-Synchronous Polar Orbit (SSPO), and Geosynchronous Transfer Orbit (GTO). In parallel, ISRO initiated the development of Geosynchronous Satellite Launch Vehicle (GSLV) with three stages, employing solid, liquid, and cryogenic propulsion modules for launching 2t class of communications satellites into GTO. Initially, a procured cryogenic stage from Russia was used for the upper stage. At the same time, technology developments needed for the cryo-engine and stage systems were undertaken.

Expansion Phase (2000s)

During the "expansion phase," extending beyond the 2000s, ISRO undertook leapfrog steps in its space program. Apart from capability in design, development, launching, and operationalization of satellite systems, ISRO expanded to undertaking interplanetary missions, reentry missions, commercial launch services, international collaboration in space technology and advanced technological missions in propulsion, and low-cost access to space. In 2008, India's first space mission to the Moon was achieved with the launch of Chandrayaan-1, which discovered lunar water in the form of ice. The Mars Orbiter Mission (MoM), which entered Mars' orbit on 24 September 2014, made India the first nation to succeed on its maiden attempt to planet Mars. ISRO launched 104 satellites in a single mission (PSLV-C37) – a world record. The cryo-engine developed for the GSLV successfully qualified through a series of ground tests, short and long duration at different levels. The cryo-stage was successfully flight tested in January 2014. The development of the next-generation launch vehicle, the GSLV MkIII, was initiated to meet the demands of launching 4t class of satellites, complete development, and operationalization. Toward future missions of human spaceflight and space science experiments, ISRO successfully conducted the Space capsule Recovery Experiment (SRE). This mission helped to master reentry technologies and recovery procedures required for the design of future reusable launch vehicles. Toward low-cost access to space, the first technology demonstration missions of the Reusable Launch Vehicle (RLV-TD) and scramjet engine were successfully accomplished.

To summarize, India entered into space activities in a humble manner, and over the years it has evolved into a major spacefaring nation through the development and demonstration of all kinds of missions, including interplanetary missions.

Dimensions of the Indian Space Program

ISRO's programs have always been driven by its vision to "harness space technology for national development while pursuing space science research and planetary exploration." All the space programs till date have aimed at developing independent access to space for which ISRO has developed the PSLV, GSLV, and GSLV MkIII launch vehicles along with the capability to build and operate Earth observation, communication, and navigation satellites. The activities of the Indian space program can be organized into the following dimensions as shown in Fig. 2:

Space Applications

A major thrust was directed toward space applications right from the very beginning of space activities in the early 1960s. Space applications are derived through synergistic use of Earth observation, communication, and navigation satellites and are complemented with ground-based observations. They play a key role in harnessing the benefits of space technology for socioeconomic security, sustainable development, disaster risk reduction, and efficient governance. Space applications and services in India are tailored to meet the evolving needs and fundamental priorities of the government. Over the past decades, Earth observation data, integrated with in situ observations and tools, have been supporting a host of applications in the areas of land and water, ocean and atmosphere, environment and

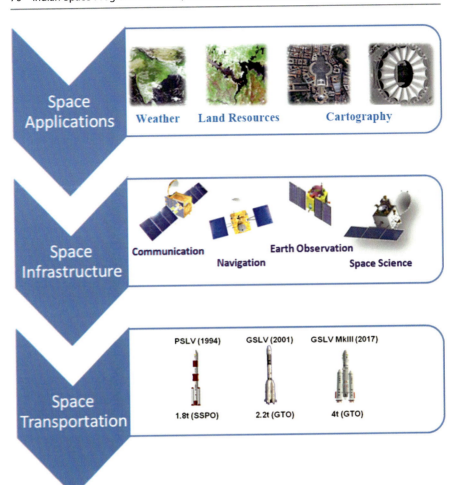

Fig. 2 Dimensions of the Indian Space Program

ecosystem, and urban and rural applications, including disaster risk reduction. Many of these applications have been effectively used and adopted by stakeholder departments for operational use. The capabilities of satellite communications and navigation have also been exploited for delivering an array of societal applications toward education, health, connectivity, skill development, and livelihood sustenance. Besides ensuring continuity of services from space, a strategy has also been drawn to strengthen applications through synergistic use of Earth observation, communication, and navigation satellites. This is planned with a specific focus on harnessing the benefits of space technology in the following broad areas, viz., (i) security of food, water, energy, health, shelter, infrastructure, and information; (ii) sustainable development; (iii) disaster risk reduction; and (iv) governance-related issues. The applications planned in these areas are in tune with the sustainable development goals, envisaged by the government. Space-based applications and tools are being used by 56 central ministries and governmental departments in planning, periodic monitoring, midcourse correction, and evaluation of developmental activities in various sectors such as agriculture, water resources, forest and environment, urban and rural planning, infrastructure development, satellite communications, and disaster management support. Many flagship programs like the Atal Mission for Rejuvenation and Urban Transformation (AMRUT), the Housing-for-all program, the National Mission for Clean Ganga, the Mahatma Gandhi National Rural Employment Guarantee (MGNREGA), etc. are also utilizing space-based tools. To facilitate these activities, customized tools to stakeholder requirements and mobile applications (Apps) are deployed and used in the current scenario. ISRO is giving further impetus to the utilization of space technology in governance and development by catalyzing the development of innovative applications and services.

Space Infrastructure

India has painstakingly developed a comprehensive space technology system over the past few decades that has played a significant role in various sectors of national development. Particularly, over the past 30 years, space systems and technologies have increasingly become a critical part in meeting the country's requirements. Space infrastructure includes spacecraft for Earth observation, communication, navigation, and space science, including associated ground segment to ensure data and services on a continued and assured basis. The emphasis has always been on self-reliance in high-technology areas with greater participation of the Indian industries and academia.

The array of Earth observation (EO) satellites that has evolved over a period of three decades with imaging capabilities in visible, infrared, thermal, and microwave regions of the electromagnetic spectrum, including hyperspectral sensors, has helped the country in realizing major operational applications. The imaging sensors have been providing spatial resolution ranging from 1 km to better than 1 m, repeat observation (temporal imaging) from 22 days to 1 day, and radiometric ranging from 7 bit to 12 bit, which has significantly helped in several usages at national level. India has a constellation of 21 EO satellites in orbit. In the coming years, the Indian EO satellites are heading toward further enhanced and improved technologies, taking cognizance of the learnings/achievements made earlier, while addressing newer observational requirements and the technological advancements including high-agility spacecraft.

Communications satellites constitute another vital segment of space infrastructure. Timely realization of multiple communications satellites to meet the national requirements in all sectors will continue to be the core objective of satellite communications program. The INSAT system is one of the largest domestic communications satellite systems in Asia Pacific region with more than 18 operational communications satellites. The INSAT system with more than 250 transponders in the C, Extended C, and Ku-bands provides services to telecommunications, television broadcasting, satellite news gathering, societal applications, weather forecasting, disaster warning, and search and rescue operations.

Satellite navigation service is an emerging satellite-based system. ISRO is committed to provide the satellite-based navigation services to meet the emerging demands of the civil aviation requirements and to meet the user requirements of the positioning, navigation, and timing based on the independent satellite navigation system. To meet the civil aviation requirements, ISRO is working jointly with the Airport Authority of India (AAI). It established the GPS-aided Geo augmented navigation (GAGAN) system with accuracy and integrity required for civil aviation applications in order to provide better air traffic management of the Indian airspace. To meet user requirements of the positioning, navigation, and timing services based on the indigenous system, ISRO has established a regional satellite navigation system called "NavIC" (Navigation with Indian Constellation) having eight satellites in orbit. Brief details on the NavIC constellation are given in Fig. 3. Expanding the navigation constellation with additional satellites and transforming NavIC system to a global system with improved indigenous time reference system will be one of the main focus areas in the coming years.

The Indian space science program encompasses research in areas like astronomy, astrophysics, planetary and Earth sciences, atmospheric sciences, and theoretical physics. Several scientific instruments have been flown on satellites especially to direct celestial X-ray and gamma-ray bursts. Major science missions undertaken by ISRO include the AstroSat, Mars Orbiter Mission, and Chandrayaan-1 and Chandrayaan-2. The AstroSat satellite is the first dedicated Indian astronomy mission aimed at studying celestial sources in X-ray and optical and UV spectral bands simultaneously. The payloads cover the energy bands of ultraviolet (near and far), limited optical and X-ray regime (0.3–100 keV). One of the unique features of AstroSat mission is that it enables the simultaneous multiwavelength observations of various astronomical objects with a single satellite. The Mars Orbiter Mission (MoM) is ISRO's first interplanetary mission to planet Mars with an orbiter designed to orbit Mars in an elliptical orbit of 372 km by 80,000 km. The MoM can be termed as a challenging technological mission considering the critical mission operation and stringent requirements on propulsion, communications, and other bus systems of the

Fig. 3 Constellation and features of NavIC

spacecraft. Chandrayaan-1, India's first mission to the Moon, was orbiting around the Moon at a height of 100 km from the lunar surface for chemical, mineralogical, and photogeologic mapping of the Moon. The spacecraft carried 11 scientific instruments built in India, the United States, the United Kingdom, Germany, Sweden, and Bulgaria. Newer missions to study the Moon, Mars, and Venus are planned in the coming years, including the study of the corona of the Sun (Aditya-L1 Mission).

To provide a platform for standalone payloads for Earth imaging and science missions within a quick turnaround time, ISRO has conceived the small satellite program. Two kinds of satellite buses have been developed, making it a more versatile platform and capable of supporting different kinds of payloads. The Indian Mini Satellite (IMS-1) bus has been developed for 100 kg class satellites which can accommodate a payload capacity of around 30 kg. The first mission of the IMS-1 series was launched successfully on 28 April 2008. YouthSat is the second mission in this series and was launched successfully on 20 April 2011. The microsatellite (Microsat) is a small satellite recently launched by ISRO which is 100 kg and derives its heritage from the IMS-1 bus. This is a technology demonstrator and a forerunner for future satellite technologies. The satellite bus is modular in design and can be fabricated and tested independent of payload. The IMS-2 bus has evolved as a standard bus for 400 kg class that can support payload capacity of around 200 kg. The IMS-2 bus development is an important milestone as it is envisaged to be a workhorse for different types of remote sensing applications in the future. The first satellite based on the IMS-2 bus was SARAL (Satellite with ARGOS and ALTIKA), a joint Indo-French mission launched onboard PSLV-C20 on 25 February 2013.

In order to maximize the impact of the capabilities acquired by the Indian space program and to foster innovation in the space sector, ISRO is following a policy of ensuring the cross-fertilization of knowledge, innovation, and ideas between space and non-space sectors and between space industry and leading research organizations and universities in the country. As part of this policy toward creating a "space-ready" generation, the available spare capacity onboard the PSLV is offered to small satellites built by students of Indian universities, which will enable exposure of the future generation to the multidisciplinary space technology. The first satellite launched in this category was ANUSAT (40 kg) from Anna University, Chennai, on 20 April 2009. As of now, ISRO has provided launch services for ten student satellites from Indian Universities.

Space Transportation System

Assured access to space is a critical goal to a nation's technological advancement, scientific discovery, security, and economic growth. Self-reliance in space transportation systems has been an important component of the guiding vision of the Indian space program in the development of space technology and its applications for societal development. The vision for space transportation systems is to carry forward the self-reliance in launch vehicle technology through increased payload capacity and reusability. ISRO also envisages to ensure low-cost access to space and progressively arrive at reliable and robust space transportation systems to enable human space activities and space exploration. India has achieved self-reliance in space transportation capability through the operationalization of Polar Satellite Launch Vehicle (PSLV) and Geosynchronous Satellite Launch Vehicle (GSLV) and Geosynchronous Satellite Launch Vehicle Mark III (GSLV MkIII) for launching satellites for Earth observation, communication, navigation, and space exploration.

Evolution of the Indian launch vehicles along with capabilities and achievements are given in Fig. 4. The PSLV has been designed and developed for catering to the need of launching remote sensing/Earth observation satellites into Sun-synchronous and low Earth orbits. The PSLV is configured as a four-stage vehicle with alternate solid and liquid propulsion stages. The first development flight of PSLV was conducted during 1993. The first successful mission of PSLV was in 1994 and orbited a 904-kg Indian Remote Sensing satellite into a Sun-synchronous orbit. PSLV created history by deploying 104 satellites in a single mission. This remarkable achievement was a new moment of pride for the scientific space community and the country. PSLV has so far completed 48 flights which include 3 developmental and 45 operational flights. The 46 flights have been successful, and PSLV has established itself as a reliable and credible launch vehicle. PSLV has successfully orbited 354 satellites of total mass over 50 t. PSLV has demonstrated end-to-end

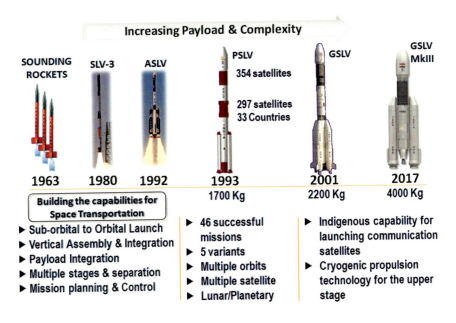

Fig. 4 Evolution of the Indian launch vehicle program

launch services by launching 297 satellites for 33 international customers including 7 dedicated commercial launches. PSLV has become the preferred launcher in the commercial launch market for the payload category up to 1500 kg. In summary, the PSLV has established itself as a workhorse vehicle for national satellites with a capacity that would enable responding quickly to commercial launch opportunities. The target for the PSLV program is to increase the frequency of launches, commercialization of activities through industry, and improvement of the global competitiveness through cost reduction measures.

The Geosynchronous Satellite Launch Vehicle is an expendable launch vehicle, designed and developed to launch 2t class of satellite to 180 km by 36,000 km geostationary transfer orbit. GSLV is a third-generation launch vehicle. It is a three-stage vehicle with four liquid strap-ons and cryogenic upper stage (CUS). GSLV uses main systems that are already proven in the PSLV in the form of solid booster and liquid-fueled Vikas engine. The first developmental flight of GSLV, i.e., GSLV-D1 carrying a payload of 1540 kg, was launched successfully on 18 April 2001. GSLV's primary payloads are 2t class of communications satellites that operate from geostationary orbits. The GSLV with indigenous cryogenic stage (GSLV-D5) was successfully launched on 5 January 2014 and has placed the GSAT-14 spacecraft with a mass 1980 kg in GTO. This was a major milestone for the Indian Space Research Organisation and a very significant demonstration of complex cryogenic technology. One of the significant achievements of the GSLV program, as a whole, is the development of complex cryogenic technologies, comprising of system design for engine and stage, special manufacturing process, new material development, testing facilities for subsystems, engine and stage and launch pad propellant filling, etc. The experience thus gained in CUS paved the way for the indigenous design and development of bigger engine and stage for the GSLV MkIII program and is also the stepping stone for the next-generation semi cryo-engine and stage development.

The Geosynchronous Launch Vehicle Mark III (GSLV MkIII) is a fourth-generation launch vehicle developed by ISRO primarily developed toward achieving indigenous launch capability to launch 4t class of communications satellites into geosynchronous transfer orbit (GTO). It is also identified as launch vehicle for crewed missions under Indian Human Spaceflight Programme. GSLV MK-III is configured with two solid strap-on motors (S200), one liquid core stage (L110), and a high thrust cryogenic upper stage (C25). The S200 solid motor is among the largest solid boosters in the world with 204.3 t of solid propellant. The liquid L110 stage uses a twin liquid engine configuration with 115.4 t of liquid propellant, while the C25 cryogenic upper stage is configured with the fully indigenous high thrust cryogenic engine (CE20) with a propellant loading of 28.35 t. The overall length of the vehicle is 43.5 m with a gross lift-off weight of 640 t and a 5-m diameter payload fairing. The first developmental mission of GSLV MkIII (GSLV MkIII-D1) was completed successfully on 5 June 2017, which launched the communications satellite (GSAT-19) into GTO orbit. So far, the GSLV MkIII vehicle has completed three missions, and all were successful.

The Human Spaceflight Program: A New Beginning in the Indian Space Program

In the early years of human spaceflight, progress was fast owing to the race for technological supremacy between the United States and the then Soviet Union. The competitive spirit drove the space engineers from these two nations to develop technologies for human spaceflight of increasing complexity including the launch vehicles, space capsules, reentry, and recovery strategies, which are all relevant to this day. While the early phase of the program till the early 1980s was focused on competitive technology development and demonstration, the next phase of the human spaceflight started incorporating science objectives with several science experiments conducted on the early Space Stations. This was followed by an era of international cooperation between the major space powers, which gave rise to the International Space Station (ISS), which has been the focus of human activity in space since 1998. ISS has been serving as a platform for evaluating human endurance in space, microgravity research, and interdisciplinary research till date. Human spaceflight capability in terms of launch capability, crew and habitat module, and safe return from orbit has been developed by the United States, the USSR, and China. From 1961 to 2018, more than 300 human spaceflight missions have been conducted by the United States, the USSR, Russia (Post 1991), and China. In recent times, there is a renewed interest in human space exploration through a globally coordinated approach synergizing the capabilities of all spacefaring nations and private players targeting exploration of the Moon, Mars, and near-Earth objects (asteroids) along with extraterrestrial resource exploitation.

Toward achieving indigenous capability in human transportation, ISRO conducted a major technology demonstration (Kumar et al. 2018) on 5 July 2018, the first in a series of tests to qualify a crew escape system, which is a critical technology relevant for human spaceflight. Launch phase of crew escape system and recovery using parachutes are shown in Fig. 5. The crew escape system is an

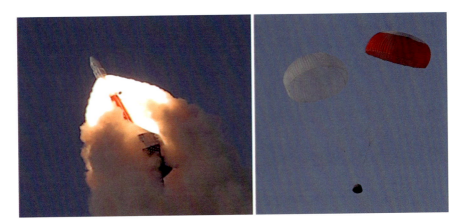

Fig. 5 Pad abort test of crew escape system

emergency escape measure designed to quickly pull the crew module along with the astronauts to a safe distance from the launch vehicle in the event of a launch abort. The crew escape system, weighing 12.6 t with the simulated crew module was lifted off using the specially developed solid motor with multiple reverse flow nozzle. In this test, the crew escape system along with the crew module was accelerated at 10 g and reached to a maximum altitude of 3 km. Recently, on 15 August 2018, the Prime Minister of India formally announced the Gaganyaan project, an Indian Human Spaceflight Programme which has given major push in this area. The Gaganyaan project is a fully autonomous 3.7-t spacecraft designed to carry a 3-member crew to orbit and safely return to the Earth after mission duration of up to 7 days. The crew module is mated to the service module, and together they are called the orbital module. The composite module of crew module and service module will have the Environmental Control and Life Support Systems (ECLSS) to ensure that conditions inside the crew module are suitable for humans to live comfortably. The ECLSS maintains a steady cabin pressure and air composition, removes carbon dioxide and other harmful gases, controls temperature and humidity, and manages parameters like fire detection and suppression, food and water management, and emergency support. Recovery and reentry aspects have been demonstrated in the Crew Module Atmospheric Reentry Experiment (CARE) that was flown onboard experimental flight of GSLV MkIII (LVM3-X) during December 2014. In this technology demonstration mission, a CARE module reentered the atmosphere at about 80 km altitude and landed in the sea, from where it was recovered by the coast guard. Many such experimental flights are planned before manned flight.

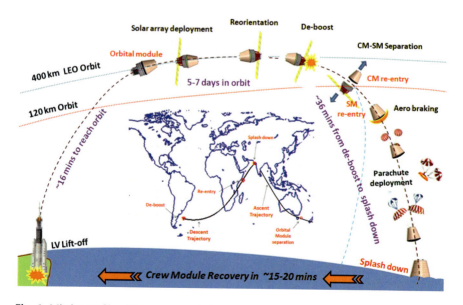

Fig. 6 Mission profile of Gaganyaan project

Based on the payload capacity, ISRO has chosen the GSLV MkIII launcher for orbiting the crew and service module. The GSLV MkIII will be human rated by incorporating improvements in the stage systems. The Launch Vehicle is also equipped with emergency mission abort and emergency escape that can be done during the action of first stage or second stage of the launch vehicle. The mission profile for Gaganyaan project is shown in Fig. 6.

Following two non-crewed orbital flight demonstrations, a crewed Gaganyaan is scheduled to be launched on the GSLV MkIII launcher by 2022.

New Initiatives Toward Low-Cost Access to Space

Currently, bringing down the cost of access to space is a primary goal of space programs around the world. After successful operationalization of the Polar Satellite Launch Vehicle (PSLV), the Geosynchronous Launch Vehicle (GSLV), and the Geosynchronous Launch Vehicle Mark III (GSLV MkIII), ISRO is in the process of developing a Reusable Launch Vehicle technologies to achieve low-cost access to space. Toward this program, a winged body configuration was conceived, which can fly at subsonic, supersonic, and hypersonic Mach number regime, reenter into Earth atmosphere, and land. ISRO is also developing technologies for air-breathing engine that uses atmospheric oxygen along with kerosene fuel to generate required thrust. A Small Satellite Launch Vehicle is being developed to provide on-demand launch at low cost.

Small Satellite Launch Vehicle

Traditionally, small satellites (below 500 kg) are launched as co-passengers along with primary satellites. In this situation the primary satellite takes the priority, and the small satellites have to be adapted to the launch schedule and orbit. This results in increased waiting time, non-optimum orbits (of primary spacecraft) that are far from ideal for the smaller spacecraft and do not always meet the unique requirements of planned missions. The small satellite industry has witnessed a rapid growth in the last few years. According to Euroconsult's recent report (Euroconsult 2019), nearly 8500 satellites will be launched worldwide over the next 10 years for government and commercial requirements. The estimated market value for small satellites manufacturing and launch is $13 billion with a potential growth rate of 13% per year. The launch of small satellites during the past decade is constrained by a limited number of launch vehicles. Innovation in miniaturization and standardization of satellite parts has trimmed the satellites' size and costs substantially. This has made feasible the building, launching, and operation of small satellite constellations. Nearly 33 small satellite launchers are under development worldwide, which are capable of offering 150–750 kg to LEO.

For catering the above requirements, ISRO has taken up the development of a cost-effective launcher, the Small Satellite Launch Vehicle (SSLV) that can offer

Fig. 7 SSLV configuration

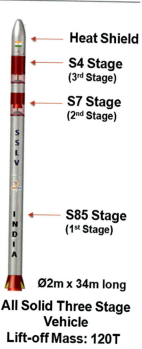

Heat Shield

S4 Stage
(3rd Stage)

S7 Stage
(2nd Stage)

S85 Stage
(1st Stage)

Ø2m x 34m long

**All Solid Three Stage
Vehicle
Lift-off Mass: 120T**

mission flexibility and on-demand launch. The SSLV is a three-stage solid propel-lant-based vehicle (S85 + S7 + S4) with a liquid velocity trimming module having payload capacity of 400–500 kg to LEO (500 km planar). The configuration of SSLV is given in Fig. 7. New technology developments for small satellite launch vehicles include the flexible nozzle driven by electromechanical actuator for solid motors, miniaturized avionics, stage separation system based on expandable tube assembly, etc. The design for all solid motors of SSLV is based on already well-proven solid motors of PSLV and GSLV launch vehicles. Propellant casting for all solid motor stages, subassembly preparation and testing, stage integration and testing, and vehicle integration and launch are planned with minimal infrastructure and man-power. The first development flight of the SSLV is expected during the second quarter of 2020.

Reusable Launch Vehicle: Technology Demonstrations

Reusability of space transportation systems, or at least parts of it, is the key capability to achieve for a significant cost reduction and quick turnaround opera-tions. Reentry technology demonstrator missions (Sivan et al. 2018) have been discussed widely in the literature. The ALFLEX, HYFLEX, and OREX of JAXA, the X-43A and X-51A of NASA, and the IXV of ISA are some examples. Many countries had embarked into RLV programs since the 1980s. A viable technology that is fully reusable and brings down launch costs is yet to emerge in the global launch market. Yet, this may be changing with the advent of the recent innovative recovery approach and reusability of the booster by the Space Exploration Technologies Corporation (SpaceX).

ISRO conceived a scale-down wing body technology demonstrator to acquire and validate the hypersonic vehicle design process as a first step toward full-scale space plane. The hypersonic experiment technology demonstrator vehicle (RLV-TD) was successfully flight tested (Sivan et al. 2018) in May 2016. The RLV-TD configuration in this mission is unique, because the double-delta wing body demonstrator was placed on top of the 11-m long and tailor-made slow-burn-rate solid booster. The mission profile for RLV-TD is given in Fig. 8.

Similar winged body flights were done by the National Aeronautics and Space Administration (NASA) and the European Space Agency (ESA). The major difference as compared to RLV-TD mission is that NASA and ESA have conducted it accommodating the winged body inside (Sivan et al. 2018) a heat shield. The RLV-TD mission was also unique because it was designed to minimize the cost and to get the maximum data in a single mission itself. In order to prepare the technology needed for the timely development of the fully reusable launch vehicle that may be required in the longer term and may be crucial for future space activity, ISRO has planned an in-flight experiment of a winged body orbital vehicle mission during the next 2–3 years. Prior to orbital mission, a series of landing experiments of the vehicle are planned to be carried out on the runway through autonomous control with a high degree of lateral accuracy, which calls for highly precise navigation system

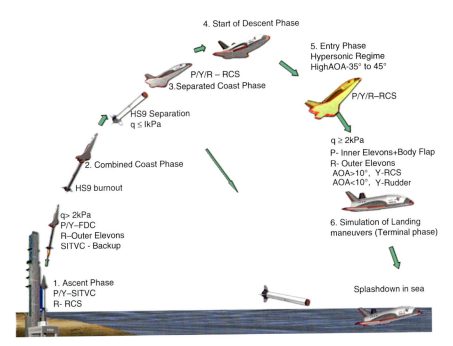

Fig. 8 Mission profile of RLV-TD

consisting of an onboard GNSS-aided Inertial Navigation System, pseudolite systems, and precise radar altimeters.

Hypersonic Air-Breathing Engine

The scramjet engine technology on air-breathing propulsion has the potential to increase the payload fraction, low-cost access to space, and launch frequency. So far very few nations (Mishra et al. 2018), namely, Russia, the United States, Australia, and India, have conducted scramjet demonstration programs. Various scramjet engine testing demonstration programs undertaken are the Kholod program (Russia), X-43A and X-51A (USA), Hyshot and HiFire (Australia), and SCRAMJET (India). The various air-breathing vehicle development programs proposed worldwide are the SPARTAN (Australia), GTX (USA), Skylon (UK), and AB-RLV (Japan). Selecting a suitable platform for flight testing of scramjet engine is important as the scramjet experiment has to be carried out in a particular flight environment governed by Mach number and dynamic pressure combination. The X43-A Hyper-X scramjet technology demonstrator was mounted on a booster (modified 1st stage of Pegasus rocket) that was air launched from the mother ship Boeing B52 aircraft. The booster was guided and controlled to achieve desired combination of Mach number

Fig. 9 Scramjet engine: technology demonstrator

and dynamic pressure for scramjet experiment. The Australian scramjet experiment Hyshot was done on a ballistic trajectory of sounding rocket.

ISRO has conducted two technology demonstration tests (Mishra et al. 2018), i.e., ATV-D01 and ATV-D02 in March 2010 and August 2016, respectively, in which a basic understanding of engine configuration and its ignition under higher Mach number have been flight demonstrated successfully. The technology demonstrator for scramjet engine is shown in Fig. 9. With regard to the overall development of a scramjet engine capable of generating positive acceleration, ISRO is working on critical technology development for hypersonic air-breathing vehicle with airframe-integrated test vehicle and demonstrated scramjet technology for long duration.

Conclusion

It was the vision of Dr. Vikram A. Sarabhai who initiated the Space Program in India in the 1960s, to utilize the potential of space technology and its applications for national development. The Indian space endeavor had a modest beginning with the launch of the sounding rocket from Thumba on 21 November 1963. Since then, in the last five decades, India has made considerable progress in the development and operationalization of satellites and space launch vehicles. With satellite constellation in LEO and GTO orbits along with ground systems, ISRO exploits the full potential of space technology for national development. The primary thrust of ISRO will be to maintain and improve national development and, at the same time, to achieve the desired self-reliance in meeting the future requirements of space infrastructure, space transportation systems, and ground systems.

References

Euroconsult (2019) Prospects of Small Satellite market – forecasts to 2028. http://www.euroconsult-ec.com/research/SS19-brochure.pdf. Accessed 10 Dec 2019

Government of India (2019) Department of Space, Indian Space Research Organization (ISRO). https://www.isro.gov.in/. Accessed 10 Dec 2019

Kumar K et al (2018) Overall mission performances of crew escape system – Pad abort test flight, ASCEND. In: National system conference 2018 by Systems Society of India, Trivandrum Chapter

Mishra P et al (2018) Mission design challenges in sounding rocket based scramjet engine test vehicle. J Aerosp Sci Technol 70(3A):195–202

Pant N (1986) Indian space program. Indian J Radio Space Phys 15:402–432

Sivan K et al (2018) An overview of reusable launch vehicle technology demonstrator. Curr Sci 114 (1):38–47

Suresh BN (2015) The Indian space launch program. In: Schrogl K-U et al (eds) Handbook of space security: policies, applications and programs. Springer, New York

Australia's Space Security Program

71

Michael Davis and Chris Schacht

Contents

Abstract

The space sector in Australia is experiencing an unprecedented level of public interest and government support. National security considerations as well as the economic benefits of a fast-growing world market for space products and services are inextricably linked as drivers for a range of government and industry initiatives outlined in this chapter. After decades of wavering government support for the space sector in Australia, there are signs that a coherent public program of capability development and R&D support, commensurate with Australia's

M. Davis (✉)
Space Industry Association of Australia, Adelaide, Australia
e-mail: mdavis@spacelaw.com.au

C. Schacht
Adelaide, Australia
e-mail: Australiaschacht@ozemail.com.au

© Springer Nature Switzerland AG 2020
K.-U. Schrogl (ed.), *Handbook of Space Security*,
https://doi.org/10.1007/978-3-030-23210-8_110

strategic and social needs, is emerging. The programs and initiatives outlined in this chapter are driven in part by a greater understanding of the national security dependencies and risks that Australia as a nation is exposed to and provide evidence that Australia is now treating space as an integral component of its role in the protection of its national security interests and in the advancement of its international responsibilities.

Introduction

The space sector in Australia is experiencing an unprecedented level of public interest and government support. National security considerations and the economic benefits of a fast-growing world market for space products and services are inextricably linked as drivers for a range of government and industry initiatives outlined in this chapter.

On the civil side, in July 2017 the Australian Government announced a review of Australia's space industry capability to ensure that Australia was able to "capitalise on the increasing opportunities within the global space industry sector." In its Interim Report in August 2017, the Expert Review Group appointed by the Government stated that:

> Australia has billions of dollars' worth of space-related existing infrastructure and facilities, a vibrant community of over 50 small companies and over 40 medium-sized companies active in the space sector ... world-leading capability including in research, satellite data analysis, radar and radio communications, and several emerging areas of strength. (Review of Australia's Space Industry Capability)

The Expert Review Group further concluded that "there is a need for a single point of contact for domestic and international partnerships, greater national coordination and strategic direction, government support to enable industry to participate in global supply chains, and a whole-of-government approach to the space sector (Ibid., pp. 68 and 69)." The stakeholder consultations carried out as part of the review revealed that there was a strong call for a space agency with the authority to coordinate Australia's space activity and to focus on the following requirements:

1. A national strategy that builds on national strengths and identifies areas of focus for Australia in the space sector
2. The coordination of domestic activities in Australia's space sector ensuring that Australia has the space capabilities to meet its strategic needs and manage risks associated with its space-related dependencies
3. A single point of contact for international engagement and partnerships with a clear point of contact within the Australian Government that can actively engage with global partners to maintain critical partnerships and broker future space opportunities
4. Support for the development of Australia's space industry capability

The Interim Report added that the space agency should be "modest, forward-looking and agile." It was not proposed that the space agency should take over the operational and regulatory activities currently being undertaken by other existing authorities. It was further said that "the future direction in space is in partnership with the private sector, as has been demonstrated through the Canadian and UK space programs."

The key recommendation for the establishment of a national space agency was quickly accepted by the Australian Government. Following on that, on 25 September 2017 at the Opening Ceremony of the International Astronautical Congress in Adelaide, the Government announced that an Australian Space Agency would be established (Space News).

This decision was seen as an important step forward for Australia and was a welcome response to the widely held view among key industry players that the development of a mature and innovative Australian space sector and a level of sovereign space capability underpinned by world class space science and technology had become an urgent national priority, taking into account:

1. The fact that the Australian Government was committed to taking advantage of innovation in science and technology to transform the national economy and strengthen security
2. The perception that Australia was missing out on opportunities enjoyed by other advanced economies in an industry sector that was enjoying growth rates exceeding 10% per annum
3. The belief that Australia could no longer assume that its vital security interests in space will continue to be maintained at virtually no cost to the Australian Government (Space Industry Association of Australia)

Australia's Dependence on Space

For over 60 years, space has played an important role in national affairs and international relations in Australia. Space-derived data and services have increasingly become embedded into the fabric of modern life providing the communications, geolocation, and timing services upon which millions of individuals and businesses rely each day. They include satellite positioning and communication services, as well as satellite earth observation and astronomy capabilities. A White Paper published by the Space Industry Association of Australia (Ibid.) concluded that without space technologies:

- Australia's ability to forecast weather and extreme weather events is immediately compromised, reducing the capacity to respond to emergencies such as bushfires and floods.
- Communications services, including Foxtel, Imparja, the National Broadband Network, and general telephony, would be significantly affected, leaving

Australia reliant on a very small number of undersea cables for connection to the outside world.

- Smartphones would not provide positioning data, with Google Maps, Uber, and any other GPS-based apps immediately useless.
- Telecommunications, critical in providing health and education support in remote communities, would suffer.
- Farmers would no longer be able to use precision agriculture, and the supply and delivery of food could no longer be coordinated through space-based navigation.
- Mining companies would lose a valuable exploration method.
- Air traffic would be disrupted.
- Financial transactions which rely on GPS satellites would slowly degrade.
- Defence would lose satellite communications, intelligence satellites, and GPS, critically affecting their ability to protect and defend Australia and its borders.

The SIAA White Paper went on to state that "...Australia has benefited greatly from access provided by partner nations such as the United States, Japan and Europe for much of its satellite data, and traditionally this has been at very little cost to the Australian Government. There are numerous examples of the economic and societal benefits generated from earth observation (EO) data in Australia, across areas such as weather forecasting, onshore and offshore mining, mitigation and management of natural disasters like bushfires and floods, water resource management, design and assessment of conservation areas, insurance assessment, and land use planning...Our public and private sectors are fundamentally dependent on EO which is entirely provided by other countries. These data and services are recognised as essential to our public and private infrastructure with numerous national reviews showing that Australian governments and industry are critically dependent on EO to maintain our economy and societal wellbeing. There is a vital national interest in maintaining the infrastructure, capabilities and international relationships necessary to secure access to satellite data sources."

The Space Industry Association of Australia document concluded that "A key issue in the development of our national space policy should therefore be the securing of long-term access, for strategic purposes, preferably from Australian territory, to foreign owned space-segment capabilities, both military and civil. Furthermore, as the geo-political environment changes, Australia needs to become a technology contributor to those partnerships, or it risks significantly rising costs or, even, loss of access. Australia would also be well-advised to consider ways to reduce its dependence on the traditional data sources, and consider its own national priorities in the development of new systems."

National Security and Space

There is an increasing awareness in Australia of the social and political implications of the country's dependence on space assets and services.

As one commentator put it, "...Australians have grown used to living in a just-in-time world for energy supply, logistics, power, heating and cooling to the point that it's only when the lights go out in Tasmania or South Australia that people realise there's a complex but imminently vulnerable interconnected system of supply and distribution that sets the rhythm of our lives. And so much of this depends on access to and control of space. While our military forces think about the implications of operating in a 'day without space', our politicians should ponder what a day or two without space would do for the quality of social harmony in Australia. If satellites go down and there are no others that can provide redundancy and resilience, how long would it take to turn our urban centres into end-of-days theme parks?" (Australian Strategic Policy Institute).

In a similar vein, another international commentator has observed:

Firstly, the sheer number of potential satellites planned will tax existing collision deconfliction and traffic management structures. On current forecasts, the number of active satellites will increase in the next decade, from approximately 1,800 to a number approaching 10,000. Within this range, a number of 'mega-constellations' are proposed, each containing thousands of satellites. Much of the explosion in numbers is forecast to occur within a fairly constrained orbital regime, within 1,000 kilometres of the surface of the earth.

Such developments may quickly overtake the current means of space surveillance, which is based around the independent verification of orbits by agencies like the US Space Surveillance Network, and in the future the US Department of Commerce. Increasingly, the best source of orbital information will be the operators themselves, which will require new ways of exchanging and verifying such data. Once exchanged, agreed norms of orbit maintenance and deconfliction measures will need to be established between operators. Additionally, the real-time monitoring and adjustment of constellations within a congested environment will drive a level of system autonomy currently unseen in current operations.

Added to this, the increasing miniaturisation of spacecraft—exemplified by the growth of 'cubesats'—will challenge the ability of existing surveillance systems to track and identify objects in orbit. The distribution of missions across multiple satellites has meant a steady reduction in object size, while modular construction techniques will make more satellites appear alike to external ground- and space-based observers. (Australian Strategic Policy Institute)

Australian Defence Policy

Since the 2016 Australian Government Defence White Paper (Australian Government, Department of Defence), there has been official recognition in Australia that access to space is important to the strategic interests of the nation. This key defence policy document proposes the continuing expansion of Australian space-based and space-enabled capabilities over the next 20 years, with space surveillance projects to be the "acquisition pipeline to strengthen Defence's awareness of space (Australian Strategic Policy Institute)." The Government's clear intention is to enhance Australian Defence sovereign space capabilities progressively through access to allied and commercial space-based capabilities (Ibid.).

As one commentator put it, ". . . a new factor driving national approaches to space is that all countries are faced with an increasingly stark choice to 'use or lose' their interests in space. Australia is acquiring at immense cost a fifth-generation-enabled defence force which, if we're ever to fight with it, must have assured access to systems that rely on space. The US alliance provides fantastic access to key space systems; however, it could benefit from increased resilience from allied systems designed with it in mind. So a defence policy for space must set out how we'll ensure that our forces have access to key systems inside our alliance with the United States and alone if necessary (Ibid.)."

The close strategic alliance between Australia and the United States features prominently in Australia's space-related capability planning. According to the Australian Strategic Policy Institute, "[a] key task is to determine how best Australia can work alongside the United States and other partners to deter threats to our critical space capabilities, and if necessary to mitigate the effects of warfare in space."

This is in stark contrast to the traditional approach in which Australia had been content to be dependent on foreign providers for space capability. According to the Australian Strategic Policy Institute, "For much of the period from the 1960s onwards, Australia adopted a supporting role in space—providing a suitable piece of real estate for ground facilities, supplying skilled personnel and managing the data coming from the satellites. However, we've not done anything significant in terms of developing sovereign capability in terms of 'the space segment' (Australian Strategic Policy Institute)."

> For defence and national security, space is a vital domain. Space is no longer a pristine global common that's a sanctuary from warfare. In the 21st century, it's recognised to be a warfighting domain that's 'contested' through the growing threat of adversaries' counter-space capability. It's also increasingly 'congested' as a result of growing space debris. Finally, space is increasingly 'competitive' as new approaches to space activities such as Space 2.0 see the acquisition of space capabilities by a broader range of state and non-state actors. Space is no longer dominated by only the major powers, and the impact of the private sector is clearly visible every time SpaceX launches—and then lands—one of its rockets.
>
> Ensuring access to space is vital to Australia's defence capability and to our approach to military tasks. Without an ability to access space capabilities—either our own, or those of an ally—our ability to fight and win information-based warfare and undertake 'multi-domain operations' is severely limited, and our adversaries are better placed to impose costs or even military defeat against our forces. (Ibid.)

The rationale for the development of defence space-related capability as set out in the White Paper and related documents is that the Australian Defence Force is reliant on space-based satellite systems to support its networked capabilities and to communicate and fight when deployed on operations. The documents note that some countries are developing capabilities to target satellites to destroy these systems or degrade their capabilities, threatening these networks. Space-based capabilities also offer potential adversaries advanced information gathering opportunities, including imagery gathering. The availability of commercial space-based systems also means that smaller countries, private interests, and non-state actors can access sensitive information about our security arrangements, such as imagery of Defence bases.

To ensure the security of its space-enabled capabilities, the Australian Government's stated policy is to strengthen Defence's space surveillance and situational awareness capabilities, including through the space surveillance radar operated jointly by Australia and the United States, and the relocation of a US optical space surveillance telescope to Australia.

It is also noted in the White Paper that "...[l]imiting the militarisation of space will also require the international community to work together to establish and manage a rules-based system – a prospect that does not seem likely in the immediate future."

In order to achieve these aims, the Australian Government has decided to strengthen its defence capability in six capability streams. One of these streams, "intelligence, surveillance, reconnaissance, space, electronic warfare, and cyber," is designed to ensure that Defence Forces have superior situational awareness.

Defence Integrated Investment Program

The Defence White Paper outlines a 10-year Integrated Investment Program under which approximately AUD195 billion was allocated in the decade to 2025–2026 for investment in new and enhanced capabilities. This includes investment in "space-related capability, including space-based and ground-based intelligence, surveillance and reconnaissance systems; and space situational awareness and command, control, communications, computer and intelligence capabilities (Australian Government)."

The Defence White Paper states that "Space-based systems for intelligence collection, communications, navigation, targeting and surveillance play a vital role in all Australian Defence Force (ADF) and coalition operations. Defence's imagery and targeting capacity will be enhanced through greater access to allied and commercial space-based capabilities, strengthened analytical capability and enhanced support systems. Enhancements to our imagery capacity will provide the basis to further develop our intelligence, surveillance and reconnaissance capabilities in the longer term, including through potential investment in space-based sensors."

The Defence White Paper further notes that:

- Satellite systems are vulnerable to space debris, which could damage or disable satellites, and advanced counter-space capabilities, such as anti-satellite missiles, which can deny, disrupt, and destroy space-based systems. It is therefore important to be able to detect and track objects in space so Defence can plan to manage the effects of any possible damage to our space-based capabilities.
- In cooperation with the United States, Australia is strengthening its space surveillance and situational awareness capabilities. At the center of this work are the establishment of the space surveillance C-band radar operated jointly by Australia and the United States and the relocation of a US optical space surveillance telescope to Australia. (Both assets will be located at the Harold E. Holt Naval Communications Station near Exmouth in Western Australia.)

- The Government intends to upgrade the Australian Defence Force's existing air defence surveillance system, including command, control and communications systems, sensors, and targeting systems, "which could be used as a foundation for development of deployed, in-theatre missile defence capabilities, should future strategic circumstances require it." The Government will also acquire new ground-based radars from around 2020 and will expand Australia's access to situational awareness information, including space-based systems.
- The Government also intends to upgrade facilities at the Harold E. Holt Communications Facility in Exmouth, Western Australia, to support enhanced space situational awareness and communications capabilities and will similarly upgrade the Jindalee Operational Radar Network and other surveillance and air defence-related facilities in northern Australia.

Other space-related initiatives flowing from the Defence White Paper include improved capability in satellite imagery analysis and space situational awareness. The 2016 Integrated Investment Program document notes the following:

Imagery
Australia's ability to collect and use imagery data will be substantially enhanced, including increasing the capacity for imagery analysis. This will be achieved primarily through additional personnel and equipment for the Australian Geospatial-Intelligence Organisation, and enhanced access to imagery, including imagery from satellites.
Australia will continue to invest in expanding access to geospatial data through both existing and new commercial and partner arrangements. This data will enhance our support to regional and global operations, and improve the resilience of our access to space-derived information, including operational imagery and targeting.
Space situational awareness
Australia's existing space situational awareness capability relies on access to comprehensive United States-sourced and processed space situational awareness information. Existing arrangements will be strengthened through the relocation of the C-band radar and optical space surveillance telescope to Australia, enhancing our access to space situational awareness information. Defence will also examine other ground-based sensors, including radar and optical systems, to develop options for expanding Australia's space situational awareness sensor coverage in the future.

Next Generation Technologies Fund

According to the 2018 Defence Industrial Capability Plan (Australian Government), the Department of Defence's nascent space programs will be developed into national capabilities over the next 10–15 years, and opportunities for industry and research institutions exist "in the design, delivery and operation of space situational awareness, satellite imagery, space operations and big data management and analytics capabilities."

The Plan states that "The Government's investment of $730 million over 10 years through the Next Generation Technologies Fund is an unprecedented opportunity to deliver high-impact future capabilities for Defence by tapping into the talent and innovation in Australia's industry and academic institutions (Ibid., paragraph 4.37)."

The Fund will support a range of collaborative projects with academia, industry (both small and large), publicly funded research agencies, Defence Science and Technology Group scientists, other areas of Defence, and Australia's allies.

Approaches to partnering between Defence and Australian innovators under the Next Generation Technologies Fund include:

- A Grand Challenge Program – bringing together government, industry, and academia to solve large-scale research problems of strategic importance to national security.
- A Defence Cooperative Research Centre Program – driving research partnerships to address high-priority defence technologies and to develop commercially viable products and solutions for future defence capabilities.
- A Small Business Exploratory Research Program – to accelerate promising science and technology of interest to the Defence, from early-stage concept development to a point where the research and technology could transition to the Defence Innovation Hub.
- A US-Australia Multidisciplinary University Research Initiative (AUSMURI) – a research network which is investing $25 million over 9 years for grants to support Australian participation in the already established US MURI program. Multidisciplinary teams of Australian university researches collaborate with US academic colleagues on high-priority projects for future defence capabilities.

Launches from Australia

In its final report delivered in March 2018, the Australian Government's Expert Reference Group recognized that space launch capability is an industry enabler and recommended the new space agency should facilitate "regulatory approval processes for small satellite launch facilities in Australia and the launch of Australian satellites overseas" and investigate opportunities to partner with appropriate international launch providers (Review of Australia's Space Industry Capability). Interest in Australia for the development of commercial launch services has since increased, and recent project announcements include the following:

Equatorial Launch Australia

Equatorial Launch Australia (ELA) has announced that it is establishing a launch site in East Arnhem Land, on the northern coast of the Australian continent (Space Connect). The company has been granted a lease from traditional owners of a 60-hectare parcel of land near Nhulunbuy in the Northern Territory. The Arnhem Space Centre project will include multiple launch sites using a variety of launch vehicles to provide suborbital and orbital access to space for commercial, research, and government organizations. The company recently signed an agreement with a US-based company with 25 years' experience working on space launch initiatives with NASA (Australian Defence Magazine).

Southern Launch

Southern Launch is a company based in South Australia aiming to provide polar earth orbit space launch capabilities (Southern Launch). It also aims to research and implement new and novel space launch vehicle guidance and control algorithms on flight hardware for customers (Southern Australian Space Industry Centre). It has announced plans for "unhindered southward launch trajectories across unpopulated areas with low density air and nautical traffic lanes" from the Whalers Way Launch Complex on the southern tip of Eyre Peninsula in South Australia (Southern Launch).

Gilmour Space Launch Services

Gilmour is an Australian launch provider whose stated aim is to develop and launch low-cost hybrid rockets for the fast-growing global small satellite industry (Gilmour Space). It has developed and tested a hybrid rocket engine and is about to conduct a series of suborbital test flights from Queensland as the next stage of its ultimate plan to provide commercial launch services from Australia. A recent report commissioned by the Government of Queensland pointed to unique opportunities for space systems, launch activities, and ground systems in Queensland (Deloitte).

Other Initiatives
- From 2014 to 2019, the Space Environment Research Centre (SERC), comprising participants from the Australian National University and RMIT University; industry partners – EOS Space Systems, Lockheed Martin, and Optus Satellite Networks; and public sector research agency, the Japanese National Institute of Information and Communications Technology, carried out a successful R&D program building on Australian expertise in the measurement, monitoring, analysis, and management of space debris in order to develop new technologies and strategies to preserve the space environment. SERC's research programs were designed to build on world-leading Australian innovations to reduce and ultimately prevent the loss of space infrastructure due to collisions between debris and satellites. Its two research objectives were to establish more efficient and effective space debris collision avoidance for active satellites by providing significant improvements in predicting the orbits of debris, allowing active satellites to maneuver in time and to maneuver space debris away from collisions using lasers on the earth. Its research outcomes included developing passive and active track sensors, development of adaptive optics astrometry capabilities, and development of high-power lasers and phased laser beam combining.

- In May 2016 the Australian Department of Defence announced that it is preparing a roadmap for a $2.3 billion next-generation satellite communications capability. The project was planned including a mix of commercial and military satcom capability. It was expected that around $2.2–2.3 billion would be allocated over the next decade for the joint project.
- In October 2016 the Australian Minister for Defence announced that the Defence Department had commenced a research program using miniature satellites in support of Defence radar capabilities and to conduct scientific experiments. Space mission "Buccaneer" was a partnership between Defence Science and Technology (DST) and the University of New South Wales to conduct calibration activities for the Jindalee Operational Radar Network (JORN) as well as undertake outer atmosphere characterization experiments.
- In June 2017 the Australian Government announced a $500 million investment to improve Australia's space-based intelligence, surveillance, and reconnaissance capabilities "to support ADF operations around the world and at home to secure our borders (Australian Government, Department of Defence)." Known as Defence Project 799, the specific aim was to enhance Australia's geospatial intelligence capabilities. The announcement said that Phase 1 of the project would provide Australia with direct and more timely access to commercial imaging satellites to support a wide range of Defence and national security activities.
- In April 2019 the Australian Government in conjunction with the Australian Space Agency laid out a 10-year roadmap for the country's space sector. The new strategy will be backed by an AUD 19.5 million national Space Infrastructure Fund to support Australia's domestic space industry. The Australian Civil Space Strategy 2019–2028 will have a staged focus across seven national priority areas. Those areas are position, navigation, and timing; earth observation, communication technologies, and services; "leapfrog" R&D; space situational awareness; robotics and automation; and access to space. Australian Government agency Geoscience Australia had previously received AUD 224.9 million for the space-based augmentation positioning initiative and AUD 36.9 million for the Digital Earth program. The period from 2021 to 2028 will focus on the remaining four priorities of leapfrog R&D, space situational awareness, robotics and automation, and access to space.
- In April 2019 the Australian Government announced AUD 55 million funding under its Cooperative Research Centres Program for a new industry and university R&D program, called the SmartSat CRC (Australian Government, Business, Cooperative Research Centres). The mission of the SmartSat CRC will be to foster the creation of next-generation space technologies and make Australia more competitive in the global space economy by supporting the next wave of growth in critical industries including agriculture, transport, logistics, communications, and mining, generating new hi-tech jobs and strengthening national defence and security. Over 90 companies, research institutions and government agencies, including the Australian Government Defence Science and Technology Group, will participate in the AUD 245 million 7-year program.

Conclusion

After decades of wavering government support for the space sector in Australia, there are signs that a coherent public program of capability development and R&D support, commensurate with Australia's strategic and social needs, is emerging. The programs and initiatives outlined in this chapter are driven in part by a greater understanding of the national security dependencies and risks that Australia as a nation is exposed to and provide evidence that Australia is now treating space as an integral component of its role in the protection of its national security interests and in the advancement of its international responsibilities.

References

Australian Defence Magazine, Space launch site for Australia wins US deal, 5 February 2019. https://www.australiandefence.com.au/news/space-launch-site-for-australia-wins-us-deal

Australian Government, Business, Cooperative Research Centres (CRC) Grants, CRC selection round outcomes. https://www.business.gov.au/assistance/cooperative-research-centres-programme/cooperative-research-centres-crcs-grants/current-crc-selection-round

Australian Government, Department of Defence, 2016 Integrated Investment Program, 25 February 2016. http://www.defence.gov.au/WhitePaper/Docs/2016-Defence-Integrated-Investment-Program.pdf

Australian Government, Department of Defence, 2016 Defence White Paper, 25 February 2016. http://www.defence.gov.au/WhitePaper/

Australian Government, Department of Defence, The Hon Christopher Pyne MP (former Minister for Defence Industry), $500 million for enhanced satellite capability, Media Release 18 June 2017. https://www.minister.defence.gov.au/minister/christopher-pyne/media-releases/500-million-enhanced-satellite-capability

Australian Government, Department of Defence, 2018 Defence Industrial Capability Plan, 23 April 2018. http://www.defence.gov.au/SPI/Industry/CapabilityPlan/Docs/DefenceIndustrialCapabilityPlan-web.pdf

Australian Strategic Policy Institute, The Strategist, Peter Jennings, Australia's future in space – building Australia's strategy for space, 14 June 2018. https://www.aspistrategist.org.au/australias-future-in-space/

Australian Strategic Policy Institute, The Strategist, Richard Harrison, Building the rewards of space – building Australia's strategy for space, 8 June 2018. https://www.aspistrategist.org.au/balancing-the-risks-and-rewards-of-space/

Australian Strategic Policy Institute, The Strategist, Malcolm Davis, Australia's space future: where to next on the final frontier? – Building Australia's strategy for space, 23 May 2018. https://www.aspistrategist.org.au/australias-space-future-where-to-next-on-the-final-frontier/

Australian Strategic Policy Institute, The Strategist, Greg Rowlands, A capability conflux: on military space, cyber & autonomous modernisation – building Australia's Strategy for Space, 23 May 2018. https://www.aspistrategist.org.au/a-capability-conflux-on-military-space-cyber-autonomous-modernisation/

Deloitte, Sky is not the limit: building Queensland's space economy, 24 February 2019. http://www.dsdmip.qld.gov.au/resources/report/space/building-qld-s-space-economy.pdf

Gilmour Space. https://www.gspacetech.com/

Review of Australia's Space Industry Capability. – report from the Expert Reference Group for the review, 29 March 2018. https://www.industry.gov.au/sites/g/files/net3906/f/June%202018/document/pdf/review_of_australias_space_industry_capability_-_report_from_the_expert_reference_group.pdf at p.68

Southern Australian Space Industry Centre. https://www.sasic.sa.gov.au/industry-and-grants/capa
   bility-directory/southern-launch
Southern Launch. https://southernlaunch.space/
Southern Launch. https://southernlaunch.space/whalers-way
Space Connect, NT launch sites one step closer to reality, 4 December 2018. https://www.spacecon
   nectonline.com.au/launch/3109-nt-launch-sites-one-step-closer-to-reality
Space Industry Association of Australia, SIAA White Paper, Advancing Australia in Space,
   21 March 2017. https://www.spaceindustry.com.au/wp-content/uploads/2019/04/SIAA-White-
   Paper-Advancing-Australia-in-Space.pdf
Space News, Australia to establish national space agency, 24 September 2017. https://spacenews.
   com/australia-to-establish-national-space-agency/

Pakistan's Space Activities

72

Ahmad Khan, Tanzeela Khalil, and Irteza Imam

Contents

Abstract

Despite political, technological, and economic constraints, Pakistan is considered an aspiring space power with a relatively modest space program compared to the larger, more successful Asian ones of China and India. Innovative leadership, smart allocation of national resources, and political will are all necessary for any

A. Khan (✉)
Department of Strategic Studies, National Defence University, Islamabad, Pakistan
e-mail: ahmad_ishaq669@yahoo.com

T. Khalil
South Asia Center, Atlantic Council, Washington, DC, USA
e-mail: tanzeela_khalil@hotmail.com

I. Imam
Department of Defence and Strategic Studies, Quaid-i-Azam University, Islamabad, Pakistan
e-mail: irttyza@gmail.com

© Springer Nature Switzerland AG 2020
K.-U. Schrogl (ed.), *Handbook of Space Security*,
https://doi.org/10.1007/978-3-030-23210-8_79

country to progress in such a high-technology field. Pakistan can utilize available resources to improve its nascent space infrastructure through collaborative efforts to gain eventual self-sufficiency for socioeconomic and strategic purposes in the South Asian region. Since its inception, space activities remained slow; however, there were substantial milestones achieved by Pakistan in space primarily for socioeconomic benefits. Likewise, Pakistan supports multilateral agreements to prevent the weaponization of space and has ratified all United Nations space treaties. Furthermore, multilateral collaboration, the utilization of available resources and public-private partnerships empowering its space program, enhances its domestic scientific and technological base and builds an indigenous space industry that can reap dividends at home and abroad. This can also benefit Pakistan's needs to mainstream its national space program for the overall national growth of the country and meet future requirements.

Introduction

On October 25, 2018, the then Minister for Information and Broadcasting, Chaudhry Fawad Hussain, declared, in a press conference after a cabinet session of the Federal Government of Pakistan, the country's intention for a first space mission by 2022 (Pakistan to Launch First Space Mission in 2022 2018). The former Information Minister announced that a Memorandum of Understanding (MOU) had been signed between a Chinese company and Pakistan's main space agency, the Space and Upper Atmosphere Research Commission (SUPARCO), for a joint manned space venture. This announcement came not only in the wake of Pakistan's recent satellite launches, of the Pakistan Remote Sensing Satellite (PRSS)-1 and the Pakistan Technology Evaluation Satellite (PakTES)-1A by China in July 2018 (Siddiqui 2018), but also in the context of Pakistan's neighboring rival, India, and its ambitious space missions (Ahsan and Khan 2019). Unlike the well-established Chinese and Indian space programs for both civilian and military purposes, Pakistan's space odyssey has been slow in keeping pace with the rest of the world due to a myriad of reasons. These range from an inadequate scientific and technological base, mostly due to technological denial from the Western countries that could propel an indigenous space program on its own (Ahsan and Khan 2019).

Pakistan's Space Program

Pakistan's National Space Agency

Soon after the prolific launch of Soviet Union's first artificial satellite, Sputnik 1, Pakistan started pursuing a space program to join the race of grasping the enormous advantages of outer space research and exploitation. The Space and Upper Atmosphere Research Commission (SUPARCO), the official space agency of the

Government of Pakistan, is in charge of the country's civil and public space program SUPARCO. Pakistan's premier space agency was established through a presidential ordinance in 1981 and its charter was approved by the National Assembly of Pakistan in 1987 during President Zia-ul-Haq's regime (1977–1988). SUPARCO is the successor to the Pakistan Space and Upper Atmosphere Research Committee, which was founded in September 1961. This committee was part of the Pakistan Atomic Energy Commission (PAEC) before becoming a separate organization in 1964. The committee was instrumental in developing and testing its Rehbar sounding rockets with support from the US National Aeronautics and Space Administration (NASA), the British National Space Centre (BNSC), and the French Centre National d'Etudes Spatiales (CNES), as well as numerous Western aerospace companies. All the major milestones achieved by Pakistan in space odyssey are due to SUPARCO (Mehmud 1989).

Since its inception in the 1960s, SUPARCO gradually expanded into satellite technology in the 1970s in the fields of remote sensing, imagery, telecommunication, atmospheric testing, and other scientific measurements, while it provided technical training and employment for its ground-based satellite stations across the country. SUPARCO continued its scientific outreach with its counterparts worldwide and was the country's main participant in numerous scientific treaties and international agreements in the United Nations (UN) and other organizations. Due to the strategic nature of space, SUPARCO, since 2000, falls under the Strategic Plans Division (SPD) umbrella of Pakistan, which is the operating arm of the country's National Command Authority (NCA). The other stakeholders partially associated with the space program are the Pakistan Telecommunication Authority (PTA), the Frequency Allocation Board (FAB), the Ministry of Information Technology (MoIT), and the Ministry of Science and Technology (Mehmud 1989).

Historical Milestones in Space by Pakistan

Pakistan's aspiration to acquire space technology can be traced back to the late 1950s. The establishment of SUPARCO in 1961 was the first step towards space exploration and space technology. Due to the lack of adequate resources and technologies, Pakistan's space program aimed at peaceful uses through the development of civilian space applications. The objectives of Pakistan's space program have largely remained confined to human resource development, telemedicines, remote sensing and geographic information system (GIS), communication satellites, and peaceful uses of outer space through support to the UN and other nation's agenda for peaceful uses of space and preventions of an arms race in outer space.

Pakistan has launched six space objects, namely are BADR-1, BADR-2, PAKSAT-1, ICUBE-1, PRSS-1, and PakTES-1A launched in 1990, 2001, 2011, 2013, and 2018 respectively. BADR-1 was launched to "test and validate the indigenously developed satellite subsystems in the space environment." In addition to this, the BADR-1 satellite was also aimed to gather real-time voice and data communications experiments between two ground systems in the ultra and very high-frequency bands. BADR-1 was launched through a Chinese rocket carrier Long March 2E. On December 10, 2001, Pakistan launched its second satellite BADR-B in space. The general function or BADR-B aimed to "acquire data on space weather" and mapping carbon and other earth resources in different parts of Pakistan. BADR-B was launched from the Baikonur Cosmodrome in Kazakhstan with the cooperation of ROSCOMOS, Russian Space Agency, through its Zenit carrier rocket. BADR-B was Pakistan's first remote sensing satellite. Pakistan's third and most successful satellite, PAKSAT-IR, was launched on August 12, 2011. It was Pakistan's first communication satellite in space exported from China. PAKSAT-IR was launched from Xichang Satellite Launch Centre through the Long March 3B rocket, and primarily it is a telecommunication satellite on which more than half of the country's TV channels are linked and broadcasting their services. Pakistan's lastest space objects launched were PRSS-1 and PakTES-1A using Long March 2C rocket (Mehdi and Su 2019).

Pakistan's acquisition of communication satellite PAKSAT-IR played a significant role in various areas of development in Pakistan. It helped Pakistan to expand its communication infrastructure to not just urban but rural areas as well. It enhanced the overall connectivity in the country as it became the preferred choice for leading mobile companies (Mehdi and Su 2019).

Launch of communication satellite PAKSAT-1R from Xichang Satellite Launch Centre into GEO is marked as the defining moment in Pakistan's space program. It replaced PAKSAT-1 and is capable of providing telecommunications and broadcast services to Pakistan. It is fitted with 12 active C-band and 18 active Ku-band transponders. The service life span of the satellite will be 15 years (Mehdi and Su 2019).

In 2016, SUPARCO and China Great Wall Industry Cooperation (CGWIC) signed an agreement for the development and launch of the Pakistan Remote Sensing System (PRSS-1). PRSS-1 was designed and built by China for Pakistan, as its first active Earth observation satellite. In July 2018, PRSS-1 and indigenously produced Pakistan Technology Evaluation Satellite (PAKTES-1A) were launched together on board Long March 2C launch vehicle from Jiuquan Satellite Launch Center. PakTES-IA remote sensing satellite has, consequently, represented the foundation of Pakistan's capability to launch its domestically build remote sensing satellite (Mehdi and Su 2019) (Fig. 1).

Space Budget

Pakistan's space program has modest financial resources to augment full-fledged space exploration activities. SUPARCO is the premier space agency in Pakistan and is primarily responsible for manufacturing, developing, and launching space objectives and other associated technologies. SUPARCO's financial support from the government remained little. SUPARCO was established in 1961 as a space science research wing of Pakistan Atomic Energy Commission (PAEC) (Moltz 2012). PAEC

Fig. 1 Pakistan's space
launches. (Source: Mehdi and
Su 2019)

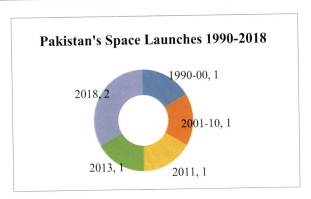

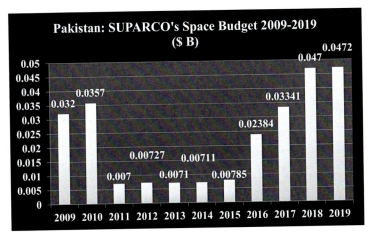

Fig. 2 SUPARCO's space budget. (Source: Pakistan Economic Survey 2009-2019)

was established in March 1956, 5 years before the establishment of SUPARCO. However, PAEC got the attention of the policymakers as it was thought that nuclear program was suitable for the socio-economic and national security purposes. Since its establishment in 1956, PAEC has now become a huge enterprise in Pakistan, serving in different fields and many corners of life including research and development (R&D), and in agriculture, industry, health sector. In the early 2000s, former President General Pervez Musharraf increased the financial budget of SUPARCO, breaking its financial stalemate. However, it is safe to say that it is still not sufficient enough to carry out an extensive space program including manufacturing its satellite indigenously and launching it from its soil. Pakistan has no dedicated launch pad for its SLVs, neither has its rocket to send satellites into space. To build an SLV and a launching pad, SUPARCO requires a tenfold budget increment for at least next 5 years. The figure shows SUPARCO's financial allocations to carry out its activities from 2009 to 2019 (Fig. 2).

Space Infrastructure in Pakistan

Ground Station Network

There are three satellite ground stations in Pakistan, located in Lahore, Islamabad, and Karachi. The ground station facility of Islamabad was developed in the 1970s and equipped with the Landsat and SPOT satellite ground receiving antennae. The remote sensing data processing facility was also established to process the raw data. The GCS for PRSS-1 satellite was established in the same facility in 2018 for TT&C and data reception. The ground station at Lahore was established in 2010 to control and operate the Paksat-1R satellite. This ground station is equipped to perform the telemetry, tracking, and control facility. The satellite ground station at Karachi was established in 2010, to perform the functions of backup station of Lahore station for Paksat-1R TT&C and operations.

Space Sector Infrastructure

Overall, Pakistan has established the following facilities related to the space sector as seen in Fig. 3 below.

Space Educational Institutions

Space education is the backbone of the Pakistan space program. There are three major space educational institutes in Pakistan providing necessary education in space applications and technological development in Pakistan. These three institutions include the Institute of Space Technology (IST) in Islamabad, the Department of Space at the University of Punjab, and the Institute of Space Science and Technology at the University of Karachi. All these institutes and departments are providing a pool of space scientists to foster Pakistan's space program. These institutes provide education at graduate, postgraduate, and PhD level in space science to foster the necessary pool of scientists in Pakistan.

The IST is the largest space education institute in Pakistan. The university was established in 2002 to solidify space science in Pakistan. It offers undergraduate and postgraduate degrees in aerospace engineering and communication systems engineering as its core discipline (About IST 2020). The department of space is the pioneer space education department in Pakistan. The University of Punjab, Lahore, established the department in 1985. The department was established to meet the rapid development in the field of space exploration, the subject of space science, telecommunications, meteorology, and related fields (Department of Space Science 2020).

The Institute of Space Science and Technology aims to achieve excellence in space science and associated technologies to bolster Pakistan's space program. The institute is the finest in Pakistan to deal with aerospace engineering. The curriculum

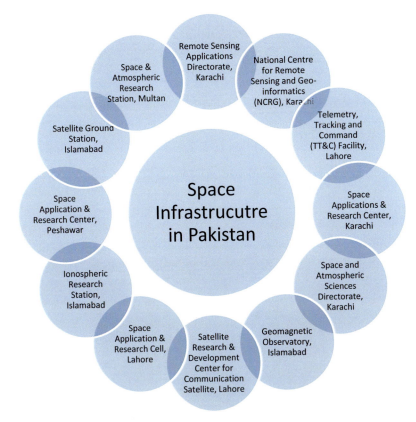

Fig. 3 Space infrastructure in Pakistan. (Source: Introduction-Facilities 2020)

offered to undergraduate and postgraduate students is designed to comply with international standards, emerging technologies, and to meet the indigenous requirements of Pakistan in space exploration (Introduction 2020).

Pakistan's Space Security and Governance

Pakistan has ratified all UN space treaties and is a party to one space-related treaty, one agreement, one convention, and one organization (see Table 1 for details). Among others, Pakistan has also supported the International Code of Conduct (ICoC). Also in 2014, the UN General Assembly adopted a Russian resolution titled "No First Placement of Weapons in Outer Space." The resolution calls for a nonbinding restriction against the first placement of weapons in outer space. The United States and other like-minded states are against the adopted resolution.

Table 1 Pakistan's participation to UN outer space treaties, international agreements, and organizations. (Source: UNCOPUOS 2020)

Treaties, conventions, agreements, organizations	UN treaties						Space related treaty, agreements, and organizations										
	OST	ARRA	LIAB	REG	MOON	NTB	BRS	ITSO	INTR	ESA	ARB	INTC	IMSO	EUTL	EUM	ITU
Pakistan	R	R	R	R	R	R	–	R	–	–	–	–	R	–	–	R

Pakistan and Russia, on the sidelines of Shanghai Cooperation Organisation (SCO) in 2019, signed a joint statement on the initiative with Russia (Khan and Sadeh 2019).

Additionally, Pakistan has been a tough proponent of disarmament and arms controls treaties. Over the years, various statements and resolutions floored by Pakistani diplomats at various international disarmament *fora* such as the UN First Committee, Conference on Disarmament (CD), etc., have called for arms control, limitation, prevention of space weaponization, and disarmament agreements at both bilateral and multilateral levels. Pakistan has been a proponent and advocate of the notion that space should be a global commons and should not be weaponized. If one observes the various official statements and resolutions tabled by Pakistan on UN governed arms control and disarmament forums such as the UN First Committee (DISEC), the CD, and UN Fourth Committee, there is a pattern that Pakistan has always been cognizant of the threats to the world peace that emanate from the militarization and weaponization of space. Pakistan has also supported the establishment of the Group of Governmental Experts (GGE) in CD to prepared expert-level recommendations to negotiate a treaty to prevent the weaponization of space and maintain the long term sustainability of activities in outer space (Khan and Khan 2019). In 2016, the then Pakistan Permanent representative in CD, Ambassador Tehmina Janjua, who later became the Foreign Secretary stated that "outer space is the common heritage of all humankind. It is in the collective interest to explore and use outer space exclusively for peaceful purposes, for the benefit of all. It should not be militarized or weaponized and turned into a realm of conflict" (Janjua 2016). She further expressed that "Weaponization of outer space is not science fiction anymore. With the ever-growing use of outer space by an increasing number of states, both for civilian and military purposes, the potential and the risk of its weaponization cannot be ruled out. Weapons in space would lead to instability, and negatively affect international and regional peace and security" (Janjua 2016).

As far as the Outer Space Treaty (OST) is concerned, Pakistan recognized that the exploration and use of outer space shall be carried out for the benefit and in the interests of all countries and shall be the province of all humanity. Pakistan endorses that the OST prohibits the placement of nuclear weapons and other weapons of mass destruction in outer space. However, Pakistan believes that there is a need to recollect thoughts on the placement of other types of weapons including conventional weapons in outer space to make space safer and sustainable (▶ Chap. 3, "Challenges to International Space Governance"). Pakistan considers that all of these concerns need to be addressed in a legally binding treaty, which is universally accepted. Pakistan is arguing for several decades in favor of developing long term sustainable guidelines for spacefaring nations to conduct peaceful activities in space. Pakistan is cognizant of the fact that the international community has continued to make efforts to prevent the weaponization of space. Pakistan has welcomed various proposals in the General Assembly which aim to address the weaponization of space with overwhelming support (Khan and Khan 2015).

The above quoted statement by Pakistan's Permanent Representative to the United Nations and CD is just one example of Pakistan's stance and commitment towards the goals set out by the idea of Prevention of an Arms Race in Space (PAROS). From the above quoted statement, Pakistan's threat perception can be seen as one of the factors driving the country's interest for a solid, effective, all-encompassing, and legally binding instrument.

Likewise, Pakistan is a proactive member of the UN Office for Outer Space Affairs (UNOOSA), Committee on Space Research (COSPAR), the UN Committee on the Peaceful Uses of Outer Space (UNCOPUOS), CD, and other multilateral forums for space security and governance (see Table 2 for details).

Pakistan's Space Strategic Priorities: Socioeconomic Benefits

The peaceful application of space capability and technology can bring about viable socioeconomic development and enhance the standards of living for the citizens of Pakistan. For instance, a communication satellite system is considered as the best service provider for teleeducation, telemedicine and channels in remote areas. Remote sensing satellites can contribute to discovering solutions for water management, land management, food security, disaster management, urban planning, and resource exploration. The PRSS-1 is a game-changer for Pakistan's space program. Remote sensing satellites can play an instrumental role in reducing the chances of intra-state conflict by providing adequate and well-informed data to deal with nontraditional threats like water scarcity and food shortage. Navigation provides free and reliable positioning on a worldwide basis. It also includes land surveying, tracking and surveillance, scientific study of earthquakes, disaster relief, and emergency services (Ahsan and Khan 2019).

Pakistan's PRSS-1 and PakTES-1A are the result of collaboration with China. PRSS-1 was purchased from China and was originally built by the China Academy of Space Technology (CAST). PakTES-1A was indigenously built by Pakistan. Since Pakistan does not have its dedicated launching facility, a Chinese Long March rocket was used to send both satellites to LEO. The primary aim of the satellites was to understand and enhance socioeconomic benefits of remote sensing systems. Another objective of sending these remote sensing satellites was to get continuous and improved data for the socioeconomic benefits, as both PRSS-1 and PAKTES-IA are Pakistan's first remote sensing satellites (Ahsan and Khan 2019).

Pakistan is extremely vulnerable to climate change in South Asia, despite being the lowest contributor to carbon emissions in the region. For example, historic floods during rainy seasons recently have hit the country, causing economic and human losses. Remote sensing satellites are necessary to monitor these climatic changes caused by natural and human activities. Likewise, Pakistan suffers from an acute shortage of freshwater. It is predicted that Pakistan will face water scarcity in the year 2025; this is a major issue for the country's national security and survival. Experts state that lack of access to water and food causes "social unrest and political

Table 2 Pakistan's participation to UN bodies for space governance. (Source: websites of these organizations)

	Multilateral forums for space governance	Agenda items	Pakistan's position
1	First Committee of the UNGA – Disarmament and International Security Committee (DISEC)	Arms control and disarmament issues in outer space	Pakistan is an active member of the First Committee of the UNGA – DISEC
2	UN Committee on the Peaceful Uses of Outer Space (UNCOPUOS)	UNCOPUOS was established in 1959 to govern the exploration and use of space for the benefit of all humanity: for peace, security, and development	Pakistan is a member of UNCOPUOS since 1973
	Standing committees (a) Scientific and Technical Subcommittee (working group on the long-term sustainability of outer space activities)	The committee discusses questions related to the scientific and technical aspects of space activities	Pakistan participates actively in the proceedings of the Scientific and Technical Subcommittee
	(b) Legal Subcommittee	Topics include the status and application of the five United Nations treaties on outer space, the definition and delimitation of outer. space, national space legislation, legal mechanisms relating to space debris mitigation, and international mechanisms for cooperation in the peaceful exploration and use of outer space	Pakistan participates actively in the proceedings of the Legal Subcommittee
3	International Telecommunication Union (ITU)	Promotes international cooperation in assigning satellite orbits, works to improve telecommunication infrastructure in the developing world	Pakistan is a member of the ITU
4	Conference on Disarmament (CD)	Proposed Prevention of an Arms Race in Space (PAROS) Treaty is one of the core issues in the permanent agenda items of the UN disarmament body	Pakistan is a member of CD. Pakistan in CD has repeatedly asked the member states to start negotiations on the PAROS treaty. Pakistan has endorsed the joint draft treaty proposal of Russia and China in CD

instability in any country." Recently, decreased access to water and food has stirred social and economic unrest in the Middle East and North Africa. Although Pakistan has not witnessed any kind of social and economic unrest similar to those of the Middle East and North Africa, these natural resource challenges must be dealt with adequately to mitigate intrastate conflict in Pakistan. To meet this challenge, Pakistan needs to build more remote sensing capabilities to improve knowledge of its freshwater supply, drainage system, and reservoirs. Water management has historically been inadequate in Pakistan, which also hampers the country's most significant economic sector – agriculture (Ahsan and Khan 2019).

In the wake of current nontraditional security challenges facing Pakistan, the PRSS-1 has ensured Pakistan to deal with them more effectively. Remote sensing capability will prove to be beneficial for addressing the peculiar challenges of water scarcity, food shortage, and climate change. Remote sensing can also help advance the understanding in the dynamics of water quantity and quality that can be used to simulate water resources management scenarios under different water quantity or quality demands. Remote sensing can also enhance the capacity of the departments and institutions related to water management in Pakistan for better evaluation of the situation and make effective strategies to deal with the imminent crisis. Furthermore, utilization of PRSS-I data, including climate forecasting to track weather-related natural disasters like floods, storms, rain, can eventually help the accelerating response, recovery, and rebuilding efforts by the relief departments. These technologies not only provide cost efficient and effective methods of water management but they also accurately monitor and predict long term trends of depletion of resources. It is high time that SUPARCO, the Ministry of Defense, and the Ministry of Science and Technology synergized their efforts to deal with these nontraditional challenges in a more cost effective manner by utilizing the available space applications. In this regard, the Government of Pakistan should also regulate and release adequate funds for the relevant ministries and commissions. The application of space technology in various domains will also help improve the economic condition of the country.

International Cooperation

Pakistan has established strong international collaborations. Pakistan is a permanent member of many international organizations, scientific committees, and United Nations bodies in conducting joint programs of research and development, trainings, and space research activities. Pakistan is collaborating with the following organizations/committees, where SUPARCO is its representative:

- Asia-Oceania Space Weather Alliance (AOSWA)
- Inter-Program Coordination Team on Space Weather (ICTSW)
- University of Massachusetts Center for Atmospheric Research (UMLCAR)
- Royal Meteorological Institute (RMI) – British Geological Survey (BGS)
- Institute of Tibetan Plateau Research (ITP), Chinese Academy of Sciences (CAS)
- European Commission (EC)

- Food and Agriculture Organization (FAO)
- United Nations Committee on the Peaceful Uses of Outer Space (UNCOPUOS)
- United Nations Economic and Social Commission for Asia and the Pacific (UNESCAP)
- Committee on Space Research (COSPAR)
- International Society for Photogrammetry and Remote Sensing (ISPRS)
- National Coordination Committee for COSPAS-SARSAT (Search & Rescue Satellite Aided Tracking System)
- American Institute of Aeronautics and Astronautics (AIAA)
- International Astronomical Federation (IAF) and International Academy of Astronautics (IAA)
- Asian Association on Remote Sensing (AARS)
- Asia-Pacific Space Cooperation Organization (APSCO)
- Inter-Islamic Network on Space Sciences and Technology (ISNET)

Pakistan's Space Vision 2047

Pakistan's Space Vision 2047 states that it needs information and data for planning, managing, and monitoring its natural resources in order to improve the quality of life across the country. The "Space Vision 2047" has envisioned an adequate deployment of remote sensing constellations in the future to reduce the chances of intrastate conflict over water and food shortages. This challenge also includes mapping and surveying of irrigated areas and proper estimates of crop yield, according to the previously mentioned challenges affecting Pakistan's agriculture sector. Also, the Space Vision 2047 maps commercial and private utilization of data from remote sensing satellites to foster economic growth. Pakistan has a humble space program that is designed to help the country better manage its landscape and deal with the rising challenges of climate change, as well as better manage the socio-economic potential of the country (Ahsan and Khan 2019). Pakistan has recently launched two satellites for remote sensing, imaging, and data collection. The Government aims to utilize these space assets to formulate strategies and mechanisms to mitigate the various environmental challenges such as water scarcity and water management. This vision outlines the following aims/goals:

(a) Launching five satellites into orbit for communication purposes. The plan envisions multiband, multifrequency communication satellites that would enable Pakistan to provide wireless communication solutions to far-flung, geographically isolated areas.
(b) Six Low Earth Orbit satellites for navigational, remote sensing, and geographic information system (GIS).
(c) Telemedicine: In this project, various hospitals of high standards will be connected to the less developed areas of Pakistan to ensure the quality treatment of the poor and needy patients.

(d) Teleeducation: This program includes broadcasting the educational materials of various universities to the students free of cost.

Challenges Ahead and Concluding Remarks

Pakistan's space program is focused on peaceful uses of outer space in achieving economic benefits for national growth. However, Pakistan's space program has insufficient budget allocations to realize the stated goals and objectives. It is not currently in a good position to develop its own commercial space technologies and therefore faces challenges in its ability to develop dual-use space assets. It will likely take decades of work in areas from education and domestic economic reform to foreign investment policies in order to create the preconditions needed to enable Pakistan to produce its own space technologies. In the interim, Pakistan will have to try to broaden its technological base through participation in the Asia-Pacific Space Cooperation Organization (APSCO) and the Asia-Pacific Regional Space Agency Forum (APRSAF), as well as to develop domestic policies to stimulate young people to study in space-related fields (Ahsan and Khan 2019).

Pakistan is an important regional player in South Asia, with significant nuclear capability. Given that its journey into space has been largely peaceful with a focus on civilian activities, Pakistan is viewed as a responsible aspiring space actor. However, the world is witnessing an era where states are cooperating and competing in land, sea, air, and space. The major space farers – the United States, Russia, and China – see their space assets as crucial for their national growth, as well as for their national security. India has been following the example of all major spacefaring states, and it has a space program with a political priority. India sees its space program as instrumental for national growth and for national security. Even though these developments posit security challenges for Pakistan, Pakistan is currently utilizing its partnership in space with China to address challenges of climate change, and water and agricultural management, by making use of space systems and applications (Ahsan and Khan 2019). Currently, Pakistan is neither capable nor inclined to militarize and weaponize outer space, despite Chinese and Indian advances in these areas. In the future, however, Pakistan can certainly achieve credible options for military uses of space assets if this is warranted and in line with the space treaties.

References

About IST: Overview (2020) Institute of Space Technology. http://www.ist.edu.pk/about. Accessed 9 Mar 2020

Ahsan A, Khan A (2019) Pakistan's journey into space. Astropolitics 17(1: Special issue on Space Power and Security Trilemma in South Asia):38–50

Department of Space Science (2020) The University of Punjab. http://pu.edu.pk/home/department/62/Department-of-Space-Science. Accessed 9 Mar 2020

Introduction-Facilities (2020) Pakistan Space and Upper Atmosphere Research Commission, http://www.suparco.gov.pk/pages/intro.asp. Accessed 27 May 2020

Introduction: Institute of Space Science and Technology (2020) University of Karachi. http://www. uok.edu.pk/research_institutes/ispa/index.php. Accessed 9 Mar 2020

Janjua T (2016) Statement by Ambassador Tehmina Janjua, permanent representative of Pakistan to the United Nations, Geneva and conference on disarmament during the thematic debate of the first committee on outer space. Pakistan Mission to United Nations

Khan Z, Khan A (2015) Chinese capabilities as global space power. Astropolitics 17(2–3):185–204

Khan Z, Khan A (2019) Space security trilemma in South Asia. Astropolitics 17(1: Special issue on Space Power and Security Trilemma in South Asia):4–22

Khan A, Sadeh E (2019) Introduction: space power and security trilemma in South Asia. Astropolitics 17(1: Special issue on Space Power and Security Trilemma in South Asia):1–4

Khan A, Ullah S (2020) Challenges to international space governance. In: Schrogl K et al. (Eds.) Handbook of Space Security: Policies, Applications and Programs. Springer, Cham

Mehdi M, Su J (2019) Pakistan space programme and international cooperation: history and prospect. Space Policy 47(1):175–180

Mehmud S (1989) Pakistan's space programme. Space Policy 5(3):217–226

Moltz C (2012) Asia's space race: national motivations, regional rivalries, and international risks. Columbia University Press, New York

Pakistan to Launch First Space Mission in 2022: Fawad Chaudhry (2018) Geo News. https://www. geo.tv/latest/216047-pakistan-to-launch-first-space-mission-in-2022-fawad-chaudhry. Accessed 25 Aug 2019

Pakistan Economic Survey (2019) Ministry of Finance, Government of Pakistan, http://www. finance.gov.pk/survey_1819.html. Accessed 27 May 2020

Siddiqui N (2018) Pakistan Launches Remote Sensing Satellite in China. Dawn, https://www.dawn. com/news/1418966/pakistan-launches-remote-sensing-satellitein-china. Accessed 9 July 2018

UNCOPUOS (2020) Status of international agreements relating to activities in outer space as at 1 January 2020. https://www.unoosa.org/oosa/en/ourwork/spacelaw/treaties/status/index.html. Accessed 29 May 2020

Space Sector Economy and Space Programs World Wide

73

Per Høyland, Estelle Godard, Marta De Oliviera, and Christina Giannopapa

Contents

Abstract

In the following chapter an overview of the space-related budgets is presented. This should provide a quantitative perspective of the overall market value and financial performance of the space activities of recent years. Accurate estimations of global space activities are complicated, due to nontransparent government space budgets in particular on defense-related programs and the lack of a stan-

P. Høyland (✉)
Department of Political Science, University of Oslo, Oslo, Norway
e-mail: perhoyl@uio.no

E. Godard
Institut d'études politiques de Paris, Paris, France
e-mail: estelle.godard@sciencespo.fr

M. De Oliviera
International Space University (ISU), Illkirch-Graffenstaden, France
e-mail: marta.oliveira@community.isunet.edu

C. Giannopapa
European Space Agency (ESA), Paris, France
e-mail: christina.giannopapa@esa.int

© Springer Nature Switzerland AG 2020
K.-U. Schrogl (ed.), *Handbook of Space Security*,
https://doi.org/10.1007/978-3-030-23210-8_72

1471

dardized approach for measuring them. A forecast for government space budgets and programs is also provided.

Introduction

Space technologies and their applications are part of our everyday life. They vary from using mobile phones, watching live TV broadcasting, making banking transactions, weather forecasting, air traffic management, etc. Its use has broaden from serving a limited number of space fearing nations driven by defense objectives into a more innovation-driven community including civilian, security, and private sector communities. Even though space activities are taking place already for more than half a century, there is no unified definition of what the space sector is, let alone space security.

In spite of being characterized by an intense international cooperation, the space sector relies above all on institutional captive markets with limited room for global competition. This is an important element in a return-on-investment analysis regarding public funding. However, it should be taken into consideration that accurate estimate of global space activities is a complicated task. This is due to the lack of a standardized approach and the lack of transparency in certain government space programs, such as defense-related programs. Additionally, the publication of financial figures by commercial companies is not uniform across the sector and varies in time.

Space remains highly oriented towards the generation of scientific, social, and/or strategic returns rather than primarily an immediate source of profit in terms of commercial return, the latter nevertheless being a major concrete feature. Thus, the value of space research and development and subsequent operational assets is first of all political and considered in a long- term economic strategy, whereby institutional funding supports the inception, then the maturing of a specific domain. This is what has justified constant public investment since the dawn of the space age. In particular institutional budgets often contribute to the start-up and development of capital-intensive and high-technology sectors such as space (OECD 2016).

As confirmed by the Organization for Economic Cooperation and Development (OECD), since the beginning of the economic crisis, the space sector has fared relatively well, in part because space is a strategic sector, often supported due to national imperatives and institutional funding, because space still supports the implementation of national objectives, and thanks to the good position of telecommunications in growing mass markets (OECD 2011, p. 29). Global government space expenditures peaked in 2012 followed by declining figures, yet this trend has changed lately. Euroconsult (2019) expect to see a growth in both defense and civil spending towards 2028. The cyclical nature of the industry in replenishing satellites as well the continuing commercial successes of many space services have contributed to the dynamism of the entire value chain. This is a fundamental element for the future forecast of space in Europe insofar as space cannot be considered a "stop and go" economy, meaning that any break in public investments would result in an immediate loss of industrial capabilities (human or otherwise) which could not be easily restarted later on.

Nowadays, a major shift in the space economy is driven by the commercial launch and satellite manufacturing industries. The market is set to grow fast due to falling costs and reduced obstacles to participate in the economy. Examples of emerging sectors in the space economy for the coming years are space tourism, asteroid mining, and in-orbit manufacturing. At the same time, exploration is brought deeper into space and existing activities related to navigation, Earth observation, and telecommunication continues to develop (ESA LTP 2017).

This chapter defines the space sector according to OECD and provides an overview of the space-related budgets worldwide. Institutional as well as commercial space activities are considered. It provides a quantitative perspective of the overall market value and financial activities over the past years complimented by a forecast of government space expenditures through 2028, as provided by Euroconsult.

The Space Sector Economy and Activities

There is no unified definition of the space sector and what it includes. According to OECD (2011), the space sector has nine main product groups of high technology: (1) aerospace, (2) computers and office machines, (3) electronics and telecommunications, (4) pharmacy, (5) scientific instruments, (6) electrical machinery, (7) chemistry, (8) nonelectrical machinery, and (9) armaments (Hatzichronoglou 1997). There is no specific "space activity classification." In the United Nations International Standard Industrial Classification (ISIC, Rev.4 released in August 2008), most parts of the space sector are included under different categories. Therefore, isolating the space sector from aerospace and defense sector remains a challenge for a number of countries.

The space sector over the years has become more commercial, and different space applications have emerged outside the traditional research and development (R&D), calling for a wider definition of space economy. This wider "space economy" can be defined using different angles. It can be defined by its products (e.g., satellites, launchers), by its services (e.g., broadcasting, imagery/data delivering), by its programmatic objectives (e.g., military, robotic space exploration, human spaceflight, Earth observation, telecommunications), by its actors/value chains (from R&D actors to users), and by its impacts (e.g., direct and indirect benefits). One drawback is that narrow definitions might ignore important aspects, such as the R&D actors (e.g., labs and universities) and the role of the military (i.e., as investor in R&D budgets and a customer for space services), or ignore scientific and space exploration programs altogether (OECD 2011, p. 16).

The OECD (2016) provides the following definition of the space economy:

The space economy is the full range of activities and the use of resources that create and provide value and benefits to human beings in the course of exploring, understanding, managing and utilising space. Hence, it includes all public and private actors involved in developing, providing and using space-related products and services, ranging from research and development, the manufacture and use of space infrastructure (ground stations, launch vehicles and satellites) to space-enabled applications (navigation equipment, satellite phones, meteorological services, etc.) and the scientific research generated by such activities. It follows that the space economy goes well beyond the space sector itself, since it also comprises the increasingly pervasive and continually changing impacts (both quantitative and qualitative) of space-derived products, services and knowledge on the economy and society.

The concept of "New Space" encapsulates the global trend of emerging investment philosophies and series of technological advancements stimulating a space industry largely driven by commercial motivations (EIB Advisory 2018). This wider space economy can be broken down in to three parameters (OECD 2016):

(i) The upstream space sector covering activities related to space infrastructures, including R&D, satellite, and launchers manufacturing, and the deployment of such infrastructures

(ii) The downstream space sector primarily related to commercial activities based on the use of data provided by space infrastructures, such as broadcasting, communication, navigation, and Earth Observation

(iii) The space-related or derived activities in other sectors

The space economy is larger than the traditional space sector (e.g., rockets and satellites). It also involves new services and product providers (e.g., geographic information systems developers, navigation equipment sellers) which are using space system's capacities to create new products. However, the unique capabilities offered by satellites (i.e., ubiquitous data, communications links, imagery) represent often only small, albeit essential, components of those new products and services (Fig. 1). Investments in manufacturing of satellites for earth observation, navigation, and telecommunication result tens of multiples in the downstream sector. Integrated

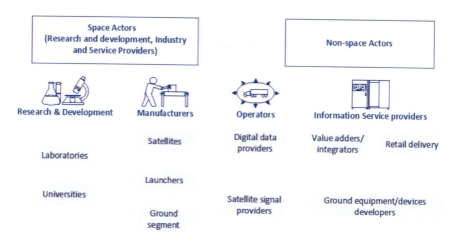

Fig. 1 The space economy's simplified value chain. (Source: OECD 2011)

application of various space technologies and often ground and airborne infrastructures broaden the space economy spectrum through value-added services sector.

Worldwide national space budgets have continued to grow in 2017 and 2018. However, it is the commercial sector that is the primary cause for the overall growth of the space economy. The total figure for the global space economy in 2017 was $383.51 billion (€339.48 billion) in government budgets and commercial revenue, an increase of 7.4% from the 2016 total of $357.18 billion (€316.20 billion) and an increase of 31.6% compared to the 2013 figure of $291.49 billion (€258.02 billion). (Exchange rate: $1 = €1.1297, average for 2017 as provided by the European Central Bank. The average figure for the year(s) in question are used throughout the chapter when figures are converted between US$ and Euros.) The majority of the increase in 2017 is attributable to growth in commercial sectors: commercial space products and services increased 8.3%, while commercial infrastructure and support industries increased by 7.5% (Space Foundation 2018). Overall, governmental spending increased from $69.99 billion in 2017 to $70.89 billion in 2018 (Euroconsult 2019). Overall, governmental spending, civilian and defense, increased by 4.8% from 2016 to 2017, although changes varied significantly from country to country, with India and China continuing their growth along with some ESA Member States. Many other space agencies, including those of the USA, Japan, Russia, and South Korea, saw a relative decrease from previous years signaling shifts in spending priorities. Such changes may be consequences of changing political climates affecting the economy and national institutional space activities. While the US figure decreased by 2.5% in 2017 compared to 2016, the total figure for non-US government space budgets increased by 16.4% (Space Foundation 2018, pp. 4–15) (Fig. 2).

The Institutional Space Sector

Figure 3 shows government space expenditure in 2018, civil and defense (Euroconsult 2019). The ESA figure includes contributions from the EU. The countries spending the most on space in 2018 were the USA (€34.71 billion),

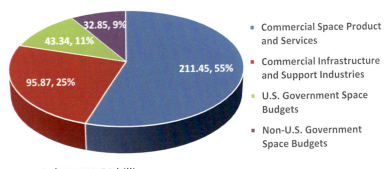

Total: $US 383.51 billion

Fig. 2 Global space activity 2017. (Source: Space Foundation 2018)

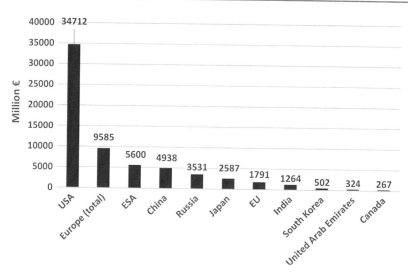

Fig. 3 Public space budgets in 2019, civil and defense. (Source: Euroconsult 2019)

China (€4.94 billion), Russia (€3.53 billion), and Japan (€2.59 billion). The total figure for Europe mounted to €9.59 billion, while ESA's budget for the same year was €5.6 billion including contributions from the EU and others. This makes ESA the second largest civilian space agency in the world, following NASA. In 2017, according to the 2018 Space Report, the total world governmental expenditure, including that of intergovernmental organizations, on space programs amounted to $76.20 billion (€53.9 billion), a figure which shows a nominal decrease of 1.78% compared to 2013. The space spending is comprised of $48.54 billion (€42.97 billion) in civil expenditures, 64% of the total, and $27.65 billion (€24.47 billion) in defense expenditures, 36% of the total, according to data gathered by the Space Foundation (2018). Of the estimated $27.65 billion (€ 24.47 billion) of defense-related space expenditure worldwide, $20.688 billion (€18.3 billion) was spent by the USA, most of it through the Department of Defence (DoD), representing a global share of 75%.

Data from 2019, gathered by Euroconsult, shows that the total figure for civilian government space expenditure mounted to $44.49 billion (€37.67 billion) in 2018, up from $42.63 billion (€36.07 billion) in 2017. In fact, civilian space budgets have increased since a low of $39.14 billion (€35.28 billion) in 2015. Defense space budgets have also increased since 2015. In 2018, the total mounted to $26.39 billion (€22.35 billion), which corresponds to about 37% of the total governmental spending for that year. This ratio varies significantly from the figures from the 2000s, where the ratio varies between 44.2% (2001) and 48.9% (2007). The last time this ratio was over 40% in 2012 (42.1%) (Euroconsult 2019). It should be noted, however, that these figures are based on accessible and public data and that differences in measurements and reporting are sources of potential errors when identifying and presenting global figures.

A prominent change in terms of the level of resources allocated to space is found in the growth of China's institutional space activities. The country's estimated total doubled over 5 years from $1.92 billion in 2008 to $4.07 billion in 2013 and has continued to increase until 2018 (Euroconsult 2019). China is now the world's second largest space country, following the USA, if one excludes the multilateral European activities undertaken by ESA. The USA has maintained its lead position with the largest budget among states, directing $22.13 billion (€18.74 billion) towards civil expenditure and $18.89 billion (€15.99 billion) towards defense expenditure. The low estimate of Russia's budget must be put into perspective, as it does not factor in the intensive military activity entailing regular classified launches or the scientific programs. China's national space budget reached €4.94 billion in 2018 and stays ahead of Russia (€3.53 billion) and Japan (€2.59 billion). The biggest governmental space expenditures in Europe are found in France (€2.67 billion), Germany (€1.82 billion), Italy (€0.95 billion), and the UK (€0.76 billion). The European Union has also become a major investor in space, with a total expenditure figure of €1.79 billion in 2018 (Euroconsult 2019). The 2019 budget for the European Space Agency (ESA), an intergovernmental organization with 22 Member States, mounts to €5.72 billion (ESA 2019). The largest contributors are France (€1.17 billion, 28.1%) and Germany (€0.93 billion, 22.2%), followed by Italy (€0.42 billion, 10.1%) and the UK (€0.37 billion, 8.8%). The total contribution from the EU, Eumetsat, and other nonstate actors is €1.53 billion (ESA 2019).

When measuring the concrete effort of countries in the space sector, it is necessary to put the figures into perspective in regard to Gross Domestic Product (GDP) (Fig. 4). However, considering the absolute numbers alone will paint only a partial

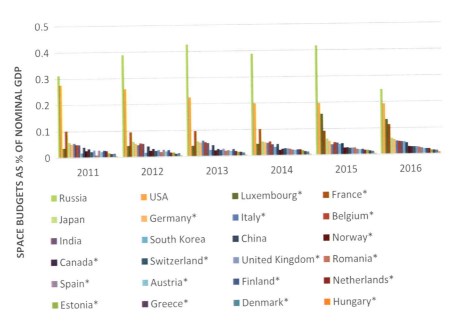

Fig. 4 Government space budgets, civil and defense, as percentage of nominal GDP in 2018. (Sources: Euroconsult 2019; IMF 2018)

picture since comparisons between countries with different economic conditions (e.g., price or wage levels) can be misleading. National space budgets as share of GDP over time indicate how space is prioritized on the national level. As illustrated in Fig. 5, this ratio has remained relatively stable for most countries, with some exceptions. Two developments are standing out in Fig. 5. First, the Russian figure decreased significantly from 2015 to 2016, from 0.22% to 0.12%. Second, the figure for Luxembourg has increased significantly from 2013 onwards. In 2016, Luxembourg's civilian space budgets as a percentage of nominal GDP was third in rank after Russia and the USA. The remaining selected countries vary between 0.02% and 0.05%, with the exception of the French figure of 0.11%.

From the values listed in the Euroconsult report, combined with population data from the International Monetary Fund (IMF), the per capita budget of Luxembourg has been calculated to be €122.50. As illustrated in Fig. 6, Luxembourg and USA (€106.10) both spend significantly more on space relative to their respective populations compared to the other countries in this overview.

Space cannot be seen in isolation from research and development (R&D) policies. Figures 7, 8, and 9 show the civil space budget in relation to the government budget allocations for R&D (GBARD). Figure 7 shows the evolution of the civil space budgets as a percentage of civil GBARD for a number of selected countries. The evolution of civil space component of public R&D shows that civil space-related R&D budgets have picked in the early to mid-1990s then decreased. Since 2007 there is a decrease with the exception of the/decrease with the exception of the UK. According to the OECD (2018), the only three space powers where the civil space budget as a part of the GBARD outreaches the OECD average (7.95%) in 2016 are the USA, Italy (9.28), and Belgium (8.39%). (Data for Russia is not available.) Since 1981 as shown in Fig. 7, the USA has been in the lead, except for 2009 when the French figure was 14.9% and the US figure was 11.9%, both well above the OECD average of 7.1 at the time. The USA spends significantly less with regards to its civil GBARD in recent years compared to the 1990s, yet the figure has increased since the lowest point in 2009. In 2009, the figure for Japan was also higher than the OECD average. The financial crisis of 2009 could be one explanatory factor behind increased focus from governments on R&D, stimulating growth.

GBAORD data are assembled by national authorities and classified by "socioeconomic" objectives on the basis of NABS 2007 (nomenclature for the analysis and comparison of scientific programs and budgets). The advantage of this is that they are reflecting government priorities.

Figure 9 shows space versus other areas of R&D expenditure according to Eurostat. This shows how space ranks as a priority within a country. The USA allocates most resources for R&D overall (€123.91 billion), followed by the combined figure for EU28 (€40.36 billion), Germany (€13.02 billion), Japan (€11.84 billion), United Kingdom (€8.50 billion), France (€6.62 billion), Italy (€5.09 billion), and Spain (€2.99 billion). For the exploration and exploitation of space, the USA spends €11.377 billion, followed by EU28 (€4.00 billion), Japan (€1.70 billion), Germany (€1.39 billion), and Italy (€929 million).

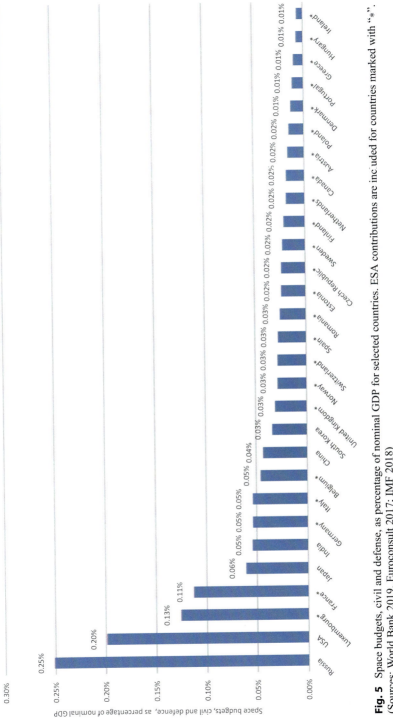

Fig. 5 Space budgets, civil and defense, as percentage of nominal GDP for selected countries. ESA contributions are included for countries marked with "*". (Sources: World Bank 2019, Euroconsult 2017; IMF 2018)

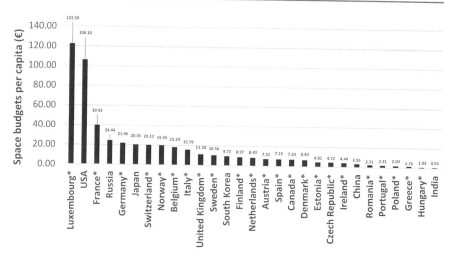

Fig. 6 Government space budgets, civil and defense, per capita in 2018. ESA contributions are included for countries marked with "*". (Sources: Euroconsult 2019; IMF 2018)

The main socioeconomic R&D objective in the EU is "industrial production and technology" (€9.69 billion), followed by "health" (€8.54 billion), "energy" (€4.66 billion), and "exploration and exploitation of space" (€4.00 billion) (Eurostat 2019).

Figure 10 shows government R&D priorities relative to each other within the same unit. Russian allocation for R&D funds suggest a very strong focus on space-related R&D relative to other objectives, while the trend in Europe is more general funds for universities and research institutions (OECD 2018).

The following tables (1, 2, and 3) give an overview of civil and defense government expenditures of recent years. Data is gathered from Euroconsult's report Government Space Programs, Benchmarks, Profiles, and Forecasts to 2028, published in July 2019.

The Commercial Space Sector

The space economy continues to grow as it has done for more than a decade, according to the Space Report 2018 published by the Space Foundation (p. 1). The greatest growth is found in the commercial space sector, where commercial space products and services make up 55.1% of the overall space activities in 2017. Combined with commercial infrastructure and support industries, the commercial sector accounts for 80% of total space activities mounting to $307.32 billion (€272.04 billion). This is a 7,4% increase compared to the 2016 figure. For commercial space products and services, this represents a growth of 8.3% compared to 2016, while commercial infrastructure and support industries grew by 7.5% (ibid.). The Space Foundation (2018, p. 5) estimates the revenue of the global space activity in 2017 to $383.51 billion (€339.48

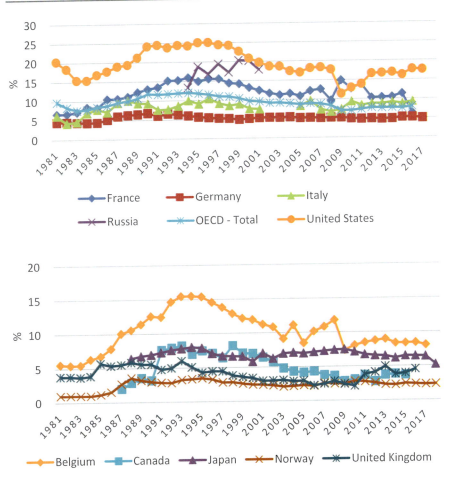

Fig. 7 Space as percentage of civil GBORD for selected countries, 1981–2017. (Source: OECD 2018)

billion), breaking the figure down in commercial space products and services ($211.45 billion/€187.17 billion), commercial infrastructure and support industries ($95.87 billion/€84.86 billion), US government space budgets ($43.34 billion/€38.36 billion), and non-US government space budgets ($32.85 billion/€29.08 billion).

For the commercial infrastructures and support industries, the highest proportion of revenues are generated in ground stations and equipment ($85.84 billion/€75.98 billion) and satellite manufacturing ($6.82 billion/€6.04 billion). The figure for the commercial satellite manufacturing represents 18.3% of the total value of the $37.30 billion (€33.02 billion) spacecraft market in 2017, representing a significant growth of 41.2% from the 2016 figures of $4.83 billion (€4.28 billion) (Space Foundation 2018). Commercial spacecraft is usually used for satellite communication, remote sensing or earth observation, commercial crew and cargo missions, launch vehicle

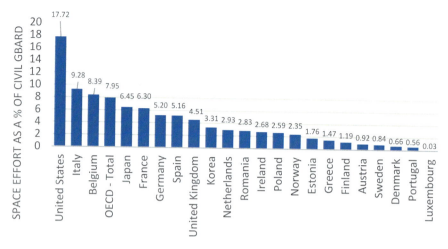

Fig. 8 Space as percentage of civil GBARD for selected countries, 2016. (Source: OECD 2018)

test missions, or other purposes such as for countries without independent access to space (Federal Aviation Administration 2018).

The total figure for governmental space budgets in 2017 mounts to approximately $76.2 billion (€67.45 billion). The USA is still in the lead with a total budget of about $43.3 billion (€38.33 billion) of which about 47.8% is channelized through the Department of Defence (DoD) and about 45.3% through NASA. The three largest governmental or inter-governmental space agencies following the USA by share of the total global figure are China (10.5%), the European Space Agency (8.6%), and Japan (4.0%) (Space Foundation 2018).

By country or territory of manufacturing, the USA and Europe are in the lead with 224 and 57 commercial spacecraft launched into orbit in 2017 (Space Foundation 2018). Combined, the USA and Europe represent 96% of the global manufacturing of commercial spacecraft, and the remaining 4% are manufactured in Argentina (1), China (6), Japan (3), and Russia (1). US launches represented 64% of commercial launches in 2017, while the European figure mounted to 24% (ibid., p. 39). The USA is heading the hierarchy of launches with 29, followed by Russia (19), China (18), Europe (11), Japan (7), and India (5) (Figs. 11, 12, and 13).

Figure 11 presents the number of launches in 2017 by country and domain. There are no defense launches in Europe but there are payloads that can be considered under that category (for instance, OPTSAT-3000 VEGA launch for the Italian Defence Ministry).

According to the Union of Concerned Scientist (2019), there are 1,957 operational satellites in orbit per November 2018. The USA is in the lead with 849 satellites followed by Russia (152) and China (284). The number of payloads manufactured and launches by country in 2017 are illustrated in Fig. 12.

Figure 13 illustrates the number of noncommercial (civilian and defense) and commercial launches from 2012 to 2017. The number of commercial

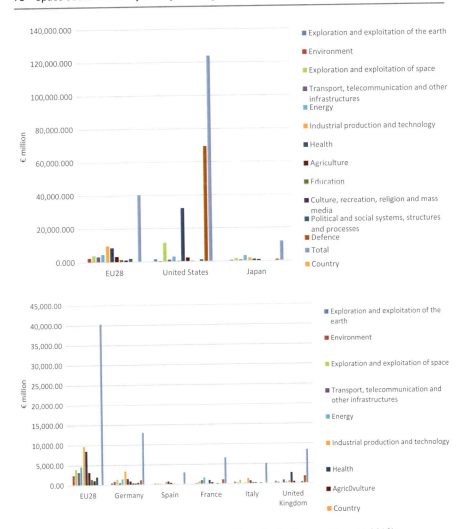

Fig. 9 Space versus other areas of R&D expenditure, 2019. (Source: Eurostat 2019)

launches has been relatively stable between 2012 and 2016, before it increased significantly from 21 to 33 from 2016 to 2017, corresponding to an increase of 57%.

Although the annual number of launches has figured between 80 and 92 since 2012, the number of spacecraft launched into orbit has increased dramatically in 2017. A total of 469 spacecraft were launched in 2017, an increase by a factor of 2.54 compared to the 2016 figure. This growth is a consequence of a significant growth in commercial payloads launched in 2017 (Fig. 14). Moore's law is also affecting the satellite industry (EIB Advisory 2018). This is reflected in the increase of small and very small (mini-, micro-, and nanoclass) payloads in clusters as 2017 was a record year for CubeSats, with 290 launched.

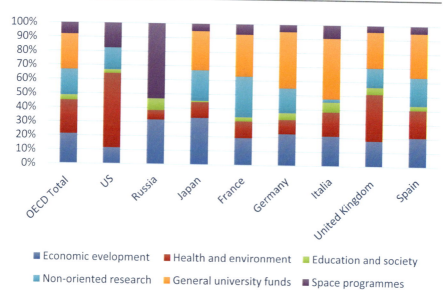

Fig. 10 Space versus other domains of government R&D priorities, 2016. (Source: OECD 2018, p. 81)

Forecasts for Government Space Budgets and Programs

The world total figure for government space budgets, civil and defense combined, continues to grow after some years of decline, posting 5 year CAGR of 5.75%. In the timeframe from 2008 to 2018, public budgets peaked in 2012 at $68.95 billion (€53.67 billion), followed by years of declining budgets towards 2015 ($62.50 billion/€56.33 billion). This decline towards 2015 can be explained by reduced defense space budgets in Russia and the USA. From 2016 onwards the total world figure has continued to increase and mounted $70.89 billion (€60.03 billion) in 2018 (Euroconsult 2019). The ceiling point is likely to meet min 2020s (Fig. 15). (Euroconsult 2019)

When breaking down the total government expenditure figure by regions domain, in our case civil or defense budgets (although it can sometimes be difficult to code the budgets as either-or), we see broad trend lines as presented in Figs. 14, 15, 16, 17, and 18. As seen in the figures, the trend is increased spending in both civil and defense sectors. This trend, for the combined figure, is expected to last until the mid-2020s.

A few insights immediately strike they eye. First, the USA spends by far the most on defense space assets (Fig. 17) relative to any other countries in the world. However, the American share of global space budgets for defense has declined gradually from 83.8% in 2008 to 68.3 in 2018. Euroconsult (2019) predicts that this share will increase again in the years to come towards 2022. Second, the share of the total figure for civilian space budgets is far more distributed among the

Table 1 Public space expenditure (civil + defense)

Countries	13	14	15	16	17	18
NORTH AMERICA	**35,782**	**35,839**	**35,594**	**37,198**	**38,152**	**41,311**
Canada	475	440	428	402	394	315
United States	35,307	35,399	35,165	36,796	37,759	40,996
ASIA	**8,899**	**9,284**	**9,967**	**10,593**	**11,013**	**11,771**
Australia	439	461	404	330	240	272
Bangladesh	25	29	63	63	67	70
Cambodia	0	0	0	0	0	4
China	4,069	4,574	4,860	5,145	5,450	5,833
India	916	870	964	1,092	1,295	1,493
Indonesia	54	75	70	73	79	205
Japan	2,907	2,585	2,824	3,005	3,047	3,056
Korea	317	485	553	643	589	593
Laos	0	0	17	17	17	17
Malaysia	10	10	10	10	10	10
Mongolia	23	50	50	53	55	5
Myanmar	0	0	0	0	0	0
New Zealand	4	4	4	7	9	9
Pakistan	11	11	26	38	33	61
Philippines	1	2	4	5	5	5
Singapore	22	34	33	18	16	29
Taiwan	59	51	39	31	35	35
Thailand	17	19	20	26	27	30
Vietnam	25	24	26	39	39	45
CIS	**10,189**	**7,357**	**5,000**	**4,278**	**4,768**	**4,355**
Azerbaijan	34	49	78	73	73	73
Belarus	21	21	22	23	23	28
Kazakhstan	284	214	22	45	47	50
Russia	9,753	6,966	4,790	4,122	4,609	4,170
Turkmenistan	67	88	74	5	5	5
Ukraine	31	18	14	10	10	28
Uzbekistan	0	0	0	0	0	2
EUROPE	**10,327**	**11,414**	**9,441**	**10,015**	**10,956**	**11,321**
ESA	4,580	4,415	3,583	4,205	4,238	4,465
Eumetsat	362	448	464	588	721	814
Austria	83	87	75	83	75	76
Belgium	279	282	216	231	248	247
Bulgaria	1	1	2	3	4	4
Czech Republic	24	25	21	23	59	59
Denmark	44	46	38	42	46	47
Estonia	4	5	4	3	6	7
European Union	1,584	2,729	2,160	1,929	2,100	2,115
Finland	72	60	48	49	55	58
France	2,975	3,072	2,391	2,782	3,055	3,158

(continued)

Table 1 (continued)

Countries	13	14	15	16	17	18
Germany	1,988	2,000	1,816	1,963	2,127	2,151
Greece	42	41	35	20	24	22
Hungary	6	6	5	9	12	12
Ireland	28	28	24	30	25	25
Italy	1,274	1,032	719	945	1,116	1,127
Lithuania	4	4	5	6	7	5
Luxembourg	25	30	90	44	60	88
Netherlands	168	195	132	159	140	171
Norway	114	103	102	107	122	125
Poland	52	56	51	70	80	90
Portugal	27	28	23	23	26	28
Romania	31	42	39	41	48	62
Slovakia	3	3	3	5	5	6
Slovenia	5	4	4	4	9	13
Spain	309	303	257	321	337	399
Sweden	157	170	173	123	129	127
Switzerland	157	184	170	186	194	202
UK	871	877	839	814	844	894
LATAM	**706**	**793**	**1,058**	**744**	**562**	**426**
Argentina	243	240	254	142	148	110
Bolivia	34	25	50	37	47	44
Brazil	260	269	255	164	173	122
Chile	5	5	5	5	5	5
Colombia	2	2	2	2	3	3
Ecuador	8	7	7	7	7	7
Mexico	104	180	244	145	16	10
Nicaragua	0	0	0	0	26	83
Paraguay	0	2	2	2	2	2
Peru	4	18	55	74	61	7
Venezuela	47	45	186	166	75	33
MIDDLE EAST AND AFRICA	**1,277**	**1,221**	**1,437**	**1,533**	**1,549**	**1,699**
Algeria	47	58	83	74	67	75
Angola	34	34	46	36	35	42
Bahrain	0	1	1	1	1	2
Egypt	31	27	14	71	100	177
Ethiopia	1	2	2	1	2	3
Gabon	0	0	0	1	0	0
Ghana	1	1	1	2	2	2
Iran	220	193	138	132	118	142
Israel	212	154	168	159	77	77
Kenya	1	1	1	1	1	1
Morocco	58	97	133	136	143	80
Nigeria	73	69	48	53	42	48

(continued)

Table 1 (continued)

Countries	13	14	15	16	17	18
Oman	0	0	0	0	0	1
Qatar	75	33	117	179	200	186
Saudi Arabia	15	15	51	130	166	165
South Africa	22	35	35	31	36	36
Tunisia	1	1	1	1	1	2
Turkey	380	350	253	136	148	276
UAE	106	151	347	389	409	383
Zimbabwe	0	0	0	0	0	1
TOTAL	67,181	65,907	62,497	64,361	67,000	70,883

Source: Euroconsult 2019

continents, although North America is still in lead in 2018 with 50.3%, followed by Europe (22.5%), Asia (18.6%), the CIS countries (Azerbaijan, Belarus, Kazakhstan, Russia, Turkmenistan, Ukraine, and Uzbekistan) (5.4%), the Middle East and Africa (2.4%), and Latin America (0.8%). Third, the space budgets for both civilian and defense purposes tends to follow the same paths; overall growth and overall decline across the board. As seen in the figures, Euroconsult (2019) foresees that government spending will continue to increase towards the mid-2020s, from the low of $62.50 billion (€56.33 billion) in 2015 to $84.59 in 2025. The decline from 2025 onwards is due to expected decreases in defense space spending, while the civilian budgets are expected to continue to grow towards 2030.

Euroconsult (2019, p. 8) predicts that the civil space budgets, overall, will grow at an average of 1.6% every year towards 2030, while defense space budgets will grow at an average yearly rate of 4.2% towards the mid-2020s followed by a negative growth. Investments in space science, exploration, and human spaceflight are drivers behind the civilian growth. The steady growth in defense budgets is predicted based on defense budget cycles in Asian, the Middle East, and the USA. The total 10-year CAGR 2019–2028 is expected to be 2.39%, civil and defense combined, in contrast to the high 11.8 CAGR of the 1999–2008 period (Euroconsult 2019, p. 12). (Compound annual growth rate (CAGR) is the average year-over-year rate of the investment over a specified period of time.)

The USA remains the largest investor in space activities, followed by China, Russia, France, and Japan. In 2018, the number of satellites (over 50 kg) launched mounted to 138, an 84% increase compared to the 75 satellites launched in 2017 (Euroconsult 2019). In the course of 2018 and 2019, Greece, Luxembourg, Australia, Zimbabwe, Turkey, and Portugal have established national space agencies.

In the area of defense space programs, funding is expected to increase until 2025. Already from 2017 to 2018, the world total for space defense budgets increased by 8.3%. The USA is dominating with 71.7% of the world total, and space defense activities are supported by the Trump administration (Euroconsult 2019, p. 15). In relative terms, Asian spending on military space activities has more than doubled over the period between 2008 and 2018, strongly driven by China along with

Table 2 Civil space expenditure

Countries	13	14	15	16	17	18
NORTH AMERICA	**18,609**	**19,494**	**19,716**	**21,068**	**21,380**	**22,398**
Canada	388	330	322	294	272	269
USA	18,222	19,163	19,394	20,774	21,108	22,129
ASIA	**6,034**	**6,443**	**6,838**	**7,397**	**7,722**	**8,278**
Australia	304	430	377	240	97	116
Bangladesh	25	29	63	63	67	70
Cambodia	0	0	0	0	0	4
China	2,392	2,665	2,788	2,994	3,441	3,699
India	829	811	904	1,030	1,227	1,406
Indonesia	53	74	69	72	65	177
Japan	1,947	1,749	1,859	2,117	1,994	2,017
Korea	317	485	553	643	589	551
Laos	0	0	17	17	17	17
Malaysia	10	10	10	10	10	10
Mongolia	23	50	50	53	55	5
New Zealand	0	0	0	2	5	5
Pakistan	11	11	26	38	33	61
Philippines	1	2	4	5	5	5
Singapore	22	34	33	18	16	26
Taiwan	59	51	39	31	35	35
Thailand	17	19	20	26	27	30
Vietnam	25	24	26	39	39	45
CIS	**5,626**	**3,778**	**2,343**	**2,095**	**2,524**	**2,391**
Azerbaijan	34	49	78	73	73	73
Belarus	21	21	22	23	23	28
Kazakhstan	284	214	22	45	47	50
Russia	5,189	3,387	2,133	1,939	2,365	2,206
Turkmenistan	67	88	74	5	5	5
Ukraine	31	18	14	10	10	28
Uzbekistan	0	0	0	0	0	2
EUROPE	**9,108**	**9,920**	**8,445**	**8,855**	**9,587**	**10,014**
ESA	4,580	4,415	3,583	4,205	4,238	4,465
Eumetsat	362	448	464	588	721	814
Austria	83	87	75	83	75	76
Belgium	279	282	216	231	248	247
Bulgaria	1	1	2	3	4	4
Czech Republic	24	25	21	23	59	59
Denmark	42	45	37	41	45	46
Estonia	4	5	4	3	6	7
European Union	1,584	2,729	2,160	1,929	2,100	2,115
Finland	72	60	48	49	55	58
France	2,432	2,346	2,047	2,251	2,373	2,517
Germany	1,897	1,898	1,695	1,799	1,873	1,964

(continued)

Table 2 (continued)

Countries	13	14	15	16	17	18
Greece	27	25	19	20	24	22
Hungary	6	6	5	9	12	12
Ireland	28	28	24	30	25	25
Italy	1,082	786	629	848	1,042	1,071
Lithuania	4	4	5	6	7	5
Luxembourg	23	27	33	28	44	56
Netherlands	165	192	129	157	138	169
Norway	100	90	92	98	113	116
Poland	52	56	51	70	80	90
Portugal	27	28	23	23	26	28
Romania	31	42	39	41	48	62
Slovakia	3	3	3	5	5	6
Slovenia	5	4	4	4	9	13
Spain	269	254	213	256	271	331
Sweden	157	170	173	123	129	127
Switzerland	157	184	170	186	194	202
UK	553	542	528	539	582	585
LATAM	**691**	**738**	**933**	**584**	**421**	**337**
Argentina	243	240	254	142	148	110
Bolivia	34	25	50	37	47	44
Brazil	260	244	196	89	104	52
Colombia	2	2	2	2	3	3
Ecuador	2	1	1	1	1	1
Mexico	104	180	244	145	16	10
Nicaragua	0	0	0	0	26	83
Paraguay	0	2	2	2	2	2
Venezuela	47	45	186	166	75	33
MIDDLE EAST AND AFRICA	**817**	**820**	**866**	**929**	**998**	**1,074**
Algeria	47	58	83	74	67	75
Angola	34	34	46	36	35	42
Bahrain	0	1	1	1	1	2
Egypt	31	27	14	23	46	57
Ethiopia	1	2	2	1	2	3
Gabon	0	0	0	1	0	0
Ghana	1	1	1	2	2	2
Iran	220	193	138	132	118	142
Israel	18	24	35	40	39	38
Morocco	3	3	3	3	3	3
Nigeria	73	69	47	52	37	41
Oman	0	0	0	0	0	1
Qatar	75	33	54	100	121	107
Saudi Arabia	15	15	51	130	166	165
South Africa	22	35	35	31	36	36

(continued)

Table 2 (continued)

Countries	13	14	15	16	17	18
Tunisia	1	1	1	1	1	2
Turkey	248	261	200	103	115	183
UAE	27	62	156	198	209	174
Zimbabwe	0	0	0	0	0	1
TOTAL	40,884	41,191	39,140	40,928	42,632	44,492

Source: Euroconsult 2019

significant Japanese budget allocations, and is expected to increase further from $3.49 billion to $5.19 billion in 2025. Europe is the fourth largest region in terms of space defense spending, totaling $1.30 billion following the USA, Asia, and CIS countries (Russia). Ariane 6 development is approaching its final phase, which may lead to reduced defense space contributions in Europe. The foreseen growth in space defense spending is mainly driven by the USA, China, India, and Japan (Euroconsult 2019, p. 15).

In terms of domains of the various programs, for launch vehicles, the total funding in 2018 corresponded to $6.71 billion (€5.68 billion), split between civil (60%), and defense (40%) domains. Euroconsult (2019) expects this figure to peak in 2019 at $7.04 billion due to the completion of the European Ariane 6 launcher foreseen in 2020, dropping to $6.02 billion in 2020.

With a 2018 total funding of $11.77 billion (€9.97 billion), Earth Observation (including metrological missions) programs remains the highest funded space application overall (Euroconsult 2019, p. 18). The figure will decrease on the short term, due to reduced funding over the US budget, yet this effect will, to a certain degree, be counterweighted by increased spending in Europe, Asia, and the Middle East and Africa. Defense spending is expected to experience a significant boost from $2.78 billion in 2021 to $3.54 in 2022.

For satellite communication, Euroconsult (2019, p. 19) anticipates an overall growth from $5.82 billion (€4.92) in 2018 to $8.29 billion in 2025. The defense figures for satellite communication make 63.3% of the total world spending in 2018. The total figure has declined from 2011 ($9.36 billion/€6.72 billion), and this is mainly due to procurement cycles of the US Department of Defence. In 2011, the defense spending corresponded to 75.6% of the overall spending. The ratio between civil and defense expenditure is expected to remain rather stable, and the growth is foreseen with the basis of more and more countries investing in satcom capabilities.

In the area of satellite navigation, funding has declined since the peak year or 2014 and the total figure of $4.67 billion (€3.52 billion). This year, the EU accounted for almost half the expenditures (45%), funding that was spent on its Galileo flagship program. Galileo is foreseen to be fully operational in the early 2020s. A GNSS constellation requires extensive funding associated with development, launches, and operations, which limits the number of countries willing to pursuit such capabilities. The 2018 GNSS spending figure for civilian and defense programs was $3.63 billion

Table 3 Defence space expenditure

Countries	13	14	15	16	17	18
NORTH AMERICA	**17,173**	**16,345**	**15,878**	**16,130**	**16,773**	**18,913**
Canada	87	109	106	108	122	46
USA	17,086	16,236	15,772	16,022	16,651	18,867
ASIA	**2,866**	**2,841**	**3,129**	**3,196**	**3,291**	**3,493**
Australia	135	31	28	90	143	156
China	1,677	1,909	2,071	2,151	2,009	2,134
India	87	60	60	62	68	87
Indonesia	1	1	1	1	14	28
Japan	961	836	965	888	1,053	1,039
Korea	0	0	0	0	0	42
New Zealand	4	4	4	4	4	4
Singapore	0	0	0	0	0	3
CIS	**4,564**	**3,579**	**2,657**	**2,183**	**2,244**	**1,964**
Russia	4,564	3,579	2,657	2,183	2,244	1,964
EUROPE	**1,219**	**1,494**	**996**	**1,160**	**1,368**	**1,307**
ESA	0	0	0	0	0	0
Eumetsat	0	0	0	0	0	0
Denmark	1	1	1	1	1	1
European Union	0	0	0	0	0	0
France	543	726	344	531	682	641
Germany	92	102	121	164	255	188
Greece	15	16	16	0	0	0
Italy	191	247	90	96	74	56
Luxembourg	3	3	58	16	16	32
Netherlands	2	2	2	2	2	2
Norway	14	13	10	10	10	10
Spain	40	49	44	65	66	69
UK	318	335	311	275	262	308
LATIN AMERICA	**15**	**55**	**125**	**161**	**141**	**89**
Brazil	0	26	59	75	69	70
Chile	5	5	5	5	5	5
Ecuador	6	6	6	6	6	6
Peru	4	18	55	74	61	7
MIDDLE EAST AND AFRICA	**461**	**402**	**571**	**604**	**551**	**626**
Egypt	0	0	0	48	54	120
Israel	194	129	133	118	39	39
Kenya	1	1	1	1	1	1
Morocco	55	94	130	133	140	77
Nigeria	0	0	1	1	5	7
Qatar	0	0	63	79	79	79
Turkey	132	89	53	33	33	93
UAE	79	89	191	191	200	210
TOTAL	26,297	24,716	23,357	23,434	24,368	26,391

Source: Euroconsult 2019

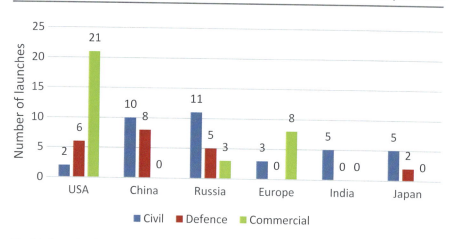

Fig. 11 Launches in 2017 by country and domain. (Source: Federal Aviation Administration 2018)

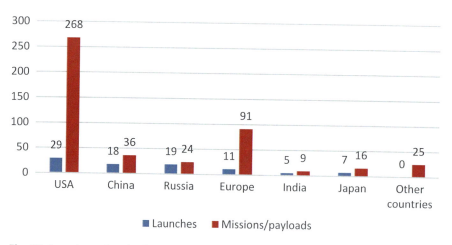

Fig. 12 Launches and payloads in 2017 ("other" countries refers to Argentina, Australia, Canada, Chile, Ecuador, Israel, Kazakhstan, South Africa, South Korea, Taiwan, Turkey, UAE, and Ukraine). (Source: Federal Aviation Administration 2018)

(€3.07 billion), and this figure is expected to grow towards, and peak at, $4.69 in 2026 (Euroconsult 2019, p. 19).

Science and exploration are funded over civilian budgets. The global expenditure mounted $7.0 billion (€5.93 billion) in 2018 and is overall expected to grow towards 2028 (expected $8.66 billion). Although about 28 countries invest in space science and exploration, the major six players account for almost all the spending (about 95%). This is, according to Euroconsult (2019, p. 18), due to the fact that these missions are highly visible thus suitable for public communication campaigns, while at the same time being ambitious.

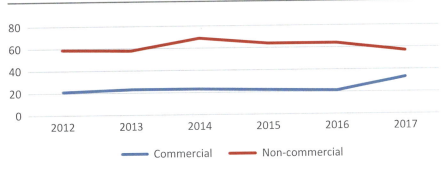

Fig. 13 Commercial and noncommercial launches 2012–2017. (Sources: Federal Aviation Administration 2012, 2013, 2014, 2016, 2017, 2018)

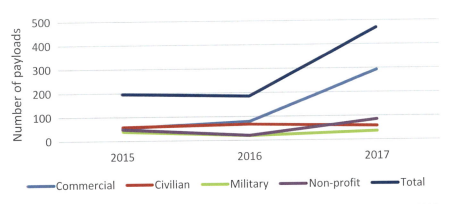

Fig. 14 Spacecraft launched 2015–2017. (Sources: Federal Aviation Administration 2016, 2017, 2018)

Human space flight is, after Earth Observation, the second single largest domain for space expenditures, totaling $11.56 billion (€9.79 billion) in 2018. The US proportion is about 77%, slightly down from 80% 2 years before, in part due to new actors entering the field including the United Arab Emirates. A significant and stable increase in spending is foreseen for the coming 10 years, mainly because of the manifestation of post-ISS activities such as the Lunar Gateway in the coming years. In 2028, Euroconsult (2019, p. 18) foresee the expenditure figures to reach $16.96 billion.

Space security, including early warning, is one of the most geographically concentrated domains in terms of expenditure, dominated by the USA, Russia, and France. As more countries express interest in space and security, and budgets increase in Asia, Europe, and Russia, the US dominance in the domain is likely to decline towards 2028. European countries, through the multilateral frameworks of ESA and the EU, have, for instance, demonstrated very keen interest in space security initiatives (including SSA for the EU) (Euroconsult 2019, p. 19). Budgets are expected to grow from $2.10 billion (€1.78 billion) in 2018, to $4.24 billion in 2023, before a decline towards the expected figure of $2.51 in 2028 (Table 4).

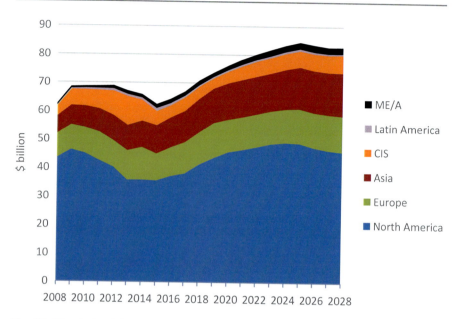

Fig. 15 Historical and forecast for government space expenditures by region, civil and defense, 2008–2028. (Source: Euroconsult 2019)

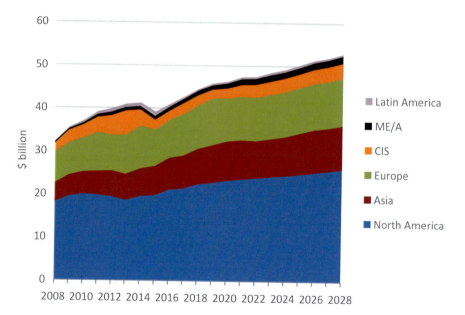

Fig. 16 Historical and forecast for civil government space expenditures by region, 2008–2028. (Source: Euroconsult 2019)

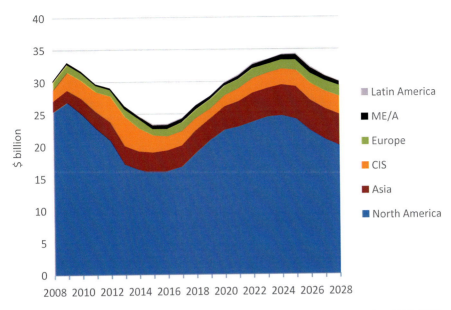

Fig. 17 Historical and forecast for defense government space expenditures by region, 2008–2028. (Source: Euroconsult 2019)

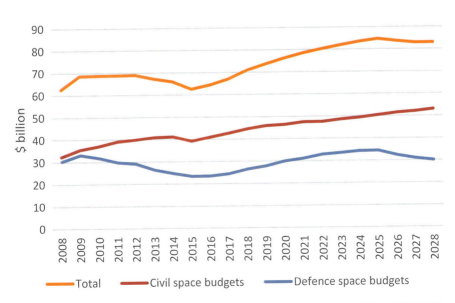

Fig. 18 Historical and forecast for civil and defense government space expenditure, 2008–2028. (Source: Euroconsult 2019)

Table 4 Exchange rates for relevant years

Year	Exchange rate (1€ = X.XXXX$)	1$ = X.XXXX€
2011	1.3920	0.718
2012	1.2848	0.778
2013	1.3281	0.753
2014	1.3285	0.753
2015	1.1095	0.901
2016	1.1069	0.903
2017	1.1297	0.885
2018	1.1810	0.8467
2019	1.1224	0.891

Source: European Central Bank

Conclusions

Accurate estimate of global space activities is a complicated task. This is due to the fact that there is lack of a standardized approach and lack of transparency in certain government space programs, such as defense-related programs. Additionally, the publication of financial figures by commercial companies is not uniform across the sector and varies in time. However, two trends stand out when investigating the space economy. First, government space budgets peaked in 2012, followed by decline, and is now increasing again and forecasts suggest that it will continue to grow in the years to come. Second, the commercial space sector constitutes the largest share of the economy with the emergence of new actors and more private investments. 2017 experienced a significant increase of commercial spacecraft launched to orbit. There also seems to be a growing interest on the recent private investments with companies such as SpaceX and Blue Origin as new drivers for the pace of the space economy. However, this must be put into perspective. While they are significant investment as compared to other private companies in the world, they still largely depend on government funding (80% for SpaceX) and do not reach government level (Fiorasco 2016). Although private space sector continues to grow, governments remain crucial actors in the space field.

References

Euroconsult (2017) Government space programs: benchmarks, profiles & forecasts to 2026. Euroconsult, Paris

Euroconsult (2019) Government space programs: benchmarks, profiles & forecasts to 2028. Euroconsult, Paris

European Commission (2018) EU budget: a €16 billion Space Programme to boost EU space leadership beyond 2020, Brussels. http://europa.eu/rapid/press-release_IP-18-4022_en.htm. Accessed 4 Mar 2019

European Investment Bank Advisory Group (2018) The Future of the European Space Sector: How to leverage Europe's Technological leadership and boost investments for space ventures. EIB, Luxembourg

European Space Agency (2017) ESA long-term plan 2018–2027. ESA, Paris. ESA/C(2017)96

European Space Agency (2019) ESA budget for 2019. http://www.esa.int/ESA_Multimedia/Images/2019/01/ESA_Budget_2019. Accessed 24 Oct 2019

Eurostat (2019) R&D expenditure. https://ec.europa.eu/eurostat/statistics-explained/index.php/R_%26_D_expenditure. Accessed 20 Oct 2019

Federal Aviation Administration (2012) The annual compendium of commercial space transportation: 2012. Federal Aviation Administration, Washington, DC

Federal Aviation Administration (2013) The annual compendium of commercial space transportation: 2012. Federal Aviation Administration, Washington, DC

Federal Aviation Administration (2014) The annual compendium of commercial space transportation: 2013. Federal Aviation Administration, Washington, DC

Federal Aviation Administration (2016) The annual compendium of commercial space transportation: 2016. Federal Aviation Administration, Washington, DC

Federal Aviation Administration (2017) The annual compendium of commercial space transportation: 2017. Federal Aviation Administration, Washington, DC

Federal Aviation Administration (2018) The annual compendium of commercial space transportation: 2018. Federal Aviation Administration, Washington, DC

Fiorasco G (2016) Open space, L'ouverture comme reponse aux defies de la filiere spatiale. Rapport au Premier ministre, Paris

Hatzichronoglou T (1997) Revision of the high technology sector and product classification. OECD Publishing, Paris

IMF (2018) World economic outlook. Washington, DC. https://www.imf.org/external/pubs/ft/weo/2018/02/weodata/index.aspx. Accessed 4 Mar 2019

Organisation for Economic Co-operation and Development (2011) Main science and technology indicators database. http://www.oecd.org/sti/msti.htm. Accessed 4 Mar 2019

Organisation for Economic Co-operation and Development (2016) Space and innovation. OECD, Paris

Organisation for Economic Co-operation and Development (2018) Main Science and Technology Indicators. Volume 2018/1. OECD, Paris

Space Foundation (2016) The space report: the authoritative guide to global space activity. The Space Foundation, Colorado Springs

Space Foundation (2018) The space report: the authoritative guide to global space activity. The Space Foundation, Colorado Springs

Union of Concerned Scientists (2019) UCS satellite database. https://www.ucsusa.org/nuclear-weapons/space-weapons/satellite-database#.XFHCsL4vw2w. Accessed 4 Mar 2019

World Bank (2019) Population estimates and projections database. https://datacatalog.worldbank.org/dataset/population-estimates-and-projections. Accessed 4 July 2019

The New Space Economy: Consequences for Space Security in Europe

74

Jean-Pierre Darnis

Contents

Abstract

The "new space economy" as a new paradigm able to revolutionize space activities shall be discussed. There is a large consensus of perceptions which tends to define this set of US-led policies and investments as a shift for all space activities. This is the reason why it triggers effects worldwide and creates a new context for European space activities. In Europe, space policies have often been characterized by a dialectic between plurality and collaboractive common frameworks. Large European Union member countries such as France, Germany, and Italy have developed a space policy on a national basis, although they traditionally collaborate through ESA, an international intergovernmental organization. More recently, the European Union, which does not share the same membership and rules as ESA, has stepped into space activities. These three layers help us define a fragmented European space panorama, which has always found it difficult to respond to the critical mass developed by US public and private policies. This is today the case with the "new space economy," where the USA is fundamentally renewing its public policy in order to create a more efficient and business-friendly approach to space, a shift which has helped create a new breed of space entrepreneurs, IT

J.-P. Darnis (✉)
Université Côte d'Azur, Istituto Affari Internazionali (IAI), Nice, France
e-mail: darnis@unice.fr

© Springer Nature Switzerland AG 2020
K.-U. Schrogl (ed.), *Handbook of Space Security*,
https://doi.org/10.1007/978-3-030-23210-8_74

tycoons investing in space with considerable financial capabilities, and tech incubator methodologies. This paradigm has renewed investments, technologies, and a new spirit of space conquest and puts enormous pressure on the European space sector, which has to renew its classic approach. Europe needs to come up with a more strategic space vision that is able to take into account not only the outcomes of new technology but also the inbred characteristics of space-based technology production and data fluxes, with key issues in terms of security.

Introduction

The "new space economy" represents a flashy and highly mediatized phenomenon. The image of the Starman piloting a Tesla roadster in space after being launched by Elon Musk Falcon Heavy rocket looks like a glimpse of a future able to create worldwide empathy among space geeks and public opinion. Behind this glamourous image, the "new space economy" describes an interesting phenomenon, which comes from a public policy reform of space and the increase of technological companies and investors in the USA. Building upon a massive capacity for public investment (space budgets, civilian, and military are roughly between 40 and 50 billion dollars a year), the USA is pushing the policy even further when Donald Trump speaks of "American dominance in space." On the European side, space has always been developed through a softer, if not blander, approach characterized by three pillars: an important European Space Agency (ESA) which has framed scientific exploration and technological research in space; some key member states such as France, Germany, and Italy which developed similar but autonomous space policies, also in the defense realm; and an emerging role from the European Commission which has fostered a "service to citizens" approach in its development of two flagship programs like Galileo and Copernicus. While for Europe security could appear as one of the services offered by space-based products, the US paradigm shift brings up not only a potential acceleration in terms of financial and technological fluxes but also the development of a more global "space security" considering the infrastructure, the data, and the products as a whole.

While European space policy can be seen as a very positive phenomenon, both in terms of outcomes and also for the political cooperative effort which has always characterized space, the post "new space economy" model illustrates the need to further intensify the cooperation framework, the goal being to develop a European way to a "global space strategy" without being pulled aside by the harvesting of data potentially coming from the USA and China.

The New Space Economy: A Push Coming from the USA

The "new space economy" is often presented as a new paradigm for space policy (Pasco 2017). It is a remarkable example of a public policy shift, which started in 2004 when the USA decided to put an end to space shuttle activities and progressively reoriented its access to space activities (Sehovic 2018). NASA became a customer buying launching capabilities from private companies rather than developing its own vehicles. Since then, the Commercial Orbital Transportation Services (COTS) program, followed by the Commercial Crew Development (CCDeV) program, has developed and shaped new bidding procedures offering new opportunities for space industries. For example, the development of SpaceX is linked to the attribution of Dragon vehicle launches within the COTS program.

The reshuffling of US space policy during the twenty-first century has initiated a movement for the entire space sector. It has fueled the interest of a new breed of entrepreneurs from the information technology industry which have renewed approaches to space (Weinzierl 2018). The "new space economy" sums up these individual and corporate investments which have taken place during the last decade. We can list a series of key characteristics which can help us to better understand the specificities of this "new space."

The first characteristic is the individual dimension of these new companies. Elon Musk (SpaceX) and Jeff Bezos (Blue Origin) represent the most popular and iconic examples of successful IT entrepreneurs who decided to invest in space technologies, with a mix of business plan and grand vision encompassing the future of the world and space exploration. Richard Branson, Robert Bigelow, and Sergei Brin can also be cited for their investment in space. These entrepreneurs are often operating not only as investors but appear to be extremely passionate and knowledgeable, somehow true space geeks, and also directly take part in the technical management of their enterprises (Lewy 2018). They also bring a specific IT business culture, with rapid innovation management processes and important investment capacities. Last but not least, they foster venture capitalism within the space sector, where space start-ups are following the trends we've already seen in the tech sector. To sum, we can see a "silicon valley" trend investing the space sector, which helps reinforce this perception of a "new space."

These trends represent a kind of inbreeding of the space sector and the tech sector. The space industry can rely on a longstanding tradition which goes back to World War II, with important public programs driving an effort toward science and exploration but also defense applications. During the last decades of the twentieth century, space applications for commercial services have grown, with satellites appearing as key infrastructures for meteorological, telecommunications, and positioning systems.

Today, satellites are key components of data retrieving, processing, and transmission. We can observe that business models that were once separate now tend to merge under the powerful umbrella of large platform companies which base their model on data harvesting and processing. This is among the reasons why OneWeb, Google, and SpaceX are planning satellite constellations. This evolution is extremely relevant as it affects both the USA and global markets. The first consequence is that US authorities are able to modify their policy for launchers and to attract new investors. Knowing they can count on the sale of a given batch of launchers enables US rocket producers to lower their price for the international commercial market. This gain in competitiveness on the US launching side represents an advantage when compared to other competitors such as the European Arianespace launcher family

(Darnis 2018). Even within the satellite market, the huge demand coming from US public authorities (civilian and military) represents a comfortable anticyclical cushion for US products. Those two factors can be considered historical parameters that can be summarized by comparing the US space budget to the European one (roughly $50 billion for the USA vs around $10 billion for Europe), underscoring a clear supremacy on the part of the USA. But this budget gap did not, until now, correspond to different systems: Europeans were more or less following the same sectors and technologies as the USA and could defend the quality of their smaller productions.

The "new space economy" indicates a different trend, a change of paradigm, with an acceleration on the US side, creating a potential integration of the space sector within the tech sector loop. This development could result in important consequences for space worldwide. On the policy side, we can observe an increasing perception of technological competition between the USA and China, and the recent announcement of the creation of a "Space Force" within the US department of defense pursues this path of a growing strategic and technological rivalry between Washington and Beijing. These trends call for a reinterpretation of global security policy, particularly in Europe.

The Transformation of Space Security

During the second part of the twentieth century, space appeared as a limited but key component of defense policies. Historically, the space race took place at the same time as nuclear dissuasion, and space technologies were developed in a parallel and cross-fertilized way: rocket technology was also useful for missiles, and satellite telecommunication and observation capabilities were integrated into the control and command chain of nuclear forces to do things like transmit orders to submarines or to define targets. Space technologies were integrated within the infrastructure of dissuasion. This is also the reason why those issues had a limited impact in Europe, with the remarkable exception of France pursuing nuclear strategic autonomy and developing some space defense tools from the mid-1970s (Nguyen 2001).

The Balkan conflicts of the 1990s represented a wake-up call for Europe: in strategic terms a conflict was taking place on European soil after decades of peace. Technically, this conflict showed that the US could dispose space-based capabilities for data gathering (Earth observation satellites) and transmission (communication satellites) that were lacking in Europe. Space technology appears to be a key enabling infrastructure for defense. This is the reason why after this conflict, several EU members decided to increase (France) or create (Italy, Germany) space capabilities for security and defense. From the 1990s up to the first decade of the twenty-first century, these "space services for defense and security" paradigm operations like Balkan conflicts or more recently the French intervention in Sahel (Serval, Epervier, and Barkhane operations) confirm the need for space capabilities to be able to project forces abroad. Indeed, space technologies have become key to provide continuity to the command, control, and information chain but also to meet the growing needs in terms of data bandwidth and also due to the expanding use of computed data-based systems such as drones. This security provided by space also had applications within the European Union with an increasing use of space-based infrastructures for security and safety applications. The launching of Galileo, an EU space infrastructure for navigation (Barbaroux 2016), is quite significant for this issue of security. One of the main reasons for this program was to provide positioning services for the citizens, with the aim of fostering the development of an important application market. This program has been strongly advocated as a civilian one, also to comply with European Union rules, which at the time carefully avoided the defense realm, a member state prerogative. Galileo has, however, created a Public Regulated Service, PRS, to be operated by public security users, including defense. Furthermore the Galileo program has seen a fierce dialectic with the USA which initially opposed this European Union initiative perceived as a competitor for their Global Positioning System (GPS). The Galileo case represents a significant shift for space security in Europe: the debate started around a program decided and built up as a "service to citizens," but it quickly triggered side effects about the inherent security applications and the foreign policy debate when dealing with USA-UE relations. To a certain extent, the Galileo case represents one of the strongest cases of European Union foreign policy action. Slowly, space policy is coming out from its "science and service to citizens" box and emerging as a potential strategic multisectoral policy.

A direct consequence of this increasing awareness about space issues has been the will to increase the capabilities that allow monitoring of space objects. The question of space debris, together with space traffic, represents a key issue for the security of systems but also the will to monitor potential hostile space objects. This is the reason why the European Union has launched a Space Surveillance and Tracking (SST) program, an initiative which was started under the umbrella of the European Space Agency in 2008 and then moved to a European Union Consortium in 2012. These institutional changes didn't accelerate a program which still seems extremely limited and relies on the pooling of national existing capabilities, but with a rather low autonomous investment capability. In the meantime, the US Department of Defense, which is responsible for space surveillance and has an impressive set of dedicated resources inherited from the Navy and Air Force space surveillance system, is planning to open space traffic management activities to private commercial firms. This announcement clearly indicates the rapid tempo of US space policy and the will, even in a DoD-only domain, to further embrace the "new space economy model," fostering the activity of private firms as a resource catalyzer. US space surveillance is extremely interesting as it reveals the evolution of the concept of space security in a post "new space economy" phase. Historically, space surveillance was, and still is, an important feature of the "space power" game played by the USA, Russia, and lately China, with a leading role of the defense administration, which aims to protect its space assets and be ready to react in case of hostile or accidental problems. This need is still strong, but the opening to the commercial market from the US DoD is a potential big bang, comparable to the NASA COTS policy which turned US policy on launchers upside down with huge consequences worldwide. The opening of space surveillance to commercial activities represents an example of a potential alliance between public authorities like the Department of Defense (DoD) and tech companies willing to retrieve and manage space data. We can already observe how global information technology platforms such as Amazon, Microsoft, and Google develop their business through contracts with the Department of Defense and other public security authorities. Incidentally, those companies are also developing cutting-edge research capabilities in sensitive sectors like quantic computing or artificial intelligence, while traditionally the DoD was the one that held a leading role in new research. These companies know how to accelerate the technological cycles, and a potential transfer of space surveillance to the private sector could enhance US space domination. This concept of "American dominance in space" was used by President Donald Trump in 2018 when he announced the launch of a "Space Force" and also corresponds to the antagonistic posture adopted by the Trump administration when dealing with China in commercial and technological relations.

Europe has to face a multilevel challenge, which is not only to keep up with US and Chinese investments but also to improve the policy level associated to space both as a security provider and a key component of EU global security.

European Space: A Resilient Model

The European space model is a federalist and somehow disparate one, much like the entire European institutional asset. It corresponds to an original buildup between sovereign countries which chose to create a mixed model while maintaining a strong grip on classic sovereignty tools (taxes, defense, security, etc.) and creating institutions for new common policy objectives. Space has a different history in Europe as its cooperation framework was consolidated in 1975 through the European Space Agency, ESA, an intergovernmental organization which has developed capacities and know-how in terms of management of science and exploration space programs. ESA works under the "juste retour" budgetary principle, meaning that each ESA member will receive a return in workload corresponding to its contribution. This rule has been, and still is, extremely useful in order to convince member states to contribute to the ESA budget, as they are certain of the return obtained. This stable mechanism has helped shape significant space activity in Europe over the entire chain of space technologies, from launchers to satellites and ground stations. Europe is part of the narrow club of space powers together with the USA, Russia, and China, countries able to master the whole set of space technologies. It is therefore remarkable to observe that even if Europe suffers a disparity when compared to the USA, with a budget ratio that is four to six times lower, a disparity that is visible in the number of launches and satellites, Europe has still been able to develop an impressive set of technologies, eventually narrowing or maintaining the gap with the USA. During the twentieth century, European space was conceived as a sectoral policy, with a strong investment in research and exploration and important efforts to develop commercial activities for launchers and satellites, benefiting from public research and development support. There is also the development of a launching facility on the European soil, with the Kourou space base. All these investments might seem difficult to justify when assessed through expenditures. Still they have allowed Europe to develop a strong scientific and industrial basis and impressive operative know-how. Europe has to rely heavily on other programs (Russian, US) only to send its astronauts in space.

The twenty-first century has seen the European Commission entering space programs. This powerful financial institution has brought new resources for the European space sector but also created a certain level of entropy for the management of programs and institutions. The "juste retour" rule represents a disparity with the European Commission which funds projects on a competitive and – officially – nonnational basis, creating difficulties for joint ESA/EC projects. As the European Commission progressively increased its action in research funding, it started to deal with space technologies and a first framework agreement with ESA started in 2004. Since then, relations between these institutions are evolving, but the differences between the two organizations' rules and membership create a series of frictions even if they both express different versions of the same intergovernmental logic. In security policy, and more precisely space security, the two organizations have different rules. ESA results are bound to its "peaceful purpose" status which prohibits defense activities (Cheli and Darnis 2004). The European Commission has insisted for a long time on the strict civilian nature of its program, with member states being extremely reluctant to have the Commission step into defense. In the 2000s, the "security" issue and the "dual-use" aspects of technologies created a rather blurred approach which was embraced by the European Commission when opening funding for "security research." Even if ESA did not strictly follow the same opening, it contributed to technological development for dual-use systems (e.g., Galileo and Copernicus). This evolution was confirmed in 2016 when the European Commission proposed a defense action plan, with a consensus allowing the European Union to invest in defense technologies. Today, we can observe that an institutional triangle for space security and defense policy is taking place between the European Commission, the European Defence Agency, and the European Space Agency, even if the outcomes of this institutional dialogue are still somewhat limited.

Since the beginning of the twenty-first century, the European space policy model has entered into a transformation phase. Before this period, it was essentially a dialectic between national agencies and ESA, and security was kept out of this collaboration, as some member states developed defense or dual-use capabilities (France, followed by Germany and Italy). In a way, we could say that during the twentieth century, there was no European space security.

Then, a hybrid period started, with the emergence of a blurred dual-use model that we can observe, for example, within the 6th framework program of the European Commission. Industry played a role in putting a budget for space and security on the Commission's agenda, as companies wished to showcase and develop space-based technologies for applied security, also with the aim of fostering potential pan-European public demand. The logic of space-based products for security services was elaborated to comply with the strong push of "services to citizens" emanating from the Commission: the global philosophy was to demonstrate that

Europe was useful in citizens' everyday life and would produce returns in terms of development of economic activity. This early 2000 mantra with a strong economic rationale was, for example, pushed in documents such as "The cost of non-Europe in the field of satellite-based systems" (Darnis et al. 2007) published by the European Parliament. Space policy analysts had to take into consideration this economic vision in order to introduce some more strategic reflections about space.

But this "space for services" approach also illustrates the resilience of the European space policy model. It is a difficult balance between public funds and commercial activities. For example, the launcher sector, with the Ariane family lately joined by Vega, has always fought to keep its international competitiveness on the commercial market while being directly subsidized through Arianespace. Lately, this model has evolved with the creation of the ArianeGroup, but the equation is still there. Europe can showcase its excellent launching technology, with an impressive rate of success both for Ariane and Vega launchers, but needs public subsidies to keep the sector alive. This is not a political problem as all space launching powers subsidize their launches. The US agencies, for example, provide batches of launches with a comfortable pricing, an operation which allows industrials to lower their price offer on the commercial market. If European Commission rules were applied to the launcher sector in the same manner as other sectors, the subsidies would be banned, an operation which would, however, undermine European interests in the international competition for access to space. This hesitation about the launching economic model in Europe reveals the lack of strategic consciousness about space. Even though important space countries such as France, Germany, and Italy might agree on the need to maintain capabilities and invest in the launcher sector, it is more difficult to share this view with a more pan-European panel which might produce contradictory considerations between general policy declarations which push for a "reliable and affordable" access to space in Europe and more economic-based thinking with criticism about the loss of a complex institutional model or countries which do not feel the need for domestic launching capability.

The US "new space economy" paradigm is pushing European institutions and companies to react. Launchers have been the first to be concerned, and the new European space vehicles Ariane 6 and Vega C represent an effort to lower costs and react to the market push. But again they are threatened by the rapidity of the transformation of US players, able not only to renew their price and public subsidies model but also to push for technological innovation such as reusable vehicles. This economic and technological competition might have a positive effect on the European players who are pushed to further evolution. It is also creating a moment for space worldwide. The popularity and the grand vision of US space tycoons, such as Elon Musk or Jeff Bezos, Donald Trump's declarations about American dominance in space, and the landing of the Change 4 Chinese probe on the dark side of the moon, are gaining relevance for the space activities worldwide. Media, public opinion, and European decision makers are caught up in debates on space policy which did not exist a few years ago. All this attention is already benefiting to the space sector, with the European Commission and European Parliament announcing budget increases.

The "new space economy" also creates a new paradigm for space activities: once isolated, they are now fully integrated into the data gathering, transmission, and processing value-added chain, a trend which is not only one of the most promising existing business models but also a key issue in terms of security and democracy, as the control of data is already affecting the economy and citizens' lives. Europe missed the information technology revolution by being unable to compete in the processors' industry and not fostering the development of large consumer-friendly IT service platforms able to play a global role. In Europe, no Microsoft, Google, or even Huawei is coming into being. Space, however, represents a sector where Europe can rely on an impressive set of in-house technology. If satellites happen to be today and tomorrow's key components of the data retrieving chain, it is important for Europe not only to maintain its capacities but also eventually to further invest in a sector which is attracting a huge interest worldwide, being able to benefit from the excellent research, technology, and industrial basis that already exists in Europe. The opportunity to maintain and invest in the whole set of space technologies can also translate into a renewed strategic vision of space security. Space can be a security provider but appears here to be a key component of a renewed global security and strategic vision: the capacity to generate and control data is not only the key to winning business models; it is also a fundamental feature to maintain the democratic order and the rule of law, meaning the continuity and the future of our societies. Space will no longer be a sector limited to extremely competent scientists and technicians but also has to be considered by society as the backbone of today and tomorrow's economic well-being and civilian life.

Those considerations mean that space be a key feature of a global European security strategy, which puts technology and data at the center of its interests. For example, we are already observing mobilization around artificial intelligence, calling for further investment and warning that existing gaps with the USA or China might jeopardize our future. But the call to invest in a sector with low capacity in Europe should not distract us from a different and potentially stronger drive: the opportunity to invest in a sector where we already have decent capabilities and where we have a large research and industrial infrastructure able to easily absorb and transform further investment. It is easier to pour money into a relatively large-sized and extremely specialized sector than to create from the scratch a whole new area. Space activities are well structured in Europe, with strong aerospace engineering and physics programs in universities, specialized master cycles, labs, start-ups, public-funded national research centers, national and European agencies, and a large industrial sector. Still, those industrial capabilities are often suffering because of the lack of demand, for example, in the launcher or satellite domain. The number of European launches is comparable to India, while China and the USA are developing an intense activity. There is a risk of marginalization but also an opportunity to maintain and potentially develop further space capabilities by filling today's research and industrial programs, and the European Union can play a decisive role in this. The road is already paved. The Copernicus flagship program is extremely promising but should be further pursued, also in order to provide a European "secure" source of Earth observation products, able to compete on the market not in terms of price and delivery, easily won by future large-sized constellations, but in terms of reliability and continuity of data, a key issue to develop important markets such as public administration as the Galileo PRS already shows. A functioning high-precision space surveillance system would also represent an impressive value added for Europe, able not only to control space and to sit at the table of future space policies with other space powers but also to take part in the development of commercial-based space awareness services. Also, space surveillance should attract more decisive funds and organization on the European side. This is a key issue in terms of security where different vetoes are keeping the European Commission from hitting the accelerator.

The European space industry does not represent the same level of flexibility and investment capability as the "new space economy" entrepreneurs: in Europe space groups are often the products of a long history, which gives continuity to the sector but sometimes represents a barrier to innovation. But those industries have shown a remarkable resilience and are able to face further technological challenges. They evolve in a composite world, between partially protected national markets, European public programs, and international competition. It would be dangerous to push for a global reform of those market structures in Europe, as they provide anticyclical capabilities to the companies. However, one key logic is missing, an increased presence of venture capital able to foster new technological developments and products for Europe's space industry. This mechanism is popping up in some member states, for example, in Italy's space agency (ASI) which has created an investment fund in partnership with venture capitalists in order to foster space-based start-ups and spin-offs. This is a limited but promising experience that could spread throughout Europe, also renewing the action of public institutions.

Conclusion

European space is today on the front line. First because it is a success story, meaning a solid science and technology sector, where Europe defends its excellence on a worldwide level. But space today is at the crossroads of different emerging policies: the reform of defense and security which no longer considers space as simply a "dimension" or as a specific outcome but more and more as a global enabler, a dimension to be included within the information and information technology sphere to create a whole new strategic approach of space. Traditionally, the strategic relations between the USA and Europe were bounded within NATO. Space was out of this scope, as space did not represent a specific transatlantic security issue and even represented as gap between Europe and the USA since space security didn't allow for dialogue to take place on some issues.

Paradoxically the fact that the European Union has stepped into space activities with its two flagship programs Galileo and Copernicus has improved the European level in terms of space security. Not only because the small size of member states did not allow strategic thinking about space, France being somehow an exception, but also because of the global foreign policy and security impact that those programs had, with Galileo opening the way. While the European Union often faces criticism about its lack of social dimension and its economic criteria, the research and innovation side is largely praised. The European Commission is providing funds and managing programs that became key for the European scientific and technologic communities, also compensating the lack of national resources. This means that the Commission plays a leadership role in terms of science and technological policy, not only because of funding but also in terms of regulatory capabilities. In the digital realm, we can observe the very relevant regulatory action performed by European Commission under the authority of Vice President Věra Jourová. This technical capacity to raise data and privacy issues and to engage large IT companies has recently created an original set of powers, while member state administrations and governments lack specialization in that domain and have agendas driven by domestic issues. We can observe a technical oriented regulatory and RD investment model emerging at the European Union level. Space is included within this panorama but suffers from institutional competition between the European Space Agency, member state agencies, and the European Commission. The European Commission is the latest player in the game and strives to treat the space sector like other technological and RD issues. The space institutional legacy and know-how lie instead in the hands of the European Space Agency, an international organization whose membership and rules differ from the European Union. As ESA represents precious know-how in terms of project management and space science, it is counterproductive to adopt an aggressive position toward ESA, for example, with a Commission aiming to unify all the space activities under its umbrella and rules. ESA can have a leadership role in research and development for space programs but can also boost its already existing capabilities in terms of Earth observation data management, also considering security issues such as the nature of the data and the need to control and eventually restrain its use.

The European Commission, on the other hand, seems better fitted to ensure the coherence of space policy with information technology and data policies and regulatory issues in a cooperative versus sovereign international technology framework. The EU already has experience with the Galileo frequency negotiations with the USA and is dealing on a day-to-day basis with international IT platforms in order to regulate their business in Europe, with a particular focus on citizens' rights and privacy but also having in mind the development of its antitrust action. Coming from a commercial policy culture, those skills today contribute to a strategic vision which shapes the European position toward non-European companies and States. This is indeed an illustration of an incisive global strategy that can be further developed and pursued within the space sector. The American "new space" example indicates a fantastic acceleration between the reforms of a very robust public demand and the breeding of space with a California-style technological culture. Europe can count on excellent capacities and resilient institutions, with a political consensus often disconnected from central organs such as the European Commission: member states still represent the center of European democracy and for space policy; they are the first and eventually the main level to be involved in key political decisions. This is a rather complicated system, where each member state evaluates its return before making a decision, a reality which helps validate the "juste retour" ESA model. The future of European space policy will have to navigate between those two rocks: member states that are often obsessed by the national justification of their investments which, for space, can only be conceived within an international cooperative framework and a European Commission which may be a potential game changer for space, with the capacity to enhance space policy within information and data global policies. Within this framework, there must be room for business models in Europe that can imitate some of the US models where public institutions are increasingly buying space capacities, leaving the private sector to deliver an "all inclusive" offer, and eventually opening public resources to commercial activities in order to enhance the system output. If we consider space surveillance as one of the hottest topics for space security, then the somewhat clumsy European consortium could support a strong reform if not a complete revision like the creation of a flagship program to develop a model able to deliver effective monitoring capabilities comparable to the US, Russian, or future Chinese ones. A key condition for the success of a future EU Space Surveillance operational program lies with member states: as surveillance information needs security management, a mechanism like the Galileo security system and procedures, including a Galileo security center, should be developed to ensure the compatibility of a EU Space Surveillance program together with MS security needs. The legacy of existing national systems, mainly radars, appears today as an obstacle as they trigger conflicts in terms of sharing of resources, investments, and ownership between Member States consortium members. A potential flagship program could put an end to these disputes, with the intent to create new capabilities from scratch. This could be developed through contracts with private partners and technical institutions, leaving defense ministers and their assets out of the loop, or proposed through a company consortium. If we agree that global space security must climb higher on the European agenda, then it is important to think in terms of new investments able to mobilize European resources and capabilities and rise up the level of Europe in the world. Space surveillance could just be the next big thing for Europe, if we move on from the current situation contaminated by legitimate, but contradictory, national interests.

References

Barbaroux P (2016) The metamorphosis of the world space economy: investigating global trends and national differences among major space nations' market structure. J Innov Econ Manag 20(2):9–35

Cheli S, Darnis JP (2004) Towards a European space strategy?, The International Spectator, 39:103–114

Darnis JP, Gasparini G, Pasco X (2007) The Cost of non-Europe in the Field of Sattelite Based Systems, European Parliament Study, Brussels, 75 p

Darnis JP (2018) Access to space, challenges and policy options for Europe, IAI Commentaries, 16/18. https://www.iai.it/it/pubblicazioni/access-space-challenges-and-policy-options-europe

Lewy Stefen W, 15/10/2018, JEFF BEZOS WANTS US ALL TO LEAVE EARTH—FOR GOOD

Nguyen A-T (2001) Les échanges technologiques entre la France et les États-Unis: les télécommunications spatiales (1960–1985). Flux 43(1):17–24

Pasco X (2017) Le nouvel age spatial, de la guerre froide au new space. CNRS, Paris, 192 p

Sehovic I (2018) The NewSpace race: the private space industry and its effect on public support for NASA funding, working paper, March 2018, 52 p, available at https://www.academia.edu/36133518/The_NewSpace_Race_The_Private_Space_Industry_and_Its_Effect_on_Public_Support_for_NASA_Funding

Weinzierl M (2018) Space, the final economic frontier. J Econ Perspect 32(2):173–192

Political Economy of Outer Space Security

75

Vasilis Zervos

Contents

Abstract

As human civilizations increasingly explore, utilize, and compete in space, the man-made security challenges are evolving and the strategies and political economic rationales become increasingly relevant for analysis. Sustainability and efficiency call for exploitation of static economies of scale and scope in space industries and services, yet the trade-offs in control, governance, and dynamic innovation point towards autonomy and oligopolistic structures with overcapacity. The economic sustainability becomes a key element of the dynamic pursue of space policies and objectives at national and partnership levels. In the latter case, specialization and its implications for the wide economy through externalities and indirect effects receive increasing attention as space becomes contested, congested, and competitive. Notwithstanding the fact that they are largely government controlled, aerospace industries play a crucial role in trading patterns. Hence, they can be considered a fiscal government spending element similar to

V. Zervos (✉)
University of Strasbourg and International Space University (ISU), Strasbourg, France
e-mail: vasilis.zervos@isunet.edu

© Springer Nature Switzerland AG 2020
K.-U. Schrogl (ed.), *Handbook of Space Security*,
https://doi.org/10.1007/978-3-030-23210-8_101

defense expenditure. The country specializations and their evolution in commercial markets and alliances are focal points in the current global trade policy paradigm shifts, affecting performance and evolution of space programs and industries. The analysis concludes with the ever-increasing role of regulation and relative power balances across nations, companies, and terrestrial-air-space systems especially for telecommunication applications.

Introduction

Outer space security historically was shaped as a concept during the post-WWII period with the space race focusing on hardware/space assets and weaponization. The maturing of relevant technologies and dissemination of space capabilities to levels whereby undergraduate engineering students are involved in designing and developing nanosatellites and space-proven software and hardware leads to the sustainability agenda evolving at a rapid pace for the future. Sustainability refers not only to manufacturing and knowledge-developing space-based applications and systems, but also to the economic model that reduces costs and augments relevant value to sustain national space industries. This has been critical since the Cold War period that led to shaping of specialization within alliances of overall space support insofar as both civilian and security capabilities are concerned. As space becomes more economically focused through time, such specialization brings forth commercial competitive pressures among allies and questions the initial specialization allocative principle. Thus, Europe, for example (but similar arguments could be made for other allies like Japan, or ex-Soviet Union countries), for several decades has been actively pursuing the objective of space autonomy and commercial competitiveness within the western alliance, despite the US leadership since the early days and the resulting security inter-alliance specialization.

This framework has implications for geopolitical dynamics and partnerships, for efficiency and duplication, but also for the space race "proxied" by national space industrial competition. The support of private-owned enterprises as the space industrial integrators has long (since the 1990s) been employed in the West as a champion of competitiveness in commercial space markets. Leaner, streamlined enterprises without the burdens of disclosure of information and profit assessment by shareholders have emerged. These enterprises are supported at large by high net worth individuals attracted by space and investing through self-developed, financed, and assessed businesses. Ambitious plans of mega-constellations of telecommunication and other application satellites are being financed and developed avoiding business risks experienced with the post-Cold War period when shareholders were exposed to opaque novel space technological operations (Iridium as a characteristic example and others). The "new space" 2.0 thus is based on private initiatives and involves massive investments and production of smaller yet effective satellites in large constellations. The public sector endorses nationally controlled initiatives in this direction as it provides a sustainability of operations and low budgetary appropriations cost not only with regard to the economic resources and organization but also with regard to industrial scale capability to replenish and mitigate against man-made and natural risks in the perilous space environment. However, the security aspects and market failures associated with space and the overcapacity are expected to lead to real challenges for a competitive market framework and pose real threats to economic and physical sustainability of space resources.

The next section focuses on the background of the security challenges in space, the trade and industrial dimensions, as well as the new space initiatives and their implications for the evolution of space systems and security aspects.

Security Dimensions and Challenges in Space

A space race whereby multiple nations and alliances are involved is ongoing with accelerated pace as new space powers and capabilities develop. Space faring nations develop space branches in the armed forces either integrating aerospace and outer space defense (Russia), or formulating a space force as a separate branch. The case of the USA as the leading space faring nation has characteristically led to a paradigm shift through the evolution of the Pentagon into a Hexagon, to incorporate by 2020 a space force and associated command (DoD 2018: 7). (The structure of the DoD is thus planned to change from one incorporating five branches of the armed forces into a six-branches one with the addition of the space force, by 2021.)

A popular space community say goes along the lines that if one was to lay down the reusable vehicles reports and projects, they would reach the moon. Despite the hyperbola of this statement, space saw limited breakthrough developments in terms of technologies since the end of the Cold War, while numerous studies for ambitious space programs were created, but few going beyond initial paper stages. Commercial considerations and applications grew substantially as the post-Cold War world enjoyed the (space) peace dividend in the form of a multinational space station (ISS), precision in positioning services (removal of selective availability from NAVSTAR/GPS), high resolution imagery and associated applications, telecommunications growth, etc. This peace dividend was the result of technologies that had developed but were reserved for military and associated security considerations. Such safeguarding of technologies and applications are explained by the principle that in the presence of rivalry, it is the relative position that matters, unlike commercial (competitive) considerations that are assumed to rely on overall improvement of positions (Zervos 2011). Thus, selective availability ensured a significant divergence between civil-signal quality and military-signal quality and associated characteristics. Historically though, there have been security threats and military considerations that boosted space technologies development, starting from the very beginning (A4/VII WWII missile) and throughout the Cold War period.

Security is, so far as its space components is concerned, classified along two major areas:

- Nature related
- Man related

Nature-related security threats incorporate elements such as near-Earth orbit threats (asteroids), or solar activity and space weather. On the other hand, man-related elements are of a strategic nature, meaning that they focus on the relative position of agents within a rivalry framework. The USA, for example, identifies the following specific strategic areas of a growing "contested and congested" space environment (see USAF 2013; DoD 2018: 5):

- Nuclear forces
- Cyberspace
- Command, control, communications, computers and intelligence, and surveillance
- Reconnaissance (C4ISR)
- Missile defense
- Joint lethality in contested environments
- Forward force maneuver and posture resilience
- Advanced autonomous systems
- Resilient and agile logistics

Space is considered a strategic domain, similar to nuclear and other WMD (weapons of mass destruction) technologies. This is because space is a major enabler and multiplier in networks (one could draw a parallel with water, electricity networks of vital importance), while at the same time, it is borderless and unregulated by treaties constraining dissemination of relevant satellite and launcher technologies, despite their inherent dual-application nature. Historically, its specific characteristics have served as both a multiplier of force and a peace-through-verification function. Dissemination of technologies, despite being unregulated by treaties, is a major element in strategic security aspects and analysis, in view of the relevant position principle. Thus, export controls exist on a global scale for trade in related items, while an illustrative example of its significance was experienced in the aftermath of the collapse of the Soviet Union, as the USA implemented policies of transfer of resources and collaboration largely to avert a chaotic dissemination of the Soviet advancements (Harford 1997). Clearly therefore, in a world whereby military alliances like NATO are critical for security, the inter-alliance specialization is of particular importance for stability, coherence, and economic (trade) profiles of participating nations. Consolidation in space and the pooling of resources to avoid duplication of R&D and enhance benefits have been dominant reasons behind the creation of space agencies like National Aeronautics and Space Administration (NASA) or the European Space Agency (ESA) at national and regional level. (The early development of space capabilities was characterized by strong competition not only at institutional but also at personality level and led to consolidation at agency level to avoid costly duplication in R&D (see Harford 1997).) Interdependence is also frequently seen as a contributing factor for club, or alliance stability, which implies specialization distributions. At the same time, however, this is an unstable mechanism as countries develop not only multidimensional evolving security policies but also as economic considerations reflect such specialization. The following section examines the systemic interdependencies by focusing on the industry and its structure-conduct and performance link to public security and defense.

However, beyond the inter-alliance specialization and resulting strategic dynamics, dissemination of space capabilities leads to a "rush for green pastures" approach. This is reflected in two dimensions, the one is the orbital level and the second is the celestial bodies' level. To avoid conflict over natural resources associated with space, the International Telecommunication Union (ITU) established since the beginning of the space-based telecommunication systems a GEO-orbit and frequency allocation mechanism that employs the principle of equality, rather than first-come. The Cold War-controlled environment has been exchanged for an environment of seemingly free commercial competition and exploration, whereby constellations of thousands of small satellites are planned and materializing. Such constellations are developed within national industrial frameworks and are compatible with security concerns as they sustain technology and production lines supporting responsiveness and ensuring rapid replenishment of assets/capabilities in case of conflict (see later section and Butler 2015). At the same time, such constellations and the required vital space they require for seamless operations lead to a contested and congested outcome across a lot more than the geostationary orbit (USAF 2013). High capacity outcomes may benefit civil users but may also exacerbate security concerns associated with earlier projects like Iridium, whereby national service providers and authorities would see restricted economic returns, while experiencing dependence for relevant telecommunication security.

Economic Background

Industry

The space industry is subject to economies of scale/scope; market failure is present at several levels through the aforementioned cost and market characteristics leading to infant industry arguments but also due to the strategic economic nature of the sector (airbus vs. boing and the launchers market analysis).

The Research & Development (R&D) intensity of the space sector is quite high compared to other manufacturing and high-tech sectors. In the United Kingdom (UK), the R&D intensity of the sector is estimated at 8.1%, which is higher than sectors such as programming and telecoms but lower than the pharmaceuticals sector. Compared with the UK average, the space industry spends 6.5 times more on R&D in value terms (HoC 2017). Other studies offer estimates for the Italian industry of R&D intensity of 14% for the space industry versus 3–4% for the space services sector (compared to 11% in aerospace; 4–5% in high-tech sectors and less than 1% in manufacturing; Graziola et al. 2011). The undoubtable relatively high R&D intensity is due to the high R&D requirements but also due to the limited production levels (compared to car manufacturing or defense manufacturing), as most programs are of a customized/limited production at national level nature.

This is clearly illustrated in the military satellites case whereby national considerations and preferential treatment are dominant. As Fig. 1 reveals, in military space, satellites are "home made" on a global scale. This phenomenon is not only due to demand side aspects whereby countries select their home industry to enhance its scale and scope economies, as well as the security factors, but also due to the supply side through export restrictions.

These export restrictions serve to safeguard technologies, as trade can rapidly diminish technological gaps limiting the scope for global trade openness in the sector. An anecdotal illustration of the strategic nature of the space sector emerged during a commercial space dispute whereby a US company filed a complaint with the US administration over European free-trade violations and associated practices in launching service provision through the heavy subsidization of the European launch vehicles. This took place during the Cold War era in 1985 when a US company (Transpace Carriers Inc.) brought a legal case against ESA and its member states to the attention of the President of the USA claiming among other things that Arianespace faced a protected home market and this was violating the US Trade Act of 1974. The case was dismissed largely on the grounds that the US public sector applied similar protective processes to its domestic space industry (using provisions related to "Buy American Act" and others):

Based on a petition filed by Transpace Carriers Inc., (TCI) the United States Trade Representative (USTR) initiated an investigation on July 9, 1984, of the European Space Agency's policies with respect to Arianespace S.A. Arianespace is a privately owned company, incorporated under the laws of France for the purpose of launching satellites. Arianespace's shareholders include the French national space agency, and aerospace companies and banks incorporated in the ESA Member States. The petitioner alleged that 1) Arianespace uses a two tier pricing policy whereby Arianespace charges a higher price to ESA Member States than to foreign customers; 2) the French national space agency (CNES) subsidises launch and range facilities, and services personnel provided to Arianespace; 3) the French national space agency subsidises the administrative and technical personnel it provides to Arianespace; and 4) Arianespace's mission insurance rates are subsidised. In addition to

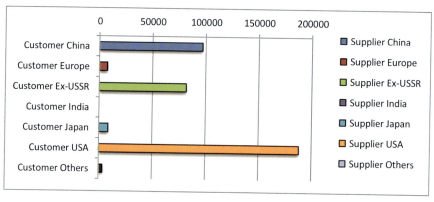

Fig. 1 Military spacecraft markets by customers and suppliers for selected areas 2013–2017 (Mass-kgs). (Source: Copyright Eurospace/Pierre Lionnet – reproduced with permission)

these allegations the U.S. also investigated three other areas: government inducements to purchasers of Arianespace's services; direct and indirect government assistance to Arianespace; and Arianespace's costs and pricing policies....Pursuant to Section 301(a) of the Trade Act of 1974, as amended (19 U.S.G. 2413(a)). I have determined that the practices of the Member States of the European Space Agency (ESA) and their instrumentalities with respect to the commercial satellite launching services of Arianespace S.A. are not unreasonable and a burden, or restriction on U.S. commerce. While Arianespace does not operate under purely commercial conditions, this is in large measure a result of the history of the launch services industry, which is marked by almost exclusive government involvement. I have determined that these conditions do not require affirmative U.S. action at this time. But because of my decision to commercialise expendable launch services in the United States, and our policies with respect to manned launch services such as the Shuttle (STS), it may become appropriate for the United States to approach other interested nations to reach an international understanding on guidelines for commercial satellite launch services at some point in the future. (Reagan 1985)

In terms of trade patterns of space goods, for spacecraft and launch vehicles, export performance is led by OECD countries and reveals upward trends as more and more countries become involved in space technologies and applications initially through trade. Figure 2 from OECD 2019 compares exporting snapshots for 2002, 2010, and 2018 illustrating relative export positions of groups of countries (constant USD) and revealing positive trends.

With reference to the breakdown of export performance by country in space goods, France appeared the leading exporter in 2018 data with nearly 28% of the total, followed by China (22%), the USA (20%), Japan and Germany at about 8% each, and Israel at nearly 6% (OECD 2019, using ITCS classification 7925 – spacecraft and spacecraft launch vehicles). For time comparison purposes, it is worth noting that the HS 880250 figures reveal that in 2017, the US exports were

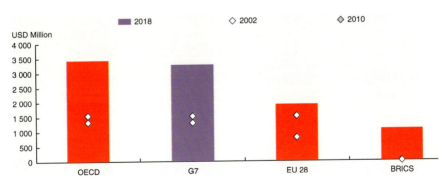

Fig. 2 Space export performance of selected areas. (Data source: OECD 2019: 31). Notes: OECD figure is based on UNCOMTRADE database and HS classification coding 880260 that is comprised of spacecraft and suborbital and spacecraft launch systems. (Direct from the UNComtrade database data is unavailable for HS 880250 (spacecraft and launch systems – without inclusion of suborbital vehicles), and there may well exist discrepancies across "as reported" data, with the values for EU28 as a group reported lower than the sum of the countries (no clear indication of consolidation), as well as for SITC 7925 ("as reported" for the selection of countries like USA, China, France, and others for 2018))

slightly higher (USD 1.48bn) than the ones of France (USD 1.72bn) highlighting the volatility of the time series data in view of the institutional nature of trading partners and limited overall values involved (Data source: UNComtrade database).

Europe and France have historically focused on an export-led model on the grounds that the "domestic" market size is too small to support an efficient size of operations of the aerospace industry in general (owing to the economies of scale and scope) compared to countries like the USA.

Trading performance is generally expected to be linked to budgetary appropriations that support and develop space technologies of domestic industries. This is owing to the fact that the space industry is subject to some traditional market failures since the early days such as:

- *Public goods*. The two characteristics of public goods, jointness of supply (zero marginal production costs) and non-excludability are found in the provision of goods, such as national security and defense based on strategic space capabilities. Prominent examples are the space-based defense applications of earth observation (EO), navigation, and telecommunications, vital elements of national defense capabilities of NATO and Warsaw Pact countries during the "Cold War." Despite variations in the supply of such goods across the NATO alliance, for the country that owns space assets with defense and security capabilities, the respective benefits accrue to all its citizens. Beyond the public goods nature of space services by nature of the service itself, there is also public good by convention. An example of this is the geo-navigation and positioning systems, whereby the USA offers a free to users signal along with its protected (encoded) military signals. This means that there is no rivalry and non-excludability (by convention) to the users (Zervos 2018).

- *Natural monopolies*. Where there is decreasing costs, production mode of a good and a sole provider reduces duplication (see later) and is also desirable for security reasons. The use of decreasing costs arguments has been used in the provision of space-based goods by "natural monopolies," such as telecommunications services prior to the recent privatization of telecommunication organizations in Europe. Prominent examples are the case of the international telecommunications satellites organization (INTELSAT) and international maritime satellites organization (INMARSAT). INTELSAT, the first civil global communications network, was created in 1964, followed by the launch of the "Early Bird" satellite. Prior to the end of the "Cold War" and the commercialization of major telecommunications service providers in Europe and Asia, it was implicitly agreed by its members that INTELSAT is a "natural monopoly" (Snow 1976, 1987a, b; Rostow 1968) and a procedure has been determined (Article XIV (d) of the Intergovernmental agreement) to establish whether other commercial systems would inflict any "significant economic harm" to its operations. The rapidly expanding size of the global telecommunications market and the large-scale privatization of telecommunications in Europe and the USA led to uncertainty surrounding the continuous non-rivalry to INTELSAT by private systems.

Following increasing competitive pressures, INTELSAT's assembly of parties (shareholding organizations, commercial, civil, and national) decided to restructure the organization and ultimately commercialize it by July 2001. Despite the growth in telecommunication service providers, the significant market concentration and network-economies associated with satellite systems signify oligopolistic markets with economies of scale and scope in operations. This is expected to extend to planned mega-constellations of satellites leading to contested and congested orbital space with potential overcapacity. Overcapacity and the ability to replenish basic operations in case of accidents or hostile operations provide first-mover advantage to space-faring nations whose industry will operate such constellations (See Commercial Space as a Space Race Catalyst). (See Rostow (1968) for an early policy-economics analysis and discussion on newly developed challenges from space-based telecommunications. Key points identified relate to avoidance of concentration of power and control to the hands of the government or the industry, while recognizing the natural monopoly challenges. The X-inefficiency (innovation factor) and power dissemination forces are arguably operating against the single-natural monopoly solution in a static decreasing-costs sector.)

- *Uncertainty, risk.* The risk associated with new space technology was and is high. As technologies for some space products mature the risk associated with such applications decreases and commercial markets are more willing to finance and support them. Telecommunications organizations and firms can thus afford to insure against a variety of risks the satellite and the launching process are subject to.
- *Externalities.* Positive production and consumption externalities, as well as spin-offs. Social cost avoidance from using, for example, geo-positioning systems like GPS on a popular basis saving on consumers' transportation costs, congestion and pollution costs, and others. Spin-offs from space technologies and production cost reduction for industries employing precision in location, but also timing information, along with technical progress associated with space goods (Zervos and Siegel 2008).

As a result of significant market failures, especially with regard to the positive externalities, public goods nature, and security considerations associated with production and control over national space assets, space-faring nations develop indigenous capabilities reflected in budgets. In view of the non-tradable nature of most space goods, international comparisons based on monetary value pose challenges of accurate reflecting relative positions and capabilities.

In this context, budget information such as OECD (2019) provides relevant estimates (Fig. 3) of relative budgetary appropriations indicating prioritization that are useful in indicating relative country positions rather than absolute numeric comparative statistics. (Clearly the downside of using percentage ratios is that small-sized budgets may well lead to low overall scale of programs and industries for smaller countries.)

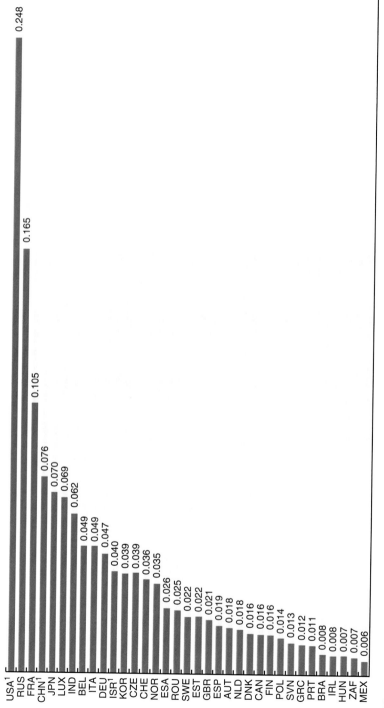

Fig. 3 Space spending as a percentage of GDP. (Source: OECD 2019: 24)

Trade Balance Effects and Atypical Patterns of the Aerospace and Defense Sector

Space versus aerospace when considering trade patterns overall performs quite interestingly. In aerospace, especially civilian commercial airliners, the dominant companies (namely Airbus and Boeing) are largely unchallenged and their export performance act as balancing forces in otherwise unbalanced trade patterns, especially with countries like China, space (military) as we saw has a different behavior. Figure 4 reveals that the USA experiences significant overall trade deficits that are persistent in recent decades.

The economics and overall rational of this global trade pattern is a point of controversy in terms of policy and trade agreements on a global scale recently, yet it is worth noticing that aerospace acts as a balancing sector with regards to specific countries such as the USA and China. Figures 5 and 6 reveal that the USA's aerospace balance has the opposite trend of the overall trade balance, mitigating somewhat overall deficits.

As Fig. 7 further illustrates, US (aerospace) exports are specifically significant with regards to China, which specializes in non-aerospace sectors in terms of export performance.

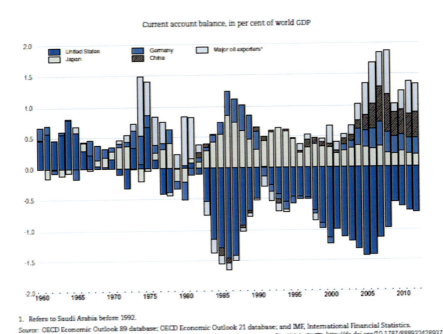

Fig. 4 Global trade patterns. (Source: OECD database)

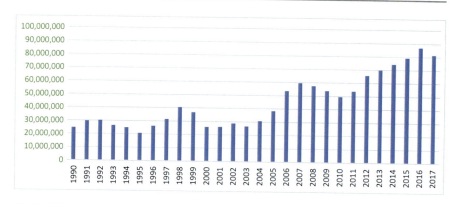

Fig. 5 US balance of trade for aerospace goods. (Source: OECD database)

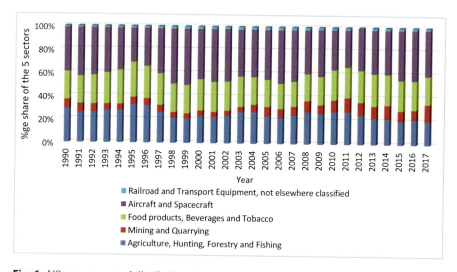

Fig. 6 US exports sectoral distribution. (Data source: OECD database)

The Chinese export performance is gradually shifting towards the aerospace sector, though the trade specialization overall is focused on alternative industries (Fig. 8).

The same is seen when considering other Western nations like France and the UK that are traditionally considered as underperformers in trade balances. By examining trade balances overall and comparing them to the sector of aerospace goods and services reveals that for major trade performers like Japan, China, the aerospace sector acts as a balancing element compared to countries such as the USA and France (Figs. 9, 10, 11, and 12).

Clearly, specialization plays a key role in trade patterns. In addition though, political aspects, such as heightened tensions and sanctions, seem to impact upon trade patterns and benefit (in relative terms) the aerospace industry. The trade pattern

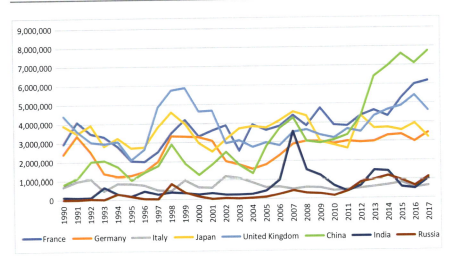

Fig. 7   US exports in aircraft and spacecraft (constant values). (Data source: OECD database)

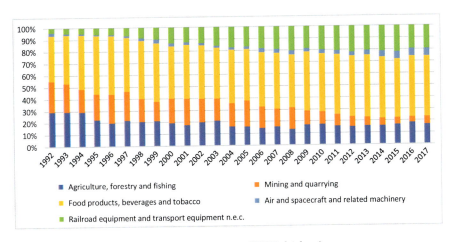

Fig. 8   Chinese exports specialization. (Data source: OECD database)

changes from the USA and France to Russia illustrate this in Figs. 13 and 14 as rising tensions and sanctions seem to result in a notable increasing aerospace share of overall export performance.

Export restrictions play a significant role in trade patterns and reveal a significant trade-off between short-run sales and maintaining of technological edge as perceived by dominant powers like the USA. Thus, the domestic "demand" for space-related security is very high in the USA, but the same can be said of other major space powers as they themselves invest in space unilaterally when it comes to critical space technologies, while they also do not release technologies on par with other (exporting) sectors. This is compatible with the dual-use, as well as the security strategic nature of space. Space capabilities are conceivably on-par in

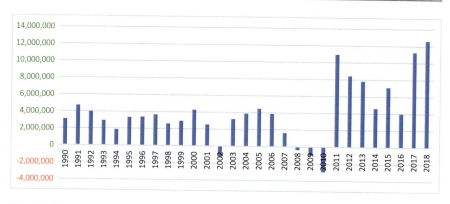

Fig. 9 Balance of trade in aerospace goods for the USA. (Data source: OECD database)

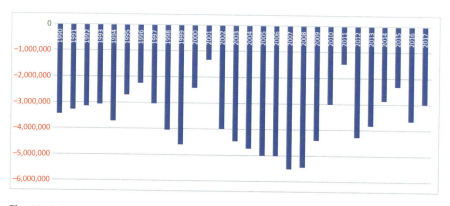

Fig. 10 Balance of trade in aerospace goods for Japan. (Data source: OECD database)

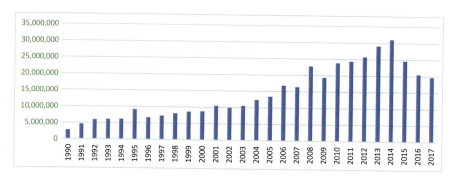

Fig. 11 Balance of trade in aerospace goods for France. (Data source: OECD database)

terms of their security special "weight" with nuclear weapons but not bound by any treaties in terms of their development. The aerospace goods and services trade patterns are clearly not applicable in the case of military satellites as seen earlier in Fig. 1.

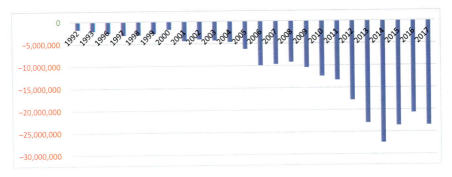

Fig. 12 Balance of trade in aerospace goods for China. (Data source: OECD database)

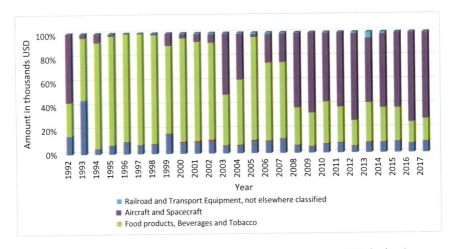

Fig. 13 Sectoral breakdown of French exports to Russia. (Data source: OECD database)

Industries subject to economies of scale and scope are long considered as strategic on the grounds that they exhibit significant industrial consolidation, frequently resulting in national champions that may collaborate in multinational institutional markets either at the industry or at the government level, while at the same time compete in commercial markets. Figure 15 captures such a framework for the space sector (not unlike the wider aerospace and defense sectors) and the resulting formation of institutional-industrial complexes.

In its simple form, such a framework is depicted by a structure whereby a national industrial champion exists (largely owing to the economies of scale and scope cost characteristics) that must also provide a level of national security in autonomous provision of security-sensitive goods. Such national champions face domestic monopolistic markets, while compete in commercial markets of an international nature.

A further element in this analysis is the nature of the "domestic" government space good or service (defense/security), as they may well constitute in the presence

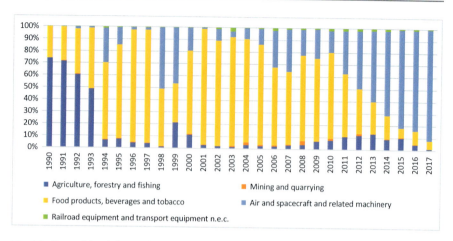

Fig. 14 Sectoral breakdown of US exports to Russia. (Data source: OECD database)

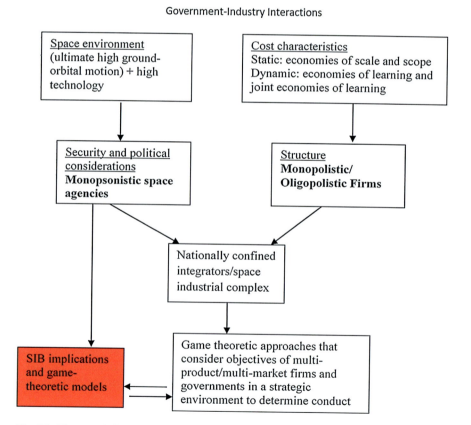

Fig. 15 The space industrial complex analytical framework. (Source: Zervos 2018)

of an alliance public goods. Public goods within an alliance or a research organization are complex goods and services. For example, "security" within NATO would refer to more than a single technological asset (ICBMs) but would require a network comprised of supportive nodes (including assets like GPS, intelligence, logistical support, tactical weapons, NATO localized basis, etc.). Thus, the main provider is seen more as an integrator and less as a vertical integrated entity. There are clear resemblances with supply chains and production technologies where "subcontractors" or lower tier suppliers exist and contribute at different levels. The distribution of alliance member contributions in the presence of economies of scale and scope is of paramount importance for the allocative efficiency.

Furthermore, following an extension of the Balassa-Samuelson framework, where a competitive tradable sector affects the relative wages of the non-tradable sector, tradable-sector specialization and development can have multiplier effects across the economy. This is not only in the presence of global supply chains but also in the case where the specialization follows a collaborative negotiated and agreed approach across partners, as is the case of alliances (NATO), or joint organizations like the ESA. Zervos (2011) shows how in strategic industries where economies of scale and scope co-exist with a tradable and governmental non-tradable sectors (defense), the non-tradable sectors are not only interconnected but can also have an unexpected impact upon the performance of the tradable sector by perverse incentives. That means that the economy with the cost advantage in the non-tradable sector may not see this advantage extended into the tradable sector, since rent-seeking behaviors prevail. This leads to an introverted focus of firms to their domestic lucrative market, rather than the more competitive global one, even though the country enjoys a theoretical cost advantage should its industry decide to capture the global market.

Wider Economy

Significant externalities and hard to measure security benefits, but also linkages of industry create an atypical environment with significant market failings, but also potential crowding-out owing to the flag-carrier and prestige impact space enjoys.

Previously examined trade balance effects thus do not take into consideration the expected greater impact of space sector manufacturing in terms of spin-offs and multiplier effects, compared to other economic activities from services and elsewhere. In addition, successes in space that have an uplifting impact on society and are associated with flag-carrier effects are hard to monetize, but potentially can have a critical impact on future generations and science and technology developments through time. Thus, it is unsurprising that there have been several attempts to quantify the economic return of space activities, either at the business-planning level for commercially minded endeavors and public-private partnerships or at the micro- (sectoral) and macroeconomic levels (Hertzfeld 2002). There are evidently significant methodological challenges to be considered, notwithstanding the "client commissioned" approach that poses additional challenges to the uncritical adoption of the results. This has been evidenced since the early US macroeconomic studies by US institutions questioning the relevant methodologies insofar as the value of the findings is concerned. Moreover, the partial nature of the analysis typically falls short of addressing the opportunity cost of appropriations devoted to space. (For a review, see Zervos (2002); regarding business studies optimism, a relevant example guiding public policy in Europe is the case of Galileo (Zervos and Siegel 2008).) On the other hand, the highly important security aspects are by-nature hard to monetize, despite the expected high value to economic activity and society.

Commercial Space as a Space Race Catalyst

The economic background indicates how the commercial markets connect seemingly unrelated monopolistic domestic markets (Zervos 1998). The conduct and performance in commercial markets therefore impacts upon specialization dynamics within security/military alliances in a dynamic environment. Thus, government support for commercial performance may well signal conflict and competitiveness at government and security-levels, especially within alliances. It is noteworthy to observe that trade specialization is thus not unrelated to alliance and geopolitical frameworks. The commercial "new space" environment and aspirations were initially experienced during the first wave of commercialization and following the end of the Cold War (peace divided applied to space) and led to sobering results (Iridium/mobile space telecommunications), following optimistic market projections of a similar nature to the market estimations with regard to the Galileo system. The newly created space companies, with origins to be found in the support by wealthy space enthusiasts like Branson, Musk, Bezos, and others, utilize capital to develop space technologies with the support of the public sector, but importantly, also seemingly rely on such support as a customer base. The commercial sector is thus government-dependent and nationally bound, perhaps more so than earlier alliances like SeaLaunch. The openness of the post-Cold War in industrial partnerships and exploration of space is therefore replaced by nationally confined considerations whereby the newly sprung enterprises compete on grounds of efficiency with publicly listed companies, focusing on innovative improvements, cost performance, and market share.

Commercial considerations are frequently evoked to equate success at the market share level (measured by market dominance and turnover), rather than a mature, government-independent industry with sound business investment (commercial) criteria. Thus, government-induced investments aiming at generating market revenue may prove to follow a nationally or geographically confined creation approach. The partnership approach utilized in other industries to avoid duplication in R&D and exploit economies of size appears heavily constrained. "Galileo, for example, is likely to have progressed at a much smoother political route, fastest pace and lower costs had the US participation been possible at contractor's level, in exchange for European industrial involvement on US relevant space projects" (Zervos and Siegel 2008).

Political Economy and Security in Space: Institutional Dimensions

Assuming that the labeled commercial space programs act as an enhancer of conflict in space, they may hinder rather than promote security even with regards to conventional space collaborative programs like NEO threats. This is owing to the fact that the impact of government programs on the commercial competitiveness of participating industry enhances transaction costs and rivalry. Thus, even though rivalry may benefit the industry through duplication of capabilities and globally increasing size and public spotlight (an impact identified since the days of Sputnik-Apollo), it may also constrain the short-run development of projects of common security interest. (In addition, such rivalry may well impact also upon collaboration in flagship programs like the ISS, as well as scientific and exploration ones.) The collapse of the Soviet Union and the resulting hegemonic position of the US diminished the rivalry perceptions between the USA and Russia allowing for a number of commercial and public collaboration, in contrast to the heightened commercial competition levels between the USA and Europe at the time. The rise of China, India, Japan, and others as they develop significant capabilities may well result in the future in relevant initiatives, but it seems that a rules of the road approach for the commercial markets and the conduct of industry/providers in these is a critical element for global sector developments in the future. Clearly, the World Trade Organization (WTO) may be a challenging environment for such developments, hence further research and novel approaches may be developed towards enhancing sectoral growth and security simultaneously.

Even though the WTO is of perhaps lesser concern to space, organizations like the ITU are critical in dealing with spectrum allocation mechanisms as they evolve into continuous liberalization from an initial equitable basis of *global commons* good are more space-focused in their agendas. Thus, the abandoning of the INTELSAT model of globalization of telecommunications and allocation through an early global partnership and governance mechanism into a private enterprise-oriented one is implemented through evolving mechanisms and licensing of orbit and spectrum allocations. The regulatory role is thus increasingly supplemented by heightened monitoring requirements requiring novel institutional arrangements as space is becoming more congested and contested in Earth orbit, but also in potential future exploration specific locations such as Langrangian points of interest, lunar hotspot ones, or Martian and other celestial bodies.

The explosion of small-satellite constellations largely associated with the advent of global telecommunication services and mobile data leads to multiple levels of contestability and congestion starting from higher atmospheric through to high orbital planes. (The term is used in this context to describe low mass satellites (less than 500Kg mass) that include what is habitually defined as nanosatellites, picosatellites.) The situation is perhaps analogous to the pre-digitization era of remote sensing satellites based on film technology that were used by the USA and the then USSR to observe the Earth and their lifetime extended to the capacity of the film that needed retrieving at the end of each satellite mission (and its end of life), thus resulting in a continuum of launches and disposable film-containing satellites. Longer lifetime, but small satellites requiring replenishment, and/or servicing are planned to support 5G and other telecommunication systems. As Fig. 16 reveals, there are planned constellations that add up to over 10,000 new smallsats all of which require their vital space in terms of physical proximity of operations and spectrum/interference topology, mitigation, disposal, and replenishment plans that are harmonized on a universal level and require situational awareness and regulatory evolution. In case of business failures, or abandonment of plans, such operational costs and legal challenges may lead to non-simple solutions requiring governments being able to support such instances when licensing and hosting relevant businesses. (It is evident that such industrial and network replenishment national industrial capacity that is economically sustainable is welcomed by the security agencies for countries like the USA. It would be significantly more costly to maintain industrial capacity without exploiting the relevant scale and scope, producing just for military/security purposes along the lines of Operationally Responsive Space (ORS), developed largely in response to the Pearl Harbor scenarios for space assets (Commission 2001).)

In addition, this new trend is placing a burden upon the ITU's role with physical and spectrum space becoming scarcer and therefore adding the dimension of placeholding, moral hazard, and speculator behaviors. In response, the ITU has tighten up the licensing conditions and implementation requirements by relevant

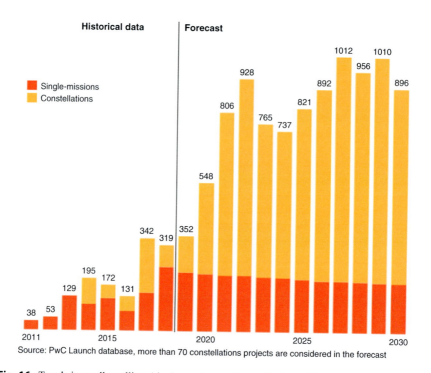

Fig. 16 Trends in small satellites (single mission and constellations). (Source: PwC 2019)

enterprises to avoid low-cost reservation of space with lengthy implementation and inactivity periods. The ITU is also faced with scarcity of spectrum as 5G systems and new technologies offer great opportunities for users but also pose challenges on the space/airspace/terrestrial architectures and allocation (as well as uses between telecommunications, science and exploration, and others). The ITU is thus planning following the WRC-19 milestone to establish harmonizing conditions for this multidimensional architecture to enhance order and efficiency in view of the congestion challenges (Fig. 17; Henry 2019; ITU 2015, 2016).

The global reach of such "utility" services raise the issue of governance of networks, safety, and security standards and control of operations. The private nature of such global utilities and the different legal regimes between users and operators are expected to exert force towards multi-stakeholder partnerships, or quasi national autonomous systems. Thus, despite potential economies of scale and scope in operations, it is likely that again there will be overcapacity and multiple systems in place augmenting harmonization and efficiency challenges. A critical challenge, though, for future security lies ahead with regards to the utilization of resources, as technical capabilities increase and disseminate with a rising number of space-faring nations coupled with challenges to the status quo of space resources utilization. The partial application of an ITU-like model for specific space cases may be an option, but it is not clear whether such an option would gather the analogous momentum at global level when dealing with a variety of resources (besides telecommunications/GEO). Moreover, space traffic management faces significant challenges, when compared with the tested aerospace/FIR applications and operability, that may though nullify as technical expertise disseminates.

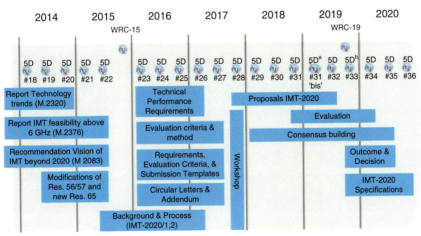

Fig. 17 ITU timeframe for regulating 5G. (Source: ITU 2016)

Conclusions

Outer space security is a multidimensional concept; the most usual direct application is found in near-Earth objects, as well as other natural threats (solar storms and others). However, as humans increasingly explore, utilize, and compete in space, the man-made security challenges are evolving and the strategies and political economic rationales become increasingly relevant for analysis. Sustainable industries and efficiency call for exploitation of static economies of scale and scope in space industries and services, yet the trade-offs in control, governance, and dynamic innovation point towards autonomy and oligopolistic structures with overcapacity. The economic sustainability becomes a key element of the dynamic pursue of space policies and objectives at national and partnership levels. In the latter case, specialization and its implications for the wide economy through externalities and indirect effects receive increasing attention as space becomes contested, congested, and competitive. Space and aerospace industries play a crucial role in trading patterns, notwithstanding the fact that they are largely government controlled, hence can be considered as a fiscal government spending element similar to defense expenditure. The country specializations and their evolution in commercial markets and alliances are focal points in the timely global trade policy paradigm shifts, affecting performance and evolution of space programs and industries. The chapter concludes with the ever-increasing role of regulation and relative power balances across nations, companies, and terrestrial-air-space systems especially for telecommunication applications.

References

Butler A (2015) USAF operationally responsive space office could oversee next SSA, Weather Sats. Aviation Week, February 12. http://aviationweek.com/space/usaf-operationally-responsive-space-office-could-oversee-next-ssa-weather-sats

Commission (2001) Report of the Commission to Assess Unites States National Security Space Management and Organization. Commission to Assess United States National Security Space Management and Organization, Washington, DC

DoD (2018) Summary of the 2018 National Defense Strategy of the United States of America. Unpublished, Department of Defense, US

Graziola G, Cefis E, Gritti P (2011) LÍndustria Spaziale Italiana nel Contesto Europeo. IL Muliko, Bologna

Harford J (1997) Korolev: how one man masterminded the soviet drive to beat America to the moon. Wiley, New York

Henry (2019) ITU wants megaconstellations to meet tougher launch milestones. Spacenews. https://spacenews.com/itu-wants-megaconstellations-to-meet-tougher-launch-milestones/?fbclid=IwAR0VpMhgrwveNmii_mNkC2a8inzFnG9Lhc_aspFLgAG_RoPcD4W6AsC5pII. Accessed July 2019

Hertzfeld HR (2002) Technology transfer in the space sector: an international perspective. J Technol Transf 27(4):307–309

HoC (2017) Space sector report. House of Commons Committee on Exiting the European Union, London

ITU (2015) Improving the dissemination of knowledge concerning the applicable regulatory procedures for small satellites, including nanosatellites and picosatellites, Resolution ITU-R 68. ITU, Geneva

ITU (2016) Workplan timeline process and deliverables for the future development of IMT. Unpublished Document, International Telecommunications Union, IMT-ADV/2 Working Group TECH, Meeting 24 June, ITU, Geneva

OECD (2019) The space economy in figures: how space contributes to the global economy. OECD Publishing, Paris. https://doi.org/10.1787/c5996201-en

PriceWaterhouseCoopers (2019) Main trends and challenges in the space sector. PWC, Paris

Reagan R (1985) Determination under Section 301 of the Trade Act of 1974, memorandum for the United States trade representative. The President of the US, White House, Washington, DC

Rostow E (1968) President's task force on communications policy, final report. Superintendent of Documents, US Government Printing Office (GPO 0-351-636), Washington, DC

Snow M (1976) Communication via satellite – a vision in retrospect. A.W. Sijthoff International Publishing Company, Boston

Snow M (1987a) National monopoly in INTELSAT: cost estimation and policy implications for a separate system issue. Telemetics Inform 4(2):133–150

Snow M (1987b) An economic issue in international telecommunications: national monopoly in commercial satellite systems. In: Macauley MM (ed) Economics and technology in space policy. R.F.F., Washington, DC

USAF (2013) Global horizons final report. United States Air Force, SAF/PA Public Release case no. 2013-0434, June, US

Zervos V (1998) Competitiveness of the European space industry, lessons from Europe's role in NATO. J Space Policy 14(1):39–47

Zervos V (2002) The economics of the European space industry. DPhil thesis, University of York

Zervos V (2011) Conflict in space. In: Braddon L, Hartley K (eds) Handbook of the economics of conflict. Edward Edgar Publishers, Cheltenham

Zervos V (2018) Vasilis Zervos: "public goods, club goods and specialization in evolving collaborative entities". In: Vliamos S, Zouboulakis M (eds) Institutionalist perspectives on development – a multidisciplinary approach. Palgrave Macmillan, Cham

Zervos V, Siegel D (2008) Technology, security and policy implications of future transatlantic partnerships in space: lessons from Galileo. Res Policy 37:1630–1642

Views on Space Security in the United Nations

76

Massimo Pellegrino

Contents

Abstract

As a forum for discussing international approaches to global challenges, the United Nations (UN) attaches great importance to space security. While no single venue addresses all aspects of space security, multiple UN bodies have become essential multilateral fora for discussing these issues. Against this backdrop, this chapter offers a definition of space security that reflects the different connotations with which this term is used within various UN entities. It then reviews and analyzes the role that relevant UN bodies and fora play in

M. Pellegrino (✉)
Vienna, Austria
e-mail: massimo.pellegrino@community.isunet.edu

© Springer Nature Switzerland AG 2020
K.-U. Schrogl (ed.), *Handbook of Space Security*,
https://doi.org/10.1007/978-3-030-23210-8_76

the international discourse on space security, with particular regard to the work being made within both international security (UNGA 1st Committee, CD, UNDC, UNODA, UNIDIR) and space (UNGA 4th Committee, COPUOS, UNOOSA) settings of the UN. The chapter concludes that the diverse initiatives discussed within the different UN entities indicate the increasing pressures facing the international community in addressing all aspects of space security; that substantial differences still exist among states over priorities, methodologies, mechanisms, and settings to tackle key space security challenges; and that the existence of political will that accommodates, rather than eliminates, these differences can prove effective in finding common ground for shared action.

Introduction

The concept of space security features prominently in current discussions and debates pertaining to both international security and disarmament, as well as peaceful uses of outer space. It has become a widely used concept, often without a common understanding. While efforts abound, little success has been achieved in developing an accurate, commonly accepted definition of space security or a consensus understanding of the concept that provides a meaningful framework for evaluating what it encompasses, which factors contribute to and detract from it, or agreed upon metrics to measure it (a positive exception is the Space Security Index project, http://spacesecurityindex.org). There is even a lack of clarity on its constituent elements: not only is there as yet no universally accepted definition of outer space under international or domestic law, but the terms security, safety, and sustainability (and stability and predictability, too) are also often used interchangeably, partly because their boundaries are blurring (Box 1).

> **Box 1: Etymology and Strict Meaning of Space Security, Safety, and Sustainability**
>
> **Space Security**
>
> Security derives from Latin *securitas -atis* (noun) or *secūrus* (adjective), composed of the privative *se-* "without" and *cură* "care" (but also "worry," "anxiety," "preoccupation," "concern," and so "threat") or *caveo* "to pay attention to," similar to Latin *sine cūrā* "without care, carefree" which led to English sinecure. The condition of "security" is about being free of threats (i.e., deliberate hostile actions). Strictly speaking, space security describes a state of affairs in which space systems are free of threats.
>
> **Space Safety**
>
> Safety derives from Latin *salūs -ūtis* (noun) or *salvus* (adjective), which literally means health and well-being (noun) or healthy, unhurt, and unharmed

(continued)

> **Box 1:** (continued)
> (adjective). The condition of "safety" is about not being harmed. Space safety describes a state of affairs in which space systems are sufficiently protected from, or not exposed to, manmade and natural hazards (i.e., unintentional actions). The difference between being protected from hazards and being free of threat is not always easily seen.
>
> **Space Sustainability**
> Sustainability derives from Latin *sustinēre* (verb), composed of *sus-*, variant of *sub-* "underneath," and *tenēre* "hold," with the meaning of "to hold up," i.e., preventing a system from collapsing. Sustainability is a concept that is generally being understood as the ability of carrying out a certain activity in the present while safeguarding the surrounding environment for the future, with particular regard to the resources therein contained, which are scarce, limited, and not always renewable. Space sustainability thus refers to a situation in which outer space will continue to be accessible to future generations on an equitable basis and its resources still available for meeting societal needs.

The rapid pace of change in both the space and security domains risks making a definition obsolete or outdated. This problem is also compounded by the different interpretations (not always publicly available) that space actors, especially spacefaring nations, attach to space security. For some actors, it is primarily linked to the potential for space weaponization and an arms race in outer space, while for other actors, space security is being concerned with the development of direct-ascent or co-orbital antisatellite (ASAT) and other counterspace capabilities (e.g., kinetic or directed energy weapons, electronic and cyber warfare, etc.); contested space operations; creation and proliferation of space debris (which also poses dangers for destructive collisions); increased congestion of (strategic) orbits; uncontrolled satellite reentries; competition for and saturation of limited natural resources such as orbital slots and radio-frequency spectrum; unintentional radio-frequency interference; the proliferation of new space technologies and dual-use capabilities (e.g., active debris removal and on-orbit satellite servicing above all); or space weather phenomena.

Achieving a mutual understanding of the different facets of space security can thus be difficult, but it is a prerequisite not only to understand how space security issues are dealt with and addressed in the most important multilateral forum for states – the United Nations (UN) – but also to develop common approaches and responses to this global challenge.

Defining Space Security

Space security is a complex term that can have several meanings dependent upon the context of its usage. Attempts to precisely define it are relatively recent and need to be contextualized (▶ Chap. 2, "Definition and Status of Space Security"; Sheehan 2015). Regardless of how one defines outer space, space security can generally be understood as being concerned with the absence of any threats to space assets. It refers to a state of affairs in which space actors are free from third party's actions and interferences preventing them from effectively accessing and using outer space. This preliminary definition is based on a general understanding of the term security applied to outer space and does not specifically take into account how the politics of space security have morphed since the beginning of the space race.

During much of the Cold War, security had a very specific connotation, since it was referred to military threats that could undermine the very existence of a state, notably nuclear strikes, conventional armed forces of adversaries, or even internal uprisings to overturn the *status quo*. Space security was primarily addressed through a military lens, in terms of both the contribution of military satellites for security purposes on earth (e.g., the maintenance of strategic stability above all) and the military threats to space systems and the risks associated with certain uses of outer space (e.g., earth-to-space and air-to-space missiles, explosive kinetic co-orbital systems, high-altitude and low outer space nuclear explosions).

Today's space and security realities have morphed however. Security threats go beyond purely military ones and embrace several aspects. Similarly, the number of players interested in space capabilities has grown. Satellites serve a wider audience than the military and are used for an ever-increasing range of non-security applications providing a wealth of socioeconomic benefits.

This expansion in space activities has resulted in a heightened focus on security and has led to new conceptions of space security that focus not only on military matters but also on how to reduce risks of any kind to the space infrastructure and ensure that space operations can be safe, secure, and sustainable in the long term (Pellegrino and Stang 2016). There is thus an alignment of interests between the space and security communities (and the general public, too) to pursue a wider space security discourse for preserving the sustainability and stability of the space environment, while ensuring protection and freedom from hazards and threats to the effective access to, and use of, outer space (Pellegrino et al. 2016). While military space actors will continue to influence the international discourse on space security, for those actors without major military reliance on space, space security is primarily related to the absence of any risks and dangers to space systems and the maintenance of outer space as stable and sustainable environment. The premise is that not only does space security concern traditional and emerging space actors but also a plethora of individuals who benefit from the services enabled by space-based capabilities.

Space Security in the UN System

The UN remains the primary multilateral forum to discuss and pursue initiatives on space security. While, as of today, there is no single venue for addressing all aspects of space security, multiple UN bodies have become essential multilateral fora for discussing these issues, each with their own specific mandate and remit (Fig. 1). These entities are spread between New York (United States), Geneva (Switzerland), and Vienna (Austria) and include the following:

- United Nations General Assembly (UNGA), particularly its First Committee (Disarmament and International Security) and Fourth Committee (Special Political and Decolonization) in New York.
- United Nations Disarmament Commission (UNDC) in New York.
- Conference on Disarmament (CD) in Geneva.
- United Nations Office for Disarmament Affairs (UNODA) in Geneva and New York.
- United Nations Institute for Disarmament Research (UNIDIR) in Geneva.
- United Nations Committee on the Peaceful Uses of Outer Space (UNCOPUOS) in Vienna.
- United Nations Office for Outer Space Affairs (UNOOSA) in Vienna.

The First Committee of the UNGA, which meets annually in October, is responsible for international security and disarmament matters and is where the Conference on Disarmament reports on its work, including on space security issues. The CD is the single multilateral disarmament negotiating forum of the international

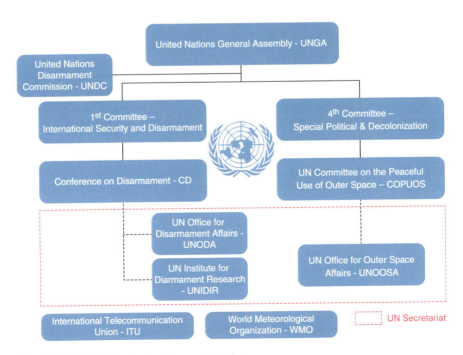

Fig. 1 Main UN bodies involved in space security

community and is supported by UNODA, which serves as its secretariat and provides support for norm-setting in the area of disarmament. The UNDC is an open-ended deliberative body and subsidiary organ of the UNGA and provides guidelines and recommendations on various disarmament-related issues, including space security. The UNIDIR is the UN's in-house think tank providing independent analysis and research on a wide range of international security issues and is well known for its annual space security conference providing a 1.5-track forum for dialogue on the policy and politics of outer space security at the multilateral level.

The Fourth Committee of the UNGA, which meets annually in November, deals with special political and decolonization issues, including outer space, and is where the UN Committee on the Peaceful Uses of Outer Space reports on its work. The UN COPUOS is the primary international forum for the development of rules and laws governing activities in outer space. It has a purely civilian focus and does not work on military or weaponization issues. Its work is supported by UNOOSA, which serves as its secretariat and promotes international cooperation in the peaceful use and exploration of space and in the utilization of space science and technology for sustainable economic and social development.

In addition to this, within the UN system, the International Telecommunication Union (ITU) and the World Meteorological Organization (WMO) are involved, too, in space security-related issues. The ITU allocates global radio-frequency spectrum and satellite orbital slots and addresses harmful radio-frequency interference. The WMO has been coordinating international efforts for space weather activities – an area of growing importance for the security community since space weather affects the continuous availability of critical space systems. Under the umbrella of the UN, the International Committee on Global Navigation Satellite Systems (ICG) has been working since 2006 to facilitate compatibility, interoperability, and transparency between all of the satellite navigation systems, including on aspects such as GNSS spectrum protection and interference detection and mitigation.

Representatives from these and other UN entities meet every year at the United Nations Inter-Agency Meeting on Outer Space Activities (UN-Space) to enhance coordination of space-related activities within the United Nations system, including on space security issues. An example of deliverable within this framework includes the UN-Space special report on the implementation of the report of the Group of Governmental Experts on Transparency and Confidence-Building Measures in Outer Space Activities (A/AC.105/1116), which was submitted to COPUOS in June 2016.

Multilateral discussions on space security take place in all of these venues which, for argument's sake, can be classified into two main categories: international security and disarmament settings and space settings. The former include the UNGA First Committee, UN Disarmament Commission, Conference on Disarmament, UN Office for Disarmament Affairs, and UN Institute for Disarmament Research; the latter include the UNGA Fourth Committee, UN Committee on the Peaceful Uses of Outer Space, and UN Office for Outer Space Affairs. The following chapters examine how space security issues are dealt with in these settings, what role each UN body plays in the international discourse on space security, which perspectives of space security are prioritized therein, and which initiatives and approaches are underway to address space security issues.

Space Security Within UN International Security and Disarmament Settings

The consideration of space security from a non-proliferation, arms control, and disarmament perspective has a long record. Attempts to protect the outer space environment and prevent indiscriminate harm to satellite systems, including one's own, have existed since the late 1950s, when the Soviet Union and the United States were engaged in negotiations on a nuclear test ban and refrained from the further testing of nuclear weapons in outer space as they recognized the incompatibility of such tests with both human spaceflights and military uses of satellites (Moltz 2014, p. 29). These discussions produced formal agreements prohibiting both nuclear tests and explosions in outer space (Article I of the Partial/Limited Test-Ban Treaty, 1963), as well as the placement and stationing of weapons of mass destruction (WMDs) in outer space or on celestial bodies (Article IV of the Outer Space Treaty, 1967).

In the decades since then, there have been multiple attempts to agree on additional norms and rules for the secure use of space, including through extending the Outer Space Treaty's ban on WMDs in outer space to weapons of any kind. These multilateral efforts have primarily included draft treaties for arms control, Transparency and Confidence-Building Measures (TCBMs), and (annual) UNGA resolutions, as well as talks in the Conference on Disarmament and in the UNGA and its First Committee under the agenda item entitled "Prevention of an Arms Race in Outer Space," commonly referred to as PAROS.

From the first discussions until today, there has been a permanent split between Soviet/Russian- and Chinese-led efforts toward a legal instrument that prohibits the placement and deployment of weapons in outer space and the views of most Western states according to which proposals for arms control measures would need to be comprehensive and precise and have proper mechanisms of monitoring, verification, and compliance (Pellegrino 2017, p. 9).

PAROS discussions and related proposals have thus become bogged down over incompatible national priorities. While the United States, for example, has an interest in defending its own freedom of action in, through, and from outer space – especially in the event that it becomes an actual war-fighting domain – China and Russia have long been concerned about the potential for space-based missile defense systems, which could undermine strategic stability and their nuclear deterrence capabilities (Hitchens 2015, p. 509).

However, the need for a re-energized dialogue on space security and the urgency of making progress, at least on those areas where there seems to be consensus, have encouraged the international community to find alternatives for moving forward, in terms of both fora in which to pursue space security dialogues and approaches for short- (e.g., non-legally binding, voluntary measures) and long-term (e.g., legally binding instruments) solutions.

The next sections review and analyze the efforts that have been made since the early 1980s within UN international security and disarmament settings to advance space security, favoring an approach based on functionality rather than a fixation on forum.

1980s and 1990s: PAROS and the Search for Various Space CBMs

Space security as an arms control and disarmament topic has a long history at the United Nations (Meyer 2011). Specific initiatives on PAROS have been discussed at the UN since August 1981, when the Soviet Union requested the Secretary-General to include the issue of space weaponization as a supplementary item in the agenda of the UN General Assembly (A/36/192). At its 36th session in December 1981, the UNGA passed two PAROS resolutions, namely, A/RES/36/97C by the Western Europe and Others Group (WEOG) and A/RES/36/99 by the Eastern European and other states. While both resolutions recognized the urgency to negotiate in the Conference on Disarmament a multilateral international agreement on PAROS, there was a substantial difference in the focus of the proposed agreements, which shows the divergence of interests and threat perceptions that has accompanied space security discussions until today. While the first resolution requested the CD to consider negotiations of an agreement to prohibit antisatellite systems, the second resolution requested the CD to start negotiating a treaty to ban the stationing of weapons of any kind in outer space.

The issue was then referred to the Conference on Disarmament. In 1982, PAROS was added as an item of the CD agenda. In 1985, after a few years of consultations, an Ad Hoc Committee of the CD on PAROS was established. Already from the first discussions in the Ad Hoc Committee, there was a rift between those states belonging to the Group of 21 and Eastern states which wanted to start negotiations on PAROS and the Western states which wanted substantive consultative work before moving to actual negotiations. Another issue for disagreement became the focus of the work of the Ad Hoc Committee. The majority of states noted that the existing legal framework governing the use of space could not prevent the risk of an arms race in outer space and should therefore be amended, while Western states were reluctant to regulate space activities further and encouraged wider compliance with existing international agreements.

With limited progress on PAROS and on negotiating new legal instruments, attention was placed on non-legally binding, voluntary measures that focus more on the degree of care with which space activities are conducted and communicated, than on which kind of orbital systems are actually deployed in outer space (Pellegrino et al. 2016). As a result, various forms of Transparency and Confidence-Building Measures (TCBMs) for outer space activates were discussed both at the CD and UNGA.

In the CD, discussions on Confidence-Building Measures (CBMs) took place in the framework of the above-mentioned Ad Hoc Committee on PAROS and focused on the transparency of pre-launch activities, rules of the road, and other measures required for monitoring purposes. In parallel, at its 45th session in December 1990, the UN General Assembly passed resolution 45/55B, requesting the Secretary-General to appoint a Group of Governmental Expert (GGE) to carry out a study on the application of CBMs in outer space. The group was made up of 10 representatives and convened four times in New York between July 1991 and July 1993. The consensus final report (A/48/305), on which some reservations were made by the US government, was submitted to the Secretary-General for transmission to the General Assembly at its 48th session in October 1993.

However, even without the pressures and expectations of legally binding arms control agreements, some divergence of views remained between those who wanted CBMs to be a stepping-stone toward legally binding treaties and those who considered them to be a valid alternative. Thus, neither did proper follow-up ensue from the work of the first GGE, nor relevant progress was made in the Ad Hoc Committee of the CD (CD/1271). Even more worrisome, the mandate of the Ad Hoc Committee on PAROS was not renewed after 10 years of work and ended in August 1994. Despite the adoption of annual UNGA resolutions on both PAROS and TCBMs and the efforts to re-establish the Ad Hoc Committee, no substantial discussions on space security cooperation took place in the late '90 and progress was prevented even due to the CD's inability to agree and implement a program of work.

2000–2015: Draft PPWT and New GGE on TCBMs – Parallel Efforts on Legally and Non-Legally Binding Instruments

With the potential for a renewal of the mandate of the Ad Hoc Committee being slim, new stimulus for discussion was offered by China and Russia. In February 2000, China submitted to the CD a Working Paper (CD/1606) on PAROS outlining, inter alia, basic elements for a new international legal instrument prohibiting the testing, deployment, and use of weapons, weapon systems, and components in outer space. In June 2002, China and Russia joined forces and submitted to the CD a joint Working Paper (CD/1679) on PAROS, outlining possible elements for a future international legal agreement to prevent the deployment of weapons in outer space as well as other threats to outer space objects.

The United Nations Institute for Disarmament Research also played an important role in facilitating dialogue through track 1.5 diplomacy and identifying potential avenues for the future. In particular, since 2002, UNIDIR has supported multilateral space security diplomacy via their annual space security conference, the report of which was for many years submitted by Canada as an official CD document.

However, it was only toward the end of the decade that new impetus was given. On 12 February 2008, after 6 years of discussions (see UN Documents CD/1769 and CD/1818), China partnered again with Russia in proposing to the CD a new PAROS legal instrument (CD/1839), namely, a draft treaty on the "Prevention of the Placement of Weapons in Outer Space, the Threat or Use of Force against Outer Space

Objects" (PPWT), an updated text of which (CD/1985) was submitted on 10 June 2014. Its declared aim is to fulfill the principle of peaceful use of outer space, extending the Outer Space Treaty's ban on WMDs in outer space to all weapons. In particular, the draft PPWT calls on adhering states not to place any weapons in outer space and not to resort to the threat or use of force against outer space objects, including through terrestrially-based weapons (Pellegrino 2017, p. 20). At present, the draft treaty does not propose any verification regime for effective monitoring and compliance, but it contains some elements for verification and resolving grievances in the form of consultative mechanisms. Critics of the PPWT, however, have noted that the proposal still suffers from a number of flaws (Pellegrino et al. 2016; see also UN Document CD/1998). For example, it ignores threats to space systems from debris-generating ASAT tests and issues related to dual-use technologies. And while prohibiting the deployment of any weapons in outer space, the proposed PPWT allows their development, production, possession, testing, use, and storage. Importantly, there are some definitional issues that need to be further addressed – the term "weapon in outer space," for example, is still an issue for disagreement given the dual-use nature of many space technologies. In its current form, the PPWT has thus been seen as a near useless arms control treaty (Pellegrino and Stang 2016). Doubts about the PPWT that have contributed to slowing international acceptance in the CD also derive from the fact that, while advancing PAROS initiatives, the proponents of the treaty are believed to develop and deploy counterspace capabilities, as well as to practice maneuvers and proximity operations that have been perceived as weapons tests (Rajeswari Pillai Rajagopalan 2017).

With limited progress on any new treaty, greater emphasis was placed once again on non-legally binding, voluntary measures. They consist of both guidelines and "rules of the road" for how to safely conduct space operations and TCBMs for how to communicate about space activities in order to enhance trust among space actors. Following discussions instigated by Russia and tabled in the UNGA's First Committee since 2005 (A/RES/60/66), in December 2010 the UNGA adopted resolution 65/68 calling for the creation of an expert group to study outer space TCBMs. The following year, Secretary-General Ban Ki-moon established the Group of Governmental Experts (GGE) on Transparency and Confidence-Building Measures in Outer Space Activities to investigate ways to improve international cooperation and reduce the risks of misunderstanding and miscalculations in outer space activities. The GGE, with representatives from 15 countries under the chairmanship of the Russian Federation, presented its Report (A/68/189) in July 2013 with recommendations for states and international organizations to implement on a voluntary basis two main types of TCBMs. The first includes information exchange on space policies and programs, space military expenditures, and registrations and orbital parameters of space objects. The second includes notifications related to outer space activities, such as launches, maneuvers, reentries, malfunctions, breakups, and emergencies. The Report also encourages states to open their space launch sites and facilities to visits, create consultative mechanisms to ensure continued dialogue, and pursue cooperation and outreach activities, including with new and non-space powers. The GGE Report and its recommendations were universally welcomed by the international space and security communities. In particular, the work of the GGE was seen by many states as a pragmatic step forward and as a model for how to quickly and effectively produce a consensus report. However, implementation of the Report's recommendations has been slow, especially by those parties which had expressed diplomatic support. A major focus going forward is how to effectively ensure compliance with the identified set of TCBMs and other recommendations contained therewith, an aspects that has more recently been jointly addressed by UNOOSA and UNODA (Pellegrino 2017).

While the GGE noted the work made in UN and non-UN fora to advance space security cooperation, the existing commitments related to non-proliferation and disarmament, and the introduction by several states of policies of not being the first state to place weapons in outer space, the Report and its recommendations do not specifically cover any issues relating to arms control in space. However, the continuing appeal of such arms control rhetoric, and a widespread concern about the potential for space weaponization, found some success in the UN and contributed to buttressing anti-Western sentiment.

Since 2014, the Russian initiative (A/RES/69/32) on "No First Placement of Weapons in Outer Space" (NFP) has indeed passed with only a few states in opposition and enjoys formal support by about 45 states as sponsors or cosponsors. The rationale behind this resolution is to establish the conditions to maintain outer space as a peaceful environment by asking states to refrain from being the first to deploy weapons in space. Its declared value lies in its politically binding nature, as it can be regarded as a TCBM that increases the level of predictability in outer space activities without needs for verification. This initiative has some similarities with the PPWT and other efforts in the CD under the PAROS agenda item. Doubts about the viability of the NFP as a measure to genuinely contribute to space security have thus been raised. For example, in addition to definitional issues about what constitutes a weapon in outer space, critics point out that the NFP initiative does not fulfill the criteria for TCBMs as developed by the GGE's consensus Report and that it may entice states to preemptively develop offensive space capabilities and place them in space once a party first breaks the agreement. Deprived of the adjective "first" – which seems to legitimize the weaponization of outer space in the event that a state introduces weapons in outer space – the NFP initiative may encounter different (and more positive) reactions. Nonetheless, the voting split in the General Assembly is worth noting when considering how the international community sees space security moving forward (during the period 2014–2018, the number of states in favor of the NFP ranged from 126 to 131, those abstaining from 40 to 48, and those in opposition from 4 to 12). These diplomatic differences are problematic and remain a critical factor in shaping the international politics and policies on the security of outer space, potentially affecting efforts to push forward and win support for other space security initiatives (Pellegrino 2017).

2015–2016: Toward a Joined-Up Approach to Space Security

The work of the 2013 GGE marked an important milestone in the context of space security cooperation, especially at the United Nations level. The GGE Report of July 2013 recommended that coordination be established between the Office for Disarmament Affairs, the Office for Outer Space Affairs, and other appropriate United Nations entities, breaking the silos that for many years had permeated the UN approach to space security. This coordination took the format of both joint ad hoc meetings of the UNGA First and Fourth Committees, as well as closer cooperation between UNODA and UNOSA, including in the framework of the UN-Space, and occasionally UNIDIR.

On 22 October 2015, the First and Fourth Committees of the UN General Assembly held a joint ad hoc meeting calling for a holistic and synergistic approach to address challenges to outer space security and sustainability. Two years later, on 12 October 2017, following adoption of resolution 71/90 of 22 December 2016, the First and Fourth Committees convened a second joint panel discussion as a contribution to the fiftieth anniversary of the Outer Space Treaty. Differently from the first meeting, this panel discussion saw a number of interventions from non-UN and non-MS representatives and brought some fresh views on the multi-faceted aspects of space security, including on commercial space and emerging technology threats to space activities (UNOOSA 2017).

While these meetings contributed to making the debate larger and reiterated the need for better overall coordination on the safety, security, and sustainability of outer space activities, they also accentuated the already extant divergences between those wishing to prioritize legally binding arms control agreements and those believing that non-discriminatory TCBMs and other conduct guidelines could provide a pragmatic way forward.

Closer cooperation between other appropriate UN entities took place, too. In particular, UNOOSA and UNODA, with contribution from UNIDIR, ITU, and WMO, prepared the UN-Space special report (A/AC.105/1116) on the role of United Nations entities in supporting member states in the implementation of outer space TCBMs, which was submitted to the COPUOS in June 2016. Another positive step in this direction was the 10th UN Workshop on Space Law held in September 2016, which saw UNOSA, UNODA, and UNIDIR partnering again (with the support of the Secure World Foundation) to discuss the contribution of space policy and law to space governance and space security.

2017–Today: New GGE on PAROS, CD Subsidiary Body 3, and UNDC WG on TCBMs – Three Parallel Initiatives for Short- and Long-Term Solutions

Although there is no evidence of a full-scale development and deployment of weapons to project force in, through, or from outer space, the latest technological developments in counterspace capabilities (Weeden and Samson 2019) orientations in military doctrines (United States Department of Defense 2019), and the continuing appeal by some states of outer space as a warfighting domain may suggest that some types of weapons can already be deployed in outer space for use in conflict. This has raised concerns in recent years over the prospect of an impending arms race in outer space and reinvigorated discussions on space security matters. The year 2017 represented an important cornerstone for the future of space security diplomacy, as states seemed to have found new source of political will to move forward.

In February 2017, the UN Secretary-General issued a report (A/72/65) prepared by UNODA on TCBMs in outer space activities, reproducing the substantive text of the special report by UN-Space (A/AC.105/1116) and incorporating updates and views received from the contributing entities and some states.

In its 2017 session from 3 to 21 April 2017, the United Nations Disarmament Commission held informal discussions on the practical implementation of outer space TCBMs with a view to taking up the issue in its 2018–2020 three-year cycle.

In February 2017, under decision CD/2090, the Conference on Disarmament established a Working Group on the "way ahead" to identify common ground for a program of work with a negotiating mandate. From 14 to 16 June 2017, the Working Group convened to address the prevention of an arms race in outer space (agenda item 3), with discussions featuring the linkages between space security and strategic stability, instruments of soft law, and legally binding instruments. Ambassador Htin Lynn of Myanmar chaired the meetings, with Ambassador Hellmut Lagos Koller of Chile co-facilitating them in his capacity as friend of the chair. While these discussions did not reconcile the divergent views of states concerning the nature of space threats and the most appropriate ways of responding to them, there was widespread agreement on the urgency of adopting immediately applicable measures. In particular, it was noted that space technology is inherently dual-use and can contribute to both maintaining and disrupting strategic stability; that, while the potential for space weaponization is a major concern, there are other dangers to space systems, such as ground-based kinetic weapons, directed energy weapons, means of electronic warfare, space debris, and space weather; that the current legal regime governing the use of space is in need of some updating and that instruments of soft law are useful supplements to legally binding provisions; that avoiding indiscriminate harm in outer space, such as that resulting from destructive ASAT tests, and prohibiting intentional creation of space debris might be an avenue for the future and form the basis for future consensus; and that there are a number of actions in space that can be perceived as hostile for which TCBMs and norms of behavior can have measurable impacts and be a faster step to move forward.

All of this work gave new impetus to space security discussions and was instrumental in laying the foundation for more substantive work, both on voluntary measures and legally binding instruments. In particular, in 2018, three separate initiatives were launched in parallel within the UN system, each with a different timeframe.

First, on 24 December 2017, the UNGA adopted resolution 72/250 establishing a new Group of Governmental Experts (GGE) on Further Practical Measures for the Prevention of an Arms Race in Outer Space. The mandate of this GGE was to explore substantial elements of a possible legally binding instrument, including on the prevention of the placement of weapons in outer space, using the draft PPWT as a the very minimum basis for discussion. Ambassador Guilherme de Aguiar Patriota, Permanent Representative of Brazil to the CD, was named the chair of the GGE, with UNIDIR acting as the technical expert. The GGE, with a membership of up to 25 representatives, met for two two-week sessions in Geneva, one from 6 to 17 August 2018 and the other one from 18 to 29 March 2019. A two-day open-ended intersessional informal consultative meeting was also held in New York on 31 January and 1 February 2019 to allow member states and the broader space community to engage in further discussions and to comment on an interim report by the chair on the work of the GGE (Patriota 2019). At its final session in March 2019, the GGE could not reach consensus on a final report. Nonetheless, discussions within the GGE have been if nothing else instrumental in identifying the pros and cons of new legal instruments and in offering a platform to discuss highly debated issues, such as definitions and verification (Meyer 2018).

Second, on 19 February 2018, through decision CD/2119, the Conference on Disarmament agreed to establish five subsidiary bodies to address individual agenda items in the absence of an overall program of work. Subsidiary Body 3 was tasked with focusing on the Prevention of an Arms Race in Outer Space. Ambassador Guilherme de Aguiar Patriota was appointed coordinator, with UNIDIR acting as the technical expert. Over the course of 2018, this Subsidiary Body met seven times and agreed on a consensus report which was submitted to and adopted by the CD plenary. This report, labeled as CD/WP.611, reviews the existing normative and institutional framework, identifies threats that raise major concerns among member states, and puts forward possible solutions, including the recourse to other UN fora to move discussions forward. Unfortunately, due to political difficulties, the CD was unable to adopt the report of Subsidiary Body 4 and therefore the final report to be sent to the UNGA. However, the CD was able to agree on and adopt a procedural report, which includes as appendixes all official documents of the Conference on Disarmament, summary records, process verbal, and reports of subsidiary bodies.

Third, at its organizational meeting on 21 February 2018, the United Nations Disarmament Commission agreed to include space security in its work program for the 2018–2020 cycle and to establish a Working Group to promote the practical implementation of outer space TCBMs with the goal of preventing an arms race in outer space. In particular, this Working Group is tasked with looking at how the recommendations contained in the consensus report of the 2013 GGE can be implemented and made operational, in an attempt to re-energize discussions over what was seen as a model for effective space security diplomacy. The UNDC Working Group held several meetings during its two-week sessions in April 2018 and April 2019, with a last meetings session being anticipated in April 2020, when it is expected to make some recommendations to give substantive effect to the 2013 GGE Report. However, even though the Working Group focuses on voluntary measures that are not legally binding, the consensus rule by which the UNDC operates and the limited track record of success that the UNDC has experienced in recent years indicate that expectations would need to be managed (Meyer 2018).

While the GGE on Further Practical Measures for PAROS and the Conference on Disarmament failed to achieve consensus on a final report and on a decision that would have re-established CD Subsidiary Body 3, respectively, all of these three separate initiatives have allowed for parallel discussions on two distinct approaches to space security – one which looks at more robust solutions through the lens of legally binding arms control agreements and the other one which looks at short-term solutions through the lens of voluntary, non-legally binding measures. By holding parallel discussions, proponents of treaties and TCBMs can still work together, potentially impacting efforts to win support for their own preferred approach.

Space Security Within UN Space Settings

The consideration of space security from the perspective of peaceful uses of outer space has a long history, too. Earliest records date back to January 1958 when, in the context of Soviet-US relations for reducing international tensions, US President Eisenhower made a proposal to use outer space only for peaceful purposes and to deny it to the purposes of war (United States Department of State 1958). On 15 March 1958, a concrete proposal on the topic was submitted by the Soviet Union for consideration by the 13th session of the UN General Assembly. Through resolution 1348 (XIII) of 13 December 1958 – the first ever UN resolution related to outer space – the UNGA established an ad hoc Committee on the Peaceful Uses of Outer Space to discuss the scientific and legal aspects of the exploration and uses of outer space. Although these issues may be not seem closely connected to security matters, they were very relevant. By advancing proposals on the peaceful uses of outer space, both superpowers reaffirmed the importance of stopping using space for the testing of missiles designed for military purposes. Not only was outer space seen as instrumental in increasing the power of new types of weapons to the detriment of the mankind but also, more importantly, controlling the uses of outer space implied controlling the development of missiles as carriers of weapons (Jacobson and Stein 1966).

In the decades since then, there have been multiple attempts in UN space settings to regulate the different uses of outer space, including on aspects pertaining to the broader perspective of space security. These multilateral efforts have included international agreements, treaties, and conventions; UNGA declarations and legal principles; UNGA resolutions; and guidelines and rules for safe and sustainable space operations; as well as talks in the Committee on the Peaceful Uses of Outer Space and its subcommittees primarily under the agenda item entitled "Long-Term Sustainability of Outer Space Activities," and in the UNGA and its Fourth Committee under the agenda item entitled "international co-operation in the peaceful uses of outer space." From the first discussions until today, space security has been seen, on the one hand, as part of the global space governance and international cooperation efforts aimed at ensuring a certain level of order and predictability in space and, on the other hand, as part of wider sustainability and development issues aimed at ensuring a safe and sustainable conduct of space activities both in space and on Earth. This is in line with both the initiative by COPUOS on the long-term sustainability of outer space activities and the strategic reflection promoted by UNOOSA in the lead-up to UNISPACE+50 and Space 2030, where space security is considered instrumental in achieving the 2030 Agenda for Sustainable Development.

By following an approach based on forum, the next paragraphs review and analyze the efforts that have been made in recent years within UN space settings to advance space security in its broader connotation.

The United Nations Committee on the Peaceful Uses of Outer Space

The Committee on the Peaceful Uses of Outer Space (COPUOS) is a permanent body of the United Nations mandated to discuss cooperative mechanisms in space activities and to develop principles governing the exploration and use of outer space. Initially established as an ad hoc committee, the UN COPUOS promotes international cooperation in the peaceful uses of outer space, encourages information sharing and the development of national space research programs, and addresses legal issues arising from the exploration of space. The Committee has a purely civilian focus and does not specifically address military activities in space, although these may be affected by its deliberations.

As of September 2019, COPUOS comprises 92 states as committee members, with over 40 intergovernmental and non-governmental organizations being granted observer status. Decisions are made by consensus and reported to the UN General Assembly for consideration and endorsement usually on the same year. The technical work of the Committee is conducted by two subcommittees, the Legal Subcommittee (LSC) and the Scientific and Technical Subcommittee (STSC), which each convene once a year and report on their work to the Committee during its annual meeting.

The work carried out in the framework of COPUOS can readily accommodate diverse aspects pertaining to space security. For instance, the five international treaties and set of legal principles governing space activities contain a number of provisions that lay down the legal foundation for space security and associated matters. Examples include the prohibition of placement of weapons of mass destruction in orbit or on celestial bodies, the liability of launching states for damages caused by space objects, the registration of space objects, the notification of space activities, the safety of spacecraft, the prevention of harmful interference with space activities, and the avoidance of harmful contamination of the space environment.

In addition to the codification of international space law, the work of the UN COPUOS in the matter of space security has been focused on the protection of space systems and on the safeguard of the space environment. Topics such as the use of nuclear power sources in space, measures for space debris mitigation and remediation, and space traffic management have all been under consideration in the Legal Subcommittee. Likewise, space debris, space weather, near-earth objects, and other aspects closely related to the safety and sustainability of space activities have been on the agenda of the Scientific and Technical Subcommittee for many years now. In particular, along with the work of the Working Group on the Long-Term Sustainability of Outer Space Activities, the development and adoption by COPUOS of a set of voluntary guidelines for space debris mitigation represents one the greatest successes of the Committee in recent years (UNOOSA 2010). Building on the IADC proposal on debris mitigation (A/AC.105/C.1/L.260), the UN COPUOS seven technical guidelines (A/AC.105/890) focus on how to safely conduct space activities while limiting the proliferation of space debris. Importantly, guideline 4 recommends that the intentional creation of long-lived debris be avoided, a rule that could also be seen as a commitment by states not to test destructive ASAT weapons (Hitchens 2015). These technical guidelines have contributed to increasing understanding of what constitutes acceptable activities (and behavior) in space, and their implementation through relevant national mechanisms would be a pragmatic way forward to reduce tensions, alter threat perception, and increase stability in space. Since 2014, under the agenda item on "ways and means of maintaining outer space for peaceful purposes," the UN COPUOS has also considered the recommendations contained in the 2013 Report by the GGE on space TCBMs, with the aim of identifying those recommendations that could ensure the safety of space operations and the long-term sustainability of outer space activities.

The UN COPUOS Working Group on the Long-Term Sustainability of Outer Space Activities

Conversely to sustainability on Earth, the topic of sustainability of space activities has been on the agenda of the United Nations only since 2007, not coincidentally a few months after China's ASAT test created thousands of long-lasting pieces of debris (Brachet 2012, 2016). At the 50th session of the UN COPUOS, then-Chair Gérard Brachet submitted a Working Paper (A/AC.105/L.268) on the future role and activities of COPUOS, suggesting that the Committee address the issue of the long-term sustainability of outer space activities and that a dedicated Working Group be set up in the STSC to develop "rules of the road" for future space operations.

In line with this proposal, an ad hoc group of experts with representatives from over 20 countries was created under the impulse of France in order to develop information exchange mechanisms and consensus-based rules of behavior for a safer and more secure space environment (see UN Document A/AC.105/C.1/L.303). The expert group worked through 2008 and 2009 to develop a document (A/AC.105/C.1/2010/CRP.3), also referred to as the "Brachet Code of Conduct," addressing a wide range of technical issues (e.g., space debris mitigation and remediation, the safety of space operations, the electromagnetic spectrum, and space weather) and putting forth a set of preliminary recommendations. While conducted informally, this work (and that of the French Delegation to COPUOS, too) led to the creation in the UN COPUOS STSC of a Working Group on the Long-Term Sustainability of Outer

Space Activities, the terms of reference of which (A/AC.105/C.1/L./307/Rev.1) were finalized the following year in June 2011.

The Working Group was tasked with preparing a consensus report containing a set of voluntary, non-legally binding best practices guidelines for implementation by all space actors to reduce risks to space activities and ensure equitable access to outer space for the benefit of all nations. In doing so, the Working Group examined the long-term sustainability of outer space activities within the wider context of sustainable development on Earth, including the contribution of space activities to the achievement of the Millennium Development Goals. In order to examine the various aspects surrounding space sustainability and prepare candidate guidelines to be then refined and agreed upon by member states at COPUOS, the Working Group relied upon four expert groups, each with a specific focus:

- Expert Group A: Sustainable space utilization supporting sustainable development on Earth.
- Expert Group B: Space debris, space operations, and tools to support collaborative space situational awareness.
- Expert Group C: Space weather.
- Expert Group D: Regulatory regimes and guidance for actors in the space arena.

Inputs from non-state actors were welcome if submitted through the relevant member states or permanent observers, both of which were also allowed to nominate non-governmental experts to the expert groups as part of their delegation (Martinez 2015). This has allowed the substantial body of knowledge and experience of the private sector and other NGOs not to be dissipated and contributed to making the guidelines relevant to all space actors and activities throughout the entire supply chain and life cycle.

The Working Group initial mandate ran until 2014, with the total number of proposed guidelines fluctuating up to 33 as inputs from expert groups, UN COPUOS member states, and WG Chair Peter Martinez were received. In June 2014, WG Chair Peter Martinez proposed to combine the 33 draft guidelines (A/AC.105/C.1/L.339) into 16 consolidated draft guidelines (A/AC.105/2014/CRP.5), to which 3 more candidate guidelines (A/AC.105/C.1/L.340) were added in October 2014.

The Committee also agreed that work should have continued in order to finalize the draft guidelines for their approval at the plenary sessions of COPUOS in 2016 and for their subsequent referral to the UNGA also in 2016. Additional 10 guidelines and alternatives were thus introduced by input from member states in 2015, and, following a process of consolidation and refinement (and compromise, too), the total number of draft guidelines was narrowed down to 29 (Martinez 2018). Since decision on the text of each guideline is reached by consensus, progress was intermittent and a first set of 12 guidelines (A/AC.105/2016/CRP.17) could be agreed upon only during the 2016 COPUOS plenary.

As the remaining guidelines were all at different stages of readiness, the mandate of the WG was extended until June 2018. In February 2018, at the 55th session of the STSC, agreement was reached on the preamble and 9 further guidelines (A/AC.105/C.1/2018/CRP.18/Rev.1), which are now ready for states and intergovernmental organizations to consider implementing on a voluntary basis (See UN Document A/AC.105/C.1/L.366).

Although the mandate of the WG expired in June 2018, with no final report by the WG being agreed upon, work has continued on the text of the remaining draft guidelines (A/AC.105/C.1/L.367) – as well as on specific procedures for reviewing, amending, and revising the text – with a view to reaching consensus on a final report and a full compendium of agreed guidelines to be then adopted by COPUOS and referred to the UN General Assembly. To expedite this process, a number of options were put forth by member states, including through extending the mandate of the Working Group by 1 year or establishing a new permanent Working Group. At its 62th session in 2019, the Committee decided to establish, under a five-year workplan, a new Working Group on the Long-Term Sustainability of Outer Space Activities under the STSC.

The set of 21 guidelines (A/74/20, Annex II) agreed to date, which were formally adopted by COPUOS in June 2019, are classified into four major categories and address a wide range of aspects pertaining to space activities, from policy and regulatory, to operational and safety, to international cooperation and capacity building, and to scientific and technical research and development. In particular, the guidelines include specific measures to ensure the safe and sustainable use of outer space for peaceful purposes. In recognizing the difference in capacity and capabilities among space actors, space-faring nations are encouraged to support emerging space actors in the implementation of the guidelines, as well as in the development of space capabilities in a manner that avoids potential harms for all parties concerned.

While voluntary and non-binding under international law, the guidelines may lead states to establish relevant national mechanisms for facilitating their implementation and to incorporate the principles therein contained in their national legislation.

Importantly, the long-term sustainability guidelines can be viewed as enablers for the implementation of existing TCBMs or even as TCBMs themselves, possibly providing the basis for more robust initiatives.

There have also been attempts by some delegations to introduce hard security issues into the work of COPUOS, but success has been limited, since those topics are seen as the prerogative of the CD and the First Committee of the UNGA. While the persistence of not contaminating the work of COPUOS with hard security issues may be seen as a limit to the ability of the body to make progress in the area of space security, it has actually allowed the Committee to exclusively focus on his core mandate and to make some tangible progress on safety and sustainability issues. However, like all voluntary measures, the success of this initiative will eventually depend on the extent to which space actors implement the agreed upon guidelines. The hope is that the awareness and positive momentum that this multilateral consensus-based process has created may facilitate future adoption of the guidelines.

The United Nations Office for Outer Space Affairs

The United Nations Office for Outer Space Affairs (UNOOSA) is the Office of the United Nations which promotes international cooperation in the peaceful uses of outer space to achieve development goals for the benefit of humankind. In addition to serving as the secretariat of the UN COPUOS, the Office is also responsible for discharging the UN Secretary-General's responsibilities and obligations under international space law, with particular regard to the implementation of the five UN space treaties, as well as information exchange and notification mechanisms.

More specifically, UNOOSA maintains the UN Register of objects launched into outer space (Article III of the Registration Convention) and disseminates information relating to outer space activities (including discovery of harmful phenomena) provided by states (Articles V and XI of the Outer Space Treaty), which are de facto treaty-based TCBMs for enhanced space security. The Office also serves as a facilitator on notifications relating to the malfunction and re-entry of nuclear-powered space objects (Principle V of the Principles Relevant to the Use of Nuclear Power Sources in Outer Space). In its resolutions 70/82 and 71/90, the UN General Assembly also encourages UNOOSA to conduct capacity-building and outreach activities associated with space security and TCBMs in outer space activities, recognizing the positive contribution that the Office can offer in these areas. Looking forward, UNOOSA could be instrumental in the implementation of a number of TCBMs, including through providing notifications on outer space activities aimed at risk reduction, disseminating pre-lunch notifications, and facilitating voluntary visits to space launch sites and facilities and to demonstrations of rockets and space technologies (Keusen 2017).

Importantly, the wider strategic review promoted by UNOOSA in the lead-up to UNISPACE+50 provided an additional opportunity to address the legal regime of outer space and global space governance, information exchange on space objects and events, and space weather services, as the thematic priorities of UNISPACE+50 have shown (see UN Document A/AC.105/2016/CRP.3). In particular, interconnection exists between the above-mentioned areas and the work of the Working Group on the Long-Term Sustainability of Outer Space Activities. With the recent tendency of merging space and development issues, and in light of the adoption of the 2030 Agenda for Sustainable Development, a safe, sustainable, and secure space environment is needed to ensure the continuous and reliable operations of space systems, including those supporting the achievement of the sustainable development goals.

Conclusions

The various initiatives being discussed in the different UN fora indicate the increasing pressures facing the international community in addressing all aspects of space security and strengthening the multilateral regime governing the use of space. They also reveal substantial differences among states over priorities, methodologies, mechanisms, and settings to address and tackle space security issues. These differences include incompatible political sensibilities and perceptions of what key space security risks are, whether the potential for space weaponization, the development of counterspace capabilities, or the proliferation of space debris; differing visions for how to address major challenges, whether by means of legally binding treaties or instruments of soft law; divergent views on the most appropriate setting in which to pursue space security diplomacy, whether in UN disarmament or space settings; and different opinions on which participants to include in the discussion (and how and when to involve them), whether to start with a like-minded core group to discuss the essence or to include, to the greatest extent practicable, as many actors as possible from the beginning (Pellegrino et al. 2016). While this proliferation of initiatives offers distinct approaches that look at both short- and long-term solutions, these proposals may end up in endless discussions, being honored only with rhetoric, or even ignored altogether. More worrisome, however, may be another challenge: deadlocks in the work of one body can be transferred over other fora, limiting progress even on those aspects on which there has traditionally been consensus.

Existing frictions among different UN bodies are also a limiting factor in pushing forward a shared agenda on space security, and so too is the long-standing division between civilian and military uses of space embedded in the UN machinery (Hitchens 2015). UNIDIR, which constitutes an invaluable forum with a frank and open dialogue on the latest challenges to the security, safety, and sustainability of outer space, could be instrumental in facilitating dialogue at all levels, both within and outside the UN, as it has always more frequently convened an ever-increasing range of established and emerging actors and stakeholders in the space and security communities with a view to exploring current trends, building bridges, and identifying potential avenues for the future. Shrinking budgets and bureaucratic politics (i.e., the promotion and protection by relevant officials of their own bureau's interests in competition with those ones of other bureau) are however hampering cooperative efforts in spite of appropriate coordination mechanisms between relevant UN entities being envisaged.

There would be great value if progress being made on space safety and sustainability in COPUOS was not seen as an end in itself, but rather as an intermediate step to create more favorable political conditions to advance discussions on hard security issues in space. The long-term sustainability guidelines are the outcome of a consensus-based process which saw the participation of all the actors engaged in arms control discussions and can be seen as a stepping-stone toward more robust international agreements.

In short, while progress has been intermittent and further work is required, especially in UN international security and disarmament settings, the future for space security diplomacy looks relatively promising, at least more than it has been in previous years. Discussions for both short- and long-term solutions will likely continue to proceed in parallel. While divergent perceptions and priorities among leading space powers still exist (and will remain), the existence of political will that accommodates, rather than eliminates, these differences can prove effective in finding common ground for future action that can be acceptable for the interests of different states. As the nature of risks to space infrastructure and services, and the available responses, are similar (although not entirely) for all space actors – whether civilian, commercial, or military – common threat perceptions may influence how states choose to cooperate and readily serve as a basis for developing common responses and finding future consensus.

References

Brachet G (2012) The origins of the "Long-term Sustainability of Outer Space Activities" initiative at UN COPUOS. Space Policy 28:161–165. https://www.academia.edu/20607722/The_origins_of_the_Long_Term_Sustainability_of_Space_Activities_inititaive_at_UNCOPUOS. Accessed 31 Aug 2019

Brachet G (2016) The security of space activities. Non-proliferation papers, no. 51, EU Non-Proliferation Consortium. https://www.nonproliferation.eu//wp-content/uploads/2018/09/the-security-of-space-activities-52.pdf. Accessed 31 Aug 2019

Hitchens T (2015) Space security-relevant international organizations: UN, ITU, and ISO. In: Schrogl K-U (et al) (eds) Handbook of space security: policies, applications and programs. Springer, New York

Jacobson H, Stein E (1966) Diplomats, scientists, and politicians: the United States and the nuclear test ban negotiations. The University of Michigan Press, Ann Arbor. https://repository.law.umich.edu/cgi/viewcontent.cgi?article=1014&context=michigan_legal_studies. Accessed 31 Aug 2019

Keusen T (2017) Safety, security and sustainability of outer space activities: UNOOSA and TCBMs looking forward. Presentation at the 2017 UNIDIR space security conference, Geneva, 20–21 April 2017. http://www.unidir.ch/files/conferences/pdfs/safety-security-and-sustainability-of-outer-space-activities-unoosa-and-tcbms-looking-forward-en-1-1244.pdf. Accessed 31 Aug 2019

Martinez P (2015) Space sustainability. In: Schrogl K-U (et al) (eds) Handbook of space security: policies, applications and programs. Springer, New York

Martinez P (2018) First fruits of the long-term sustainability discussions in UN COPUOS. Paper presented at the 69th International Astronautical Congress (IAC), IAC-18-E3.4.1, Bremen, 1–5 Oct 2018

Meyer P (2011) The CD and PAROS: a short history. United Nations Institute for Disarmament Research, Geneva. http://www.unidir.org/files/publications/pdfs/the-conference-on-disarmament-and-the-prevention-of-an-arms-race-in-outer-space-370.pdf. Accessed 31 Aug 2019

Meyer P (2018) Diplomacy: the missing ingredient in space security. Simons papers in security and development, no. 67/2018, School for International Studies, Simon Fraser University, Vancouver. http://summit.sfu.ca/system/files/iritems1/18290/SimonsWorkingPaper67.pdf. Accessed 31 Aug 2019

Moltz J (2014) Crowded orbits: conflict and cooperation in space. Columbia University Press, New York

Patriota G (2019) Interim Report by the Chair of the Group of Governmental Experts on Further Practical Measures for the Prevention of an Arms Race in Outer Space. https://s3.amazonaws.com/unoda-web/wp-content/uploads/2019/02/oral-report-chair-gge-paros-2019-01-31.pdf. Accessed 31 August 2019

Pellegrino M (2017) UNIDIR space security conference report 2017. United Nations Institute for Disarmament Research, Geneva. http://unidir.org/files/publications/pdfs/unidir-space-security-2017-en-685.pdf. Accessed 31 Aug 2019

Pellegrino M, Stang G (2016) Space security for Europe. European Union Institute for Security Studies, Paris. https://www.iss.europa.eu/sites/default/files/EUISSFiles/Report_29_0.pdf. Accessed 31 Aug 2019

Pellegrino M et al (2016) Security in space: challenges to international cooperation and options for moving forward. Paper presented at the 67th International Astronautical Congress (IAC), IAC-

16-E3,4,12,x35460, Guadalajara, 26–30 September 2016. https://swfound.org/media/205876/manuscript-security-in-space-iac-2016.pdf. Accessed 31 Aug 2019

Rajagopalan RP (2017) Space security, deterrence and strategic stability. Presentation at the 2017 UNIDIR Space Security Conference, Geneva, 20–21 April 2017. http://www.unidir.ch/files/conferences/pdfs/space-security-deterrence-and-strategic-stability-en-1-1224.pdf. Accessed 31 Aug 2019

Sheehan M (2015) Defining space security. In: Schrogl K-U (et al) (eds) Handbook of space security: policies, applications and programs. Springer, New York

United States Department of Defense (2019) Missile Defense Review. The United States Department of Defense, Washington, DC. https://media.defense.gov/2019/Jan/17/2002080666/-1/-1/1/2019-MISSILE-DEFENSE-REVIEW.PDF. Accessed 31 August 2019

United States Department of State (1958) The Department of State bulletin, Vol. XXXVIII (No. 970, 27 January 1958, pp. 122–127) and Vol. XXXVIII (No. 976, 10 March 1958, pp. 376–380), US Government Printing Office, Washington, DC. https://catalog.hathitrust.org/Record/000598610. Accessed 31 August 2019

UNOOSA (2010) Space Debris Mitigation Guidelines of the Committee on the Peaceful Uses of Outer Space. United Nations, Vienna. https://www.unoosa.org/pdf/publications/st_space_49E.pdf. Accessed 31 August 2019

UNOOSA (2017) Joint panel discussion of the First and Fourth Committees on possible challenges to space security and sustainability. https://www.unoosa.org/documents/pdf/gajointpanel/Co-Chair_Summary_C1-C4_Joint_Panel_Discussion_Final_2.pdf. Accessed 31 August 2019

Weeden B, Samson V (eds) (2019) Global counterspace capabilities: an open source assessment. The Secure World Foundation, Washington, DC. https://swfound.org/media/206408/swf_global_counterspace_april2019_web.pdf. Accessed 31 Aug 2019

Further Readings

Brachet G (2007) Collective security in space: a key factor for sustainable long-term use of space. In: Logsdon J et al (eds) Collective security in space: European perspectives. Space Policy Institute, Elliott School of International Affairs, The George Washington University, Washington, DC

Hersch M, Steer C (eds) (2020) War and Peace in Outer Space: Ethics, Law and Policy. Oxford University Press, Oxford. (in progress)

Rajagopalan RP (2018) Space governance. Oxford research encyclopedia of planetary science. https://doi.org/10.1093/acrefore/9780190647926.013.107. Accessed 31 Aug 2019

The Role of COSPAR for Space Security and Planetary Protection

Leslie I. Tennen

Contents

L. I. Tennen (✉)
Law Offices of Sterns and Tennen, Glendale, AZ, USA
e-mail: Ltennen@astrolaw.com

© Springer Nature Switzerland AG 2020
K.-U. Schrogl (ed.), *Handbook of Space Security*,
https://doi.org/10.1007/978-3-030-23210-8_145

Abstract

The environment of the Earth can be at risk of potential contamination from extraterrestrial material brought back to this planet by robotic and human explorers. Similarly, spacecraft sent to explore celestial bodies can carry microbes that can contaminate an extraterrestrial environment. Contamination presents significant implications for space security. The Committee on Space Research (COSPAR) of the International Science Council (ISC) has maintained a Planetary Protection Policy (PPP) to address both forward and back contamination. States have implemented the COSPAR PPP as the recognized standard to comply with international treaty obligations in regard to contamination from biological matter.

Introduction

How did life begin? Are we alone in the universe? These questions are among the most essential, eternal, and enduring mysteries of science. With the advent of the space age, mankind acquired new capabilities to search for alien life. We have the technology to venture to distant worlds, conduct experiments in the pursuit of evidence of life, and even bring samples from celestial bodies back to our home planet. However, extraterrestrial matter brought back to the Earth could potentially harbor deleterious material that could jeopardize the safety and security of the entire globe. This issue of "back contamination" is not confined to the realm of science fiction but has been a significant concern since the days of the Apollo program.

Just as the environment of the Earth is in need of protection from potential harmful contamination from extraterrestrial sources, so it is also necessary to protect celestial environments from harmful contamination from terrestrial matter. The discovery of evidence of life on another celestial body would forever change humanity's view of mankind's place in the cosmos. The significance of such a discovery cannot be overstated, nor can it be completely comprehended in advance; therefore it is of the utmost importance that the scientific integrity of such a discovery be assured. Forward contamination of a celestial environment from terrestrial biological materials could jeopardize the scientific integrity of subsequent investigation and also could harmfully interfere with future exploration and experimentation. This directly impacts the risk of conflict between nations and threatens the safeguarding of space for peaceful endeavors.

In its most basic terms, forward contamination controls seek to ensure that any evidence of life found on a celestial body is indigenous and not of terrestrial origin. Similarly, back contamination requirements are a way to answer the question of how will we know that extraterrestrial matter brought back to the Earth will not be dangerous. The policy of planetary protection can be summarized as follows:

The conduct of scientific investigations of possible extraterrestrial life forms, precursors, and remnants must not be jeopardized.

The Earth must be protected from the potential hazard posed by extraterrestrial matter carried by a spacecraft returning from a celestial body (Coustenis et al. 2019)

The Committee on Space Research

The international diplomatic and scientific communities have acted to prevent both forward and back contamination. The United Nations has been the leading organization for the development of legal principles to protect celestial environments. The Committee on Space Research (COSPAR) has been the preeminent scientific forum for the articulation of specific policies and guidelines for planetary protection (the Planetary Protection Policy or PPP).

COSPAR was formed in 1958 by the International Council of Scientific Unions (ICSU), now the International Science Council (ISC). The membership of COSPAR is comprised of national scientific institutions and international scientific unions. The organizational structure consists of eight commissions which represent every scientific discipline involved in space research. In addition, there are ten panels which are designed to deal with issues of interest to specific segments of the space research community. The prevention of forward and back contamination is the charge of the Panel on Planetary Protection (PP Panel).

The PP Panel is appointed by the COSPAR Bureau and is led by a chair and two vice-chairs, one of which is reserved for a representative of the United Nations Office for Outer Space Affairs (UNOOSA). There is an equal number of representatives from national or international authorities and representatives from COSPAR Scientific Commission B, Space Studies of the Earth-Moon System, Planets, and Small Bodies of the Solar System, and Commission F, Life Sciences as Related to Space.

The competence of the COSPAR PP Panel is limited to forward and back contamination from biological sources. As such, it is primarily concerned with microbes and other potential organic contaminants. The COSPAR PP Panel does not otherwise consider protecting celestial bodies, unique environments, or historical sites. Nor does the COSPAR Planetary Protection Policy apply to protecting the Earth from man-made space debris or defending the planet from the impact of large asteroids or comets.

Foundations and Development of the Planetary Protection Policy

Initial Activities of the Diplomatic Community

In 1957, the United Nations General Assembly took the first formal steps to regulate mankind's movement into the cosmos by the creation of the *Ad Hoc* Committee on the Peaceful Uses of Outer Space (COPUOS), which became a permanent committee the following year. COPUOS conducts its work through two subcommittees, one of which is devoted to scientific and technical matters, while the other considers legal issues. Membership in the Committee has expanded over the years to include all states actively launching and performing missions in space. Currently 95 nations are members of COPUOS, and in addition a number of international organizations, including both intergovernmental and nongovernmental entities, have observer status in the Committee. COPUOS and its subcommittees function on the basis of consensus.

Initial Scientific Policies

The Planetary Protection Policy began as a means of self-regulation by the international scientific community and pre-dated any applicable international law. In 1958, within months of the launch of Sputnik I, the ICSU formed the *Ad Hoc* Committee on Contamination by Extraterrestrial Exploration (CETEX), which considered celestial bodies to be scientific preserves. CETEX identified four primary objectives:

(i) Freedom of exploration of celestial bodies, subject to limitations such as planetary quarantine requirements.
(ii) Disclosure to COSPAR of information concerning activities and experiments.
(iii) Only experiments which are likely to yield useful scientific data should be conducted.
(iv) Nuclear explosions should not occur near the surface of celestial bodies (Tennen 2003/2004).

Policies of Planetary Protection

In March 1962, Chairman Khrushchev of the Soviet Union drew the attention of the diplomatic community to the issue of protecting planetary environments. The Chairman wrote a letter to President Kennedy about what he described as "heavenly matters" and proposed that:

> in carrying out experiments in outer space, *no one should create obstacles* to the study and use of space for peaceful purposes by other States...any experiments in outer space which may *hinder the exploration* of space by other countries should be the subject of preliminary discussion and of an agreement... (emphasis added)

Khrushchev inextricably linked protection of celestial environments to the right of states to conduct activities in the exploration and use of outer space. His focus, however, was not on the intrinsic value of preserving pristine celestial environments. Rather Khrushchev asserted and sought to protect the right of states to conduct activities in space without hindrance created by the activities of other states. He proposed that there should be a right to prior consent over activities and experiments of other states (Jakhu and Pelton 2017).

These and other issues were subject to debate and discussion in COPUOS, which recognized that the problem of preventing potentially harmful interference in the peaceful uses of outer space was of urgent concern. On November 22, 1963, COPUOS unanimously approved the Declaration of Legal Principles Governing the Activities of States in the Exploration and Use of Outer Space. The General Assembly of the United Nations thereafter adopted the Declaration of Principles as Resolution 1962. Planetary protection concerns were addressed in Paragraph 6:

> In the exploration and use of outer space, States shall be guided by the principle of co-operation and mutual assistance and shall conduct all their activities in outer space with *due regard for the corresponding interests of other States.* (emphasis added)

The Declaration further provided that a state which has reason to believe its activities or experiments may cause harmful interference with the activities of another state shall undertake appropriate consultations with such second state. Similarly, if a state believes its activities may be interfered with by the activities of another state, it could request such other state to participate in consultations. This first legal planetary protection standard approved by the community of nations confirmed the complex interplay between preservation of celestial environments from harmful contamination and due regard for the right of states to conduct activities on other worlds without harmful interference.

Scientific Regulation Tthrough Planetary Quarantine Requirements

Following the work of CETEX, the ICSU established the COSPAR Consultative Group on Potentially Harmful Effects of Space Experiments, which released a comprehensive Planetary Protection Policy in 1964. This policy established "planetary quarantine requirements" (PQR) which adopted a probabilistic approach to protecting pristine celestial environments (COSPAR 1964). That is, the PQR established limitations expressed as a probability that a spacecraft could contaminate a celestial environment. The probability of contamination (P(c)) by any mission was determined by the formula:

$$P(c) = mi(0) \cdot P(vt) \cdot P(uv) \cdot P(a) \cdot P(sa) \cdot P(r) \cdot P(g)$$

where

mi(O) Initial microbial burden (at launch, after decontamination)
P(vt) Probability of surviving space vacuum-temperature
P(uv) Probability of surviving ultraviolet space radiation
P(a) Probability of arriving at planet
P(sa) Probability of surviving atmospheric entry
P(r) Probability of release
P(g) Probability of growth

The PQR set a probability limit for an accidental planetary impact by an unsterilized flyby or orbiting spacecraft to be 3×10^{-5} or less ($< 1/300,000$). Spacecraft which were intended to penetrate the atmosphere or land on the surface of a planet were subject to a much higher standard. For these spacecraft, the probability limit for contamination by a single viable terrestrial organism aboard the craft was to be less than 1×10^{-4} ($<1/10,000$). These P(c) limits were to apply for an initial period of planetary exploration of 10 years, later extended to 30 years (Phillips 1975).

In 1966 the P(c) limit for landing or atmospheric penetration spacecraft was reduced to 1×10^{-3} (1/1000). This probability limit was an aggregate for all missions from all nations, and individual countries were allocated specific portions of the overall probability limits. The recipient states, in turn, apportioned their share among the various missions each conducted. The overall P(c) limits were divided as follows (Meltzer 2012):

USA	4.4×10^{-4}
USSR	4.4×10^{-4}
All others	1.2×10^{-4}
Total	1×10^{-3}

The United States, in turn, distributed its national allotment to the following interplanetary missions (Sterns and Tennen 2019):

Mariner Mars	7.1×10^{-5}
Pioneer Jupiter	6.4×10^{-5}
Mariner Venus	7×10^{-5}
Viking	2×10^{-4} (1×10^{-4} for each of two missions)

Compliance with the PQR generally required missions to engage in active decontamination techniques to reduce the presence of microbes on the spacecraft at launch. States employ a variety of techniques for this purpose. Prevention and limitation of the initial microbial burden are achieved by assembly and testing of a spacecraft in biologically controlled cleanrooms. The bioburden is reduced by cleaning with various solvents and by dry heat, plasma, or ionizing radiation. Barrier systems such as purging, filters, and seals are utilized to avoid possible re-contamination of the spacecraft. However, absolute sterilization of a spacecraft is not considered possible (Meltzer 2012).

Planetary Protection and the Emergence of Binding International Law

The legal regulation of activities in space entered a new phase when COPUOS reached consensus on the text of the Treaty on Principles Governing the Activities of States in the Exploration and Use of Outer Space, including the Moon and Other Celestial Bodies (OST), which entered into force in October 1967. Article I of the OST recognizes the right of states to conduct peaceful activities on all areas of the Moon and other celestial bodies. Article IX of the treaty incorporated the due regard and consultation provisions of GA Res. 1962, the Declaration of Principles. In addition, and significantly, Article IX went further and explicitly addressed planetary protection and provided that in pursuing studies and conducting exploration of celestial bodies, states shall:

> avoid their harmful contamination and also adverse changes to the environment of the Earth resulting from the introduction of extraterrestrial matter and, where necessary, shall adopt appropriate measures for this purpose.

Article IX is very broad in its terms and encompasses not just the concepts of due regard for, and the prevention of, harmful interference with activities of other states but also the inherent value in the preservation of natural celestial environments from harmful contamination. Nevertheless, there is no international consensus on the definitions of "harmful contamination" or "interference." Nor have the interests of other states that shall be given due regard been identified, other than avoidance of harmful contamination. However, the COSPAR PPP represents a consensus that, at a minimum, harmful contamination includes the introduction of biological matter from the Earth into at least certain celestial environments.

Revisions to the PQR

The COSPAR Planetary Protection Policy has not been static but rather has been subject to continuing review and revision. Within COSPAR is a recognized process by which changes to the policy can be considered, evaluated, and examined by pertinent stakeholders, including scientific commissions of COSPAR and members of the international planetary science community (Coustenis et al. 2019). As part of this process, the probabilistic approach on which the PQR was based was criticized for the inherent difficulty of assigning specific numerical values to criteria with largely unknown properties in order to determine the probability of contamination for a mission. Recognizing this difficulty in application of the PQR, COSPAR in 1969 limited the policy to missions to Mars and other planets deemed important in the search for extraterrestrial life (Stabekis 2002).

This change in the application of the policy was a significant departure from manner in which the risk of contamination to a celestial environment was viewed. The PQR proceeded from the original perspective articulated by CETEX that celestial bodies were scientific preserves, and as such the PQR were presumed to be applicable to all interplanetary missions. The revisions to the PPP, however, limited the applicability of the decontamination requirements to only a small subset of interplanetary missions.

Subsequent revisions to the policy reinforced this change in perspective. For example, in 1978, the probability that terrestrial organisms could grow in extraterrestrial environments was re-evaluated and deemed to be virtually zero for most celestial bodies (Space Science Board 1978). As a result, requirements that missions to those bodies engage in active decontamination techniques were largely eliminated. The movement away from the probabilistic approach turned protection of celestial environments from a comprehensive blanket policy to one of restricted application. That is, PQR became the exception, not the norm, and active decontamination techniques would not be required for missions to alien worlds deemed too hostile for terrestrial microbes to survive and replicate.

The Moon Agreement and Elaboration of Legal Regulation

In 1979 COPUOS achieved consensus on a second treaty which addressed forward and back contamination, the Moon Agreement (MA), which provides, in Article 7.1:

> In exploring and using the Moon, States Parties **shall take measures to prevent the disruption of the existing balance of its environment** whether by introducing adverse changes in that environment, by its harmful contamination through the introduction of extra-environmental matter or otherwise. States Parties shall also take measures to avoid harmfully affecting the environment of the Earth through the introduction of extraterrestrial matter or otherwise. (emphasis added)

The obligations of states under the MA are more extensive than required by the OST. While the MA prohibits harmful contamination, as does the OST, the MA also prohibits states from disrupting the existing balance of the lunar environment. (Pursuant to Article 1.1 the provisions of the MA apply to other celestial bodies in the Solar System unless specific legal norms enter into force for such objects.) The MA makes it clear that environmental disruption could occur by several means, including but not limited to harmful contamination by the introduction of extra-environmental biological and other matter.

Article 7 contains a further departure from the OST in paragraph 3, which provides that states shall notify the Secretary General of areas of the Moon with special scientific interest, so that consideration shall be given to the designation of an area as an "international scientific preserve." Such areas shall be subject to "special protective arrangements to be agreed upon in consultation with the competent bodies of the United Nations."

The Preamble to the MA reflects that the treaty was formulated, in part, to define and develop the provisions of the OST, thereby providing some clarification as to the meaning of Article IX. However, the MA has not received widespread acceptance by the community of nations. While the OST has been signed or ratified by more than 130 states, only 18 states have ratified the MA, and just 4 more have signed the instrument. These include India and members of ESA including Belgium, France, the Netherlands, and Austria. These states must ensure that any mission to celestial bodies in which they participate is conducted in compliance with their international obligations, including the MA (Jakhu and Pelton 2017).

Transformation of the PQR

The derogation of the PQR continued into the 1980s, when the requirements were replaced with a new approach that provided the imposition of planetary protection constraints would be dependent upon the nature of the mission and the target body or bodies to be explored. These guidelines completely eliminated the overall probability of contamination restrictions and did not require any decontamination techniques or documentation for target bodies which were deemed not to be of biological interest in the search for life, including the Moon. The planetary protection classification for missions to other target bodies was to be determined on a case-by-case basis (Stabekis 2002).

Subsequent revisions continued to erode and limit the application of the PPP. In addition to the prerequisite conditions that a target body be both biologically interesting and deemed to provide an environment in which terrestrial microbes could survive and replicate, a third condition was added: that the mission spacecraft must include experiments intended to detect evidence of extraterrestrial life. An exception to this condition has been carved out for missions to areas of a celestial body which are designated as a "special region," that is, an area which possesses properties which are known to be conducive to sustaining life in terms of temperature range and the presence of water. Spacecraft intended to land or explore or create (even temporarily by the operation of spacecraft hardware) a special region must achieve Viking-level decontamination, even where the mission does not include life detection experiments. However, the designation of an area as a special region under the PPP is not to be considered as the equivalent to an international scientific preserve pursuant to Article 7.3 of the MA. Special regions are expressly defined in relation to the search for evidence of indigenous life, while the designation as a scientific preserve is not limited to this purpose.

Current COSPAR Planetary Protection Policy

Categorization of Target Bodies

The overarching policy is stated by COSPAR as follows (COSPAR 2017):

The conduct of scientific investigations of possible extraterrestrial life forms, precursors, and remnants must not be jeopardized. In addition, the Earth must be protected from the potential hazard posed by extraterrestrial matter carried by a spacecraft returning from an interplanetary mission. Therefore, for certain space mission/target planet combinations, controls on contamination shall be imposed in accordance with issuances implementing this policy

The implementation of the policy divides celestial bodies and mission types into four categories, with a fifth category designated for the return of extraterrestrial materials to the Earth. The concern for contamination and the concomitant contamination controls vary by category.

Categories I and II

Category I consists of flyby, orbiter, and landing missions to target bodies which are not considered to be of significant interest in the search for life, such as Io and undifferentiated, metamorphosed asteroids. No specific PPP requirements are imposed for such mission/target body combinations.

Category II consists of flyby, orbiter, and landing missions to most of the moons in the solar system, including the Moon of the Earth, the planets – including Pluto – other than Mars, carbonaceous chondrite asteroids, comets, and some Kuiper belt objects. These target bodies are considered to be of "significant interest relative to the process of chemical evolution and the origin of life, but where there is only a remote chance that contamination carried by a spacecraft could compromise future investigations." The term "remote" in this context "implies the absence of environments where terrestrial organisms could *survive and replicate*... (emphasis added)." Category II missions are subject to only simple documentation requirements, consisting of "a short planetary protection plan" identifying possible impact sites, brief pre- and post-launch analyses of impact strategies, and a post-encounter or mission report providing the location of any impact which occurred.

Category III: Flyby and Orbiter Mission to Mars, Europa, and Enceladus

Category III target bodies are of both significant interest in the search for evidence of life and carry a significant chance that contamination could compromise future investigations. Since these missions are not intended to impact the target body, the policy allows for the spacecraft to satisfy a requirement of either orbital lifetime parameters or bioburden reduction. Orbital lifetime parameters require a probability of greater than or equal to 99% that the spacecraft will not impact the target body within 20 years of launch and greater than or equal to 95% on non-impact within 50 years. The bioburden reduction requirements limit the total number of spores on and in the spacecraft to less than or equal to 5×10^5 (Spores are defined as aerobic microorganisms that survive a heat shock of 80 °C for 15 min and cultured on Tryptic-Soy-Agar at 32 °C for 72 h).

Category IV: Lander Missions to Mars, Europa, and Enceladus

Category IV for landing craft is divided into three subcategories, depending on the mission type and objectives. Subcategory IV(a) applies to spacecraft which do not carry instruments designed to search for extant life. These spacecraft must meet bioburden limits of a total of less than or equal to 3×10^5 spores and an average of not more than 300 spores per square meter.

Category IV(b) missions are designed to search for extant extraterrestrial life. Such missions are required to meet bioburden reduction levels four orders of magnitude more stringent than for Category IV(a) spacecraft, that is, a total of 30 spores, or the "levels of bioburden reduction driven by the nature and sensitivity of the particular life detection experiments." However, if the life detection experiment hardware is cleaned to this level and the spacecraft is designed to ensure that the hardware cannot be exposed to possible recontamination, the remainder of the spacecraft can satisfy the PPP by meeting Category IV(a) limits.

Category IV(c) is for missions to special regions. Spacecraft landing within a special region must comply with Category IV(b) limits. If the special region is accessed through horizontal or vertical mobility, either the entire spacecraft or the subsystems which will encounter the special region must reach that level of cleanliness, and in the event of the latter, the subsystems must be designed to prevent recontamination after the bioburden is reduced.

Category V Sample Return Missions

The four mission type/target body categories described above are concerned with forward contamination. The issue of back contamination is addressed in Category V, which applies to sample return missions. These sample return missions are classified as either restricted or nonrestricted. Missions to the Moon or Venus are nonrestricted Earth return and do not require any specific decontamination measures to be utilized. However, missions returning samples from Mars or Europa are restricted Earth return and warrant the highest level of caution to protect the Earth from potentially catastrophic contamination. (The importance of protective measures from potential back contamination is undeniably demonstrated by the Covid-19 health crisis which is devastating the globe as this chapter is in preparation.)

The PPP requires that the outbound leg of a restricted Earth return mission follow Category IV(b) requirements, to guard against "false positives" in experiments searching for extraterrestrial life. Unless the samples are to be sterilized prior to return to the Earth, the samples are to be kept in closed containers and a means provided to "break the chain of contact" with the target body. The PPP further provides that mission reviews and approvals are to be conducted at three stages: prior to launch from the Earth, after sample collection and before maneuvers to enter a biased Earth return trajectory, and prior to commitment to Earth reentry. In addition, unsterilized samples are either to be sterilized or subject to a program of life detection and biohazard testing, as an absolute precondition to the controlled distribution of any portion of the sample.

In 2019, in response to a request from JAXA for guidance, the COSPAR PP Panel recommended that sample return missions to Phobos and Deimos be designated as nonrestricted Earth return (Coustenis et al. 2019). Samples to be returned from other small bodies in the solar system, which are not otherwise categorized, are to be designated as restricted or nonrestricted on a case-by-case basis. The criteria for this determination are set out in the PPP, as a series of questions:

1. Does the preponderance of scientific evidence indicate that there was never liquid water in or on the target body?
2. Does the preponderance of scientific evidence indicate that metabolically useful energy sources were never present?
3. Does the preponderance of scientific evidence indicate that there was never sufficient organic matter (or CO_2 or carbonates and an appropriate source of reducing equivalents) in or on the target body to support life?
4. Does the preponderance of scientific evidence indicate that subsequent to the disappearance of liquid water, the target body has been subjected to extreme temperatures (i.e., $>160\ °C$)?
5. Does the preponderance of scientific evidence indicate that there is or was sufficient radiation for biological sterilization of terrestrial life forms?
6. Does the preponderance of scientific evidence indicate that there has been a natural influx to the Earth, e.g., via meteorites, of material equivalent to a sample returned from the target body?

For containment procedures to be necessary ("restricted Earth return"), an answer of "no" or "uncertain" needs to be returned to all six questions.

The sample return missions which have been launched since Apollo have all been nonrestricted, as these missions to comets, asteroids, or Phobos were not intended to return samples from a body deemed capable of supporting life. This situation will soon change as NASA is preparing to launch the first restricted Earth return mission, Mars 2020. This mission is planned to land on the red planet in 2021 and includes the capability to collect 40 separate samples of Martian material for return to the Earth. Mars 2020 is not itself a sample return mission but only a precursor to an as of yet not fully defined future mission that will return the collected samples. Nevertheless, these samples are subject to the standards of restricted Earth return, and the protocols to be developed for containing, isolating, transporting, and investigating any such samples are not without current controversy (MSR Science Planning Group Report 2019).

Human Missions to Mars

Robotic sample return missions such as Mars 2020 are a stepping stone to the eventual human exploration of the planet. Application of the COSPAR PPP is not limited to robotic missions, although the introduction of crews into the equation considerably complicates matters. The PPP recognizes that it will not be possible for human explorers to operate in entirely closed systems, nor can exposure with Martian materials be avoided. Moreover, the limitations on the initial microbial burden transported by robotic spacecraft are inapposite and unsuitable for human missions, which "will carry microbial populations that will vary in both kind and quantity..." Accordingly, the PPP identifies specific implementation guidelines for human missions to Mars, including:

- The development of a comprehensive protocol for addressing forward and backward contamination concerns during all phases of a mission.
- Continuous monitoring and evaluation of microbes carried by a mission.
- The designation of a member of the crew with primary responsibility for implementing planetary protection provisions.
- A quarantine capability for the crew during and after the mission in case of potential contact with a Martian life form.
- Robotic precursors to evaluate whether or not a site should be characterized as a special region prior to crew access.
- Planetary protection requirements for human missions should be based on a conservative approach recognizing the lack of specific knowledge of the Martian environment, possible life, and the utility and efficacy of human support systems.
- Planetary protection requirements for subsequent missions should not be relaxed without scientific justification and consensus.

The development of the necessary policies and procedures will be able to draw upon the experience and precedent of the Apollo program, in particular in regard to the Lunar Receiving Laboratory. The astronauts of Apollo 11, 12, and 14 were subject to quarantine upon their return from the Moon, as were the lunar materials they brought back to the Earth, until it was determined that they had not been exposed to harmful extraterrestrial contamination. It is known that the microbe *Microbispora* survived the harsh conditions of the ill-fated reentry of Columbia, (McLeana et al. 2006), and the possibility exists that terrestrial organisms can mutate in space and present new dangers. The lessons of Apollo can be instructive from technical, operational, and management perspectives, as issues were encountered in each area that resulted in certain breaks in the biological barrier that must not be repeated with Martian materials (Mangus and Larsen 2004).

Implementation of the COSPAR PPP by Space Agencies and Authorities

The COSPAR PPP requirements do not have the force of binding law as they are neither a treaty nor a formal intergovernmental agreement. The specific requirements set forth in the PPP are more in the nature of guidelines rather than compulsory obligations, and COSPAR does not direct the manner in which competent national authorities and international agencies can implement the policy. States which launch or register a spacecraft are internationally responsible and liable for the mission pursuant to the OST and have the duty to comply with Article IX to prevent the harmful contamination of celestial environments. COPUOS has recognized the PPP as the international reference standard for guiding compliance with Article IX of the OST in relation to contamination from biological sources (COPUOS 2017), but the PPP is not necessarily the only method by which states can comply with the treaty. Nevertheless, space active states generally have sought to satisfy the specific requirements articulated by COSPAR, and several agencies have incorporated the substantive provisions of the PPP into their formal governing documentation.

European Space Agency

The Council of the European Space Agency (ESA) has adopted the "ESA Planetary Protection Policy." The ESA PPP is implemented by specific "ESA Planetary Protection Requirements," which are based on and consistent with the COSPAR PPP, and set forth the Agency's overall planetary protection management responsibilities. These include technical requirements for robotic and human missions, requirements related to procedures, and descriptions of necessary documentation (Kminek 2017). In addition, this document contains internal ESA organizational descriptions and requirements. Neither the ESA PPP nor the ESA PPR have been released to the public. However, specific standards have been published by the European Cooperation for Space Standardization (ECSS) in "Space Sustainability – Planetary Protection" which are to be followed by ESA member states in their own projects. The ESA PPP and corresponding implementation standards apply to spaceflight missions conducted by ESA and to contributions to ESA spaceflight missions. Significantly, the ESA PPP also applies to ESA contributions to non-ESA spaceflight missions, such as missions conducted by other states, agencies, or the private sector which are launched by ESA or from ESA facilities.

The United States

NASA has adopted NASA Policy Directive NPD 8020.7G, "Biological Contamination Control for Outbound and Inbound Planetary Spacecraft," which mirrors the COSPAR policy statement.

The NPD is implemented by NASA Procedural Requirements NPR 8020.12D, "Planetary Protection Provisions for Robotic Extraterrestrial Missions," which are based on and are in conformity with the COSPAR PPP. Pursuant to the NASA policy, the agency has established a Planetary Protection Officer, with responsibility to categorize missions and oversee compliance with the Planetary Protection Policy.

In 2017, NASA promulgated NASA Interim Directive NID 8020.109A, "Planetary Protection Provisions for Robotic Extraterrestrial Missions." Section 2.2 of this

NID concerns NASA participation in non-NASA or non-US Missions, which assigns sole responsibility for planetary protection categorization and certification of compliance to the lead and launching organizations. Subsection 2.2.2 provides:

> NASA shall provide hardware, services, data, funding, and other resources to non-NASA missions (including but not limited to resources agreements) only if the recipient organization(s), whether governmental or private entity, demonstrate adherence to appropriate policies, regulations, and laws regarding planetary protection that are generally consistent with the COSPAR Planetary Protection Policy and Guidelines

Japan

Japan has launched several deep space explorations, including the Hayabusa 2 asteroid sample return mission currently in progress, without an overarching Planetary Protection Policy document in place. Nevertheless, individual projects implemented the COSPAR PPP by the adoption of conforming design standards and by achieving agreement with the COSPAR Planetary Protection Panel. Japan steadily has been increasing its space exploration activities, and in December 2018 JAXA established a planetary protection organization. Shortly thereafter, the agency issued a formal Planetary Protection Policy and associated requirements which are compliant with the COSPAR PPP, JAXA Management Requirement JMR-014, "Planetary Protection Program Standards." These standards specify administrative, technical, and procedural aspects of planetary protection for both forward and back contamination. These standards apply to missions conducted by JAXA, parties who participate in missions conducted by JAXA, and JAXA's participation in missions hosted by other organizations.

Russia

The Russian Federal Law "On space activities" Decree No. 5663-1, Article 6, designates the State Corporation Roscosmos as the body authorized to ensure the safety of space activities. Article 4.1 provides that Roscosmos is responsible for ensuring that space activities are carried out in accordance with the principle of environmental protection. In order to ensure "strategic and ecological security," Article 4.2 prohibits the creation of "harmful contamination of outer space which leads to unfavorable changes of the environment, . . ." This prohibition appears to be more limited than is Article IX of the OST, which expressly includes "the Moon and other celestial bodies" as part of outer space. The omission of specific reference to the Moon and other celestial bodies in this subsection of the Russian law indicates that it is directed only to the outer space environment, as another subsection expressly prohibits the use of the Moon and other celestial bodies for military purposes. In addition, the express direction that harmful contamination includes the deliberate elimination or destruction of space objects in outer space implies objects in Earth orbit. Nevertheless, Article 26.3 affirms the duty of Russia to fulfill its obligations under the OST.

Russian Federal Laws "on technical regulation" No. 184-FZ, and "on sanitary and epidemiological well-being of the population" No. 52-FZ, address safety in general, including biological safety, but do not contain specific rules dedicated or directly addressed to issues of forward or back contamination. Nor apparently has Roscosmos publicly issued any official documents related to policies, procedures, or standards related to the biological protection of the Earth from the risks associated with space missions (Dobrokhotsky et al. 2012). However, missions such as Phobos-Grunt sample return were reviewed and approved as being in compliance with the COSPAR PPP, although the reliability of such certification can be questioned (Sterns and Tennen 2019).

Israel

Israel has conducted a lunar mission and is planning future missions but does not have a publicly available formal Planetary Protection Policy governing document. The Israel IL lunar landing craft Beresheet was launched on a private American rocket from a NASA launch facility in Florida. As such, per NASA NID 8020.109A, the craft was subject to the NASA policy and requirements as set forth in NPD 8020.7G and NPR 8020.12D. Beresheet crashed on the lunar surface while attempting to land. Several months later reports began to circulate that a payload supplied by a third party had secretly encased some tardigrades, a primitive form of life, into some components that were carried to the Moon. A deliberate failure to disclose the presence of living organisms on a spacecraft launched to a celestial body would violate US law and policy (Johnson et al. 2019).

China/India/UAE

China, India, and the UAE are each conducting missions to celestial bodies but do not have publicly available formal Planetary Protection Policy governing documents. The Chinese have a long history of space exploration, and have participated in COSPAR since 1993. China has sent four spacecraft to the Moon, the most recent of which is the Chang'e-4 that soft landed on the lunar far side in January, 2019, and continues to operate together with the Yutu-2 rover. China also sent a Martian orbiter, Yinghuo-1, and a microgravity grinding tool as payloads on the Phobos-Grunt mission. The Chinese plan to launch an orbiter and rover, Tianwen 1, to Mars in July 2020. Although an internal coordination mechanism has been implemented to formulate a relevant work plan for this mission, the Chinese are deferring formalizing a planetary protection policy pending further practice and research work.

The Indian Space Research Organization launched the Mars Orbiter Mission and the Chandrayaan mission to the Moon, and the agency followed the COSPAR policy for both missions. ISRO currently is planning additional lunar and planetary missions and is in the process of preparing a formal Planetary Protection Policy. The UAE is preparing for the launch of the "Hope Probe" to Mars. The Emirates have embarked upon an ambitious program to become a significant participant in space activities. The UAE Space Agency is in the process of developing internal policies and procedures, including legal regulation; however it has not been publicly announced whether these policies, procedure, and regulations will include planetary protection considerations.

Lacuna in the Planetary Protection Policy

The COSPAR PPP began as a comprehensive policy that recognized pristine celestial environments to be scientific preserves and established a Planetary Quarantine Policy applicable to all solar system bodies within our technological reach. Revisions to the policy have narrowed the scope and excluded celestial objects based on scientific explorations and results, new perspectives and approaches to planetary protection, and conclusions drawn regarding the biological interest of particular alien environments vis-a-vis the impact of the introduction of terrestrial organic matter and the ability vel non of terrestrial microbes to survive and replicate in such environments. As the COSPAR PP Panel reviews and re-evaluates the policy, consideration should be given to filling the gaps that remain, including the following:

Expand Target Bodies in Categories III and IV

The more recent revisions to the PPP have included Europa and Enceladus in the list of target bodies subject to heightened protective measures. Six additional potential ocean worlds have been identified which may support alien life: Titan, Ganymede, Triton, Dione, Callisto, and Pluto. In addition to Mars, recent reports have indicated that Venus also may have been a habitable planet for much of the history of the solar system, and life may be present today in her clouds (Wall 2018; NASA 2016). Even Mercury, which long was considered to be completely uninhabitable, has been discovered to have ice deposits in shielded craters just like the Moon (NASA 2012). These discoveries expand the universe of extraterrestrial environments which are of potential interest in understanding the process of chemical evolution or the origin of life in the solar system, yet few of these target bodies currently are categorized so as to require active protective measures. Nevertheless, each category of the PPP notes that additional bodies can be added to any category on a case-by-case basis.

Discoveries of extremophiles on this planet confirm that we must be prepared to encounter life as we do not know it or may not even be able to recognize. Indeed, it was only just over 40 years ago that cellular organisms were discovered to be composed of three distinct types of organisms – archaea, bacteria, and eukarya

(Woese and Fox 1977). The gaps in our knowledge are great, and assumptions must be continuously tested. We do know that the basic building blocks of life are not unique to this planet and that once life takes hold, it grabs on tenaciously. It is not possible to conclude that life, or its precursors, could not have been or is not present in environments previously considered inhospitable if not antithetical to life without first conducting specific and thorough scientific investigations.

Missions to Phobos and Deimos Can Increase the Risk of Contaminating Mars

The categorization of target bodies in the COSPAR PPP has a consequence which may have been unintended but nevertheless places certain celestial objects at increased and unnecessary risk of contamination. Specifically, a focus on a target body fails to adequately consider other bodies that may be within its relative vicinity. This gap within the PPP is illustrated by the Phobos-Grunt mission, which was conducted by the Russian Space Agency in 2011. The mission was intended to land a spacecraft on Phobos, which orbits less than 6000 km above the surface of Mars, collect samples of surface dirt and rocks, and return them to the Earth. The landing craft included a "bioshield module" provided by the Planetary Society, an American public advocacy organization. Called the "Living Interplanetary Flight Experiment (LIFE)" Project, the bioshield held 11 types of organisms from all three domains of life. These organisms ranged from the "mundane" to the "bizarre" and included extremophiles that were resistant to radiation, desiccation, salt, and heat. The Life Project sought to test the transpermia hypothesis, that is, whether a living organism could survive inside a meteorite and travel from planet to planet.

Phobos is considered to have an environment with a remote chance that terrestrial organisms could survive or replicate and therefore did not require stringent decontamination according to the COSPAR PPP. However, the Life Project BioShield enclosure was specifically designed to enhance and promote the survivability and replication of the organisms housed therein, negating the rationale which underlies Category II classifications.

There is no question but that the intentional introduction of potentially contaminating organisms within 6,000 km of Mars placed the planet at increased and unnecessary risk of despoiling the environment. Missions to Mars are difficult at best, and sample return missions are inherently more complex and challenging than a flyby, orbiting, or landing mission. The possibilities for a spacecraft off-nominal occurrence on or near Phobos are numerous, especially while engaging in arrival, landing, and return maneuvers, any one of which could result in an unintended and uncontrolled encounter with the Martian environment. Such accidental encounters could occur anywhere on the planet, including in special regions.

The launch of the LIFE Project failed to adequately consider the risk to the Martian environment and neither represented the best practice of states nor established good precedent. The mission ended when the spacecraft failed to leave

Earth orbit and fell back into the atmosphere and burned up. As a matter of due regard, the PPP should strictly scrutinize any proposed experiment that increases the risk of harmful contamination to a celestial environment, even if it is not the intended target body. Justification for such a mission must demand that the scientific results sought to be attained are of a very high significance and are not available by any other alternative means (Sterns and Tennen 2019).

Challenges to Planetary Protection by Private Sector Commercial Projects

Additional gaps in the COSPAR PPP are presented by the emergence of New Space commercial ventures. The OST provides in Article VI that states are internationally responsible for the activities of their nationals in space and that nongovernmental entities must be authorized and continuously supervised by their appropriate state. Neither ESA, NASA, nor JAXA authorize private sector activities, and as noted above, the agencies have adopted policies which require adherence with their planetary protection policies for any missions in which they participate. This includes commercial missions; however private sector activities without ESA, NASA, or JAXA participation are not subject to those agencies' internal policy requirement documents. Several states have adopted or are in the process of drafting national licensing regimes for the authorization and supervision of New Space activities (Jakhu and Pelton 2017). However, to date none of these national regimes has included specific reference to, or a requirement to comply with, the COSPAR PPP or similar strictures (Babb et al. 2018).

The COSPAR PPP does not address the implications and ramifications of the discovery of evidence of alien life. The scientific community presumably would take all necessary steps to protect and preserve the evidence and proceed with additional experimentation and exploration with the utmost care. The motivations of the private sector, however, are not necessarily congruent with scientific investigation. Therefore, it is imperative that the Planetary Protection Policy considers the adoption of rules and procedures for the private sector to follow in the event of discovery of evidence of alien life, or the remnants or precursors thereof, or other discoveries of scientific interest.

Policies will need to be developed regarding the disclosure of a discovery of evidence of extraterrestrial life. This would be consistent with the provisions of Article 5.3 of the MA which obligates states to disclose the discovery of any indication of organic life. This disclosure is to be made to the Secretary General, the international scientific community, and the public at large. Private sector ventures may be hesitant to make such a disclosure, especially if there is the possibility of obtaining a financial opportunity from intellectual property rights derived from a non-terrestrial microbe (Long 2016).

The area in proximity to a discovery of extraterrestrial life, and other discoveries of scientific interest, must also be protected and preserved. In addition, it must be determined whether commercial activities should be prohibited from conducting resource extraction and other activities from areas of special scientific interest, such as special regions on Mars, pending further scientific investigation (Sterns and Tennen 2019).

Commercial activities present additional issues from a planetary protection perspective, such as the need for the scientific community to be informed of and have the opportunity to express any concerns about proposed missions prior to the issuance of licenses or other authorizations. Private sector missions should be required to disclose an inventory of hazardous materials carried on a spacecraft and provide pre- and post-mission environmental impact assessments for activities conducted in situ.

The focus of the COSPAR PPP on biological contamination is only one aspect of the impact New Space ventures will have on celestial environments. The regulation of the private sector in space will need to consider planetary environmental protection more broadly, such as what requirements, if any, should be imposed on commercial ventures to restore the surface and subsurface of celestial bodies and remove private spacecraft and other objects at the end of their operations. Moreover, historic sites and other unique locations of special interest should be protected from being despoiled in the name of profit.

Whether or not an area has particular historical or scientific interest, appropriate regulation of the private sector will need to consider limitations and prohibitions on materials that can be transported on board a spacecraft and deposited on a celestial object. Commercial ventures have already begun to market the "service" of lunar "burial" of human remains and to carry almost unlimited personal objects on board a landing craft to establish a permanent monument on the Moon. In addition to the obvious environmental impact, erecting shrines to vanity and littering the lunar surface with useless detritus violate any number of provisions of international law, including the non-appropriation principle in Article II of the OST. These commercial ventures harken back to the concerns expressed by Khrushchev in 1962 about activities that create obstacles and hinder the activities of others. The creation of obstacles and hindrances is inherently provocative and destabilizing to space security.

Conclusion

The protection of celestial environments presents scientific, legal, ethical, moral, aesthetic, and philosophical considerations (Rummel et al. 2012). COSPAR has been the leading scientific organization to articulate and develop a policy of planetary protection, which has been recognized as the international standard for compliance with Article IX of the OST regarding protecting against forward and back contamination from biological sources. The COSPAR PPP promotes space security by furthering the interests of science and enhancing the essential goal of international space law to safeguard space, including the Moon and other celestial bodies, for peaceful exploration and use by all states.

References

Babb RJ, Erb H, Howard D (2018) Cost reduction solutions in regard to PP for commercial companies, IAF Paper No. IAC-18-F1.2.3

COSPAR Planetary Protection Policy, Space Research Today, No. 200, p. 12 (2017). https://cosparhq.cnes.fr/assets/uploads/2019/12/PPPolicyDecember-2017.pdf. Last accessed February 29 2020

Coustenis A, Kminek G, Hedman N, Ammannito E, Deshevaya E, Doran PT, Grasset O, Hayes A, Lei L, Nakamura A, Prieto-Ballesteros O, Raulin F, Rettberg P, Sreekumar P, Tsuneta S, Viso M, Zaitsev M, Zorzano-Mier P (2019) The COSPAR panel on planetary protection role, structure and activities, https://cosparhq.cnes.fr/assets/uploads/2019/07/PPP_SRT-Article_Role-Structure_Aug-2019.pdf. Last accessed February 29 2020

Dobrokhotsky ON, Dyatlov IA, Orlov OI, Novikova ND, Hamidullina NM, Deshevaya EA (2012) Ensuring biosafety in the study of samples of extraterrestrial origin by an example of preparation for the "Phobos-Grunt" Mission. https://cyberleninka.ru/article/n/obespechenie-biologicheskoy-bezopasnosti-pri-issledovanii-materialov-vnezemnogo-proishozhdeniya-na-primere-podgotovki-ekspeditsii/viewer (in Russian). Last accessed February 29 2020

Hofmann M, Retberg P, Williamson M (eds) (2010) Protecting the environment of celestial bodies: the need for policy and guidelines jaxa jmr-014, the planetary protection standards. http://sma.jaxa.jp/TechDoc/Docs/JAXA-JMR-014.pdf (in Japanese). Last accessed February 29 2020

Jakhu RS, Pelton JN (eds) (2017) Global space governance: an International study 389, 392, 453–54

Johnson CD, Porras D, Hearsay CM and O'Sullivan S (2019) The curious case of the transgressing tardigrades (part 2). https://www.thespacereview.com/article/3786/1. Last accessed February 29 2020

Kminek G (2017) Planetary Protection at ESA. http://pposs.org/wp-content/uploads/2017/03/7.-PPOSS-PP-at-ESA-G.-Kminek.pdf. Last accessed February 29 2020

Long GA (2016) The meaning of life and close encounters of the commercial kind, 2015 proceedings of the international institute of space law 175

Mangus S, Larsen W (2004) Lunar receiving laboratory project history https://www.lpi.usra.edu/lunar/documents/lunarReceivingLabCr2004_208938.pdf. Last accessed February 29 2020

McLeana RJC, Welsha AK, Casasanto VA (2006) Microbial survival in space shuttle crash. Icarus 181(1):323–325

Meltzer M (2012) When biospheres collide: a history of nasa's planetary protection programs 82 http://www.nasa.gov/connect/ebooks/when_biospheres_collide_detail.html. Last accessed February 29 2020

MSR Science Planning Group (2019) Science-driven contamination control issues associated with the receiving and initial processing of the msr samples. https://mepag.jpl.nasa.gov/reports.cfm. Last accessed February 29 2020

NASA (2012) Messenger finds new evidence for water ice at mercury's poles https://www.nasa.gov/mission_pages/messenger/media/PressConf20121129.html. Last accessed February 29 2020

NASA (2016) NASA climate modeling suggests venus may have been habitable. https://www.nasa.gov/feature/goddard/2016/nasa-climate-modeling-suggests-venus-may-have-been-habitable. Last accessed February 29 2020

Phillips CR (1975) The planetary quarantine program: origins and achievements 37. https://ntrs.nasa.gov/archive/nasa/casi.ntrs.nasa.gov/19750006598.pdf. Last accessed February 29 2020

Rummel JD, Race MS, Horneck G (2012) Ethical considerations for planetary protection in space exploration: a workshop. Astrobiology 12(11):1017–1023

Space Science Board Committee on Planetary Biology and Chemical Evolution (1978) Recommendations on Quarantine Policy for Mars, Jupiter, Saturn, Uranus, Neptune and Titan 27–28 (Appendix C)

Stabekis P (2002) History and processing of changes, in report, COSPAR/IAU Workshop on planetary protection (Appendix C)

Sterns PM, Tennen LI (2019) Lacuna in the updated planetary protection policy and international law, 23 life sciences in space research 10 https://www.sciencedirect.com/science/article/pii/S2214552418301007. Last accessed February 29 2020

Tennen LI, Evolution of the planetary protection policy: conflict of science and jurisprudence? Proceedings of the 45th colloquium on the law of outer space 466 (2003) and Advances in Space Research 34 (2004): 2354–2362

Wall M (2018) Life on venus? Why its not an absurd thought. https://www.space.com/40304-venus-clouds-alien-life-search.html. Last accessed February 29 2020

Woese CR, Fox GE (1977) Phylogenetic structure of the prokaryotic domain: the primary kingdoms. Proc Natl Acad Sci U S A 74:5088–5090. https://www.ncbi.nlm.nih.gov/pubmed/270744. Last accessed February 29 2020

Additional Reading

Compendium on mechanisms adopted in relation to non-legally binding United Nations instruments on outer space, Submission by Japan, UN Doc. A/AC.105/C.2/2019/CRP.16 (2 April 2019). http://www.unoosa.org/res/oosadoc/data/documents/2019/aac_105c_22019crp/aac_105c_22019crp_16_0_html/AC105_C2_2019_CRP16E.pdf. Last accessed February 29 2020

European Cooperation for Space Standardization (ECSS) (1 August 2019) Space Sustainability – Planetary Protection, ECSS-U-ST-20C. https://ecss.nl/standard/ecss-u-st-20c-space-sustainability-planetary-protection. Last accessed February 29 2020

Hofmann M, Retberg P, Williamson M (eds) (2010) Protecting the environment of celestial bodies: the need for policy and guidelines jaxa jmr-014, the planetary protection standards. http://sma.jaxa.jp/TechDoc/Docs/JAXA-JMR-014.pdf (in Japanese). Last accessed February 29 2020

NASA Interim Directive NID 8020.109A, Planetary Protection Provisions for Robotic Extraterrestrial Missions. https://nodis3.gsfc.nasa.gov/OPD_docs/NID_8020_109A.pdf. Last accessed February 29 2020

NASA Policy Directive NPD 8020.7G, Biological Contamination Control for Outbound and Inbound Planetary Spacecraft. https://nodis3.gsfc.nasa.gov/displayDir.cfm?t=NPD&c=8020&s=7G. Last accessed February 29 2020

NASA Procedural Requirements NPR 8020.12D, planetary protection Provisions For Robotic Extraterrestrial Missions, https://nodis3.gsfc.nasa.gov/displayDir.cfm?t=NPR&c=8020&s=12D. Last accessed February 29 2020

Russian Federal Law "On space activities" Decree No. 5663-1 http://www.unoosa.org/oosa/en/ourwork/spacelaw/nationalspacelaw/russian_federation/decree_5663-1_E.html. Last accessed February 29 2020

Index